Natural Capital Accounting and Valuation of Ecosystem Services, Karnataka State, India

T. V. Ramachandra · Bharath Haridas Aithal · Bharath Setturu · S. Vinay · K. S. Asulabha · V. Sincy

Natural Capital Accounting and Valuation of Ecosystem Services, Karnataka State, India

Ecosystem Services

Springer

T. V. Ramachandra
Energy and Wetlands Research Group
Centre for Ecological Sciences
Indian Institute of Science
Bengaluru, Karnataka, India

Bharath Haridas Aithal
Ranbir and Chitra Gupta School
of Infrastructure Design and Management
Indian Institute of Technology Kharagpur
Kharagpur, West Bengal, India

Bharath Setturu
Energy and Wetlands Research Group
Centre for Ecological Sciences
Indian Institute of Science
Bengaluru, Karnataka, India

S. Vinay
Energy and Wetlands Research Group
Centre for Ecological Sciences
Indian Institute of Science
Bengaluru, Karnataka, India

K. S. Asulabha
Energy and Wetlands Research Group
Centre for Ecological Sciences
Indian Institute of Science
Bengaluru, Karnataka, India

V. Sincy
Energy and Wetlands Research Group
Centre for Ecological Sciences
Indian Institute of Science
Bengaluru, Karnataka, India

ISBN 978-981-97-2404-8 ISBN 978-981-97-2405-5 (eBook)
https://doi.org/10.1007/978-981-97-2405-5

The work was supported by the United Nations Environment Programme (UNEP) as part of the international, EU-funded Natural Capital Accounting and Valuation of Ecosystem Services (NCAVES) project.

This Springer imprint is published by the registered company Springer Nature Singapore Pte Ltd.
The registered company address is: 152 Beach Road, #21-01/04 Gateway East, Singapore 189721, Singapore

Paper in this product is recyclable.

Preface

Humans depend on ecosystems for their basic needs, such as food, fuel, minerals, water, and air. All forms of interaction between ecosystems and people, including in situ and remote interactions, are often referred to as ecosystem services. The supply of an ecosystem service is associated with an ecosystem structure or process or a combination of ecosystem structures and processes that reflect the biological, chemical, and physical interactions among ecosystem components. Ecosystem services are broadly categorized as (i) provisioning services, which are those ecosystem services representing the contributions to benefits that are extracted or harvested from ecosystems; (ii) regulating and maintenance services are those ecosystem services resulting from the ability of ecosystems to regulate biological processes and to influence climate, hydrological, and biochemical cycles, and thereby maintain environmental conditions beneficial to individuals and society, (iii) cultural services are the experiential and intangible services related to the perceived or actual qualities of ecosystems whose existence and functioning contribute to a range of cultural benefits.

India is attempting to accelerate economic growth and relax environmental laws, and there is tremendous pressure to divert natural systems to other uses. Hence, there is a pressing need to undertake the natural capital accounting and valuation of the ecosystem services, especially intangible benefits, provided by ecosystems. The value of all ecosystem services, including the degradation costs, needs to be understood to develop appropriate policies for the conservation and sustainable management of ecosystems. Scientific efforts during the past decade have refined the understanding of ecosystem function and demonstrated the links between functions and the provision of ecosystem services. This knowledge needs to be communicated effectively to decision-makers and the public, which will lead to developing policies that adequately consider the trade-offs between the conservation of ecosystems and natural resources and economic growth.

There is an urgent policy need for more comprehensive assessments of the natural capital of ecosystems, which will aid in comparing these aggregate values with the opportunity cost of this land. Policymakers need such information to gain support for conservation funding, engage local communities, and develop market-based instruments for conservation, which necessitates accounting for the natural capital found in ecosystems and incorporating their economic worth into the measurement of the wealth of a region.

Ecosystem accounts make the value of ecosystem services visible, allowing them to be internalized into decision-making. Accounting of ecosystem services enables an assessment of trade-offs between economic development and environmental conservation and restoration, resulting in better-informed decisions. It also strengthens the economic case for conserving forests in states in India and developing countries where there can be tremendous pressure to relax forest laws and divert forests to non-forest uses without proper consideration of the sustainability of such actions.

United Nations (UNSD and UNEP) developed the System of Environmental-Economic Accounting (SEEA) framework for natural capital accounting and valuation of ecosystem services (NCAVES) by integrating economic and environmental data and providing a comprehensive and multipurpose view of the interrelationships between the economy and the environment and the stocks and changes in stocks of environmental assets, as they bring benefits to humanity. Karnataka state was chosen by UNSD (SSFA/2019/1502), to pilot the compilation of selected ecosystem accounts in physical and monetary terms based on policy priorities and to contribute to policy mainstreaming.

The research focussing on natural capital accounting and valuation of ecosystem services was carried out by the Energy and Wetlands Research Group at CES, Indian Institute of Science, in collaboration with the United Nations Environment Programme (UNEP), United Nations Statistics Division (UNSD), the Ministry of Statistics and Programme Implementation (MoSPI), Government of India and the ENVIS division, The Ministry of Environment Forests and Climate Change (MoEFCC), Government of India and Ranbir and Chitra Gupta School of Infrastructure Design and Management (RCG SIDM), and Indian Institute of Technology Kharagpur (IIT-KGP) as part of the international, EU-funded Natural Capital Accounting and Valuation of Ecosystem Services (NCAVES) project.

The valuation of ecosystem services (forests, agriculture, and aquatic) is implemented district-wise for Karnataka State, India, as per the validated statistical framework for natural capital accounting—SEEA: System of Environmental-Economic Accounting (SEEA.un.org). According to SEEA protocol, ecosystem services are defined as the contributions of ecosystems to the benefits that are used in economic and other human activities. The valuation of ecosystem services (VES) provides an unbiased framework to value unaccounted ecosystem benefits and helps in developing meaningful policy interventions. The approach allows for adjusted regional or national accounts, which reflect the output of ecosystem services as well as the depletion of natural resources and the degradation costs (externalized costs of the loss of ecosystem services) of ecosystems in economic terms. In this perspective, the current publication, focusing on the natural capital accounting and valuation of ecosystem services in Karnataka, India, will help raise awareness and provide a quantitative tool to evaluate the sustainability of policies.

Bengaluru, India	T. V. Ramachandra
Kharagpur, India	Bharath Haridas Aithal
Bengaluru, India	Bharath Setturu
Bengaluru, India	S. Vinay
Bengaluru, India	K. S. Asulabha
Bengaluru, India	V. Sincy

Acknowledgments

The valuation of ecosystem services of terrestrial (forests and agriculture) and aquatic ecosystems is implemented district-wise for Karnataka State, India, as per the validated protocol—System of Environmental-Economic Accounting (SEEA EA; SEEA.un.org) protocol, which constitutes the statistical framework for natural capital accounting. According to this protocol, ecosystem services are defined as the contributions of ecosystems to the benefits that are used in economic and other human activities. The valuation of ecosystem services (VES) allows for adjusted national accounts that reflect the output of ecosystem services as well as the depletion of natural resources and the degradation costs (externalized costs of the loss of ecosystem services) of ecosystems in economic terms, which will help raise awareness and provide a quantitative tool to evaluate the sustainability of policies. It provides an unbiased and dependable national framework to value so far unaccounted ecosystem benefits and helps develop meaningful policy interventions.

The research focussing on natural capital accounting and valuation of ecosystem services in Karnataka State, India, was commissioned by the United Nations Environment Programme (UNEP) as part of the international, EU-funded Natural Capital Accounting and Valuation of Ecosystem Services (NCAVES) project. The NCAVES project was carried out as a collaboration between UNEP, the United Nations Statistics Division (UNSD), the Ministry of Statistics and Programme Implementation (MoSPI), Government of India and ENVIS Division, The Ministry of Environment Forests and Climate Change (MoEF&CC), Government of India.

Acknowledgments go to the European Union for funding the NCAVES Project and the Delegation of the European Union to India for supporting its implementation in Karnataka State, India, and the UNSD and UNEP for leading the NCAVES Project globally and supporting its implementation and management in Karnataka State, India.

We are grateful to the review inputs from Dr. William Speller of UNEP, Dr. Bram Edens of UNSD, and the members of the UN Technical Committee on the SEEA EA and its working groups.

We appreciate the contributions of Dr. P. Bhanumati and Mr. Rakesh Kumar Maurya of the Ministry of Statistics and Programme Implementation (MoSPI),

Government of India (GoI); Dr. Anandi Subramnyam, Dr. James Mathew, and Mr. Kumar Rajnish of the Ministry of Environment, Forests, and Climate Change (MoEFCC), GoI in developing the ecosystem services account. Thanks to (i) Dr. Prabhuraj, Director, Karnataka State Remote Sensing Centre (KSRSAC), Government of Karnataka (GoK), Bengaluru, for providing spatial data of the district administrative boundaries, stream and river network, population data; (ii) Dr. Hemanth Kumar, Executive Secretary, Karnataka State Council for Science and Technology (KSCST), GoK for providing spatial data related to geology, lithology, etc; (iii) Director, National Remote Sensing Centre, Department of Space, GoI, Hyderabad; and (iv) Director, Department of Agriculture, GoK for providing soil health data—district-wise for Karnataka.

We acknowledge the efforts of Mr. Karthik Naik and Mr. Vinayka Bhatta (for assisting in field data collection and district-wise remote sensing data analyses—extent and fragmentation analyses), Ms. Madhumita Dey (for assistance in the computation of Land surface temperature analyses), Mr. D. R. Rakesh (soil quality analyses), Mr. M. Chandan (for the assistance in agent-based modelling and geo-visualisation), Ms. Harita (for help in compiling data related to biodiversity), Ms. Minsa (for compiling data related to fauna), Dr. G. R. Rao, Mr. Vishnu Mukri, and Mr. Shrikanth Naik (for assistance in field data collection related to flora and fauna in five districts of Western Ghats), and Mr. Vrijulal and Dr. M. D. Subash Chandran (for verification of data—flora and fauna). Ms. V. Sincy, Ms. K. S. Asulabha, Ms. H. Deepthi, Ms. G. Saranya, and Mr. Sudarshan Bhat assisted with the ecosystems' economic valuation and compiling the accounts.

The work was carried out using (i) field data; (ii) collateral data compiled KSRSAC, GoK; KSCST, GoK; Department of Agriculture, GoK, Karnataka Forest Department, GoK; (iii) spatial data (Landsat series available in the public domain); and (iv) spatial data (IRS LISS Data) procured from the National Remote Sensing Centre, GoI, Hyderabad.

The views, thoughts, and opinions expressed in the publication are of the authors and do not necessarily reflect the views of either the publisher, funding agencies, or the employer. The designations employed and the presentation of material including on any map in this work do not imply the expression of any opinion whatsoever on the part of the publisher, funding agencies, or the employer concerning the legal status of any country, territory, city or area or its authorities, or concerning the delimitation of its frontiers or boundaries.

Introduction

Ecosystems (terrestrial and aquatic) are vital for the survival of humans on the earth. Ecosystems and the biological diversity within them provide a number of goods (food, fodder, etc.) and services (remediation, clean air, water, etc.) essential to sustain life with economic prosperity among other aspects of human welfare. The well-being of humans is entirely dependent on the health of the ecosystems. Humans, directly and indirectly, are dependent on the goods and services provided by the ecosystems in their surroundings to meet their everyday needs. These goods and services are generated as a consequence of the biotic and abiotic interactions that take place within the ecosystem. This implies that any change in the nature of their components would affect their capacity to generate the said goods and services. Ecosystem services can be understood as the benefits (often indirect) humans obtain from ecosystems. Forest ecosystems, in particular, owing to their abiotic and biotic diversity and abundance, foster goods and services that are of high economic importance. The direct benefits are mostly in the form of goods and include food, timber, fuelwood, etc., while the indirect benefits include air purification, flood mitigation, carbon sequestration, pollination, cultural amenities, etc. These goods and services are often not given any economic importance as most of them do not directly enter the market and, hence, do not possess a market value. This has led to the development of policies focussing on the economic benefits of any given activity without taking into consideration the environmental loading caused by the same. Taking into account the natural capital found in the form of forest ecosystems among others in such a way that their economic worth is being added to the measurement of the wealth of a nation is of importance. Valuation of ecosystem goods and services is a step taken towards the same with the agenda of assigning an economic value to the various goods and services provided by an ecosystem.

Human beings are dependent on diverse ecosystems to meet their basic needs, such as food, fuel, minerals, water, and air. In developing countries, on average, 40% of the Gross Domestic Product (GDP) is contributed by agriculture, with nearly 70–80% of the labour force engaged in agricultural or resource-based activities. This massive dependency on natural resources and overexploitation over the years have led to their degradation and depletion owing to the unsustainable practices involved in their extraction. Unplanned development activities coupled with the increasing population have put tremendous pressure on environmental resources and increased concerns over environmental degradation. An increased surge in the developmental and technological activities after the industrial revolution and opening up of markets over the last two decades have led to an increase in the quantities of chemicals contributing to damage caused in the natural ecosystems. Over the past 50 years, rapid environmental changes have resulted in a substantial and largely irreversible loss in the diversity of life on earth. Unsustainable utilisation of land and other natural resources persists, despite the increasing understanding of the impacts that human activities have on the environment. The linkage between the health of the environment and the sustenance of humankind makes it imperative to maintain a balance between the carrying capacity of the environment and the availability of natural resources by ensuring their prudent extraction. Conservation of natural ecosystems has long-term benefits for humans in utilitarian terms through their provision of food, timber, minerals, and a variety of valuable resources that have provided the backbone for economic development. Going beyond the utilitarian values, it has also been a source for the maintenance of gene pools, biodiversity, and other potentially useful factors that are of indirect use to humans. Hence, the ecosystems' intrinsic values and rights, regardless of human needs, therefore, should be taken into account apart from considering it a resource to be exploited for human settlement, food, and energy production. The dilemma associated with rapid land use change for accommodating the growing demand for the use of natural resources is that the more we change the land to meet our needs for food, water, fuel, shelter, etc., the more we impact and possibly degrade the ecosystems. Hence, it necessitates a structured economic development that optimises the economic and social benefits without jeopardising the likely potential for similar benefits in the future. Sustainable development of a region requires a synoptic ecosystem approach that relates to the dynamics of natural variability and the effects of human interventions on key indicators of biodiversity and productivity. The ecosystem service approach offers a way to understand and deal with the negative feedback loop that is ultimately created when ecosystems are used up to meet the needs of society. The ecosystem services framework allows the analyst to capture the full range of environmental impacts more systematically by linking the ecological effects to the changes in human welfare. The benefits from ecosystems include provisioning services such as food and water; regulating services such as regulation of floods, drought, land degradation, and disease; and cultural services such as recreational, spiritual, aesthetic, and science.

There is a pressing need to undertake the natural capital account and valuation of the ecosystem services, especially intangible benefits, provided by ecosystems. The value of all ecosystem services, including the externalized costs of their loss, needs to be understood for developing appropriate policies toward the conservation and sustainable management of ecosystems. Scientific efforts during the past decade have refined the understanding of ecosystem function and demonstrated the links between functions and the provision of ecosystem services, and this knowledge needs to be communicated effectively to decision-makers and the public, which will lead to the development of policies focussing on the economic benefits of any given activity with consideration of the environmental loading caused by the same.

There is an urgent policy need for more comprehensive assessments of the natural capital of ecosystems, which will aid in comparing these aggregate values with the opportunity cost of this land. Policymakers need such information to gain support for conservation funding but also to engage local communities and develop market-based instruments for conservation. This necessitates accounting for the natural capital found in the forest ecosystems (among others) and incorporation of their economic worth added to the measurement of the wealth of a region. Valuation of ecosystem goods and services is a step taken towards the same with the agenda of assigning an economic value to the various goods and services provided by an ecosystem.

The purpose of environmental-economic accounting has been the derivation of adjusted measures of value added and wealth that take into account the cost of using up environmental assets. This is considered in ecosystem accounting by measuring ecosystem degradation as reflecting the loss of future flows of ecosystem services.

The ecological interaction among biotic and abiotic factors results in the flow of goods and services essential for human survival and can be measured as tangible and intangible. The direct benefits are primarily in the form of goods such as food, timber, and fuelwood while the indirect services include air purification, flood mitigation, carbon sequestration, pollination, the prevention of soil erosion, pest control, regulating local and global climate, ameliorate weather events, regulate the hydrological cycle, climate change mitigation, control of disease vectors, and local ecosystem resilience and stability, cultural amenities, etc. and provide a vast store of genetic information much of which has yet to be uncovered. These goods and services, directly and indirectly, support livelihoods but are often not assigned any economic significance as most of them do not directly enter the market and, hence, do not possess a market value.

Ecosystems deliver goods and services to humankind and, therefore, have an economic value defined as their contribution to human welfare (financial, social, hydrological, environmental, and health). Ecosystem values are broadly classified into use values and non-use values, based on the benefits derived from the ecosystems to sustain the present and future generations. Use value refers to a value arising from an actual use made of a given resource, whereas the non-use value refers to one that is available independent of the ecosystem's interaction with humans. The use value is further classified into direct, non-direct, and option use values. Direct use value is derived from direct interactions with the ecosystem that may be consumptive or extractive by nature, like timber extraction, fruits, and fishing. Indirect use value is understood as the value derived from the functions served by the ecosystem. It does not have a direct market value but can be analysed using several techniques available that use proxy variables. It includes functions like nutrient removal, prevention of downstream flooding, and microclimate regulation. The option value refers to that value people put on a particular ecosystem owing to having the option of using the resources available in the future, etc.

The non-use value is classified into existence, bequest, altruistic, and option values. It deals with the value provided by an ecosystem without any interaction with humans and necessarily entails intangible benefits. Existence value is derived simply from the satisfaction of knowing that a particular ecosystem continues to exist irrespective of the benefits gained from it and is often also understood as the intrinsic value held by an ecosystem. The bequest value is associated with the knowledge that ecosystems and their services will be passed on to future generations so that they can enjoy them as well, which also happens to be the underlying theme of sustainable development. Altruistic value is the value derived from knowing that people can enjoy the goods and services provided by ecosystems. Finally, option value is also a part of the non-use value where the value is derived from the option of using a particular service/function in the future.

Most resource management decisions are influenced by ecosystem goods entering markets; as a result, the non-marketed benefits are often lost or degraded. Both renewable resources (water supply, air quality, etc.) and non-renewable resources (mineral deposits, some soil nutrients, fossil fuels, etc.) are capital assets and provide the backbone for numerous economic activities that account for the development of a region. Hitherto, traditional national accounts have not included resource depletion or degradation measures. The Gross Domestic Product (GDP), a measure of the current economic well-being of a region based on the market-exchange of material well-being, will therefore not be representative of the decline in these assets (wealth) in correspondence to the positive gain in the economy that is contributed through the process. Hence, GDP cannot be a true measure of the country's sustained economic wealth and cannot be a proxy for understanding its future economic well-being. Quantitative evidence on the economic value of such assets is thereby necessary for most of these services and goods provided by them are not traded in the markets and hence do not have a market value. The monetary valuation of the various services and goods provided by the ecosystems can help in building a better understanding of their influence on the economic well-being and can further facilitate information-driven decisions and policy reforms that align with the Sustainable Development Goals (SDGs), a successor to the Millennium Development Goals (MDGs) identified by the United Nations Development Programme (UNDP). Environmental accounting systems seek to set out a region's environmental, social, and economic assets and can be used to assess whether economic development is consistent with sustainable development or to help ensure optimal use of natural resources and environment. Recent efforts, especially the System of Environmental-Economic Accounting, Central Framework, and Experimental Ecosystem Accounting (SEEA CF and SEEA EEA), aim to extend and integrate the national accounts for environmental and ecosystem assets.

The stock of ecosystem resources (biotic and abiotic) that provide goods and services is referred to as natural capital. The services provided by these ecosystems support the economy and society. However, the contributions by the ecosystems or natural capital are often undervalued while making economic decisions. This necessitates accounting of natural resources to minimize overexploitation, mismanagement, etc. The process of accounting stocks and flows of goods and services (natural resources) in a given ecosystem of a region is often termed as natural capital accounting.

United Nations Statistical Division (UNSD) has proposed a System of Environmental-Economic Accounting—Experimental Ecosystem Accounting (SEEA—EEA) framework for ecosystem accounting. A key feature of ecosystem accounting, as per SEEA, is its capacity to integrate spatially referenced data about ecosystems, i.e. data about the location, size, and condition of ecosystems within a given area and how these are changing over time. The ecosystem accounting framework related to ecosystem services concerns (i) the supply of ecosystem services to users; and (ii) the contribution of ecosystem services to benefits (i.e., the goods and services ultimately used and enjoyed by people and society). Ecosystem services are structured into three broad categories:

- *Provisioning services* are those ecosystem services representing the contributions to benefits that are extracted or harvested from ecosystems.
- *Regulating services* are those ecosystem services resulting from the ability of ecosystems to regulate biological processes and to influence climate, hydrological, and biochemical cycles, and hereby maintain environmental conditions beneficial to individuals and society.
- *Cultural services* are the experiential and intangible services related to the perceived or actual qualities of ecosystems whose existence and functioning contribute to a range of cultural benefits.

Together, these ecosystem goods and services contribute to the Total Ecosystem Supply Value (TESV) of the natural resource. The key stages in the valuation of ecosystem services are: (i) natural capital accounting—establishing the environmental baseline through assessment of ecosystem extent and condition accounts, (ii) quantification of goods and services, (iii) valuation of ecosystem services, (iv) identify and provide qualitative assessment of the potential impacts of policy options on ecosystem services, (iv) quantify the impacts of policy options on specific ecosystem services and assess the effects on human welfare, and (v) value the changes in ecosystem services.

The ecosystem accounts have been developed as per the protocol of the System of Environmental-Economic Accounting—Experimental Ecosystem Accounts (SEEA—EEA) considering the following:

- **Ecosystem extent** accounts record the total area of each ecosystem, which is classified by type within an ecosystem accounting area and, over time, in a specified area (district, Karnataka State, India).
- **Ecosystem condition** accounts record the condition of ecosystem assets in terms of selected characteristics at specific points in time and, over time, record the changes to their condition. These changes are mainly due to human activities with economic interventions.
- **Ecosystem services** accounts record the supply of ecosystem services by ecosystem assets and the use of those services by economic units, including households. Ecosystem services accounts are presented both in physical and monetary units, using techniques for the valuation of ecosystem services.

- **Ecosystem monetary asset** accounts record information on stocks and changes in stocks (additions and reductions) of ecosystem assets. This includes accounting for ecosystem degradation and enhancement.

Ecosystem accounting entails the estimation of the extent of different ecosystem assets (EAs) in the study region (district and federal state) using remote sensing data (acquired through space-borne sensors at regular intervals and available since the 1970s). This spatial data helps in compiling ecosystems extent account and provides an underlying infrastructure for measuring ecosystem conditions and modeling ecosystem services. The analyses of temporal data helped in assessing the extent of changes in an ecosystem over time.

The outline of this book is as follows: Chap. 1 presents ecosystem extent, and Chap. 2 assesses the condition of terrestrial (forests and agriculture) ecosystems. Chapter 3 focuses on the valuation of terrestrial ecosystem services. Chapter 4 quantifies the impacts of policy options on specific ecosystem services and assesses the effects on human welfare. Chapter 5 presents the worth of aquatic (inland wetlands and estuaries) ecosystems.

Contents

Ecosystem Extent Account for Karnataka State, India ... 1
1 Introduction ... 3
2 Study Region: KARNATAKA STATE, INDIA ... 5
3 Data Sources and Methods ... 24
3.1 Classifying Ecosystem Types ... 24
3.2 Remote Sensing (RS) Data ... 25
3.3 Methods ... 28
3.4 Ecosystem Extent Quantification Through Land Use Analyses ... 31
4 Results ... 33
5 Conclusions ... 40
References ... 41

Ecosystem Condition Account for Karnataka State, India ... 43
1 Introduction ... 46
1.1 Fragmentation of Forest Ecosystem ... 48
1.2 Land Surface Temperature Response to Land-Use Changes ... 50
1.3 Vegetation Dynamics in Karnataka ... 52
1.4 Biotic Elements of Flora and Fauna ... 54
1.5 Characteristics of Soil ... 55
2 Data Sources ... 56
3 Methods ... 62
3.1 Forest Fragmentation ... 62
3.2 Land Surface Temperature ... 64
3.3 Vegetation Structure in Karnataka ... 65
3.4 Characteristics of Soil ... 68
3.5 Ecosystem Condition Indicator Account ... 69
4 Results ... 72
4.1 Forest Fragmentation ... 72
4.2 Land Surface Temperature ... 78

4.3 Ecosystem Condition Indicator of Landscape Level Characteristics Based on Land Surface Temperature (LST) 80
4.4 Vegetation (AGB: Above Ground Biomass, BGB: Below Ground Biomass, NPP: Net Primary Productivity), Distribution of Flora and Fauna 84
4.5 Soil–Characteristics 105
5 Ecosystem Condition Based on Abiotic Ecosystem Characteristics 168
6 Ecosystem Condition Index (ECI) 168
7 Conclusion 235
References 239

Valuation of Ecosystem Services, Karnataka State, India 247
1 Introduction 250
1.1 Ecosystem Services 252
1.2 Ecosystem Services Accounting—Physical as well as Monetary (Based on Residual Value Techniques) 254
2 Data 255
3 Method 266
3.1 Valuation of Forest Ecosystem Services 267
3.2 Valuation of Services from Agriculture Ecosystems 300
3.3 Total Ecosystem Supply Value [TESV] 304
3.4 Net Present Value (NPV) of Ecosystem Assets 304
4 Results and Discussion 305
4.1 Assessment of Ecosystem Extent Over Time 305
4.2 Valuation of the Ecosystem Services 312
4.3 Valuation of the Forest Ecosystem Services—Forest Circle Wise, Karnataka 321
5 Cultural Services from Forest Ecosystems 376
6 Total Provisioning, Regulating, and Cultural Services of Forest Ecosystems in Karnataka 377
6.1 Valuation of Agriculture (Croplands and Horticulture) Ecosystem Services 385
6.2 TESV—Total Ecosystem Supply Value, GEP—Gross Ecosystem Product, Karnataka [District-Wise] State, India 430
6.3 Net Present Value (NPV) of Ecosystem Assets 447
7 Conclusion 449
8 Recommendations 451
Annexure A Valuation of Forest Ecosystem Services (Provisioning, Regulating and Cultural) at the District Level 452
References 466

Scenario-Based Assessment of Policy Interventions in Karnataka State, India 473
1 Introduction 477
2 Review of Land-Use Change Models 480
2.1 Cellular Automata Model 480
2.2 Analytical Hierarchical Process 482
2.3 Agent-Based Model 483
3 Data 485
4 Method 488
4.1 Scenario-Based Prediction 494
5 Results and Discussion 499
5.1 Assessment of Historical Land-Use Transitions and Simulation 499
5.2 Scenario-Based Prediction and Validation 500
5.3 Ecosystem Services with the Simulated Land Uses (of 2033) 515
6 Conclusions and Recommendations 529
6.1 Recommendations 540
References 543

Natural Capital Accounting and Valuation of Aquatic Ecosystem Services in Karnataka, India 547
1 Introduction 548
2 Ecosystem Services from Wetlands 551
3 Approaches of Ecosystem Service Valuation 554
4 Materials and Methods 558
4.1 Study Area 558
5 Water Sample Collection and Water Quality Analysis 570
6 Valuation of Ecosystem Service from Wetlands 571
7 Data and Source: Freshwater Ecosystem in Karnataka 571
8 Data and Source: Estuaries in Karnataka 571
9 Method 605
10 Results and Discussion 612
10.1 Water Quality of Lakes 612
11 Ecosystem Services Provided by Microalgae 615
12 Ecosystem Services Provided by Fish 620
13 Ecosystem Services Provided by Macrophyte 624
14 Ecosystem Services from Wetlands in Karnataka 628
15 Total Ecosystem Supply Value (TESV) and Net Present Value (NPV) of Freshwater Wetlands in Karnataka 636
16 Estuarine Ecosystem in Karnataka 639
16.1 Provisioning Services 639
17 Regulating Services 650
18 Cultural Services 654
19 Ecosystem Services of Estuaries in the West Coast, Karnataka, India (Districtwise Analyses) 654

20 Total Ecosystem Supply Value (TESV) 657
21 Net Present Value (NPV) 660
22 Framework for Accounting Aquatic (Inland, Estuaries) Ecosystems 660
22.1 Accounting of Aquatic Ecosystem in Karnataka, India 662
23 Conclusion 665
Appendix 667
References 673

About the Authors

Dr. T. V. Ramachandra FIE, FIEE (UK), FNIE obtained Ph.D. in Ecology and Energy from the Indian Institute of Science. At present, Coordinator of Energy and Wetlands Research Group (EWRG) and Convener of Environmental Information System (ENVIS) at Centre for Ecological Sciences (CES). During the past 25 years, he has established an active school of research in the area of energy and environment (http://wgbis.ces.iisc.ac.in/energy). He is a member of Karnataka State Wetland Authority (since 2018), Kerala Wetland Authority (since 2022), National Wetland Committee, GoI., and Karnataka State Audit Advisory Committee. He was a Member of the Karnataka State-level Environment Expert Appraisal Committee (2007–2010), appointed by the Ministry of Environment and Forests, Government of India, and a member of the Western Ghats task force appointed by the Government of Karnataka. Apart from this TVR is serving in many committees of NGT (National Green Tribunal) related to the wetlands of Bangalore.

TVR is a recipient of Johny Biosphere Award for Ecology and Environment (2004), Satish Dhawan Young Scientist Award, 2007 of the Karnataka State Government, ENVIS Award (Thrice: 2004, 2014 and 2015) of the Ministry of Environment, Forests, and Climate Change, GoI, Namma Bangalorean, 2016 award (Namma Bengaluru Foundation) and Rotary Exemplars 2017—Citizen Extraordinaire award (Rotary Club of Bangalore), **Karnataka State Parisara Award, 2017–2018** (Environment Award, Government of Karnataka), Khadri Achuathan Memorial Samvahana

Award (Environment) 2018, on World Communicators Day by PRCI, research advisor at the Nanyang Academy of Sciences, Singapore (2019), and Eminent Engineer, Institution of Engineers, Karnataka (2019).

TVR features in the recent global ranking of top 2% of scientists (developed by Stanford University) with ranking in Environment: 0.1887, Energy: 0.2138, Enabling and strategic technologies: 0.2642. He is a honorary fellow, Karnataka Science and Technology Academy (2022), *Elected Fellow* of National Institute of Ecology (2011), Indian Association of Hydrologists (India; 2006), Indian Social Science Academy (ISSA), Institution of Engineers (IE, India; 2003), the Institution of Electrical Engineers (IEE, UK; 2005), and a Senior Member, IEEE (USA; 2000) and Association of Energy Engineers (USA; 2000).

TVR's research interests are in the area of aquatic ecosystems, biodiversity, ecological modeling, Western Ghats ecology, energy systems, renewable energy, energy conservation, energy planning, geoinformatics, environmental engineering education research, and curriculum development at the tertiary level. He has published over 352 research papers in the reputed peer-reviewed international and national journals, 75 book chapters, 333 papers in the international and national symposiums as well as 19 books. In addition, he has delivered a number of plenary lectures at national and international conferences. Publication "Milking diatoms for energy" is seminal work in biofuel research evident from reports in Scientific American, BBC, national dailies, etc.

He has guided 178 students for Master's dissertation and nineteen students for Doctoral degrees. He teaches courses (i) environmental management and (ii) Principles and application of GIS and remote sensing for masters and doctoral students. TVR has travelled widely across the country for field research and also for delivering lectures (1500+) at Schools, Colleges, Research Institutes, etc. He has taken initiatives through biennial symposium (popular as Lake series, conducted 12 events since 1998), Environment management course for in-service working professionals (18 sessions) training programmes and workshops for capacity building at various levels. Publications are available at https://www.researchgate.net/profile/T_V_Ramachandra/publications, https://scholar.google.co.in/citations?user=Woh1fa8AAAAJ&hl=en

Co-ordinator, Energy and Wetlands Research Group, Centre for Ecological Sciences [CES]

Convenor, Environmental Information System [ENVIS], CES TE 15, Indian Institute of Science

Associate Faculty, Centre for Sustainable Technologies (astra)

Centre for infrastructure, Sustainable Transportation and Urban Planning [CiSTUP]

Indian Institute of Science, Bengaluru, Karnataka, India

Web URL: http://wgbis.ces.iisc.ernet.in/energy; http://wgbis.ces.iisc.ernet.in; http://wgbis.ces.iisc.ac.in/energy

E-mail: tvr@iisc.ac.in; energy.ces@iisc.ac.in

Dr. Bharath Haridas Aithal is currently an Associate Professor in Ranbir and Chitra Gupta School of Infrastructure Design and Management at the Indian Institute of Technology Kharagpur. He obtained his Ph.D. from Indian Institute of Science. His area of interest is spatial pattern analysis, Urban growth modelling, natural disasters, geoinformatics, landscape modelling, urban planning, open-source GIS, and digital image processing. He has published over 50 research papers in reputed peer-reviewed international and national journals, 100 papers in the international and national symposiums as well as three books and six book chapters to his credit. https://sites.google.com/view/bharathhaithal/home

Bharath Setturu is currently an assistant professor at Chankya University (School of Mathematics and Natural Sciences) and a post-doctoral fellow at Energy and Wetlands Research Group (EWRG), Centre for Ecological Sciences, Indian Institute of Science, Bengaluru. His research contributions are as follows: 40 research papers in peer-reviewed international journals, five book chapters, 55 papers in national and international conference proceedings, 50+ technical reports, and one book. His work has received a large appreciation across the globe. He is a recipient of "Sahyadri Ecologist (2016 and 2020)" award. He has received the best research paper awards during NSRS-2019 Conference at Bengaluru, CCIVA-2019 Conference at IIT-Kharaghpur, BDCC-2018 Conference at IIT-Kharaghpur, and WEIZ-2017 Conference at Kolkata by IIT-Kharaghpur. He has also received "Dr. Parameshwaran Research Grant" in the year 2016.

S. Vinay is an Associate Professor in the Department of Civil Engineering, Alva's Institute of Engineering and technology, Moodbidri, Karnataka, India. He has received his Bachelors in Civil Engineering from R. V. College of Engineering, Bengaluru, India; Masters in Geoinformatics from Karnataka State Remote Sensing Centre—VTU RRC, Bengaluru, India; and Ph.D. from Visvesvaraya Technological University, Belagavi, India. He was a post-doctoral fellow at RCGSIDM, IIT Kharagpur, India. Since 2013 he is associated as a researcher at EWRG, CES, IISc. His research interest areas are Environmental flows, Hydrology, applications of Remote Sensing and GIS in disaster risk assessment, visualisaiton of landscape, spatial modelling for forecasting, Climate, and Environment aspects. He has published over 100 articles as journals, conferences, book chapters, and conference proceedings. He has been a reviewer in national and international journals and conferences.

K. S. Asulabha is currently a research scholar in Ecoinformatics working in the Microalgae spatial decision support system at the Energy and Wetlands Research Group, CES TE15, Indian Institute of Science, Bengaluru. Her areas of interest are monitoring aquatic ecosystems, water quality assessment, wetland ecosystem evaluation, soil quality assessment, algal culturing, algae, fish, and zooplankton documentation, and open-source GIS. She has published over seven research papers in reputed peer-reviewed international and national journals, 19 papers in international and national symposiums, one strategy paper, 20 technical reports, and two book chapters.

V. Sincy is currently a Research Scholar in Ecoinformatics working in the Ichthyofauna spatial decision support system at the Energy and Wetlands Research Group, CES TE15, Indian Institute of Science, Bengaluru. Her areas of interest are wetland ecosystem evaluation, monitoring of aquatic ecosystems, water quality assessment, soil quality assessment, algal culturing in wastewater, fish, algae, and zooplankton, and open-source GIS. She has published over 6 research papers in reputed peer-reviewed international and national journals, 19 papers in international and national symposiums, one strategy paper, 21 technical reports, and two book chapters.

Ecosystem Extent Account for Karnataka State, India

Abstract Ecosystems (terrestrial and aquatic) are vital for the survival of humans on the Earth. Ecosystems provide an array of goods (food, fodder, etc.) and services (remediation, clean air, water, etc.) that humans depend upon to sustain their lives. The stock of ecosystem resources (biotic and abiotic) that provide goods and services is referred to as natural capital. The services provided by these ecosystems support the economy and society. However, the contributions by the ecosystems or natural capital are often undervalued while making economic decisions. This necessitates accounting of natural resources to minimize overexploitation, mismanagement, etc. The process of accounting stocks and flows of goods and services (natural resources) in a given ecosystem of a region is often termed natural capital accounting. In natural capital accounting, ecosystems are assets that provide ecosystem services to people.

A key feature of ecosystem accounting is its capacity to integrate spatially referenced data about ecosystems, which includes data about the location, size, and condition of ecosystems within a given area and how these are changing over time as per the official international framework for natural capital accounting (The System of Environmental-Economic Accounting, SEEA).

Ecosystem accounting entailed estimation of the extent of different ecosystem assets (EAs) in the study region (district and federal state) using temporal remote sensing data (acquired through space-borne sensors at regular intervals and available since 1970's. This spatial data helps in compiling ecosystem extent accounts and provides an underlying infrastructure for measuring ecosystem conditions and modeling ecosystem services. The analyses of temporal remote sensing data helped in assessing the extent of changes of an ecosystem over time. Ecosystem extent accounts record the total area of each ecosystem, which is classified by type within an ecosystem accounting area and, over time, in a specified area (Karnataka State, India).

The objective of this chapter is to assess the extent of ecosystems for select districts in Karnataka State through temporal remote sensing data with collateral data. This involves the assessment of land use dynamics using temporal remote sensing (RS) data. The latest RS data (2018/19) was analysed using supervised classification techniques, and compared with the land uses of the earlier years to understand the extent of ecosystems, extent of degradation, habitat loss, and deforestation rates.

T. V. Ramachandra et al., *Natural Capital Accounting and Valuation of Ecosystem Services, Karnataka State, India*, https://doi.org/10.1007/978-981-97-2405-5_1

Ecosystem extent through land use analyses reveals that the agriculture ecosystem, with a spatial extent of 127,962 sq.km (66.72%) constitutes the major asset in Karnataka, India. This is followed by plantations (21,325 sq.km., 11.123%), evergreen forests (10,887 sq.km, 5.68%), moist deciduous forests (7892 sq.km, 4.12%), dry deciduous forests (4280 sq.km., 2.23%) and scrub grasslands (4906 sq.km, 2.56%). The State has 5.7% area under evergreen forest cover (2019) compared to 6.5% (2005) and 4% moist deciduous forests compared to 5.2%. The loss of evergreen and deciduous forest cover across the State articulates the anthropogenic pressure, which signifies an increase in monoculture and built-up areas. The State had 1.4% of the area under built-up (2005) compared to 3% in 2019.

District-wise analyses reveal that the evergreen forest cover has lost from 68 to 29% in Uttara Kannada district, and the area under human habitation has increased during the last four decades, evident from the increase of built-up area from 0.38 to 5.2% (1973–2018).

LU analyses of Shimoga highlight the loss of forest cover from 43.83% (1973) to 34% (2018) with the increase of built-up from 0.63% (1973) to 2.35% (2018), plantations (9–30%), industrial and cascaded developmental activities.

Temporal LU analyses highlight that the major portion of Mysore district is under agriculture (66.57% in 2019), and a loss of moist deciduous forest cover from 3.8 to 2.2% is noticed. The built-up cover has increased from 0.3 to 5% due to the growth in Mysore city and its sub-urbans.

The land use analysis of Belgaum district shows loss of evergreen forest cover from 7.75 to 5.64% from 1989 to 2019.

Land use analyses portray changes in deciduous forest cover and an increase in scrub forest cover over four decades in Chamarajanagar district. Recurring fires in the Bandipur region have transformed large tracts of forests from deciduous to scrub cover. The region has shown an increase in agriculture by 37–43%.

The spatio-temporal land use analysis of Kodagu district highlights the loss of evergreen cover and the increase in built-up and monoculture plantation areas. The region had an evergreen cover of 40.47–24.17% by 2019. The increase in resorts, buildings, and other infrastructure developments has resulted in an increase in built-up cover from 0.42 to 2.34%.

The Davangere district has undergone changes in land use in built-up cover from 0.47% to 3.15%, horticulture area from 0.89 to 2.5%, and agriculture area from 83 to 79% from 1989 to 2019. The non-forest cover has increased to 85.31%.

List of Abbreviatons

ADTR	Anshi-Dandeli Tiger Reserve
ESR	Ecologically Sensitive Regions
ETM+	Enhanced Thematic Mapper Plus
FCC	False Colour Composite
GHG	Green House Gas

GIS	Geographic Information System
GLP	Global Land Project
GPS	Global Positioning System
IGBP	International Geosphere-Biosphere Programme
IHDP	International Human Dimensions Programme on Global Environmental Change
IPCC	Intergovernmental Panel on Climate Change
IRS	Indian Remote Sensing
LANDSAT	Land Remote-Sensing Satellite (System)
LISS 4 MX	Linear Imaging Self-Scanning Sensor 4 Multispectral
LULC	Land Use Land Cover
MoEFCC	Ministry of Environment and Forests and Climate Change
MSS	Multi-Spectral Scanner
NDVI	Normalized Difference Vegetation Index
NIR	Near Infrared
NRSC	National Remote Sensing Agency
QGIS	Quantum GIS
RS	Remote Sensing
SOI	Survey of India
TM	Thematic Mapper
USGS	United States Geological Survey

1 Introduction

Ecosystems (terrestrial and aquatic) are vital for the survival of humans on Earth. Ecosystems provide an array of goods (food, fodder, etc.) and services (remediation, clean air, water, etc.) that humans depend upon to sustain their lives. The stock of ecosystem resources (biotic and abiotic) that provide goods and services is referred to as natural capital. The services provided by these ecosystems support the economy and society. However, the contributions by the ecosystems or natural capital are often undervalued while making economic decisions. This necessitates the accounting of natural resources to minimize overexploitation, mismanagement, etc. The process of accounting stocks and flows of goods and services (natural resources) in a given ecosystem of a region is often termed natural capital accounting. In natural capital accounting, ecosystems are assets that provide ecosystem services to people.

Landscapes are composed of many dynamic components (landscape elements/ ecosystems) with their dynamics. A natural landscape has complex ecological, economic, and cultural qualities on which humans and other life forms depend directly. Landscape is the heterogeneous land area of interacting systems that forms an interconnected system called ecosystems (Forman and Gordron 1986). The functional aspects (interaction of spatial elements, cycling of water and nutrients, bio-geo-chemical cycles) of an ecosystem depend on its structure (size, shape, and configuration) and constituent's spatial patterns (linear, regular, aggregated). The status of a particular landscape is derived from land use land cover [LULC] information. Temporal land use and land cover information of a region provides a base for accounting the natural resources availability and their utilization. The information pertaining to LULC provides a framework for decision-making towards sustainable natural resources management. The analysis of the LULC change addresses issues like climate change, deforestation, soil erosion by water and wind, salinization etc.

Land use, Land cover [LULC] dynamics: Land cover [LC] relates to the discernible Earth surface expressions, such as vegetation or non-vegetation (soil, water or anthropogenic features), indicating the extent of Earth's physical state in terms of the natural environment (Lambin et al. 2001; Ramachandra et al. 2012). Variations in topography, vegetation cover, and other physical characteristics of the land surface influence surface-atmosphere fluxes of sensible heat, latent heat, and momentum of heated air particulates caused by conduction, convection and radiation, influencing weather and climate. Land use [LU] expresses human uses of the landscape, e.g., for residential, commercial, or agricultural purposes. Land cover changes induced by human and natural processes play a significant role at global and regional scale patterns of the climate and biogeochemistry of the Earth system. Land use information is vital for regional planning and management activities and has been considered essential for modeling and understanding the Earth as a system (Ramachandra and Shruthi 2007).

Land use changes alter the homogeneous landscape into a heterogeneous mosaic of patches. The LULC changes are due to natural as well as human-induced alterations. These changes are highly dynamic and characterized by a heterogeneous landscape facilitating socioeconomic–environmental interactions. Natural events such as weather, flooding, fire, climate fluctuations, and ecosystem dynamics initiate changes in land cover. Globally, land cover is altered principally by direct anthropogenic use such as agriculture, livestock raising, forest harvesting & management, population change, urbanisation, and other developmental activities (Meyer 1995). Natural disturbances tend to alter forest landscape patterns differently from anthropogenic impacts (Mladenoff 1993), human-induced impacts are quantified as more deleterious effects between patches as compared with natural changes (Hudak et al. 2007). The undisturbed (or wilderness) areas represent only 46% of the Earth's land surface (Mittermeier et al. 2003). Forests covered about 50% of the Earth's land area 8000 years ago, as opposed to 30% today (Ball 2001).

Land use categories in a region depend on the agroecological aspects. Hence, in districts located in hot dry arid or hot dry semi-arid and arid, do not have all categories of land use (especially forest classes). Hence, land use/land cover categories were classified based on the agroecological and agro-climatic zones, using a supervised classification technique based on a Gaussian maximum likelihood algorithm with training data (collected from the field using GPS).

A key feature of ecosystem accounting as per the official international framework for natural capital accounting (The System of Environmental-Economic Accounting, SEEA) is its capacity to integrate spatially referenced data of ecosystems (i.e., data about the location, size, and condition of ecosystems) within a given area and account how these are changing.

Ecosystem accounting entails estimation of the extent of different ecosystem assets (EAs) in the study region (district and federal state) using remote sensing data (acquired through space-borne sensors at regular intervals and available since the 1970s. The spatial information helped to compile an ecosystem extent account and provided an underlying infrastructure for measuring ecosystem conditions and modeling ecosystem services. The temporal data analyses help assess the extent of changes in an ecosystem over time. Ecosystem extent accounts record the total area of each ecosystem, which is classified by type within an ecosystem accounting area and, over time, in a specified region (district, Karnataka State, India),

The objective of the current chapter is to compile ecosystem extent accounts of Karnataka State across districts using temporal remote sensing data, collateral data and field data (collected using pre-calibrated global positioning system from districts, chosen based on agro-climatic zones) in Karnataka State.

2 Study Region: KARNATAKA STATE, INDIA

Karnataka is one of the four southern states of Peninsular India and came into existence with the States Re-organisation Act (1956, Nov 1). Extending 760 km N–S (11°34′ N and 18°27′ N) and 420 km E–W (74°3′ E and 78° 34′ E), Karnataka has a spatial extent of 1,91,846 sq.km, which accounts for 5.8% of India's geographical area (Fig. 1).

Karnataka is bounded by the Arabian Sea and the Laccadive Sea on the west, Goa on the north-west, Maharashtra on the north, Telangana on the north-east, Andhra

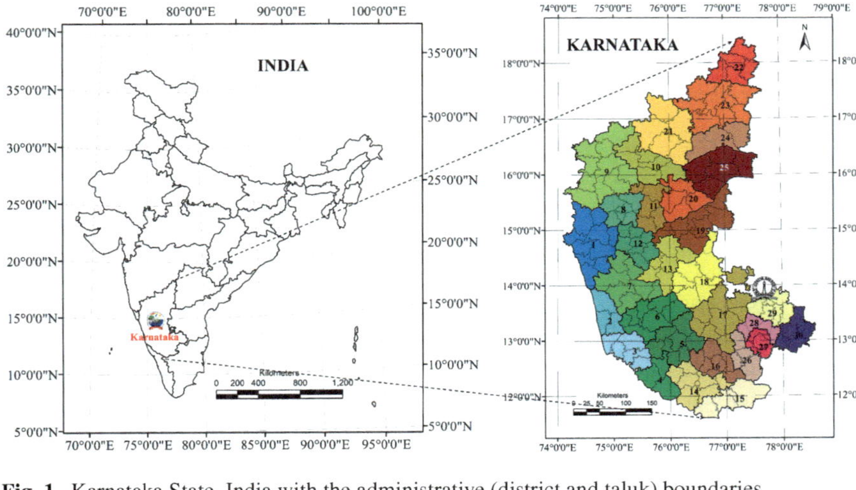

Fig. 1 Karnataka State, India with the administrative (district and taluk) boundaries

Pradesh on the east, Tamil Nadu on the south-east and Kerala on the south-west. It is situated on a tableland where the Western and Eastern Ghats' ranges converge into the Nilgiri hill complex.

Karnataka is divided into 30 Districts, which consist 178 Sub-districts (taluks), with 367 towns and 27,397 villages (Table 1 and Fig. 1) according to the Census of India.

Table 1 Karnataka administrative divisions (Census 2011)

Sl. no	District	Area km^2	Taluks	City/Town	Villages
1	Uttara Kannada	10,306	11	21	1243
2	Udupi	3573	3	21	233
3	Dakshina Kannada	4850	5	42	331
4	Kodagu	4105	3	5	291
5	Hassan	6821	8	14	2418
6	Chikkamagaluru	7214	7	9	1022
7	Shimoga	8479	7	9	1444
8	Dharwad	4258	6	6	361
9	Belagavi	13,392	10	34	1263
10	Bagalkot	6567	7	15	613
11	Gadag	4658	5	9	322
12	Haveri	4821	7	10	696
13	Davanagere	5919	6	6	800
14	Mysore	6321	7	20	1199
15	Chamarajanagar	5636	4	5	428
16	Mandya	4946	7	9	1368
17	Tumkur	10,600	10	12	2582
18	Chitradurga	8436	6	9	948
19	Bellary	8457	7	13	522
20	Koppal	5578	4	6	595
21	Vijayapura	10,965	5	6	679
22	Bidar	5446	5	8	595
23	Kalburgi	10,507	7	13	871
24	Yadgir	5282	3	7	487
25	Raichur	8468	5	9	815
26	Ramanagara	3524	4	6	820
27	Bengaluru (Urban)	2193	4	19	562
28	Bengaluru (Rural)	2298	4	8	957
29	Chikkaballapura	4245	6	8	1324
30	Kolar	3981	5	8	1608

Belagavi district is the largest district with a land area of 13,392 sq.km, and Bengaluru urban district has the smallest area of 2193 sq.km. Tumkur and Hassan have the largest number of villages, i.e., 2582 and 2418 respectively.

Forest Resources: Karnataka State has 3.83 million ha of recorded forest cover, covering about 20% of its geographical spread. Having been endowed with the most magnificent forests in the country, it harbours the Western Ghats region, one of the 36 global priority hotspots for conservation, resulting in a significant variety of flora and fauna, a significant part of them being endemic and threatened species. The forest ecosystem of Karnataka is unique and highly diverse. It forms an important component of the natural resources of the environment. Different forest ecosystems result from the interplay of topographic, climatic and edaphic differences influenced by altitude and the distance from the sea. Forest types include tropical evergreen, semi-evergreen, moist deciduous, dry deciduous, thorny scrubs, sholas, and coastal mangroves, which account for the second-largest land use after agriculture. The total forest cover in the State is 43,356.47 sq.km (2016–17). i.e., about 22.61% of the State's geographical area is under forest cover. Of the total forests, reserve forest constitutes 15.48%, protected forest constitutes 1.85%, village forest constitutes 0.03%, unclassified forest constitutes 5.23%, and private forest constitutes 0.03%.

The State's forest resources are under severe pressure, with a drastic fall in dense forest cover area between 2001 and 2015. The State's forest cover has slightly declined compared to the country's forest cover during the period. Increased deforestation and degradation of environmental resources have serious implications for the ecosystem's production and resilience. The loss of forest cover seriously threatens the environment, sustainable development, and the livelihoods of millions of people in the State. Forest resources significantly contribute to the State's Gross Domestic Product (GDP) by being a major source of timber, medicinal plants, NTFPs, grazing, recreational activities, carbon sequestration, watershed provisions, etc. The State has formed 4467 Biodiversity Management Committees at Grama Pachayat level as per the provisions of the Biological Diversity Act of 2002, (BDA 2002, Government of India) to protect and develop biodiversity. Biodiversity heritage sites (such as the 400-year-old tamarind grove at Nallur, Devanahalli taluk) are being protected to conserve and develop unique genetic bio-diversity.

Karnataka has a repository of rich biodiversity with more than 1.2 lakh known species, including 4,500 flowering plants, 800 fishes, 600 birds, 160 reptiles, 120 mammals and 1,493 medicinal plants. 50% of Western Ghats' biodiversity is present in Karnataka. These forests of Karnataka support a wide range of flora and fauna (biodiversity) through a network of well-connected and protected Wildlife Sanctuaries and National Parks. The State has 5 National Parks and 30 Wildlife Sanctuaries covering an area of 9,586.02 sq km. Apart from the National Parks and Sanctuaries, the State has 15 conservation reserves and 1 community reserve comprising 652.369 sq km. All these areas form 23.59% of the total forest area. These are spread over evergreen forests to scrub forests, representing diverse ecosystems with rare and endangered species of plants, animals, and birds. The State has been actively formulating and implementing various programmes to develop forests and protect its natural environment. Among the Forest Department's schemes concerning wildlife and national parks, Long Term Measures to mitigate Man-Animal Conflict incurred an expenditure of 24.80%, Project Tiger 30.40%, Integrated Development of Wildlife Habitats 2.47%, Nature conservation activities attracted 13.38% and Rs. 27.50 crores of total expenditure were incurred towards Voluntary Rehabilitation of families from Tiger Reserves and National parks during 2016–17.

Topography: Karnataka comprises varied topographical structures that include high mountains, plateaus, residual hills, and coastal plains. It is enclosed by chains of mountains in its west, east, and south. The Western Ghats generally exhibit a narrow coastal plain followed to the east by small and short plateaus at different altitudes, then suddenly rising to great heights, followed by the gentle east and east-north-west sloping plateau. The entire landscape of the State rests on undulating terrain, broken up by various mountain ranges and deep ravines, though it mainly consists of a plateau with elevations ranging between < 0 m and >1900 m AMSL and slopes between 0 and 65° (Fig. 2). The thin strip coastal plains facing the Arabian Sea along the west coast districts of Uttara Kannada, Udupi, and Dakshina Kannada have flat slopes with elevations ranging just over 50 m and slopes less than 10°. There are a few high peaks in Western and Eastern Ghats systems with more than 1,500 m altitudes. Among the tallest peaks of Karnataka are the Mullayyana Giri (1,925 m), Bababudangiri (Chandradrona Parvata 1,894 m), and the Kudremukh (1,895 m) from the Chikmagalur district and the Pushpagiri (1,908 m) in the Kodagu district. These Ghats have highly undulating terrain with slopes ranging over 30°. The Deccan plains

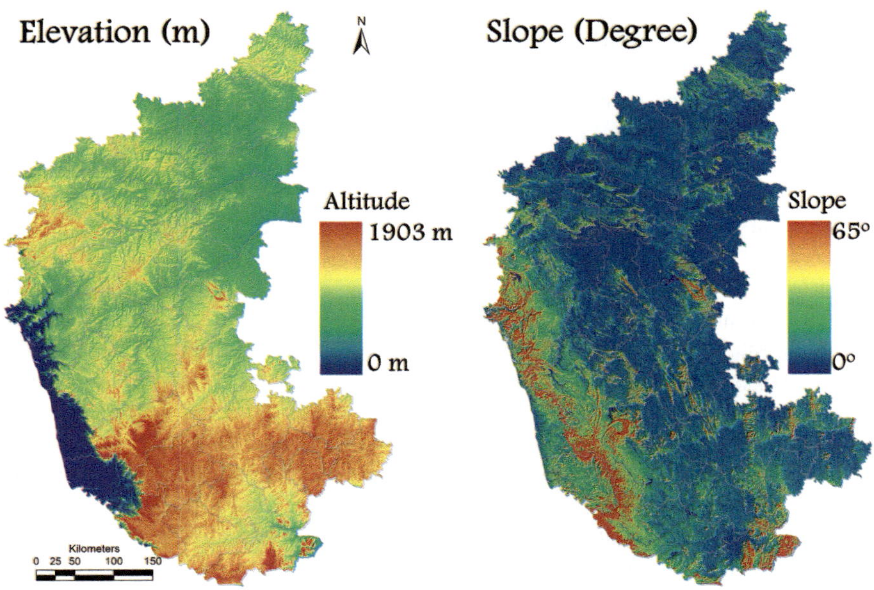

Fig. 2 Karnataka—topography

Table 2 Agroecological zone and distinct feature

Agroecological zone	Geographical area covered	Distinct feature
Karnataka plateau, arid	Northern part of Karnataka, Northwest part of Karnataka, Southern parts of Karnataka, Eastern part of Karnataka	90 days length of growing period (LGP)
Karnataka plateau, moist semi-arid		90–120 days LGP
Karnataka plateau, hot dry semi-arid	The northern part of Karnataka	120–150 days LGP
Karnataka plateau, hot moist semi-arid	Southern parts of Karnataka	150–180 days LGP
Karnataka plateau, hot dry sub-humid	The coastal part of Karnataka	180–210 days LGP
Western Ghats, hot moist sub humid	Hilly region like the Western Ghats	210–240 days LGP
West coast plain, hot humid	The coastal part of Karnataka	240–270 days LGP

expanding to the east are flat with slopes less than 10° with altitudes ranging over 700 m.

Agroecological Zones: Regions with similar geographic, edaphic, and meteorological characteristics and duration of crop growing period (LGP: Length of the growing period) are grouped and referred to as agro-climatic zones. The State is divided into seven agroecological zones at the macro level based on physiography, soil, bio-climate, and details in Table 2. Agroecological zones, district-wise and at the decentralized level grid-wise, are presented in Fig. 3.

Based on physiography, soil types, crops, and cropping pattern, the State has been divided into ten agro-climatic zones (North-eastern transition zone, North-eastern dry zone, Northern dry zone, Central dry zone, Eastern dry zone, Southern dry zone, Southern transition zone, Northern transition zone, Hilly zone, Coastal zone) are depicted in Fig. 4 (district-wise and at grid-level).

- Coastal zone includes districts like Dakshina Kannada, Udupi, Uttara Kannada;
- The hilly zone includes districts like Belgaum, Shivmogga (Shimoga), Chikkamagaluru, Madikeri, Kodagu, Hassan;
- The northeastern transition zone includes Bidar and parts of Gulbarga;
- Northeastern dry zone include Gulbarga, Yadgiri, and parts of Raichur;
- The northern dry zone includes Bellary, Bijapur, Dharwad, Raichur;
- The central dry zone includes Chitradurga, Tumkur, and some parts of Hassan and Chikkamagaluru;
- The eastern dry zone includes Bengaluru, Kolar, Ramanagaram, Bengaluru rural;
- The southern transition zone includes Hassan, Shimoga, and parts of Mysore;
- The northern transition zone includes Belgaum and Dharwad.

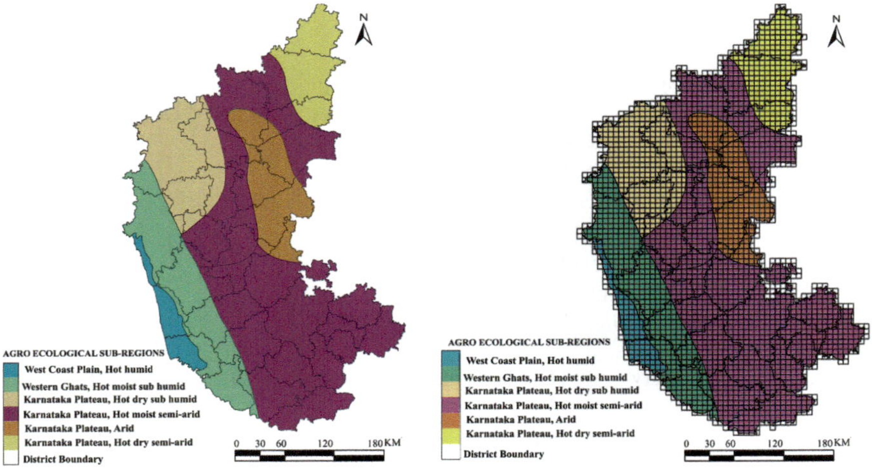

Fig. 3 Agro ecological sub-regions in Karnataka

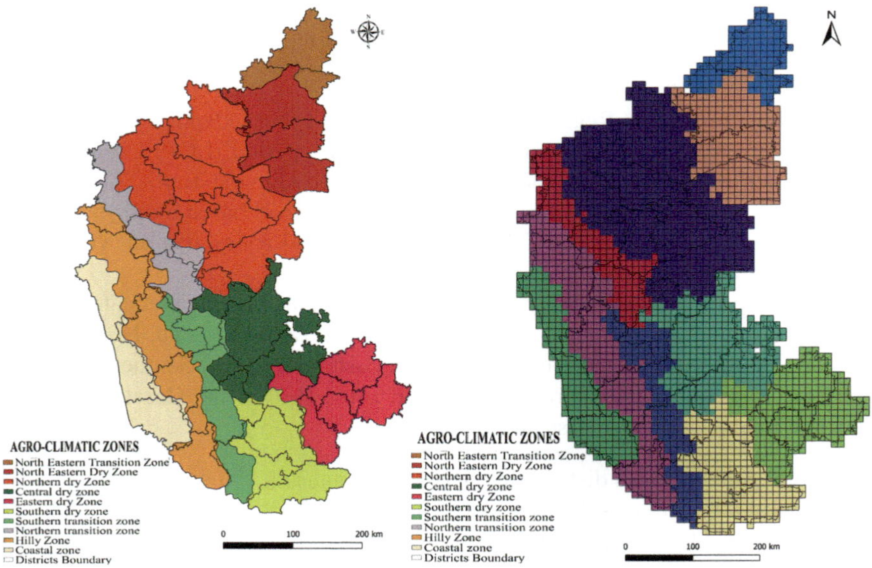

Fig. 4 Agro-climatic zones in Karnataka

Water Resource: Karnataka has seven river systems and tributaries flowing through the State. The major river basins of Karnataka are Krishna (59.48%), Cauvery (17.99%), West Flowing Rivers (12.76%), North Pennar (3.64%), Godavari (2.31%), South Pennar (2.29%), Palar (1.56%). Karnataka has 26 east-flowing rivers and 10 west-flowing rivers. The west-flowing rivers of Karnataka provide 60% of the State's inland water resources. Figure 5 depicts various water bodies of Karnataka categorised under rivers (rivers, reservoirs, and estuaries) and lakes.

The State has over 12 highly productive estuaries along its west coast, namely Kali, Bedti, Aghanashini, Sharavathi, Venkatapura, Chakra, Varahi, Netravati, Barpole, Payaswini, Sita, Souparnika, and numerous small creeks. Based on the water availability in the region, the rulers/administrators in the past have created interconnected lakes to sustain the regional water requirements as a result, the transition zone and the Deccan traps have a large number of lakes. Karnataka has over 39,000 water

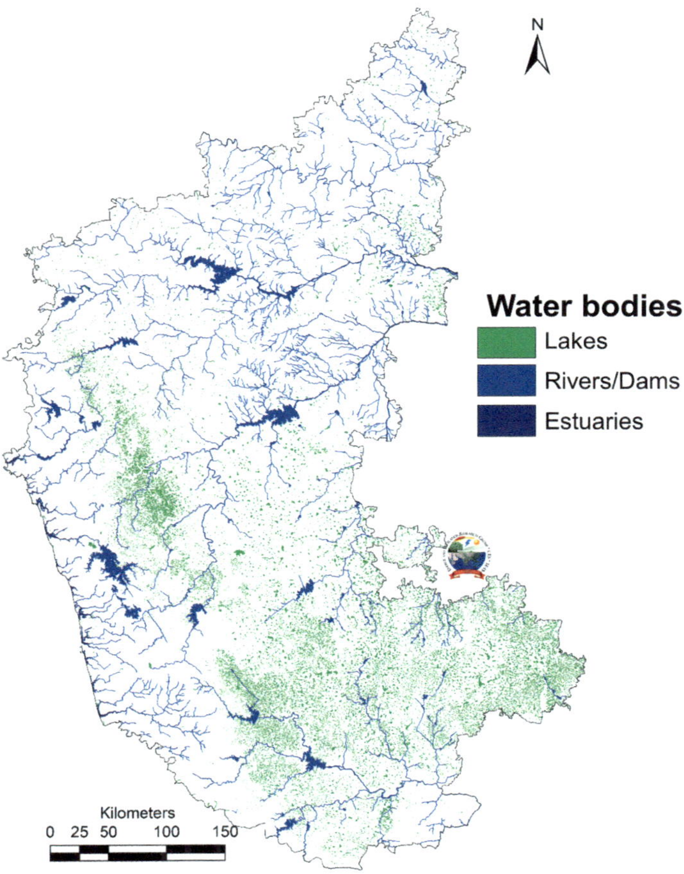

Fig. 5 Water resources of Karnataka

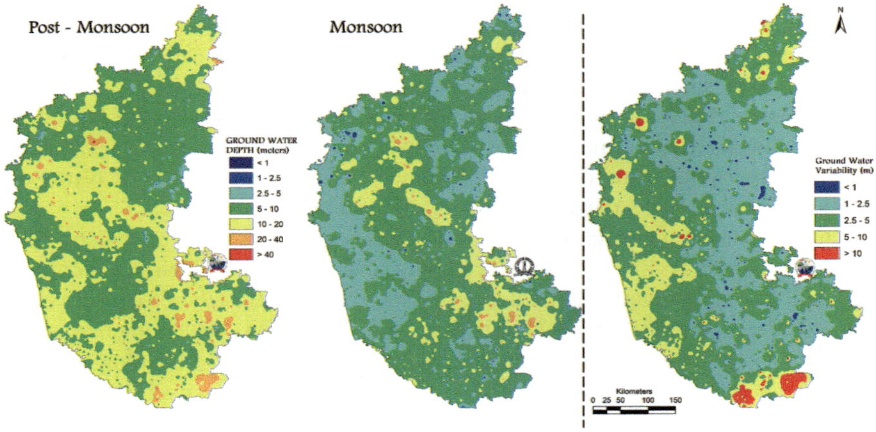

Fig. 6 Ground water dynamics

bodies with areas ranging from 2500 sq. m to more than ten sq.km. In the recent past, dams have been constructed in the State to encourage irrigation and power generation. Karnataka has more than 15 major reservoirs namely Alamatti, Bhadra, Varahi, Hemavarhi, Kabini, KRS, Harangi, Ghataprabha, Malaprabha, Narayanapura, Supa, Linganamakki, Kadra, Gersoppa, Chakra, Mani, Kodasalli, Renuka Sagar, Vanivilasa Sagara, Basavaragara.

Groundwater: The State is covered by peninsular gneisses, granites, schists, and basalts, along with sedimentaries of Kaldagi and Bhima groups. The recent alluvium is restricted to coastal areas and stream courses. The extent of weathering and fracturing primarily controls the water-bearing and yield characteristics in hard rock. In limestone areas, solution cavities impart secondary porosity. The yield of tube wells tapping hard rocks is as high as 50 m^3/hr. The tube wells in sedimentaries can yield up to 15 m^3/hr. Groundwater depth (Fig. 6) in most parts of the State goes beyond 10 m (BGL) in the post-monsoon season, while in the monsoon, groundwater depth raises to less than 10 m (BGL). Topography, lithology and soil play a major role in holding the groundwater. At the coasts, the groundwater depth varies over 2.5 between post-monsoon and monsoon, while in the Deccan plains, the groundwater depth variations are less than 2.5 m. Very high depths were observed in cities such as Bangalore, Belagavi, etc.

Geology: Four main types of geological formations (Fig. 7) are: (i) The Archean complex made up of Dharwad schists and granitic gneisses: These cover around 60% of the area of the State and consist of gneisses, granites, and charnockite rocks. Some of the minerals found in this region are (i) dolomite, limestone, gabbro, quartzite, pyroxenite, manganese and iron ores, and metabasalt, (ii) The Proterozoic non-fossiliferous sedimentary formations of the Kaladgi and Bhima series: The Kaladgi series has horizontal rocks that run for 160 km in the districts of Belgaum,

Raichur, Dharwad and Bijapur districts. The Bhima series that is present on either side of the Bhima River consists of rocks containing sandstone, limestone, and shale, and this is present in the Gulbarga and Bijapur districts, (iii) The Deccan trappean and inter trappean deposits: This is a part of the Deccan traps which were formed by the accumulation of basaltic lava. This is made up of greyish to black augite-basalt, and (iv) The tertiary and recent laterites and alluvial deposits: Laterite capping are found over the Deccan traps and were formed after the cessation of volcanic activity in the early tertiary period. These are found in many districts in the Deccan plateau and in the coast.

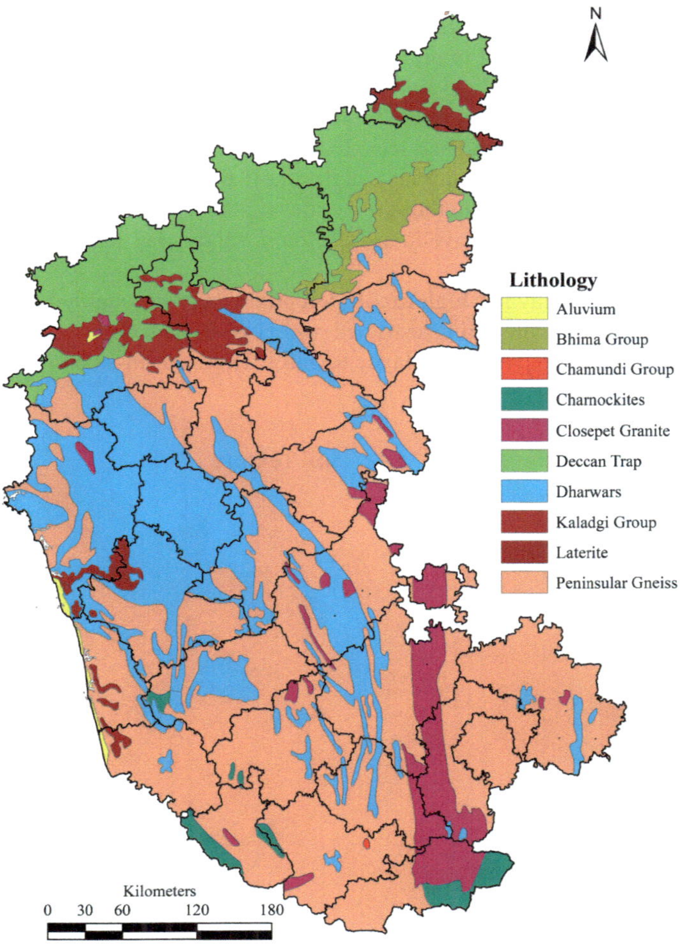

Fig. 7 Geology of Karnataka

Soil: Soil can be grouped into nine categories: red sandy soils, red loamy soils, shallow black soils, medium black soils, deep black soils, mixed red and black soils, laterite soils, laterite gravelly soils, and coastal alluvium of the orders Alfisols, Andisols, Aridisols, Entisols, Histosols, Inceptisols, Mollisols, Oxisols, Spodosols, Ultisols, Vertisols. Figure 8 describes soils based on their texture; most of Karnataka is dominated by loamy and fine soils towards the Ghats and Deccan plains. The other major soil types (based on soil texture) are Clayey, Clayey Sandy, Loamy Sandy, Sandy, which are spread across the State. Based on the soil texture, the water holding capacities vary and are categorised under Soil Hydrological Group, namely A, B, C and D (Fig. 8).

1. Group A: Sandy: Low runoff potential when thoroughly wet, Less than 10% clay and more than 90% sand or gravel and have gravel or sand textures
2. Group B: Sandy Loamy and Loam: Moderately low runoff potential when thoroughly wet, Water transmission through the soil is unimpeded, 10% and 20% clay and 50% to 90% sand and have loamy sand or sandy loam textures
3. Group C: Clay Loam: Moderately high runoff potential when thoroughly wet, Water transmission through the soil is somewhat restricted. 20 and 40% clay and less than 50% sand and have loam, silt loam, sandy clay loam, clay loam, and silty clay loam textures.
4. Group D: Clay: Clay has high runoff potential when thoroughly wet, and water movement through the soil is restricted or very restricted. Greater than 40% clay, less than 50% sand, and have clayey textures. High shrink-swell potential.

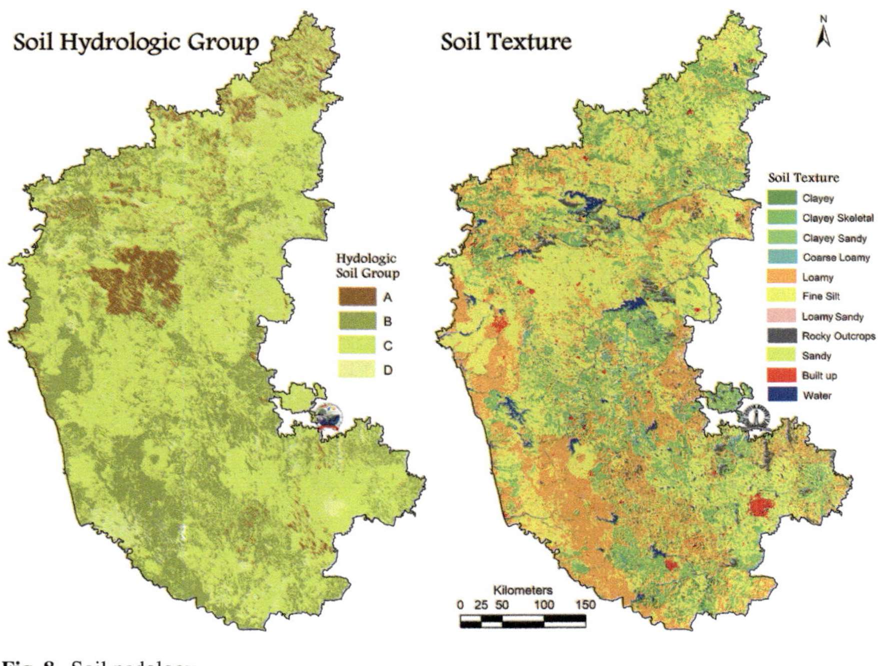

Fig. 8 Soil pedology

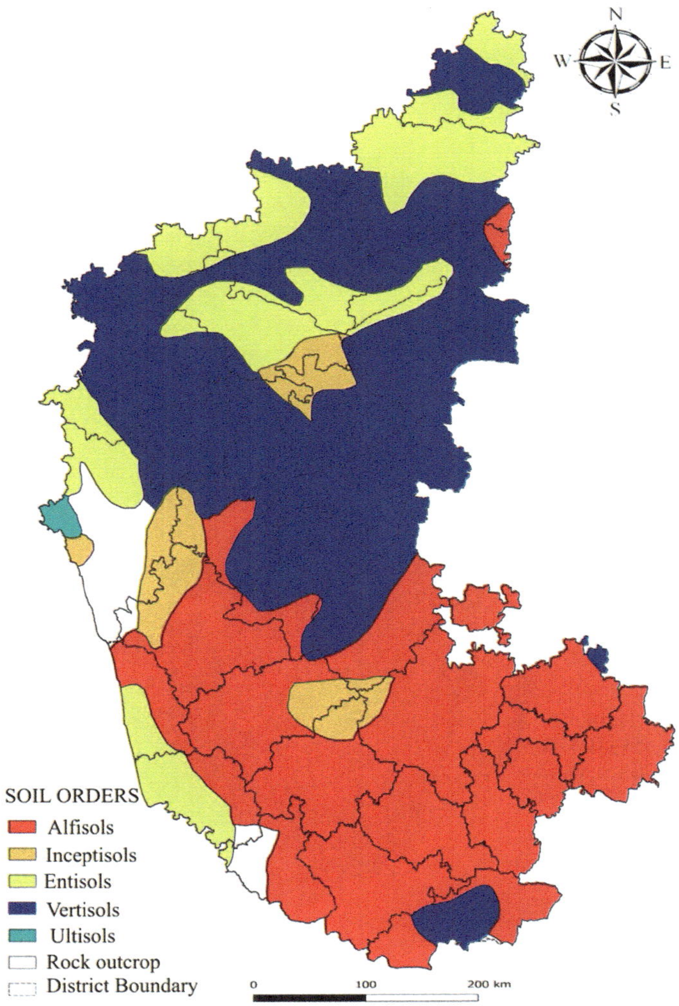

Fig. 9 Soil order in Karnataka

The soils of Karnataka state were classified as per soil Taxonomy and traditional classification. The soil of Karnataka belongs to 7 orders, 12 suborders, 27 great groups, 47 subgroups, and 96 soil families. Of the total area of Karnataka, 27% is covered by Alfisols, 25% by Inceptisols, 16% by Entisols, 15% by Vertisols, 8% by Ultisols, 5% by Aridisols and 1% by Mollisols. Southern parts of Karnataka are mainly of Alfisols. Vertisols cover central parts of Karnataka, the Northern region of Karnataka is of Entisols. Figure 9 depicts the soil order in Karnataka. Figure 10 illustrates the Soil Classification, which highlights that in north region consists of

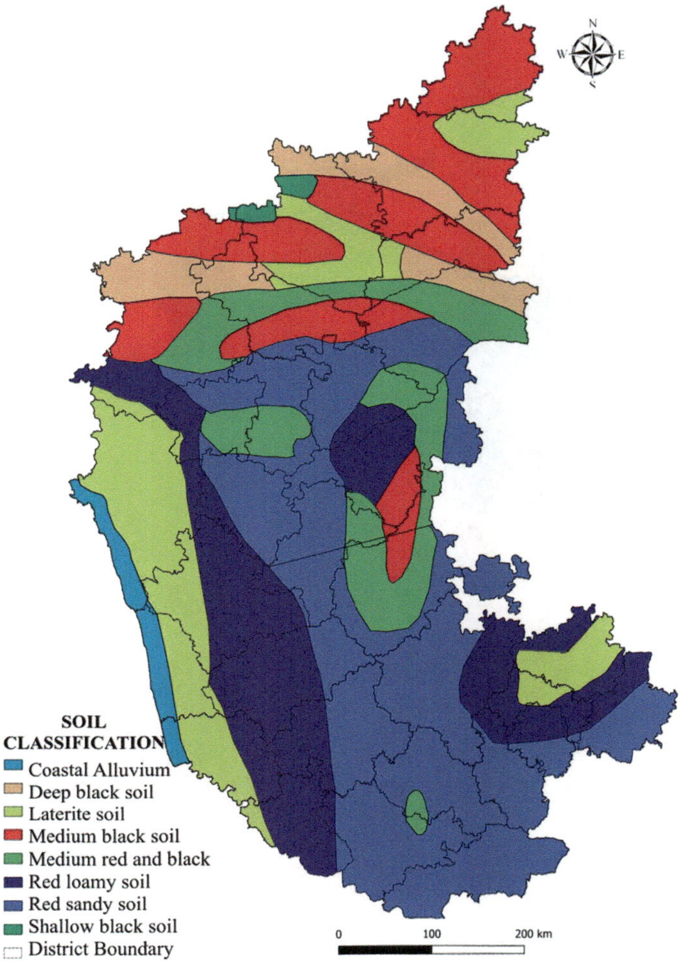

Fig. 10 Soil classification in Karnataka

black soil, the southern and central part is mainly of red soil and the coastal region of Karnataka is mainly of coastal alluvium soil.

Soil depth: Figure 11 depicts the soil depth across the Karnataka region. The Karnataka region's southern and central part has a deep and moderately deep soil of about 75–100 cm; the Northwestern and northern part of Karnataka has a shallow soil depth ranging < 10–25 cm. In some regions, it is moderately shallow.

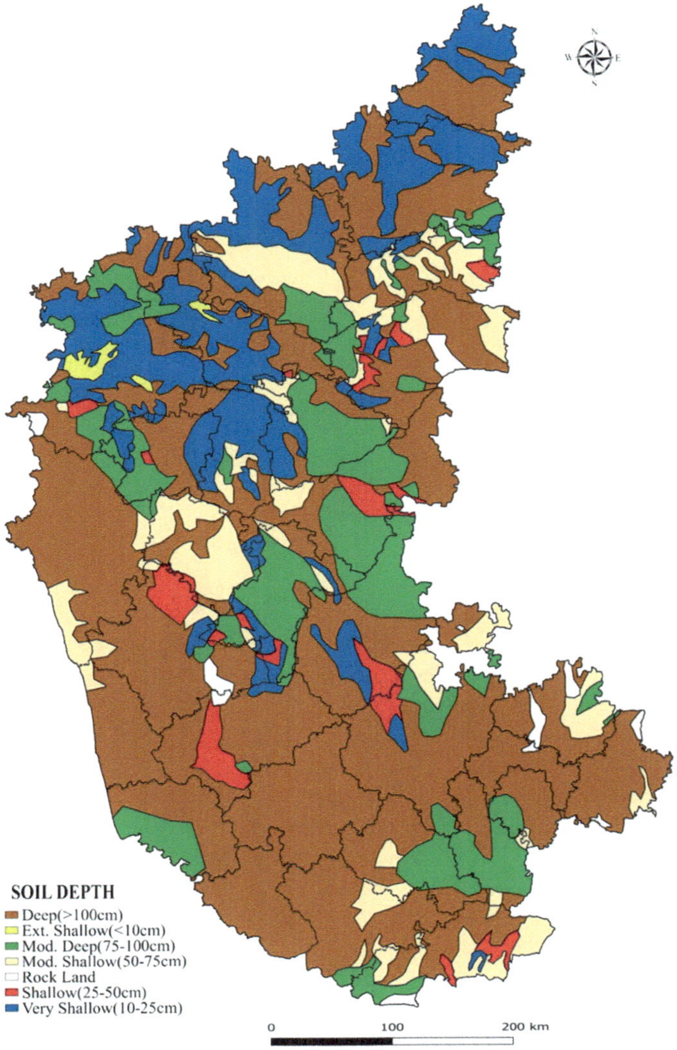

Fig. 11 Soil depth in Karnataka

Soil Gravelliness: Figure 12 depicts the Soil Gravelliness in Karnataka state, which indicates that districts Gulbarga, Bidar, Tumkur, Bengaluru, Bijapur has slight or moderately gravelly soil (<15 to about 35%) and Mangalore, Karwar, Hassan, Kolar, Bellary has strongly gravelly soil (35–80%).

Climate: Karnataka experiences hot and moist summer, cool and dry winters due to the tropical monsoon climate. The State has diverse climatic patterns due to the influence of geographical location, oceans, physical features, vegetation, and

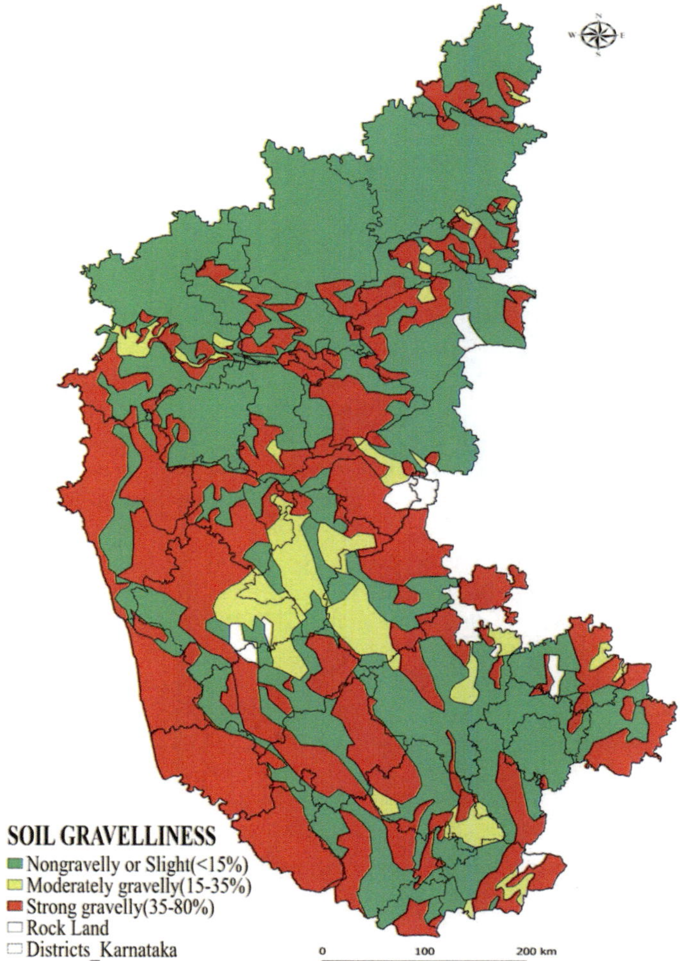

Fig. 12 Soil Gravelliness status

monsoon winds. Karnataka has a total of four climates. The winter season prevails from January to February. During the winter season, temperature, humidity, and rainfall are low. The average temperature is between 25 and 27 °C. The minimum temperature recorded in Belagavi is 6.7 °C. There is a low temperature in other northern districts, i.e., Vijayapura, Kalaburagi, and Bidar. This is followed by the summer season from March to May. The summer season is very hot, dry, and humid.

The temperature rises uniformly throughout the State and reaches the maximum from April to May. Raichur has recorded a temperature of 45.6 °C, and it is the hottest place in Karnataka. The temperature is high even in the other districts of the northern

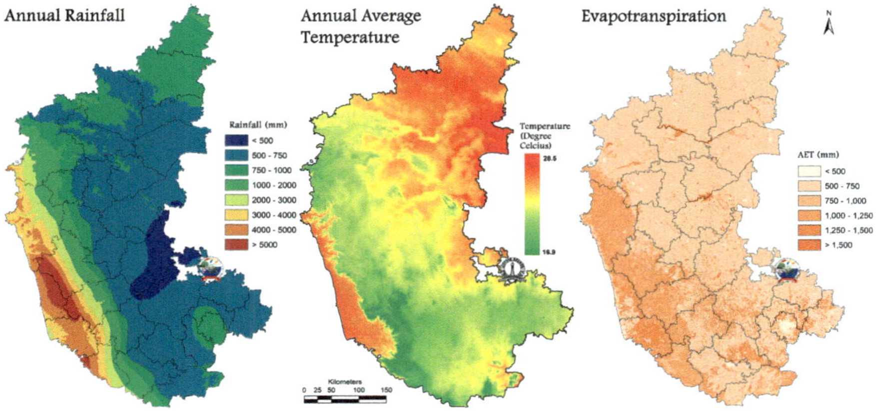

Fig. 13 Climate in Karnataka

plains. They are then followed by the monsoon season (from June to September) and, sub-sequentially, the post-monsoon (that lasts for the three months of October to December). The Western Ghats receive the highest rainfall over 3000 mm, due to the moisture-laden winds blowing from the Arabian Sea. This is followed by coastal plains with rainfall ranging between 2000 and 3000 mm. This was followed by the transition zones between the Western Ghats and the Deccan plains, wherein the rainfall ranges between 1000 and 3000 mm. The Deccan plains receive rainfall up to 1000 mm. Lowest rainfall was observed in the district of Chitradurga, with precipitation less than 500 mm. Southwest monsoon caters to over 70% of the rain received in the State (Fig. 13). Retreating Monsoon Season is also called the North-East Monsoon season. Due to the decrease in temperature and pressure, the winds blow from the north-east to the south-west. These winds are dry because they blow from the land. The winds that blow from the Bay of Bengal bring a little rain to the southern maidan's districts. Besides, due to the cyclones coming from the Bay of Bengal during November–December, rainfall occurs. During this season, the State receives about 12% of the precipitation.

The long-term annual average temperature in the State of Karnataka is depicted in Fig. 13. The temperature in Karnataka varies between 16.9 and 28.5 °C. The transition zones between the Ghats and the Deccan plains have lower temperatures ranging less than 18 °C. The Coastal Plains and the northeast and eastern parts of the Deccan Plains have higher average annual temperatures over 25 °C.

Demography: The population dynamics of Karnataka are depicted in Table 3 and Fig. 14. Taluk level population were collated from the District at a glance and Census of India. Karnataka is the eighth largest State by population, forming 5.1% of India's population, with a growth rate of 15.7% from 2001 to 2011. The State's total population is 6.1 crores, of which male (31,057,742) and female (30,072,962), respectively, with a density of 319 persons per sq km. The population's growth rates in rural and urban areas are 6.49 and 27.16%, respectively. Bangalore Urban District accommodates 11.59% of the state population, which exhibited the highest growth rate in the urban population at 46.68%. The State has a birth rate of 2.2%, a death rate of 0.72%, and an infant mortality rate of 5.5%, and a maternal mortality rate of 0.195%. The total fertility rate of the State is 2.2. The State has a 75.6% literacy rate.

Table 3 Population dynamics in 2001, 2011 and 2021

District	2001	2011	2021*
Bagalkot	1,651,892	1,889,752	2,162,334
Bangalore rural	850,968	990,923	1,155,649
Bangalore urban	6,629,636	9,621,551	14,066,760
Belgaum	4,214,505	4,779,661	5,425,978
Bellary	2,027,140	2,452,595	2,975,288
Bidar	1,502,373	1,703,300	1,932,227
Bijapur	1,806,918	2,177,331	2,627,733
Chamarajanagar	965,462	1,020,791	1,079,330
Chikballapur	1,149,007	1,255,104	1,371,243
Chikmagalur	1,140,905	1,137,961	1,136,720
Chitradurga	1,517,896	1,659,456	1,815,242
Dakshina Kannada	1,897,730	2,089,649	2,302,443
Davangere	1,790,952	1,945,497	2,116,812
Dharwar	1,604,253	1,847,023	2,131,178
Gadag	971,835	1,064,570	1,166,583
Gulbarga	2,174,742	2,566,326	3,035,650
Hassan	1,721,669	1,776,421	1,835,717
Haveri	1,439,116	1,597,668	1,773,991
Kodagu	548,561	554,519	560,631
Kolar	1,387,062	1,536,401	1,702,729
Koppal	1,196,089	1,389,920	1,616,467
Mandya	1,755,212	1,805,769	1,859,496
Mysore	2,641,027	3,001,127	3,427,465
Raichur	1,669,762	1,928,812	2,233,133
Ramanagara	1,030,546	1,082,636	1,138,947
Shimoga	1,642,545	1,752,753	1,875,975
Tumkur	2,584,711	2,678,980	2,786,076
Udupi	1,112,243	1,177,361	1,246,320
Uttara Kannada	1,353,644	1,437,169	1,528,709
Yadgir	956,180	1,174,271	1,442,157

*Predicted population

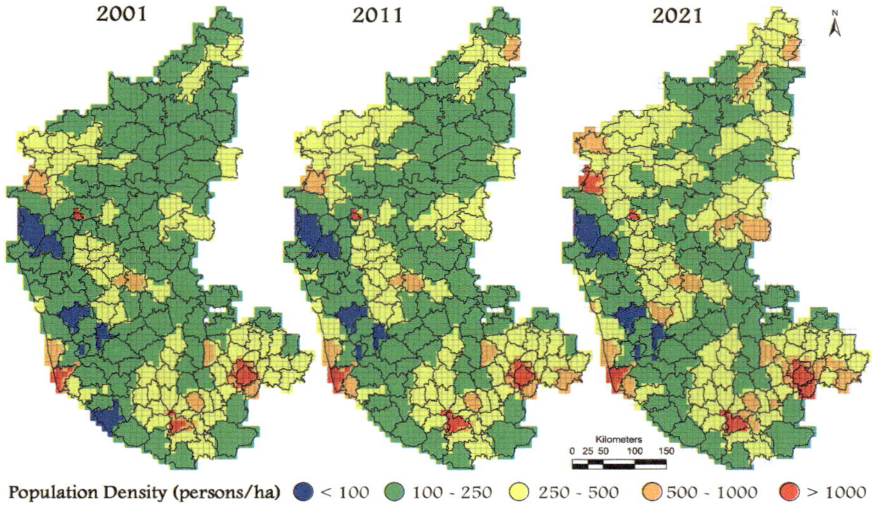

Fig. 14 Population dynamics-temporal population density across taluks in Karnataka

Livestock: Livestock is a prime sector supporting enormous Indian rural populations and contributes to India's economy. The livestock sector offers self-employment opportunities to the people by contributing 16% to small farm households' income against an average of 14% for all rural households. This sector contributes 4.11% GDP through various products such as milk and milk products, eggs, meat products, raw materials, etc. This sector also provides non-edible by-products besides essential proteins and nutritious products. The 20th Livestock census was carried out in 2017, which shows an increase of 12% in poultry and significantly contributes to the Animal Husbandry sector with higher exports and imports. But, indigenous species of cattle, sheep, and pigs declined between 2012 and 2017 while the exotic crossbred population increased, reflecting the market's influence. The union government has substantially supported this sector by providing monetary benefits of 23,538.298 lakh rupees for 2016–17 under the National Livestock Mission (NLM), which commenced in 2014–15. The total benefit of 62,592.157 lakh rupees was reported for the year 2014–17 under various schemes such as Rural Backyard beneficiaries assisted; distribution of Hand driven chaff cutter; distribution of Power-driven chaff cutter; fodder seed procurement & distribution in quintals; fodder production from the forest (in hectare); health cover and Entrepreneurship Development & Employment Generation (EDEG), etc. Karnataka's Animal Husbandry, Poultry, and Dairying is one of the most successful initiatives undertaken to increase farmer's income through well-developed vast infrastructure based on a cooperative network. This network covers milk procurement, processing, and marketing and promotes dairy farming and veterinary services. The daily procurement of milk in the State is over 6.35 million tons, which results in farmers getting an income of over Rs. 5000 crores per year, supporting more than 25 lakh rural households. Women play a significant role in the generation of income through dairy farming. Table 4 lists district-wise livestock density, and livestock density are presented spatially in Figs. 15 (cattle and buffalo) and 16 (sheep and goat), respectively.

Table 4 Livestock density (animals/sq.km)

District	Cattle	Buffalo	Goat	Sheep	Poultry
Bagalkot	33.7	30.7	81.2	46.3	79.0
Bangalore rural	55.1	7.6	40.4	28.8	1102.5
Bangalore urban	71.0	5.2	39.4	22.6	1111.2
Belgaum	45.5	68.8	60.2	37.9	84.7
Bellary	37.1	14.4	80.0	20.4	260.6
Bidar	50.7	32.8	14.9	32.8	31.3
Bijapur	18.7	12.2	26.4	26.9	18.6
Chamarajanagar	43.2	4.2	23.4	17.5	27.1
Chikballapur	39.6	6.0	61.9	21.7	43.4
Chikmagalur	45.2	9.8	9.1	4.6	23.5
Chitradurga	42.0	23.7	100.7	44.0	513.9
Dakshina Kannada	53.0	0.8	0.1	5.3	118.6
Davangere	63.9	33.3	84.6	22.5	126.2
Dharwar	46.9	18.0	16.3	14.1	21.7
Gadag	25.7	9.8	32.7	20.3	8.6
Gulbarga	46.3	9.7	7.5	29.8	23.2
Hassan	90.7	20.7	21.8	13.8	239.4
Haveri	61.6	20.4	49.5	25.5	39.9
Kodagu	23.8	3.7	0.4	2.0	38.7
Kolar	56.8	11.6	109.5	21.9	1064.1
Koppal	28.0	9.0	58.3	17.2	14.8
Mandya	56.4	23.6	53.8	40.4	138.3
Mysore	92.0	9.3	39.2	28.2	467.0
Raichur	21.4	9.1	37.6	14.8	15.4
Ramanagara	54.7	9.0	37.6	24.9	64.0
Shimoga	68.1	18.1	5.9	8.2	44.9
Tumkur	46.8	13.7	76.9	23.6	69.8
Udupi	70.9	2.4	0.0	2.0	45.4
Uttara Kannada	37.9	9.3	0.5	0.9	62.1
Yadgir	57.0	13.9	83.6	46.6	44.0

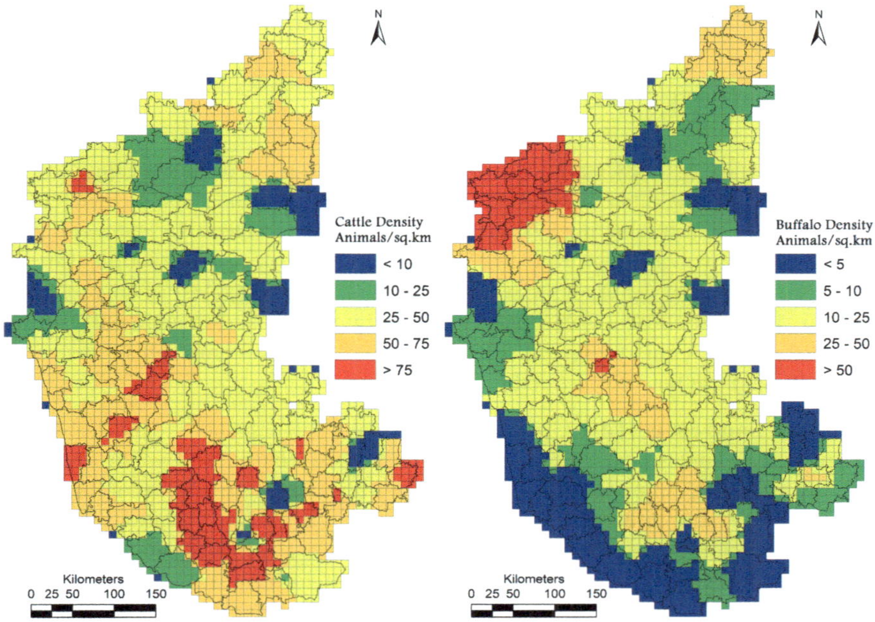

Fig. 15 Livestock density—Cattle and Buffalo

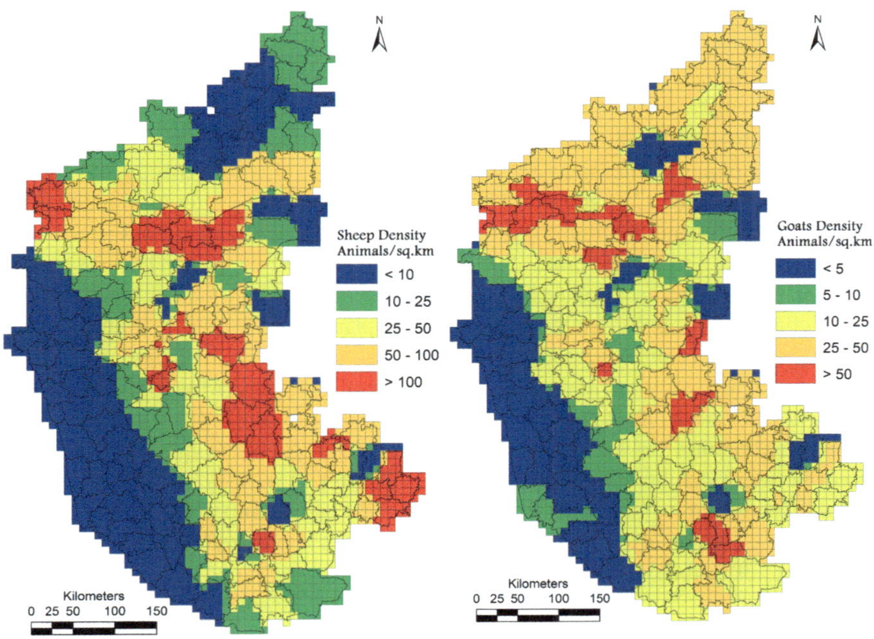

Fig. 16 Livestock density—Sheep and goat

3 Data Sources and Methods

3.1 Classifying Ecosystem Types

A natural landscape has complex ecological, economic, and cultural qualities on which humans and other life forms depend directly. A landscape is a heterogeneous land area of interacting systems, forming an interconnected system called an ecosystem (Forman and Gordron 1986). The functional aspects (interaction of spatial elements, cycling of water and nutrients, bio-geo-chemical cycles) of an ecosystem depend on its structure (size, shape, and configuration) and constituent's spatial patterns (linear, regular, aggregated). The extent of ecosystems can be described based on land use land cover [LULC] information.

Land use, Land cover [LULC] dynamics: Land cover [LC] relates to the discernible Earth surface expressions, such as vegetation or non-vegetation (soil, water, or anthropogenic features), indicating the extent of Earth's physical State in terms of the natural environment (Lambin et al. 2001; Ramachandra et al. 2012). Variations in topography, vegetation cover, and other physical characteristics of the land surface does influence surface-atmosphere fluxes of sensible heat, latent heat, and momentum of heated air particulates caused by conduction, convection, and radiation, influencing weather and climate. Land use [LU] is an expression of the landscape's human uses, e.g., for residential, commercial, or agricultural purposes, etc. Land cover changes induced by human and natural processes play a major role at global as well as at regional-scale patterns of the climate and biogeochemistry of the Earth system. Land use information is vital for regional planning and management activities and has been considered essential for modeling and understanding the Earth as a system (Ramachandra and Shruthi 2007).

Land-use changes alter the homogeneous landscape into a heterogeneous mosaic of patches. The LULC changes are due to natural as well as human-induced alterations. These changes are highly dynamic and characterized by a heterogeneous landscape facilitating socioeconomic–environmental interactions. Natural events such as weather, flooding, fire, climate fluctuations, and ecosystem dynamics initiate land cover changes. Globally, land cover is altered principally by direct anthropogenic use such as agriculture, livestock raising, forest harvesting & management, population change, urbanisation, and other developmental activities (Meyer 1995). Natural disturbances tend to alter forest landscape patterns differently from anthropogenic impacts (Mladenoff et al. 1993). Human-induced impacts are quantified as more effect between patches than natural changes (Hudak et al. 2007). The undisturbed (or wilderness) areas represent only 46% of the Earth's land surface (Mittermeier et al. 2003). Forests covered about 50% of the Earth's land area 8000 years ago, as opposed to 30% today (Ball 2001).

Land use/land cover categories were classified based on the agroecological and agro-climatic zones, using a supervised classification technique based on the Gaussian maximum likelihood algorithm with training data (collected from the field using GPS).

3.2 *Remote Sensing (RS) Data*

Detection of land cover changes relies on an accurate interpretation of baseline conditions and changes in surface spectral properties over time. RS data used in the study are Landsat MSS (1973), TM (1989, 1999), IRS LISS-IV MX (2010, 2019), Landsat ETM^+ (2013), and Google Earth (http://earth.google.com). The Landsat data is cost-effective, with high spatial resolution, and freely downloadable from public domains like GLCF (http://glcfapp.glcf.umd.edu:8080/esdi/index.jsp) and USGS (http://glovis.usgs.gov/). Cloud-free data availability poses a significant challenge in land-use analyses using remote sensing optical data. Cloud-free spatial data were downloaded for the respective districts from the public domain while ensuring a decade's data representativeness.

Districts such as Uttara Kannada and Shimoga, located in the Western Ghats region, are known for good forest cover. However, these districts witnessed large-scale land-use changes due to the construction of reservoirs (dams) in the 80's and market interventions (with globalisation in 1990s). Hence land-use analyses for the districts located in the Western Ghats were done for the period of 1970s.

IRS P6 LISS-IV (Indian Remote Sensing Satellite, part of the Indian Space Programme) data concurring with the field data collection period were purchased from the National Remote Sensing Centre, Hyderabad (http://www.nrsc.gov.in). Table 5 provides the details of IRS data procured from the National Remote Sensing Centre (NRSC), Hyderabad.

Ancillary data: Ancillary data include cadastral revenue maps (1:6000), the Survey of India (SOI), topographic maps (1:50,000 and 1:250,000 scales), vegetation map of South India developed by the French Institute (1986) of scale 1:250,000. Topographic maps provided ground control points (GCP's) to rectify remote sensing data and scanned paper maps. The vegetation map of South India (1986) of scale 1:250,000 (Pascal 1986) was digitized to identify various forest cover types and classify RS data of 1980's. Other ancillary data includes land cover maps, administration boundary data, transportation data (road network), etc. Pre-calibrated **GPS** (Global Positioning System—**Garmin GPS)** units were used for field data collection, required for RS data classification, and validation.

Table 5 Details of RS data for Karnataka State (2019) with respective path and row

S. no	District	Satellite data path & row
1	Bagalkot	146,048, 146,049, 145,049
2	Belagavi	146,048, 146,049
3	Bellary	145,049, 145,050, 144,049, 144,050
4	Bengaluru Rural	144,051
5	Bengaluru Urban	144,051
6	Bidar	145,057, 145,058
7	Chamarajanagar	144,052, 144,051
8	Chikballapur	144,050, 144,051
9	Chikkamagaluru	145,050, 145,051
10	Chitradurga	145,050, 145,051, 144,050, 144,051
11	Dakshina Kannada	146,051, 145,051
12	Davanagere	145,050
13	Dharwad	146,049, 146,050
14	Gadag	145,049, 145,050, 146,049, 146,050
15	Hassan	145,051, 144,051
16	Haveri	146,050, 145,050
17	Kalaburagi	145,048, 144,048

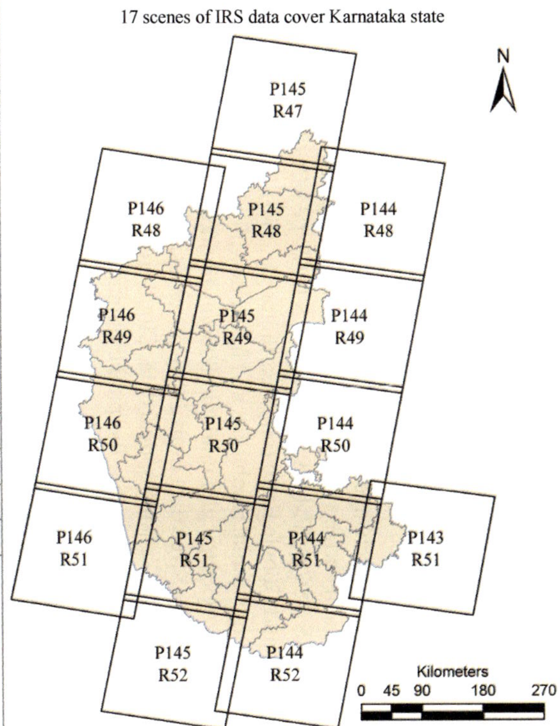

(continued)

Table 5 (continued)

S. no	District	Satellite data path & row
18	Kodagu	145,051, 145,052
19	Kolar	143,051
20	Koppal	145,051
21	Mandya	144,051, 144,052
22	Mysuru	144,051, 144,052, 145,051, 145,052
23	Raichur	145,049, 144,049
24	Ramanagara	144,051
25	Shivamogga	146,050, 146,051, 145,050, 145,051
26	Tumakuru	144,050, 144,051, 145,051
27	Udupi	146,050, 146,051, 145,051
28	Uttara Kannada	146,049, 146,050
29	Vijayapura	146,048, 146,049, 145,048, 145,049
30	Yadgir	145,048, 145,049, 144,048, 144,049

3.3 Methods

Pre-processing of data: Remote sensing data obtained were geo-referenced, rectified, and cropped, corresponding to the study area. Geo-registration of remote sensing data (Landsat data) has been done using ground control points collected from the field using pre-calibrated GPS (Global Positioning System) and also from known points (such as road intersections, etc.) collected from geo-referenced topographic maps published by the Survey of India. In the correction process, numerous GCPs are located in terms of their two image coordinates; on the distorted image, their ground coordinates are typically measured from a map or located in the field in terms of UTM coordinates or latitude and longitude. The Landsat satellite 1973 images, having a spatial resolution of 57.5 × 57.5 m (nominal resolution), were resampled to 28.5 m, comparable to the 1989–2010 data, which are 28.5 × 28.5 m (nominal resolution). The spatiotemporal change detection process involves determining the changes associated with land use and land cover properties with reference to geo-registered multi-temporal remote sensing data. Vector data of the district, taluk and village boundaries, drainage network, water bodies (lakes, ponds) were digitized from the Survey of India topographic maps and cadastral maps. Population census and taluk-wise village boundaries were collected from the Directorate of Census Operations, Bangalore region (http://censuskarnataka.gov.in).

Land cover analysis: Land cover [LC] relates to the discernible Earth surface expressions, such as vegetation or non-vegetation (soil, water or anthropogenic features), indicating the extent of Earth's physical state in terms of the natural environment (Lambin et al. 2001, Ramachandra et al. 2012). Land cover analysis essentially involves delineating the region under vegetation and non-vegetation, which is done through the computation of vegetation indices NDVI (Normalized Difference Vegetation Index). Among all land cover mapping techniques, NDVI is the most widely accepted and applied (Weismiller et al. 1977; Nelson 1983; Ramachandra et al. 2009). NDVI is calculated using visible Red and NIR bands reflected by vegetation. Healthy vegetation absorbs most of the visible light that hits it and reflects a large portion of the near-infrared light. Sparse vegetation reflects more visible light and less near-infrared light. NDVI for a given pixel always results in a number ranging from minus one (–1) to plus one (+1). Very low values of NDVI (–0.1 and below) correspond to soil or barren rock, sand, or built-up areas. Zero indicates the water bodies. Moderate values represent low-density vegetation (0.1–0.3), while high values indicate thick canopy vegetation (0.6–0.9). The outcome of NDVI (for the latest time period) was verified through field investigation and virtual spatila data—Google Earth (http://earth.google.com). NDVI was calculated using Eq. (1).

$$\mathbf{NDVI} = {}^{(\mathbf{NIR} - \mathbf{R})}/{(\mathbf{NIR} + \mathbf{R})} \tag{1}$$

Land use analysis: Land use [LU] is an expression of human uses of the landscape, e.g. for residential, commercial, or agricultural purposes, etc. Land cover changes

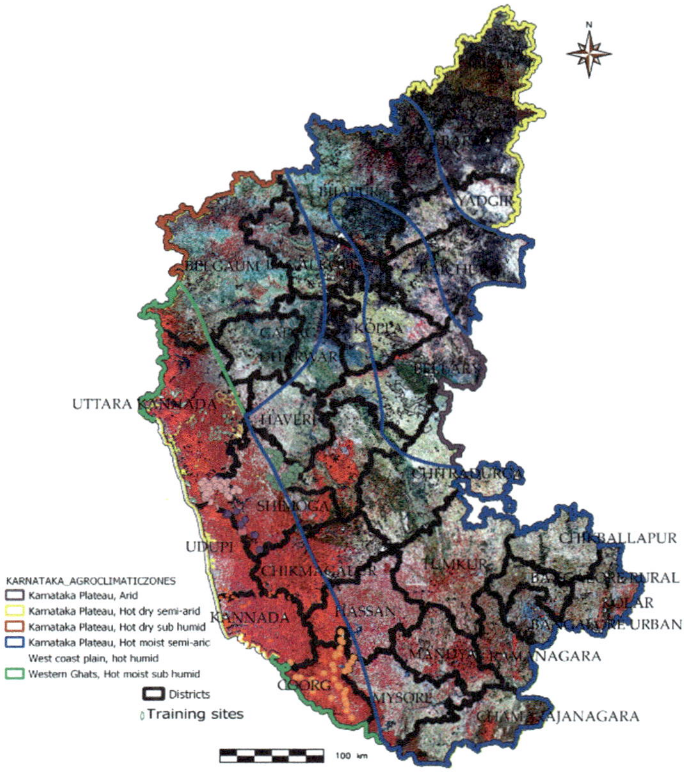

Fig. 17 False-color composite with the digitized training polygons

induced by human and natural processes play a major role in global and regional scale patterns of Earth's climate and biogeochemistry.

Land use analysis involved

(i) downloading Landsat data of 2005 and (ii) procuring IRS MSS data for Karnataka State from NRSC,
(ii) geo-rectification of scenes,
(iii) generating False Color Composite (FCC) of remote sensing data (bands–green, red, and NIR, depending on the data LANDSAT/IRS). This composite image helps in locating heterogeneous patches in the landscape (Fig. 16),
(iv) selecting training polygons by covering 15% of the study area (polygons are uniformly distributed over the entire study area), and loading these training polygons coordinates into pre-calibrated GPS,
(v) collecting attribute information corresponding to these training polygons from online portals (Bhuvan, Google Earth) and validation from field using pre-calibrated GPS (Fig. 17)

(vi) corresponding attribute data (land-use types) for these polygons (training data—corresponding to various land uses and vegetation types) were collected from ten districts (Fig. 17) from October 2019 to February 2020 using a pre-calibrated handheld global positioning system (GPS). GPS helped in locating respective training polygons in the field,

(vii) 60% of the training data has been used for classification, while the balance is used for validation or accuracy assessment.

Data analyses involved (i) downloading Landsat data of 1985 and 2005, and (ii) procured IRS MSS data for Karnataka State from NRSC, (iii) geo-rectification of scenes, (iv) generating FCC with NIR, R, and G (Fig. 17), (iv) digitizing training polygons corresponding to heterogeneous patches spread across the study region (Fig. 18), (v) collecting attribute information corresponding to these training polygons from online portals (Bhuvan, Google Earth) and validation from field using pre-calibrated GPS (Fig. 19).

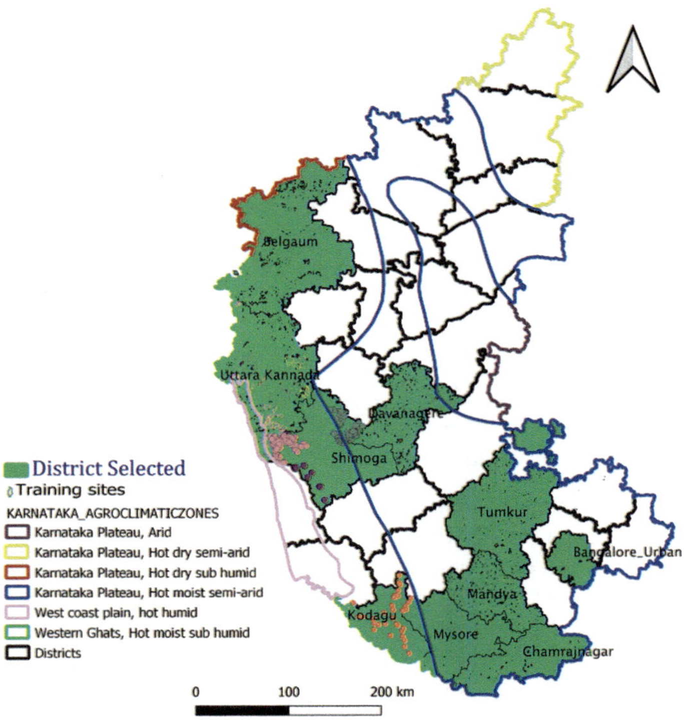

Fig. 18 Districts with training polygons (collected from the selected districts representing agro-climatic zones)

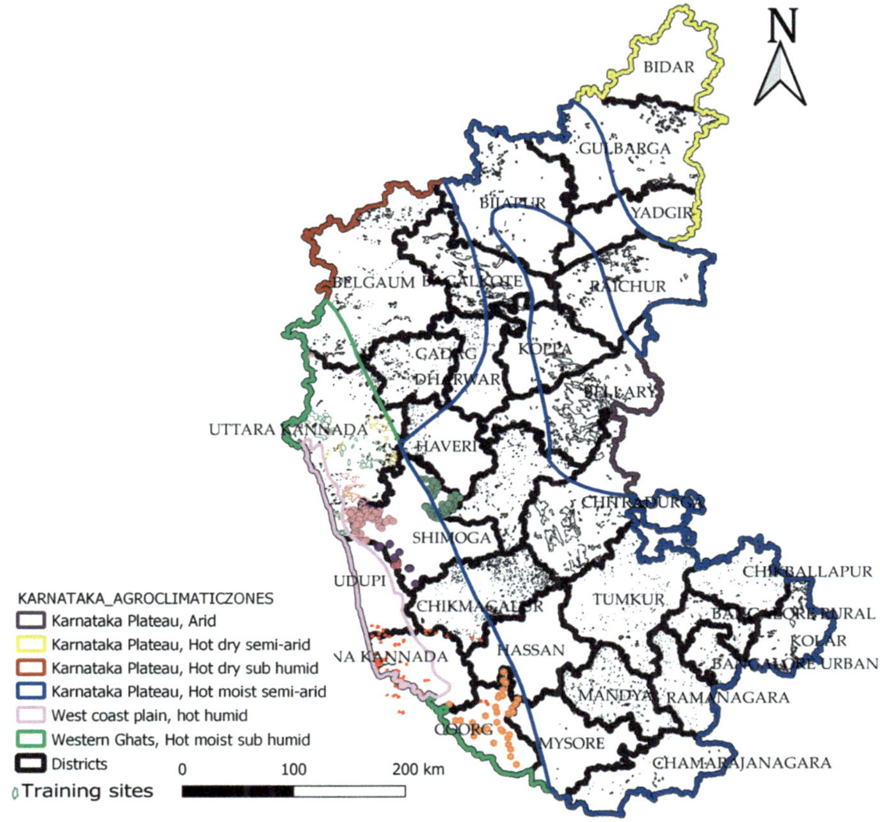

Fig. 19 Training data across agro-climatic zones in Karnataka for classifying remote sensing data

3.4 *Ecosystem Extent Quantification Through Land Use Analyses*

The land use analysis was done using a supervised classification technique based on a Gaussian maximum likelihood algorithm with training data (collected from the field using GPS). The method followed for LULC dynamics is represented in Fig. 20.

The Maximum Likelihood algorithm has been widely applied as an appropriate and efficient classifier to extract information from remote sensing data. This approach quantitatively evaluated the variance and covariance of the category spectral response patterns when classifying an unknown pixel of remote sensing data, assuming the distribution of data points to be Gaussian. The statistical probability of a given pixel value being a member of a particular class is computed. After evaluating each category's probability, the pixel is assigned to the most likely class (highest probability value). **GRASS GIS** ***(Geographical Resources Analysis Support System)*** software

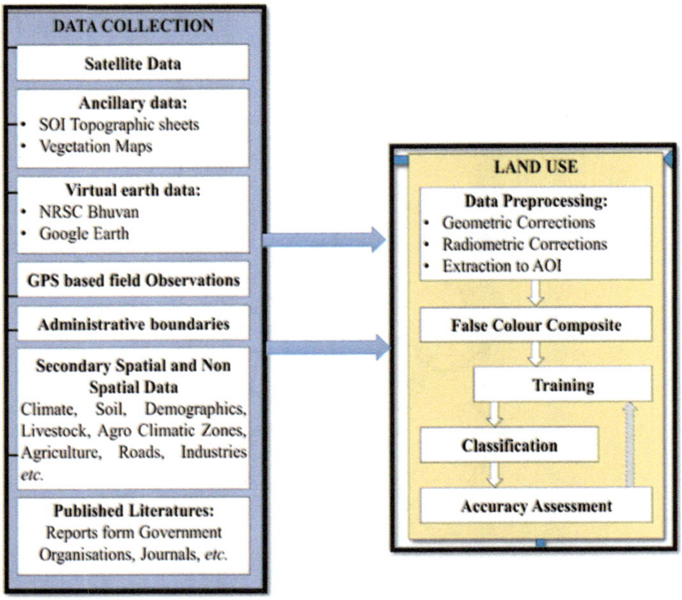

Fig. 20 Method followed for assessing the spatial extent of ecosystems and dynamics (LU changes)

is used for the analysis. It is free and open-source software with robust support for processing vector and raster files, accessible at http://wgbis.ces.iisc.ernet.in/grass/index.php. Temporal remote sensing data have been classified through supervised classification techniques by using available multi-temporal "ground truth" information. Earlier time data were classified using the training polygon along with attribute details compiled from the historical published topographic maps, French Institute vegetation maps, revenue maps, land records available from local administrative authorities, etc.

Accuracy assessments of the classified information have been done through an error matrix (also referred to as confusion matrix), and computation of Kappa (κ) statistics and overall (producer and user) accuracies. This is done to evaluate the quality of the information derived from remotely sensed data considering reference pixels. **Kappa** statistic compares two or more matrices and weighs cells in an error matrix according to the magnitude of misclassification (Lillisand and Keifer 1987; Liu et al. 2007). The producer accuracy measures omission errors through correctly classified pixels in a particular category as a percentage of the total number of pixels belonging to that category in the image. User accuracy (UA) measures commission errors, using the number of correctly classified pixels to the total number of pixels assigned to a particular category.

4 Results

Land use dynamics of Karnataka have been analyzed by classifying temporal (1985, 2005, 2019) Remote Sensing (RS) data into nine different land use categories through a supervised classification technique based on the Gaussian maximum likelihood algorithm. Training data (ground-truthing data, ground controlling points) required for classifying remote sensing data were collected from the districts representing agroclimatic zones using Pre-calibrated Global Positioning System (GPS) supplemented with the training data compiled from virtual portals (Google Earth, Bhuvan) and Collateral data (the Survey of India topographic maps, Vegetation maps, etc.). This also aided in enhancing the accuracy of the land use analysis and in the verification of classified (remote sensing data) information. Secondary Data collection involves a collection of ancillary data such as French Institute Puducherry vegetation maps (Pascal 1986), Geographical Survey of India topographic land use maps (http://www.portal.gsi.gov.in), Biodiversity portal (http://indiabiodiversity.org/), and Virtual earth data such as Google Earth (http://earth.google.com), Bhuvan (http://bhuvan.nrsc.gov.in). The Secondary data provide additional information along with the field data for data pre-processing and classification. Data pre-processing was carried out through Geo-rectification and Radiometric correction of RS data. GPS-based field data, along with Google earth (https://earth.google.com), and Bhuvan (https://bhuvan.nrsc.gov.in) are used for geo-reference the RS data. The supervised classifier based on the Gaussian maximum likelihood classifier (GMLC) scheme was adopted for land use analysis to categorise land uses under nine different categories using GRASS GIS (Geographical Analysis Support System). GRASS is a free and open-source geospatial software with robust functionalities for processing vector and raster data available at (http://wgbis.ces.iisc.ernet.in/grass/). The training data (60%) collected has been used for classification, while the balance is used for accuracy assessment to validate the classification. The test samples are then used to create an error matrix (also referred to as confusion matrix), Kappa (κ) statistics, and overall (producer and user) accuracies to assess the classification accuracies.

Temporal land use analyses reveal the decline of forest cover in Karnataka from 1985 to 2019 (Fig. 21). Currently, 15% of the State's geographical area is under forest compared to 21% in 1985. Large-scale developmental activities such as the construction of a series of reservoirs and dams, the creation of special economic zones and townships, and land conversion for built-up areas have led to the loss of large tracts of forests. The abrupt land-use conversion has resulted in a loss of productive agricultural lands near cities such as Bengaluru, Mysore, Hubli-Dharwad, Shimoga, etc. The districts such as Kodagu, Uttara Kannada, Bengaluru, Shimoga, Belgaum, Dakshina Kannada, and Chikmagalur have been experiencing large-scale land cover due to unplanned developmental activities. Post-1990s, the State witnessed large-scale land-use transitions due to industrialization, urbanization, increased horticulture crops, conversion from agriculture to market-based crops (higher economic), etc. The forest cover is now confined to major conservation reserves such as protected areas, national parks, and wildlife sanctuaries. The built-up cover has increased from

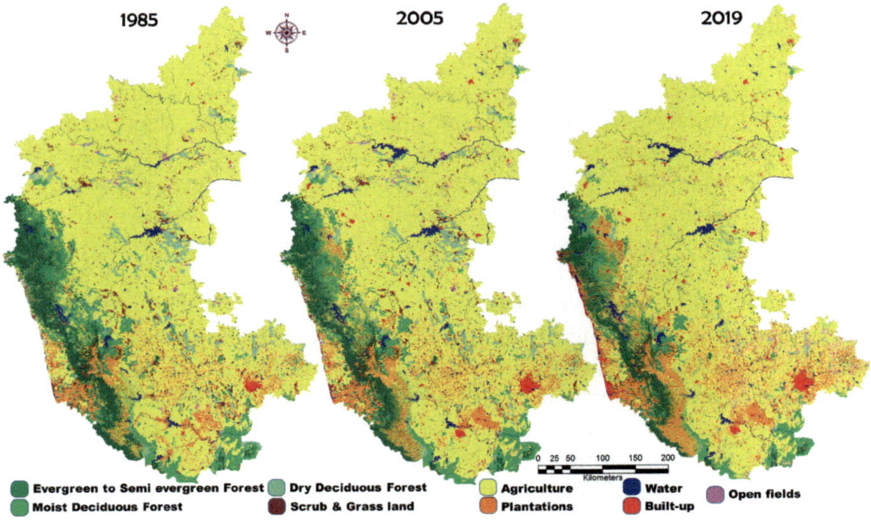

Fig. 21 Land use dynamics in Karnataka

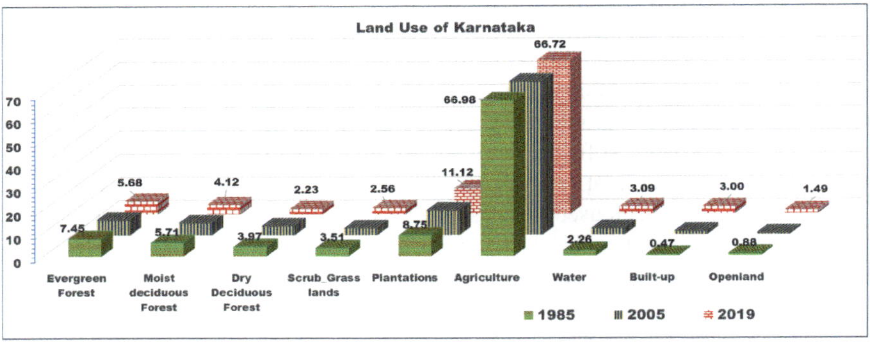

Fig. 22 Land use of Karnataka from 1985–2019

0.47% to 3% from 1985 to 2019, causing an impact on agriculture, forests, and lakes (Fig. 22), necessitating sustainable land-use policies to arrest deforestation and abrupt land conversions.

The built-up area has increased from 0.5% (1985) to 3% (2019), and the horticulture area has increased from 8.8% (1985) to 11.1% (2019). Category-wise land-use dynamics are presented in Table 6. Natural forests show the decline evident from the decrease of evergreen forests from 7.5% (1985) to 5.7% (by 2019), moist deciduous forests from 5.7% (1985) to 4.1% (by 2019) and dry deciduous forests from 4.0% (1985) to 2.2% (2019). Accuracy assessment of the classified remote sensing data was done through the computation of category-wise accuracies and Kappa statistics.

Table 6 Ecosystem extent—**Karnataka State** (units in hectares and percentage)—based on temporal remote sensing data analyses

Year	Units	Land use categories									
		Built-up	Crop land	Horticulture	Fallow land	Evergreen forest	Moist deciduous forest	Dry deciduous forest	Scrub grass lands	Water	Total area
1985	sq.km	904	128,468	16,790	1678	14,293	10,960	7622	6733	4344	191,791
	%	0.5	67.0	8.8	0.9	7.5	5.7	4.0	3.5	2.3	100
2005	sq.km	2666	127,196	20,209	1185	12,445	9900	7410	5604	5177	191,791
	%	1.4	66.3	10.5	0.6	6.5	5.2	3.9	2.9	2.7	21.13
2019	sq.km	5748	127,962	21,325	2854	10,888	7892	4281	4907	5934	191,791
	%	3.0	66.7	11.1	1.5	5.7	4.1	2.2	2.6	3.1	17.68
LU changes during 1985–2019											
1985	sq.km	904	128,468	16,790	1678	14,293	10,960	7622	6733	4344	191,791
2019	sq.km	5748	127,962	21,325	2854	10,888	7892	4281	4907	5934	191,791
Net change of extent (during 1985–2019)											
Extent	sq.km	4844	–505	4536	1175	–3405	–3068	–3341	–1826	1590	0
	%	535.8	–0.4	27.0	70.0	–23.8	–28.0	–43.8	–27.1	36.6	

Table 7 lists category-wise accuracy, which indicates that overall accuracy ranges from 86.84% (1985), 90.08% (2005) to 89.39% (2019), and Kappa ranges from 0.85 (1985) to 0.88 (2019). LU transitions across ecosystems from 2009 to 2019 are listed in Table 8.

Ecosystem extent accounts for land use—land cover (LULC) category-wise for Karnataka, which is provided in Table 9. The new urban agglomerations were noticed across the cities and major towns such as Bengaluru, Mangalore, Hubli, Hassan, Mysore, etc. Large-scale monoculture plantations of Eucalyptus, Rubber, Acacia, Teak, and Arecanut have increased and now cover 12% of the State. These abrupt changes result in an imbalance of ecosystem services, affecting the hydrologic regime and natural resource availability. The coastal and Tier–1 cities such as Bangalore and Tier–2 cities such as Mysore, Hubli-Dharwad, and Belgaum are experiencing a loss of agricultural areas in the sub-urban regions with new layouts and satellite towns.

Table 7 Accuracy assessment

Year	1985		2005		2019	
Category	PA	UA	PA	UA	PA	UA
Built-up	93.97	95.19	92.58	88.01	98.99	96.27
Agriculture	94.01	94.71	97.85	100.00	93.97	93.15
Horticulture	63.98	69.40	90.34	81.38	93.13	88.27
Openland	84.58	76.09	98.14	100.00	91.34	90.88
Evergreen forest	91.77	47.95	88.53	52.43	79.46	100.00
Moist deciduous forest	94.89	75.42	67.05	86.23	66.95	83.99
Dry deciduous forest	69.93	69.70	77.77	79.76	93.80	87.63
Scrub_grass lands	66.95	64.54	80.24	92.82	39.21	81.23
Water	91.45	79.95	91.60	100.00	97.62	86.39
Overall accuracy	86.84		90.08		89.39	
Kappa	0.85		0.87		0.88	

Table 8 Transitions across LU categories during 1985 to 2019—Karnataka State (Extent in hectares and percentage)

1985	2019 Land use categories										
	Units	Built-up	Agriculture	Horticulture	Fallow land	Evergreen forest	Moist deciduous forest	Dry deciduous forest	Scrub/grass lands	Water	Total (Opening stock) 1985
Built-up	sq.km	859	20	15	2	1	3	1	2	2	904
	%	95.0	2.2	1.7	0.2	0.2	0.4	0.1	0.2	0.3	
Agriculture	sq.km	3169	114,150	4050	4529	69	119	117	254	2008	128,468
	%	2.5	88.9	3.2	3.5	0.1	0.1	0.1	0.2	1.6	
Plantations	sq.km	775	3344	11,661	185	243	258	93	108	122	16,790
	%	4.6	19.9	69.5	1.1	1.4	1.5	0.6	0.6	0.7	
Openland	sq.km	50	843	32	710	1	2	12	13	15	1678
	%	3.0	50.2	1.9	42.3	0.1	0.1	0.7	0.8	0.9	
Evergreen forest	sq.km	175	416	2372	151	9097	1402	211	286	182	14,293
	%	1.2	2.9	16.6	1.1	63.6	9.8	1.5	2.0	1.3	
Moist deciduous forest	sq.km	190	1973	1648	485	388	5581	470	164	61	10,960
	%	1.7	18.0	15.0	4.4	0.4	54.1	4.3	1.5	0.6	
Dry deciduous forest	sq.km	85	3374	779	458	68	419	2306	87	47	7622
	%	1.1	44.3	10.2	6.0	0.9	5.5	30.3	1.1	0.6	

(continued)

Table 8 (continued)

1985	2019 Land use categories										
	Units	Built-up	Agriculture	Horticulture	Fallow land	Evergreen forest	Moist deciduous forest	Dry deciduous forest	Scrub/grass lands	Water	Total (Opening stock) 1985
Scrub_grass lands	sq.km	327	3056	701	438	136	113	72	1787	103	6733
	%	4.9	45.4	10.4	6.5	2.0	1.7	1.1	26.5	1.5	
Water	sq.km	93	734	114	35	14	16	7	9	3321	4344
	%	2.1	16.9	2.6	0.8	0.3	0.4	0.2	0.2	76.5	
Closing stock, 2019	sq.km	5725	127,910	21,371	6994	10,018	7914	3288	2710	5862	191,791
	%	3.0	66.7	11.1	3.6	5.2	4.1	1.7	1.4	3.1	191,791

Table 9 Ecosystem extent account for land use—land cover (LULC) in Karnataka State, India

Level-1	Level-2	Karnataka				
		Opening stock 1985	Additions to stock	Reduction in stock	Closing stock 2019	Net change (in %) during 1985–2019
Built-up land	Built-up	904	4866	45	5725	533.1
	Urban					
	Rural					
	Mining					
	Sub-total 1	904	4866	45	5725	533.1
Agricultural land	Horticulture	16,790	9711	5129	21,371	27.3
	Cropland	128,468	13,760	14,317	127,910	-0.4
	Fallow land	1678	6284	968	6994	316.7
	Sub-total 2	146,936	29,754	20,414	156,275	6.4
Forests	Evergreen/ Semi-evergreen	14,293	921	5196	10,018	-29.9
	Moist deciduous	10,960	2333	5379	7914	-27.8
	Dry deciduous	7622	981	5316	3288	-56.9
	Scrub forest	6733	922	4946	2710	-59.8
	Forest plantation					
	Swamp/ Mangroves					
	Sub-total 4	39,607	5158	20,836	23,929	-39.6
Grass/ Grazing	Grass/Grazing					
	Sub-total 5					
Snow and glacier	Snow and glacier					
	Sub-total 6					
Wetlands/ water bodies	Inland wetland					
	Coastal wetland					
	River/stream/ canals					
	Waterbodies	4344	2541	1023	5862	35.0
	Sub-total 7	4344	2541	1023	5862	35.0
Grand total		191,791	42,319	42,319	191,791	

5 Conclusions

Land use information of a region provides a base for accounting for the availability and utilization of natural resources. The information about land use and land cover (LULC) provides a framework for decision-making towards sustainable natural resources management.

Land-use changes alter the homogeneous landscape into a heterogeneous mosaic of patches. The LULC changes are due to natural as well as human-induced alterations. These changes are highly dynamic and characterized by a heterogeneous landscape facilitating socioeconomic–environmental interactions. The LULC changes analysis helps mitigate the impacts due to climate changes, deforestation, soil erosion, etc.

Monitoring LULC plays a vital role at the local/regional and global level in understanding the dynamics associated with the Earth. Monitoring and managing natural resources requires timely, synoptic, and repetitive coverage over large areas across various spatial scales that help assess the temporal and spatial changes. Remote sensing data acquired at regular intervals through space-borne sensors since the 1970's provides an up to date, reliable spatial data, which are useful for LULC analysis. Geographic Information System (GIS) helps compile, analyze, and manage spatial data with attribute information. Remote sensing data with a better spectral and spatial resolution (Multi-Spectral data, Hyperspectral data, etc.) and GIS technologies play an essential role in evaluating the natural resource dynamics for the management and planning of activities such as land-use development, natural resource exploitation, and engineering projects.

The System of Environmental-Economic Accounting—Experimental Ecosystem Accounting (SEEA EEA), through the consultative process, has developed the protocol for ecosystem extent accounts as part of the ongoing initiative of "Natural Capital Accounting and Valuation of Ecosystem Services" of the United Nations Statistics Division (UNSD), UN Environment, and the Secretariat of the Convention on Biological Diversity. The protocol serves as a common starting point for ecosystem accounting. It organizes information on the extent of different ecosystem types within a region (federal state, country). The pilot study for Karnataka State has been implemented through an internationally agreed method (SEEA EEA) during 2020–21.

The purpose of this work is to demonstrate the low-cost techniques for combining temporal remote sensing (RS) data with collateral data (field data with the data available from the government sources) using Geographic Information System (GIS) and develop relatively accurate temporal land-use maps for the ten representative districts chosen based on agro-climatic zones and scaling up the approach for the entire Karnataka State. The availability of qualitative and quantitative information on land-use dynamics in a region helps formulate policies, planning, and decision-making toward the prudent management of natural resources.

Ecosystem extent accounts for ten selected districts in Karnataka representing agro-climatic zones in Karnataka, which have been compiled using various data sources. All districts show a decline in the spatial extent of forest ecosystems.

Ecosystem extent in Karnataka State, India, through land-use analyses, reveals that the agriculture ecosystem, with a spatial extent of 127,962 sq.km (66.72%) constitutes the major asset in the State. This is followed by plantations (21,325 sq.km, 11.123%), evergreen forests (10,887 sq.km, 5.68%), moist deciduous forests (7892 sq.km, 4.12%), dry deciduous forests (4280 sq.km, 2.23%) and scrub grasslands (4906 sq.km, 2.56%). The State has a 5.7% area under evergreen forest cover (2019) compared to 6.5% (2005) and 4% moist deciduous forests compared to 5.2%. The loss of evergreen and deciduous forest cover across the state articulates the anthropogenic pressure, signifying an increase in monoculture and built-up areas. The state had 1.4% of the area under built-up (2005) compared to 3% in 2019.

Changes in the landscape structure (through large-scale land use changes) have altered the functional abilities of an ecosystem, evident from the altered hydrological regime, barren hilltops, lowered water retention capabilities of catchments, disappearing perennial streams, frequent flooding and drought, recurring instances of human–animal conflicts, declined ecosystem goods, etc. This necessitates the restoration of native forests in the region to ensure water and food security apart from the livelihood of the local people.

In conclusion, the adopted method allows effective use of spatial data acquired through space-borne sensors while setting a guideline for high-resolution analysis of landscape dynamics over the past five decades.

References

Ball JB (2001) Global forest resources: history and dynamics. In Evans J (ed) The forests handbook, vol 1, pp 3–22, pp 418. Oxford: Blackwell Sci

BDA (2002) The biological diversity act 2002, Government of India. https://www.indiacode.nic.in/bitstream/123456789/2046/4/a2003-18.pdf

Census (2011) The census of India. https://censusindia.gov.in/

FAO (1995) Forest resources assessment 1990. Global Synthesis. FAO, Rome

Forman RTT, Gordron M (1986) Landscape ecology. JohnWiley & Sons, New York. ISBN 0-471-87037-4

Hudak AT, Morgan P, Bobbitt M, Lentile L (2007) Characterizing stand-replacing harvest and fire disturbance patches in a forested landscape: a case study from Cooney Ridge Montana. In Wulder MA, Franklin SE (Eds) Understanding Forest disturbance and spatial pattern: remote sensing and gis approaches. CRC Press (Taylor and Francis), Boca Raton, FL, 246, pp 209–231

Lambin EF, Turner BL II, Geist HJ, Agbola S, Angelsen A, Bruce JW, Coomes O, Dirzo R, Fischer G, Folke C, George PS, Homewood K, Im-bernon J, Leemans R, Li X, Moran EF, Morti-more M, Ramakrishnan PS, Richards JF, Skånes H, Steffen W, Stone GD, Svedin U, Veldkamp T, Vogel C, Xu J (2001) The causes of land-use and land-cover change: Moving beyond the myths. Glob Environ Chang: Hum Policy Dimens 11:261–269

Lillesand TM, Keifer RW (1987) Remote sensing and image interpretation. John Willey and Sons, NewYork

Liu C, Frazier P, Kumar L (2007) Comparative assessment of the measures of thematic classification accuracy. Remote Sens Environ 107:606–616

Meyer WB (1995) Past and present land-use and land-cover in the U.S.A. Consequences:24–33

Mittermeier R, Mittermeier CG, Gil PR, Pilgrim J, Fonseca G (2003) Wilderness: Earth's last wild places. Univ. Chicago Press, Chicago, p 576

Mladenoff DJ, White MA, Pastor J, Crow TR (1993) Comparing spatial pattern in unaltered old-growth and disturbed forest landscapes. Ecol Appl 3:294–306

Nelson RF (1983) Detecting forest canopy change due to insect activity using Landsat MSS. Photogram Eng Remote Sensing 49:1303–1314

Pascal JP (1986) Explanatory booklet of vegetation maps 1, 2 and 3. Institut Français de Pondichéry, India. pp 88, http://hal.archives-ouvertes.fr/hal-00444285

Ramachandra TV, Shruthi BV (2007) Spatial mapping of renewable energy potential. Renew Sustain Energy Rev 11(7):1460–1480

Ramachandra TV, Setturu B, Bharath HA (2012) Peri-urban to urban landscape patterns elucidation through spatial metrics. Int J Eng Res Dev 2(12):58–81

Ramachandra TV, Kumar U, Diwakar PG, Joshi NV (2009) Land cover Assessment using À Trous Wavelet fusion and K-Nearest Neighbour classification. In: Proceedings of the 25th annual in-house symposium on space science and technology, 29 - 30 January 2009, ISRO—IISc Space Technology Cell, Indian Institute of Science, Bangalore

SEEA Experimental Ecosystem Accounting and the Central Framework. https://seea.un.org/ecosystem-accounting

United Nations (2012) System of environmental-economic accounting for water. UN. Series F No. 100 (ST/ESA/SER.F/100)

United Nations, European Commission, Food and Agricultural Organization of the United Nations, International Monetary Fund, Organisation for Economic Co-operation and Development, The World Bank (2014a) System of environmental-economic accounting 2012—Central Framework. United Nations, New York

United Nations, European Commission, Food and Agricultural Organization of the United Nations, Organisation for Economic Co-operation and Development, The World Bank (2014b) System of environmental-economic accounting 2012—Experimental ecosystem accounting. United Nations, New York

Weismiller RA, Kristof SJ, Scholtz DK (1977) Change detection in coastal zone environments. Photogram Eng Remote Sensing 43:1533–1539

Ecosystem Condition Account for Karnataka State, India

Abstract Ecosystem accounting entails estimation of the extent of different ecosystem assets (EAs) in the study region (district and federal state) using remote sensing data (acquired through space-borne sensors at regular intervals and available since the 1970s. The multi-resolution spatial data helps in assessing ecosystem extent accounts and provides an underlying infrastructure for measuring ecosystem conditions and modeling ecosystem services. The analyses of temporal data help in assessing the extent of changes in an ecosystem over time.

Ecosystem condition accounts record the condition of ecosystem assets in terms of selected characteristics at specific points in time and, over time, record the changes to their condition. These changes are mainly due to human activities with economic interventions. Ecosystem condition accounts provide a structured approach for describing ecosystem assets' characteristics by recording and aggregating data describing the dynamics of characteristics of ecosystem assets characteristics (SEEA EA). Ecosystem quality is measured in terms of its abiotic and biotic characteristics with respect to an ecosystem's composition, structure, and function, which in turn reflect multiple values across a range of temporal and spatial scales indicating ecological integrity.

An ecosystem condition account was developed considering (i) characteristics of abiotic components of an ecosystem (physical and chemical status of soil), (ii) biotic ecosystem characteristics considering compositional state (occurrence of flora and fauna species chosen as per IUCN categories of threats), structural state (vegetation–above-ground biomass, below-ground biomass of forest ecosystems), functional state (NPP: net primary productivity in forest ecosystems) and (iii) landscape-level characteristics (fragmentation of forest landscapes, LST: the land surface temperature at landscape levels) for 2005 and 2020.

Landscape-level characteristics (fragmentation of forest landscapes, LST: land surface temperature at landscape levels) for 2005 and 2019 were assessed considering remote sensing data. The fragmentation analysis reveals that the spatial extent of contiguous interior forests constituted only 11,335 sq.km (5.91%) in 2019. The state has witnessed a decline in interior forests by 63.7% from 1985 to 2019. Very-density forests constitute 11,334 sq.km (5.91%), and medium-density forests are about 12,869 sq.km (6.71%). Analyses of fragmentation of forest ecosystem reveal

T. V. Ramachandra et al., *Natural Capital Accounting and Valuation of Ecosystem Services, Karnataka State, India*, https://doi.org/10.1007/978-981-97-2405-5_2

that districts, even ecologically fragile regions–Western Ghats have been experiencing fragmentation of forests evident from the decline of contiguous intact forests to the extent of 54% (Chikkamagalur), 57.6% (Dakshina Kannada), 44% (Udupi), 41% (Uttara Kannada) and 35% (Shimoga) during 2005 to 2019.

Land surface temperature (LST) estimations district-wise for Karnataka considering thermal bands–remote sensing data for 2005 and 2019 indicate a considerable increase in temperature, which could be attributed to land cover changes in the region with the decline of vegetation cover. The higher temperature can be seen especially in non-forest areas due to intensified built-up activities. The temperature rise can be attributed to high deforestation with unplanned urbanisation and industrialisation. Temporal analysis of LST indicates a rise in mean temperature over the periods. The highest temperature of 48.46 °C was found in the hot, dry semi-arid region, and the lowest was found in the hot humid of 32.46 °C. About 821,600 hectares had temperatures less than 30 °C (high) in 2005 compared to 485,566 hectares in 2019. Regions under the medium category (with temperatures 30–35 °C) have increased from 2898 sq.km (2005) to 3666 sq.km (2019). The results highlight the role of biophysical variables in maintaining regional climate. The increase in temperature can induce a higher loss of species, imbalance in the ecosystem, and vegetation die-off. Proper land-use planning and restoration of degraded land through afforestation with native species would help mitigate the increase in temperature.

Biotic ecosystem characteristics considering compositional state (occurrence of endemic flora and fauna species chosen as per IUCN categories of threats) show there has been a decline in the number of flora species (Critically endangered, Endangered, Vulnerable, Near-threatened) from 3614 species (2005) to 3374 and a similar trend in the occurrence of endemic fauna species from 4243 (2005) to 3923 (2020) species.

The structural state of the forest ecosystem was done through an assessment of vegetation characteristics with the computation of above-ground biomass (AGB), and below-ground biomass (BGB) of forest ecosystems, revealing a decline of 23.4% of AGB from 987,686 (2005) to 732,828 (2019) Gg and similar trend of decrease in BGB from 191,537 (2005) to 146,566 (2019) Gg. The total biomass (AGB + BGB) of forests is about 1013.7 Tg (Tera gram) with stored carbon of 506.8 Tg (in 1985), which is now reduced to 678 Tg and 339 Tg, respectively (2019). The temporal decline of AGB values in Kodagu, Shimoga, Uttara Kannada, and Dakshina Kannada districts is due to anthropogenic pressure. The Mysore, Chamrajnagara, and Bellary districts also reflect a decline in AGB values from 2005 to 2019.

Karnataka Forest Department has implemented monoculture plantations in the state to protect the land under greening initiatives and sustain market demand for timber. The AGB and BGB were accounted to understand the role of plantations in arresting land degradation. The AGB for forest and plantations accounted for 1056.90 Tg (in 1985), which is now reduced to 732.83 Tg (2019).

Changes in the land cover have impacted the functional state of the forest ecosystem, which was assessed through the computation of NPP (net primary productivity), indicating 15.5% decline in NPP from 17,323 Gg/yr (2005) to 14,635 Gg/yr (2019).

Abiotic characteristics of the ecosystem are assessed through the physical and chemical status of soil in agricultural lands. Soil health assessment indicates that 43.7% of soil in Karnataka shows good soil quality, where 48.6% of soil in Karnataka shows medium-quality soil, while low-quality soil is found in 7.6% of the area in Karnataka. Among the chemical parameters,

- 10.6% of cropland has high potassium nutrients, where 85.3% of the area is medium while low potassium is 4% of the total area.
- Nitrogen nutrient is found low in 95.5% of the cropland soil, while 4.5% of the area shows medium quantity.
- Phosphorus nutrient availability in the soil is high in 9.9% of the area, whereas 85.8% of the area is in medium status while low availability is in 4.1% of the state area.
- Zinc nutrients in the soil are medium (sufficient) in 90.5% of the area, while they are low (deficient) in 9.5% of croplands in Karnataka.
- Iron nutrient availability in the soil is medium (sufficient) in 98.8% of the area, while 1.2% of the cropland in the State is low (deficient).
- Organic carbon in the soil is medium (sufficient) in 88.25% of the area, while it is found deficient in 11.75% of the area of the state.
- Boron is in medium (sufficient) in 57% of the cropland soil, while 43% of the area have low (deficient).
- Sulphur nutrient is high in 54.7% of the area, medium in 43.58%, and low in 1.6% of the area of the state.
- Copper availability in the soil is low (deficient) in 86.8% of the area while medium (sufficient) in 13.2% of the area.
- Manganese nutrients in the soil are low (deficient) in 0.15% of the area and medium (sufficient) in 99.58% of the area of the state.

List of Abbreviations

ADTR	Anshi-Dandeli Tiger Reserve
ESR	Ecologically Sensitive Regions
ETM	Enhanced Thematic Mapper Plus
FCC	False Colour Composite
GHG	Green House Gas
GIS	Geographic Information System
GLP	Global Land Project
GPS	Global Positioning System
IGBP	International Geosphere-Biosphere Programme
IHDP	International Human Dimensions Programme on Global Environmental Change
IPCC	Intergovernmental Panel on Climate Change

IRS	Indian Remote Sensing
LANDSAT	Land Remote-Sensing Satellite (System)
LISS 4 MX	Linear Imaging Self-Scanning Sensor 4 Multispectral
LULC	Land Use Land Cover
MoEFCC	Ministry of Environment and Forests and Climate Change
MSS	Multi-Spectral Scanner
NDVI	Normalized Difference Vegetation Index
NIR	Near Infrared
NRSC	National Remote Sensing Agency
QGIS	Quantum GIS
RS	Remote Sensing
SOI	Survey of India
TM	Thematic Mapper
USGS	United States Geological Survey

1 Introduction

Ecosystem condition accounts provide a structured approach for describing ecosystem assets' characteristics by recording and aggregating data describing the dynamics of characteristics of ecosystem assets characteristics (SEEA EA). Ecosystem quality is measured in terms of its abiotic and biotic characteristics concerning an ecosystem's composition, structure, and function, which, in turn, reflect multiple values across a range of temporal and spatial scales, indicating ecological integrity. **Ecosystem condition** accounts record the condition of ecosystem assets in terms of selected characteristics at specific points in time and, over time, record the changes to their condition. These changes are mainly due to human activities with economic interventions. The ecosystem condition typology as per SEEA are listed in Table 1.

The current study's objective is to compile ecosystem conditions for Karnataka State based on the SEEA EA framework (United Nations et al. 2020). The focus is on (i) compiling accounts for the ten selected districts (representative of agro-climatic zones) and Karnataka State (District wise), and (ii) computing ecosystem condition index at the district level for Karnataka State, India, considering (i) abiotic characteristics (physical and chemical state of soil), (ii) biotic characteristics (compositional, structural and functional) and (iii) landscape-level characteristics (land surface temperature and fragmentation of forest ecosystems).

Table 1 The SEEA Ecosystem Condition Typology (SEEA ECT)

Ecosystem condition	ECT groups	ECT classes
	Abiotic ecosystem characteristics	1. Physical state characteristics (**soil structure**, water availability)
		2. Chemical state characteristics (**soil nutrient levels**, water quality, air pollutant concentrations)
	Biotic ecosystem characteristics	3. Compositional state characteristics (including **species-based indicators**)
		4. Structural state characteristics (including **vegetation, biomass**, food chains)
		5. Functional state characteristics (including **ecosystem processes**, disturbance regimes)
	Landscape level characteristics	6. Landscape and seascape characteristics (including **landscape diversity, connectivity, fragmentation**, embedded semi-natural elements in farmland)

This involves,

(a) **Fragmentation of forests**

(i) Land use dynamics analyses (which reveals the extent of an ecosystem at different interval of time and also helps in assessing the rate of change);
(ii) Fragmentation analyses through landscape metrics;
(iii) Align and integrate the various types of maps by overlaying them in an integrated GIS database;
(iv) Compile a set of ecosystem extent and condition accounts that describe a range of ecosystem characteristics by ecosystem type (i.e. land cover types), and how they change over time.

(b) **Land surface temperature responses to land use changes**

(i) Understand the LU dynamics in select districts;
(ii) Biophysical parameters of earth to sub-pixel thermal variations;
(iii) Evaluation of temperature variations across forest and non-forest cover regions.

(c) **Assessment of vegetation dynamics**

(i) Understanding spatial patterns of land use dynamics in Karnataka State, India;
(ii) Quantification of the carbon emissions;
(iii) Structural state characteristics (including **vegetation, biomass**)

(d) **Biotic elements–Flora and fauna in Karnataka**
(e) **Physical and Chemical status of soil**

(i) Assessment of physical characteristics of soil across agro-climatic zones;
(ii) Evaluation of chemical and nutrient status of soil.

1.1 Fragmentation of Forest Ecosystem

The landscape is a mosaic of heterogeneous interacting dynamic elements, which are the manifestation of the natural and anthropogenic processes. The structure of a landscape (size, shape, and configuration) influences functional aspects such as biogeochemical cycling, hydrologic regime, etc. The interactions among the landscape elements result in the flow of nutrients, minerals, and energy, which contribute to the functioning of the landscape. Forest ecosystems constitute a key component of the global carbon cycle, accounting for over two-thirds of net primary production on land by converting solar energy into biomass through photosynthesis (MEA 2005; Roy et al. 2001; Ramachandra et al. 2013, 2016a, b). Forest ecosystems offer timber, non-timber forest products (NTFP), medicinal resources, fuelwood, etc., and provide recreational value (Kindstrand et al. 2008). They aid as biodiversity repository (Li et al. 2009), restrain soil erosion (Nandy et al. 2011), prevent landslides as tree roots bind the soil, regulate air humidity, temperature and mitigate global warming (Cabral et al. 2010) by absorbing 30% of fossil fuel CO_2 emissions (Pan et al. 2011a, b). The goods and services provided by forested landscapes are vital to human socio-economic development (DeFries et al 2004) and survival (Ramachandra et al. 2013). In recent times, large-scale land use and land cover changes (LULC) have altered the ecosystem structure, influencing the respective ecosystem's goods and services. These disturbances have resulted in the fragmentation of forests with a mosaic of natural patches surrounded by other land uses (Ramachandra and Kumar 2011). A host of anthropogenic activities (tree logging, conversion of forest land for agriculture, intense agriculture practices, forest fire, and unplanned infrastructure development) has contributed to the disruption of forests' contiguity in predominantly natural landscapes (Buskirk et al. 2000; Boogaert et al. 2004). An alteration in forest structure through fragmentation of forests has influenced the functional abilities evident from the decline of water yield, carbon sequestration potential, biodiversity, etc. (Diaz et al. 2006; Ramachandra and Kumar 2011).

The fragmentation process involves altering the structure and composition of native forests through the division of contiguous forests into smaller non-contiguous fragments with a sharp increase in edges (Riitters et al. 2002; Broadbent et al. 2008). This process will have detrimental effects such as disruption in biogeochemical cycling, nutrient and water cycling, ecological processes (Fahrig 2003; Holway 2005), easier access, and further land-use changes (Bennett and Saunders 2010; Holway 2005; Ramachandra et al. 2012a). The edge effect may even perish large trees within 300 m of the forest edge and are replaced by densely spaced short-lived pioneers (Laurance 1999), resulting in the decline of forest biomass (Harper et al. 2005). The negative impacts of edges include alterations in plant and animal community composition with the changes in diversity (Cagnolo et al. 2006), seed dispersion, pollination, predation, fire susceptibility, altered microclimate, and increased carbon emissions (Laurance et al. 2002). Consequences of edge effects (Carolina 1995; Asner et al. 2006) include abiotic changes such as microclimate (Kampichler et al. 2012), water availability, etc. Figure 1 explains the process of fragmentation with

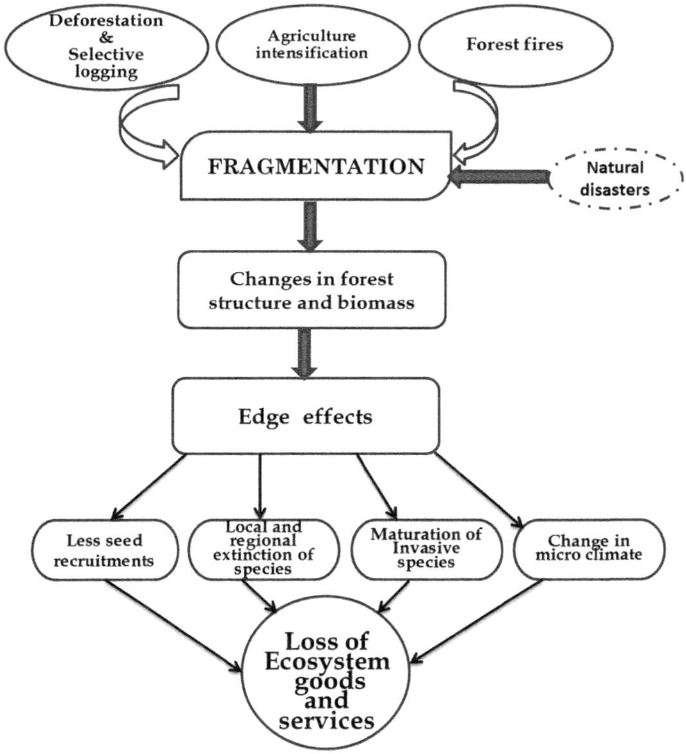

Fig. 1 Fragmentation of forests and its consequence

drivers. The earlier stages of forest fragmentation lead to forest composition changes due to the transition of evergreen forests to semi-evergreen, leading to alterations in microclimate, habitat, etc. (Ramachandra et al. 2013). During the succession stage, edges will become more prominent with higher light availability and soil moisture loss, directly impacting seed dispersal, seed viability, species distribution, propagation of invasive species, etc. (Ramachandra and Kumar 2011). These cascading effects have the potential to disrupt many basic ecological processes over larger areas seriously. The changes in forest structure will result in abrupt effects and disturb the natural ecosystem's productivity of goods and services. The analyses of the extent of forest fragmentation will provide insights into complex dynamic interactions, which will help in adopting appropriate location-specific conservation measures to mitigate disruptions in ecological processes.

Quantification of forest fragmentation: Land use (LU) changes driven by anthropogenic activities alter the structure of a landscape, which adversely influences functional aspects of ecosystems. Land use patterns are the collective result of local geophysical (elevation, slope, rainfall, etc.) interactions and agro-climatic indicators, demographic variation, market forces, and related development policies (Munroe et al. 2004; Ameztegui et al. 2010). Numerous studies focusing on deforestation at landscape levels have explored the spatial patterns and interactions among the geophysical elements of the landscape and its dynamics (Nelson and Geoghegan 2002; Alix-Garcia et al. 2005; Echeverria et al. 2008; Ramachandra and Kumar 2011; Ramachandra et al. 2013). These changes measured at a temporal scale help monitor the ecosystems and aid in implementing location-specific mitigation measures. Spatial data acquired remotely through space-borne sensors (remote sensing data) at regular intervals helps assess the temporal changes in spatial patterns (Ramachandra et al. 2014a). Remote sensing data (RS) with geographic information systems (GIS) have made significant contributions in assessing the spatiotemporal patterns and processes of forest ecosystems (Ramachandra et al. 2012a, 2016a, b) in the criteria-based decision-making and selection of the best alternative. The availability of remote sensing data with improvements in resolutions (spatial, spectral, and temporal resolutions) enables the creation of land use, land cover maps (Chen et al. 2015), and innovative analytical techniques, help to monitor changes in a cost-effective way (Bharath et al. 2012a). Changes in forested landscapes of the Western Ghats have been more sudden since the middle of the last century due to industrialization policy impetus consequent to globalization. Quantifying forest fragmentation in an ecologically fragile region (such as the Western Ghats) will help formulate appropriate mitigation measures for biodiversity conservation.

1.2 Land Surface Temperature Response to Land-Use Changes

Landscape patterns play an important role in ecological processes, and the structure of a landscape, shape, and configuration are responsible for its functional aspects such as biogeochemical cycling and hydrologic regimes (Chen et al. 2014; Ramachandra et al. 2016a, b). The landscape transition can be natural or anthropogenic. Several studies across the globe have proved that anthropogenic activities are altering the global climate (Peng et al. 2011; Newbold et al. 2015; Grizzetti et al. 2017; Szymura et al. 2018). The abrupt environmental changes at local, regional, and global scales threaten humans' survival and other biotas. Land use land cover (LULC) is an essential landscape parameter, reflecting its change characteristics (Ganasri et al. 2015). LC refers to the observed physical cover on the earth's surface, whereas LU expresses arrangements, activities, and inputs people undertake in a certain LC. LULC change is governed by crucial drivers such as population, urban expansion, agricultural production, and land management policies, altering a landscape into a heterogeneous mosaic of patches (Ramachandra et al. 2012a, b, c). LULC alterations will affect the range of functionalities of the landscape, such as nutrient cycling, biogeochemical cycle, hydrologic cycling, carbon sequestration potential, water retention capability, etc. (Ramachandra et al. 2018a, b; Ramachandra and Bharath 2019a, b). The rampant LULC changes in a region will affect energy exchanges between the biophysical properties and the atmosphere. A comprehensive understanding of these changes is necessary to evaluate critical drivers of changes and frame policies for sustainable development (Ramachandra et al. 2017a, b). Land surface temperature (LST) is a prime climate variable for climate change across the local and global scales. It also indicates the energy balance of land surface processes (James and Mundia 2014). LST changes can be directly linked with associate LULC changes to account for the adverse effect on the environment, energy balance, and biogeochemical interactions.

Global warming and consequent changes in the climate have given momentum to research to understand the causes of LULC. LST is sensitive to vegetation, canopy layer, and soil moisture. Hence, it is used to detect climate alterations with LULC changes (Mallick et al. 2008). The unplanned developmental activities will intensify environmental degradation, induce disasters, and disrupt sustainable economic growth (Ramachandra et al. 2019a, b). LST changes are essential in most environmental processes and affect microclimate through associated biophysical changes. An increase in surface temperature has been observed due to the conversion of pervious or vegetation surfaces to impervious surfaces. These changes, in turn, affect the absorption of solar radiation, surface temperature, evaporation rates, and wind turbulence and alter the conditions of the near-surface atmosphere (Aithal et al. 2014). Thus, understanding the linkages between LST and landscape dynamics helps in evaluating human–environment interactions, which aids in the planning and management of natural resources with persistent knowledge of landscape at a temporal scale (Bharath et al. 2013a, b). The advancement of spatial technologies has aided in understanding LULC changes, precise evaluation of LST, and environmental consequences. The temporal satellite remote sensing data of the thermal band can capture the variation in energy across various landscape features. Numerous studies have tried to capture LST using split-window algorithms in Landsat 8 using bands 10 and 11 (Li et al. 2013; Bunai et al. 2018). Moderate Resolution Imaging Spectral radiometer (MODIS) data has advantages in LST estimation due to its global coverage, high radiometric resolution, and dynamic ranges, making it efficient for monitoring and estimating land surface temperature over a large area (Wan et al. 2002). The satellite-derived LST of a larger region will provide information on different types of Earth surface processes, surface-atmosphere interactions, and evapotranspiration (Anderson et al. 2011). The Karnataka state of India is experiencing LULC changes due to increased anthropogenic pressure from the 1990's. In this regard, the current research tries to understand the LULC changes at a temporal scale from 1985–2019. It explores the correlation between the earth's biophysical parameters and subpixel thermal variations based on agro-climatic zones of Karnataka. It includes evaluation of temperature variations across the forest and non-forest cover regions.

1.3 Vegetation Dynamics in Karnataka

Vegetation constitutes the fundamental element in the earth system biota and helps in mitigating carbon levels in the atmosphere. Autotrophic organisms uptake carbon dioxide (CO_2) during photosynthesis, transforming the sun's energy into a chemical carbohydrate molecule, converting carbon in the atmosphere to fuel and structural materials for living organisms. Post industrialization era witnessed an increasing GHG footprint, of which 72% constitute CO_2. Escalating anthropogenic greenhouse gas (GHG) emissions (400 parts per million from 280 ppm CO_2 emissions of the pre-industrial era) have contributed to global warming (Blunden 2017) with changes in the climate, which has given impetus to the decline of ecosystem productivity and sustenance of water, affecting the livelihood of people.

Forest ecosystems are large repositories of terrestrial carbon and play a crucial role in the carbon cycle (C-cycle) through sequestration of atmospheric carbon in the above-ground biomass (AGB), below-ground biomass (BGB), and soil organic carbon (SOC). Forests and soil have been playing a vital role in the carbon cycle, evidenced by the sequestration of about 30% of annual global anthropogenic CO_2 emissions (2 petagrams (Pg) of carbon per year) from the atmosphere. The forests hold large quantities of carbon per unit area packed down through photosynthesis, which gets released with fragile ecosystems' mismanagement due to unplanned developmental activities with anthropogenic pressures (Randerson et al. 2005). The world's forests' annual carbon sequestration has been estimated at 2.4 Gigaton C (Pan et al. 2011a, b). Soil stores about 2–3 times more carbon in organic form than forest woody biomass (Scharlemann et al. 2014). Carbon stored in soils as soil organic carbon (SOC) and the studies focusing on soil's potential in sequestering carbon are scanty and received relatively limited attention from the policy community compared to carbon storage in the above-ground wood biomass (Batjes 1996; Lal 2004; Scharlemann et al. 2014). SOC constitutes the largest terrestrial carbon pool globally, with an estimated 700–3000 PgC (1 PgC = $1*10^{15}$gC) (Chhabra et al. 2003). SOC content in the soils varies based on climate, moisture, physiography, soil type, elevation, terrestrial vegetation type, density, and extent. However, with the inappropriate land-use changes due to mismanagement, the soil becomes a source of greenhouse gas emissions (CO_2, CH_4). This necessitates prudent land management in agricultural practices and restoration of eroded and degraded forest soils to improve the soil carbon pool (Lal 1999). The total soil C pool in the top 1 m has been estimated as 2011 PgC (IPCC 2000), which constitutes 4.1 times the biotic pool and three times the atmospheric carbon. Mean SOC is estimated as 92.1 t/ha in littoral and swamp forests, 37.5 t/ha in tropical dry deciduous forest regions (for top 50 cm soil depth) in India (Chhabra et al. 2003). The total SOC in Indian forests is estimated as 4.13 PgC (top 50 cm soil depth) and 6.81 PgC (top 1 m), which highlights the need for the protection of soil with the appropriate conservation strategies to mitigate greenhouse gas emissions and associated climate changes. Incessant burning of fossil fuel (Mundaca 2017), escalated industrial activities (Liu and Bae 2018), higher deforestation (Bharath et al. 2013a, b), and land degradation (Ramachandra et al. 2018a; b) highlight the extent of anthropogenic-induced global warming. This unrestrained increase in global atmospheric carbon since the dawn of the industrial revolution and implications such as changes in the climate on water and food security have driven the attention of policymakers across the globe to focus on the Earth's carbon stocks and flows. Large-scale land use land cover changes (LULC) altering the integrity of forests, soil, and aquatic ecosystems with the associated emissions have been contributing to higher greenhouse gas (GHG) footprint. LULC changes have eroded the sequestration capability directly and disturbed the amount of vegetation residues (organic matter) returned to the soil (Pongratz et al. 2014). LULC changes have been posing a more significant threat by altering their potential of sequestration, escalating vegetation die-off, and increasing instances of wildfire (Gallardo et al. 2016) and have contributed to about one-third of all anthropogenic carbon (Houghton and Hackler 1999). LULC change induced deforestation, resulting in 90% of net carbon emissions across the globe and acting as a source of 20% of annual greenhouse gas emissions into the atmosphere (IPCC 2007). This has prioritized the need to understand LULC changes with the associated decline of biomass and carbon storage for framing international policy strategies to reduce greenhouse gas emissions by reducing the abrupt LULC changes. LULC changes and their impacts vary across the regions, necessitating region-specific management (Ramachandra et al. 2019a, b) in contrast to global policy and regulation. Agriculture, energy production, industrial activities, waste mismanagement, and transportation are the major carbon-emitting sectors to be accounted for carbon budgeting as mismanagement in these sectors has contributed to a higher quantum of greenhouse gas emissions (Singh et al. 2008; Ali et al. 2011; Yu et al. 2014; Abbas et al. 2018).

Advancements in Geoinformatics (GIS technologies) coupled with the availability of multi-resolution spatial data at regular intervals acquired through spaceborne sensors (remote sensing data) with field data have aided in the land use land cover mapping, quantification of above-ground biomass (AGB), below-ground biomass (BGB) and soil carbon. Remote sensing with continuous data support has been helpful in the quantification of carbon footprint through measurement of carbon stock and emissions, which vary with the climate, land use practices, and changes in the land cover and land uses (Dadhwal et al. 2009; Ramachandra and Bharath 2019a, b).

Afforestation with the location-specific endemic species of vegetation, arresting the deforestation process through improved regulatory mechanisms, the transition to energy-efficient devices, and environmentally sound technologies are potential approaches for sequestering carbon and mitigating carbon emissions. Plants (trees, grasses, herbs) take up atmospheric carbon dioxide during photosynthesis and are stored as carbon in biomass (trunks, branches, foliage, roots) and soils. Storing carbon in forests or through plantations in the form of standing biomass constitutes a potential carbon capture and storage (CCS) option (Ni et al. 2016). The global potential of carbon sequestration through plants was estimated as 5–15 Gt C/year, which depends on land-use practices, climate, etc. (Scholz and Hasse 2008). REDD and REDD+ initiative (Reducing Emission from Deforestation and Degradation) developed by Parties to the United Nations Framework Convention on Climate Change (UNFCCC) is an efficient strategy to promote conservation while reducing greenhouse gas emissions due to deforestation and forest degradation (accounting for 11% of global carbon emissions). Mitigation of impacts of the changes in climate and stabilizing global average temperatures within two degrees Celsius entails reducing emissions from the forest sector and other sector mitigation actions. REDD+ provides financial incentives to developing countries to reduce emissions from forested lands and invest in low-carbon paths to sustainable development through increasing forest cover, lessening nationwide land degradation, arresting deforestation, mitigating carbon emissions, and reducing degradation of various geographical regions (Sheikh et al. 2011). As per the Kyoto Protocol obligations, the carbon credit payment scheme is another initiative to curb carbon towards carbon sequestration through effective management. The scheme allocated credits according to the actual amount of carbon sequestered by the trees as modest land-based ($77.91 per hectare per year) and tree-based ($0.2 per m^3 per year) to minimize the abandonment or degradation of forests (Guthrie and Kumareswaran 2009). These international initiatives towards mitigating carbon dioxide emissions through improved forestry activities necessitate understanding spatial and temporal carbon dynamics.

1.4 Biotic Elements of Flora and Fauna

Karnataka has 5164 species of flora in which the most abundant species found are *Mangifera indica, Cassia fistula, Azadirachta indica, Ficus racemose* and *F. religiosa.* Altogether 294 families were found with Fabaceae, Poaceae constituting the most dominant families. The prime endemic flora includes *Hemidesmus indicus, Erythrina indica, Flacourtia montana, Garcinia gummi_gutta, Holigarna graham, Hopea ponga, Knema attenuata, Saraca asoca, Syzygium mundagam* etc.

Fauna of the state covers birds (812 species), fishes (428), mammals (249), amphibians (165), reptiles (247), and butterflies (438), etc., as per the review. The endemic fauna includes *Macaca silenus, Vandeleuria nilagirica, Viverra civettina, Fejervarya nilagirica, Indirana beddomii, Indirana gundia, Nasikabatrachus sahyadrensis, Turdoides subrufa, Garra bicornuta, Puntius cauveriensis, Tor mussullah, Mabuya beddomii, Trimeresurus malabaricus* etc. The grid-wise spatial distribution reveals that the highest is in Uttara Kannada, and the lowest is in the Koppal and Raichur districts.

Globally, establishing protected areas (PA) has expanded measures for conservation, and according to the International Union for the Conservation of Nature (IUCN), nearly 15.2% of the global land surface is now under some form of protection. Mysore Elephant Conservation Reserve is a critical habitat for elephants and provides a rich resource base. Karnataka has a primary elephant corridor connecting northern and southern portions as a Mysore elephant reserve boundary. Mysore Elephant Reserve has 6072 elephants as per the Karnataka state census (2012) estimations over an area of 6,463 sq.km (Baskaran et al. 2011). Mysore Elephant Reserve connects Kerala, Karnataka, and Tamilnadu state forests. The instances of human-elephant are widely observed due to discontinuity and disturbances with the fragmentation of forests.

1.5 Characteristics of Soil

Soil is the basis of all forms of life on planet Earth. All living organisms absorb their food and nutrients from the soil; the decay of all dead organisms and plants occurs in the soil. Hence the soil can be described as the source and sink of all nutrients. Soil is a reservoir of nutrients in one form or another and differs from the parent material and the morphological, physical, chemical and biological properties.

Characteristics of soil: it is necessary to know the characteristics of the soil to understand the basic properties of soil.

- Chemical properties: this refers to the soil's ability to react with water, pesticides, and fertilisers, which will alter the properties of the soil. The main elements present in the soil with high concentrations are oxygen, silicon, aluminium, iron, calcium, sodium, potassium and magnesium. Some chemical properties are discussed below:
  - Soil pH: it is defined as the negative logarithmic concentration of hydrogen ion present in soil. Soil acidification is a natural process, which is increased due to the decomposition of acids and acid-forming substances from the atmosphere. Alkalinity occurs where there is a high degree of base saturation.
  - Ion exchange: the reversible process in which anions and cations are exchanged between solids and liquid phases when they are in close contact with each other. It is an important factor for all processes occurring in soil. Some chemical and physical processes connected with ion exchange are weathering of minerals, nutrient absorption, swelling and shrinkage of clay, leaching of soluble salts, etc.
  - Electrical conductivity: it is the measure of the amount of salts present in the soil. It is a reciprocal of resistance and it is measured with a conductivity meter.

Constituents of soil: soil composition will vary depending on the origin, time, and the natural process involved in the formation. The proportion may also vary between the horizons. The major constituents in the topsoil are water, soil air, minerals, and organic matter.

Mineral and organic components: the soil is made up of mineral and organic matter and contain pore spaces filled by water and air. The minerals may include residues from the parent material, as the products of weathering. Organic matter includes dead plants and animals which are decomposed partially or fully. Table 2 lists soil characteristics ranges (chemical and nutrients) used for assessing the health of agricultural soil.

Table 2 Ranges of soil characteristics (chemical and nutrients)

Values	Range		
	Low	*Medium*	*High*
Potassium (Kg/ha)	<118	118–280	>280
Nitrogen (Kg/ha)	<280	580–560	>560
Phosphorus (Kg/ha)	<10	10–25	>25
	Acidic	*Neutral*	*Alkaline*
pH	<6.5	6.5–7.5	>7.5
Micro nutrients	*Low*	*Medium*	*High*
	Deficient[a]	*Sufficient*[a]	
Zinc (ppm)	<0.5	0.5–1.2	>1.2
Iron (ppm)	<6.3	>6.3	>20
Organic carbon (%)	<0.5	>0.5	>5

[a]As per the department of agriculture categorization (for soil health monitoring)

2 Data Sources

Table 3 lists the ecosystem condition indicators chosen for the pilot study in Karnataka State, India, with details of the method adopted and data source.

Table 3 Ecosystem condition indicators with details of the method adopted and data source

Ecosystem condition indicator	Method	Data source	Reference or target condition/ state variable	Scope
Fragmentation of ecosystems	Ecosystem extent assessment Computation of P_f and P_{ff} Figures 2 and 3 explain the method adopted	Land use data based on temporal remote sensing data classification	P_f and P_{ff}–details in Figs. 1 and 3 and Table 7	Forests

(continued)

(continued)

Ecosystem condition indicator	Method	Data source	Reference or target condition/ state variable	Scope
Land surface temperature	MODIS Land surface temperature (LST) and emissivity data has been retrieved from Land Processes Distributed Active Archive Center (LPDAAC), which has been re-projected from sinusoidal projection to UTM Zone 43 projection with WGS 84 datum using MODIS Re-Projection Tools. To investigate temporal patterns of temperature across the study region 2005, 2012 and 2018 of May month data were used. LST was computed by multiplying DN (Digital numbers) with a scale factor of 0.02 and then converted into degree Celsius Figure 4 depicts the protocol for estimating LST	MODIS Land surface temperature (LST) and emissivity data, 1 km resolution and LU data	LST compared with the ground measurements State variable–LST of the reference year	All Ecosystems
Vegetation	(i) Assessment of land use dynamics using temporal remote sensing data; (ii) Field data collection to classify remote sensing data, (iii) Quantification of AGB through field measurements of girth and height and sampling of the locations through transect-based quadrat	Remote sensing data Field data–AGB, BGB, and NPP estimations	AGB, BGB, and NPP	Karnataka–District wise

(continued)

(continued)

Ecosystem condition indicator	Method	Data source	Reference or target condition/ state variable	Scope
Biotic Elements Flora, Fauna	Transect-based quadrat method for mapping flora and fauna Data from published literature, Biodiversity portals, Karnataka Biodiversity Board–published reports Karnataka Forest Department Publications of EMPRI–Environmental Management Policy Research Institute	Field investigations Data from the Karnataka Biodiversity Board (KBB)	Threat status–as per IUCN and KBB	All districts–forests
Soil	Soil physical and chemical parameters–macro variables from NSSSLUP, GoK, and soil nutrients at block levels (village-wise for the entire Karnataka) Figure 5 depicts the procedure adopted for assessing soil health	Details are provided next	Soil characteristics in 1998 compared with the latest field data	All Ecosystems

Physical and chemical characteristics of Soil: District-Wise for Karnataka

Physical and chemical characteristics of soil at macro level with data source are listed in Table 4.

Table 4 Macro level data–physical and chemical characteristics of soil

Sl.No.	Parameter	Data source and resolution	Explanation
1	Geology, River Basin, Agro-ecological sub division, Agro climatic zones	National Bureau of Soil Survey and Land Use Planning (NBSS&LUP) (1:250,000 scale) (https://nbsslup.in/), 2000	Geological formation of soil based on parent rock
2			Major rivers serve water for various purposes
3			Agro-Ecological sub-division is based on climatic conditions and the length of the growing period of crops
4			Agro-climatic zones are based on soil types, crops, cropping patterns, and physiography

(continued)

(continued)

Sl.No.	Parameter	Data source and resolution	Explanation
5	Soil orders	NBSS&LUP (1:250,000 scale), 2000	Classified based on soil taxonomy
6	Soil classification		Traditional classification of soil
7	Soil erosion	NBSS&LUP (1:250,000 scale), 2000	Soil erosion results from burning, excessive grazing, forest cutting, tillage, etc
8	Soil depth	NBSS&LUP (1:250,000 scale), 2000	Depth of a soil in which nutrients are available and can hold water
9	Soil gravelliness	NBSS&LUP (1:250,000 scale), 2000	Gravel is the term used for coarse fragments between 2 mm and 7.5 cm in diameter. It influences the soil moisture storage capacity, infiltration and plant growth
10	Soil moisture	https://water.karnataka.gov.in, 2000	Soil moisture is the water stored in the soil and which is influenced by precipitation, temperature, soil characteristics, etc
11	Soil organic carbon	https://water.karnataka.gov.in, 2000	It is the total amount of carbon content present in the soil in the form of organic matter
12	Soil slope	NBSS&LUP (1:250,000 scale), 2000	It is the inclination of the surface of the land. It influences soil formation depth, and affects land use and land development
13	Soil texture	NBSS&LUP (1:250,000 scale), 2000	Soil texture refers to the relative proportion of soil primary particles of sand, silt, and clay. It is classified as Sandy soil, loamy soil, and clayey soil
14	Soil pH	https://water.karnataka.gov.in	It is the concentration of hydrogen ions present in the soil. It ranges from 1–14. If less than 6.5, it is acidic, between 6.5 and 7.5 neutral and greater than 7.5 alkaline
15	Nitrogen status	NBSS&LUP (1:250,000 scale)	It is the major nutrient required for plant growth. It is measured in Kg/ha
16	Iron status	NBSS&LUP (1:250,000 scale)	It is the minor nutrient required for plant growth. It is measured in parts per million (ppm)
17	Zinc status	NBSS&LUP (1:250,000 scale), 2000	It is the minor nutrient required for plant growth. It is measured in parts per million (ppm)
18	Potassium status	NBSS&LUP (1:250,000 scale), 2000	It is the major nutrient required for plant growth. It is measured in Kg/ha

(continued)

(continued)

Sl.No.	Parameter	Data source and resolution	Explanation
19	Phosphorus status	NBSS&LUP (1:250,000 scale), 2000	It is the major nutrient required for plant growth. It is measured in Kg/ha

Micro Level Data: Data corresponding to **241,373 sampling plots** (sampled during 2018–2020) across villages and districts in Karnataka were compiled from the Department of Agriculture at the respective district headquarters. Micro-level data details are given in Tables 5 and 6.

Table 5 Micro level soil data

Parameters (physical and chemical):	Soil pH, Electrical Conductivity, Organic Carbon, Nitrogen Availability, Phosphorus Availability, Potassium Availability, Sulphur Availability
Microelements	Zinc, Iron, Copper, Manganese and Boron
Data source	Soil data (details collected in each sample plot, village wise -during 2018–2020) from the respective district Department of Agriculture

Table 6 No of samples, and source of micro level soil data

Sl.No.	District name	Source	No. of sample points
1	Bagalkot	Soil details collected in each sample plot, and village wise from the Department of Agriculture, Bagalkot https://soilhealth.dac.gov.in/PublicReports/GridFormNSVW/Baglkot	20,581
2	Ballari	https://soilhealth.dac.gov.in/PublicReports/GridFormNSVW/Ballari	2304
3	Belgaum	Soil details were collected in each sample plot, and village wise from the Department of Agriculture	15,450
4	Bengaluru Rural	https://soilhealth.dac.gov.in/PublicReports/GridFormNSVW/BenagalureRural	362
5	Bengaluru Urban	Soil details were collected in each sample plot, and village wise from the Department of Agriculture	288
6	Bidar	https://soilhealth.dac.gov.in/PublicReports/GridFormNSVW/Bidar	6363
7	Chamarajanagar	https://soilhealth.dac.gov.in/PublicReports/GridFormNSVW/Chamrajnagar	49
8	Chikballapur	https://soilhealth.dac.gov.in/PublicReports/GridFormNSVW/Chikaballapur	5129
9	Chikmagalur	Soil details were collected in each sample plot, and village wise from the Department of Agriculture	19,357

(continued)

(continued)

Sl.No.	District name	Source	No. of sample points
10	Chitradurga	Soil details were collected in each sample plot, and village wise from the Department of Agriculture	27,731
11	Dakshina Kannada	Soil details were collected in each sample plot, and village wise from the Department of Agriculture	27,253
12	Davangere	Soil details were collected in each sample plot, and village wise from the Department of Agriculture	39,815
13	Dharwad	https://soilhealth.dac.gov.in/PublicReports/GridFormNSVW/Dharwad	382
14	Gadag	https://soilhealth.dac.gov.in/PublicReports/GridFormNSVW/Gadag	7
15	Hassan	https://soilhealth.dac.gov.in/PublicReports/GridFormNSVW/Hassan	4477
16	Haveri	https://soilhealth.dac.gov.in/PublicReports/GridFormNSVW/Havari	7813
17	Kalaburagi	https://soilhealth.dac.gov.in/PublicReports/GridFormNSVW/Kalaburgi	804
18	Kodagu	https://soilhealth.dac.gov.in/PublicReports/GridFormNSVW/Kodagu	2611
19	Kolar	Soil details collected in each sample plot, village wise from the Department of Agriculture	8593
20	Koppal	https://soilhealth.dac.gov.in/PublicReports/GridFormNSVW/Koppal	5755
21	Mandya	Soil details collected in each sample plot, and village wise from the Department of Agriculture, Bagalkot	787
22	Mysuru	Soil details were collected in each sample plot, and village wise from the Department of Agriculture	1245
23	Raichur	https://soilhealth.dac.gov.in/PublicReports/GridFormNSVW/Raichur	30
24	Ramanagara	https://soilhealth.dac.gov.in/PublicReports/GridFormNSVW/Ramanagara	234
25	Shimoga	Soil details were collected in each sample plot, and village wise from the Department of Agriculture	5954
26	Tumakuru	Soil details were collected in each sample plot, and village wise from the Department of Agriculture	3957
27	Udupi	https://soilhealth.dac.gov.in/PublicReports/GridFormNSVW/Udupi	1191
28	Uttara Kannada	Soil details collected in each sample plot, village wise from the Department of Agriculture	31,166
29	Vijayapura	https://soilhealth.dac.gov.in/PublicReports/GridFormNSVW/Vijayapura	1673
30	Yadgir	https://soilhealth.dac.gov.in/PublicReports/GridFormNSVW/Yadgir	12
	Total–sampled locations		241,373

3 Methods

3.1 Forest Fragmentation

Figure 2 elucidates the protocol for analysing the fragmentation of forests using land-use data. Forest fragmentation categories (at pixel level) are computed through P_f (the ratio of pixels that are forested to the total non-water pixels in the window) and P_{ff} (the proportion of all adjacent (cardinal directions only) pixel pairs that include at least one forest pixel, for which both pixels are forested). In other words, the P_f is simply the proportion of non-water pixels within the moving window with a specified size that is forest, and P_{ff} is the ratio of the number of pixel pairs in cardinal directions that are both forests divided by the number of pixel pairs in cardinal directions where either one or both are forested. P_{ff} estimates the conditional probability that given a pixel of the forest, its neighbour is also the forest. A moving window with the size of 5×5 pixels (Fig. 3) was used for the fragmentation analysis to maintain a fair representation of the proportion (P_f) of pixels in the window and to maintain the interior forest at an appropriate level, due to the fact that the outcomes of the model are scale-dependent and threshold dependent (Riitters et al. 2000). The spatial maps of forest fragmentation components are thus derived by computing 'P_f' and 'P_{ff}' using the sliding kernel (of size 5×5) based on the total extent of the forest and its occurrence as adjacent pixels. A pixel is categorized based on the type of fragmentation, and the result of the kernel is stored at the location of the center pixel (in the derived map), which represents between-pixel fragmentation around the corresponding forest location. Details of levels of fragmentation with discriminant criteria of each component are listed in Table 7 based on two indices, P_f and P_{ff}. Depending on these indices values, the analysis will derive different fragmentation components to assess the health of a region's forest. Forest pixels that comprise a small forested area surrounded by non-forested land cover are referred as **Patch** forests. Forest pixels that define the boundary between interior forest and large non-forested land cover features are **Edge** forests. **Perforated** forest refers to pixels that are the boundary between interior forests and relatively small clearings (perforations) within the forested landscape. The **interior** forest is the forest pixels that are reasonably far away from the forest, non-forest boundary. Interior forested areas are surrounded by more thick forested areas. Transitional types depend on non-forest and edge patterns. These are in between edge types and non-forest types. If higher pixels are non-forest then they will be tending to non-forest cover with a higher degree of edge. The water bodies or river courses are non-fragmenting features because they act as natural corridors in the forested landscape. Non-forested areas, including buildings, roads, agricultural fields, and barren land, and developed land, are considered fragmenting features.

The computation of P_f and P_{ff} is explained in Fig. 3, considering a 5×5 kernel with 25 pixels shown. Forest pixels are shaded (15 pixels), non-forest pixels are not

Fig. 2 Method for fragmentation analysis considering LU

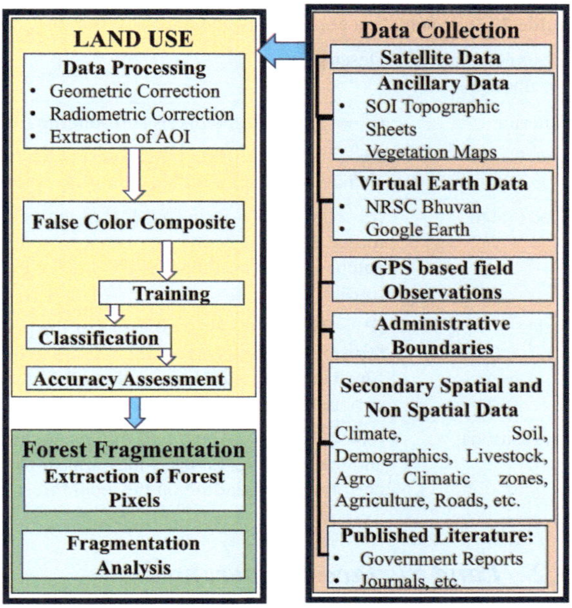

Fig. 3 KERNEL (5 × 5) for computation of P_f, P_{ff} values

shaded, and P_f is 15/25 = 0. 6. Considering pairs of forest pixels in cardinal directions, the total number of adjacent pixel pairs is 40, and of these, 40 pairs include at least one forested pixel. Eight of those 40 pairs are **forest-forest** pairs, so P_{ff} equals 8/40 = 0.2.

Table 7 Fragmentation components and their description

Fragmentation component	Description
Interior	($P_f = 1$), All of the pixels surrounding the center pixel are forest
Patch	($P_f < 0.4$), Pixel is part of a forest patch on a non-forest background, such as a small wooded lot within a built-up region
Perforated	($P_f > 0.6$ and $P_f – P_{ff} > 0$), Most of the pixels in the surrounding area are forested, but the center pixel appears to be part of the inside edge of a forest patch, such as would occur if a small clearing was made within a patch of forest
Edge	($P_f > 0.6$ and $P_f - P_{ff} < 0$), Most of the pixels in the surrounding area are forested, but the center pixel appears to be part of the outside edge of the forest, such as would occur along the boundary of a large built-up area or agricultural field
Transitional	($0.4 < P_f < 0.6$), About half of the cells in the surrounding area are forested and the center forest pixel may appear to be part of a patch, edge, or perforation depending on the local forest pattern

3.2 *Land Surface Temperature*

The method adopted for the analysis is illustrated in Fig. 4. LU analysis involved (i) generation of False Color Composite (FCC) of remote sensing data (bands–green, red, and NIR). This composite image helps in locating heterogeneous patches in the landscape, (ii) selection of training polygons by covering 15% of the study area (polygons are uniformly distributed over the entire study area), (iii) 60% of the training data has been used for classification, while the balance is used for validation

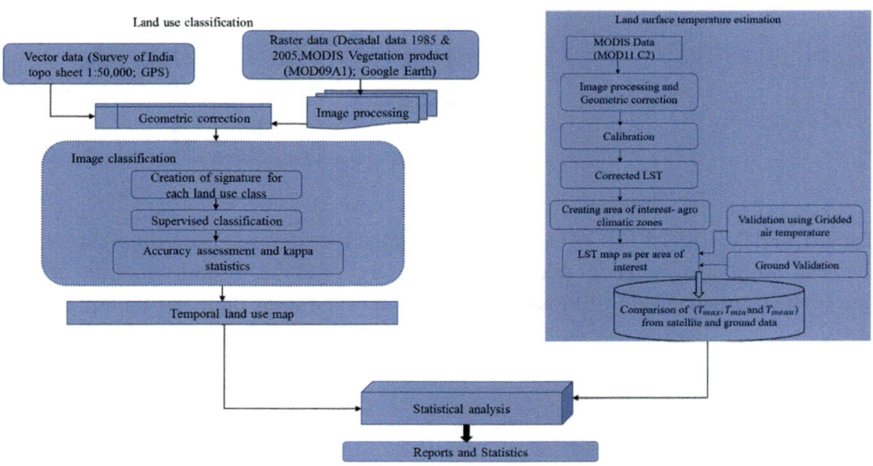

Fig. 4 Method adopted for LST

or accuracy assessment. (iv) LU analysis was done using a supervised classification technique based on the Gaussian maximum likelihood algorithm through GRASS GIS (Geographical Resources Analysis Support System). A free and open-source software accessible at http://wgbis.ces.iisc.ac.in/grass/index.php.

MODIS Land surface temperature (LST) and emissivity data have been retrieved from the Land Process Distributed Active Archive Center (LPDAAC), which has been re-projected from sinusoidal projection UTM Zone 43 projection with WGS 84 datum using MODIS Re-Projection Tools. To investigate temporal patterns of temperature across the study region 2005, 2012, and 2018 of May month data were used. LST was computed by multiplying DN (Digital numbers) with a scale factor of 0.02 and then converted into degrees Celsius. Validation of land surface temperature was done using the ground data of 0.5 degree from Princeton University, which gives air temperature above the 2 m ground level. Finally, statistical analysis was performed to understand the relationship between various LU classes and LST.

3.3 Vegetation Structure in Karnataka

The protocol adopted to assess the vegetation in Karnataka is presented in Fig. 5. The research involved (i) assessment of land-use dynamics using temporal remote sensing data; (ii) field data collection to classify remote sensing data, (iii) quantification of AGB through field measurements of girth and height and sampling of the locations through transect-based quadrat; (iv) data mining.

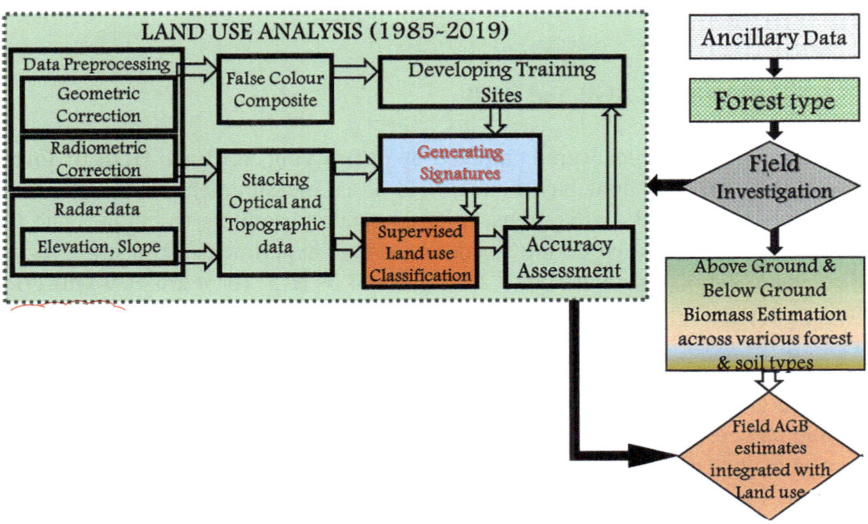

Fig. 5 Method adopted for ABG and BGB budgeting for Karnataka

3.3.1 Land Use Dynamics

Spatial patterns of land use dynamics assessment using temporal remote sensing data: The remote sensing data of Landsat series for 1985, 2005, and 2019 (downloaded from the public domain http://landsat.org) were analysed through a supervised classifier based on Gaussian maximum likelihood algorithm using GRASS GIS software (http://ces.iisc.ernet.in/grass). The training data collected from the field was used to classify the remote sensing data of 2019, coinciding with the field data collection period. Remote sensing data of earlier period were classified using collateral data compiled from various sources such as Karnataka Forest Department reports (http://aranya.gov.in), vegetation map of South India of 1:250,000, the French Institute of India (http://www.ifpindia.org). The process of remote sensing data classification involved (i) the creation of a false-color composite (FCC) using three bands (R, G and NIR) of LANDSAT data, which helped in the selection of training sites through the identification of heterogeneous patches corresponding to diverse landscape elements, (ii) collection of attribute data from the field corresponding to these training polygons using pre-calibrated GPS (Global positioning system) and virtual data (Google Earth–http://earth.google.com, Bhuvan http://bhuvan.nrsc.gov.in), (iii) classification of RS data for land use categories through Gaussian Maximum likelihood algorithm using training data and (iv) accuracy assessment through the computation of error matrix (confusion matrix) and kappa statistics. The Gaussian maximum likelihood classifier (GMLC) is proved to be an efficient supervised classification technique for deriving 8 different LU categories from RS data using free and open-source GRASS GIS (Geographical Analysis Support System–http://wgbis.ces.iisc.ernet.in/grass/). The training data (60%) collected has been used for classification, while the balance used for accuracy assessment and validation (Lillesand et al. 2015).

3.3.2 Estimation of AGB and BGB

The vegetation dynamics were computed by quantifying AGB and BGB of **forest ecosystems**. This was done based on (i) field-based measurements collected across the forests of Karnataka using transect-based quadrat sampling techniques and (ii) published literature based on the standard biomass experiments. The study region (Karnataka State) was divided into 2597 grids of 5' × 5' (or 9 km × 9 km) grids corresponding to 5' × 5' grids of 1:50,000 topographic maps of the Survey of India. Select grids corresponding to agro-climatic zones were chosen for biomass and carbon estimation through field investigations. The basal area, height, vegetation type (evergreen, deciduous, semi-evergreen, moist deciduous, scrub forests), diversity, biomass, etc. were computed based on the collected field data. The field estimations were done across the forest types with about 424 transects in Uttara Kannada, Shimoga, Chikmagalur, Kodagu, Dakshina Kannada, Udupi, and Dharwad districts.

Table 8 Biomass and sequestered carbon based on forest types

Index	Forest type	Equation	Quantification
Biomass (T/Ha)	Evergreen	(Forestcover) × 485.67	Above-ground biomass content
	Deciduous	(Forestcover) × 258.12	
	Scrub	(Forestcover) × 74.25	
	Plantations	(Extent) × 45.25	
Annual Increment in Biomass (T/Ha)	Evergreen	(Forestcover) × 10.48	Incremental growth in biomass (Ramachandra et al. 2000a; Pandey et al. 2011; Devagiri et al. 2013; Do et al. 2018)
	Deciduous	(Forestcover) × 13.82	
	Scrub	(Forestcover) × 5.4	
	Plantations	(Extent) × 1.4	
Net annual Biomass productivity (T/Ha)	Evergreen	(Forestcover) × 3.6	Used to compute the annual availability of woody biomass in the region. (Ramachandra et al. 2000b)
	Deciduous	(Forestcover) × 3.9	
	Scrub	(Forestcover) × 0.5	
	Deciduous	(Forestcover) × 58	
	Scrub	(Forestcover) × 44	
	Agriculture	(Extent) × 2.43	
	Plantations	(Extent) × 55	
Annual Increment of soil carbon (T/Ha)	All	(Cover) × 2.5	Annual increment of carbon stored in the soil

The number of quadrats per transect varied between 3 and 5 depending on the occurrence of species in the sampling locality. The biomass was estimated using GBH (girth at breast height) for the trees >30 cm. The transect data and standard literature data were used for biomass quantification. The biomass, annual increment in biomass of various forest types, and productivity have been computed using field data integrated with the information compiled from the literature, which is listed in Tables 8 and 9, respectively. The probable relationship of biomass with the vegetation cover has been evaluated through multivariate regression analysis across coastal, Sahyadri, and plain regions of Karnataka. The carbon is stored in the soil (SOC) as soil organic matter in both organic and inorganic forms. SOC is calculated based on the field estimations in the top 30 cm soil for different forests (Table 10) and mean soil carbon reported in the literature (Ravindranath et al. 1997; Nair et al. 2009; Ramachandra and Bharath 2019a, b).

Table 9 Above ground Biomass for various forest types and plantations

Sno	Forest type	Standing Biomass (T/ha)	Source
1	Dense Evergreen to Semi evergreen	486–834	Field-based transect cum quadrat method; (Rai 1986; Ramachandra et al. 2000a, b; Ramachandra et al. 2010; Rao et al. 2013; Murthy et al. 2015; Ramachandra and Bharath 2019a, b)
2	Low evergreen	226	
3	Dense Deciduous	258	
4	Degraded Deciduous	130	
5	Savanna Woodlands	75–90	
6	Thorn degraded	40	
7	Littoral and swamp	215	
8	Plantations	45–126	

Table 10 Soil carbon storage in different forest types and agriculture filed

Sno	Forest types	Mean SOC in top 30 cm (t/ha)	Source
1	Tropical Wet Evergreen Forest	132.8	Swamy (1992), Ravindranath et al. (1997), Ravindranath and Ostwald (2007), Kaul et al. (2010), Murthy et al. (2015), Ramachandra and Bharath (2019a, b)
2	Tropical Semi Evergreen Forest	171.75	
3	Tropical Moist Deciduous Forest	57.14	
4	Littoral and Swamp Forest	34.9	
5	Tropical Dry Deciduous Forest	58	
6	Tropical Thorn Forest	44	
7	Tropical Dry Evergreen Forest	33	
8	Agriculture Fields	4	
9	Plantations	**55**	

3.4 Characteristics of Soil

Analysis of soil property in Karnataka outlined in Fig. 6 involved (i). data (macro level, 1:50,000, for 2000) collection from NBSSLUP and (ii) district-wise micro-level data (plot-level soil micro and macronutrient) from the agriculture department offices at the respective district headquarters; II. These point data were converted into spatial data, with georeferencing and digitizing the vector layer. The point data of soil parameters were converted to spatial data covering unsampled plots) through the interpolation (Inverse Distance Weighting) method.

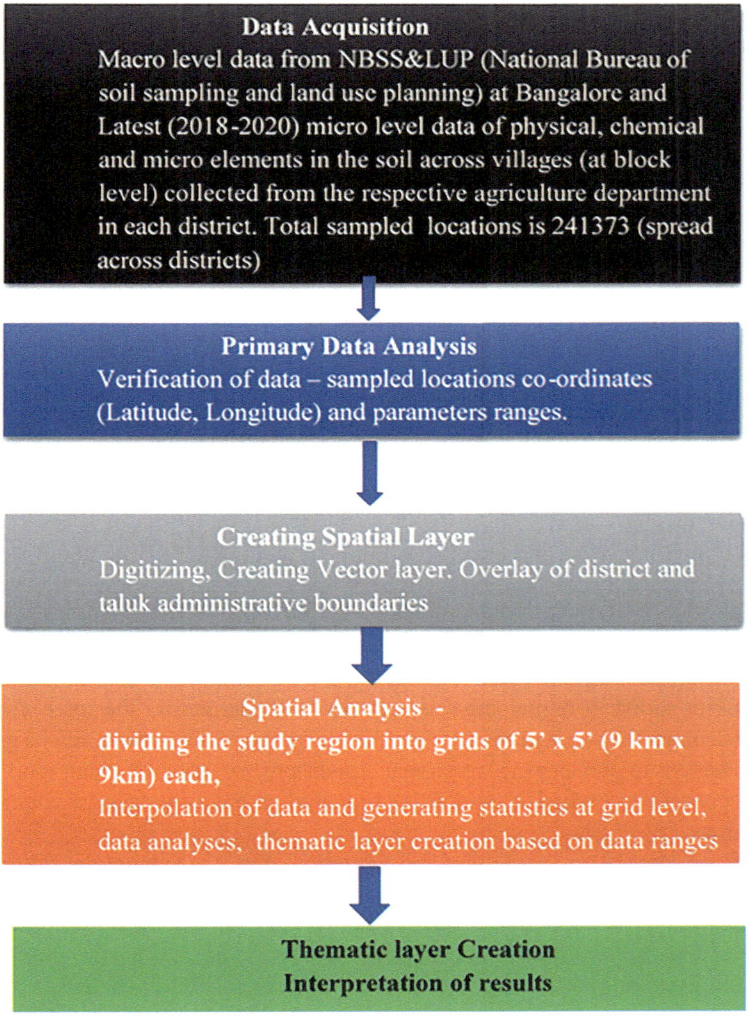

Fig. 6 Method adopted for soil analyses

3.5 *Ecosystem Condition Indicator Account*

3.5.1 Analyses at Disaggregated Levels (Districts and Grids)

The condition indicator account for Karnataka State was developed at disaggregated levels, and the work involved (i) dividing the study region into 9 km × 9 km or 5' × 5' grids, which are comparable to the grids in the Survey of India topographic maps of scale 1:50,000. These grids also overlap with the panchayat spatial extent, which is the administrative unit for decentralized governance as per Panchayath Raj Act, GoI (The Constitution (Seventy-Third Amendment) Act, 1992, Government of India). Study Region was divided into 5' × 5' (9 km × 9 km) grids. These grids correspond to.

(i) Grids of the Survey of India topographic maps of scale 1:50,000
(ii) Each district was divided into 5' × 5' grids. This helps in aggregating various baseline data at the grid level for visualisation
(iii) the spatial extent of each grid corresponding to a panchayat, an administrative unit for the decentralised level of governance
(iv) most of the administrative records are available at 1:50,000. This aids in the compilation of collateral data (vegetation maps of the Karnataka Forest Department)
(v) Appropriate spatial unit for field sampling across agro-climatic zones and collection of primary data collections. This approach aided in the analysis to account for micro-level changes, which is the optimum sample size (which defines the sampling intensity) for recording the data on different inventory parameters.
(vi) Representative grids across agro-climatic regions for detailed field sampling (attribute data compilation corresponding to training polygons for classifying remote sensing data, vegetation sampling, biodiversity inventorying, etc.)
(vii) The extent of ecosystems (forests, plantations, agriculture, water bodies, etc.) is computed at grid level.
(viii) Grids aided in organising the data on condition across the three condition accounts (high, medium and low) based on the relative share across grids.
(ix) Helped in developing reference conditions of past, present, and future ecosystem conditions, which is compared to measure relative change over time
(x) Ecosystem asset accounting and Valuation of ecosystem goods and services.
(xi) As per the Biodiversity Act, 2002 (GoI), in line with GEF guidelines, each panchayat is supposed to have a biodiversity management committee. The information available at panchayat level (corresponding to grids) would empower decentralised governance mechanism –BMC and help in the prudent management of natural resources.

This also aided in computing the opening stock and closing stock at three levels–grids were grouped based on the relative proportion of abiotic, biotic, and landscape parameters as high, medium, and low, as indicated in Table 11.

At the landscape level, the analyses of fragmentation of forests have been carried out at grid levels over a temporal scale. Interior forests depict the contiguity and health of a forest ecosystem, patch forests signify the disturbances, and edge denotes the anthropogenic pressure on forests. Area (proportion of grids corresponding to interior forest cover or contiguous forest) is considered a high indicator of ecosystem condition (reference level), areas under patch and the transitional forest cover are considered as a medium, while grids corresponding to the perforated and edge forest types are considered as low.

Table 11 Classification of ecosystem conditions at three levels

SNO		Variable	Condition		
			High	Medium	Low
1	Landscape Level	Fragmentation	Area of Interior forest type	Area of patch and transitional types	Area of Perforated and Edge types
		LST	< 30 °C	30–35 °C	>35 °C
2	**Biotic-Compositional State**	**Flora**	Occurrence of endemic species and threat status based on IUCN > mean + standard deviation	Occurrence of endemic species (Mean + standard deviation) < **medium** < mean value of species occurrence	Occurrence of endemic species < mean
		Fauna	IUCN CATEGORIES (CR; Critically Endangered; VU: vulnerable)	EN: Endangered	NT: Near Threatened
3	**Biotic–Structural State**	**AGB**	>AGB_Mean + 2SD	AGB_Mean < Medium < AGB_Mean + 2SD	<AGB_Mean
		BGB	>BGB_Mean + 2SD	BGB_Mean < Medium < BGB_Mean + 2SD	<BGB_Mean
4	**Biotic–Functional State**	**NPP_Annual**	>NPPA_Mean + 2SD	NPPA_Mean < Medium < NPPA_Mean + 2SD	<NPPA_Mean

The LST of Karnataka has been analyzed at a temporal scale for each grids. The LST data has been classified into three categories to depict the condition. The regions (proportion of grids) with temperatures less than 30 °C have been grouped under the high category, further confirming that these regions have interior forests. Grids are grouped as medium, with a temperature range from 30 to 35 °C and low conditions for girds with LST >35 °C which denotes the highly urbanized regions with little or no forest cover.

The above-ground biomass (AGB) and below-ground biomass (BGB) of the forest ecosystem are quantified for 2005 and 2019, which are the prime measures to assess the structural state of the ecosystem and greenhouse gas inventory. The study region is divided into ecosystem condition reference levels–high, medium, and low depending on the proportion of AGB considering AGB > ($\mu + 2\sigma$: Mean + 2 standard deviation) for high, grids covering (values within μ and $< \mu + 2\sigma$) AGB values for medium category, while low category includes grids having $<\mu$ AGB. A similar strategy was adopted for BGB and NPP.

4 Results

4.1 *Forest Fragmentation*

Karnataka State: Figure 7 highlights the extent of fragmentation of forests in the State. The fragmentation analysis reveals that the spatial extent of contiguous interior forests constituted only 11,335 km^2 (5.91%) in 2019. The state has witnessed a decline in interior forests by 63.7% from 1985 to 2019. Table 12 lists various levels

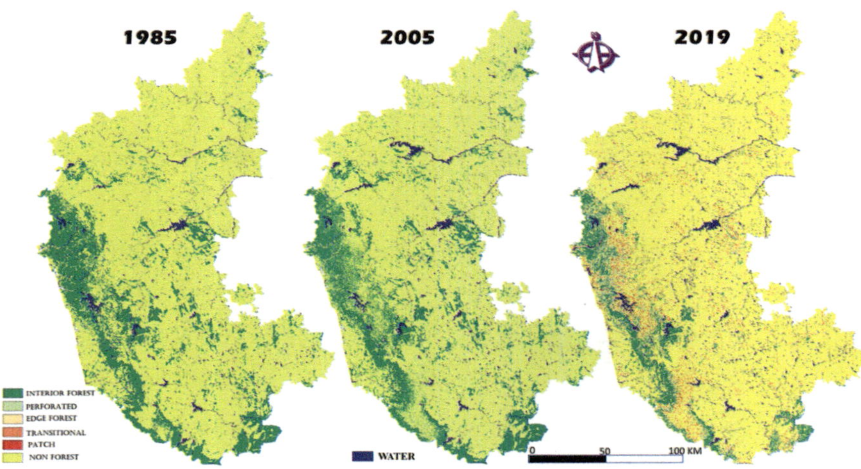

Fig. 7 Forest fragmentation in Karnataka

Table 12 Details of fragmentation of forests in Karnataka

Year	Units	Fragmentation metrics							
		Interior	Patch	Transitional	Edge	Perforated	Non forests	Water	Total
1985	sq.km	31,224	1247	211	3184	1189	150,057	4680	191,791
	%	16.3	0.7	0.1	1.7	0.6	78.2	2.4	
2005	sq.km	24,607	1170	2359	6655	1151	150,671	5178	
	%	12.8	0.6	1.2	3.5	0.6	78.6	2.7	
2019	sq.km	11,335	2839	2071	7365	595	161,661	5926	
	%	5.9	1.5	1.1	3.8	0.3	84.3	3.1	
Changes during 1985 to 2019									
1989	sq.km	31,224	1247	211	3184	1189	150,057	4680	191,791
2019	sq.km	11,335	2839	2071	7365	595	161,661	5926	191,791
Net changes during 1985 to 2019									
	sq.km	−19,889	1592	1860	4181	−595	11,603	1247	
	%	−63.7	127.7	881.8	131.3	−50.0	7.7	26.6	

of forest fragmentations. Forest Survey of India classifies the condition of forests as very dense forests, medium dense forests, open forests. Hence, forest condition was reclassified as per FSI categories depending on the level of fragmentation. Figure 8 depicts forest cover as very dense forests and medium dense forests in Karnataka. Very density forests constitute 11,334 km^2 (5.91%), and medium-density forests are about 12,869 km^2 (6.71%). Table 13 lists the spatial extent of forests (classifications as per the Forest Survey of India) based on various fragmentation levels from 1985 to 2019.

Ecosystem condition indicators (SEEA EA): At the landscape level, the analyses of fragmentation of forests have been carried out at grid levels over a temporal scale. Interior forests depict the forest ecosystem's contiguity and health; patch forests signify disturbances, and edge denotes the anthropogenic pressure on forests. Area (proportion of grids corresponding to interior forest cover or contiguous forest) is considered a high indicator of ecosystem condition, areas under patch and transitional forest cover are considered as medium, while grids corresponding to the perforated and edge types are considered as low. Table 14 lists ecosystem condition indicators of landscape-level characteristics based on the fragmentation of forests in Karnataka.

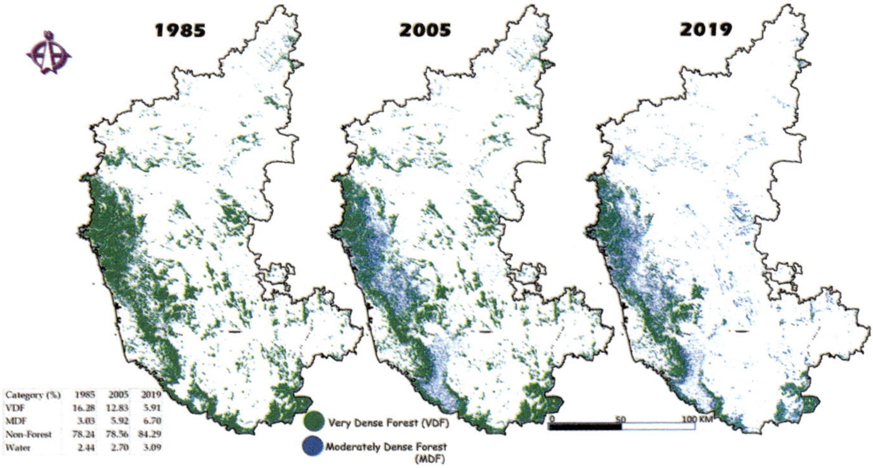

Fig. 8 Forest cover (depending on the level of fragmentation) in Karnataka

Table 13 Spatial extent of forests (as per Forest Survey of India categories) under various levels of fragmentation

Year	Units	Forest types based on fragmentation metrics				
		Very dense forest	Moderately dense forest	Non-forest	Water	Total area
1985	sq.km	31,224	5830	150,057	4680	191,791
	%	16.3	3.0	78.2	2.4	
2005	sq.km	24,607	11,335	150,671	5178	
	%	12.8	5.9	78.6	2.7	
2019	sq.km	11,335	12,869	161,661	5926	
	%	5.9	6.7	84.3	3.1	
Changes during 1985 to 2019						
1985	sq.km	31,224	5830	150,057	4680	191,791
2019	sq.km	11,335	12,869	161,661	5926	191,791
Net changes during 1985 to 2019						
sq.km		−19,889	7039	11,603	1247	
%		−63.7	120.7	7.7	26.6	

Table 14 Ecosystem Condition Indicators: Landscape-level Characteristics of forest ecosystem based on the fragmentation of forests in Karnataka (district-wise)

Karnataka State (district wise)						Ecosystem: Forests			
Districts		Opening stock 2005				Closing stock 2019			
	Units	Total	High	Medium	Low	Total	High	Medium	Low
Bagalkot	Ha	52,882	34,350	4954	13,578	14,150	1042	8807	4301
	%		65.0	9.4	25.7		7.4	62.2	30.4
Bangalore-Rural	Ha	8581	5887	719	1975	3444	532	1414	1498
	%		68.6	8.4	23.0		15.4	41.1	43.5
Bangalore-Urban	Ha	15,821	14,220	383	1218	24,776	3013	13,956	7807
	%		89.9	2.4	7.7		12.2	56.3	31.5
Belgaum	Ha	242,026	191,044	12,686	38,296	241,622	81,580	66,825	93,217
	%		78.9	5.2	15.8		33.8	27.7	38.6
Bellary	Ha	103,257	76,914	8296	18,047	41,214	5587	16,732	18,895
	%		74.5	8.0	17.5		13.6	40.6	45.8
Bidar	Ha	25,224	11,187	5334	8703	8122	1016	4237	2869
	%		44.4	21.1	34.5		12.5	52.2	35.3
Bijapur	Ha	15,365	6457	2820	6088	7908	1287	3116	3505
	%		42.0	18.4	39.6		16.3	39.4	44.3
Chamarajanagar	Ha	287,779	272,557	2928	12,294	250,530	200,521	13,040	36,969
	%		94.7	1.0	4.3		80.0	5.2	14.8
Chikkaballapura	Ha	48,618	37,519	2597	8502	24,414	4197	9856	10,361
	%		77.2	5.3	17.5		17.2	40.4	42.4
Chikkamagaluru	Ha	251,681	209,068	9923	32,690	181,681	95,678	17,392	68,611
	%		83.1	3.9	13.0		52.7	9.6	37.8
Chitradurga	Ha	101,026	81,607	4833	14,586	30,047	5123	10,222	14,702

(continued)

Table 14 (continued)

Karnataka State (district wise)						Ecosystem: Forests			
Districts		Opening stock 2005				Closing stock 2019			
	Units	Total	High	Medium	Low	Total	High	Medium	Low
	%		80.8	4.8	14.4		17.0	34.0	48.9
Dakshina Kannada	Ha	196,726	162,817	7032	26,877	115,575	69,113	10,489	35,973
	%		82.8	3.6	13.7		59.8	9.1	31.1
Davanagere	Ha	111,383	97,292	3367	10,724	77,746	50,803	8464	18,478
	%		87.3	3.0	9.6		65.3	10.9	23.8
Dharwad	Ha	38,297	30,194	1972	6131	27,467	10,811	4557	12,099
	%		78.8	5.1	16.0		39.4	16.6	44.0
Gadag	Ha	32,317	24,644	1875	5798	7894	217	5515	2162
	%		76.3	5.8	17.9		2.7	69.9	27.4
Hassan	Ha	105,449	38,881	37,413	29,155	56,491	8400	26,502	21,589
	%		36.9	35.5	27.6		14.9	46.9	38.2
Haveri	Ha	39,153	29,229	2269	7655	16,033	4997	4927	6109
	%		74.7	5.8	19.6		31.2	30.7	38.1
Kalburgi	Ha	39,678	24,177	5559	9942	10,678	2308	4375	3995
	%		60.9	14.0	25.1		21.6	41.0	37.4
Kodagu	Ha	219,596	93,764	60,034	65,798	186,845	80,734	46,730	59,381
	%		42.7	27.3	30.0		43.2	25.0	31.8
Kolara	Ha	32,253	23,335	2205	6713	10,118	2022	4033	4063
	%		72.3	6.8	20.8		20.0	39.9	40.2
Koppal	Ha	31,361	21,571	2932	6858	11,515	1134	6604	3777
	%		68.8	9.3	21.9		9.8	57.4	32.8

(continued)

Table 14 (continued)

Karnataka State (district wise)						Ecosystem: Forests			
Districts		Opening stock 2005				Closing stock 2019			
	Units	Total	High	Medium	Low	Total	High	Medium	Low
Mandya	Ha	38,598	29,800	2278	6520	37,665	11,244	20,129	6292
	%		77.2	5.9	16.9		29.9	53.4	16.7
Mysuru	Ha	108,087	97,804	2245	8038	103,591	75,591	13,208	14,793
	%		90.5	2.1	7.4		73.0	12.7	14.3
Raichur	Ha	39,030	25,440	3812	9778	19,116	2434	8460	8222
	%		65.2	9.8	25.1		12.7	44.3	43.0
Ramanagara	Ha	86,657	73,756	2775	10,126	36,182	6584	9849	19,749
	%		85.1	3.2	11.7		18.2	27.2	54.6
Shimoga	Ha	362,116	147,013	74,997	140,106	179,200	95,226	31,270	52,704
	%		40.6	20.7	38.7		53.1	17.4	29.4
Tumkur	Ha	125,077	102,859	5466	16,752	124,528	39,074	44,449	41,005
	%		82.2	4.4	13.4		31.4	35.7	32.9
Udupi	Ha	119,965	99,580	4242	16,143	83,450	55,526	6207	21,717
	%		83.0	3.5	13.5		66.5	7.4	26.0
Uttara Kannada	Ha	689,000	378,655	74,366	235,979	486,781	222,849	65,553	198,379
	%		55.0	10.8	34.2		45.8	13.5	40.8
Yadgir	Ha	27,159	19,057	2583	5519	7455	678	4067	2710
	%		70.2	9.5	20.3		9.1	54.6	36.4

Note high = aggregation of grids overing interior forests, medium = aggregation of grids of patch and transitional forests, and low = grids corresponding to perforated and edge forests

4.2 Land Surface Temperature

Karnataka State: The temporal LST results further highlight that the loss of forest cover has increased temperature gradient across the state (Fig. 9). The higher temperature can be seen especially in non-forest areas due to intensified built-up activities. The temperature rise can be attributed to high deforestation with unplanned urbanisation and industrialisation. Table 15 illustrates the temperature gradient across the agro-climatic zones. Urban areas show an increase in temperature (maximum) as population explosion exerts pressure on local resources such as open lands and forest cover. Temporal analysis of LST indicates a rise in mean temperature over the periods. The highest temperature of 48.46 °C was found in the hot, dry semi-arid region, and the lowest was found in the hot humid of 32.46 °C. Figure 10 confirms that LST (satellite temperature) are comparable to ground measurement across the

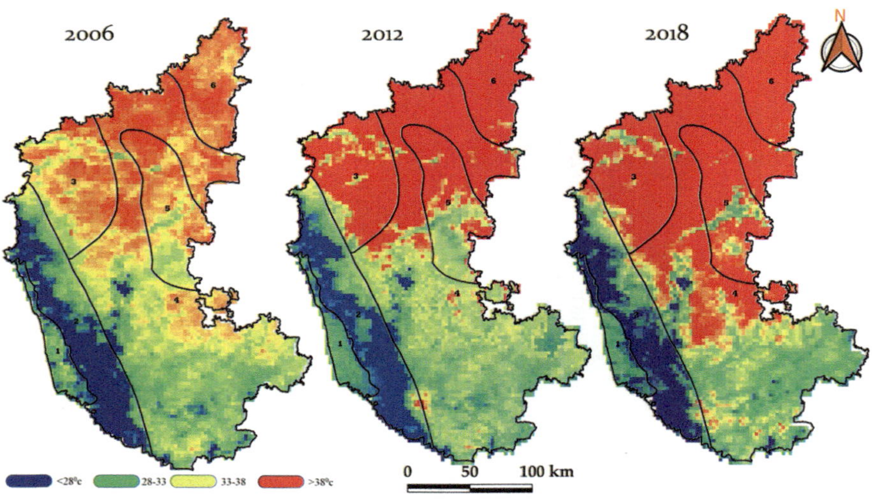

Fig. 9 LST dynamics of Karnataka from 2006 to 2018

Table 15 LST across the agro-climatic zones

Regions	2006		2012		2018	
	Min (°C)	Max (°C)	Min (°C)	Max (°C)	Min (°C)	Max (°C)
Hot dry semi-arid (1)	38.55	41.92	40.35	43.5	43.27	48.46
Arid (2)	33.57	38.08	29.39	33.88	37.21	41.25
Hot moist semi-arid (3)	26.91	41.58	26.49	38.7	28.78	44.57
Hot moist sub humid (4)	22.85	32.38	21.53	27.3	27.11	38.31
Hot humid (5)	26.83	29.9	25.77	28.9	29.89	32.46
Hot dry sub humid (6)	32.49	35.92	29.33	39.08	38.11	45.68

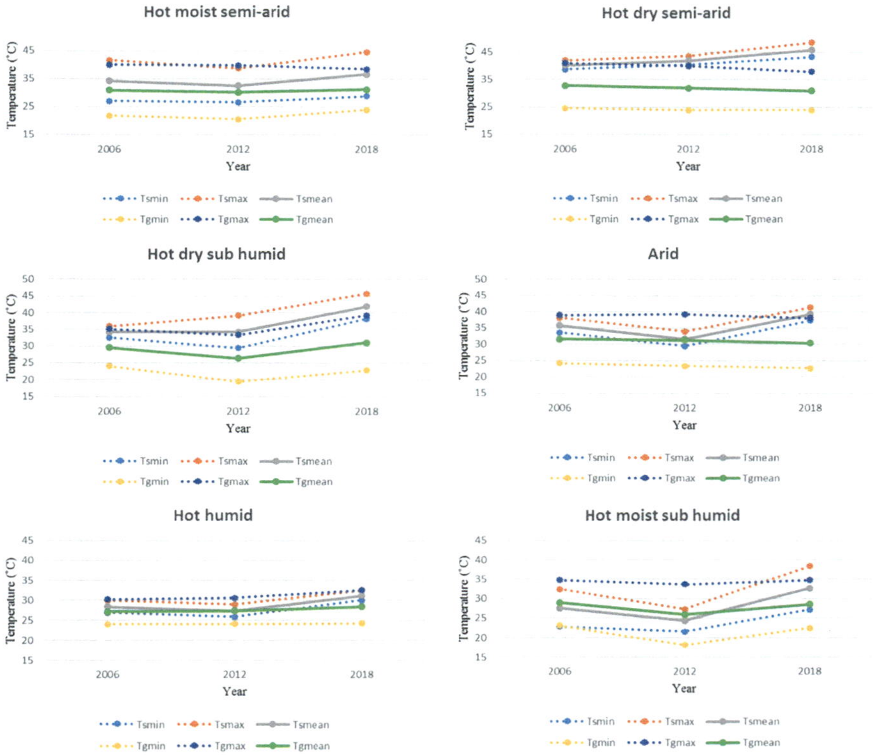

Fig. 10 LST across agro climatic zones from 2006–2018. Where Tsmin = Satellite minimum temperature, Tsmax = Satellite maximum temperature, Tsmean = Satellite mean temperature, Tgmin = Ground minimum temperature, Tgmax = Ground maximum temperature Tgmean = Ground mean temperature

agro climatic zones. The ground air temperature was measured at a distance of 2 m above ground while the satellite gives a temperature of the land surface, which may be 2–3 °C higher.

One km gradient has been used for sampling across the agro-climatic zones from forest and non-forest categories to assess the temperature variations across various land uses. The two samples were taken for each agro-climatic zone except Hot Moist semi-arid (4 samples) due to larger coverage. The correlation analysis shows (Fig. 11) that forest decreases with the increase in temperature, indicating an indirect relationship. This indicates that the forest holds a strong co-relationship of 0.8 (r = 0.94) with temperature. It was observed that an increase in built-up areas, led to an increase in the temperature, which shows a direct relationship. Also, non-forest cover holds a strong co-relationship of 0.5 (r = 0.74) with temperature. The gradients depict as the non-forest cover has increased in the region, depicting a rise in temperature. The results highlight the role of biophysical variables in maintaining

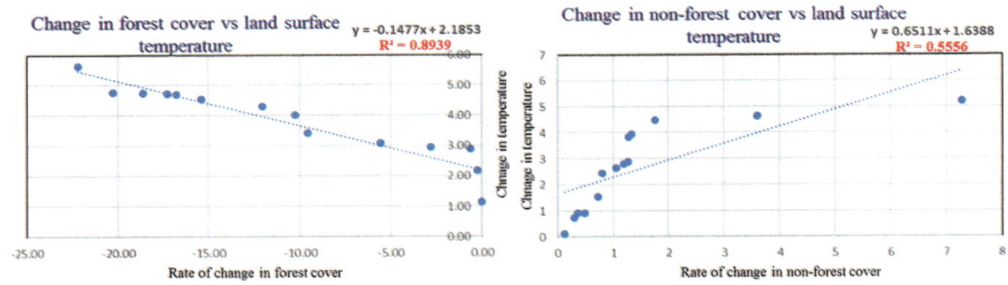

Fig. 11 Correlation among forest, non-forest LU with LST

regional climate. The increase in temperature can induce a higher loss of species, imbalance in the ecosystem, and vegetation die-off. It is suggested that curbing the unplanned developments and increasing forest cover with native species through stakeholder's active involvement would help in the conservation and prevent further degradation of the landscape. Effective forest management with forest regeneration activities would help in reducing LST and its ill effects.

The general trend in the ambient temperature of Karnataka further clarifies that there was a fundamental drift of temperature rise in recent years, especially during the last decade. Therefore, there is an urgent need to mitigate these issues and possible drought conditions. Proper land-use planning and restoration of degraded land through afforestation with native species would help reduce the increase in temperature.

4.3 *Ecosystem Condition Indicator of Landscape Level Characteristics Based on Land Surface Temperature (LST)*

The LST data has been classified into three categories to depict the condition. Table 16 lists ecosystem indicators based on landscape-level characteristics considering land surface temperature, district-wise, for Karnataka state. The regions (proportion of grids) having temperature less than 30 °C have been grouped under the high category, which further confirms the regions of interior forests compared with fragmentation analysis. Grids are grouped as mediums with a temperature range of 30–35 °C and as low conditions for grids with temperature >35 °C which denotes highly urbanized regions with little or no forest cover.

Table 16 Ecosystem Condition Indicators based on Landscape level Characteristics considering land Surface Temperature in Karnataka State (district wise)

Karnataka state (district wise)						Scope: landscape level			
Districts		Opening stock 2005				Closing stock 2019			
	Units	Total	High(<30 °C)	Medium(30–35 °C)	Low(>35 °C)	Total	High (<30 °C)	Medium (30–35 °C)	Low (>35 °C)
Bagalkot	Ha	688,140	747	8870	678,523	688,140	0	11,498	676,642
	%		0.11	1.29	98.60		0.00	1.67	98.33
Bangalore-Rural	Ha	251,101	0	3780	247,321	251,100	0	6596	244,504
	%		0.00	1.51	98.49		0.00	2.63	97.37
Bangalore-Urban	Ha	226,800	0	8771	218,029	226,800	0	25,953	200,847
	%		0.00	3.87	96.13		0.00	11.44	88.56
Belgaum	Ha	1,547,100	11,643	68,806	1,466,651	1,547,100	567	117,563	1,428,970
	%		0.75	4.45	94.80		0.04	7.60	92.36
Bellary	Ha	939,240	1493	48,260	889,487	939,240	54	24,604	914,582
	%		0.16	5.14	94.70		0.01	2.62	97.37
Bidar	Ha	682,740	0	981	681,759	682,740	0	4831	677,909
	%		0.00	0.14	99.86		0.00	0.71	99.29
Bijapur	Ha	1,132,920	306	5001	1,127,613	1,132,920	0	5883	1,127,037
	%		0.03	0.44	99.53		0.00	0.52	99.48
Chamarajanagar	Ha	688,500	38,759	114,849	534,892	688,500	38,839	191,068	458,593
	%		5.63	16.68	77.69		5.64	27.75	66.61
Chikkaballapura	Ha	518,400	0	666	517,734	518,400	0	3860	514,540
	%		0.00	0.13	99.87		0.00	0.74	99.26
Chikkamagaluru	Ha	704,255	67,002	346,041	291,212	704,255	66,422	356,566	281,267
	%		9.51	49.14	41.35		9.43	50.63	39.94

(continued)

Table 16 (continued)

Karnataka state (district wise)						Scope: landscape level			
Districts		Opening stock 2005				Closing stock 2019			
	Units	Total	High(<30 °C)	Medium(30 –35 °C)	Low(>35 °C)	Total	High (<30 °C)	Medium (30–35 °C)	Low (>35 °C)
Chitradurga	Ha	914,760	0	2303	912,457	914,760	0	1179	913,581
	%		0.00	0.25	99.75		0.00	0.13	99.87
Dakshina Kannada	Ha	534,410	28,343	364,955	141,112	534,329	24,974	448,162	61,193
	%		5.30	68.29	26.41		4.67	83.87	11.45
Davanagere	Ha	574,470	20,308	76,853	477,309	574,470	657	106,547	467,266
	%		3.54	13.38	83.09		0.11	18.55	81.34
Dharwad	Ha	413,100	0	0	413,100	413,100	0	0	413,100
	%		0.00	0.00	100.00		0.00	0.00	100.00
Gadag	Ha	461,700	0	0	461,700	461,700	0	599	461,101
	%		0.00	0.00	100.00		0.00	0.13	99.87
Hassan	Ha	655,201	29,516	119,804	505,881	655,200	33,623	220,410	401,167
	%		4.50	18.29	77.21		5.13	33.64	61.23
Haveri	Ha	502,200	0	4405	497,795	502,200	0	8391	493,809
	%		0.00	0.88	99.12		0.00	1.67	98.33
Kalburgi	Ha	1,214,730	0	90	1,214,640	1,214,649	0	2763	1,211,886
	%		0.00	0.01	99.99		0.00	0.23	99.77
Kodagu	Ha	485,730	255,520	209,138	21,072	485,568	255,084	223,207	7277
	%		52.61	43.06	4.34		52.53	45.97	1.50
Kolar	Ha	471,960	0	5040	466,920	471,960	0	23,322	448,638
	%		0.00	1.07	98.93		0.00	4.94	95.06

(continued)

Table 16 (continued)

Karnataka state (district wise)						Scope: landscape level			
Districts		Opening stock 2005				Closing stock 2019			
	Units	Total	High(<30 °C)	Medium(30 –35 °C)	Low(>35 °C)	Total	High (<30 °C)	Medium (30–35 °C)	Low (>35 °C)
Koppal	Ha	541,894	9353	39,167	493,374	541,890	594	11,729	529,567
	%		1.73	7.23	91.05		0.11	2.16	97.73
Mandya	Ha	510,300	1251	148,425	360,624	510,228	0	125,883	384,345
	%		0.25	29.09	70.67		0.00	24.67	75.33
Mysore	Ha	671,760	46,777	108,979	516,004	671,679	24,624	168,063	478,992
	%		6.96	16.22	76.81		3.67	25.02	71.31
Raichur	Ha	923,400	3374	96,935	823,091	923,400	81	14,445	908,874
	%		0.37	10.50	89.14		0.01	1.56	98.43
Ramanagara	Ha	356,400	0	49,059	307,341	356,400	0	31,565	324,835
	%		0.00	13.77	86.23		0.00	8.86	91.14
Shimoga	Ha	858,613	56,216	335,484	466,913	858,600	13,925	537,995	306,680
	%		6.55	39.07	54.38		1.62	62.66	35.72
Tumkur	Ha	1,296,000	0	18,022	1,277,978	1,296,000	0	25,941	1,270,059
	%		0.00	1.39	98.61		0.00	2.00	98.00
Udupi	Ha	359,082	32,335	131,358	195,389	358,920	19,948	186,694	152,278
	%		9.00	36.58	54.41		5.56	52.02	42.43
Uttara Kannada	Ha	1,060,593	218,657	563,911	278,025	1,060,503	6154	740,898	313,451
	%		20.62	53.17	26.21		0.58	69.86	29.56
Yadgir	Ha	591,300	0	18,345	572,955	591,300	0	39,975	551,325
	%		0.00	3.10	96.90		0.00	6.76	93.24

Note High = aggregation of grids having temperature <30 °C, medium = aggregation of grids having temperature (30–35 °C) and low = aggregation of grids having temperature >35 °C

4.4 Vegetation (AGB: Above Ground Biomass, BGB: Below Ground Biomass, NPP: Net Primary Productivity), Distribution of Flora and Fauna

4.4.1 Structure of Vegetation in Forest Ecosystems of Karnataka

This has been done through assessment of above-ground biomass (AGB) and below-ground biomass (BGB) and Net Primary Productivity (NPP) based on (i) field-based measurements across various types of forests in Karnataka using transect based quadrat sampling techniques and (ii) published literature based on the standard biomass experiments. The analyses of the above-ground biomass show that the grids in the Western Ghats part of Karnataka have higher AGB >1000 Gg (Giga gram). The grids of evergreen forested areas represent the greater values of biomass compared to the other forest types. The total AGB of forests is about 1013.7 Tg (Tera gram) with stored carbon of 506.8 Tg (in 1985) which is now reduced to 678 Tg and 339 Tg, respectively (2019). The temporal decline of AGB values in Kodagu, Shimoga, Uttara Kannada, and Dakshina Kannada districts is due to anthropogenic pressure (Fig. 12). The Mysore, Chamrajnagara and Bellary districts also reflect a decline in AGB values from 2005 to 2019.

Temporal BGB highlights the decline from 275 Tg (1985) to 180 Tg (2019). The grids consisting of evergreen forests (of Western Ghats) show higher values of >600 Gg SOC, while other regions are with relatively lower values (Fig. 13).

Karnataka Forest Department has implemented monoculture plantations in the state to protect the land under greening initiatives and sustain market demand for timber. The AGB, BGB were accounted to understand the role of plantations in arresting land degradations. The total AGB and BGB have been estimated as a sum

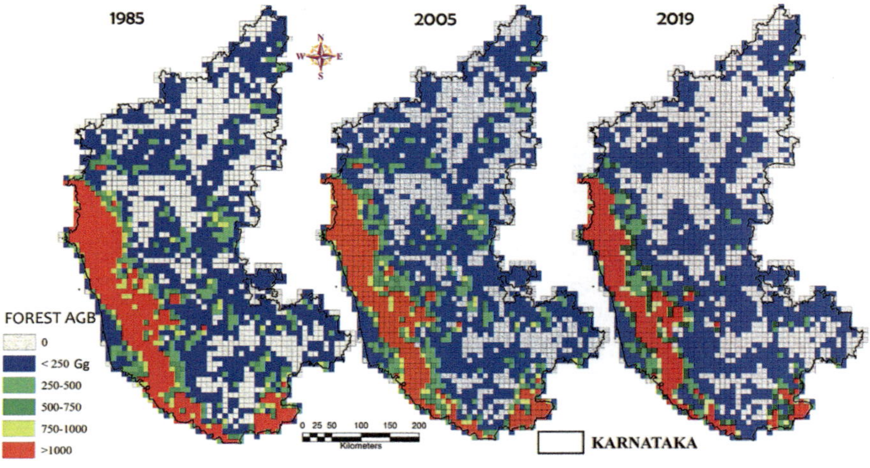

Fig. 12 Temporal AGB in forest areas of Karnataka

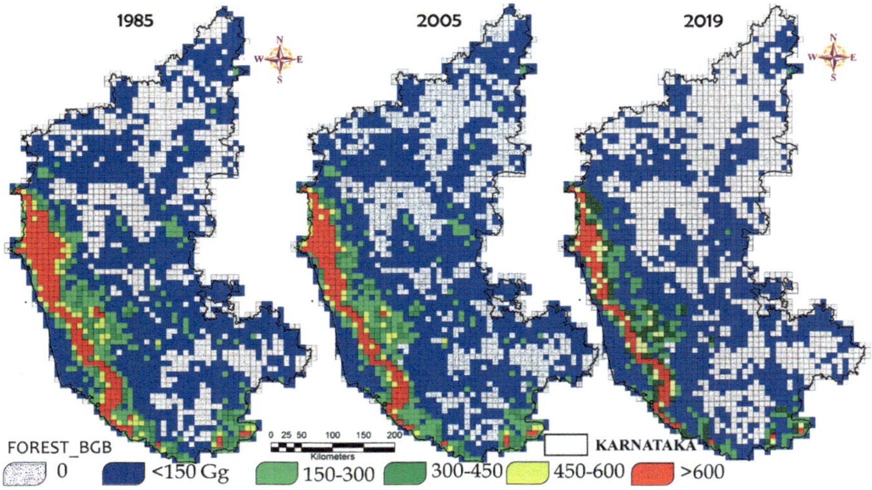

Fig. 13 BGB across the forests of Karnataka from 1985 to 2019

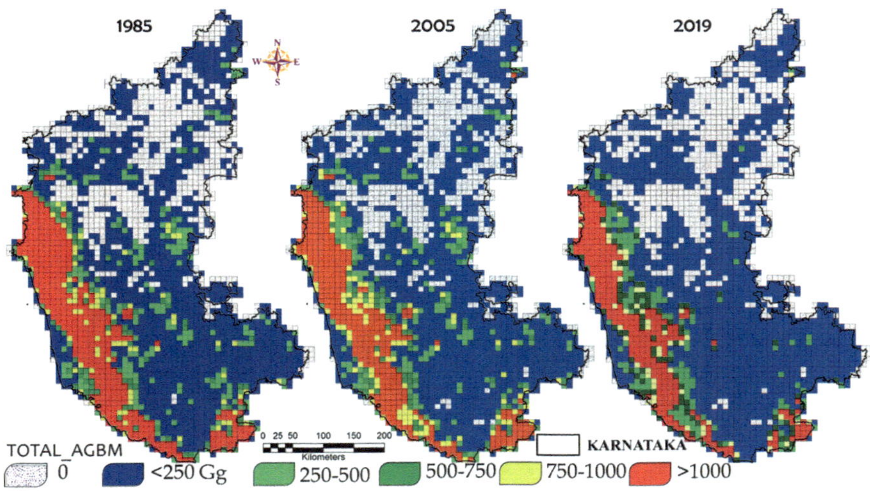

Fig. 14 Total AGB of Karnataka from 1985 to 2019

of forest and forest plantation biomass. Figure 14 shows the AGB for forest and plantations accounted for 1056.90 Tg (in 1985), which is now reduced to 732.83 Tg (2019). Figure 15 shows that BGB from forest plantation and agriculture areas across the state accounted for 275.43 (1985), now reduced to 180.54 Tg. The plantations, though not shown any significant contribution of ecosystem services compared to the forest, have supported carbon sequestration. The Uttara Kannada grids have significant AGB and BGB values.

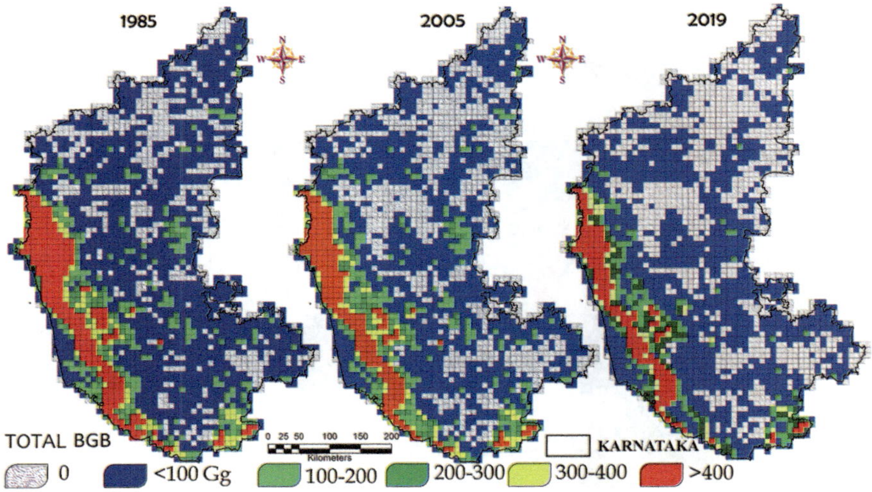

Fig. 15 Total BGB in Karnataka during 1985 to 2019

Total AGB and BGB from forests are about 782.1 Tg (1985), which is reduced to 519.36 Tg (2019) due to LU conversions (Fig. 16). The grids covering Bellary, Mysore, Chamarajanagar, Uttara Kannada, and Kodagu districts have witnessed higher transitions in AGB and BGB dynamics. Figure 17 depicts the changes in NPP in Karnataka from 1985 to 2019. Figure 18 illustrates the distribution of flora and fauna in Karnataka.

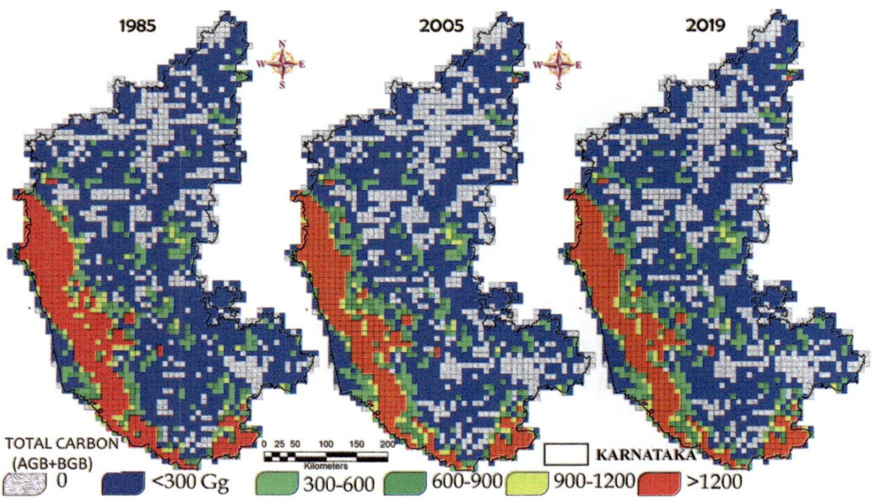

Fig. 16 Total tree biomass AGB & BGB of Karnataka from 1985 to 2019

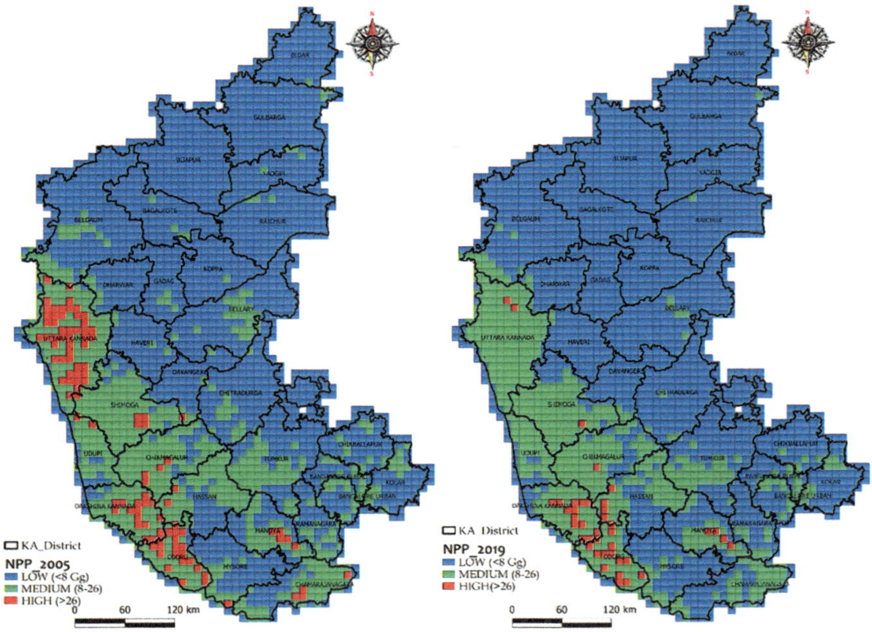

Fig. 17 NPP in forest ecosystems in Karnataka

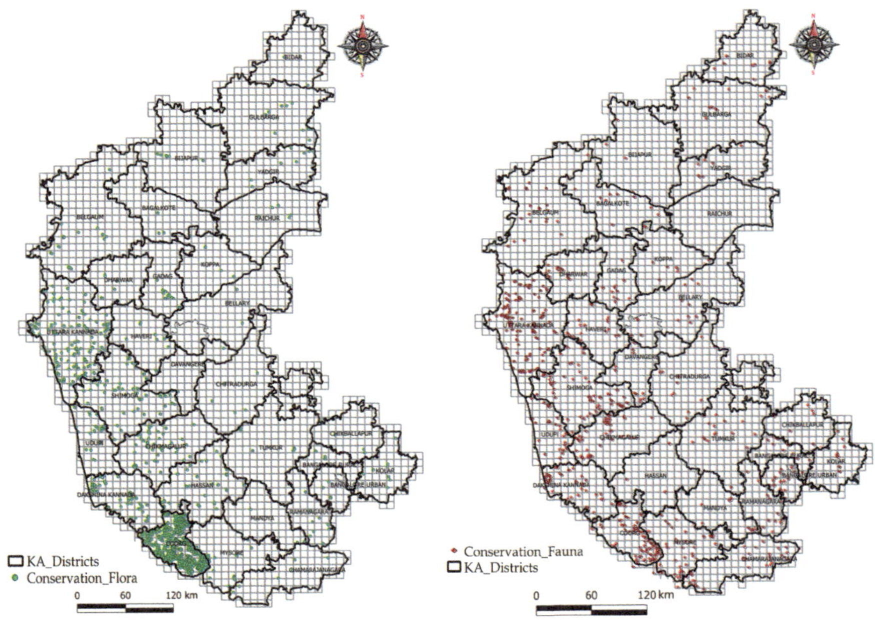

Fig. 18 Flora and fauna species distribution in Karnataka

Ecosystem Condition Indicator based on Biotic ecosystem characteristics considering structural state characteristics (AGB, BGB): The above-ground biomass, below-ground biomass (AGB, BGB) of the forest ecosystem are quantified for the years 2005, and 2019, which are the prime measures to assess the structural state of an ecosystem. The study region is divided into high, medium, and low depending on the proportion of AGB considering AGB > ($\mu + 2\sigma$: Mean+2 standard deviation) for high, grids covering (values within μ and <$\mu + 2\sigma$) AGB values for medium category, while low category includes grids having <μ AGB. A similar strategy was adopted for BGB. Tables 17 and 18 list ecosystem indicators for Karnataka State (district-wise) based on AGB and BGB, respectively.

Ecosystem Condition Indicator based on Biotic ecosystem characteristics considering Compositional state (species-based indicators): Flora (endemic taxa), Fauna (endemic taxa and threat categories based on IUCN, considered–critically endangered, endangered, vulnerable, near threatened, least concern categories).

Ecosystem condition as per SEEA Ecosystem Condition Indicator based on flora and fauna are given in Tables 19 and 20, respectively. Reference levels are computed based on the occurrence of endemic species and threat status based on IUCN as high (>(mean + standard deviation)), medium [Occurrence of endemic species (Mean + standard deviation) < **medium** < mean value of species occurrence] and low (Occurrence of endemic species < mean). Similarly, for fauna as high (species–IUCN categories–CR; Critically Endangered; VU: vulnerable), medium (EN: Endangered) and low (NT: Near Threatened).

Ecosystem Condition Indicator based on Biotic ecosystem characteristics considering functional state characteristics (NPP): The net primary productivity (NPP) of the forest ecosystem is quantified for 2005, and 2019. The study region is divided into high, medium, and low depending on the proportion of NPP, considering NPP > ($\mu + 2\sigma$: Mean + 2 standard deviation) for high, grids covering (values within μ and < $\mu + 2\sigma$) NPP values for medium category, while low category includes grids having < μ NPP to derive the reference state as per SEEA. Table 21 lists ecosystem indicators based on NPP in Karnataka State (district-wise).

Table 17 Ecosystem Condition Indicator based on Biotic ecosystem characteristics considering structural state characteristics (standing biomass–AGB)

Standing Biomass (AGB)						Karnataka State (districtwise)						
Districts	Units	Total	Opening–2005			Total	Closing–2019			2005	2019	
		2005	High	Medium	Low	2019	High	Medium	Low	AGB_Gg	AGB_Gg	% Net Change
Bagalkot	Ha	688,500	0	0	688,500	688,500	0	0	688,500	6363	1643	−74.2
	%		0.0	0.0	100.0		0.0	0.0	100.0			
Bangalore (Rural)	Ha	251,100	0	8100	243,000	251,100	0	0	251,100	2913	1988	−31.7
	%		0.0	3.2	96.8		0.0	0.0	100.0			
Bangalore (Urban)	Ha	226,800	0	8100	218,700	226,800	0	0	226,800	2297	1744	−24.1
	%		0.0	3.6	96.4		0.0	0.0	100.0			
Belgaum	Ha	1,547,100	48,600	162,000	1,336,500	1,547,100	40,500	121,500	1,385,100	46,196	34,730	−24.8
	%		3.1	10.5	86.4		2.6	7.9	89.5			
Bellary	Ha	939,600	0	72,900	866,700	939,600	0	0	939,600	14,999	5401	−64.0
	%		0.0	7.8	92.2		0.0	0.0	100.0			
Bidar	Ha	688,500	0	0	688,500	688,500	0	0	688,500	2231	713	−68.0
	%		0.0	0.0	100.0		0.0	0.0	100.0			
Bijapur	Ha	1,134,000	0	0	1,134,000	1,134,000	0	0	1,134,000	1198	560	−53.2
	%		0.0	0.0	100.0		0.0	0.0	100.0			
Chamarajnagar	Ha	688,500	56,700	388,800	243,000	688,500	32,400	315,900	340,200	76,832	52,383	−31.8
	%		8.2	56.5	35.3		4.7	45.9	49.4			
Chikkaballapura	Ha	518,400	0	8100	510,300	518,400	0	0	518,400	5895	3425	−41.9
	%		0.0	1.6	98.4		0.0	0.0	100.0			
Chikmagalur	Ha	704,700	81,000	348,300	275,400	704,700	72,900	332,100	299,700	77,233	70,965	−8.1
	%		11.5	49.4	39.1		10.3	47.1	42.5			
Chitradurga	Ha	915,300	0	56,700	858,600	915,300	0	16,200	899,100	13,325	6084	−54.3

(continued)

Table 17 (continued)

Standing Biomass (AGB)						Karnataka State (districtwise)						
Districts	Units	Total	Opening–2005			Total	Closing–2019			2005	2019	
		2005	High	Medium	Low	2019	High	Medium	Low	AGB_Gg	AGB_Gg	% Net Change
	%		0.0	6.2	93.8		0.0	1.8	98.2			
Dakshina Kannada	Ha	558,900	121,500	234,900	202,500	558,900	121,500	218,700	218,700	77,191	73,802	−4.4
	%		21.7	42.0	36.2		21.7	39.1	39.1			
Davangere	Ha	575,100	0	40,500	534,600	575,100	0	24,300	550,800	9529	4648	−51.2
	%		0.0	7.0	93.0		0.0	4.2	95.8			
Dharwad	Ha	413,100	0	32,400	380,700	413,100	0	24,300	388,800	5212	3894	−25.3
	%		0.0	7.8	92.2		0.0	5.9	94.1			
Gadag	Ha	461,700	0	24,300	437,400	461,700	0	0	461,700	3482	876	−74.8
	%		0.0	5.3	94.7		0.0	0.0	100.0			
Hassan	Ha	656,100	32,400	89,100	534,600	656,100	16,200	48,600	591,300	28,794	21,400	−25.7
	%		4.9	13.6	81.5		2.5	7.4	90.1			
Haveri	Ha	502,200	0	40,500	461,700	502,200	0	16,200	486,000	5928	2672	−54.9
	%		0.0	8.1	91.9		0.0	3.2	96.8			
Kalburgi	Ha	1,215,000	0	16,200	1,198,800	1,215,000	0	8100	1,206,900	4414	1638	−62.9
	%		0.0	1.3	98.7		0.0	0.7	99.3			
Kodagu	Ha	486,000	113,400	332,100	40,500	486,000	81,000	234,900	170,100	78,795	60,045	−23.8
	%		23.3	68.3	8.3		16.7	48.3	35.0			
Kolar	Ha	494,100	0	16,200	477,900	494,100	0	8100	486,000	6575	3518	−46.5
	%		0.0	3.3	96.7		0.0	1.6	98.4			
Koppal	Ha	542,700	0	8100	534,600	542,700	0	0	542,700	2636	942	−64.3
	%		0.0	1.5	98.5		0.0	0.0	100.0			

(continued)

Table 17 (continued)

Standing Biomass (AGB)						Karnataka State (districtwise)						
Districts	Units	Total	Opening–2005			Total	Closing–2019			2005	2019	
		2005	High	Medium	Low	2019	High	Medium	Low	AGB_Gg	AGB_Gg	% Net Change
Mandya	Ha	510,300	0	24,300	486,000	510,300	0	8100	502,200	8999	6809	−24.3
	%		0.0	4.8	95.2		0.0	1.6	98.4			
Mysore	Ha	672,300	16,200	105,300	550,800	672,300	8100	97,200	567,000	26,204	21,167	−19.2
	%		2.4	15.7	81.9		1.2	14.5	84.3			
Raichur	Ha	923,400	0	8100	915,300	923,400	0	0	923,400	3392	1645	−51.5
	%		0.0	0.9	99.1		0.0	0.0	100.0			
Ramanagara	Ha	356,400	0	89,100	267,300	356,400	0	64,800	291,600	15,452	10,969	−29.0
	%		0.0	25.0	75.0		0.0	18.2	81.8			
Shimoga	Ha	858,600	121,500	486,000	251,100	858,600	72,900	405,000	380,700	105,221	81,112	−22.9
	%		14.2	56.6	29.2		8.5	47.2	44.3			
Tumkur	Ha	1,296,000	0	32,400	1,263,600	1,296,000	0	8100	1,287,900	14,940	8344	−44.1
	%		0.0	2.5	97.5		0.0	0.6	99.4			
Udupi	Ha	405,000	89,100	113,400	202,500	405,000	81,000	121,500	202,500	49,099	46,526	−5.2
	%		22.0	28.0	50.0		20.0	30.0	50.0			
Uttara Kannada	Ha	1,174,500	542,700	413,100	218,700	1,174,500	388,800	494,100	291,600	258,687	202,018	−21.9
	%		46.2	35.2	18.6		33.1	42.1	24.8			
Yadgir	Ha	591,300	0	16,200	575,100	591,300	0	0	591,300	3654	1165	−68.1
	%		0.0	2.7	97.3		0.0	0.0	100.0			

Note AGB > ($\mu + 2\sigma$: Mean + 2 standard deviation) for high, grids covering (values within μ and < $\mu + 2\sigma$) AGB values for Medium category, while low category includes grids having <μ AGB

Table 18 Ecosystem Condition Indicator based on Biotic ecosystem characteristics considering structural state characteristics (BGB) of vegetation for Karnataka State, India

Below Ground Biomass (BGB)						Karnataka State (district wise)						
Districts	Units	Total	Opening–2005			Total	Closing–2019			2005	2019	Net Change
		2005	High	Medium	Low	2019	High	Medium	Low	BGB_Gg	BGB_Gg	%
Bagalkot	Ha	688,500	0	0	688,500	688,500	0	0	688,500	1273	329	−74.2
Bagalkot	%		0.0	0.0	100.0		0.0	0.0	100.0			
Bangalore (Rural)	Ha	251,100	0	8100	243,000	251,100	0	0	251,100	583	398	−31.7
Bangalore (Rural)	%		0.0	3.2	96.8		0.0	0.0	100.0			
Bangalore (Urban)	Ha	226,800	0	8100	218,700	226,800	0	0	226,800	459	349	−24.1
Bangalore (Urban)	%		0.0	3.6	96.4		0.0	0.0	100.0			
Belgaum	Ha	1,547,100	48,600	162,000	1,336,500	1,547,100	40,500	121,500	1,385,100	9239	6946	−24.8
Belgaum	%		3.1	10.5	86.4		2.6	7.9	89.5			
Bellary	Ha	939,600	0	72,900	866,700	939,600	0	0	939,600	3000	1080	−64.0
Bellary	%		0.0	7.8	92.2		0.0	0.0	100.0			
Bidar	Ha	688,500	0	0	688,500	688,500	0	0	688,500	446	143	−68.0
Bidar	%		0.0	0.0	100.0		0.0	0.0	100.0			
Bijapur	Ha	1,134,000	0	0	1,134,000	1,134,000	0	0	1,134,000	240	112	−53.2
Bijapur	%		0.0	0.0	100.0		0.0	0.0	100.0			
Chamarajnagar	Ha	688,500	56,700	388,800	243,000	688,500	32,400	315,900	340,200	15,366	10,477	−31.8
Chamarajnagar	%		8.2	56.5	35.3		4.7	45.9	49.4			
Chikkaballapura	Ha	518,400	0	8100	510,300	518,400	0	0	518,400	1179	685	−41.9
Chikkaballapura	%		0.0	1.6	98.4		0.0	0.0	100.0			

(continued)

Table 18 (continued)

Below Ground Biomass (BGB)						Karnataka State (district wise)						
Districts	Units	Total	Opening–2005			Total	Closing–2019			2005	2019	Net Change
		2005	High	Medium	Low	2019	High	Medium	Low	BGB_Gg	BGB_Gg	%
Chikmagalur	Ha	704,700	81,000	348,300	275,400	704,700	72,900	332,100	299,700	15,447	14,193	−8.1
Chikmagalur	%		11.5	49.4	39.1		10.3	47.1	42.5			
Chitradurga	Ha	915,300	0	56,700	858,600	915,300	0	16,200	899,100	2665	1217	−54.3
Chitradurga	%		0.0	6.2	93.8		0.0	1.8	98.2			
Dakshina Kannada	Ha	558,900	121,500	234,900	202,500	558,900	121,500	218,700	218,700	15,438	14,760	−4.4
Dakshina Kannada	%		21.7	42.0	36.2		21.7	39.1	39.1			
Davangere	Ha	575,100	0	40,500	534,600	575,100	0	24,300	550,800	1906	930	−51.2
Davangere	%		0.0	7.0	93.0		0.0	4.2	95.8			
Dharwad	Ha	413,100	0	32,400	380,700	413,100	0	24,300	388,800	1042	779	−25.3
Dharwad	%		0.0	7.8	92.2		0.0	5.9	94.1			
Gadag	Ha	461,700	0	24,300	437,400	461,700	0	0	461,700	696	175	−74.8
Gadag	%		0.0	5.3	94.7		0.0	0.0	100.0			
Hassan	Ha	656,100	32,400	89,100	534,600	656,100	16,200	48,600	591,300	5759	4280	−25.7
Hassan	%		4.9	13.6	81.5		2.5	7.4	90.1			
Haveri	Ha	502,200	0	40,500	461,700	502,200	0	16,200	486,000	1186	534	−54.9
Haveri	%		0.0	8.1	91.9		0.0	3.2	96.8			
Kalburgi	Ha	1,215,000	0	16,200	1,198,800	1,215,000	0	8100	1,206,900	883	328	−62.9
Kalburgi	%		0.0	1.3	98.7		0.0	0.7	99.3			
Kodagu	Ha	486,000	113,400	332,100	40,500	486,000	81,000	234,900	170,100	15,759	12,009	−23.8
Kodagu	%		23.3	68.3	8.3		16.7	48.3	35.0			

(continued)

Table 18 (continued)

Below Ground Biomass (BGB)						Karnataka State (district wise)						
Districts	Units	Total	Opening–2005			Total	Closing–2019			2005	2019	Net Change
		2005	High	Medium	Low	2019	High	Medium	Low	BGB_Gg	BGB_Gg	%
Kolar	Ha	494,100	0	16,200	477,900	494,100	0	8100	486,000	1315	704	−46.5
Kolar	%		0.0	3.3	96.7		0.0	1.6	98.4			
Koppal	Ha	542,700	0	8100	534,600	542,700	0	0	542,700	527	188	−64.3
Koppal	%		0.0	1.5	98.5		0.0	0.0	100.0			
Mandya	Ha	510,300	0	24,300	486,000	510,300	0	8100	502,200	1800	1362	−24.3
Mandya	%		0.0	4.8	95.2		0.0	1.6	98.4			
Mysore	Ha	672,300	16,200	105,300	550,800	672,300	8100	97,200	567,000	5241	4233	−19.2
Mysore	%		2.4	15.7	81.9		1.2	14.5	84.3			
Raichur	Ha	923,400	0	8100	915,300	923,400	0	0	923,400	678	329	−51.5
Raichur	%		0.0	0.9	99.1		0.0	0.0	100.0			
Ramanagara	Ha	356,400	0	89,100	267,300	356,400	0	64,800	291,600	3090	2194	−29.0
Ramanagara	%		0.0	25.0	75.0		0.0	18.2	81.8			
Shimoga	Ha	858,600	121,500	486,000	251,100	858,600	72,900	405,000	380,700	21,044	16,222	−22.9
Shimoga	%		14.2	56.6	29.2		8.5	47.2	44.3			
Tumkur	Ha	1,296,000	0	32,400	1,263,600	1,296,000	0	8100	1,287,900	2988	1669	−44.1
Tumkur	%		0.0	2.5	97.5		0.0	0.6	99.4			
Udupi	Ha	405,000	89,100	113,400	202,500	405,000	81,000	121,500	202,500	9820	9305	−5.2
Udupi	%		22.0	28.0	50.0		20.0	30.0	50.0			
Uttara Kannada	Ha	1,174,500	542,700	413,100	218,700	1,174,500	388,800	494,100	291,600	51,737	40,404	−21.9
Uttara Kannada	%		46.2	35.2	18.6		33.1	42.1	24.8			
Yadgir	Ha	591,300	0	16,200	575,100	591,300	0	0	591,300	731	233	−68.1
Yadgir	%		0.0	2.7	97.3		0.0	0.0	100.0			

Note BGB > ($\mu + 2\sigma$: Mean + 2 standard deviation) for high, grids covering (values within μ and < $\mu + 2\sigma$) BGB values for Medium category, while low category includes grids having <μ BGB

Table 19 Ecosystem conditions based on compositional state characteristics (**species-based indicators**–endemic species of flora as per IUCN)

Flora			Karnataka state (district wise)									
				Opening–2005				Closing–2020			2005	2020
	District	Units	Total	High	Med	Low	Total	High	Med	Low	Flora Sp	Flora Sp
1	Bagalkot	Ha	688,500	0	105,300	583,200	688,500	0	105,300	583,200	10	10
		%		0.0	15.3	84.7		0.0	15.3	84.7		
2	Bangalore (Rural)	Ha	251,100	0	48,600	202,500	251,100	0	48,600	202,500	11	11
		%		0.0	19.4	80.6		0.0	19.4	80.6		
3	Bangalore (Urban)	Ha	226,800	0	72,900	153,900	226,800	0	72,900	153,900	21	21
		%		0.0	32.1	67.9		0.0	32.1	67.9		
4	Belgaum	Ha	1,547,100	755,461	8100	783,539	1,547,100	170,100	8100	1,368,900	67	48
		%		48.8	0.5	50.6		11.0	0.5	88.5		
5	Bellary	Ha	939,600	0	72,900	866,700	939,600	0	72,900	866,700	6	6
		%		0.0	7.8	92.2		0.0	7.8	92.2		
6	Bidar	Ha	688,500	0	137,700	550,800	688,500	0	137,700	550,800	18	18
		%		0.0	20.0	80.0		0.0	20.0	80.0		
7	Bijapur	Ha	1,134,000	0	56,700	1,077,300	1,134,000	0	56,700	1,077,300	18	18
		%		0.0	5.0	95.0		0.0	5.0	95.0		
8	Chamarajanagar	Ha	688,500	115,018	121,500	451,982	688,500	97,200	121,500	469,800	16	15
		%		16.7	17.6	65.6		14.1	17.6	68.2		
9	Chikkaballapura	Ha	518,400	0	121,500	396,900	518,400	0	121,500	396,900	19	19
		%		0.0	23.4	76.6		0.0	23.4	76.6		

(continued)

Table 19 (continued)

Flora							Karnataka state (district wise)					
				Opening–2005				Closing–2020			2005	2020
	District	Units	Total	High	Med	Low	Total	High	Med	Low	Flora Sp	Flora Sp
10	Chikkamagaluru	Ha	704,700	587,729	16,200	100,771	704,700	372,600	16,200	315,900	145	104
		%		83.4	2.3	14.3		52.9	2.3	44.8		
11	Chitradurga	Ha	915,300	0	153,900	761,400	915,300	0	153,900	761,400	25	25
		%		0.0	16.8	83.2		0.0	16.8	83.2		
12	Dakshina Kannada	Ha	558,900	526,895	16,200	15,805	558,900	380,700	16,200	162,000	187	166
		%		94.3	2.9	2.8		68.1	2.9	29.0		
13	Davanagere	Ha	575,100	0	56,700	518,400	575,100	0	56,700	518,400	8	8
		%		0.0	9.9	90.1		0.0	9.9	90.1		
14	Dharwad	Ha	413,100	113,575	16,200	283,325	413,100	56,700	16,200	340,200	22	17
		%		27.5	3.9	68.6		13.7	3.9	82.4		
15	Gadag	Ha	461,700	0	64,800	396,900	461,700	0	64,800	396,900	14	14
		%		0.0	14.0	86.0		0.0	14.0	86.0		
16	Hassan	Ha	656,100	562,391	64,800	28,909	656,100	226,800	64,800	364,500	87	63
		%		85.7	9.9	4.4		34.6	9.9	55.6		
17	Haveri	Ha	502,200	271,624	32,400	198,176	502,200	113,400	32,400	356,400	35	24
		%		54.1	6.5	39.5		22.6	6.5	71.0		
18	Kalburgi	Ha	1,215,000	0	170,100	1,044,900	1,215,000	0	170,100	1,044,900	33	33
		%		0.0	14.0	86.0		0.0	14.0	86.0		
19	Kodagu	Ha	486,000	456,244	8100	21,656	486,000	461,700	8100	16,200	1898	1890
		%		93.9	1.7	4.5		95.0	1.7	3.3		

(continued)

Table 19 (continued)

Flora							Karnataka state (district wise)					
				Opening–2005				Closing–2020			2005	2020
	District	Units	Total	High	Med	Low	Total	High	Med	Low	Flora Sp	Flora Sp
20	Kolar	Ha	494,100	0	129,600	364,500	494,100	0	129,600	364,500	26	26
		%		0.0	26.2	73.8		0.0	26.2	73.8		
21	Koppal	Ha	542,700	0	97,200	445,500	542,700	0	97,200	445,500	13	13
		%		0.0	17.9	82.1		0.0	17.9	82.1		
22	Mandya	Ha	510,300	0	72,900	437,400	510,300	0	72,900	437,400	11	11
		%		0.0	14.3	85.7		0.0	14.3	85.7		
23	Mysuru	Ha	672,300	160,710	32,400	479,190	672,300	129,600	32,400	510,300	178	137
		%		23.9	4.8	71.3		19.3	4.8	75.9		
24	Raichur	Ha	923,400	0	81,000	842,400	923,400	0	81,000	842,400	11	11
		%		0.0	8.8	91.2		0.0	8.8	91.2		
25	Ramanagara	Ha	356,400	0	121,500	234,900	356,400	0	121,500	234,900	22	22
		%		0.0	34.1	65.9		0.0	34.1	65.9		
26	Shimoga	Ha	858,600	334,171	24,300	500,129	858,600	437,400	24,300	396,900	119	117
		%		38.9	2.8	58.2		50.9	2.8	46.2		
27	Tumkur	Ha	1,296,000	21,229	145,800	1,128,971	1,296,000	8100	145,800	1,142,100	28	26
		%		1.6	11.3	87.1		0.6	11.3	88.1		
28	Udupi	Ha	405,000	212,203	24,300	168,497	405,000	170,100	24,300	210,600	80	66
		%		52.4	6.0	41.6		42.0	6.0	52.0		

(continued)

Table 19 (continued)

Flora				Karnataka state (district wise)								
				Opening–2005				Closing–2020			2005	2020
	District	Units	Total	High	Med	Low	Total	High	Med	Low	Flora Sp	Flora Sp
29	Uttara Kannada	Ha	1,174,500	1,030,714	64,800	78,986	1,174,500	858,600	64,800	251,100	465	414
		%		87.8	5.5	6.7		73.1	5.5	21.4		
30	Yadgir	Ha	591,300	0	129,600	461,700	591,300	0	129,600	461,700	20	20
		%		0.0	21.9	78.1		0.0	21.9	78.1		

High: grids corresponding to occurrence of species > $\mu + \sigma$; medium: within $\mu + 2\sigma$ and μ; low: grids having <μ

Table 20 Ecosystem conditions based on Compositional state characteristics (endemic species of fauna as per IUCN categories)

Fauna							Karnataka state (district wise)					
				Opening–2005				Closing–2020			2005	2020
	DIST	Units	Total	High	Medium	Low	Total	High	Medium	Low	Fauna	Fauna
1	Bagalkot	Ha	688,500	0	56,700	631,800	688,500	0	56,700	631,800	30	30
		%		0.0	8.2	91.8		0.0	8.2	91.8		
2	Bangalore (Rural)	Ha	251,100	0	16,200	234,900	251,100	0	16,200	234,900	39	39
		%		0.0	6.5	93.5		0.0	6.5	93.5		
3	Bangalore (Urban)	Ha	226,800	0	48,600	178,200	226,800	0	48,600	178,200	142	142
		%		0.0	21.4	78.6		0.0	21.4	78.6		
4	Belgaum	Ha	1,547,100	179,872	24,300	1,342,928	1,547,100	40,500	24,300	1,482,300	89	67
		%		11.6	1.6	86.8		2.6	1.6	95.8		
5	Bellary	Ha	931,501	0	80,302	851,199	931,501	0	81,000	850,501	54	54
		%		0.0	8.6	91.4		0.0	8.6	90.5		
6	Bidar	Ha	688,500	0	40,500	648,000	688,500	0	40,500	648,000	65	65
		%		0.0	5.9	94.1		0.0	5.9	94.1		
7	Bijapur	Ha	1,134,000	0	16,200	1,117,800	1,134,000	0	16,200	1,117,800	19	19
		%		0.0	1.4	98.6		0.0	1.4	98.6		
8	Chamarajanagar	Ha	688,500	258,790	48,600	381,110	688,500	218,700	48,600	421,200	122	118
		%		37.6	7.1	55.4		31.8	7.1	61.2		
9	Chikkaballapura	Ha	518,400	0	40,500	477,900	518,400	0	40,500	477,900	39	39
		%		0.0	7.8	92.2		0.0	7.8	92.2		
10	Chikkamagaluru	Ha	704,700	357,748	64,800	282,152	704,700	226,800	64,800	413,100	412	345
		%		50.8	9.2	40.0		32.2	9.2	58.6		

(continued)

Table 20 (continued)

Fauna							Karnataka state (district wise)					
				Opening–2005				Closing–2020			2005	2020
	DIST	Units	Total	High	Medium	Low	Total	High	Medium	Low	Fauna	Fauna
11	Chitradurga	Ha	915,300	0	24,300	891,000	915,300	0	24,300	891,000	39	39
		%		0.0	2.7	97.3		0.0	2.7	97.3		
12	Dakshina Kannada	Ha	558,900	336,316	72,900	149,684	558,900	243,000	72,900	243,000	211	163
		%		60.2	13.0	26.8		43.5	13.0	43.5		
13	Davanagere	Ha	575,100	0	48,600	526,500	575,100	0	48,600	526,500	77	77
		%		0.0	8.5	91.5		0.0	8.5	91.5		
14	Dharwad	Ha	413,100	0	81,000	332,100	413,100	0	81,000	332,100	44	44
		%		0.0	19.6	80.4		0.0	19.6	80.4		
15	Gadag	Ha	453,601	0	63,663	389,938	453,601	0	64,800	388,801	40	41
		%		0.0	14.0	86.0		0.0	14.0	84.2		
16	Hassan	Ha	656,100	140,598	40,500	475,002	656,100	56,700	40,500	558,900	48	26
		%		21.4	6.2	72.4		8.6	6.2	85.2		
17	Haveri	Ha	502,200	19,402	24,300	458,498	502,200	8100	24,300	469,800	73	54
		%		3.9	4.8	91.3		1.6	4.8	93.5		
18	Kalburgi	Ha	1,215,000	0	16,200	1,198,800	1,215,000	0	16,200	1,198,800	65	65
		%		0.0	1.3	98.7		0.0	1.3	98.7		
19	Kodagu	Ha	486,000	272,145	64,800	149,055	486,000	275,400	64,800	145,800	374	374
		%		56.0	13.3	30.7		56.7	13.3	30.0		
20	Kolara	Ha	494,100	0	32,400	461,700	494,100	0	32,400	461,700	39	39
		%		0.0	6.6	93.4		0.0	6.6	93.4		

(continued)

Table 20 (continued)

Fauna							Karnataka state (district wise)					
				Opening–2005				Closing–2020			2005	2020
	DIST	Units	Total	High	Medium	Low	Total	High	Medium	Low	Fauna	Fauna
21	Koppal	Ha	542,700	0	8100	534,600	542,700	0	8100	534,600	38	38
		%		0.0	1.5	98.5		0.0	1.5	98.5		
22	Mandya	Ha	510,300	0	81,000	429,300	510,300	0	81,000	429,300	53	53
		%		0.0	15.9	84.1		0.0	15.9	84.1		
23	Mysuru	Ha	672,300	150,665	81,000	440,635	672,300	121,500	81,000	469,800	269	244
		%		22.4	12.0	65.5		18.1	12.0	69.9		
24	Raichur	Ha	923,400	0	56,700	866,700	923,400	0	56,700	866,700	6	6
		%		0.0	6.1	93.9		0.0	6.1	93.9		
25	Ramanagara	Ha	356,400	227,317	8100	120,983	356,400	48,600	8100	299,700	41	34
		%		63.8	2.3	33.9		13.6	2.3	84.1		
26	Shimoga	Ha	858,600	241,346	32,400	584,854	858,600	315,900	32,400	510,300	543	542
		%		28.1	3.8	68.1		36.8	3.8	59.4		
27	Tumkur	Ha	1,296,000	0	81,000	1,215,000	1,296,000	0	81,000	1,215,000	70	70
		%		0.0	6.3	93.8		0.0	6.3	93.8		
28	Udupi	Ha	405,000	242,518	24,300	138,182	405,000	194,400	24,300	186,300	237	194
		%		59.9	6.0	34.1		48.0	6.0	46.0		
29	Uttara Kannada	Ha	1,174,500	661,213	129,600	383,687	1,174,500	550,800	129,600	494,100	934	881
		%		56.3	11.0	32.7		46.9	11.0	42.1		
30	Yadgir	Ha	591,300	0	8100	583,200	591,300	0	8100	583,200	31	31
				0.0	1.4	98.6		0.0	1.4	98.6		

Notes High = grids having occurrence of fauna species belonging to (CR; Critically Endangered; VU: vulnerable) categories, Medium: grids with the occurrence of fauna species of IUCN endangered category, Low: grids with the occurrence of fauna species belong to NT: Near threatened category of IUCN

Table 21 Net primary productivity (NPP) in forest ecosystems in Karnataka State

Net annual Biomass productivity (deadwood, fallen leaves, etc.)						Karnataka State (district wise)			Ecosystem: Forests		
			Opening 2005			Closing 2019			NPP 2005	NPP 2019	Net changes
S.No	District	Units	High	Medium	Low	High	Medium	Low	Gg/Yr	Gg/Yr	
1	Bagalkot	Ha	0	32,400	656,100	0	0	688,500	165.1	42.8	−74.1
1	Bagalkot	%	0.0	4.7	95.3	0.0	0.0	100.0			
2	Bangalore (Rural)	Ha	0	89,100	162,000	0	72,900	178,200	217.8	199.6	−8.4
2	Bangalore (Rural)	%	0.0	35.5	64.5	0.0	29.0	71.0			
3	Bangalore (Urban)	Ha	0	97,200	129,600	0	81,000	145,800	205.2	185.6	−9.5
3	Bangalore (Urban)	%	0.0	42.9	57.1	0.0	35.7	64.3			
4	Belgaum	Ha	0	234,900	1,312,200	0	137,700	1,409,400	565.0	374.5	−33.7
4	Belgaum	%	0.0	15.2	84.8	0.0	8.9	91.1			
5	Bellary	Ha	0	162,000	777,600	0	40,500	899,100	409.5	169.3	−58.7
5	Bellary	%	0.0	17.2	82.8	0.0	4.3	95.7			
6	Bidar	Ha	0	8100	680,400	0	0	688,500	48.9	19.3	−60.4
6	Bidar	%	0.0	1.2	98.8	0.0	0.0	100.0			
7	Bijapur	Ha	0	0	1,134,000	0	0	1,134,000	12.8	6.3	−50.8
7	Bijapur	%	0.0	0.0	100.0	0.0	0.0	100.0			
8	Chamarajnagar	Ha	32,400	388,800	267,300	0	324,000	364,500	1034.6	704.8	−31.9
8	Chamarajnagar	%	4.7	56.5	38.8	0.0	47.1	52.9			
9	Chikkaballapura	Ha	0	89,100	429,300	0	32,400	486,000	224.5	161.1	−28.2
9	Chikkaballapura	%	0.0	17.2	82.8	0.0	6.3	93.8			
10	Chikmagalur	Ha	32,400	558,900	113,400	8100	550,800	145,800	1426.9	1361.0	−4.6
10	Chikmagalur	%	4.6	79.3	16.1	1.1	78.2	20.7			
11	Chitradurga	Ha	0	137,700	777,600	0	32,400	882,900	404.2	256.2	−36.6

(continued)

Table 21 (continued)

Net annual Biomass productivity (deadwood, fallen leaves, etc.)						Karnataka State (district wise)			Ecosystem: Forests		
			Opening 2005			Closing 2019			NPP 2005	NPP 2019	Net changes
S.No	District	Units	High	Medium	Low	High	Medium	Low	Gg/Yr	Gg/Yr	
11	Chitradurga	%	0.0	15.0	85.0	0.0	3.5	96.5			
12	Dakshina Kannada	Ha	129,600	348,300	81,000	105,300	340,200	113,400	1361.3	1296.1	−4.8
12	Dakshina Kannada	%	23.2	62.3	14.5	18.8	60.9	20.3			
13	Davangere	Ha	8100	48,600	518,400	0	32,400	542,700	206.7	114.1	−44.8
13	Davangere	%	1.4	8.5	90.1	0.0	5.6	94.4			
14	Dharwad	Ha	0	48,600	364,500	0	24,300	388,800	98.7	79.3	−19.7
14	Dharwad	%	0.0	11.8	88.2	0.0	5.9	94.1			
15	Gadag	Ha	0	32,400	429,300	0	0	461,700	103.9	28.5	−72.6
15	Gadag	%	0.0	7.0	93.0	0.0	0.0	100.0			
16	Hassan	Ha	16,200	429,300	210,600	32,400	380,700	243,000	973.6	929.3	−4.5
16	Hassan	%	2.5	65.4	32.1	4.9	58.0	37.0			
17	Haveri	Ha	0	48,600	453,600	0	16,200	486,000	115.7	62.8	−45.7
17	Haveri	%	0.0	9.7	90.3	0.0	3.2	96.8			
18	Kalburgi	Ha	0	24,300	1,190,700	0	8100	1,206,900	82.4	29.4	−64.3
18	Kalburgi	%	0.0	2.0	98.0	0.0	0.7	99.3			
19	Kodagu	Ha	162,000	267,300	56,700	121,500	307,800	56,700	1301.2	1282.8	−1.4
19	Kodagu	%	33.3	55.0	11.7	25.0	63.3	11.7			
20	Kolar	Ha	0	113,400	380,700	0	81,000	413,100	297.2	241.6	−18.7
20	Kolar	%	0.0	23.0	77.0	0.0	16.4	83.6			
21	Koppal	Ha	0	8100	534,600	0	0	542,700	38.7	13.1	−66.0
21	Koppal	%	0.0	1.5	98.5	0.0	0.0	100.0			

(continued)

Table 21 (continued)

Net annual Biomass productivity (deadwood, fallen leaves, etc.)						Karnataka State (district wise)			Ecosystem: Forests		
			Opening 2005			Closing 2019			NPP 2005	NPP 2019	Net changes
S.No	District	Units	High	Medium	Low	High	Medium	Low	Gg/Yr	Gg/Yr	
22	Mandya	Ha	40,500	218,700	251,100	32,400	202,500	275,400	649.9	608.0	−6.5
22	Mandya	%	7.9	42.9	49.2	6.3	39.7	54.0			
23	Mysore	Ha	8100	153,900	510,300	0	145,800	526,500	490.4	424.4	−13.5
23	Mysore	%	1.2	22.9	75.9	0.0	21.7	78.3			
24	Raichur	Ha	0	24,300	899,100	0	8100	915,300	70.1	33.2	−52.6
24	Raichur	%	0.0	2.6	97.4	0.0	0.9	99.1			
25	Ramanagara	Ha	0	129,600	226,800	0	72,900	283,500	321.8	241.7	−24.9
25	Ramanagara	%	0.0	36.4	63.6	0.0	20.5	79.5			
26	Shimoga	Ha	40,500	688,500	129,600	8100	648,000	202,500	1832.9	1636.1	−10.7
26	Shimoga	%	4.7	80.2	15.1	0.9	75.5	23.6			
27	Tumkur	Ha	0	405,000	891,000	0	299,700	996,300	871.9	716.5	−17.8
27	Tumkur	%	0.0	31.3	68.8	0.0	23.1	76.9			
28	Udupi	Ha	16,200	267,300	121,500	0	275,400	129,600	747.1	710.2	−4.9
28	Udupi	%	4.0	66.0	30.0	0.0	68.0	32.0			
29	Uttara Kannada	Ha	380,700	648,000	145,800	16,200	996,300	162,000	2952.6	2669.0	−9.6
29	Uttara Kannada	%	32.4	55.2	12.4	1.4	84.8	13.8			
30	Yadgir	Ha	0	32,400	558,900	0	0	591,300	92.4	38.4	−58.4
30	Yadgir	%	0.0	5.5	94.5	0.0	5.5	94.5			

Note High: Grids with NPP > ($\mu + 2\sigma$: Mean + 2 standard deviation), Medium: grids covering (values within μ and < $\mu + 2\sigma$) NPP values, while low category includes grids having <μ NPP

4.5 Soil–Characteristics

Soil quality (physical and chemical) characteristics of soil district-wise for Karnataka State, India was assessed based on the data compiled from the National Bureau of Soil Survey and Land Use Planning (NBSSLUP) at the macro level.

Soil erosion: Southern districts like Hassan, Tumkur, Mandya, Chikmagalur have slight to moderate soil erosion status; Districts like Dharwad, Bengaluru, Kodagu, Belgaum, Bijapur has severe soil erosion status. Figure 19 depicts the soil erosion status (1998) in Karnataka state.

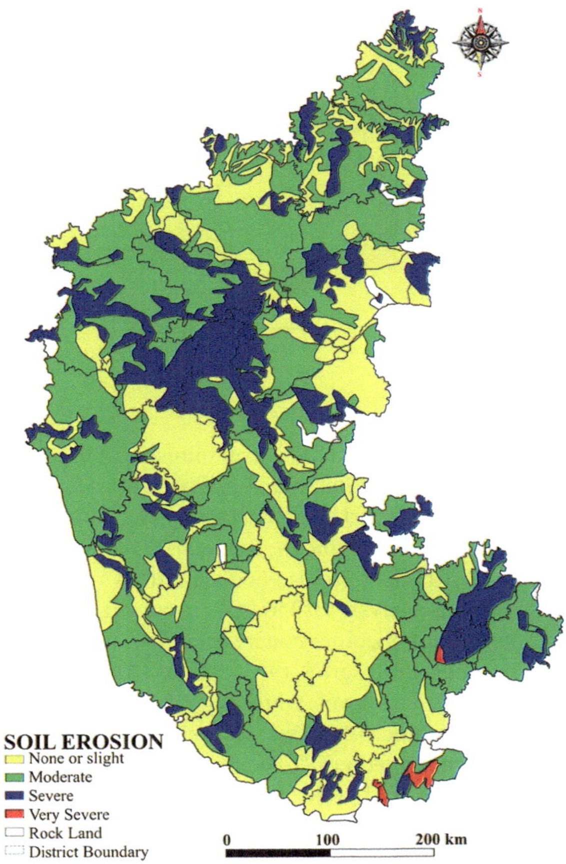

Fig. 19 Soil erosion status in Karnataka

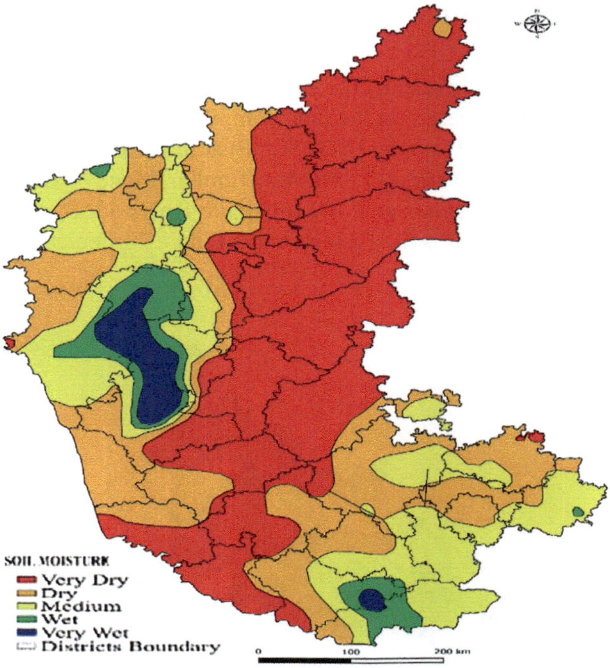

Fig. 20 Soil moisture in Karnataka

Soil Moisture: Figure 20 depicts the soil moisture status (during 1998) in Karnataka state. Districts Bidar, Gulbarga, Raichur, Hassan, Chakamagaluru has very dry soil. Districts–Belgaum, Uttara Kannada, Mangalore, parts of Mandya, Bengaluru, has dry soil. Mysore, Shimoga, Belgaum has medium to wet soil.

Potassium: Table 22 lists district-wise concentrations of potassium from 1998 and 2020. The macro-level analysis (1:250,000) shows higher potassium levels in 1028 grids (83,268 km^2 area), medium concentration in 1460 grids (118,260 km^2), and very low concentration in 94 grids (9612 km^2). Figure 21 depicts the potassium concentration at the district level, while Fig. 22 highlights the variation at the grid level during 1998. The micro-level analysis shows higher potassium levels in 274 grids (22,194 km^2 area), medium concentration in 2195 grids (177,795 km^2), and very low concentration in 103 grids (10,365 km^2). Changes in potassium status levels in the soil of Karnataka, district wise is given in Fig. 23. Changes in potassium concentrations are evident during the period 1998–2020. For example, in Dakshina Kannada 3483 km^2 area of medium concentration and 1863 km^2 area of low concentration in the year 1998 has become 1215 km^2 of medium and 4131 km^2 of low concentration, which highlights the decline in nutrient levels due to lack of use of

Table 22 District wise potassium levels in the soil during 1998 and 2020

Sl.No.	District Name	1998						2020					
		Low		Medium		High		Low		Medium		High	
		Grids	Area	Grids	Area	Grids	Area	Grids	Area	Grids	Area	Grids	Area
1	Bagalkot		0	27	2187	58	4698		0	37	2997	48	3888
2	Bangalore Urban		0	28	2268	4	324		0	31	2511	1	81
3	Bangalore (Rural)		0	27	2187	3	243		0	26	2106	4	324
4	Belgaum		0	153	12,393	27	2187		0	173	14,013	16	1296
5	Bellary	10	810	7	567	109	8829		0	116	9396		0
6	Bidar		0		0	85	6885		0	75	6075	10	810
7	Bijapur	6	486	57	4617	76	6156		0	41	3321	97	7857
8	Chamarajanagar		0	77	6237	8	648		0	83	6723	1	81
9	Chikballapur		0	60	4860	4	324		0	60	4860	3	243
10	Chikkamagaluru	6	486	62	5022	19	1539	4	324	79	6399	4	324
11	Chitradurga		0	26	2106	87	7047	14	1134	99	8019		0
12	Dakshina Kannada	23	1863	43	3483		0	51	4131	15	1215		0
13	Davangere		0	40	3240	31	2511		0	71	5751		0
14	Dharwad		0	22	1782	29	2349		0	41	3321	10	810
15	Gadag		0	22	1782	35	2835		0	54	4374	3	243
16	Gulbarga		0	22	1782	129	10,449		0	109	8829	42	3402
17	Hassan		0	29	2349	52	4212		0	76	6156	5	405
18	Haveri		0	42	3402	20	1620		0	62	5022		0
19	Kodagu	18	1458	38	3078	3	243		0	54	4374	5	405

(continued)

Table 22 (continued)

Sl.No.	District Name	1998						2020					
		Low		Medium		High		Low		Medium		High	
		Grids	Area	Grids	Area	Grids	Area	Grids	Area	Grids	Area	Grids	Area
20	Kolar		0	55	4455	4	324		0	59	4779		0
21	Koppal		0		0	67	5427		0	43	3483	24	1944
22	Mandya		0	51	4131	12	972		0	63	5103		0
23	Mysuru		0	65	5265	18	1458		0	82	6642		0
24	Raichur		0	22	1782	91	7371		0	112	9072	1	81
25	Ramanagara		0	27	2187	15	1215		0	42	3402		0
26	Shimoga		0	106	8586		0		0	106	8586		0
27	Tumkur		0	148	11,988	10	810		0	154	12,474		0
28	Udupi	8	648	44	3564		0	33	2673	19	1539		0
29	Uttara Kannada	23	1863	119	9639		0	1	81	140	11,340		0
30	Yadgir		0	41	3321	32	2592		0	73	5913		0
	Total	**94**	**9612**	**1460**	**118,260**	**1028**	**83,268**	**103**	**10,363**	**2195**	**177,795**	**274**	**22,194**

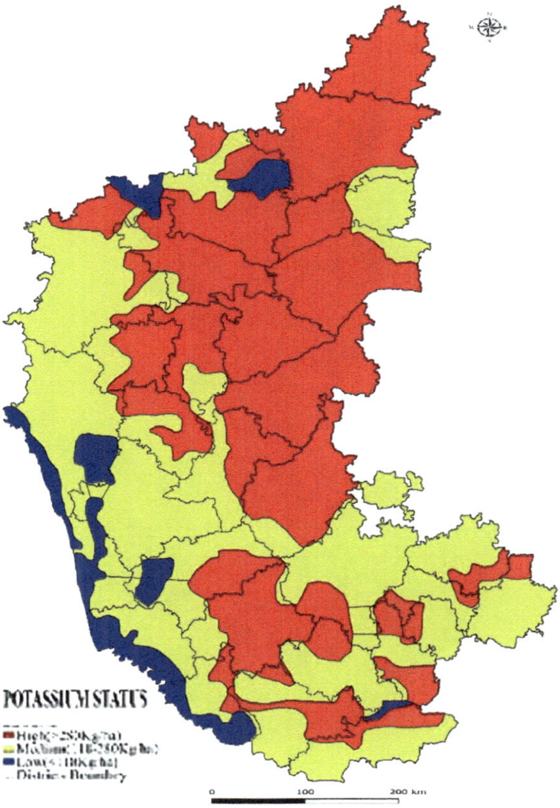

Fig. 21 Potassium status in soil of Karnataka (1998), district wise (1:250,000 scale)

organic manure and a similar trend is observed in all districts. A study on the deficiency of potassium in soils and crops of India was carried out earlier (Naidu et al. 2011) and found that potassium removal crops are more cultivated than potassium addition crops by farmers and insufficient use of NPK fertilizers are contributing to large scale potassium mining; they also found that depletion of potassium in red, laterite and shallow black soils. Figure 24 depicts the potassium status during 2020 at the taluk level, while Fig. 25 depicts the potassium status at grid levels.

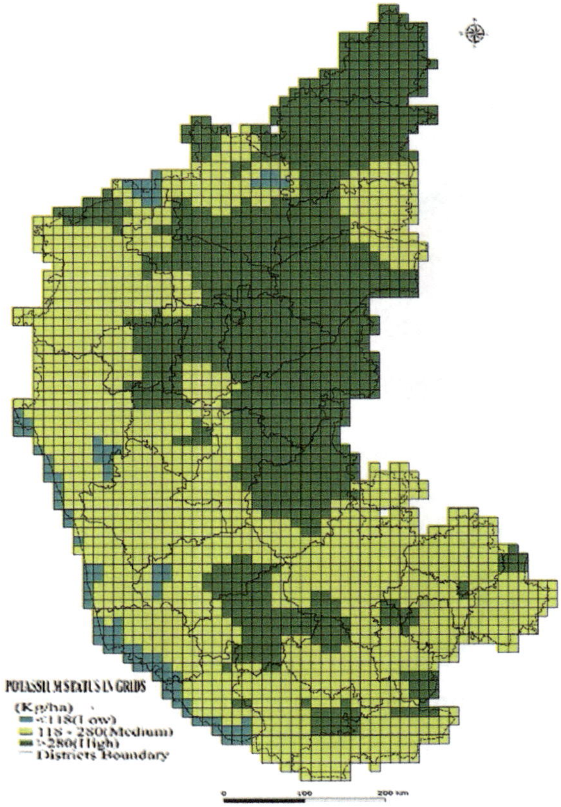

Fig. 22 Potassium status in soil of Karnataka (1998), Grid wise (1:250,000 scale)

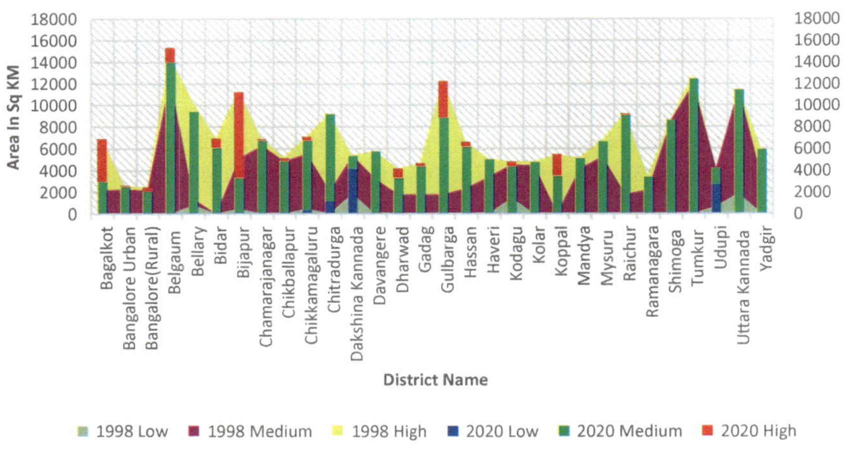

Fig. 23 Change in Potassium status during 1998 to 2020

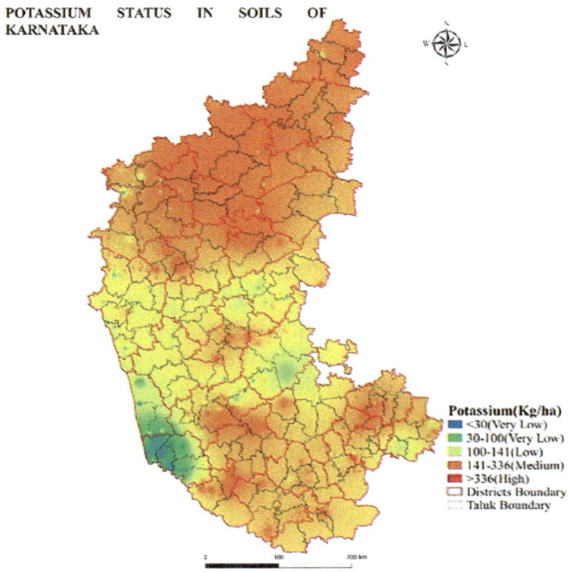

Fig. 24 Potassium status in soil of Karnataka (2020), taluk wise

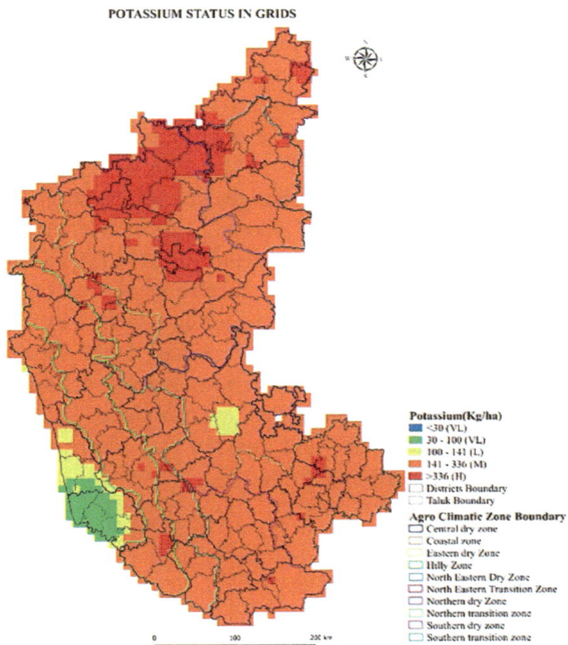

Fig. 25 Potassium status in soil of Karnataka (2020), Grid and agro-climatic zone wise

Nitrogen: Table 23 lists district-wise concentrations of nitrogen during 1998 and 2020. The macro level analysis (1:250,000) shows that higher nitrogen levels in 1074 grids (86,994 km^2 area), medium concentration in 872 grids (70,632 km^2), and very low concentration in 552 grids (44,712 Km^2). Figure 26 depicts the nitrogen concentration at the district level, while Fig. 27 highlights the variation at the grid level during 1998. The micro level analysis shows higher nitrogen levels in 0 grids, medium concentration in 118 grids (9558 km^2), and very low concentration in 2454 grids (198,774 km^2)—Fig. 28 highlights changes in nitrogen status levels in the soil of Karnataka. Figure 29 depicts the nitrogen status during 2020 at taluk level, while Fig. 30 depicts the nitrogen status at the grid level. The changes in concentration are evident during the 1998–2020 period. For example, in Belgaum 7776 km^2 area of high concentration, 5184 km^2 area of medium concentration, and 2430 km^2 area of low concentration in the year 1998 has become 972 km^2 of medium and 14,337 km^2 of low concentration, which highlights the decline in nutrient levels due to lack of rotation of crops in the district and a similar trend is observed in all districts. Mapping of major nutrients in Malaprabha right bank command area of Karnataka (Ravikumar et al. 2007) show high potassium in soil and low nitrogen concentration and suggested to use of organic manure to avoid soil mining for nitrogen.

Phosphorus: Table 24 lists district-wise concentrations of phosphorus during 1998 and 2020. The macro-level analysis (1:250,000) shows medium concentration in 244 grids (19,764 Km^2), and very low concentration in 2419 grids (195,939 Km^2). Figure 31 depicts the phosphorus concentration at the district level during 1998, while Fig. 32 highlights the variation at grid levels. The micro-level analysis shows higher phosphorus levels in 255 grids (20,655 km^2), medium concentration in 2209 grids (178,929 km^2), and very low concentration in 108 grids (8748 km^2). Figure 33 indicates the changes in phosphorus levels district-wise in the soil of Karnataka. Figure 34 depicts the phosphorus status during 2020 at the taluk level, while Fig. 35 depicts the phosphorus status at grid levels. The changes in concentration are evident during the period 1998–2020; for example, in Chikkamagaluru, 810 km^2 area of medium concentration and 6237 km^2 area of low concentration (in 1998) and now 3483 km^2 of high concentration, 3483 km^2 of medium and 81km^2 of low concentration, which highlights incline in nutrient levels due to usage of high fertilizers in the district and a similar trend is observed in all districts. This is contrary to the earlier report of phosphorus status in soils of the eastern dry zone, Karnataka (Chandrakala et al. 2017), which indicated phosphorus and nitrogen deficiency and recommended the management of phosphorus in both irrigated and rain-fed agriculture apart from reducing phosphorus losses and minimizing pollution.

Table 23 District wise nitrogen levels in soil during 1998 and 2020

Sl.No.	District name	1998						2020					
		Low		Medium		High		Low		Medium		High	
		Grids	Area	Grids	Area	Grids	Area	Grids	Area	Grids	Area	Grids	Area
1	Bagalkot	31	2511	49	3969	5	405	52	4212	33	2673		0
2	Bangalore Urban	6	486	17	1377	9	729	32	2592		0		0
3	Bangalore (Rural)	22	1782	8	648		0	30	2430		0		0
4	Belgaum	30	2430	64	5184	96	7776	177	14,337	12	972		0
5	Bellary	27	2187	46	3726	43	3483	116	9396		0		0
6	Bidar	0	0	5	405	80	6480	85	6885		0		0
7	Bijapur	0	0	46	3726	9	729	137	11,097	1	81		0
8	Chamarajanagar	0	0	21	1701	64	5184	83	6723	1	81		0
9	Chikballapur	64	5184	0	0	0	0	63	5103		0		0
10	Chikkamagaluru	5	405	9	729	73	5913	85	6885	2	162		0
11	Chitradurga	33	2673	72	5832	8	648	113	9153		0		0
12	Dakshina Kannada	0	0	0	0	66	5346	62	5022	4	324		0
13	Davangere	2	162	40	3240	29	2349	70	5670	1	81		0
14	Dharwad		0	22	1782	29	2349	51	4131		0		0
15	Gadag	13	1053	34	2754	10	810	57	4617		0		0
16	Gulbarga	25	2025	86	6966	40	3240	151	12,231		0		0
17	Hassan		0	40	3240	41	3321	39	3159	42	3402		0
18	Haveri	4	324	15	1215	43	3483	62	5022		0		0
19	Kodagu	0	0	5	405	54	4374	55	4455	4	324		0

(continued)

Table 23 (continued)

Sl.No.	District name	1998						2020					
		Low		Medium		High		Low		Medium		High	
		Grids	Area	Grids	Area	Grids	Area	Grids	Area	Grids	Area	Grids	Area
20	Kolar	51	4131	8	648	0	0	59	4779		0		0
21	Koppal	12	972	53	4293	2	162	67	5427		0		0
22	Mandya	18	1458	38	3078	7	567	63	5103		0		0
23	Mysuru	24	1944	39	3159	20	1620	82	6642		0		0
24	Raichur	54	4374	50	4050	9	729	113	9153		0		0
25	Ramanagara	8	648	30	2430	4	324	42	3402		0		0
26	Shimoga	0	0	3	243	103	8343	102	8262	4	324		0
27	Tumkur	62	5022	60	4860	36	2916	140	11,340	14	1134		0
28	Udupi	0	0	0	0	52	4212	52	4212		0		0
29	Uttara Kannada	0	0	0	0	142	11,502	141	11,421		0		0
30	Yadgir	61	4941	12	972	0	0	73	5913		0		0
	Total	**552**	**44,712**	**872**	**70,632**	**1074**	**86,994**	**2454**	**198,774**	**118**	**9558**		**0**

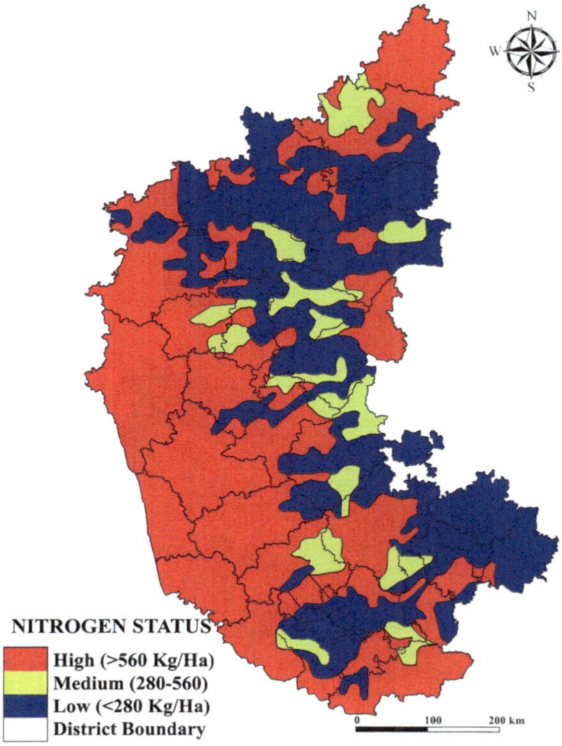

Fig. 26 Nitrogen status in soil of Karnataka (1998), district wise (1:250,000 scale)

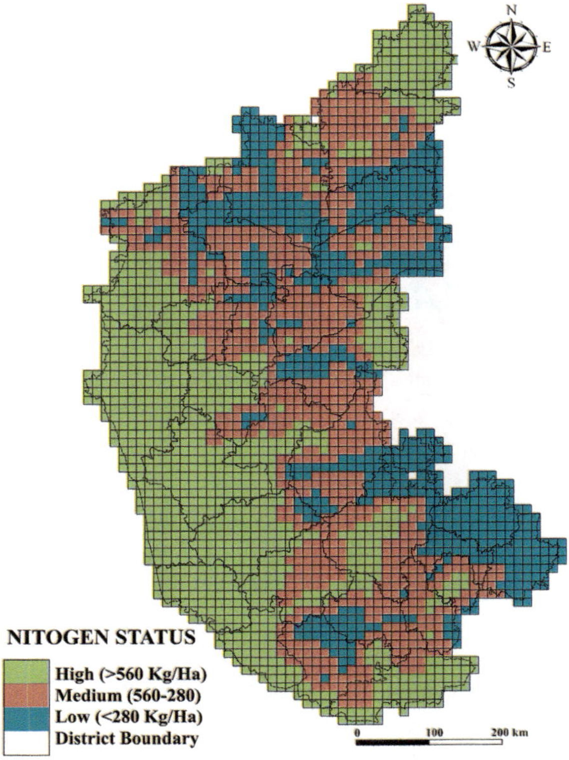

Fig. 27 Nitrogen status in soil of Karnataka (1998),Grid wise (1:250,000 scale)

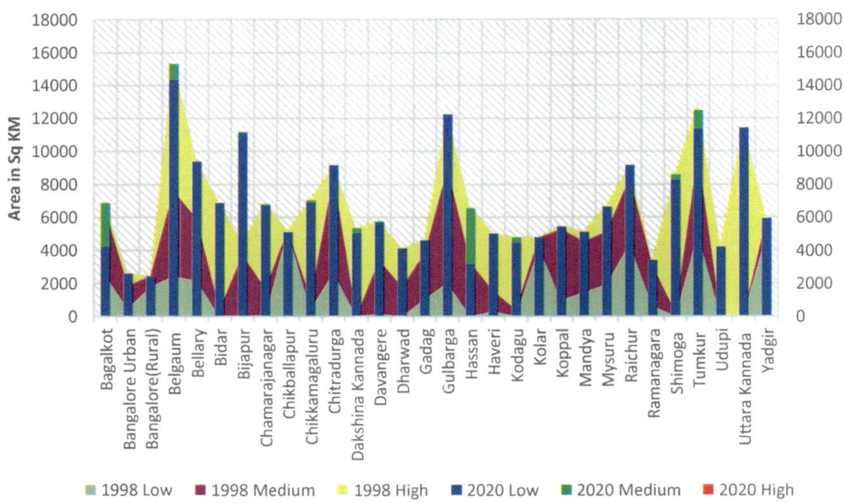

Fig. 28 Variations in Nitrogen in soil (district wise) during 1998 to 2020

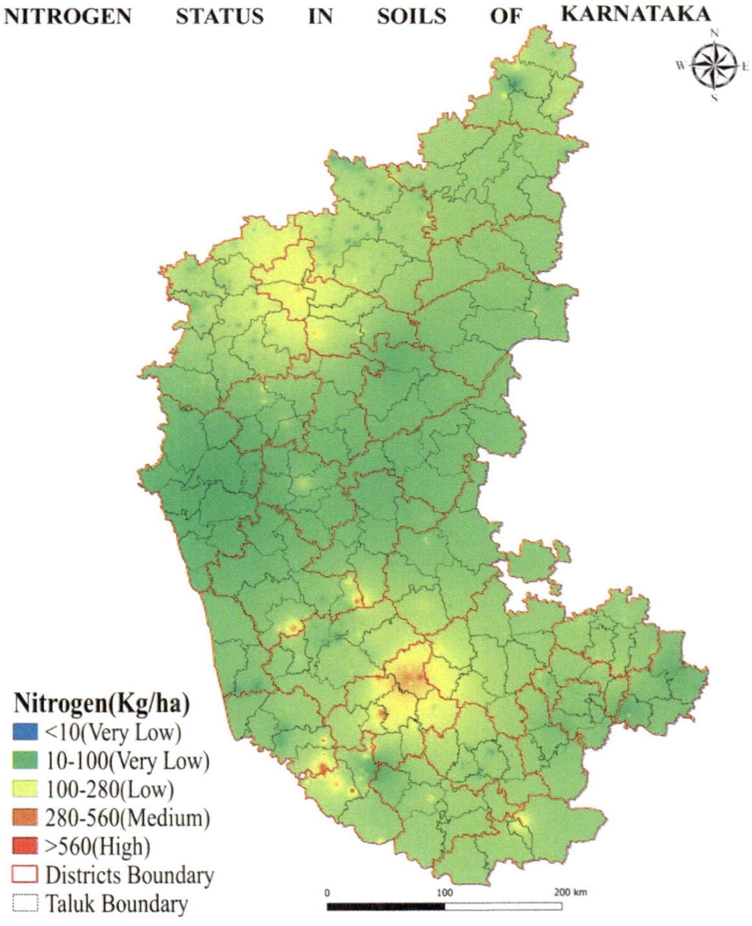

Fig. 29 Nitrogen status in soil of Karnataka (2020), taluk wise

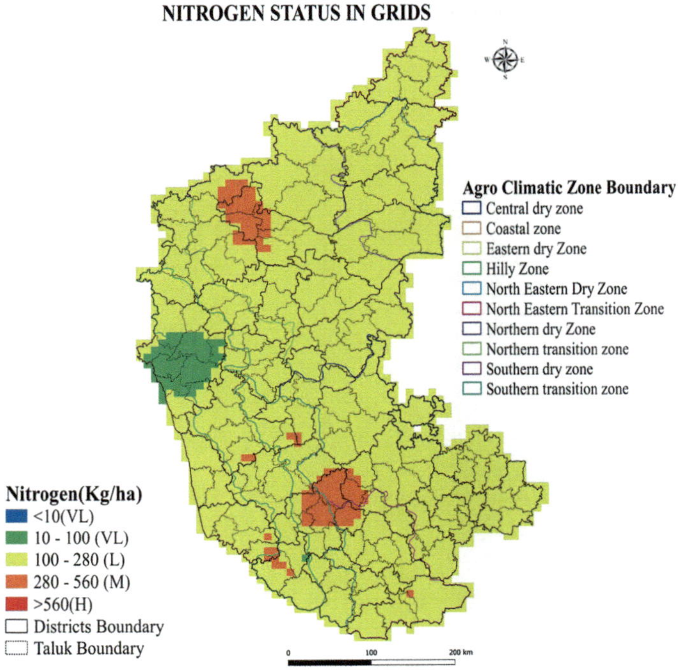

Fig. 30 Nitrogen status in soil of Karnataka (2020), Grid and agro-climatic zone wise

Table 24 District-wise phosphorus levels in the soil during 1998 and 2020

Sl.No.	District name	1998						2020					
		Low		Medium		High		Low		Medium		High	
		Grids	Area	Grids	Area	Grids	Area	Grids	Area	Grids	Area	Grids	Area
1	Bagalkot	81	6561	4	324		0	1	81	77	6237	7	567
2	Bangalore Urban	23	1863	9	729		0		0	32	2592		0
3	Bangalore (Rural)	23	1863	7	567		0		0	30	2430		0
4	Belgaum	163	13,203	28	2268		0	6	486	183	14,823		0
5	Bellary	104	8424	12	972		0	1	81	111	8991	4	324
6	Bidar	52	4212	33	2673		0	55	4455	30	2430		0
7	Bijapur	206	16,686	6	486		0		0	137	11,097	1	81
8	Chamarajanagar	85	6885	0	0		0		0	83	6723	1	81
9	Chikballapur	58	4698	6	486		0		0	63	5103		0
10	Chikkamagaluru	77	6237	10	810		0	1	81	43	3483	43	3483
11	Chitradurga	104	8424	9	729		0		0	97	7857	16	1296
12	Dakshina Kannada	67	5427	0	0		0	31	2511	34	2754	1	81
13	Davangere	61	4941	10	810		0		0	33	2673	38	3078
14	Dharwad	51	4131	0	0		0		0	51	4131		0
15	Gadag	57	4617	0	0		0		0	53	4293	4	324
16	Gulbarga	138	11,178	14	1134		0	1	81	150	12,150		0
17	Hassan	70	5670	11	891		0		0	47	3807	34	2754
18	Haveri	62	5022	0	0		0		0	59	4779	3	243
19	Kodagu	53	4293	7	567		0	1	81	58	4698		0

(continued)

Table 24 (continued)

Sl.No.	District name	1998						2020					
		Low		Medium		High		Low		Medium		High	
		Grids	Area	Grids	Area	Grids	Area	Grids	Area	Grids	Area	Grids	Area
20	Kolar	54	4374	6	486		0	4	324	55	4455		0
21	Koppal	67	5427	0	0		0		0	31	2511	36	2916
22	Mandya	42	3402	21	1701		0		0	55	4455	8	648
23	Mysuru	73	5913	10	810		0		0	82	6642		0
24	Raichur	113	9153		0		0		0	108	8748	5	405
25	Ramanagara	40	3240	2	162		0		0	42	3402		0
26	Shimoga	87	7047	19	1539		0	1	81	105	8505		0
27	Tumkur	142	11,502	18	1458		0		0	153	12,393	1	81
28	Udupi	50	4050	2	162		0	6	486	46	3726		0
29	Uttara Kannada	143	11,583		0		0		0	88	7128	53	4293
30	Yadgir	73	5913		0		0		0	73	5913		0
	Total	**2419**	**195,939**	**244**	**19,764**		**0**	**108**	**8748**	**2209**	**178,929**	**255**	**20,655**

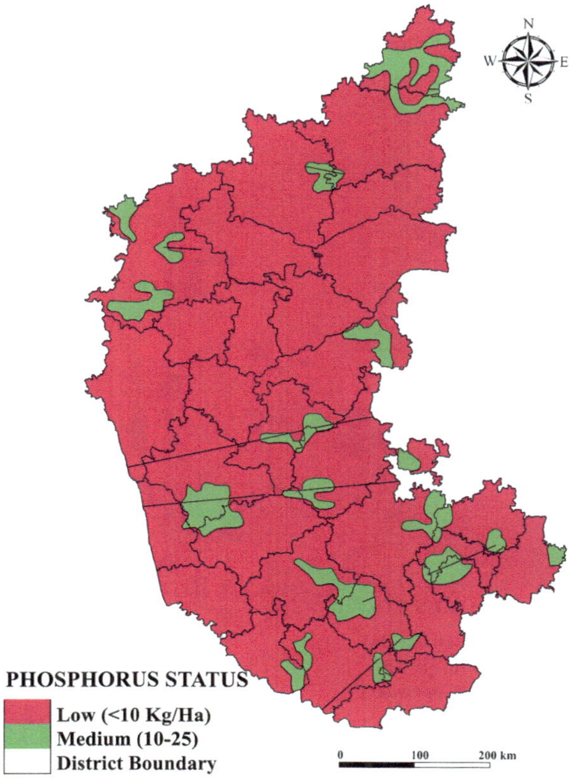

Fig. 31 Phosphorus status in soil of Karnataka (1998), district wise (1:250,000 scale)

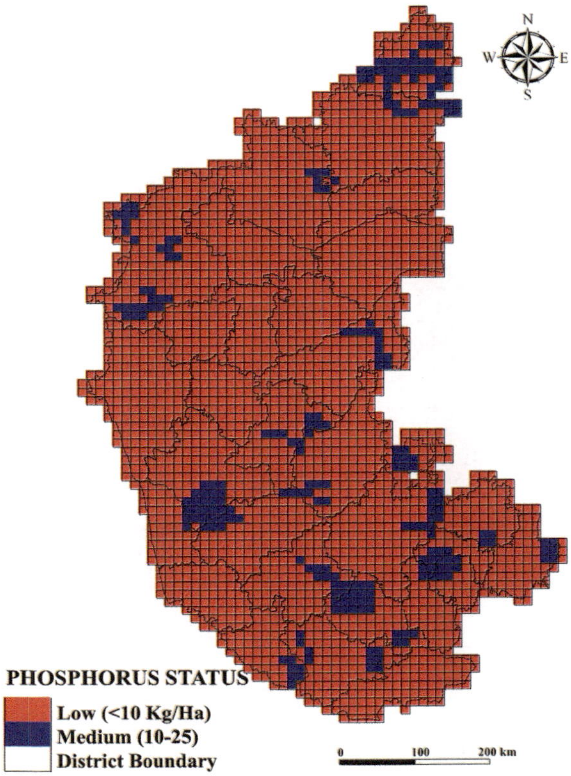

Fig. 32 Phosphorus status in soil of Karnataka (1998), Grid wise (1:250,000 scale)

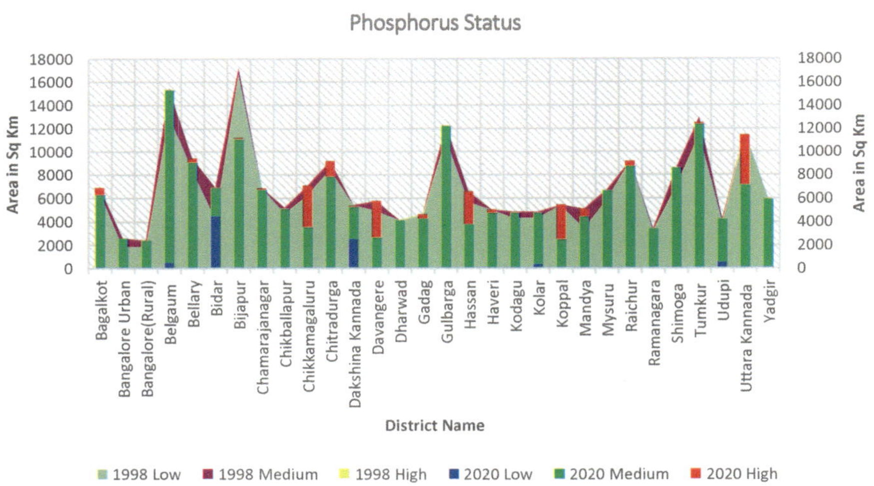

Fig. 33 Changes in Phosphorus status during 1998 to 2020 (district wise)

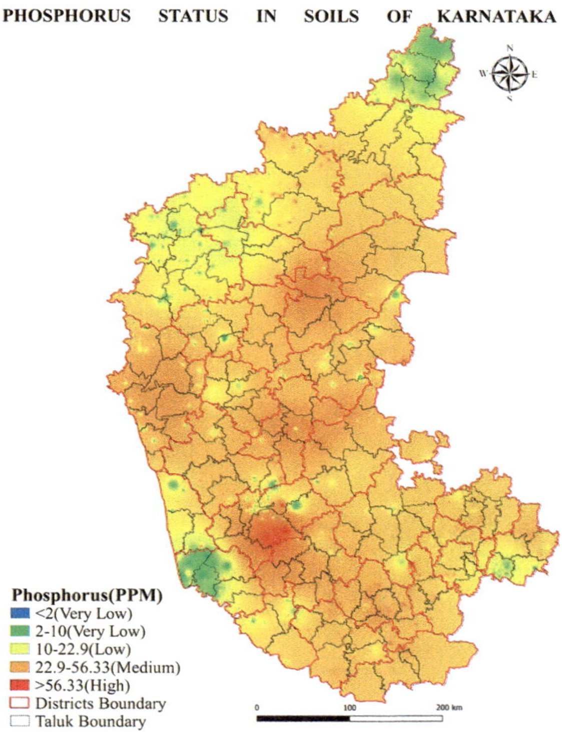

Fig. 34 Phosphorus status in soil of Karnataka (2020), taluk wise

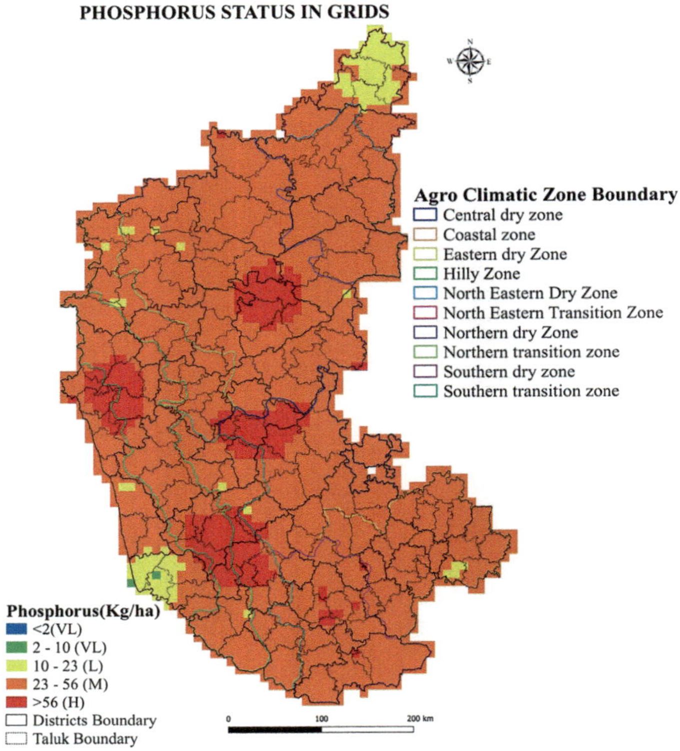

Fig. 35 Phosphorus status in soil of Karnataka (2020), Grid and agro-climatic zone wise

Zinc: Table 25 lists district-wise concentrations of zinc during 1998 and 2020. The macro-level analysis (1:250,000) shows zinc deficiency in 1994 grids (161,514 Km^2) and sufficiency in 588 grids (47,628 Km^2). Figure 36 depicts the zinc concentration at the district level, while Fig. 37 highlights the variation at the grid level during 1998. The micro-level analysis shows zinc deficiency levels in 243 grids (19,683 Km^2), and sufficiency levels in 2329 grids (188,649 Km^2). Figure 38 highlights changes in zinc levels in the soil of Karnataka. The changes in concentration are evident during the period 1998–2020; for example, in the Tumkur district 12,312 km^2 area had deficient, and 486 km^2 had sufficient status in 1998. Now, 4941 km^2 is deficient, and 7533 km^2 has sufficient status. Figure 39 depicts the zinc status at taluk levels during 2020, while Fig. 40 depicts the zinc status at grid levels. A study on chemical properties and available zinc in soils under Bhadra command, Karnataka (Ranjitha et al. 2019) reported the sufficiency of zinc due to the regular addition of $ZnSO_4$ in every season to paddy, it was also influenced by pH, organic carbon and $CaCO_3$ contents in soil.

Table 25 District wise zinc levels in soil during 1998 and 2020

Sl.No.	District name	1998				2020			
		Deficient		Sufficient		Deficient		Sufficient	
		Grid	Area	Grid	Area	Grid	Area	Grid	Area
1	Bagalkot	85	6885		0	29	2349	56	4536
2	Bangalore Urban	5	405	27	2187	3	243	29	2349
3	Bangalore (Rural)	21	1701	9	729	8	648	22	1782
4	Belgaum	138	11,178	52	4212	39	3159	150	12,150
5	Bellary	113	9153	3	243		0	116	9396
6	Bidar	85	6885		0	1	81	84	6804
7	Bijapur	139	11,259		0	11	891	127	10,287
8	Chamarajnagar	59	4779	26	2106		0	84	6804
9	Chikballapur	64	5184		0	21	1701	42	3402
10	Chikmagalur	19	1539	68	5508	4	324	83	6723
11	Chitradurga	109	8829	4	324	48	3888	65	5265
12	Dakshina Kannada	59	4779	7	567		0	66	5346
13	Davangere	50	4050	21	1701	2	162	69	5589
14	Dharwad	25	2025	26	2106		0	51	4131
15	Gadag	57	4617		0		0	57	4617
16	Gulbarga	151	12,231		0	1	81	150	12,150

(continued)

Table 25 (continued)

Sl.No.	District name	1998				2020			
		Deficient		Sufficient		Deficient		Sufficient	
		Grid	Area	Grid	Area	Grid	Area	Grid	Area
17	Hassan	55	4455	26	2106	7	567	74	5994
18	Haveri	35	2835	27	2187	3	243	59	4779
19	Kodagu	13	1053	46	3726		0	59	4779
20	Kolar	55	4455	4	324	1	81	58	4698
21	Koppal	67	5427		0		0	67	5427
22	Mandya	50	4050	13	1053		0	63	5103
23	Mysore	58	4698	25	2025		0	82	6642
24	Raichur	113	9153		0		0	113	9153
25	Ramanagara	9	729	33	2673	1	81	41	3321
26	Shimoga	64	5184	42	3402	3	243	103	8343
27	Tumkur	152	12,312	6	486	61	4941	93	7533
28	Udupi	51	4131	1	81		0	52	4212
29	Uttara Kannada	20	1620	122	9882		0	141	11,421
30	Yadgir	73	5913		0		0	73	5913
	Total	**1994**	**161,514**	**588**	**47,628**	**243**	**19,683**	**2329**	**188,649**

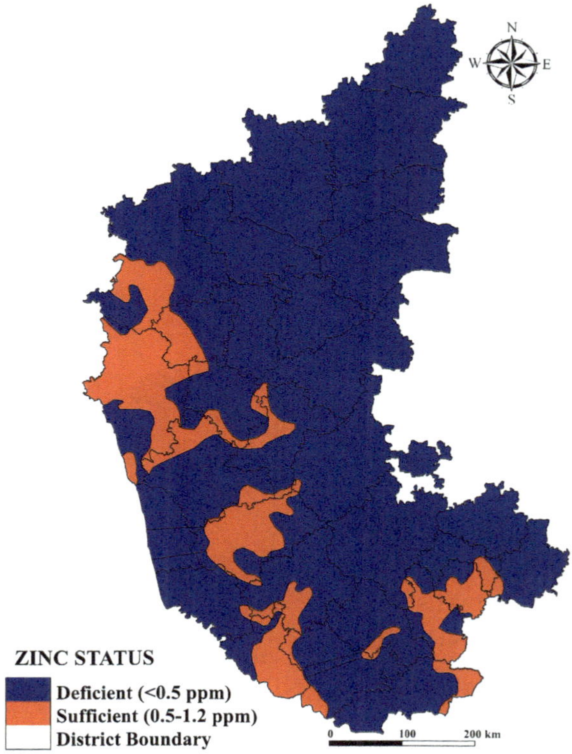

Fig. 36 Zinc status in soil of Karnataka (1998), district wise (1:250,000 scale)

Fig. 37 Zinc status in soil of Karnataka (1998), Grid wise (1:250,000 scale)

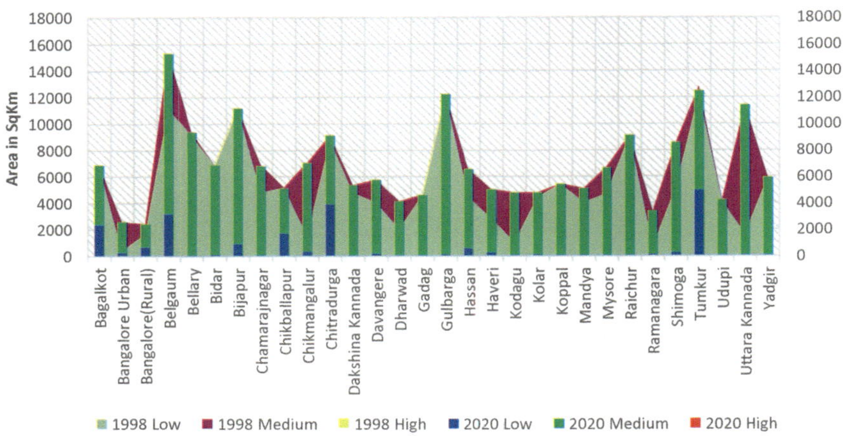

Fig. 38 Change in Zinc status during 1998 to 2020

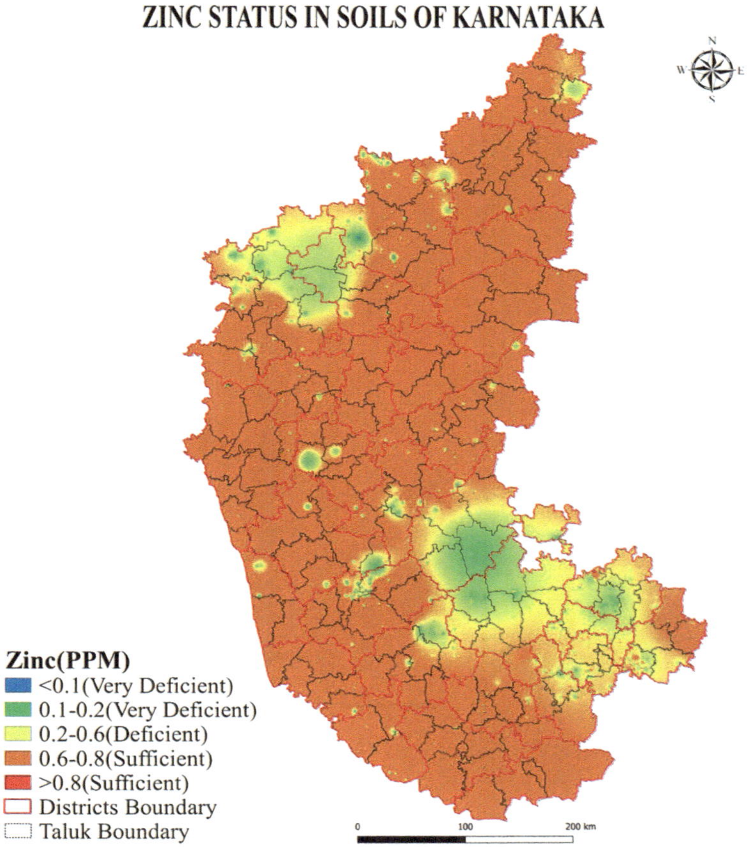

Fig. 39 Zinc status in soil of Karnataka (2020), taluk wise

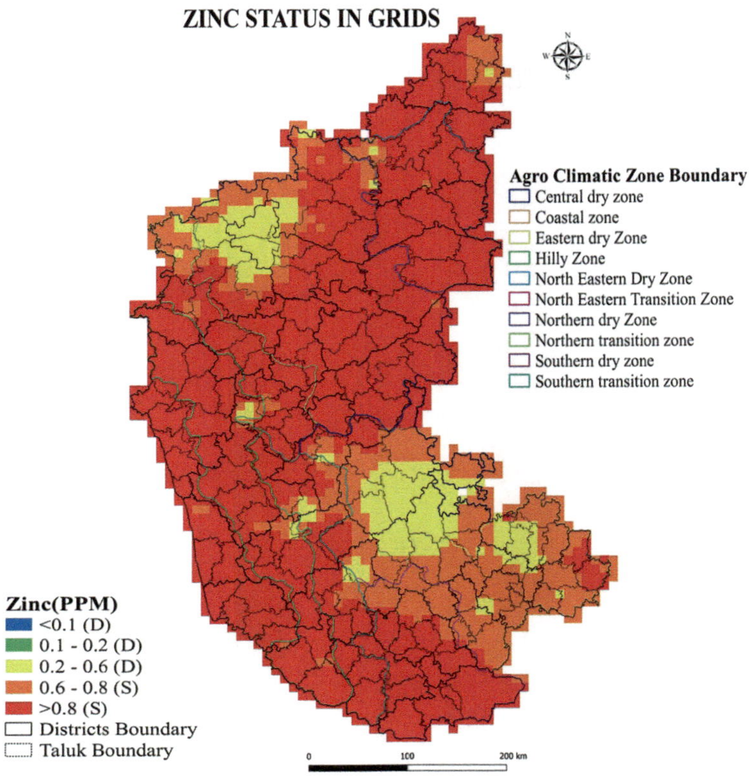

Fig. 40 Zinc status in soil of Karnataka (2020), Grid and agro-climatic zone wise

Iron: Table 26 lists district wise concentration of iron in the soil during 1998 and 2020. The macro level analysis (1:250,000) shows iron deficiency levels in 613 grids (49,653 km^2), and sufficiency levels in 1969 grids (159,489 km^2). Figure 41 depicts the iron concentration at the district level, while Fig. 42 highlights the variation at the grid level during 1998. The micro level analysis shows iron deficiency levels in 29 grids (2349 km^2) and sufficiency levels in 2543 grids (205,983 km^2). Figure 43 depicts changes in iron levels district-wise in the soil of Karnataka from 1998 to 2020. The changes in concentration are evident during the period 1998–2020; for example, in Raichur district 8424 km^2 area was deficient and 729 km^2 area which was sufficient in the year 1998, has become 0 km^2 of area deficient, and 9153 km^2 of area sufficient, which highlights improvements in nutrient levels. A similar trend is observed in all districts. Figure 44 depicts the iron status during 2020 at the taluk level, while Fig. 45 depicts the iron status at the grid level. Analysis of heavy metal concentration in soils of Bijapur, Karnataka, was carried out by Pujar et al. (2012) and higher values of nutrients were due to anthropogenic inputs, either due to the application of fertilizers or industrial and mining activities.

Table 26 District wise Iron levels in soil during 1998 and 2020

Sl.No.	District name	1998				2020			
		Deficient		Sufficient		Deficient		Sufficient	
		Grid	Area	Grid	Area	Grid	Area	Grid	Area
1	Bagalkot	47	3807	38	3078	19	1539	66	5346
2	Bangalore Urban	1	81	31	2511		0	32	2592
3	Bangalore (Rural)		0	30	2430		0	30	2430
4	Belgaum	108	8748	82	6642	9	729	180	14,580
5	Bellary	46	3726	70	5670		0	116	9396
6	Bidar	18	1458	67	5427	1	81	84	6804
7	Bijapur	66	5346	73	5913		0	138	11,178
8	Chamarajnagar	10	810	75	6075		0	84	6804
9	Chikballapur		0	64	5184		0	63	5103
10	Chikmagalur		0	87	7047		0	87	7047
11	Chitradurga	17	1377	96	7776		0	113	9153
12	Dakshina Kannada		0	66	5346		0	66	5346
13	Davangere		0	71	5751		0	71	5751
14	Dharwad	15	1215	36	2916		0	51	4131
15	Gadag	49	3969	8	648		0	57	4617
16	Gulbarga	58	4698	93	7533		0	151	12,231
17	Hassan		0	81	6561		0	81	6561
18	Haveri	13	1053	49	3969		0	62	5022
19	Kodagu		0	59	4779		0	59	4779
20	Kolar		0	59	4779		0	59	4779

(continued)

Table 26 (continued)

Sl.No.	District name	1998				2020			
		Deficient		Sufficient		Deficient		Sufficient	
		Grid	Area	Grid	Area	Grid	Area	Grid	Area
21	Koppal	19	1539	48	3888		0	67	5427
22	Mandya	1	81	62	5022		0	63	5103
23	Mysore	1	81	82	6642		0	82	6642
24	Raichur	104	8424	9	729		0	113	9153
25	Ramanagara		0	42	3402		0	42	3402
26	Shimoga		0	106	8586		0	106	8586
27	Tumkur	3	243	155	12,555		0	154	12,474
28	Udupi		0	52	4212		0	52	4212
29	Uttara Kannada		0	142	11,502		0	141	11,421
30	Yadgir	37	2997	36	2916		0	73	5913
	Total	**613**	**49,653**	**1969**	**159,489**	**29**	**2349**	**2543**	**205,983**

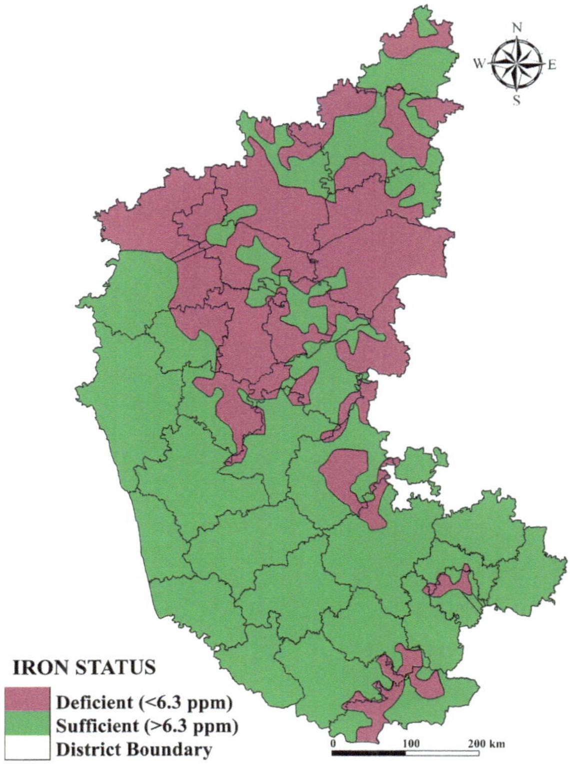

Fig. 41 Iron in soil of Karnataka (1998), district wise (1:250,000 scale)

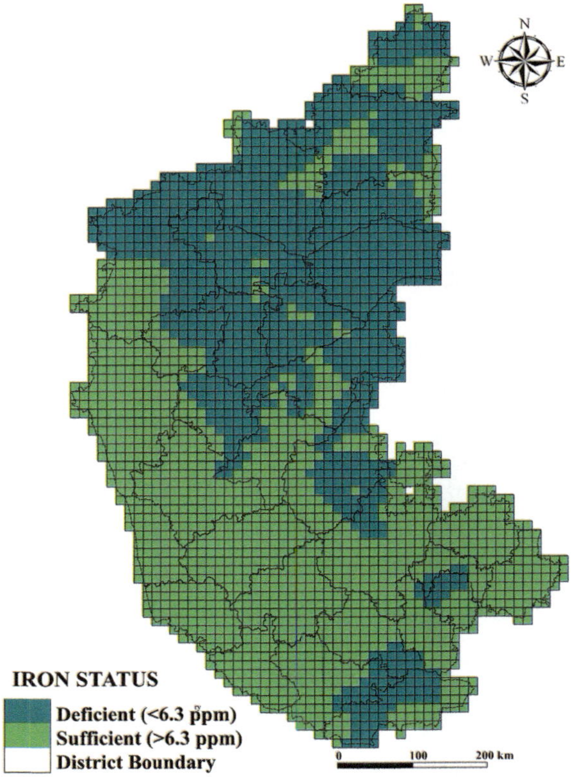

Fig. 42 Iron in soil of Karnataka (1998), Grid wise (1:250,000 scale)

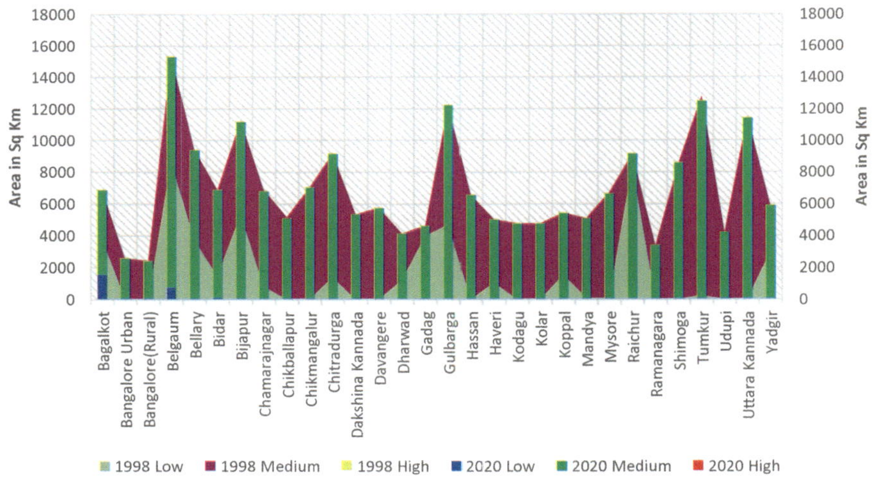

Fig. 43 Changes in iron content in soil of Karnataka (district wise) during 1998 to 2020

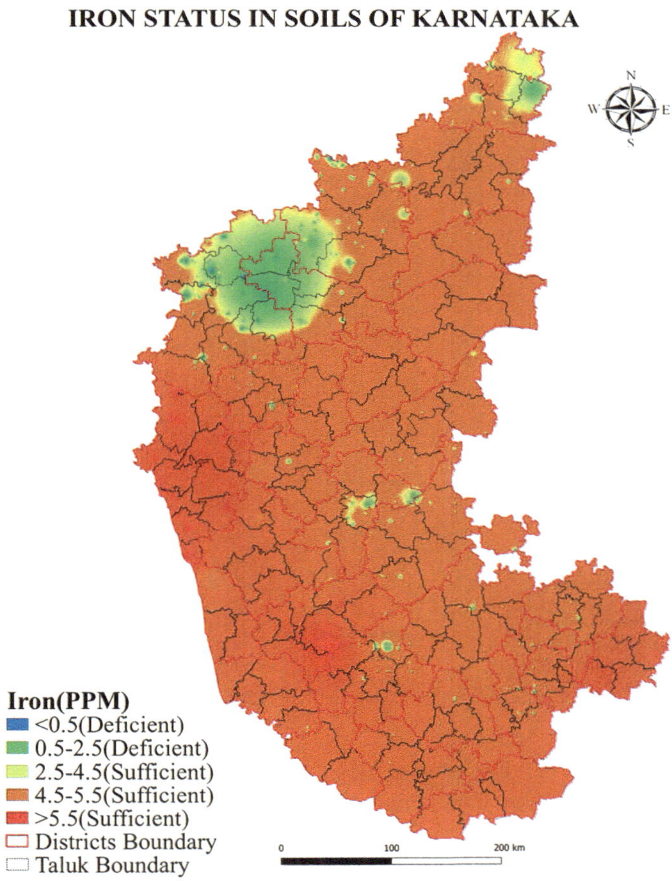

Fig. 44 Iron status in soil of Karnataka (2020), taluk wise

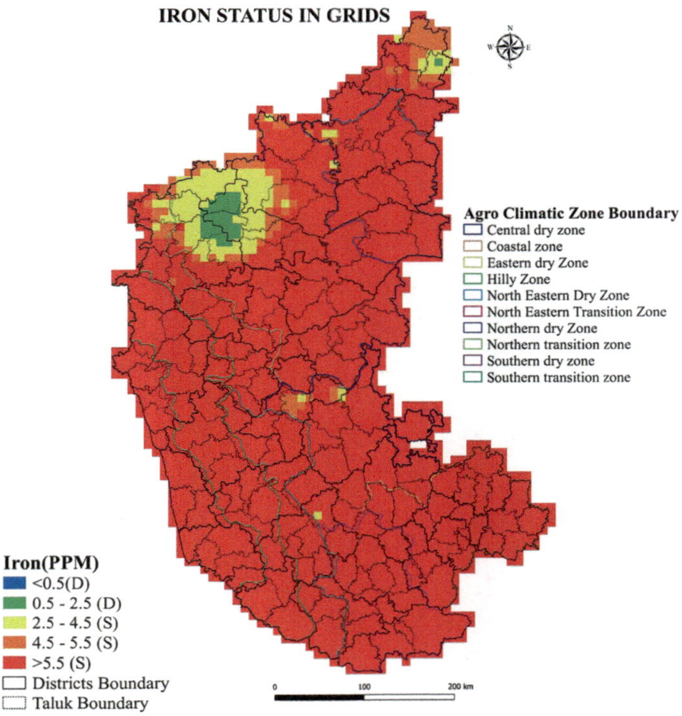

Fig. 45 Iron status in soil of Karnataka (2020), Grid and agro-climatic zone wise

pH: Table 27 lists the district-wise pH of soil during 1998 and 2020. The macro-level analysis (1:250,000) shows higher acidic levels in 578 grids (46,818 km^2), neutral levels in 1897 grids (153,657 km^2), and alkaline levels in 107 grids (8667 km^2). Figure 46 depicts the pH concentration at the district level, while Fig. 47 highlights pH variations at grid levels during 1998. The micro-level analysis shows acidic pH levels in 741 grids (60,021 km^2), neutral levels in 1831 grids (148,311 km^2), and no alkalinity grids. Figure 48 depicts district-wise variations in pH in the soil of Karnataka. Figure 49 depicts the pH status during 2020 at the taluk level, while Fig. 50 depicts the pH status at the grid level. The changes in pH levels are evident during the period 1998–2020; for example, in Chitradurga district, 113 km^2 area of neutral levels concentration in the year 1998 became acidic (6561 km^2), and 2592 km^2 of neutral level concentration, which highlights the decline in pH levels due to usage of high fertilizers in the district which increases the acidity of the soil and similar trend is observed in all districts.

Table 27 District wise variations in pH in the soil during 1998 and 2020

Sl.No.	District name	1998						2020					
		Acidic		Neutral		Alkaline		Acidic		Neutral		Alkaline	
		Grid	Area	Grid	Area	Grid	Area	Grid	Area	Grid	Area	Grid	Area
1	Bagalkot		0	83	6723	2	162		0	85	6885	0	0
2	Bangalore Urban		0	32	2592		0	9	729	23	1863	0	0
3	Bangalore (Rural)	10	810	20	1620		0	4	324	26	2106	0	0
4	Belgaum	32	2592	157	12,717	1	81	10	810	179	14,499	0	0
5	Bellary	1	81	115	9315		0	2	162	114	9234	0	0
6	Bidar		0	85	6885		0	1	81	84	6804	0	0
7	Bijapur	2	162	113	9153	24	1944		0	138	11,178	0	0
8	Chamarajanagar	2	162	83	6723		0	11	891	73	5913	0	0
9	Chikballapur	2	162	62	5022		0	1	81	62	5022	0	0
10	Chikkamagaluru	53	4293	34	2754		0	36	2916	51	4131	0	0
11	Chitradurga		0	113	9153		0	81	6561	32	2592	0	0
12	Dakshina Kannada	66	5346		0		0	66	5346		0	0	0
13	Davangere		0	71	5751		0	5	405	66	5346	0	0
14	Dharwad	1	81	35	2835	15	1215	6	486	45	3645	0	0
15	Gadag		0	47	3807	10	810		0	57	4617	0	0
16	Gulbarga	1	81	149	12,069	1	81		0	151	12,231	0	0
17	Hassan	29	2349	52	4212		0	44	3564	37	2997	0	0
18	Haveri	1	81	61	4941		0		0	62	5022	0	0
19	Kodagu	55	4455	4	324		0	57	4617	2	162	0	0

(continued)

Table 27 (continued)

Sl.No.	District name	1998						2020					
		Acidic		Neutral		Alkaline		Acidic		Neutral		Alkaline	
		Grid	Area	Grid	Area	Grid	Area	Grid	Area	Grid	Area	Grid	Area
20	Kolar		0	59	4779		0	55	4455	4	324	0	0
21	Koppal		0	65	5265	2	162		0	67	5427	0	0
22	Mandya		0	63	5103		0		0	63	5103	0	0
23	Mysuru	6	486	77	6237		0	15	1215	67	5427	0	0
24	Raichur		0	63	5103	50	4050		0	113	9153	0	0
25	Ramanagara	14	1134	28	2268		0	1	81	41	3321	0	0
26	Shimoga	97	7857	9	729		0	29	2349	77	6237	0	0
27	Tumkur	20	1620	138	11,178		0	134	10,854	20	1620	0	0
28	Udupi	52	4212		0		0	33	2673	19	1539	0	0
29	Uttara Kannada	134	10,854	8	648		0	141	11,421		0	0	0
30	Yadgir		0	71	5751	2	162		0	73	5913	0	0
	Total	**578**	**46,818**	**1897**	**153,657**	**107**	**8667**	**741**	**60,021**	**1831**	**148,311**	**0**	**0**

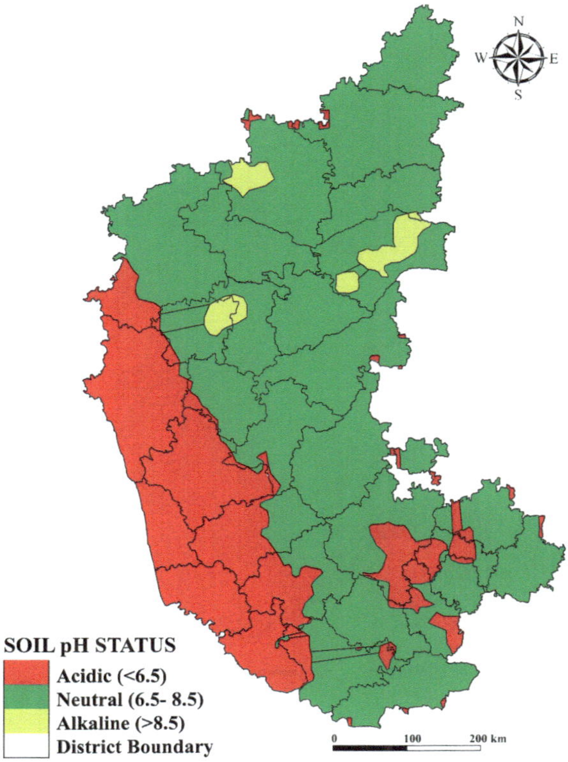

Fig. 46 pH in soil of Karnataka (1998), district wise (1:250,000 scale)

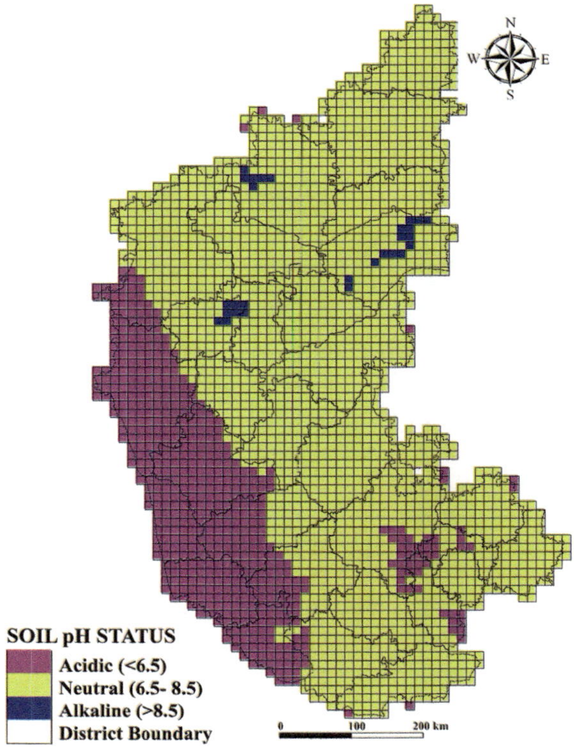

Fig. 47 pH in soil of Karnataka (1998), grid wise (1:250,000 scale)

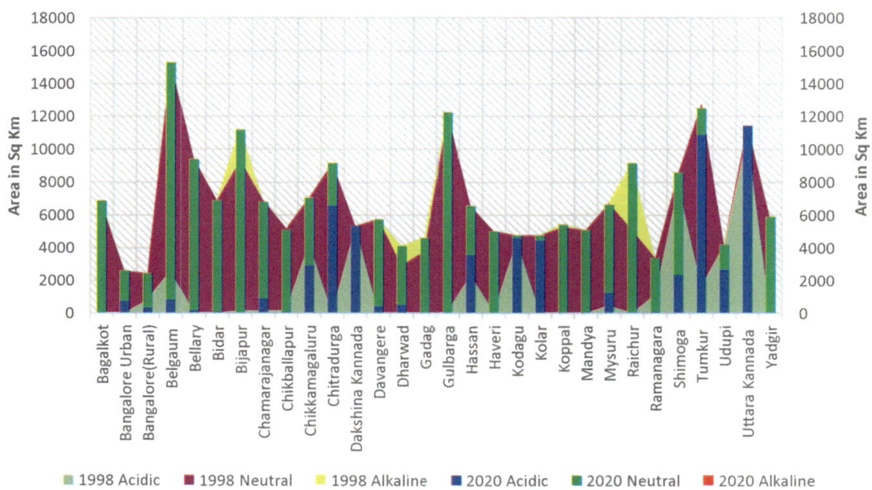

Fig. 48 Variations in pH in soil of Karnataka during 1998 to 2020

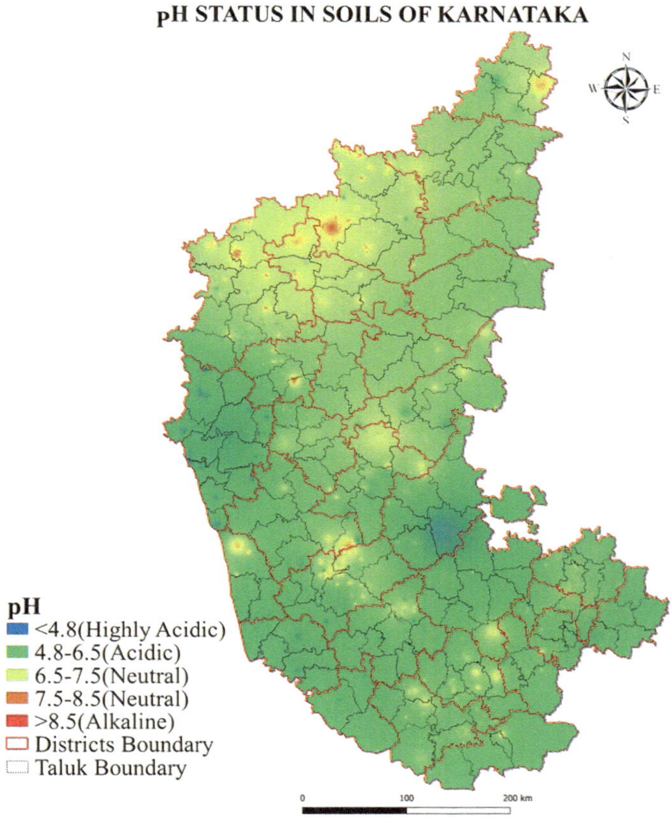

Fig. 49 pH status in soil of Karnataka (2020), taluk wise

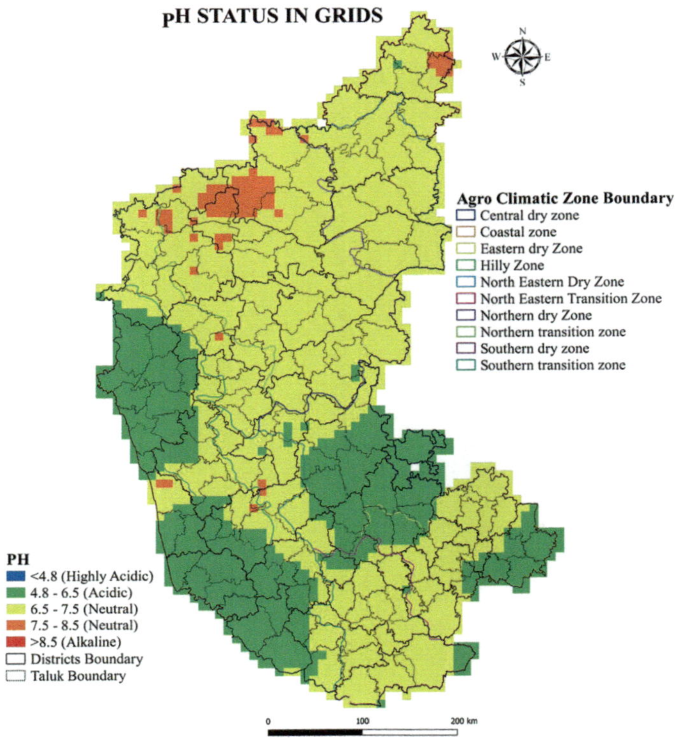

Fig. 50 pH status in soil of Karnataka (2020), Grid and agro-climatic zone wise

Organic Carbon: Table 28 lists district-wise organic carbon content in the soil during 1998 and 2020. The macro-level analysis (1:250,000) shows organic carbon deficiency in 1254 grids (101,574 km^2) and sufficiency levels in 1338 grids (108,378 km^2). Figure 51 depicts the organic carbon concentration at district levels, while Fig. 52 highlights the variation at the grid level during 1998. The micro-level analysis shows that organic carbon has low (deficiency) levels in 302 grids (24,462 km^2), and sufficiency levels in 2270 grids (183,870 km^2). Figure 53 depicts variations in organic carbon levels district-wise in the soil of Karnataka. Figure 54 depicts the organic carbon status during 2020 at the taluk level, while Fig. 55 depicts the organic carbon status at the grid level. The changes in concentration are evident during the period 1998–2020 for example, in Chamrajnagar district 6642 km^2 area was deficient and 243 km^2 area, which was sufficient in the year 1998 has become 0 km^2 of area deficient, and 6804 km^2 of area sufficient and similar trend is observed in all districts. This is contrary to a study of soil organic carbon (Koppad and Janagowdar 2017) in Bhatkal taluk of Uttar Kannada district, Karnataka, which found that soil organic carbon is less in agriculture than in forest soil.

Table 28 District wise organic carbon levels in soil during 1998 and 2020

Sl.No.	District name	1998				2020			
		Deficient		Sufficient		Deficient		Sufficient	
		Grid	Area	Grid	Area	Grid	Area	Grid	Area
1	Bagalkot	9	729	76	6156	12	972	73	5913
2	Bangalore Urban	18	1458	14	1134	28	2268	4	324
3	Bangalore (Rural)	28	2268	2	162	13	1053	17	1377
4	Belgaum	6	486	185	14,985	28	2268	161	13,041
5	Bellary	3	243	113	9153	9	729	107	8667
6	Bidar	11	891	74	5994	11	891	74	5994
7	Bijapur	126	10,206	13	1053	10	810	128	10,368
8	Chamarajnagar	82	6642	3	243		0	84	6804
9	Chikballapur	64	5184		0	45	3645	18	1458
10	Chikmagalur	24	1944	63	5103		0	87	7047
11	Chitradurga	106	8586	7	567	1	81	112	9072
12	Dakshina Kannada		0	67	5427		0	66	5346
13	Davangere	30	2430	41	3321	17	1377	54	4374
14	Dharwad	18	1458	33	2673		0	51	4131
15	Gadag	41	3321	16	1296	4	324	53	4293
16	Gulbarga	101	8181	51	4131		0	151	12,231
17	Hassan	37	2997	44	3564	1	81	80	6480
18	Haveri	20	1620	42	3402	14	1134	48	3888
19	Kodagu		0	60	4860	4	324	55	4455
20	Kolar	60	4860		0	59	4779	0	0

(continued)

Table 28 (continued)

Sl.No.	District name	1998				2020			
		Deficient		Sufficient		Deficient		Sufficient	
		Grid	Area	Grid	Area	Grid	Area	Grid	Area
21	Koppal	49	3969	18	1458	29	2349	38	3078
22	Mandya	9	729	54	4374		0	63	5103
23	Mysore	51	4131	32	2592	3	243	79	6399
24	Raichur	84	6804	29	2349	3	243	110	8910
25	Ramanagara	35	2835	7	567	4	324	38	3078
26	Shimoga		0	106	8586	3	243	103	8343
27	Tumkur	144	11,664	16	1296	3	243	151	12,231
28	Udupi		0	52	4212	1	81	51	4131
29	Uttara Kannada	40	3240	105	8505		0	141	11,421
30	Yadgir	58	4698	15	1215		0	73	5913
	Total	**1254**	**101,574**	**1338**	**108,378**	**302**	**24,462**	**2270**	**183,870**

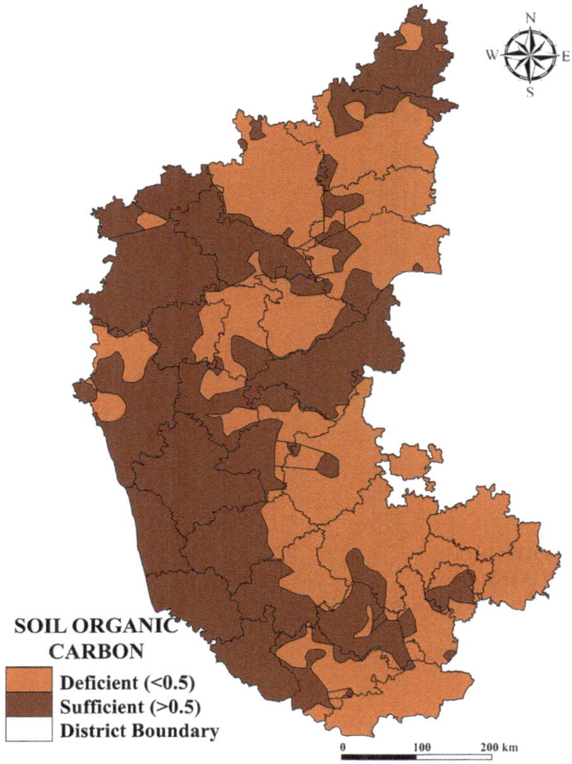

Fig. 51 Organic carbon in soil of Karnataka (1998), district wise (1:250,000 scale)

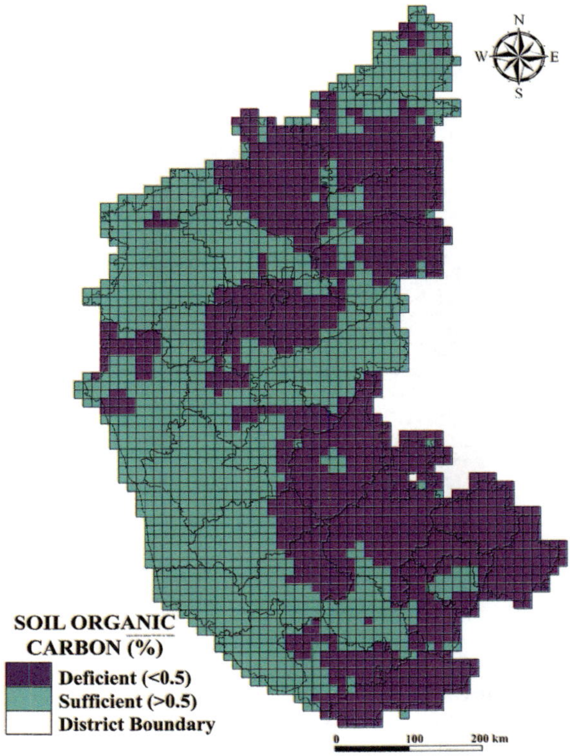

Fig. 52 Organic carbon in soil of Karnataka (1998), Grid level (1:250,000 scale)

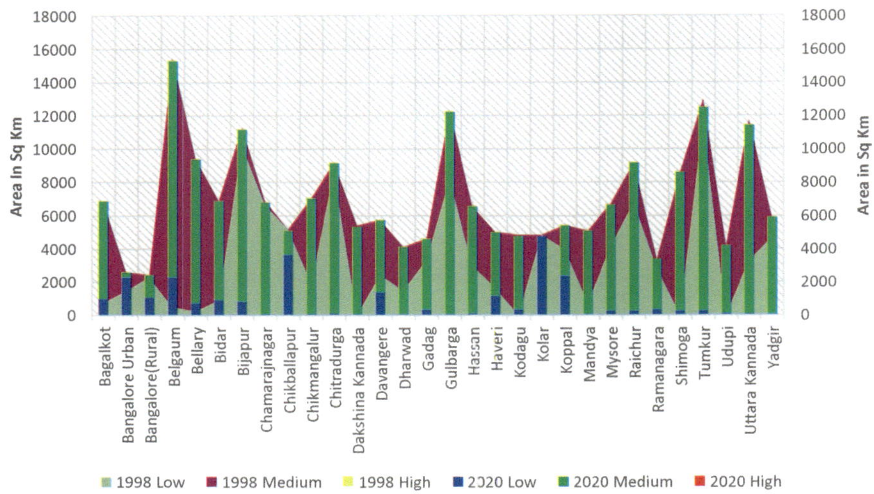

Fig. 53 Variations in organic carbon in soil (district wise), Karnataka during 1998 to 2020

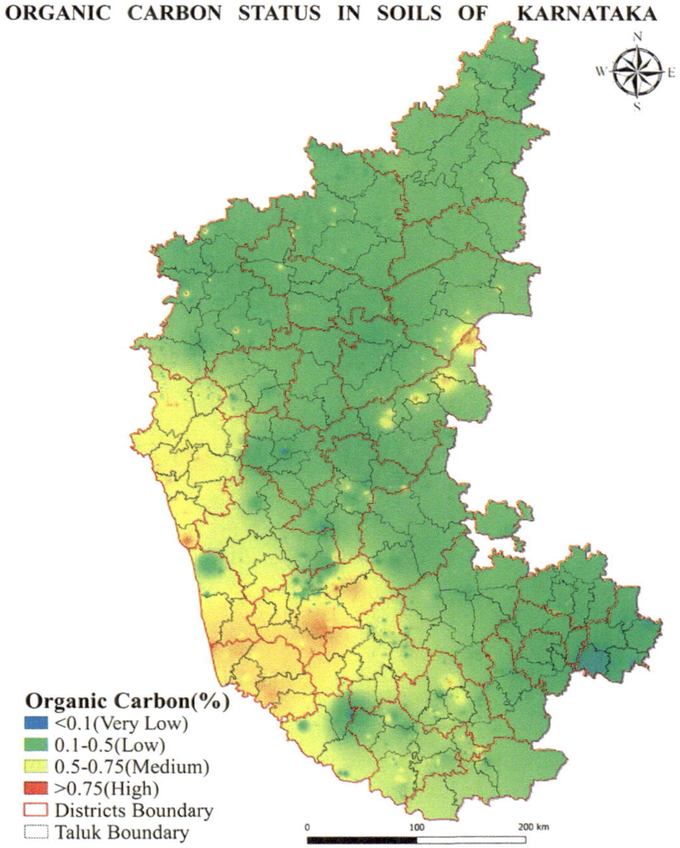

Fig. 54 Organic carbon in soil of Karnataka (2020), taluk wise

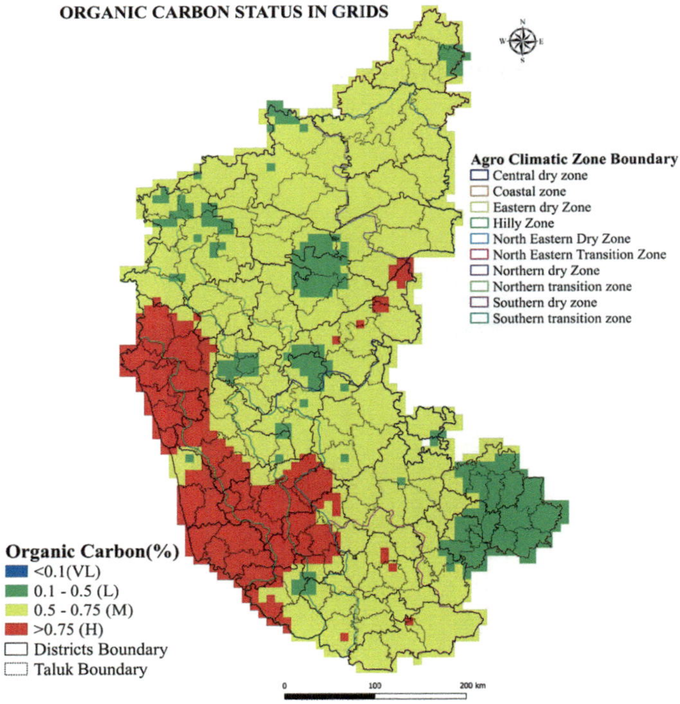

Fig. 55 Organic carbon in soil of Karnataka (2020), Grid and agro-climatic zone wise

Boron: Table 29 lists the district-wise concentration of boron during 2020. The macro-level analysis shows boron deficiency in 1104 grids (89,424 km^2) and sufficiency in 1468 grids (183,870 km^2). Figure 56 depicts the boron status during 2020 at the taluk level, while Fig. 57 depicts the boron status at grid levels. Figure 58 illustrates district-wise boron content in the soil of Karnataka. Districts such as Uttara Kannada, and Belgaum are deficient in boron, while Tumkur, Bagalkot, and Chitradurga have sufficient boron quantity in the soil. The distribution of plant-available boron in major soil types of Karnataka was investigated (Anitha et al. 2015) and reported deficiency of boron in rice-based cropping systems of coastal and hill zones experiencing high rainfall due to leaching of highly soluble boron by rainwater.

Table 29 District-wise Boron status in the soil during 2020

Sl.No.	District name	Deficient		Sufficient	
		Grid	Area	Grid	Area
1	Bagalkot	10	810	75	6075
2	Bangalore Urban	26	2106	6	486
3	Bangalore (Rural)	9	729	21	1701
4	Belgaum	122	9882	67	5427
5	Bellary	10	810	106	8586
6	Bidar	56	4536	29	2349
7	Bijapur	28	2268	110	8910
8	Chamarajnagar	32	2592	52	4212
9	Chikballapur		0	63	5103
10	Chikmagalur	80	6480	7	567
11	Chitradurga	19	1539	94	7614
12	Dakshina Kannada	63	5103	3	243
13	Davangere	47	3807	24	1944
14	Dharwad	51	4131		0
15	Gadag	29	2349	28	2268
16	Gulbarga	68	5508	83	6723
17	Hassan	15	1215	66	5346
18	Haveri	62	5022		0
19	Kodagu	13	1053	46	3726
20	Kolar	4	324	55	4455
21	Koppal		0	67	5427

(continued)

Table 29 (continued)

Sl.No.	District name	Deficient		Sufficient	
		Grid	Area	Grid	Area
22	Mandya	2	162	61	4941
23	Mysore	11	891	71	5751
24	Raichur	12	972	101	8181
25	Ramanagara	18	1458	24	1944
26	Shimoga	91	7371	15	1215
27	Tumkur	5	405	149	12,069
28	Udupi	52	4212		0
29	Uttara Kannada	141	11,421		0
30	Yadgir	28	2268	45	3645
	Grand total	1104	89,424	1468	118,908

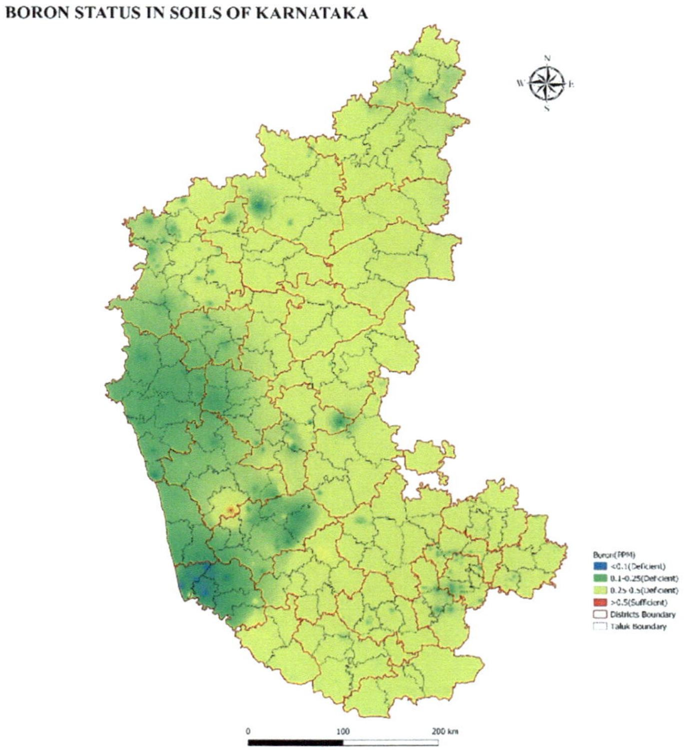

Fig. 56 Boron status in soil of Karnataka (2020), taluk wise

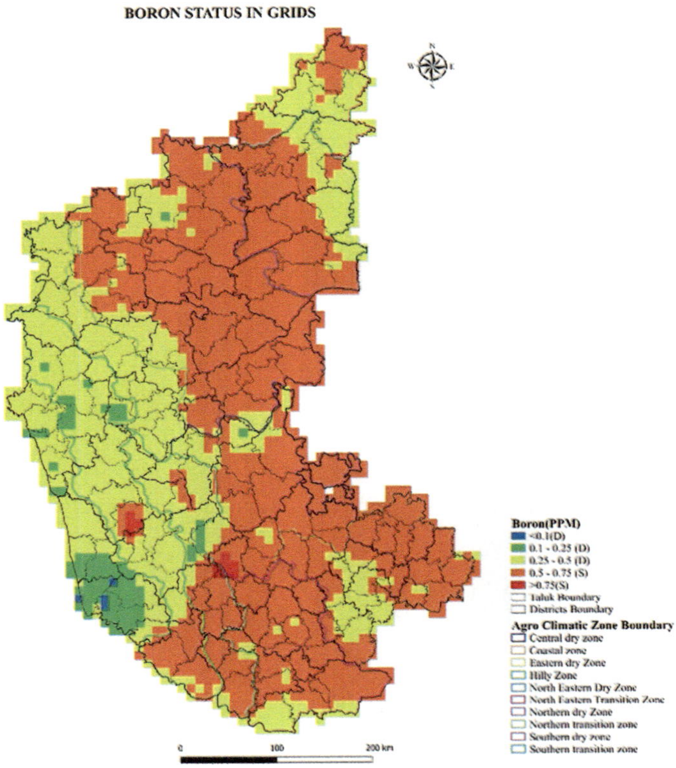

Fig. 57 Boron status in soil of Karnataka (2020), Grid and agro-climatic zone wise

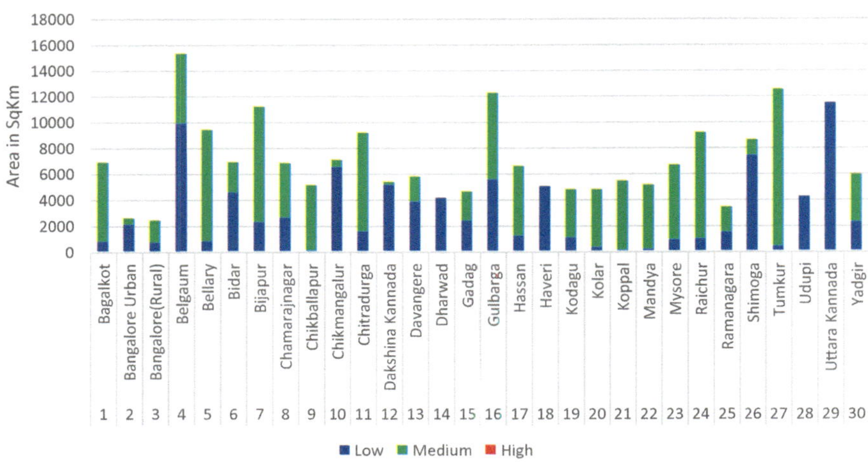

Fig. 58 Boron status in districts of Karnataka during 2020

Table 30 District wise Copper status in soil during 2020

Sl.No.	District name	Deficient		Sufficient	
		Grid	Area	Grid	Area
1	Bagalkot	83	6723	2	162
2	Bangalore Urban	32	2592		0
3	Bangalore (Rural)	30	2430		0
4	Belgaum	180	14,580	9	729
5	Bellary	116	9396		0
6	Bidar	85	6885		0
7	Bijapur	138	11,178		0
8	Chamarajnagar	84	6804		0
9	Chikballapur	63	5103		0
10	Chikmagalur	15	1215	72	5832
11	Chitradurga	113	9153		0
12	Dakshina Kannada	60	4860	6	486
13	Davangere	71	5751		0
14	Dharwad	40	3240	11	891
15	Gadag	55	4455	2	162
16	Gulbarga	151	12,231		0
17	Hassan	50	4050	31	2511
18	Haveri	59	4779	3	243
19	Kodagu	57	4617	2	162
20	Kolar	59	4779		0
21	Koppal	45	3645	22	1782
22	Mandya	63	5103		0
23	Mysore	82	6642		0
24	Raichur	112	9072	1	81
25	Ramanagara	42	3402		0
26	Shimoga	71	5751	35	2835
27	Tumkur	154	12,474		0
28	Udupi	51	4131	1	81
29	Uttara Kannada	1	81	140	11,340
30	Yadgir	73	5913		0
	Grand total	2235	181,035	337	27,297

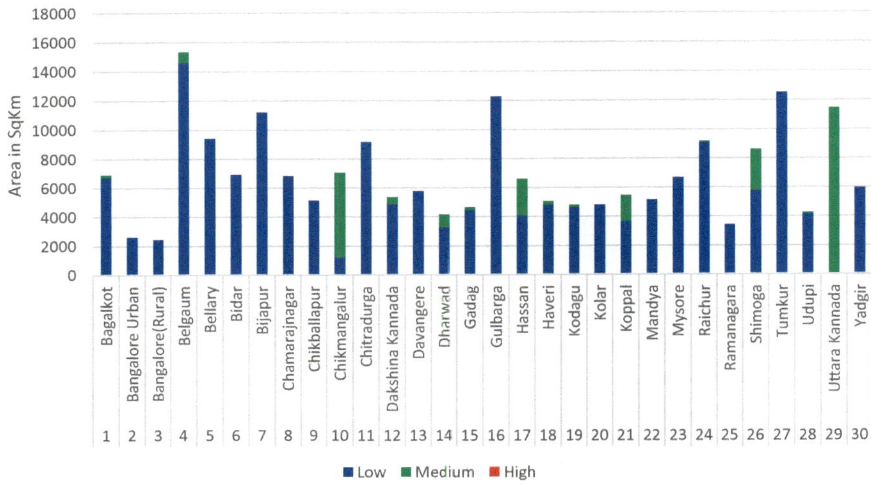

Fig. 59 Copper in soil of Karnataka (district wise) during 2020

Copper: Table 30 lists the district-wise concentration of copper during 2020. The micro-level analysis shows copper deficiency in 2235 grids (181,035 km^2) and sufficiency levels in 337 grids (27,297 km^2). Figure 59 depicts copper levels in the soil (district-wise) of Karnataka, highlighting that most parts are low (deficient) in the copper nutrient. Gulbarga, Chitradurga, and Bijapur districts soil is low (deficient) in copper. Uttara Kannada and Chikamagalur districts have the medium (sufficient) quantity of copper in the soil. Figure 60 depicts the copper status in the soil during 2020 at taluk levels, while Fig. 61 depicts the copper status at the grid level. A study of micronutrients with soil properties of tomato growing soils of Chikkaballapura, Karnataka (Prasad et al. 2020) indicates that copper was medium in only 5.3% of soil samples while the rest, 94.6% had high copper nutrients.

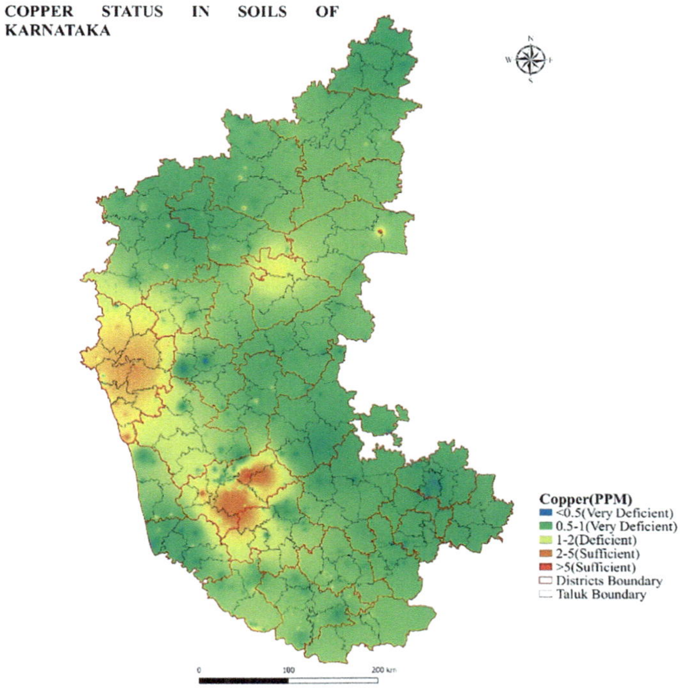

Fig. 60 Copper status in soil of Karnataka (2020), taluk wise

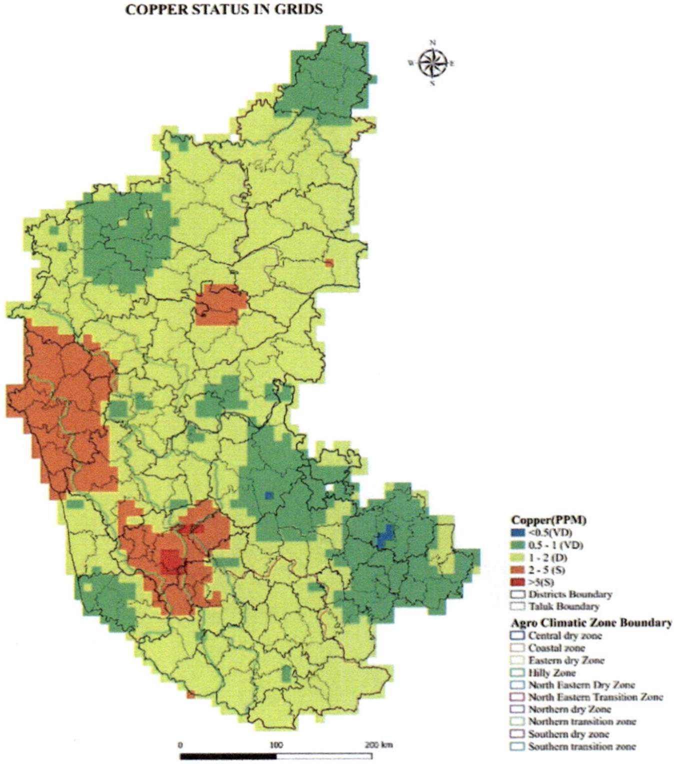

Fig. 61 Copper status in soil of Karnataka (2020), Grid and agro-climatic zone wise

Electrical Conductivity: Table 31 lists district-wise electrical conductivity in the soil of Karnataka during 2020. The micro-level analysis shows that electrical conductivity is low in 2557 grids (207,117 km^2), medium levels in 14 grids (1134 km^2), and a high level is found in 1 grid (81 km^2). Figure 62 depicts the variation in electrical conductivity in the soil (district-wise) of Karnataka. The districts like Bagalkot and parts of Belgaum had moderate conductivity, and one grid in Bagalkot had high conductivity. Most of Karnataka is low in electrical conductivity. Figure 63 depicts the electrical conductivity at the taluk level, while Fig. 64 shows electrical conductivity variations at grid levels. The soil fertility assessment in the micro-watershed of Hassan, Karnataka (Biradar et al. 2020) reveals normal electrical conductivity levels.

Table 31 District-wise electrical conductivity status in the soil during 2020

Sl.No.	District name	Low		Medium		High		Very High	
		Grid	Area	Grid	Area	Grid	Area	Grid	Area
1	Bagalkot	74	5994	10	810	1	81		0
2	Bangalore Urban	32	2592		0		0		0
3	Bangalore (Rural)	30	2430		0		0		0
4	Belgaum	186	15,066	3	243		0		0
5	Bellary	116	9396		0		0		0
6	Bidar	85	6885		0		0		0
7	Bijapur	138	11,178		0		0		0
8	Chamarajnagar	84	6804		0		0		0
9	Chikballapur	63	5103		0		0		0
10	Chikmagalur	87	7047		0		0		0
11	Chitradurga	113	9153		0		0		0
12	Dakshina Kannada	65	5265	1	81		0		0
13	Davangere	71	5751		0		0		0
14	Dharwad	51	4131		0		0		0
15	Gadag	57	4617		0		0		0
16	Gulbarga	151	12,231		0		0		0
17	Hassan	81	6561		0		0		0
18	Haveri	62	5022		0		0		0
19	Kodagu	59	4779		0		0		0
20	Kolar	59	4779		0		0		0
21	Koppal	67	5427		0		0		0
22	Mandya	63	5103		0		0		0
23	Mysore	82	6642		0		0		0
24	Raichur	113	9153		0		0		0
25	Ramanagara	42	3402		0		0		0
26	Shimoga	106	8586		0		0		0
27	Tumkur	154	12,474		0		0		0
28	Udupi	52	4212		0		0		0
29	Uttara Kannada	141	11,421		0		0		0
30	Yadgir	73	5913		0		0		0
	Grand total	2557	207,117	14	1134	1	81		0

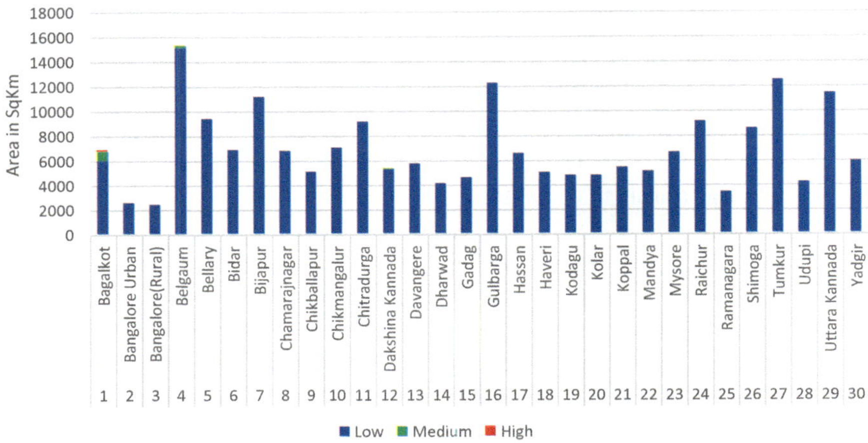

Fig. 62 Electrical conductivity in soil (district wise), Karnataka during 2020

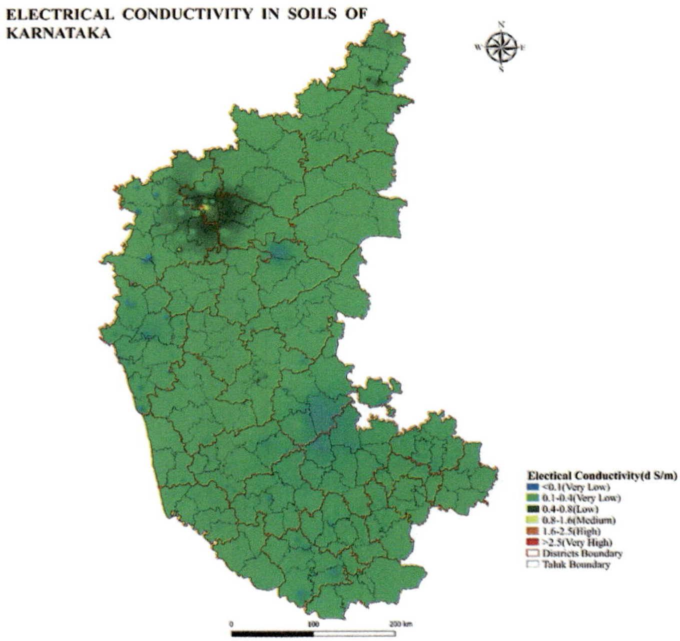

Fig. 63 Electrical conductivity status in soil of Karnataka (2020), taluk wise

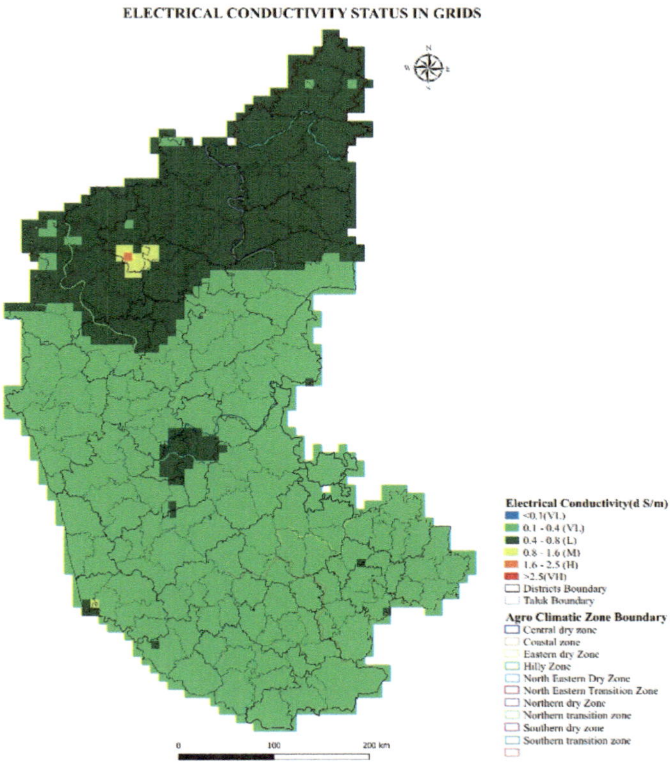

Fig. 64 Electrical conductivity status in the soil of Karnataka (2020), Grid and agro-climatic zone wise

Manganese: Table 32 lists district-wise manganese concentration in the soil of Karnataka during 2020. The micro level analysis shows manganese deficiency levels in 4 grids (324 km^2), and sufficiency levels in 2568 grids (208,008 km^2). Figure 65 depicts manganese in the soil (district-wise) of Karnataka. Some parts of Dakshina Kannada are deficient in manganese, and all other districts of Karnataka have enough manganese in the soil. Figure 66 depicts the manganese status during 2020 at the taluk level, while Fig. 67 illustrates the manganese in soil at the grid level. Evaluation of the soil nutrient index of Siddapura catchment, Karnataka (Sindhu et al. 2022) shows sufficient manganese availability in the soil.

Table 32 District-wise manganese status in soil during 2020

Sl.No.	District name	Deficient		Sufficient	
		Grid	Area	Grid	Area
1	Bagalkot		0	85	6885
2	Bangalore Urban		0	32	2592
3	Bangalore (Rural)		0	30	2430
4	Belgaum		0	189	15,309
5	Bellary		0	116	9396
6	Bidar		0	85	6885
7	Bijapur		0	138	11,178
8	Chamarajnagar		0	84	6804
9	Chikballapur		0	63	5103
10	Chikmagalur		0	87	7047
11	Chitradurga		0	113	9153
12	Dakshina Kannada	4	324	62	5022
13	Davangere		0	71	5751
14	Dharwad		0	51	4131
15	Gadag		0	57	4617
16	Gulbarga		0	151	12,231
17	Hassan		0	81	6561
18	Haveri		0	62	5022
19	Kodagu		0	59	4779
20	Kolar		0	59	4779
21	Koppal		0	67	5427
22	Mandya		0	63	5103
23	Mysore		0	82	6642
24	Raichur		0	113	9153
25	Ramanagara		0	42	3402
26	Shimoga		0	106	8586
27	Tumkur		0	154	12,474
28	Udupi		0	52	4212
29	Uttara Kannada		0	141	11,421
30	Yadgir		0	73	5913
	Grand total	4	324	2568	208,008

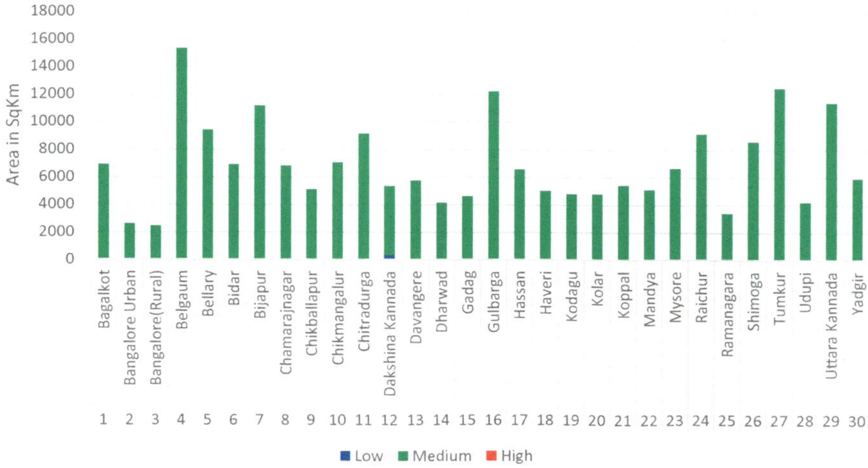

Fig. 65 Manganese status in area during 2020

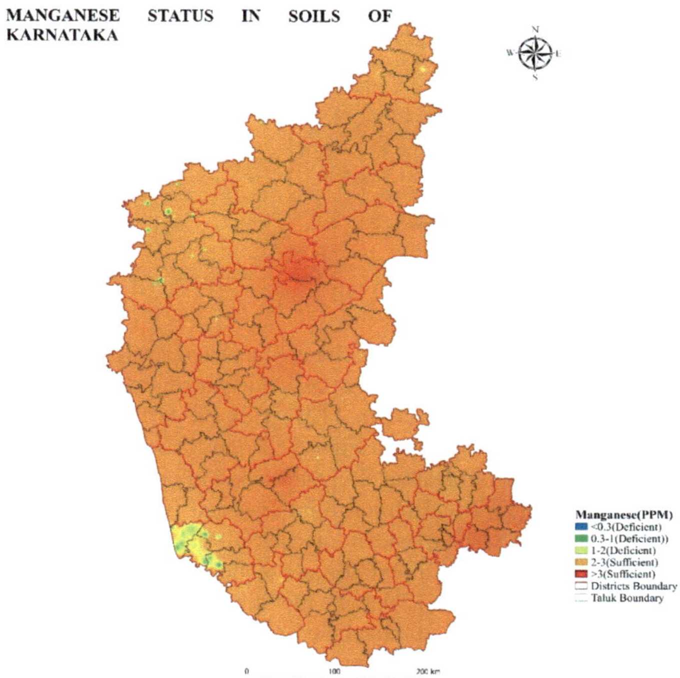

Fig. 66 Manganese status in soil of Karnataka (2020), taluk wise

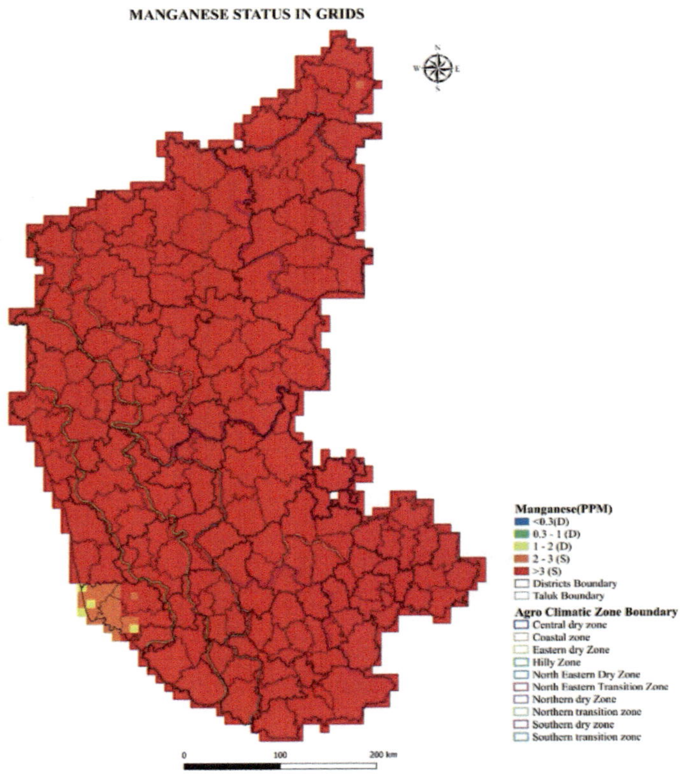

Fig. 67 Manganese status in soil of Karnataka (2020), Grid and agro-climatic zone wise

Sulphur: Table 33 lists district-wise Sulphur status in the soil of Karnataka during 2020. The micro-level analysis shows that Sulphur concentration levels are low in 43 grids (3483 km^2), medium levels in 1121 grids (90,801 km^2), and high level is found in 1408 grids (114,048 km^2). Figure 68 depicts sulphur levels in the soil (district-wise) of Karnataka. Bagalkot, Belgaum and Gulbarga districts have a high concentration of Sulphur, while Tumkur, Shimoga, and Bengaluru districts have a moderate concentration of Sulphur, and in some parts of Bidar and Dakshina Kannada soil have a low concentration of Sulphur. Figure 69 depicts the Sulphur status during 2020 at the taluk level, while Fig. 70 depicts the Sulphur status at the grid level. Soil fertility status of the Pannur micro watershed of Manvi taluk, Raichur, Karnataka (Meenakshi et al. 2018) reveals low to medium Sulphur concentration in most of the samples collected, and the nutrients decreased with the depth of soil.

Table 33 District-wise sulphur status in soil during 2020

Sl.No.	District name	Low		Medium		High	
		Grid	Area	Grid	Area	Grid	Area
1	Bagalkot		0		0	85	6885
2	Bangalore Urban		0	32	2592		0
3	Bangalore (Rural)		0	30	2430		0
4	Belgaum		0	2	162	187	15,147
5	Bellary		0	26	2106	90	7290
6	Bidar	23	1863	56	4536	6	486
7	Bijapur		0		0	138	11,178
8	Chamarajnagar		0	65	5265	19	1539
9	Chikballapur		0	63	5103		0
10	Chikmagalur		0	50	4050	37	2997
11	Chitradurga		0	74	5994	39	3159
12	Dakshina Kannada	14	1134	47	3807	5	405
13	Davangere		0	17	1377	54	4374
14	Dharwad		0	1	81	50	4050
15	Gadag		0	5	405	52	4212
16	Gulbarga		0	4	324	147	11,907
17	Hassan		0	18	1458	63	5103
18	Haveri		0	44	3564	18	1458
19	Kodagu		0	58	4698	1	81
20	Kolar		0	59	4779		0
21	Koppal		0	16	1296	51	4131
22	Mandya		0	16	1296	47	3807
23	Mysore		0	44	3564	38	3078
24	Raichur		0		0	113	9153
25	Ramanagara		0	39	3159	3	243
26	Shimoga		0	93	7533	13	1053
27	Tumkur		0	140	11,340	14	1134
28	Udupi	6	486	45	3645	1	81
29	Uttara Kannada		0	77	6237	64	5184
30	Yadgir		0		0	73	5913
	Grand total	43	3483	1121	90,801	1408	114,048

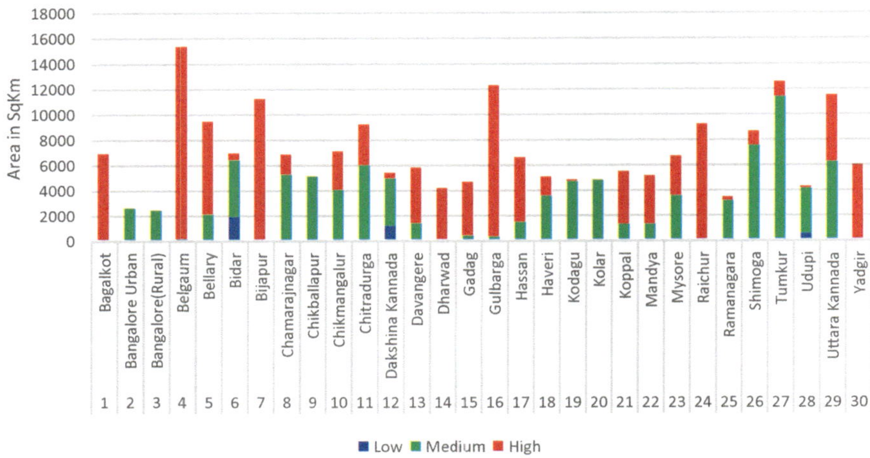

Fig. 68 Sulphur is soil (district wise) of Karnataka during 2020

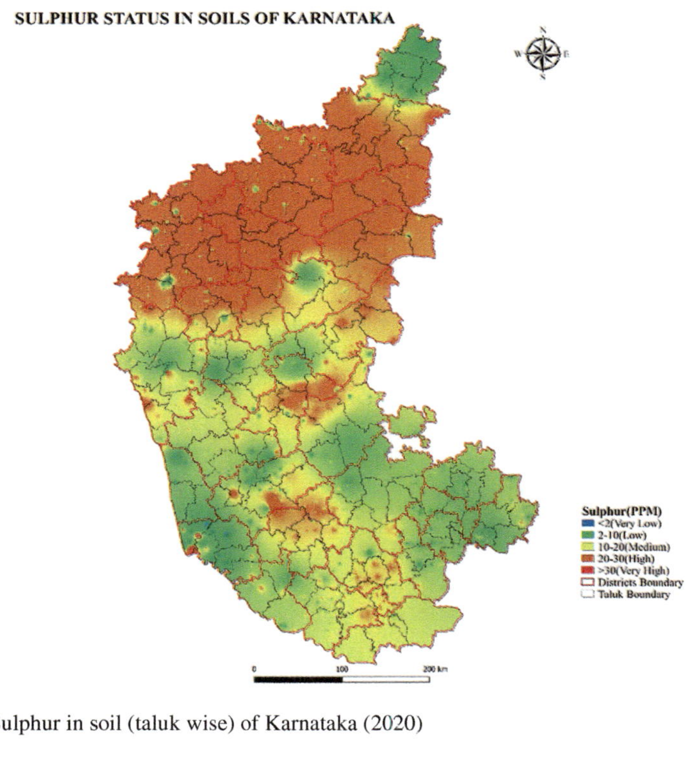

Fig. 69 Sulphur in soil (taluk wise) of Karnataka (2020)

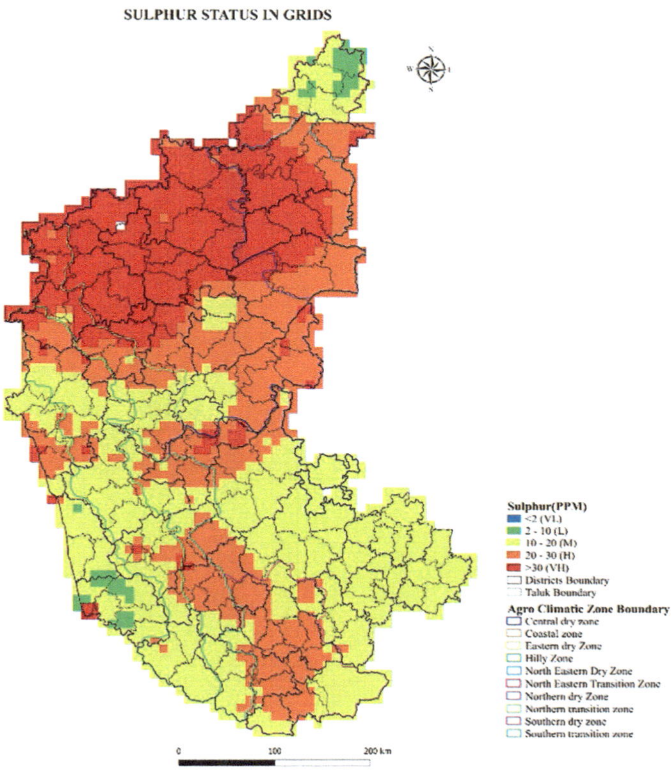

Fig. 70 Sulphur status in the soil of Karnataka (2020), Grid and agro-climatic zone wise

Nutrient status: Table 34 lists district-wise variations in nutrient concentrations (1998).

Table 34 District-wise nutrient status during 1998

Sl.No.	District name	Potassium (Kg/ha)	Nitrogen (Kg/ha)	Phosphorus (Kg/ha)	Zinc (ppm)	Iron (ppm)	pH	Organic carbon (%)
1	Bagalkot	199–281	80–560	10–20	0.4	6.2–6.4	7–7.8	0.4–0.6
2	Ballari	240–281	279–561	10–20	0.4–0.625	6.2–6.4	6.4–7	0.4–0.6
3	Belagavi	117–281	279–560	10–20	0.4–0.85	6.2–6.4	6.4–7.8	0.4–0.6
4	Bengaluru Rural	199–281	279–420	10.-20	0.4–0.8	6.3–6.4	6.4–7	0.4–0.6
5	Bengaluru Urban	199–281	279–561	10–20	0.4–0.85	6.2–6.4	6.7–7	0.4–0.6
6	Bidar	281	420–561	10–20	0.4	6.2–6.4	7	0.4–0.6
7	Chamarajanagar	199–281	420–561	10	0.4–0.85	6.2–6.4	6.4–7	0.4–0.6
8	Chikballapur	199–281	279	10–20	0.4	6.4	6.4–7	0.4
9	Chikkamagaluru	117–281	279–561	10–20	0.4–0.85	6.4	6.4–7	0.4–0.6
10	Chitradurga	199–281	279–561	10–20	0.4–0.625	6.2–6.4	6.7–7	0.4–0.6
11	Dakshina Kannada	117–199	561	10	0.4–0.625	6.4	6.4	0.6
12	Davangere	119–281	279–561	10–20	0.4–0.85	6.3–6.4	6.7–7	0.4–0.6
13	Dharwad	199–281	420–561	10	0.4–0.85	6.2–6.4	6.4–8.6	0.4–0.6
14	Gadag	199–281	279–561	10	0.4	7–8.6	6.2–6.4	0.4–0.6
15	Hassan	199–281	349–569	10–20	0.4–0.85	6.4	6.4–7	0.4–0.6
16	Haveri	199–281	279–581	10	0.4–0.85	6.2–6.4	6.4–7	0.4–0.6
17	Kalaburagi	199–281	279–561	10–20	0.4	6.2–6.4	6.4–8.6	0.4–0.6
18	Kodagu	117–281	420–561	10–20	0.4–0.85	6.4	6.4–6.7	0.6
19	Kolar	199–281	279–420	10.20	0.4–0.85	6.4	6.7–7	0.4
20	Koppal	281	279–561	10	0.4	6.2–6.4	7–7.8	0.4–0.6
21	Mandya	158–581	279–561	10–20	0.4–0.625	6.2–6.4	6.7–7	0.4–0.6

(continued)

Table 34 (continued)

Sl.No.	District name	Potassium (Kg/ha)	Nitrogen (Kg/ha)	Phosphorus (Kg/ha)	Zinc (ppm)	Iron (ppm)	pH	Organic carbon (%)
22	Mysuru	158–581	279–561	10–20	0.4–0.85	6.2–6.4	6.4–7	0.4–0.6
23	Raichur	199–281	279–561	10	0.4	6.2–6.3	7.8.6	0.4–0.6
24	Ramanagara	199–281	279–561	10–20	0.4–0.85	6.3–6.4	6.4–7	0.4–0.6
25	Shivamogga	158–199	420–561	10–20	0.4–0.85	6.3–6.4	6.4–6.7	0.6
26	Tumakuru	199–281	279–561	10–20	0.4–0.625	6.2–6.4	6.4–7	0.4–0.6
27	Udupi	117–199	561	10–20	0.4–0.625	6.4	6.4	0.6
28	Uttara Kannada	117–240	561	10	0.4–0.85	6.4	6.4–7	0.4–0.6
29	Vijayapura	117–281	279–561	10–20	0.4	6.2–6.4	6.4–8.6	0.4–0.6
30	Yadgir	199–281	279–420	10	0.4	6.2–6.4	7–7.8	0.4–0.6

5 Ecosystem Condition Based on Abiotic Ecosystem Characteristics

Considering Physical state characteristics (**soil pH and EC**) and chemical state characteristics (**soil nutrient levels**, water quality, air pollutant concentrations).

Soil data for 2005 were for districts compiled from published literature for some districts, while the data of 1998 was adjusted for a few districts (4 districts, data from 2005 was unavailable) using 2005 data of neighbouring districts and the base level of 1998, in consultation with the expert from the Agriculture department. Micro nutrients are grouped as sufficient and deficient categories from a crop-land health status point of view, which were translated to SEEA reference levels as–medium (sufficient) and low (deficient) in consultation with the experts from the agriculture sector. Table 35 lists district-wise ecosystem conditions based on abiotic characteristics considering physical state characteristics (pH and EC) and chemical state characteristics (soil nutrient levels) of soil for 2005 and 2019 in Karnataka.

6 Ecosystem Condition Index (ECI)

The ecosystem condition index for Karnataka State (district-wise) is computed considering ecosystem condition variables–abiotic (soil physical and chemical status), biotic ecosystem (compositional, structural, and functional variables), and landscape-level (fragmentation and LST). The aggregation is performed by assigning weights as 0.2 for abiotic, 0.3 for biotic (0.1 for compositional, 0.1 for structural, and 0.1 for functional), and 0.5 for landscape-level characteristics. Table 36 lists the ecosystem condition index computed for all districts in Karnataka for 2005 and 2020 (Fig. 71).

Table 35 Ecosystem conditions based on abiotic characteristics considering Soil in Karnataka State (district wise), India–physical state characteristics (EC and pH) **and** Chemical state characteristics (**soil nutrient levels**)

Soil condition based on physical and chemical parameters					Ecosystem: agricultural lands						
					Opening 2005				Closing 2020		
District	ECT	Parameter	Units	Total	High	Medium	Low	Total	High	Medium	Low
Bagalkot	Soil–chemical (macro nutrients)	K	sq.km	6885	4698	2187	0	6885	3888	2997	0
			%		68.2	31.8	0.0		56.5	43.5	0.0
		N	sq.km	6885	405	3969	2511	6885	0	2673	4212
			%		5.9	57.6	36.5		0.0	38.8	61.2
		P	sq.km	6885	0	324	6561	6885	567	6237	81
			%		0.0	4.7	95.3		8.2	90.6	1.2
		Zn	sq.km	6885	0	0	6885	6885	0	4536	2349
			%		0.0	0.0	100.0		0.0	65.9	34.1
		Fe	sq.km	6885	0	3078	3807	6885	0	5346	1539
			%		0.0	44.7	55.3		0.0	77.6	22.4
		OC	sq.km	6885	0	6156	729	6885	0	5913	972
			%		0.0	89.4	10.6		0.0	85.9	14.1
	Soil–chemical (micro nutrients)	B	sq.km	6885	0	6075	810	6885	0	6075	810
			%		0.0	88.2	11.8		0.0	88.2	11.8
		Cu	sq.km	6885	0	162	6723	6885	0	162	6723
			%		0.0	2.4	97.6		0.0	2.4	97.6
		Mn	sq.km	6885	0	6885	0	6885	0	6885	0
			%		0.0	100.0	0.0		0.0	100.0	0.0
		S	sq.km	6885	6885	0	0	6885	6885	0	0
			%		100.0	0.0	0.0		100.0	0.0	0.0

(continued)

Table 35 (continued)

Soil condition based on physical and chemical parameters					Ecosystem: agricultural lands							
					Opening 2005				Closing 2020			
District	ECT	Parameter	Units	Total	High	Medium	Low	Total	High	Medium	Low
	Soil physical	EC	sq.km	6885	5994	810	81	6885	5994	810	81
			%		87.1	11.8	1.2		87.1	11.8	1.2
		pH	sq.km	6885	0	6723	162	6885	0	6885	0
			%		0.0	97.6	2.4		0.0	100.0	0.0
Bangalore (Rural)	Soil -chemical (macro nutrients)	K	sq.km	2430	243	2187	0	2430	324	2106	0
			%		10.0	90.0	0.0		13.3	86.7	0.0
		N	sq.km	2430	0	648	1782	2430	0	0	2430
			%		0.0	26.7	73.3		0.0	0.0	100.0
		P	sq.km	2430	0	567	1863	2430	0	2430	0
			%		0.0	23.3	76.7		0.0	100.0	0.0
		Zn	sq.km	2430	0	729	1701	2430	0	1782	648
			%		0.0	30.0	70.0		0.0	73.3	26.7
		Fe	sq.km	2430	0	2430	0	2430	0	2430	0
			%		0.0	100.0	0.0		0.0	100.0	0.0
		OC	sq.km	2430	0	162	2268	2430	0	1377	1053
			%		0.0	6.7	93.3		0.0	56.7	43.3
	Soil–chemical (micro nutrients)	B	sq.km	2430	0	1701	729	2430	0	1701	729
			%		0.0	70.0	30.0		0.0	70.0	30.0
		Cu	sq.km	2430	0	0	2430	2430	0	0	2430
			%		0.0	0.0	100.0		0.0	0.0	100.0

(continued)

Table 35 (continued)

Soil condition based on physical and chemical parameters				Ecosystem: agricultural lands							
					Opening 2005				Closing 2020		
District	ECT	Parameter	Units	Total	High	Medium	Low	Total	High	Medium	Low
		Mn	sq.km	2430	0	2430	0	2430	0	2430	0
			%		0.0	100.0	0.0		0.0	100.0	0.0
		S	sq.km	2430	0	2430	0	2430	0	2430	0
			%		0.0	100.0	0.0		0.0	100.0	0.0
	Soil physical	EC	sq.km	2430	2430	0	0	2430	2430	0	0
			%		100.0	0.0	0.0		100.0	0.0	0.0
		pH	sq.km	2430	810	1620	0	2430	324	2106	0
			%		33.3	66.7	0.0		13.3	86.7	0.0
Bangalore (Urban)	Soil–chemical (macro nutrients)	K	sq.km	2592	324	2268	0	2592	81	2511	0
			%		12.5	87.5	0.0		3.1	96.9	0.0
		N	sq.km	2592	729	1377	486	2592	0	0	2592
			%		28.1	53.1	18.8		0.0	0.0	100.0
		P	sq.km	2592	0	729	1863	2592	0	2592	0
			%		0.0	28.1	71.9		0.0	100.0	0.0
		Zn	sq.km	2592	0	2187	405	2592	0	2349	243
			%		0.0	84.4	15.6		0.0	90.6	9.4
		Fe	sq.km	2592	0	2511	81	2592	0	2592	0
			%		0.0	96.9	3.1		0.0	100.0	0.0
		OC	sq.km	2592	0	1134	1458	2592	0	324	2268
			%		0.0	43.8	56.3		0.0	12.5	87.5

(continued)

Table 35 (continued)

Soil condition based on physical and chemical parameters					Ecosystem: agricultural lands						
					Opening 2005				Closing 2020		
District	ECT	Parameter	Units	Total	High	Medium	Low	Total	High	Medium	Low
	Soil–chemical (micro nutrients)	B	sq.km	2592	0	486	2106	2592	0	486	2106
			%		0.0	18.8	81.3		0.0	18.8	81.3
		Cu	sq.km	2592	0	0	2592	2592	0	0	2592
			%		0.0	0.0	100.0		0.0	0.0	100.0
		Mn	sq.km	2592	0	2592	0	2592	0	2592	0
			%		0.0	100.0	0.0		0.0	100.0	0.0
		S	sq.km	2592	0	2592	0	2592	0	2592	0
			%		0.0	100.0	0.0		0.0	100.0	0.0
	Soil physical	EC	sq.km	2592	2592	0	0	2592	2592	0	0
			%		100.0	0.0	0.0		100.0	0.0	0.0
		pH	sq.km	2592	0	2592	0	2592	729	1863	0
			%		0.0	100.0	0.0		28.1	71.9	0.0
Belgaum	Soil–chemical (macro nutrients)	K	sq.km	15,309	2187	12,393	729	15,309	1296	14,013	0
			%		14.3	81.0	4.8		8.5	91.5	0.0
		N	sq.km	15,309	7776	5103	2430	15,309	0	972	14,337
			%		50.8	33.3	15.9		0.0	6.3	93.7
		P	sq.km	15,309	0	2268	13,041	15,309	0	14,823	486
			%		0.0	14.8	85.2		0.0	96.8	3.2
		Zn	sq.km	15,309	0	12,150	3159	15,309	0	12,150	3159
			%		0.0	79.4	20.6		0.0	79.4	20.6

(continued)

Table 35 (continued)

Soil condition based on physical and chemical parameters				Ecosystem: agricultural lands							
					Opening 2005				Closing 2020		
District	ECT	Parameter	Units	Total	High	Medium	Low	Total	High	Medium	Low
		Fe	sq.km	15,309	0	14,580	729	15,309	0	14,580	729
			%		0.0	95.2	4.8		0.0	95.2	4.8
		OC	sq.km	15,309	0	13,041	2268	15,309	0	13,041	2268
			%		0.0	85.2	14.8		0.0	85.2	14.8
	Soil–chemical (micro nutrients)	B	sq.km	15,309	0	5427	9882	15,309	0	5427	9882
			%		0.0	35.4	64.6		0.0	35.4	64.6
		Cu	sq.km	15,309	0	729	14,580	15,309	0	729	14,580
			%		0.0	4.8	95.2		0.0	4.8	95.2
		Mn	sq.km	15,309	0	15,309	0	15,309	0	15,309	0
			%		0.0	100.0	0.0		0.0	100.0	0.0
		S	sq.km	15,309	15,147	162	0	15,309	15,147	162	0
			%		98.9	1.1	0.0		98.9	1.1	0.0
	Soil physical	EC	sq.km	15,309	15,066	243	0	15,309	15,066	243	0
			%		98.4	1.6	0.0		98.4	1.6	0.0
		pH	sq.km	15,309	2592	12,636	81	15,309	810	14,499	0
			%		16.9	82.5	0.5		5.3	94.7	0.0
Bellary	Soil–chemical (macro nutrients)	K	sq.km	9396	8019	567	810	9396	0	9396	0
			%		85.3	6.0	8.6		0.0	100.0	0.0
		N	sq.km	9396	3483	3726	2187	9396	0	0	9396
			%		37.1	39.7	23.3		0.0	0.0	100.0

(continued)

Table 35 (continued)

Soil condition based on physical and chemical parameters					Ecosystem: agricultural lands						
					Opening 2005				Closing 2020		
District	ECT	Parameter	Units	Total	High	Medium	Low	Total	High	Medium	Low
		P	sq.km	9396	0	972	8424	9396	324	8991	81
			%		0.0	10.3	89.7		3.4	95.7	0.9
		Zn	sq.km	9396	0	243	9153	9396	0	9396	0
			%		0.0	2.6	97.4		0.0	100.0	0.0
		Fe	sq.km	9396	0	5670	3726	9396	0	9396	0
			%		0.0	60.3	39.7		0.0	100.0	0.0
		OC	sq.km	9396	0	9153	243	9396	0	8667	729
			%		0.0	97.4	2.6		0.0	92.2	7.8
	Soil–chemical (micro nutrients)	B	sq.km	9396	0	8586	810	9396	0	8586	810
			%		0.0	91.4	8.6		0.0	91.4	8.6
		Cu	sq.km	9396	0	0	9396	9396	0	0	9396
			%		0.0	0.0	100.0		0.0	0.0	100.0
		Mn	sq.km	9396	0	9396	0	9396	0	9396	0
			%		0.0	100.0	0.0		0.0	100.0	0.0
		S	sq.km	9396	7290	2106	0	9396	7290	2106	0
			%		77.6	22.4	0.0		77.6	22.4	0.0
	Soil physical	EC	sq.km	9396	9396	0	0	9396	9396	0	0
			%		100.0	0.0	0.0		100.0	0.0	0.0
		pH	sq.km	9396	81	9315	0	9396	162	9234	0
			%		0.9	99.1	0.0		1.7	98.3	0.0

(continued)

Table 35 (continued)

Soil condition based on physical and chemical parameters				Ecosystem: agricultural lands							
				Opening 2005				Closing 2020			
District	ECT	Parameter	Units	Total	High	Medium	Low	Total	High	Medium	Low
Bidar	Soil–chemical (macro nutrients)	K	sq.km	6885	6885	0	0	6885	810	6075	0
			%		100.0	0.0	0.0		11.8	88.2	0.0
		N	sq.km	6885	6480	405	0	6885	0	0	6885
			%		94.1	5.9	0.0		0.0	0.0	100.0
		P	sq.km	6885	0	2673	4212	6885	0	2430	4455
			%		0.0	38.8	61.2		0.0	35.3	64.7
		Zn	sq.km	6885	0	0	6885	6885	0	6804	81
			%		0.0	0.0	100.0		0.0	98.8	1.2
		Fe	sq.km	6885	0	5427	1458	6885	0	6804	81
			%		0.0	78.8	21.2		0.0	98.8	1.2
		OC	sq.km	6885	0	5994	891	6885	0	5994	891
			%		0.0	87.1	12.9		0.0	87.1	12.9
	Soil–chemical (micro nutrients)	B	sq.km	6885	0	2349	4536	6885	0	2349	4536
			%		0.0	34.1	65.9		0.0	34.1	65.9
		Cu	sq.km	6885	0	0	6885	6885	0	0	6885
			%		0.0	0.0	100.0		0.0	0.0	100.0
		Mn	sq.km	6885	0	6885	0	6885	0	6885	0
			%		0.0	100.0	0.0		0.0	100.0	0.0
		S	sq.km	6885	486	4536	1863	6885	486	4536	1863
			%		7.1	65.9	27.1		7.1	65.9	27.1

(continued)

Table 35 (continued)

Soil condition based on physical and chemical parameters				Ecosystem: agricultural lands							
					Opening 2005				Closing 2020		
District	ECT	Parameter	Units	Total	High	Medium	Low	Total	High	Medium	Low
	Soil physical	EC	sq.km	6885	6885	0	0	6885	6885	0	0
			%		100.0	0.0	0.0		100.0	0.0	0.0
		pH	sq.km	6885	0	6885	0	6885	81	6804	0
			%		0.0	100.0	0.0		1.2	98.8	0.0
Bijapur	Soil–chemical (macro nutrients)	K	sq.km	11,178	6156	4536	486	11,178	7857	3321	0
			%		55.1	40.6	4.3	100.0	70.3	29.7	0.0
		N	sq.km	11,178	729	10,449	0	11,178	0	81	11,097
			%		6.5	93.5	0.0	100.0	0.0	0.7	99.3
		P	sq.km	11,178	0	486	10,692	11,178	81	11,097	0
			%		0.0	4.3	95.7	100.0	0.7	99.3	0.0
		Zn	sq.km	11,178	0	0	11,178	11,178	0	10,287	891
			%		0.0	0.0	100.0	100.0	0.0	92.0	8.0
		Fe	sq.km	11,178	0	5913	5265	11,178	0	11,178	0
			%		0.0	52.9	47.1	100.0	0.0	100.0	0.0
		OC	sq.km	11,178	0	1053	10,125	11,178	0	10,368	810
			%		0.0	9.4	90.6	100.0	0.0	92.8	7.2
	Soil–chemical (micro nutrients)	B	sq.km	11,178	0	8910	2268	11,178	0	8910	2268
			%		0.0	79.7	20.3	100.0	0.0	79.7	20.3
		Cu	sq.km	11,178	0	0	11,178	11,178	0	0	11,178
			%		0.0	0.0	100.0	100.0	0.0	0.0	100.0

(continued)

Table 35 (continued)

Soil condition based on physical and chemical parameters				Ecosystem: agricultural lands							
					Opening 2005				Closing 2020		
District	ECT	Parameter	Units	Total	High	Medium	Low	Total	High	Medium	Low
		Mn	sq.km	11,178	0	11,178	0	11,178	0	11,178	0
			%		0.0	100.0	0.0	100.0	0.0	100.0	0.0
		S	sq.km	11,178	11,178	0	0	11,178	11,178	0	0
			%		100.0	0.0	0.0	100.0	100.0	0.0	0.0
	Soil Physical	EC	sq.km	11,178	11,178	0	0	11,178	11,178	0	0
			%		100.0	0.0	0.0	100.0	100.0	0.0	0.0
		pH	sq.km	11,178	162	9153	1863	11,178	0	11,178	0
			%		1.4	81.9	16.7	100.0	0.0	100.0	0.0
Chamarajanagar	Soil–Chemical (macro nutrients)	K	sq.km	6885	648	6237	0	6804	81	6723	0
			%		9.4	90.6	0.0		1.2	98.8	0.0
		N	sq.km	6885	5184	1701	0	6804	0	81	6723
			%		75.3	24.7	0.0		0.0	1.2	98.8
		P	sq.km	6885	0	0	6885	6804	81	6723	0
			%		0.0	0.0	100.0		1.2	98.8	0.0
		Zn	sq.km	6885	0	2106	4779	6804	0	6804	0
			%		0.0	30.6	69.4		0.0	100.0	0.0
		Fe	sq.km	6885	0	6075	810	6804	0	6804	0
			%		0.0	88.2	11.8		0.0	100.0	0.0
		OC	sq.km	6885	0	243	6642	6804	0	6804	0
			%		0.0	3.5	96.5		0.0	100.0	0.0

(continued)

Table 35 (continued)

Soil condition based on physical and chemical parameters					Ecosystem: agricultural lands						
					Opening 2005				Closing 2020		
District	ECT	Parameter	Units	Total	High	Medium	Low	Total	High	Medium	Low
	Soil–chemical (micro nutrients)	B	sq.km	6804	0	4212	2592	6804	0	4212	2592
			%		0.0	61.9	38.1	100.0	0.0	61.9	38.1
		Cu	sq.km	6804	0	0	6804	6804	0	0	6804
			%		0.0	0.0	100.0		0.0	0.0	100.0
		Mn	sq.km	6804	0	6804	0	6804	0	6804	0
			%		0.0	100.0	0.0		0.0	100.0	0.0
		S	sq.km	6804	1539	5265	0	6804	1539	5265	0
			%		22.6	77.4	0.0		22.6	77.4	0.0
	Soil physical	EC	sq.km	6804	6804	0	0	6804	6804	0	0
			%		100.0	0.0	0.0		100.0	0.0	0.0
		pH	sq.km	6804	891	5913	0	6804	891	5913	0
			%		13.1	86.9	0.0		13.1	86.9	0.0
Chikballapur	Soil–chemical (macro nutrients)	K	sq.km	5184	324	4860	0	5103	243	4860	0
			%		6.3	93.8	0.0		4.8	95.2	0.0
		N	sq.km	5184	0	0	5184	5103	0	0	5103
			%		0.0	0.0	100.0		0.0	0.0	100.0
		P	sq.km	5184	0	486	4698	5103	0	5103	0
			%		0.0	9.4	90.6		0.0	100.0	0.0
		Zn	sq.km	5184	0	0	5184	5103	0	3402	1701
			%		0.0	0.0	100.0		0.0	66.7	33.3

(continued)

Table 35 (continued)

Soil condition based on physical and chemical parameters					Ecosystem: agricultural lands						
					Opening 2005				Closing 2020		
District	ECT	Parameter	Units	Total	High	Medium	Low	Total	High	Medium	Low
		Fe	sq.km	5184	0	5184	0	5103	0	5103	0
			%		0.0	100.0	0.0		0.0	100.0	0.0
		OC	sq.km	5184	0	0	5184	5103	0	1458	3645
			%		0.0	0.0	100.0		0.0	28.6	71.4
	Soil–chemical (micro nutrients)	B	sq.km	5103	0	5103	0	5103	0	5103	0
			%		0.0	100.0	0.0		0.0	100.0	0.0
		Cu	sq.km	5103	0	0	5103	5103	0	0	5103
			%		0.0	0.0	100.0		0.0	0.0	100.0
		Mn	sq.km	5103	0	5103	0	5103	0	5103	0
			%		0.0	100.0	0.0		0.0	100.0	0.0
		S	sq.km	5103	0	5103	0	5103	0	5103	0
			%		0.0	100.0	0.0		0.0	100.0	0.0
	Soil physical	EC	sq.km	5103	5103	0	0	5103	5103	0	0
			%		100.0	0.0	0.0		100.0	0.0	0.0
		pH	sq.km	5184	162	5022	0	5103	81	5022	0
			%		3.1	96.9	0.0		1.6	98.4	0.0
Chikkamagaluru	Soil–chemical (macro nutrients)	K	sq.km	7047	1539	5022	486	7047	324	6399	324
			%		21.8	71.3	6.9		4.6	90.8	4.6
		N	sq.km	7047	5913	729	405	7047	0	162	6885
			%		83.9	10.3	5.7		0.0	2.3	97.7

(continued)

Table 35 (continued)

Soil condition based on physical and chemical parameters					Ecosystem: agricultural lands						
					Opening 2005				Closing 2020		
District	ECT	Parameter	Units	Total	High	Medium	Low	Total	High	Medium	Low
		P	sq.km	7047	0	810	6237	7047	3483	3483	81
			%		0.0	11.5	88.5		49.4	49.4	1.1
		Zn	sq.km	7047	0	5508	1539	7047	0	6723	324
			%		0.0	78.2	21.8		0.0	95.4	4.6
		Fe	sq.km	7047	0	7047	0	7047	0	7047	0
			%		0.0	100.0	0.0		0.0	100.0	0.0
		OC	sq.km	7047	0	5103	1944	7047	0	7047	0
			%		0.0	72.4	27.6		0.0	100.0	0.0
	Soil–chemical (micro nutrients)	B	sq.km	0	0	0	0	7047	0	567	6480
			%						0.0	8.0	92.0
		Cu	sq.km	7047	0	5832	1215	7047	0	5832	1215
			%		0.0	82.8	17.2		0.0	82.8	17.2
		Mn	sq.km	7047	0	7047	0	7047	0	7047	0
			%		0.0	100.0	0.0		0.0	100.0	0.0
		S	sq.km	7047	2997	4050	0	7047	2997	4050	0
			%		42.5	57.5	0.0		42.5	57.5	0.0
	Soil physical	EC	sq.km	7047	7047	0	0	7047	7047	0	0
			%		100.0	0.0	0.0		100.0	0.0	0.0
		pH	sq.km	7047	2916	4131	0	7047	2916	4131	0
			%		41.4	58.6	0.0		41.4	58.6	0.0

(continued)

Table 35 (continued)

Soil condition based on physical and chemical parameters					Ecosystem: agricultural lands						
					Opening 2005				Closing 2020		
District	ECT	Parameter	Units	Total	High	Medium	Low	Total	High	Medium	Low
Chitradurga	Soil–chemical (macro nutrients)	K	sq.km	9153	7047	2106	0	9153	0	8019	1134
			%		77.0	23.0	0.0		0.0	87.6	12.4
		N	sq.km	9153	648	5832	2673	9153	0	0	9153
			%		7.1	63.7	29.2		0.0	0.0	100.0
		P	sq.km	9153	0	729	8424	9153	1296	7857	0
			%		0.0	8.0	92.0		14.2	85.8	0.0
		Zn	sq.km	9153	0	324	8829	9153	0	5265	3888
			%		0.0	3.5	96.5		0.0	57.5	42.5
		Fe	sq.km	9153	0	7776	1377	9153	0	9153	0
			%		0.0	85.0	15.0		0.0	100.0	0.0
		OC	sq.km	9153	0	567	8586	9153	0	9072	81
			%		0.0	6.2	93.8		0.0	99.1	0.9
	Soil–chemical (micro nutrients)	B	sq.km	9153	0	7614	1539	9153	0	7614	1539
			%		0.0	83.2	16.8		0.0	83.2	16.8
		Cu	sq.km	9153	0	0	9153	9153	0	0	9153
			%		0.0	0.0	100.0		0.0	0.0	100.0
		Mn	sq.km	9153	0	9153	0	9153	0	9153	0
			%		0.0	100.0	0.0		0.0	100.0	0.0
		S	sq.km	9153	3159	5994	0	9153	3159	5994	0
			%		34.5	65.5	0.0		34.5	65.5	0.0

(continued)

Table 35 (continued)

Soil condition based on physical and chemical parameters					Ecosystem: agricultural lands						
					Opening 2005				Closing 2020		
District	ECT	Parameter	Units	Total	High	Medium	Low	Total	High	Medium	Low
	Soil physical	EC	sq.km	9153	9153	0	0	9153	9153	0	0
			%		100.0	0.0	0.0		100.0	0.0	0.0
		pH	sq.km	9153	0	9153	0	9153	6561	2592	0
			%		0.0	100.0	0.0		71.7	28.3	0.0
Dakshina Kannada	Soil–chemical (macro nutrients)	K	sq.km	5346	0	3483	1863	5346	0	1215	4131
			%		0.0	65.2	34.8		0.0	22.7	77.3
		N	sq.km	5346	5346	0	0	5346	0	324	5022
			%		100.0	0.0	0.0		0.0	6.1	93.9
		P	sq.km	5427	0	0	5427	5346	81	2754	2511
			%		0.0	0.0	100.0		1.5	51.5	47.0
		Zn	sq.km	5346	0	567	4779	5346	0	5346	0
			%		0.0	10.6	89.4		0.0	100.0	0.0
		Fe	sq.km	5346	0	5346	0	5346	0	5346	0
			%		0.0	100.0	0.0		0.0	100.0	0.0
		OC	sq.km	5427	0	5427	0	5346	0	5346	0
			%		0.0	100.0	0.0		0.0	100.0	0.0
	Soil–chemical (micro nutrients)	B	sq.km	5346	0	243	5103	5346	0	243	5103
			%		0.0	4.5	95.5		0.0	4.5	95.5
		Cu	sq.km	5346	0	486	4860	5346	0	486	4860
			%		0.0	9.1	90.9		0.0	9.1	90.9

(continued)

Table 35 (continued)

Soil condition based on physical and chemical parameters					Ecosystem: agricultural lands						
					Opening 2005				Closing 2020		
District	ECT	Parameter	Units	Total	High	Medium	Low	Total	High	Medium	Low
		Mn	sq.km	5346	0	5022	324	5346	0	5022	324
			%		0.0	93.9	6.1		0.0	93.9	6.1
		S	sq.km	5346	405	3807	1134	5346	405	3807	1134
			%		7.6	71.2	21.2		7.6	71.2	21.2
	Soil physical	EC	sq.km	5346	5265	81	0	5346	5265	81	0
			%		98.5	1.5	0.0		98.5	1.5	0.0
		pH	sq.km	5346	5346	0	0	5346	5346	0	0
			%		100.0	0.0	0.0		100.0	0.0	0.0
Davangere	Soil–chemical (macro nutrients)	K	sq.km	5751	2511	3240	0	5751	0	5751	0
			%		43.7	56.3	0.0		0.0	100.0	0.0
		N	sq.km	5751	2349	3240	162	5751	0	81	5670
			%		40.8	56.3	2.8		0.0	1.4	98.6
		P	sq.km	5751	0	810	4941	5751	3078	2673	0
			%		0.0	14.1	85.9		53.5	46.5	0.0
		Zn	sq.km	5751	0	1701	4050	5751	0	5589	162
			%		0.0	29.6	70.4		0.0	97.2	2.8
		Fe	sq.km	5751	0	5751	0	5751	0	5751	0
			%		0.0	100.0	0.0		0.0	100.0	0.0
		OC	sq.km	5751	0	3321	2430	5751	0	4374	1377
			%		0.0	57.7	42.3		0.0	76.1	23.9

(continued)

Table 35 (continued)

Soil condition based on physical and chemical parameters					Ecosystem: agricultural lands						
					Opening 2005				Closing 2020		
District	ECT	Parameter	Units	Total	High	Medium	Low	Total	High	Medium	Low
	Soil–chemical (micro nutrients)	B	sq.km	5751	0	1944	3807	5751	0	1944	3807
			%		0.0	33.8	66.2		0.0	33.8	66.2
		Cu	sq.km	5751	0	0	5751	5751	0	0	5751
			%		0.0	0.0	100.0		0.0	0.0	100.0
		Mn	sq.km	5751	0	5751	0	5751	0	5751	0
			%		0.0	100.0	0.0		0.0	100.0	0.0
		S	sq.km	5751	4374	1377	0	5751	4374	1377	0
			%		76.1	23.9	0.0		76.1	23.9	0.0
	Soil physical	EC	sq.km	5751	5751	0	0	5751	5751	0	0
			%		100.0	0.0	0.0		100.0	0.0	0.0
		pH	sq.km	5751	0	5751	0	5751	405	5346	0
			%		0.0	100.0	0.0		7.0	93.0	0.0
Dharwad	Soil–chemical (macro nutrients)	K	sq.km	4131	2349	1782	0	4131	810	3321	0
			%		56.9	43.1	0.0		19.6	80.4	0.0
		N	sq.km	4131	2349	1782	0	4131	0	0	4131
			%		56.9	43.1	0.0		0.0	0.0	100.0
		P	sq.km	4131	0	0	4131	4131	0	4131	0
			%		0.0	0.0	100.0		0.0	100.0	0.0
		Zn	sq.km	4131	0	2106	2025	4131	0	4131	0
			%		0.0	51.0	49.0		0.0	100.0	0.0

(continued)

Table 35 (continued)

Soil condition based on physical and chemical parameters					Ecosystem: agricultural lands						
					Opening 2005				Closing 2020		
District	ECT	Parameter	Units	Total	High	Medium	Low	Total	High	Medium	Low
		Fe	sq.km	4131	0	2916	1215	4131	0	4131	0
			%		0.0	70.6	29.4		0.0	100.0	0.0
		OC	sq.km	4131	0	2673	1458	4131	0	4131	0
			%		0.0	64.7	35.3		0.0	100.0	0.0
	Soil–chemical (micro nutrients)	B	sq.km	4131	0	0	4131	4131	0	0	4131
			%		0.0	0.0	100.0		0.0	0.0	100.0
		Cu	sq.km	4131	0	891	3240	4131	0	891	3240
			%		0.0	21.6	78.4		0.0	21.6	78.4
		Mn	sq.km	4131	0	4131	0	4131	0	4131	0
			%		0.0	100.0	0.0		0.0	100.0	0.0
		S	sq.km	4131	4050	81	0	4131	4050	81	0
			%		98.0	2.0	0.0		98.0	2.0	0.0
	Soil physical	EC	sq.km	4131	4131	0	0	4131	4131	0	0
			%		100.0	0.0	0.0		100.0	0.0	0.0
		pH	sq.km	4131	81	2835	1215	4131	486	3645	0
			%		2.0	68.6	29.4		11.8	88.2	0.0
Gadag	Soil–chemical (macro nutrients)	K	sq.km	4617	2835	1782	0	4617	243	4374	0
			%		61.4	38.6	0.0		5.3	94.7	0.0
		N	sq.km	4617	810	2754	1053	4617	0	0	4617
			%		17.5	59.6	22.8		0.0	0.0	100.0

(continued)

Table 35 (continued)

Soil condition based on physical and chemical parameters					Ecosystem: agricultural lands						
					Opening 2005				Closing 2020		
District	ECT	Parameter	Units	Total	High	Medium	Low	Total	High	Medium	Low
		P	sq.km	4617	0	0	4617	4617	324	4293	0
			%		0.0	0.0	100.0		7.0	93.0	0.0
		Zn	sq.km	4617	0	0	4617	4617	0	4617	0
			%		0.0	0.0	100.0		0.0	100.0	0.0
		Fe	sq.km	4617	0	648	3969	4617	0	4617	0
			%		0.0	14.0	86.0		0.0	100.0	0.0
		OC	sq.km	4617	0	1296	3321	4617	0	4293	324
			%		0.0	28.1	71.9		0.0	93.0	7.0
	Soil–chemical (micro nutrients)	B	sq.km	4617	0	2268	2349	4617	0	2268	2349
			%		0.0	49.1	50.9		0.0	49.1	50.9
		Cu	sq.km	4617	0	162	4455	4617	0	162	4455
			%		0.0	3.5	96.5		0.0	3.5	96.5
		Mn	sq.km	4617	0	4617	0	4617	0	4617	0
			%		0.0	100.0	0.0		0.0	100.0	0.0
		S	sq.km	4617	4212	405	0	4617	4212	405	0
			%		91.2	8.8	0.0		91.2	8.8	0.0
	Soil physical	EC	sq.km	4617	4617	0	0	4617	4617	0	0
			%		100.0	0.0	0.0		100.0	0.0	0.0
		pH	sq.km	4617	0	3807	810	4617	0	4617	0
			%		0.0	82.5	17.5		0.0	100.0	0.0

(continued)

Table 35 (continued)

Soil condition based on physical and chemical parameters					Ecosystem: agricultural lands						
					Opening 2005				Closing 2020		
District	ECT	Parameter	Units	Total	High	Medium	Low	Total	High	Medium	Low
Gulbarga	Soil–chemical (macro nutrients)	K	sq.km	12,231	10,449	1782	0	12,231	3402	8829	0
			%		85.4	14.6	0.0		27.8	72.2	0.0
		N	sq.km	12,231	3240	6966	2025	12,231	0	0	12,231
			%		26.5	57.0	16.6		0.0	0.0	100.0
		P	sq.km	12,312	0	1134	11,178	12,231	0	12,150	81
			%		0.0	9.2	90.8		0.0	99.3	0.7
		Zn	sq.km	12,231	0	0	12,231	12,231	0	12,150	81
			%		0.0	0.0	100.0		0.0	99.3	0.7
		Fe	sq.km	12,231	0	7533	4698	12,231	0	12,231	0
			%		0.0	61.6	38.4		0.0	100.0	0.0
		OC	sq.km	12,312	0	4131	8181	12,231	0	12,231	0
			%		0.0	33.6	66.4		0.0	100.0	0.0
	Soil–chemical (micro nutrients)	B	sq.km	12,231	0	6723	5508	12,231	0	6723	5508
			%		0.0	55.0	45.0		0.0	55.0	45.0
		Cu	sq.km	12,231	0	0	12,231	12,231	0	0	12,231
			%		0.0	0.0	100.0		0.0	0.0	100.0
		Mn	sq.km	12,231	0	12,231	0	12,231	0	12,231	0
			%		0.0	100.0	0.0		0.0	100.0	0.0
		S	sq.km	12,231	11,907	324	0	12,231	11,907	324	0
			%		97.4	2.6	0.0		97.4	2.6	0.0

(continued)

Table 35 (continued)

Soil condition based on physical and chemical parameters					Ecosystem: agricultural lands						
					Opening 2005				Closing 2020		
District	ECT	Parameter	Units	Total	High	Medium	Low	Total	High	Medium	Low
	Soil physical	EC	sq.km	12,231	12,231	0	0	12,231	12,231	0	0
			%		100.0	0.0	0.0		100.0	0.0	0.0
		pH	sq.km	12,231	81	12,069	81	12,231	0	12,231	0
			%		0.7	98.7	0.7		0.0	100.0	0.0
Hassan	Soil–chemical (macro nutrients)	K	sq.km	6561	4212	2349	0	6561	405	6156	0
			%		64.2	35.8	0.0		6.2	93.8	0.0
		N	sq.km	6561	3321	3240	0	6561	0	3402	3159
			%		50.6	49.4	0.0		0.0	51.9	48.1
		P	sq.km	6561	0	891	5670	6561	2754	3807	0
			%		0.0	13.6	86.4		42.0	58.0	0.0
		Zn	sq.km	6561	0	2106	4455	6561	0	5994	567
			%		0.0	32.1	67.9		0.0	91.4	8.6
		Fe	sq.km	6561	0	6561	0	6561	0	6561	0
			%		0.0	100.0	0.0		0.0	100.0	0.0
		OC	sq.km	6561	0	3564	2997	6561	0	6480	81
			%		0.0	54.3	45.7		0.0	98.8	1.2
	Soil–chemical (micro nutrients)	B	sq.km	6561	0	5346	1215	6561	0	5346	1215
			%		0.0	81.5	18.5		0.0	81.5	18.5
		Cu	sq.km	6561	0	2511	4050	6561	0	2511	4050
			%		0.0	38.3	61.7		0.0	38.3	61.7

(continued)

Table 35 (continued)

Soil condition based on physical and chemical parameters				Ecosystem: agricultural lands							
				Opening 2005				Closing 2020			
District	ECT	Parameter	Units	Total	High	Medium	Low	Total	High	Medium	Low
		Mn	sq.km	6561	0	6561	0	6561	0	6561	0
			%		0.0	100.0	0.0		0.0	100.0	0.0
		S	sq.km	6561	5103	1458	0	6561	5103	1458	0
			%		77.8	22.2	0.0		77.8	22.2	0.0
	Soil physical	EC	sq.km	6561	6561	0	0	6561	6561	0	0
			%		100.0	0.0	0.0		100.0	0.0	0.0
		pH	sq.km	6561	2349	4212	0	6561	3564	2997	0
			%		35.8	64.2	0.0		54.3	45.7	0.0
Haveri	Soil–chemical (macro nutrients)	K	sq.km	5022	1620	3402	0	5022	0	5022	0
			%		32.3	67.7	0.0		0.0	100.0	0.0
		N	sq.km	5022	3483	1215	324	5022	0	0	5022
			%		69.4	24.2	6.5		0.0	0.0	100.0
		P	sq.km	5022	0	0	5022	5022	243	4779	0
			%		0.0	0.0	100.0		4.8	95.2	0.0
		Zn	sq.km	5022	0	2187	2835	5022	0	4779	243
			%		0.0	43.5	56.5		0.0	95.2	4.8
		Fe	sq.km	5022	0	3969	1053	5022	0	5022	0
			%		0.0	79.0	21.0		0.0	100.0	0.0
		OC	sq.km	5022	0	3402	1620	5022	0	3888	1134
			%		0.0	67.7	32.3		0.0	77.4	22.6

(continued)

Table 35 (continued)

Soil condition based on physical and chemical parameters					Ecosystem: agricultural lands						
					Opening 2005				Closing 2020		
District	ECT	Parameter	Units	Total	High	Medium	Low	Total	High	Medium	Low
	Soil–chemical (micro nutrients)	B	sq.km	5022	0	0	5022	5022	0	0	5022
			%		0.0	0.0	100.0		0.0	0.0	100.0
		Cu	sq.km	5022	0	243	4779	5022	0	243	4779
			%		0.0	4.8	95.2		0.0	4.8	95.2
		Mn	sq.km	5022	0	5022	0	5022	0	5022	0
			%		0.0	100.0	0.0		0.0	100.0	0.0
		S	sq.km	5022	1458	3564	0	5022	1458	3564	0
			%		29.0	71.0	0.0		29.0	71.0	0.0
	Soil physical	EC	sq.km	5022	5022	0	0	5022	5022	0	0
			%		100.0	0.0	0.0		100.0	0.0	0.0
		pH	sq.km	5022	81	4941	0	5022	0	5022	0
			%		1.6	98.4	0.0		0.0	100.0	0.0
Kodagu	Soil–chemical (macro nutrients)	K	sq.km	4779	243	3078	1458	4779	405	4374	0
			%		5.1	64.4	30.5		8.5	91.5	0.0
		N	sq.km	4779	4374	405	0	4779	0	324	4455
			%		91.5	8.5	0.0		0.0	6.8	93.2
		P	sq.km	4779	0	567	4212	4779	0	4698	81
			%		0.0	11.9	88.1		0.0	98.3	1.7
		Zn	sq.km	4779	0	3726	1053	4779	0	4779	0
			%		0.0	78.0	22.0		0.0	100.0	0.0

(continued)

Table 35 (continued)

Soil condition based on physical and chemical parameters					Ecosystem: agricultural lands						
					Opening 2005				Closing 2020		
District	ECT	Parameter	Units	Total	High	Medium	Low	Total	High	Medium	Low
		Fe	sq.km	4779	0	4779	0	4779	0	4779	0
			%		0.0	100.0	0.0		0.0	100.0	0.0
		OC	sq.km	4779	0	4779	0	4779	0	4455	324
			%		0.0	100.0	0.0		0.0	93.2	6.8
	Soil–chemical (micro nutrients)	B	sq.km	4779	0	3726	1053	4779	0	3726	1053
			%		0.0	78.0	22.0		0.0	78.0	22.0
		Cu	sq.km	4779	0	162	4617	4779	0	162	4617
			%		0.0	3.4	96.6		0.0	3.4	96.6
		Mn	sq.km	4779	0	4779	0	4779	0	4779	0
			%		0.0	100.0	0.0		0.0	100.0	0.0
		S	sq.km	4779	81	4698	0	4779	81	4698	0
			%		1.7	98.3	0.0		1.7	98.3	0.0
	Soil physical	EC	sq.km	4779	4779	0	0	4779	4779	0	0
			%		100.0	0.0	0.0		100.0	0.0	0.0
		pH	sq.km	4779	4455	324	0	4779	4617	162	0
			%		93.2	6.8	0.0		96.6	3.4	0.0
Kolar	Soil–chemical (macro nutrients)	K	sq.km	4779	324	4455	0	4779	0	4779	0
			%		6.8	93.2	0.0		0.0	100.0	0.0
		N	sq.km	4779	0	648	4131	4779	0	0	4779
			%		0.0	13.6	86.4		0.0	0.0	100.0

(continued)

Table 35 (continued)

Soil condition based on physical and chemical parameters					Ecosystem: agricultural lands						
					Opening 2005				Closing 2020		
District	ECT	Parameter	Units	Total	High	Medium	Low	Total	High	Medium	Low
		P	sq.km	4779	0	486	4293	4779	0	4455	324
			%		0.0	10.2	89.8		0.0	93.2	6.8
		Zn	sq.km	4779	0	324	4455	4779	0	4698	81
			%		0.0	6.8	93.2		0.0	98.3	1.7
		Fe	sq.km	4779	0	4779	0	4779	0	4779	0
			%		0.0	100.0	0.0		0.0	100.0	0.0
		OC	sq.km	4779	0	0	4779	4779	0	0	4779
			%		0.0	0.0	100.0		0.0	0.0	100.0
	Soil–chemical (micro nutrients)	B	sq.km	4779	0	4455	324	4779	0	4455	324
			%		0.0	93.2	6.8		0.0	93.2	6.8
		Cu	sq.km	4779	0	0	4779	4779	0	0	4779
			%		0.0	0.0	100.0		0.0	0.0	100.0
		Mn	sq.km	4779	0	4779	0	4779	0	4779	0
			%		0.0	100.0	0.0		0.0	100.0	0.0
		S	sq.km	4779	0	4779	0	4779	0	4779	0
			%		0.0	100.0	0.0		0.0	100.0	0.0
	Soil physical	EC	sq.km	4779	4779	0	0	4779	4779	0	0
			%		100.0	0.0	0.0		100.0	0.0	0.0
		pH	sq.km	4779	0	4779	0	4779	4455	324	0
			%		0.0	100.0	0.0		93.2	6.8	0.0

(continued)

Table 35 (continued)

Soil condition based on physical and chemical parameters					Ecosystem: agricultural lands						
					Opening 2005				Closing 2020		
District	ECT	Parameter	Units	Total	High	Medium	Low	Total	High	Medium	Low
Koppal	Soil–chemical (macro nutrients)	K	sq.km	5427	5427	0	0	5427	1944	3483	0
			%		100.0	0.0	0.0		35.8	64.2	0.0
		N	sq.km	5427	162	4293	972	5427	0	0	5427
			%		3.0	79.1	17.9		0.0	0.0	100.0
		P	sq.km	5427	0	0	5427	5427	2916	2511	0
			%		0.0	0.0	100.0		53.7	46.3	0.0
		Zn	sq.km	5427	0	0	5427	5427	0	5427	0
			%		0.0	0.0	100.0		0.0	100.0	0.0
		Fe	sq.km	5427	0	3888	1539	5427	0	5427	0
			%		0.0	71.6	28.4		0.0	100.0	0.0
		OC	sq.km	5427	0	1458	3969	5427	0	3078	2349
			%		0.0	26.9	73.1		0.0	56.7	43.3
	Soil–chemical (micro nutrients)	B	sq.km	5427	0	5427	0	5427	0	5427	0
			%		0.0	100.0	0.0		0.0	100.0	0.0
		Cu	sq.km	5427	0	1782	3645	5427	0	1782	3645
			%		0.0	32.8	67.2		0.0	32.8	67.2
		Mn	sq.km	5427	0	5427	0	5427	0	5427	0
			%		0.0	100.0	0.0		0.0	100.0	0.0
		S	sq.km	5427	4131	1296	0	5427	4131	1296	0
			%		76.1	23.9	0.0		76.1	23.9	0.0

(continued)

Table 35 (continued)

Soil condition based on physical and chemical parameters					Ecosystem: agricultural lands						
					Opening 2005				Closing 2020		
District	ECT	Parameter	Units	Total	High	Medium	Low	Total	High	Medium	Low
	Soil physical	EC	sq.km	5427	5427	0	0	5427	5427	0	0
			%		100.0	0.0	0.0		100.0	0.0	0.0
		pH	sq.km	5427	0	5265	162	5427	0	5427	0
			%		0.0	97.0	3.0		0.0	100.0	0.0
Mandya	Soil–chemical (macro nutrients)	K	sq.km	5103	972	4131	0	5103	0	5103	0
			%		19.0	81.0	0.0		0.0	100.0	0.0
		N	sq.km	5103	567	3078	1458	5103	0	0	5103
			%		11.1	60.3	28.6		0.0	0.0	100.0
		P	sq.km	5103	0	1701	3402	5103	648	4455	0
			%		0.0	33.3	66.7		12.7	87.3	0.0
		Zn	sq.km	5103	0	1053	4050	5103	0	5103	0
			%		0.0	20.6	79.4		0.0	100.0	0.0
		Fe	sq.km	5103	0	5022	81	5103	0	5103	0
			%		0.0	98.4	1.6		0.0	100.0	0.0
		OC	sq.km	5103	0	4374	729	5103	0	5103	0
			%		0.0	85.7	14.3		0.0	100.0	0.0
	Soil–chemical (micro nutrients)	B	sq.km	5103	0	4941	162	5103	0	4941	162
			%		0.0	96.8	3.2		0.0	96.8	3.2
		Cu	sq.km	5103	0	0	5103	5103	0	0	5103
			%		0.0	0.0	100.0		0.0	0.0	100.0

(continued)

Table 35 (continued)

Soil condition based on physical and chemical parameters					Ecosystem: agricultural lands						
					Opening 2005				Closing 2020		
District	ECT	Parameter	Units	Total	High	Medium	Low	Total	High	Medium	Low
		Mn	sq.km	5103	0	5103	0	5103	0	5103	0
			%		0.0	100.0	0.0		0.0	100.0	0.0
		S	sq.km	5103	3807	1296	0	5103	3807	1296	0
			%		74.6	25.4	0.0		74.6	25.4	0.0
	Soil physical	EC	sq.km	5103	5103	0	0	5103	5103	0	0
			%		100.0	0.0	0.0		100.0	0.0	0.0
		pH	sq.km	5103	0	5103	0	5103	0	5103	0
			%		0.0	100.0	0.0		0.0	100.0	0.0
Mysore	Soil–chemical (macro nutrients)	K	sq.km	6723	1458	5265	0	6642	0	6642	0
			%		21.7	78.3	0.0		0.0	100.0	0.0
		N	sq.km	6723	1620	3159	1944	6642	0	0	6642
			%		24.1	47.0	28.9		0.0	0.0	100.0
		P	sq.km	6723	0	810	5913	6642	0	6642	0
			%		0.0	12.0	88.0		0.0	100.0	0.0
		Zn	sq.km	6642	0	6642	0	6642	0	6642	0
			%		0.0	100.0	0.0		0.0	100.0	0.0
		Fe	sq.km	6642	0	6642	0	6642	0	6642	0
			%		0.0	100.0	0.0		0.0	100.0	0.0
		OC	sq.km	6642	0	6399	243	6642	0	6399	243
			%		0.0	96.3	3.7		0.0	96.3	3.7

(continued)

Table 35 (continued)

Soil condition based on physical and chemical parameters					Ecosystem: agricultural lands						
					Opening 2005				Closing 2020		
District	ECT	Parameter	Units	Total	High	Medium	Low	Total	High	Medium	Low
	Soil–chemical (micro nutrients)	B	sq.km	6642	0	5751	891	6642	0	5751	891
			%		0.0	86.6	13.4		0.0	86.6	13.4
		Cu	sq.km	6642	0	0	6642	6642	0	0	6642
			%		0.0	0.0	100.0		0.0	0.0	100.0
		Mn	sq.km	6642	0	6642	0	6642	0	6642	0
			%		0.0	100.0	0.0		0.0	100.0	0.0
		S	sq.km	6642	3078	3564	0	6642	3078	3564	0
			%		46.3	53.7	0.0		46.3	53.7	0.0
	Soil physical	EC	sq.km	6642	6642	0	0	6642	6642	0	0
			%		100.0	0.0	0.0		100.0	0.0	0.0
		pH	sq.km	6642	486	6156	0	6642	1215	5427	0
			%		7.3	92.7	0.0		18.3	81.7	0.0
Raichur	Soil–chemical (macro nutrients)	K	sq.km	9153	7371	1782	0	9153	81	9072	0
			%		80.5	19.5	0.0		0.9	99.1	0.0
		N	sq.km	9153	729	4050	4374	9153	0	0	9153
			%		8.0	44.2	47.8		0.0	0.0	100.0
		P	sq.km	9153	0	0	9153	9153	405	8748	0
			%		0.0	0.0	100.0		4.4	95.6	0.0
		Zn	sq.km	9153	0	0	9153	9153	0	9153	0
			%		0.0	0.0	100.0		0.0	100.0	0.0

(continued)

Table 35 (continued)

Soil condition based on physical and chemical parameters					Ecosystem: agricultural lands						
					Opening 2005				Closing 2020		
District	ECT	Parameter	Units	Total	High	Medium	Low	Total	High	Medium	Low
		Fe	sq.km	9153	0	729	8424	9153	0	9153	0
			%		0.0	8.0	92.0		0.0	100.0	0.0
		OC	sq.km	9153	0	2349	6804	9153	0	8910	243
			%		0.0	25.7	74.3		0.0	97.3	2.7
	Soil–chemical (micro nutrients)	B	sq.km	9153	0	8181	972	9153	0	8181	972
			%		0.0	89.4	10.6		0.0	89.4	10.6
		Cu	sq.km	9153	0	81	9072	9153	0	81	9072
			%		0.0	0.9	99.1		0.0	0.9	99.1
		Mn	sq.km	9153	0	9153	0	9153	0	9153	0
			%		0.0	100.0	0.0		0.0	100.0	0.0
		S	sq.km	9153	9153	0	0	9153	9153	0	0
			%		100.0	0.0	0.0		100.0	0.0	0.0
	Soil physical	EC	sq.km	9153	9153	0	0	9153	9153	0	0
			%		100.0	0.0	0.0		100.0	0.0	0.0
		pH	sq.km	9153	0	5103	4050	9153	0	9153	0
			%		0.0	55.8	44.2		0.0	100.0	0.0
Ramanagara	Soil–chemical (macro nutrients)	K	sq.km	3402	1215	2187	0	3402	0	3402	0
			%		35.7	64.3	0.0		0.0	100.0	0.0
		N	sq.km	3402	324	2430	648	3402	0	0	3402
			%		9.5	71.4	19.0		0.0	0.0	100.0

(continued)

Table 35 (continued)

Soil condition based on physical and chemical parameters					Ecosystem: agricultural lands						
					Opening 2005				Closing 2020		
District	ECT	Parameter	Units	Total	High	Medium	Low	Total	High	Medium	Low
		P	sq.km	3402	0	162	3240	3402	0	3402	0
			%		0.0	4.8	95.2		0.0	100.0	0.0
		Zn	sq.km	3402	0	2673	729	3402	0	3321	81
			%		0.0	78.6	21.4		0.0	97.6	2.4
		Fe	sq.km	3402	0	3402	0	3402	0	3402	0
			%		0.0	100.0	0.0		0.0	100.0	0.0
		OC	sq.km	3402	0	567	2835	3402	0	3078	324
			%		0.0	16.7	83.3		0.0	90.5	9.5
	Soil–chemical (micro nutrients)	B	sq.km	3402	0	1944	1458	3402	0	1944	1458
			%		0.0	57.1	42.9		0.0	57.1	42.9
		Cu	sq.km	3402	0	0	3402	3402	0	0	3402
			%		0.0	0.0	100.0		0.0	0.0	100.0
		Mn	sq.km	3402	0	3402	0	3402	0	3402	0
			%		0.0	100.0	0.0		0.0	100.0	0.0
		S	sq.km	3402	243	3159	0	3402	243	3159	0
			%		7.1	92.9	0.0		7.1	92.9	0.0
	Soil physical	EC	sq.km	3402	3402	0	0	3402	3402	0	0
			%		100.0	0.0	0.0		100.0	0.0	0.0
		pH	sq.km	3402	1134	2268	0	3402	81	3321	0
			%		33.3	66.7	0.0		2.4	97.6	0.0

(continued)

Table 35 (continued)

Soil condition based on physical and chemical parameters					Ecosystem: agricultural lands						
					Opening 2005				Closing 2020		
District	ECT	Parameter	Units	Total	High	Medium	Low	Total	High	Medium	Low
Shimoga	Soil–chemical (macro nutrients)	K	sq.km	8586	0	8586	0	8586	0	8586	0
			%		0.0	100.0	0.0		0.0	100.0	0.0
		N	sq.km	8586	8343	243	0	8586	0	324	8262
			%		97.2	2.8	0.0		0.0	3.8	96.2
		P	sq.km	8586	0	1539	7047	8586	0	8505	81
			%		0.0	17.9	82.1		0.0	99.1	0.9
		Zn	sq.km	8586	0	3402	5184	8586	0	8343	243
			%		0.0	39.6	60.4		0.0	97.2	2.8
		Fe	sq.km	8586	0	8586	0	8586	0	8586	0
			%		0.0	100.0	0.0		0.0	100.0	0.0
		OC	sq.km	8586	0	8586	0	8586	0	8343	243
			%		0.0	100.0	0.0		0.0	97.2	2.8
	Soil–chemical (micro nutrients)	B	sq.km	8586	0	1215	7371	8586	0	1215	7371
			%		0.0	14.2	85.8		0.0	14.2	85.8
		Cu	sq.km	8586	0	2835	5751	8586	0	2835	5751
			%		0.0	33.0	67.0		0.0	33.0	67.0
		Mn	sq.km	8586	0	8586	0	8586	0	8586	0
			%		0.0	100.0	0.0		0.0	100.0	0.0
		S	sq.km	8586	1053	7533	0	8586	1053	7533	0
			%		12.3	87.7	0.0		12.3	87.7	0.0

(continued)

Table 35 (continued)

Soil condition based on physical and chemical parameters				Ecosystem: agricultural lands							
					Opening 2005				Closing 2020		
District	ECT	Parameter	Units	Total	High	Medium	Low	Total	High	Medium	Low
	Soil physical	EC	sq.km	8586	8586	0	0	8586	8586	0	0
			%		100.0	0.0	0.0		100.0	0.0	0.0
		pH	sq.km	8586	7857	729	0	8586	2349	6237	0
			%		91.5	8.5	0.0		27.4	72.6	0.0
Tumkur	Soil–chemical (macro nutrients)	K	sq.km	12,474	810	11,664	0	12,474	0	12,474	0
			%		6.5	93.5	0.0		0.0	100.0	0.0
		N	sq.km	12,474	2916	4536	5022	12,474	0	1134	11,340
			%		23.4	36.4	40.3		0.0	9.1	90.9
		P	sq.km	12,474	0	1458	11,016	12,474	81	12,393	0
			%		0.0	11.7	88.3		0.6	99.4	0.0
		Zn	sq.km	12,474	0	486	11,988	12,474	0	7533	4941
			%		0.0	3.9	96.1		0.0	60.4	39.6
		Fe	sq.km	12,474	0	12,231	243	12,474	0	12,474	0
			%		0.0	98.1	1.9		0.0	100.0	0.0
		OC	sq.km	12,474	0	1296	11,178	12,474	0	12,231	243
			%		0.0	10.4	89.6		0.0	98.1	1.9
	Soil–chemical (micro nutrients)	B	sq.km	12,474	0	12,069	405	12,474	0	12,069	405
			%		0.0	96.8	3.2		0.0	96.8	3.2
		Cu	sq.km	12,474	0	0	12,474	12,474	0	0	12,474
			%		0.0	0.0	100.0		0.0	0.0	100.0

(continued)

Table 35 (continued)

Soil condition based on physical and chemical parameters				Ecosystem: agricultural lands							
					Opening 2005				Closing 2020		
District	ECT	Parameter	Units	Total	High	Medium	Low	Total	High	Medium	Low
		Mn	sq.km	12,474	0	12,474	0	12,474	0	12,474	0
			%		0.0	100.0	0.0		0.0	100.0	0.0
		S	sq.km	12,474	1134	11,340	0	12,474	1134	11,340	0
			%		9.1	90.9	0.0		9.1	90.9	0.0
	Soil physical	EC	sq.km	12,474	12,474	0	0	12,474	12,474	0	0
			%		100.0	0.0	0.0		100.0	0.0	0.0
		pH	sq.km	12,474	1620	10,854	0	12,474	10,854	1620	0
			%		13.0	87.0	0.0		87.0	13.0	0.0
Udupi	Soil–chemical (macro nutrients)	K	sq.km	4212	0	3564	648	4212	0	1539	2673
			%		0.0	84.6	15.4		0.0	36.5	63.5
		N	sq.km	4212	4212	0	0	4212	0	0	4212
			%		100.0	0.0	0.0		0.0	0.0	100.0
		P	sq.km	4212	0	162	4050	4212	0	3726	486
			%		0.0	3.8	96.2		0.0	88.5	11.5
		Zn	sq.km	4212	0	81	4131	4212	0	4212	0
			%		0.0	1.9	98.1		0.0	100.0	0.0
		Fe	sq.km	4212	0	4212	0	4212	0	4212	0
			%		0.0	100.0	0.0		0.0	100.0	0.0
		OC	sq.km	4212	0	4212	0	4212	0	4131	81
			%		0.0	100.0	0.0		0.0	98.1	1.9

(continued)

Table 35 (continued)

Soil condition based on physical and chemical parameters					Ecosystem: agricultural lands						
					Opening 2005				Closing 2020		
District	ECT	Parameter	Units	Total	High	Medium	Low	Total	High	Medium	Low
	Soil–chemical (micro nutrients)	B	sq.km	4212	0	0	4212	4212	0	0	4212
			%		0.0	0.0	100.0		0.0	0.0	100.0
		Cu	sq.km	4212	0	81	4131	4212	0	81	4131
			%		0.0	1.9	98.1		0.0	1.9	98.1
		Mn	sq.km	4212	0	4212	0	4212	0	4212	0
			%		0.0	100.0	0.0		0.0	100.0	0.0
		S	sq.km	4212	81	3645	486	4212	81	3645	486
			%		1.9	86.5	11.5		1.9	86.5	11.5
	Soil physical	EC	sq.km	4212	4212	0	0	4212	4212	0	0
			%		100.0	0.0	0.0		100.0	0.0	0.0
		pH	sq.km	4212	4212	0	0	4212	2673	1539	0
			%		100.0	0.0	0.0		63.5	36.5	0.0
Uttara Kannada	Soil–chemical (macro nutrients)	K	sq.km	11,421	0	9639	1782	11,421	0	11,340	81
			%		0.0	84.4	15.6		0.0	99.3	0.7
		N	sq.km	11,421	11,421	0	0	11,421	0	0	11,421
			%		100.0	0.0	0.0		0.0	0.0	100.0
		P	sq.km	11,421	0	0	11,421	11,421	4293	7128	0
			%		0.0	0.0	100.0		37.6	62.4	0.0
		Zn	sq.km	11,421	0	9801	1620	11,421	0	11,421	0
			%		0.0	85.8	14.2		0.0	100.0	0.0

(continued)

Table 35 (continued)

Soil condition based on physical and chemical parameters				Ecosystem: agricultural lands							
					Opening 2005				Closing 2020		
District	ECT	Parameter	Units	Total	High	Medium	Low	Total	High	Medium	Low
		Fe	sq.km	11,421	0	11,421	0	11,421	0	11,421	0
			%		0.0	100.0	0.0		0.0	100.0	0.0
		OC	sq.km	11,745	0	8505	3240	11,421	0	11,421	0
			%		0.0	72.4	27.6		0.0	100.0	0.0
	Soil–chemical (micro nutrients)	B	sq.km	11,421	0	0	11,421	11,421	0	0	11,421
			%		0.0	0.0	100.0		0.0	0.0	100.0
		Cu	sq.km	11,421	0	11,340	81	11,421	0	11,340	81
			%		0.0	99.3	0.7		0.0	99.3	0.7
		Mn	sq.km	11,421	0	11,421	0	11,421	0	11,421	0
			%		0.0	100.0	0.0		0.0	100.0	0.0
		S	sq.km	11,421	5184	6237	0	11,421	5184	6237	0
			%		45.4	54.6	0.0		45.4	54.6	0.0
	Soil physical	EC	sq.km	11,421	11,421	0	0	11,421	11,421	0	0
			%		100.0	0.0	0.0		100.0	0.0	0.0
		pH	sq.km	11,502	10,854	648	0	11,421	11,421	0	0
			%		94.4	5.6	0.0		100.0	0.0	0.0
Yadgir	Soil–chemical (macro nutrients)	K	sq.km	5913	2592	3321	0	5913	0	5913	0
			%		43.8	56.2	0.0		0.0	100.0	0.0
		N	sq.km	5913	0	972	4941	5913	0	0	5913
			%		0.0	16.4	83.6		0.0	0.0	100.0

(continued)

Table 35 (continued)

Soil condition based on physical and chemical parameters					Ecosystem: agricultural lands						
					Opening 2005				Closing 2020		
District	ECT	Parameter	Units	Total	High	Medium	Low	Total	High	Medium	Low
		P	sq.km	5913	0	0	5913	5913	0	5913	0
			%		0.0	0.0	100.0		0.0	100.0	0.0
		Zn	sq.km	5913	0	0	5913	5913	0	5913	0
			%		0.0	0.0	100.0		0.0	100.0	0.0
		Fe	sq.km	5913	0	2916	2997	5913	0	5913	0
			%		0.0	49.3	50.7		0.0	100.0	0.0
		OC	sq.km	5913	0	1215	4698	5913	0	5913	0
			%		0.0	20.5	79.5		0.0	100.0	0.0
	Soil–chemical (micro nutrients)	B	sq.km	5913	0	3645	2268	5913	0	3645	2268
			%		0.0	61.6	38.4		0.0	61.6	38.4
		Cu	sq.km	5913	0	0	5913	5913	0	0	5913
			%		0.0	0.0	100.0		0.0	0.0	100.0
		Mn	sq.km	5913	0	5913	0	5913	0	5913	0
			%		0.0	100.0	0.0		0.0	100.0	0.0
		S	sq.km	5913	5913	0	0	5913	5913	0	0
			%		100.0	0.0	0.0		100.0	0.0	0.0
	Soil physical	EC	sq.km	5913	5913	0	0	5913	5913	0	0
			%		100.0	0.0	0.0		100.0	0.0	0.0
		pH	sq.km	5913	0	5751	162	5913	0	5913	0
			%		0.0	97.3	2.7		0.0	100.0	0.0

Note N: Nitrogen, P: Phosphorous, K: Potash, OC: Organic carbon, Zn: Zinc, Fe: Iron, B: Boron, Cu: Copper, Mn: Manganese, S: Sulphur, EC: Electrical conductivity

Table 36 Ecosystem condition index account–Karnataka state (district wise)

Districts	SEEA-EA conditions	Indicator	Parameter		Opening–2005 (%)			Closing–2020 (%)		
				Weight	High	Medium	Low	High	Medium	Low
Bagalkot	Abiotic ecosystem	Soil	K	0.017	68.2	31.8	0.0	56.5	43.5	0.0
		Soil	N	0.017	5.9	57.6	36.5	0.0	38.8	61.2
			P	0.017	0.0	4.7	95.3	8.2	90.6	1.2
			OC	0.017	0.0	89.4	10.6	0.0	85.9	14.1
			S	0.017	100.0	0.0	0.0	100.0	0.0	0.0
			Zn	0.017	0.0	0.0	100.0	0.0	65.9	34.1
			Fe	0.017	0.0	44.7	55.3	0.0	77.6	22.4
			B	0.017	0.0	88.2	11.8	0.0	88.2	11.8
			Cu	0.017	0.0	2.4	97.6	0.0	2.4	97.6
			Mn	0.017	0.0	100.0	0.0	0.0	100.0	0.0
			EC	0.017	87.1	11.8	1.2	87.1	11.8	1.2
			pH	0.017	0.0	97.6	2.4	0.0	100.0	0.0
	Biotic–compositional state	Flora		0.05	0.0	15.3	84.7	0.0	15.3	84.7
		Fauna		0.05	0.0	8.2	91.8	0.0	8.2	91.8
	Biotic–structural state	AGB		0.05	0.0	0.0	100.0	0.0	0.0	100.0
		BGB		0.05	0.0	0.0	100.0	0.0	0.0	100.0
	Biotic–functional state	NPP		0.10	0.0	4.7	95.3	0.0	0.0	100.0
	Landscape level	Fragmentation		0.25	65.0	9.4	25.7	7.4	62.2	30.4
		LST		0.25	0.1	1.3	98.6	0.0	1.7	98.3

(continued)

Table 36 (continued)

Districts	SEEA-EA conditions	Indicator	Parameter		Opening–2005 (%)			Closing–2020 (%)		
				Weight	High	Medium	Low	High	Medium	Low
Ecosystem condition account			Index	1.00	20.6	13.1	66.3	6.0	28.9	65.1
Bangalore (Rural)	Abiotic ecosystem	Soil	K	0.017	10.0	90.0	0.0	13.3	86.7	0.0
		Soil	N	0.017	0.0	26.7	73.3	0.0	0.0	100.0
			P	0.017	0.0	23.3	76.7	0.0	100.0	0.0
			OC	0.017	0.0	6.7	93.3	0.0	56.7	43.3
			S	0.017	0.0	100.0	0.0	0.0	100.0	0.0
			Zn	0.017	0.0	30.0	70.0	0.0	73.3	26.7
			Fe	0.017	0.0	100.0	0.0	0.0	100.0	0.0
			B	0.017	0.0	70.0	30.0	0.0	70.0	30.0
			Cu	0.017	0.0	0.0	100.0	0.0	0.0	100.0
			Mn	0.017	0.0	100.0	0.0	0.0	100.0	0.0
			EC	0.017	100.0	0.0	0.0	100.0	0.0	0.0
			pH	0.017	33.3	66.7	0.0	13.3	86.7	0.0
	Biotic–compositional state	Flora		0.05	0.0	19.4	80.6	0.0	19.4	80.6
		Fauna		0.05	0.0	6.5	93.5	0.0	6.5	93.5
	Biotic–structural state	AGB		0.05	0.0	3.2	96.8	0.0	0.0	100.0
		BGB		0.05	0.0	3.2	96.8	0.0	0.0	100.0
	Biotic–functional state	NPP		0.10	0.0	35.5	64.5	0.0	29.0	71.0
	Landscape level	Fragmentation		0.25	68.6	8.4	23.0	15.4	41.1	43.5
		LST		0.25	0.0	1.5	98.5	0.0	2.6	97.4
Ecosystem condition account			Index	1.00	19.5	17.9	62.6	6.0	28.0	66.0

(continued)

Table 36 (continued)

Districts	SEEA-EA conditions	Indicator	Parameter		Opening–2005 (%)			Closing–2020 (%)		
				Weight	High	Medium	Low	High	Medium	Low
Bangalore (Urban)	Abiotic ecosystem	Soil	K	0.017	12.5	87.5	0.0	3.1	96.9	0.0
		Soil	N	0.017	28.1	53.1	18.8	0.0	0.0	100.0
			P	0.017	0.0	28.1	71.9	0.0	100.0	0.0
			OC	0.017	0.0	43.8	56.3	0.0	12.5	87.5
			S	0.017	0.0	100.0	0.0	0.0	100.0	0.0
			Zn	0.017	0.0	84.4	15.6	0.0	90.6	9.4
			Fe	0.017	0.0	96.9	3.1	0.0	100.0	0.0
			B	0.017	0.0	18.8	81.3	0.0	18.8	81.3
			Cu	0.017	0.0	0.0	100.0	0.0	0.0	100.0
			Mn	0.017	0.0	100.0	0.0	0.0	100.0	0.0
			EC	0.017	100.0	0.0	0.0	100.0	0.0	0.0
			pH	0.017	0.0	100.0	0.0	28.1	71.9	0.0
	Biotic–compositional state	Flora		0.05	0.0	32.1	67.9	0.0	32.1	67.9
		Fauna		0.05	0.0	21.4	78.6	0.0	21.4	78.6
	Biotic–structural state	AGB		0.05	0.0	3.6	96.4	0.0	0.0	100.0
		BGB		0.05	0.0	3.6	96.4	0.0	0.0	100.0
	Biotic–functional state	NPP		0.10	0.0	42.9	57.1	0.0	35.7	64.3
	Landscape level	Fragmentation		0.25	89.9	2.4	7.7	12.2	56.3	31.5
		LST		0.25	0.0	3.9	96.1	0.0	11.4	88.6
Ecosystem condition account			Index	1.00	24.8	20.8	54.4	5.2	34.7	60.1

(continued)

Table 36 (continued)

Districts	SEEA-EA conditions	Indicator	Parameter		Opening–2005 (%)			Closing–2020 (%)		
				Weight	High	Medium	Low	High	Medium	Low
Belgaum	Abiotic ecosystem	Soil	K	0.017	14.3	81.0	4.8	8.5	91.5	0.0
		Soil	N	0.017	50.8	33.3	15.9	0.0	6.3	93.7
			P	0.017	0.0	14.8	85.2	0.0	96.8	3.2
			OC	0.017	0.0	85.2	14.8	0.0	85.2	14.8
			S	0.017	98.9	1.1	0.0	98.9	1.1	0.0
			Zn	0.017	0.0	79.4	20.6	0.0	79.4	20.6
			Fe	0.017	0.0	95.2	4.8	0.0	95.2	4.8
			B	0.017	0.0	35.4	64.6	0.0	35.4	64.6
			Cu	0.017	0.0	4.8	95.2	0.0	4.8	95.2
			Mn	0.017	0.0	100.0	0.0	0.0	100.0	0.0
			EC	0.017	98.4	1.6	0.0	98.4	1.6	0.0
			pH	0.017	16.9	82.5	0.5	5.3	94.7	0.0
	Biotic–compositional state	Flora		0.05	48.8	0.5	50.6	11.0	0.5	88.5
		Fauna		0.05	11.6	1.6	86.8	2.6	1.6	95.8
	Biotic–structural state	AGB		0.05	3.1	10.5	86.4	2.6	7.9	89.5
		BGB		0.05	3.1	10.5	86.4	2.6	7.9	89.5
	Biotic–functional state	NPP		0.10	0.0	15.2	84.8	0.0	8.9	91.1
	Landscape level	Fragmentation		0.25	78.9	5.2	15.8	33.8	27.7	38.6
		LST		0.25	0.8	4.4	94.8	0.0	7.6	92.4

(continued)

Table 36 (continued)

Districts	SEEA-EA conditions	Indicator	Parameter		Opening–2005 (%)			Closing–2020 (%)		
				Weight	High	Medium	Low	High	Medium	Low
Ecosystem condition account			Index	1.00	27.9	15.3	56.8	12.9	22.1	65.0
Bellary	Abiotic ecosystem	Soil	K	0.017	85.3	6.0	8.6	0.0	100.0	0.0
		Soil	N	0.017	37.1	39.7	23.3	0.0	0.0	100.0
			P	0.017	0.0	10.3	89.7	3.4	95.7	0.9
			OC	0.017	0.0	97.4	2.6	0.0	92.2	7.8
			S	0.017	77.6	22.4	0.0	77.6	22.4	0.0
			Zn	0.017	0.0	2.6	97.4	0.0	100.0	0.0
			Fe	0.017	0.0	60.3	39.7	0.0	100.0	0.0
			B	0.017	0.0	91.4	8.6	0.0	91.4	8.6
			Cu	0.017	0.0	0.0	100.0	0.0	0.0	100.0
			Mn	0.017	0.0	100.0	0.0	0.0	100.0	0.0
			EC	0.017	100.0	0.0	0.0	100.0	0.0	0.0
			pH	0.017	0.9	99.1	0.0	1.7	98.3	0.0
	Biotic–compositional state	Flora		0.05	0.0	7.8	92.2	0.0	7.8	92.2
		Fauna		0.05	0.0	8.6	91.4	0.0	8.6	90.5
	Biotic–structural state	AGB		0.05	0.0	7.8	92.2	0.0	0.0	100.0
		BGB		0.05	0.0	7.8	92.2	0.0	0.0	100.0
	Biotic–functional state	NPP		0.10	0.0	17.2	82.8	0.0	4.3	95.7
	Landscape level	Fragmentation		0.25	74.5	8.0	17.5	13.6	40.6	45.8
		LST		0.25	0.2	5.1	94.7	0.0	2.6	97.4

(continued)

Table 36 (continued)

Districts	SEEA-EA conditions	Indicator	Parameter		Opening–2005 (%)			Closing–2020 (%)		
				Weight	High	Medium	Low	High	Medium	Low
Ecosystem condition account			Index	1.00	23.7	15.4	60.9	6.4	25.4	68.1
Bidar	Abiotic ecosystem	Soil	K	0.017	100.0	0.0	0.0	11.8	88.2	0.0
		Soil	N	0.017	94.1	5.9	0.0	0.0	0.0	100.0
			P	0.017	0.0	38.8	61.2	0.0	35.3	64.7
			OC	0.017	0.0	87.1	12.9	0.0	87.1	12.9
			S	0.017	7.1	65.9	27.1	7.1	65.9	27.1
			Zn	0.017	0.0	0.0	100.0	0.0	98.8	1.2
			Fe	0.017	0.0	78.8	21.2	0.0	98.8	1.2
			B	0.017	0.0	34.1	65.9	0.0	34.1	65.9
			Cu	0.017	0.0	0.0	100.0	0.0	0.0	100.0
			Mn	0.017	0.0	100.0	0.0	0.0	100.0	0.0
			EC	0.017	100.0	0.0	0.0	100.0	0.0	0.0
			pH	0.017	0.0	100.0	0.0	1.2	98.8	0.0
	Biotic–compositional state	Flora		0.05	0.0	20.0	80.0	0.0	20.0	80.0
		Fauna		0.05	0.0	5.9	94.1	0.0	5.9	94.1
	Biotic–structural state	AGB		0.05	0.0	0.0	100.0	0.0	0.0	100.0
		BGB		0.05	0.0	0.0	100.0	0.0	0.0	100.0
	Biotic–functional state	NPP		0.10	0.0	1.2	98.8	0.0	0.0	100.0
	Landscape level	Fragmentation		0.25	44.4	21.1	34.5	12.5	52.2	35.3
		LST		0.25	0.0	0.1	99.9	0.0	0.7	99.3

(continued)

Table 36 (continued)

Districts	SEEA-EA conditions	Indicator	Parameter		Opening–2005 (%)			Closing–2020 (%)		
				Weight	High	Medium	Low	High	Medium	Low
Ecosystem condition account			Index	1.00	16.1	15.2	68.6	5.1	26.3	68.6
Bijapur	Abiotic ecosystem	Soil	K	0.017	55.1	40.6	4.3	70.3	29.7	0.0
		Soil	N	0.017	6.5	93.5	0.0	0.0	0.7	99.3
			P	0.017	0.0	4.3	95.7	0.7	99.3	0.0
			OC	0.017	0.0	9.4	90.6	0.0	92.8	7.2
			S	0.017	100.0	0.0	0.0	100.0	0.0	0.0
			Zn	0.017	0.0	0.0	100.0	0.0	92.0	8.0
			Fe	0.017	0.0	52.9	47.1	0.0	100.0	0.0
			B	0.017	0.0	79.7	20.3	0.0	79.7	20.3
			Cu	0.017	0.0	0.0	100.0	0.0	0.0	100.0
			Mn	0.017	0.0	100.0	0.0	0.0	100.0	0.0
			EC	0.017	100.0	0.0	0.0	100.0	0.0	0.0
			pH	0.017	1.4	81.9	16.7	0.0	100.0	0.0
	Biotic–compositional state	Flora		0.05	0.0	5.0	95.0	0.0	5.0	95.0
		Fauna		0.05	0.0	1.4	98.6	0.0	1.4	98.6
	Biotic–structural state	AGB		0.05	0.0	0.0	100.0	0.0	0.0	100.0
		BGB		0.05	0.0	0.0	100.0	0.0	0.0	100.0
	Biotic–functional state	NPP		0.10	0.0	0.0	100.0	0.0	0.0	100.0
	Landscape level	Fragmentation		0.25	42.0	18.4	39.6	16.3	39.4	44.3
		LST		0.25	0.0	0.4	99.5	0.0	0.5	99.5

(continued)

Table 36 (continued)

Districts	SEEA-EA conditions	Indicator	Parameter		Opening–2005 (%)			Closing–2020 (%)		
				Weight	High	Medium	Low	High	Medium	Low
Ecosystem condition account			Index	1.00	14.9	12.7	72.4	8.6	21.9	69.5
Chamarajanagar	Abiotic ecosystem	Soil	K	0.017	9.4	90.6	0.0	1.2	98.8	0.0
		Soil	N	0.017	75.3	24.7	0.0	0.0	1.2	98.8
			P	0.017	0.0	0.0	100.0	1.2	98.8	0.0
			OC	0.017	0.0	3.5	96.5	0.0	100.0	0.0
			S	0.017	22.6	77.4	0.0	22.6	77.4	0.0
			Zn	0.017	0.0	30.6	69.4	0.0	100.0	0.0
			Fe	0.017	0.0	88.2	11.8	0.0	100.0	0.0
			B	0.017	0.0	61.9	38.1	0.0	61.9	38.1
			Cu	0.017	0.0	0.0	100.0	0.0	0.0	100.0
			Mn	0.017	0.0	100.0	0.0	0.0	100.0	0.0
			EC	0.017	100.0	0.0	0.0	100.0	0.0	0.0
			pH	0.017	13.1	86.9	0.0	13.1	86.9	0.0
	Biotic–compositional state	Flora		0.05	16.7	17.6	65.6	14.1	17.6	68.2
		Fauna		0.05	37.6	7.1	55.4	31.8	7.1	61.2
	Biotic–structural state	AGB		0.05	8.2	56.5	35.3	4.7	45.9	49.4
		BGB		0.05	8.2	56.5	35.3	4.7	45.9	49.4
	Biotic–functional state	NPP		0.10	4.7	56.5	38.8	0.0	47.1	52.9
	Landscape level	Fragmentation		0.25	94.7	1.0	4.3	80.0	5.2	14.8
		LST		0.25	5.6	16.7	77.7	5.6	27.8	66.6

(continued)

Table 36 (continued)

Districts	SEEA-EA conditions	Indicator	Parameter		Opening–2005 (%)			Closing–2020 (%)		
				Weight	High	Medium	Low	High	Medium	Low
Ecosystem condition account			Index	1.00	32.8	26.4	40.9	26.5	32.5	41.0
Chikballapur	Abiotic ecosystem	Soil	K	0.017	6.3	93.8	0.0	4.8	95.2	0.0
		Soil	N	0.017	0.0	0.0	100.0	0.0	0.0	100.0
			P	0.017	0.0	9.4	90.6	0.0	100.0	0.0
			OC	0.017	0.0	0.0	100.0	0.0	28.6	71.4
			S	0.017	0.0	100.0	0.0	0.0	100.0	0.0
			Zn	0.017	0.0	0.0	100.0	0.0	66.7	33.3
			Fe	0.017	0.0	100.0	0.0	0.0	100.0	0.0
			B	0.017	0.0	100.0	0.0	0.0	100.0	0.0
			Cu	0.017	0.0	0.0	100.0	0.0	0.0	100.0
			Mn	0.017	0.0	100.0	0.0	0.0	100.0	0.0
			EC	0.017	100.0	0.0	0.0	100.0	0.0	0.0
			pH	0.017	3.1	96.9	0.0	1.6	98.4	0.0
	Biotic–compositional state	Flora		0.05	0.0	23.4	76.6	0.0	23.4	76.6
		Fauna		0.05	0.0	7.8	92.2	0.0	7.8	92.2
	Biotic–structural state	AGB		0.05	0.0	1.6	98.4	0.0	0.0	100.0
		BGB		0.05	0.0	1.6	98.4	0.0	0.0	100.0
	Biotic–functional state	NPP		0.10	0.0	17.2	82.8	0.0	6.3	93.8
	Landscape level	Fragmentation		0.25	77.2	5.3	17.5	17.2	40.4	42.4
		LST		0.25	0.0	0.1	99.9	0.0	0.7	99.3

(continued)

Table 36 (continued)

Districts	SEEA-EA conditions	Indicator	Parameter		Opening–2005 (%)			Closing–2020 (%)		
				Weight	High	Medium	Low	High	Medium	Low
Ecosystem condition account			Index	1.00	21.1	14.8	64.1	6.1	25.6	68.3
Chikkamagaluru	Abiotic ecosystem	Soil	K	0.017	21.8	71.3	6.9	4.6	90.8	4.6
		Soil	N	0.017	83.9	10.3	5.7	0.0	2.3	97.7
			P	0.017	0.0	11.5	88.5	49.4	49.4	1.1
			OC	0.017	0.0	72.4	27.6	0.0	100.0	0.0
			S	0.017	42.5	57.5	0.0	42.5	57.5	0.0
			Zn	0.017	0.0	78.2	21.8	0.0	95.4	4.6
			Fe	0.017	0.0	100.0	0.0	0.0	100.0	0.0
			B	0.017	0.0	8.0	92.0	0.0	8.0	92.0
			Cu	0.017	0.0	82.8	17.2	0.0	82.8	17.2
			Mn	0.017	0.0	100.0	0.0	0.0	100.0	0.0
			EC	0.017	100.0	0.0	0.0	100.0	0.0	0.0
			pH	0.017	41.4	58.6	0.0	41.4	58.6	0.0
	Biotic–compositional state	Flora		0.05	83.4	2.3	14.3	52.9	2.3	44.8
		Fauna		0.05	50.8	9.2	40.0	32.2	9.2	58.6
	Biotic–structural state	AGB		0.05	11.5	49.4	39.1	10.3	47.1	42.5
		BGB		0.05	11.5	49.4	39.1	10.3	47.1	42.5
	Biotic–functional state	NPP		0.10	4.6	79.3	16.1	1.1	78.2	20.7
	Landscape level	Fragmentation		0.25	83.1	3.9	13.0	52.7	9.6	37.8
		LST		0.25	9.5	49.1	41.4	9.4	50.6	39.9

(continued)

Table 36 (continued)

Districts	SEEA-EA conditions	Indicator	Parameter		Opening–2005 (%)			Closing–2020 (%)		
				Weight	High	Medium	Low	High	Medium	Low
Ecosystem condition Account			Index	1.00	36.3	37.6	26.1	24.9	40.6	34.5
Chitradurga	Abiotic ecosystem	Soil	K	0.017	77.0	23.0	0.0	0.0	87.6	12.4
		Soil	N	0.017	7.1	63.7	29.2	0.0	0.0	100.0
			P	0.017	0.0	8.0	92.0	14.2	85.8	0.0
			OC	0.017	0.0	6.2	93.8	0.0	99.1	0.9
			S	0.017	34.5	65.5	0.0	34.5	65.5	0.0
			Zn	0.017	0.0	3.5	96.5	0.0	57.5	42.5
			Fe	0.017	0.0	85.0	15.0	0.0	100.0	0.0
			B	0.017	0.0	83.2	16.8	0.0	83.2	16.8
			Cu	0.017	0.0	0.0	100.0	0.0	0.0	100.0
			Mn	0.017	0.0	100.0	0.0	0.0	100.0	0.0
			EC	0.017	100.0	0.0	0.0	100.0	0.0	0.0
			pH	0.017	0.0	100.0	0.0	71.7	28.3	0.0
	Biotic–compositional state	Flora		0.05	0.0	16.8	83.2	0.0	16.8	83.2
		Fauna		0.05	0.0	2.7	97.3	0.0	2.7	97.3
	Biotic–structural state	AGB		0.05	0.0	6.2	93.8	0.0	1.8	98.2
		BGB		0.05	0.0	6.2	93.8	0.0	1.8	98.2
	Biotic–functional state	NPP		0.10	0.0	15.0	85.0	0.0	3.5	96.5
	Landscape level	Fragmentation		0.25	80.8	4.8	14.4	17.0	34.0	48.9
		LST		0.25	0.0	0.3	99.7	0.0	0.1	99.9

(continued)

Table 36 (continued)

Districts	SEEA-EA conditions	Indicator	Parameter		Opening–2005 (%)			Closing–2020 (%)		
				Weight	High	Medium	Low	High	Medium	Low
Ecosystem condition account			Index	1.00	23.8	13.3	62.8	7.9	21.8	70.2
Dakshina Kannada	Abiotic ecosystem	Soil	K	0.017	0.0	65.2	34.8	0.0	22.7	77.3
		Soil	N	0.017	100.0	0.0	0.0	0.0	6.1	93.9
			P	0.017	0.0	0.0	100.0	1.5	51.5	47.0
			OC	0.017	0.0	100.0	0.0	0.0	100.0	0.0
			S	0.017	7.6	71.2	21.2	7.6	71.2	21.2
			Zn	0.017	0.0	10.6	89.4	0.0	100.0	0.0
			Fe	0.017	0.0	100.0	0.0	0.0	100.0	0.0
			B	0.017	0.0	4.5	95.5	0.0	4.5	95.5
			Cu	0.017	0.0	9.1	90.9	0.0	9.1	90.9
			Mn	0.017	0.0	93.9	6.1	0.0	93.9	6.1
			EC	0.017	98.5	1.5	0.0	98.5	1.5	0.0
			pH	0.017	100.0	0.0	0.0	100.0	0.0	0.0
	Biotic–compositional state	Flora		0.05	94.3	2.9	2.8	68.1	2.9	29.0
		Fauna		0.05	60.2	13.0	26.8	43.5	13.0	43.5
	Biotic–structural state	AGB		0.05	21.7	42.0	36.2	21.7	39.1	39.1
		BGB		0.05	21.7	42.0	36.2	21.7	39.1	39.1
	Biotic–functional state	NPP		0.10	23.2	62.3	14.5	18.8	60.9	20.3
	Landscape level	Fragmentation		0.25	82.8	3.6	13.7	59.8	9.1	31.1
		LST		0.25	5.3	68.3	26.4	4.7	83.9	11.5

(continued)

Table 36 (continued)

Districts	SEEA-EA conditions	Indicator	Parameter		Opening–2005 (%)			Closing–2020 (%)		
				Weight	High	Medium	Low	High	Medium	Low
Ecosystem condition account			Index	1.00	39.3	36.8	23.9	29.2	43.4	27.4
Davangere	Abiotic ecosystem	Soil	K	0.017	43.7	56.3	0.0	0.0	100.0	0.0
		Soil	N	0.017	40.8	56.3	2.8	0.0	1.4	98.6
			P	0.017	0.0	14.1	85.9	53.5	46.5	0.0
			OC	0.017	0.0	57.7	42.3	0.0	76.1	23.9
			S	0.017	76.1	23.9	0.0	76.1	23.9	0.0
			Zn	0.017	0.0	29.6	70.4	0.0	97.2	2.8
			Fe	0.017	0.0	100.0	0.0	0.0	100.0	0.0
			B	0.017	0.0	33.8	66.2	0.0	33.8	66.2
			Cu	0.017	0.0	0.0	100.0	0.0	0.0	100.0
			Mn	0.017	0.0	100.0	0.0	0.0	100.0	0.0
			EC	0.017	100.0	0.0	0.0	100.0	0.0	0.0
			pH	0.017	0.0	100.0	0.0	7.0	93.0	0.0
	Biotic–compositional state	Flora		0.05	0.0	9.9	90.1	0.0	9.9	90.1
		Fauna		0.05	0.0	8.5	91.5	0.0	8.5	91.5
	Biotic–structural state	AGB		0.05	0.0	7.0	93.0	0.0	4.2	95.8
		BGB		0.05	0.0	7.0	93.0	0.0	4.2	95.8
	Biotic–functional state	NPP		0.10	1.4	8.5	90.1	0.0	5.6	94.4
	Landscape level	Fragmentation		0.25	87.3	3.0	9.6	65.3	10.9	23.8
		LST		0.25	3.5	13.4	83.1	0.1	18.5	81.3

(continued)

Table 36 (continued)

Districts	SEEA-EA conditions	Indicator	Parameter		Opening–2005 (%)			Closing–2020 (%)		
				Weight	High	Medium	Low	High	Medium	Low
Ecosystem condition account			Index	1.00	27.2	16.1	56.7	20.3	20.5	59.2
Dharwad	Abiotic ecosystem	Soil	K	0.017	56.9	43.1	0.0	19.6	80.4	0.0
		Soil	N	0.017	56.9	43.1	0.0	0.0	0.0	100.0
			P	0.017	0.0	0.0	100.0	0.0	100.0	0.0
			OC	0.017	0.0	64.7	35.3	0.0	100.0	0.0
			S	0.017	98.0	2.0	0.0	98.0	2.0	0.0
			Zn	0.017	0.0	51.0	49.0	0.0	100.0	0.0
			Fe	0.017	0.0	70.6	29.4	0.0	100.0	0.0
			B	0.017	0.0	0.0	100.0	0.0	0.0	100.0
			Cu	0.017	0.0	21.6	78.4	0.0	21.6	78.4
			Mn	0.017	0.0	100.0	0.0	0.0	100.0	0.0
			EC	0.017	100.0	0.0	0.0	100.0	0.0	0.0
			pH	0.017	2.0	68.6	29.4	11.8	88.2	0.0
	Biotic–compositional state	Flora		0.05	27.5	3.9	68.6	13.7	3.9	82.4
		Fauna		0.05	0.0	19.6	80.4	0.0	19.6	80.4
	Biotic–structural state	AGB		0.05	0.0	7.8	92.2	0.0	5.9	94.1
		BGB		0.05	0.0	7.8	92.2	0.0	5.9	94.1
	Biotic–functional state	NPP		0.10	0.0	11.8	88.2	0.0	5.9	94.1
	Landscape level	Fragmentation		0.25	78.8	5.1	16.0	39.4	16.6	44.0
		LST		0.25	0.0	0.0	100.0	0.0	0.0	100.0

(continued)

Table 36 (continued)

Districts	SEEA-EA conditions	Indicator	Parameter		Opening–2005 (%)			Closing–2020 (%)		
				Weight	High	Medium	Low	High	Medium	Low
Ecosystem condition account			Index	0.98	25.4	13.1	61.5	14.1	18.2	67.6
Gadag	Abiotic ecosystem	Soil	K	0.017	61.4	38.6	0.0	5.3	94.7	0.0
		Soil	N	0.017	17.5	59.6	22.8	0.0	0.0	100.0
			P	0.017	0.0	0.0	100.0	7.0	93.0	0.0
			OC	0.017	0.0	28.1	71.9	0.0	93.0	7.0
			S	0.017	91.2	8.8	0.0	91.2	8.8	0.0
			Zn	0.017	0.0	0.0	100.0	0.0	100.0	0.0
			Fe	0.017	0.0	14.0	86.0	0.0	100.0	0.0
			B	0.017	0.0	49.1	50.9	0.0	49.1	50.9
			Cu	0.017	0.0	3.5	96.5	0.0	3.5	96.5
			Mn	0.017	0.0	100.0	0.0	0.0	100.0	0.0
			EC	0.017	100.0	0.0	0.0	100.0	0.0	0.0
			pH	0.017	0.0	82.5	17.5	0.0	100.0	0.0
	Biotic–compositional state	Flora		0.05	0.0	14.0	86.0	0.0	14.0	86.0
		Fauna		0.05	0.0	14.0	86.0	0.0	14.0	84.2
	Biotic–structural state	AGB		0.05	0.0	5.3	94.7	0.0	0.0	100.0
		BGB		0.05	0.0	5.3	94.7	0.0	0.0	100.0
	Biotic–functional state	NPP		0.10	0.0	7.0	93.0	0.0	0.0	100.0
	Landscape level	Fragmentation		0.25	76.3	5.8	17.9	2.7	69.9	27.4
		LST		0.25	0.0	0.0	100.0	0.0	0.1	99.9

(continued)

Table 36 (continued)

Districts	SEEA-EA conditions	Indicator	Parameter		Opening–2005 (%)			Closing–2020 (%)		
				Weight	High	Medium	Low	High	Medium	Low
Ecosystem condition account			Index	1.00	23.6	10.5	65.9	4.1	31.3	64.6
Kalburgi	Abiotic ecosystem	Soil	K	0.017	85.4	14.6	0.0	27.8	72.2	0.0
		Soil	N	0.017	26.5	57.0	16.6	0.0	0.0	100.0
			P	0.017	0.0	9.2	90.8	0.0	99.3	0.7
			OC	0.017	0.0	33.6	66.4	0.0	100.0	0.0
			S	0.017	97.4	2.6	0.0	97.4	2.6	0.0
			Zn	0.017	0.0	0.0	100.0	0.0	99.3	0.7
			Fe	0.017	0.0	61.6	38.4	0.0	100.0	0.0
			B	0.017	0.0	55.0	45.0	0.0	55.0	45.0
			Cu	0.017	0.0	0.0	100.0	0.0	0.0	100.0
			Mn	0.017	0.0	100.0	0.0	0.0	100.0	0.0
			EC	0.017	100.0	0.0	0.0	100.0	0.0	0.0
			pH	0.017	0.7	98.7	0.7	0.0	100.0	0.0
	Biotic–compositional state	Flora		0.05	0.0	14.0	86.0	0.0	14.0	86.0
		Fauna		0.05	0.0	1.3	98.7	0.0	1.3	98.7
	Biotic–structural state	AGB		0.05	0.0	1.3	98.7	0.0	0.7	99.3
		BGB		0.05	0.0	1.3	98.7	0.0	0.7	99.3
	Biotic–functional state	NPP		0.10	0.0	2.0	98.0	0.0	0.7	99.3
	Landscape level	Fragmentation		0.25	60.9	14.0	25.1	21.6	41.0	37.4
		LST		0.25	0.0	0.0	100.0	0.0	0.2	99.8

(continued)

Table 36 (continued)

Districts	SEEA-EA conditions	Indicator	Parameter		Opening–2005 (%)			Closing–2020 (%)		
				Weight	High	Medium	Low	High	Medium	Low
Ecosystem condition account			Index	1.00	20.4	11.8	67.8	9.2	23.3	67.5
Hassan	Abiotic ecosystem	Soil	K	0.017	64.2	35.8	0.0	6.2	93.8	0.0
		Soil	N	0.017	50.6	49.4	0.0	0.0	51.9	48.1
			P	0.017	0.0	13.6	86.4	42.0	58.0	0.0
			OC	0.017	0.0	54.3	45.7	0.0	98.8	1.2
			S	0.017	77.8	22.2	0.0	77.8	22.2	0.0
			Zn	0.017	0.0	32.1	67.9	0.0	91.4	8.6
			Fe	0.017	0.0	100.0	0.0	0.0	100.0	0.0
			B	0.017	0.0	81.5	18.5	0.0	81.5	18.5
			Cu	0.017	0.0	38.3	61.7	0.0	38.3	61.7
			Mn	0.017	0.0	100.0	0.0	0.0	100.0	0.0
			EC	0.017	100.0	0.0	0.0	100.0	0.0	0.0
			pH	0.017	35.8	64.2	0.0	54.3	45.7	0.0
	Biotic–compositional state	Flora		0.05	85.7	9.9	4.4	34.6	9.9	55.6
		Fauna		0.05	21.4	6.2	72.4	8.6	6.2	85.2
	Biotic–structural state	AGB		0.05	4.9	13.6	81.5	2.5	7.4	90.1
		BGB		0.05	4.9	13.6	81.5	2.5	7.4	90.1
	Biotic–functional state	NPP		0.10	2.5	65.4	32.1	4.9	58.0	37.0
	Landscape level	Fragmentation		0.25	36.9	35.5	27.6	14.9	46.9	38.2
		LST		0.25	4.5	18.3	77.2	5.1	33.6	61.2

(continued)

Table 36 (continued)

Districts	SEEA-EA conditions	Indicator	Parameter		Opening–2005 (%)			Closing–2020 (%)		
				Weight	High	Medium	Low	High	Medium	Low
Ecosystem condition account			Index	1.00	21.9	32.0	46.1	12.6	40.5	46.9
Haveri	Abiotic ecosystem	Soil	K	0.017	32.3	67.7	0.0	0.0	100.0	0.0
		Soil	N	0.017	69.4	24.2	6.5	0.0	0.0	100.0
			P	0.017	0.0	0.0	100.0	4.8	95.2	0.0
			OC	0.017	0.0	67.7	32.3	0.0	77.4	22.6
			S	0.017	29.0	71.0	0.0	29.0	71.0	0.0
			Zn	0.017	0.0	43.5	56.5	0.0	95.2	4.8
			Fe	0.017	0.0	79.0	21.0	0.0	100.0	0.0
			B	0.017	0.0	0.0	100.0	0.0	0.0	100.0
			Cu	0.017	0.0	4.8	95.2	0.0	4.8	95.2
			Mn	0.017	0.0	100.0	0.0	0.0	100.0	0.0
			EC	0.017	100.0	0.0	0.0	100.0	0.0	0.0
			pH	0.017	1.6	98.4	0.0	0.0	100.0	0.0
	Biotic–compositional state	Flora		0.05	54.1	6.5	39.5	22.6	6.5	71.0
		Fauna		0.05	3.9	4.8	91.3	1.6	4.8	93.5
	Biotic–structural state	AGB		0.05	0.0	8.1	91.9	0.0	3.2	96.8
		BGB		0.05	0.0	8.1	91.9	0.0	3.2	96.8
	Biotic–functional state	NPP		0.10	0.0	9.7	90.3	0.0	3.2	96.8
	Landscape level	Fragmentation		0.25	74.7	5.8	19.6	31.2	30.7	38.1
		LST		0.25	0.0	0.9	99.1	0.0	1.7	98.3

(continued)

Table 36 (continued)

Districts	SEEA-EA conditions	Indicator	Parameter		Opening–2005 (%)			Closing–2020 (%)		
				Weight	High	Medium	Low	High	Medium	Low
Ecosystem condition account			Index	1.00	25.4	13.3	61.3	11.2	21.7	67.1
Kodagu	Abiotic ecosystem	Soil	K	0.017	5.1	64.4	30.5	8.5	91.5	0.0
		Soil	N	0.017	91.5	8.5	0.0	0.0	6.8	93.2
			P	0.017	0.0	11.9	88.1	0.0	98.3	1.7
			OC	0.017	0.0	100.0	0.0	0.0	93.2	6.8
			S	0.017	1.7	98.3	0.0	1.7	98.3	0.0
			Zn	0.017	0.0	78.0	22.0	0.0	100.0	0.0
			Fe	0.017	0.0	100.0	0.0	0.0	100.0	0.0
			B	0.017	0.0	78.0	22.0	0.0	78.0	22.0
			Cu	0.017	0.0	3.4	96.6	0.0	3.4	96.6
			Mn	0.017	0.0	100.0	0.0	0.0	100.0	0.0
			EC	0.017	100.0	0.0	0.0	100.0	0.0	0.0
			pH	0.017	93.2	6.8	0.0	96.6	3.4	0.0
	Biotic–compositional state	Flora		0.05	93.9	1.7	4.5	95.0	1.7	3.3
		Fauna		0.05	56.0	13.3	30.7	56.7	13.3	30.0
	Biotic–structural state	AGB		0.05	23.3	68.3	8.3	16.7	48.3	35.0
		BGB		0.05	23.3	68.3	8.3	16.7	48.3	35.0
	Biotic–functional state	NPP		0.10	33.3	55.0	11.7	25.0	63.3	11.7
	Landscape level	Fragmentation		0.25	42.7	27.3	30.0	43.2	25.0	31.8
		LST		0.25	52.6	43.1	4.3	52.5	46.0	1.5

(continued)

Table 36 (continued)

Districts	SEEA-EA conditions	Indicator	Parameter		Opening–2005 (%)			Closing–2020 (%)		
				Weight	High	Medium	Low	High	Medium	Low
Ecosystem condition account			Index	1.00	41.8	41.5	16.7	39.1	42.5	18.3
Kolar	Abiotic ecosystem	Soil	K	0.017	6.8	93.2	0.0	0.0	100.0	0.0
		Soil	N	0.017	0.0	13.6	86.4	0.0	0.0	100.0
			P	0.017	0.0	10.2	89.8	0.0	93.2	6.8
			OC	0.017	0.0	0.0	100.0	0.0	0.0	100.0
			S	0.017	0.0	100.0	0.0	0.0	100.0	0.0
			Zn	0.017	0.0	6.8	93.2	0.0	98.3	1.7
			Fe	0.017	0.0	100.0	0.0	0.0	100.0	0.0
			B	0.017	0.0	93.2	6.8	0.0	93.2	6.8
			Cu	0.017	0.0	0.0	100.0	0.0	0.0	100.0
			Mn	0.017	0.0	100.0	0.0	0.0	100.0	0.0
			EC	0.017	100.0	0.0	0.0	100.0	0.0	0.0
			pH	0.017	0.0	100.0	0.0	93.2	6.8	0.0
	Biotic–compositional state	Flora		0.05	0.0	26.2	73.8	0.0	26.2	73.8
		Fauna		0.05	0.0	6.6	93.4	0.0	6.6	93.4
	Biotic–structural state	AGB		0.05	0.0	3.3	96.7	0.0	1.6	98.4
		BGB		0.05	0.0	3.3	96.7	0.0	1.6	98.4
	Biotic–functional state	NPP		0.10	0.0	23.0	77.0	0.0	16.4	83.6
	Landscape level	Fragmentation		0.25	72.3	6.8	20.8	20.0	39.9	40.2
		LST		0.25	0.0	1.1	98.9	0.0	4.9	95.1

(continued)

Table 36 (continued)

Districts	SEEA-EA conditions	Indicator	Parameter		Opening–2005 (%)			Closing–2020 (%)		
				Weight	High	Medium	Low	High	Medium	Low
Ecosystem condition account			Index	1.00	19.9	16.5	63.6	8.2	26.2	65.6
Koppal	Abiotic ecosystem	Soil	K	0.017	100.0	0.0	0.0	35.8	64.2	0.0
		Soil	N	0.017	3.0	79.1	17.9	0.0	0.0	100.0
			P	0.017	0.0	0.0	100.0	53.7	46.3	0.0
			OC	0.017	0.0	26.9	73.1	0.0	56.7	43.3
			S	0.017	76.1	23.9	0.0	76.1	23.9	0.0
			Zn	0.017	0.0	0.0	100.0	0.0	100.0	0.0
			Fe	0.017	0.0	71.6	28.4	0.0	100.0	0.0
			B	0.017	0.0	100.0	0.0	0.0	100.0	0.0
			Cu	0.017	0.0	32.8	67.2	0.0	32.8	67.2
			Mn	0.017	0.0	100.0	0.0	0.0	100.0	0.0
			EC	0.017	100.0	0.0	0.0	100.0	0.0	0.0
			pH	0.017	0.0	97.0	3.0	0.0	100.0	0.0
	Biotic–compositional state	Flora		0.05	0.0	17.9	82.1	0.0	17.9	82.1
		Fauna		0.05	0.0	1.5	98.5	0.0	1.5	98.5
	Biotic–structural state	AGB		0.05	0.0	1.5	98.5	0.0	0.0	100.0
		BGB		0.05	0.0	1.5	98.5	0.0	0.0	100.0
	Biotic–functional state	NPP		0.10	0.0	1.5	98.5	0.0	0.0	100.0
	Landscape level	Fragmentation		0.25	68.8	9.3	21.9	9.8	57.4	32.8
		LST		0.25	1.7	7.2	91.0	0.1	2.2	97.7

(continued)

Table 36 (continued)

Districts	SEEA-EA conditions	Indicator	Parameter		Opening–2005 (%)			Closing–2020 (%)		
				Weight	High	Medium	Low	High	Medium	Low
Ecosystem condition account			Index	1.00	22.3	14.3	63.5	6.9	27.9	65.2
Mandya	Abiotic ecosystem	Soil	K	0.017	19.0	81.0	0.0	0.0	100.0	0.0
		Soil	N	0.017	11.1	60.3	28.6	0.0	0.0	100.0
			P	0.017	0.0	33.3	66.7	12.7	87.3	0.0
			OC	0.017	0.0	85.7	14.3	0.0	100.0	0.0
			S	0.017	74.6	25.4	0.0	74.6	25.4	0.0
			Zn	0.017	0.0	20.6	79.4	0.0	100.0	0.0
			Fe	0.017	0.0	98.4	1.6	0.0	100.0	0.0
			B	0.017	0.0	96.8	3.2	0.0	96.8	3.2
			Cu	0.017	0.0	0.0	100.0	0.0	0.0	100.0
			Mn	0.017	0.0	100.0	0.0	0.0	100.0	0.0
			EC	0.017	100.0	0.0	0.0	100.0	0.0	0.0
			pH	0.017	0.0	100.0	0.0	0.0	100.0	0.0
	Biotic–compositional state	Flora		0.05	0.0	14.3	85.7	0.0	14.3	85.7
		Fauna		0.05	0.0	15.9	84.1	0.0	15.9	84.1
	Biotic–structural state	AGB		0.05	0.0	4.8	95.2	0.0	1.6	98.4
		BGB		0.05	0.0	4.8	95.2	0.0	1.6	98.4
	Biotic–functional state	NPP		0.10	7.9	42.9	49.2	6.3	39.7	54.0
	Landscape level	Fragmentation		0.25	77.2	5.9	16.9	29.9	53.4	16.7
		LST		0.25	0.2	29.1	70.7	0.0	24.7	75.3

(continued)

Table 36 (continued)

Districts	SEEA-EA conditions	Indicator	Parameter		Opening–2005 (%)			Closing–2020 (%)		
				Weight	High	Medium	Low	High	Medium	Low
Ecosystem condition account			Index	1.00	23.6	26.7	49.7	11.2	38.7	50.1
Mysore	Abiotic ecosystem	Soil	K	0.017	21.7	78.3	0.0	0.0	100.0	0.0
		Soil	N	0.017	24.1	47.0	28.9	0.0	0.0	100.0
			P	0.017	0.0	12.0	88.0	0.0	100.0	0.0
			OC	0.017	0.0	96.3	3.7	0.0	96.3	3.7
			S	0.017	46.3	53.7	0.0	46.3	53.7	0.0
			Zn	0.017	0.0	100.0	0.0	0.0	100.0	0.0
			Fe	0.017	0.0	100.0	0.0	0.0	100.0	0.0
			B	0.017	0.0	86.6	13.4	0.0	86.6	13.4
			Cu	0.017	0.0	0.0	100.0	0.0	0.0	100.0
			Mn	0.017	0.0	100.0	0.0	0.0	100.0	0.0
			EC	0.017	100.0	0.0	0.0	100.0	0.0	0.0
			pH	0.017	7.3	92.7	0.0	18.3	81.7	0.0
	Biotic–compositional state	Flora		0.05	23.9	4.8	71.3	19.3	4.8	75.9
		Fauna		0.05	22.4	12.0	65.5	18.1	12.0	69.9
	Biotic–structural state	AGB		0.05	2.4	15.7	81.9	1.2	14.5	84.3
		BGB		0.05	2.4	15.7	81.9	1.2	14.5	84.3
	Biotic–functional state	NPP		0.10	1.2	22.9	75.9	0.0	21.7	78.3
	Landscape level	Fragmentation		0.25	90.5	2.1	7.4	73.0	12.7	14.3
		LST		0.25	7.0	16.2	76.8	3.7	25.0	71.3

(continued)

Table 36 (continued)

Districts	SEEA-EA conditions	Indicator	Parameter		Opening–2005 (%)			Closing–2020 (%)		
				Weight	High	Medium	Low	High	Medium	Low
Ecosystem condition account			Index	1.00	30.4	22.1	47.6	23.9	27.5	48.6
Raichur	Abiotic ecosystem	Soil	K	0.017	80.5	19.5	0.0	0.9	99.1	0.0
		Soil	N	0.017	8.0	44.2	47.8	0.0	0.0	100.0
			P	0.017	0.0	0.0	100.0	4.4	95.6	0.0
			OC	0.017	0.0	25.7	74.3	0.0	97.3	2.7
			S	0.017	100.0	0.0	0.0	100.0	0.0	0.0
			Zn	0.017	0.0	0.0	100.0	0.0	100.0	0.0
			Fe	0.017	0.0	8.0	92.0	0.0	100.0	0.0
			B	0.017	0.0	89.4	10.6	0.0	89.4	10.6
			Cu	0.017	0.0	0.9	99.1	0.0	0.9	99.1
			Mn	0.017	0.0	100.0	0.0	0.0	100.0	0.0
			EC	0.017	100.0	0.0	0.0	100.0	0.0	0.0
			pH	0.017	0.0	55.8	44.2	0.0	100.0	0.0
	Biotic–compositional state	Flora		0.05	0.0	8.8	91.2	0.0	8.8	91.2
		Fauna		0.05	0.0	6.1	93.9	0.0	6.1	93.9
	Biotic–structural state	AGB		0.05	0.0	0.9	99.1	0.0	0.0	100.0
		BGB		0.05	0.0	0.9	99.1	0.0	0.0	100.0
	Biotic–functional state	NPP		0.10	0.0	2.6	97.4	0.0	0.9	99.1
	Landscape level	Fragmentation		0.25	65.2	9.8	25.1	12.7	44.3	43.0
		LST		0.25	0.4	10.5	89.1	0.0	1.6	98.4

(continued)

Table 36 (continued)

Districts	SEEA-EA conditions	Indicator	Parameter		Opening–2005 (%)			Closing–2020 (%)		
				Weight	High	Medium	Low	High	Medium	Low
Ecosystem condition account			Index	1.00	21.2	11.9	66.9	6.6	25.3	68.1
Ramanagara	Abiotic ecosystem	Soil	K	0.017	35.7	64.3	0.0	0.0	100.0	0.0
		Soil	N	0.017	9.5	71.4	19.0	0.0	0.0	100.0
			P	0.017	0.0	4.8	95.2	0.0	100.0	0.0
			OC	0.017	0.0	16.7	83.3	0.0	90.5	9.5
			S	0.017	7.1	92.9	0.0	7.1	92.9	0.0
			Zn	0.017	0.0	78.6	21.4	0.0	97.6	2.4
			Fe	0.017	0.0	100.0	0.0	0.0	100.0	0.0
			B	0.017	0.0	57.1	42.9	0.0	57.1	42.9
			Cu	0.017	0.0	0.0	100.0	0.0	0.0	100.0
			Mn	0.017	0.0	100.0	0.0	0.0	100.0	0.0
			EC	0.017	100.0	0.0	0.0	100.0	0.0	0.0
			pH	0.017	33.3	66.7	0.0	2.4	97.6	0.0
	Biotic–compositional state	Flora		0.05	0.0	34.1	65.9	0.0	34.1	65.9
		Fauna		0.05	63.8	2.3	33.9	13.6	2.3	84.1
	Biotic–structural state	AGB		0.05	0.0	25.0	75.0	0.0	18.2	81.8
		BGB		0.05	0.0	25.0	75.0	0.0	18.2	81.8
	Biotic–functional state	NPP		0.10	0.0	36.4	63.6	0.0	20.5	79.5
	Landscape level	Fragmentation		0.25	85.1	3.2	11.7	18.2	27.2	54.6
		LST		0.25	0.0	13.8	86.2	0.0	8.9	91.1

(continued)

Table 36 (continued)

Districts	SEEA-EA conditions	Indicator	Parameter		Opening–2005 (%)			Closing–2020 (%)		
				Weight	High	Medium	Low	High	Medium	Low
Ecosystem condition account			Index	1.00	27.6	23.1	49.4	7.1	28.6	64.3
Shimoga	Abiotic ecosystem	Soil	K	0.017	0.0	100.0	0.0	0.0	100.0	0.0
		Soil	N	0.017	97.2	2.8	0.0	0.0	3.8	96.2
			P	0.017	0.0	17.9	82.1	0.0	99.1	0.9
			OC	0.017	0.0	100.0	0.0	0.0	97.2	2.8
			S	0.017	12.3	87.7	0.0	12.3	87.7	0.0
			Zn	0.017	0.0	39.6	60.4	0.0	97.2	2.8
			Fe	0.017	0.0	100.0	0.0	0.0	100.0	0.0
			B	0.017	0.0	14.2	85.8	0.0	14.2	85.8
			Cu	0.017	0.0	33.0	67.0	0.0	33.0	67.0
			Mn	0.017	0.0	100.0	0.0	0.0	100.0	0.0
			EC	0.017	100.0	0.0	0.0	100.0	0.0	0.0
			pH	0.017	91.5	8.5	0.0	27.4	72.6	0.0
	Biotic–compositional state	Flora		0.05	38.9	2.8	58.2	50.9	2.8	46.2
		Fauna		0.05	28.1	3.8	68.1	36.8	3.8	59.4
	Biotic–structural state	AGB		0.05	14.2	56.6	29.2	8.5	47.2	44.3
		BGB		0.05	14.2	56.6	29.2	8.5	47.2	44.3
	Biotic–functional state	NPP		0.10	4.7	80.2	15.1	0.9	75.5	23.6
	Landscape level	Fragmentation		0.25	40.6	20.7	38.7	53.1	17.4	29.4
		LST		0.25	6.5	39.1	54.4	1.6	62.7	35.7

(continued)

Table 36 (continued)

Districts	SEEA-EA conditions	Indicator	Parameter		Opening–2005 (%)			Closing–2020 (%)		
				Weight	High	Medium	Low	High	Medium	Low
Ecosystem condition account			Index	1.00	22.0	39.0	38.9	21.3	46.0	32.6
Tumkur	Abiotic ecosystem	Soil	K	0.017	6.5	93.5	0.0	0.0	100.0	0.0
		Soil	N	0.017	23.4	36.4	40.3	0.0	9.1	90.9
			P	0.017	0.0	11.7	88.3	0.6	99.4	0.0
			OC	0.017	0.0	10.4	89.6	0.0	98.1	1.9
			S	0.017	9.1	90.9	0.0	9.1	90.9	0.0
			Zn	0.017	0.0	3.9	96.1	0.0	60.4	39.6
			Fe	0.017	0.0	98.1	1.9	0.0	100.0	0.0
			B	0.017	0.0	96.8	3.2	0.0	96.8	3.2
			Cu	0.017	0.0	0.0	100.0	0.0	0.0	100.0
			Mn	0.017	0.0	100.0	0.0	0.0	100.0	0.0
			EC	0.017	100.0	0.0	0.0	100.0	0.0	0.0
			pH	0.017	13.0	87.0	0.0	87.0	13.0	0.0
	Biotic–compositional state	Flora		0.05	1.6	11.3	87.1	0.6	11.3	88.1
		Fauna		0.05	0.0	6.3	93.8	0.0	6.3	93.8
	Biotic–structural state	AGB		0.05	0.0	2.5	97.5	0.0	0.6	99.4
		BGB		0.05	0.0	2.5	97.5	0.0	0.6	99.4
	Biotic–functional state	NPP		0.10	0.0	31.3	68.8	0.0	23.1	76.9
	Landscape level	Fragmentation		0.25	82.2	4.4	13.4	31.4	35.7	32.9
		LST		0.25	0.0	1.4	98.6	0.0	2.0	98.0

(continued)

Table 36 (continued)

Districts	SEEA-EA conditions	Indicator	Parameter		Opening–2005 (%)			Closing–2020 (%)		
				Weight	High	Medium	Low	High	Medium	Low
Ecosystem condition account			Index	1.00	23.2	16.2	60.7	11.2	25.5	63.4
Udupi	Abiotic ecosystem	Soil	K	0.017	0.0	84.6	15.4	0.0	36.5	63.5
		Soil	N	0.017	100.0	0.0	0.0	0.0	0.0	100.0
			P	0.017	0.0	3.8	96.2	0.0	88.5	11.5
			OC	0.017	0.0	100.0	0.0	0.0	98.1	1.9
			S	0.017	1.9	86.5	11.5	1.9	86.5	11.5
			Zn	0.017	0.0	1.9	98.1	0.0	100.0	0.0
			Fe	0.017	0.0	100.0	0.0	0.0	100.0	0.0
			B	0.017	0.0	0.0	100.0	0.0	0.0	100.0
			Cu	0.017	0.0	1.9	98.1	0.0	1.9	98.1
			Mn	0.017	0.0	100.0	0.0	0.0	100.0	0.0
			EC	0.017	100.0	0.0	0.0	100.0	0.0	0.0
			pH	0.017	100.0	0.0	0.0	63.5	36.5	0.0
	Biotic–compositional state	Flora		0.05	52.4	6.0	41.6	42.0	6.0	52.0
		Fauna		0.05	59.9	6.0	34.1	48.0	6.0	46.0
	Biotic–structural state	AGB		0.05	22.0	28.0	50.0	20.0	30.0	50.0
		BGB		0.05	22.0	28.0	50.0	20.0	30.0	50.0
	Biotic–functional state	NPP		0.10	4.0	66.0	30.0	0.0	68.0	32.0
	Landscape level	Fragmentation		0.25	83.0	3.5	13.5	66.5	7.4	26.0
		LST		0.25	9.0	36.6	54.4	5.6	52.0	42.4

(continued)

Table 36 (continued)

Districts	SEEA-EA conditions	Indicator	Parameter		Opening–2005 (%)			Closing–2020 (%)		
				Weight	High	Medium	Low	High	Medium	Low
Ecosystem condition Account			Index	1.00	36.2	28.0	35.7	27.3	36.1	36.7
Uttara Kannada	Abiotic ecosystem	Soil	K	0.017	0.0	84.4	15.6	0.0	99.3	0.7
		Soil	N	0.017	100.0	0.0	0.0	0.0	0.0	100.0
			P	0.017	0.0	0.0	100.0	37.6	62.4	0.0
			OC	0.017	0.0	72.4	27.6	0.0	100.0	0.0
			S	0.017	45.4	54.6	0.0	45.4	54.6	0.0
			Zn	0.017	0.0	85.8	14.2	0.0	100.0	0.0
			Fe	0.017	0.0	100.0	0.0	0.0	100.0	0.0
			B	0.017	0.0	0.0	100.0	0.0	0.0	100.0
			Cu	0.017	0.0	99.3	0.7	0.0	99.3	0.7
			Mn	0.017	0.0	100.0	0.0	0.0	100.0	0.0
			EC	0.017	100.0	0.0	0.0	100.0	0.0	0.0
			pH	0.017	94.4	5.6	0.0	100.0	0.0	0.0
	Biotic–compositional state	Flora		0.05	87.8	5.5	6.7	73.1	5.5	21.4
		Fauna		0.05	56.3	11.0	32.7	46.9	11.0	42.1
	Biotic–structural state	AGB		0.05	46.2	35.2	18.6	33.1	42.1	24.8
		BGB		0.05	46.2	35.2	18.6	33.1	42.1	24.8
	Biotic–functional state	NPP		0.10	32.4	55.2	12.4	1.4	84.8	13.8
	Landscape level	Fragmentation		0.25	55.0	10.8	34.2	45.8	13.5	40.8
		LST		0.25	20.6	53.2	26.2	0.6	69.9	29.6

(continued)

Table 36 (continued)

Districts	SEEA-EA conditions	Indicator	Parameter		Opening–2005 (%)			Closing–2020 (%)		
				Weight	High	Medium	Low	High	Medium	Low
Ecosystem condition account			Index	1.00	39.6	35.9	24.5	25.8	46.3	28.0
Yadgir	Abiotic ecosystem	Soil	K	0.017	43.8	56.2	0.0	0.0	100.0	0.0
		Soil	N	0.017	0.0	16.4	83.6	0.0	0.0	100.0
			P	0.017	0.0	0.0	100.0	0.0	100.0	0.0
			OC	0.017	0.0	20.5	79.5	0.0	100.0	0.0
			S	0.017	100.0	0.0	0.0	100.0	0.0	0.0
			Zn	0.017	0.0	0.0	100.0	0.0	100.0	0.0
			Fe	0.017	0.0	49.3	50.7	0.0	100.0	0.0
			B	0.017	0.0	61.6	38.4	0.0	61.6	38.4
			Cu	0.017	0.0	0.0	100.0	0.0	0.0	100.0
			Mn	0.017	0.0	100.0	0.0	0.0	100.0	0.0
			EC	0.017	100.0	0.0	0.0	100.0	0.0	0.0
			pH	0.017	0.0	97.3	2.7	0.0	100.0	0.0
	Biotic–compositional state	Flora		0.05	0.0	21.9	78.1	0.0	21.9	78.1
		Fauna		0.05	0.0	1.4	98.6	0.0	1.4	98.6
	Biotic–structural state	AGB		0.05	0.0	2.7	97.3	0.0	0.0	100.0
		BGB		0.05	0.0	2.7	97.3	0.0	0.0	100.0
	Biotic–functional state	NPP		0.10	0.0	5.5	94.5	0.0	5.5	94.5
	Landscape level	Fragmentation		0.25	70.2	9.5	20.3	9.1	54.6	36.4
		LST		0.25	0.0	3.1	96.9	0.0	6.8	93.2
Ecosystem condition account			Index	1.00	21.6	11.8	66.6	5.6	29.7	64.7

Note N: Nitrogen, P: Phosphorous, K: Potash, OC: Organic carbon, Zn: Zinc, Fe: Iron, B: Boron, Cu: Copper, Mn: Manganese, S: Sulphur, EC: Electrical conductivity, AGB: Above ground biomass, BGB: Below ground Biomass, NPP: Net Primary Productivity, LST: Land Surface Temperature

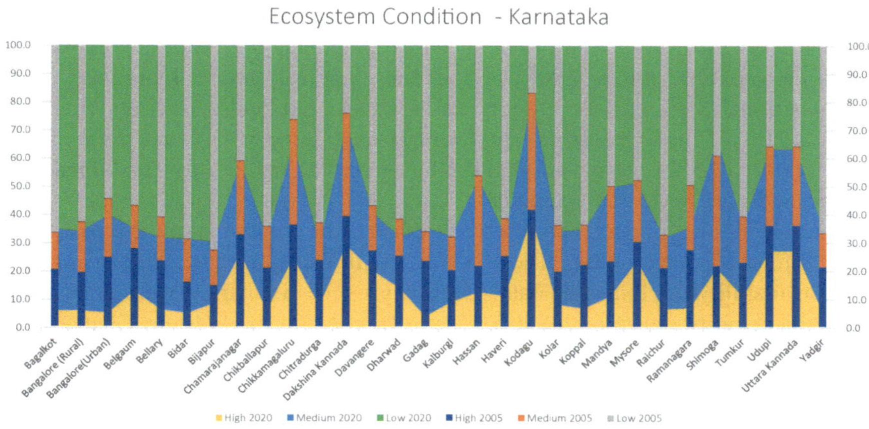

Fig. 71 Ecosystem condition index–district-wise for Karnataka State

7 Conclusion

Ecosystem condition accounts provide a structured approach for describing ecosystem assets' characteristics by recording and aggregating data relating the dynamics of characteristics of ecosystem assets characteristics (SEEA EA). Ecosystem quality is measured in terms of its abiotic and biotic features with respect to an ecosystem's composition, structure, and function, reflecting multiple values across a range of temporal and spatial scales indicating ecological integrity.

Ecosystem condition account was developed considering (i) abiotic ecosystem characteristics (physical and chemical status of soil), (ii) biotic ecosystem characteristics considering compositional state (occurrence of flora and fauna species chosen as per IUCN categories of threats), structural state (vegetation – above-ground biomass, below-ground biomass of forest ecosystems), functional state (NPP: net primary productivity in forest ecosystems) and (iii) landscape-level characteristics (fragmentation of forest landscapes, LST: the land surface temperature at landscape levels) for 2005 and 2020. The ecosystem condition indicator account was developed for Karnataka State at disaggregated levels by dividing the study region into 9 km × 9 km or 5' × 5' grids, comparable to the grids in the Survey of India topographic maps of scale 1:50,000. This also aided in computing the opening stock (2005) and closing stock at three levels – grids were grouped based on the relative proportion of abiotic, biotic, and landscape parameters as high, medium and low (reference levels). The ecosystem condition index for Karnataka State (district wise) is computed considering all these ecosystem condition variables–abiotic (soil physical and chemical status), biotic ecosystem (compositional, structural and functional variables) and landscape-level (fragmentation and LST). The aggregation is performed by assigning weights as 0.2 for abiotic, 0.3 for biotic (0.1 for compositional, 0.1 for structural and

0.1 for functional) and 0.5 for landscape-level characteristics. The ecosystem condition index for Karnataka State (district wise) is computed considering ecosystem condition variables–abiotic (soil physical and chemical status), biotic ecosystem (compositional, structural, and functional variables) and landscape-level (fragmentation and LST). The aggregation is performed by assigning weights as 0.2 for abiotic, 0.3 for biotic (0.1 for compositional, 0.1 for structural, and 0.1 for functional) and 0.5 for landscape-level characteristics.

Landscape-level characteristics were assessed using temporal remote sensing data (optical remote sensing data). The constraint with the optical remote sensing data is the availability of cloud-free data. The analyses for 10 districts were done based on the availability (year) of cloud-free data. Districts in the Western Ghats region witnessed large-scale land-use changes in the eighties with the construction of dam/ reservoir for hydroelectric projects. Hence an attempt is made to analyse spatial data of 70's for these districts to understand the role of developmental projects in landscape dynamics.

Ecosystem condition variables were compiled district-wise for Karnataka State, India for 2005 and 2019/2020 (depending on the data availability).

Abiotic characteristics of ecosystems in Karnataka were captured through the soil data available with the department of agriculture, Government of Karnataka for 1998–2000, and at block levels (to assess the soil health of croplands) for 2020. Soil quality data were updated for 2005 by compiling data from published literatures of University of Agriculture, National Bureau of soil survey and land use planning, etc.

Biotic ecosystem characteristics considering compositional state (occurrence of flora and fauna species chosen as per IUCN categories of threats) was assessed based on the field data collected from 10 districts (primary data) using GPS (Global positioning system) and the data compiled from biodiversity portal of Karnataka Biodiversity Board and Karnataka forest Department. IUCN categories of compiled flora and fauna data were included by referring online portal of IUCN (https://www.iucnredlist.org/). Data for 2005 were compiled from published literatures from academia.

Structural state (vegetation–above ground biomass, below ground biomass of forest ecosystems), and functional state (NPP: net primary productivity in forest ecosystems) of ecosystems were based on the field data collected from select districts across agro-climatic zones and extrapolation to the entire state based on the land uses using remote sensing data and preliminary analyses across districts.

Landscape-level characteristics (fragmentation of forest landscapes, LST: land surface temperature at landscape levels) for 2005 and 2019 were assessed considering remote sensing data.

Analyses of fragmentation of forest ecosystem reveal that districts, even ecologically fragile regions–Western Ghats have been experiencing fragmentation of forests evident from the decline of contiguous intact forests to the extent of 54% (Chikkamagalur), 57.6% (Dakshina Kannada), 44% (Udupi), 41% (Uttara Kannada) and 35% (Shimoga) during 2005 to 2019. The decline in ecosystem condition has altered ecosystem process affecting services.

Land surface temperature estimations district-wise for Karnataka considering thermal bands–remote sensing data for 2005 and 2019 indicate a considerable increase in temperature, which could be attributed to land cover changes in the region with the decline of vegetation cover. About 821,600 hectares had temperatures less than 30 °C (high) in 2005 compared to 485,566 hectares in 2019. Regions under medium category (with temperatures 30-35 °C) have increased from 2898 sq.km (2005) to 3666 sq.km (2019).

Biotic ecosystem characteristics considering compositional state (occurrence of endemic flora and fauna species chosen as per IUCN categories of threats) show there has been a decline in the number of flora species (Critically endangered, Endangered, Vulnerable, Near-threatened) from 3614 species (2005) to 3374 and a similar trend in the occurrence of endemic fauna species from 4243 (2005) to 3923 (2020) species.

Structural state of forest ecosystem was done through assessment of vegetation characteristics with the computation of above-ground biomass (AGB), below-ground biomass (BGB) of forest ecosystems, reveals a decline of 23.4% of AGB from 987,686 (2005) to 732,828 (2019) Gg and similar trend of decrease in BGB from 191,537 (2005) to 146,566 (2019) Gg.

Changes in the land cover have impacted the functional state of the forest ecosystem, which was assessed through the computation of NPP (net primary productivity), indicate 15.5% decline in NPP from 17,323 Gg/yr (2005) to 14,635 Gg/yr (2019.

Abiotic characteristics of the ecosystem are assessed through the physical and chemical status of soil in agricultural lands. Soil health assessment indicates that 43.7% of soil in Karnataka shows good soil quality, where 48.6% of soil in Karnataka shows medium quality soil while low quality of soil is found in 7.6% area in Karnataka. Among the chemical parameters,

- 10.6% of cropland has high potassium nutrient where 85.3% of area is medium while low potassium in 4% of the total area.
- Nitrogen nutrient is found low in 95.5% of the cropland soil while 4.5% of the area shows medium quantity.
- Phosphorus nutrient availability in the soil is high in 9.9% of the area, where 85.8% of the area is in medium status while low availability is in 4.1% of the area of the state.
- Zinc nutrient in the soil is medium (sufficient) in 90.5% of the area while it is low (deficient) in 9.5% of croplands in Karnataka.
- Iron nutrient availability in the soil is medium (sufficient) in 98.8% area while 1.2% of the cropland in the State are low (deficient).
- Organic carbon in the soil is medium (sufficient) in 88.25% of the area, while it is found deficient in 11.75% of the area of the state.
- Boron is in medium (sufficient) in 57% of the cropland soil while 43% of area have low (deficient).
- Sulphur nutrient is high in 54.7% of the area, where it is medium in 43.58% and it is low in 1.6% of the area of the state.
- Copper availability in the soil is low (deficient) in 86.8% of the area while it is medium (sufficient) in 13.2% of area.

- Manganese nutrient in the soil is low (deficient) in 0.15% of the area and medium (sufficient) in 99.58% of the area of the state.

Ecosystem Condition index computed district-wise considering reference levels of ecosystem condition variables (abiotic, biotic–vegetation, structural and functional, and landscape characteristics) highlighting land-use-related changes. Kodagu district had maximum value of 41.8 (high), 41.5 (medium) during 2005 has now 39.5 (high), indicating a decline in ecosystem integrity. Similar trends are observed in all districts in Karnataka and are listed in Table 37.

Table 37 Ecosystem index across districts in Karnataka for 2005 and 2020

District		High 2005	Medium 2005	Low 2005	High 2020	Medium 2020	Low 2020
Bagalkot		20.6	13.1	66.3	6.0	28.9	65.1
Bangalore (Rural)		19.5	17.9	62.6	6.0	28.0	66.0
Bangalore (Urban)		24.8	20.8	54.4	5.2	34.7	60.1
Belgaum		27.9	15.3	56.8	12.9	22.1	65.0
Bellary		23.7	15.4	60.9	6.4	25.4	68.1
Bidar		16.1	15.2	68.6	5.1	26.3	68.6
Bijapur		**14.9**	12.7	**72.4**	8.6	21.9	69.5
Chamarajanagar		32.8	26.4	40.9	26.5	32.5	41.0
Chikballapur		21.1	14.8	64.1	6.1	25.6	68.3
Chikkamagaluru		36.3	37.6	26.1	24.9	40.6	34.5
Chitradurga		23.8	13.3	62.8	7.9	21.8	**70.2**
Dakshina Kannada		39.3	36.8	23.9	29.2	43.4	27.4
Davangere		27.2	16.1	56.7	20.3	20.5	59.2
Dharwad		25.4	13.1	61.5	14.1	18.2	67.6
Gadag		23.6	**10.5**	65.9	4.1	31.3	64.6
Kalburgi		20.4	11.8	67.8	9.2	23.3	67.5
Hassan		21.9	32.0	46.1	12.6	40.5	46.9
Haveri		25.4	13.3	61.3	11.2	21.7	67.1
Kodagu		**41.8**	**41.5**	**16.7**	**39.1**	42.5	**18.3**
Kolar		19.9	16.5	63.6	8.2	26.2	65.6
Koppal		22.3	14.3	63.5	6.9	27.9	65.2
Mandya		23.6	26.7	49.7	11.2	38.7	50.1
Mysore		30.4	22.1	47.6	23.9	27.5	48.6
Raichur		21.2	11.9	66.9	6.6	25.3	68.1
Ramanagara		27.6	23.1	49.4	7.1	28.6	64.3

(continued)

Table 37 (continued)

District		High 2005	Medium 2005	Low 2005	High 2020	Medium 2020	Low 2020
Shimoga		22.0	39.0	38.9	21.3	**46.0**	32.6
Tumkur		23.2	16.2	60.7	11.2	25.5	63.4
Udupi		36.2	28.0	35.7	27.3	36.1	36.7
Uttara Kannada		36.2	28.0	35.7	27.3	36.1	36.7
Yadgir		21.6	11.8	66.6	5.6	29.7	64.7
	Max	41.8	41.5	72.4	39.1	46.0	70.2
	Min	14.9	10.5	16.7	4.1	18.2	18.3
	avg	25.7	20.5	53.8	13.7	29.9	56.4
	Std. dev	6.7	9.2	14.7	9.4	7.6	15.0

References

Forest Fragmentation

Alix-Garcia J, Janvry AD, Sadoulet E (2005) A tale of two communities: explaining deforestation in Mexico. World Dev 33(2):219–235

Ameztegui A, Brotons L, Coll L (2010) Land-use changes as major drivers of mountain pine (Pinus uncinata Ram.) expansion in the Pyrenees. Global Ecol Biogeogr 19(5):632–641

Asner GP, Broadbent EN, Oliveira PJC, Keller M, Knapp DE, Silva JN (2006) Condition and fate of logged forests in the Brazilian Amazon. Proc Natl Acad Sci 103:12947–12950

Baskaran N, Varma S, Sar CK Sukumar R (2011) Current status of Asian elephants in India. Gajah 35(1–2):47–54

Bennett A, Saunders D (2010) Habitat fragmentation and landscape change. In Sodhi NS, Ehrlich PR (Eds), Conservation biology for all. Oxford University Press, UK.

Bharath S, Bharath H Aithal, Rajan KS, Ramachandra TV (2012a) Cost effective mapping, monitoring and visualisation of spatial patterns of urbanisation using FOSS. FOSS 4G, 10/2012

Bharath HA, Bharath S, Durgappa SD, Ramachandra TV (2012b) Empirical patterns of the influence of spatial resolution of remote sensing data on landscape metrics. Int J Eng Res Appl (IJERA) 2(3):767–775

Boogaert J, Ceulemans R, Salvador-Van Eysenrode D (2004) Decision tree algorithm for detection of spatial processes in landscape transformation. Environ Manag 33:62–73

Broadbent E, Asner GP, Keller M, Knapp D, Oliveira P, Silva J (2008) Forest fragmentation and edge effects from deforestation and selective logging in the Brazilian Amazon. Biol Conserv 140:142–155

Biradar B, Jayadeva HM, Channakeshava S, Geetha KN, Sannagoudar MS, Pavan AS, Prakash KN (2020) Assessment of soil fertility through GIS techniques and thematic mapping in micro-watershed of Hassan, Karnataka. J Pharmacognosy Phytochem 9(4):3218–3228

Buskirk S, William H, Federick R, Smith W, Richard Knight L (2000) An overview of forest fragmentation in the Southern Rocky Mountains. In: Knight RL, Frederick Smith W, Steven Buskirk W, William Romme H, William Baker Boulder L (eds) Forest fragmentation in the Southern Rocky Mountains. University Press of Colorado, pp 3–14

Cabral AIR, Vasconcelos MJ, Oom D, Sardinha R (2010) Spatial dynamics and quantification of deforestation in the central-plateau woodlands of Angola (1990–2009). Appl Geogr 31:1185–1193
Cagnolo L, Cabido M, Valladares G (2006) Plant species richness in the Chaco Serrano woodland from central Argentina: ecological traits and habitat fragmentation effects. Biol Conserv 132:510–519
Carolina M (1995) Edge effects in fragmented forests: implications for conservation. Tree 10(2)
Chen J, Chen J, Liao A, Cao X, Chen L, Chen X, He C, Han G, Peng S, Lu M, Tong X, Zhang W (2015) China's global land cover mapping at 30 m resolution: A POK-based operational approach ISPRS. J Photogramm Remote Sens 103:7–27
DeFries R, Foley J, Asner G (2004) Land-use choices: balancing human needs and ecosystem function. Front Ecol Environ 2(5):249–257
Diaz S, Fargione J, Chapin F, Tilman D (2006) Biodiversity loss threatens human well-being. PLoS Biol 4(8) 277:1300–1306
Echeverria C, Coomes DA, Hall M, Newton AC (2008) Spatially explicit models to analyze forest loss and fragmentation between 1976 and 2020 in southern Chile. Ecol Model 212(3):439–449
Fahrig L (2003) Effects of habitat fragmentation on biodiversity. Annu Rev Ecol Evol Syst 34:487–515
Gong C, Chen J, Yu S (2011) Spatiotemporal dynamics of urban forest conversion through model urbanization in Shenzhen, China. Int J Remote Sens 32(24):9071–9092
Gong C, Yu S, Joesting H, Chen J (2013) Determining socioeconomic drivers of urban forest fragmentation with historical remote sensing images. Landsc Urban Plan 117:57–65. ISSN 0169-2046. https://doi.org/10.1016/j.landurbplan.2013.04.009
Gupta RK, Prasad TS, Krishna Rao PV, Bala Manikavelu PM (2000) Problems in upscaling of high resolution remote sensing data to coarse spatial resolution over land surface. Adv Space Res 26(7):1111–1121. ISSN 0273-1177. https://doi.org/10.1016/S0273-1177(99)01127-8
Harper KA, MacDonald SE, Burton PJ, Chen J, Brosofske KD, Saunders SC, Euskirchen ES, Roberts D, Jaiteh MS, Esseen PA (2005) Edge influence on forest structure and composition in fragmented landscapes. Conserv Biol 19:768–782
Holway D (2005) Edge effects of an invasive species across a natural ecological boundary. Biol Cons 121:561–567
Kampichler C, van Turnhout CAM, Devictor V, van der Jeugd HP (2012) Large-scale changes in community composition: determining land use and climate change signals. PLoS ONE 7(4):e35272. https://doi.org/10.1371/journal.pone.0035272
Kindstrand C, Norman J, Boman M, Mattsson L (2008) Attitude towards various forest functions: a comparison between private forest owners and forest officers. Scand J Forest Res 23:133–136
Kitron U, Clennon JA, Cecere MC, Gürtler RE, King CH, Vazquez-Prokopec G (2006) Upscale or downscale: applications of fine scale remotely sensed data to Chagas disease in Argentina and schistosomiasis in Kenya. Geospat Health 1(1):49–58
Kuèas A, Trakimas G, Balèiauskas L, Vaitkus G (2011) Multi-scale analysis of forest fragmentation in Lithuania. Balt for 17(1):128–135
Laurance WF (1999) Reflections on the tropical deforestation crisis. Biol Conserv 91:109–117
Laurance WF, Lovejoy TE, Vasconcelos HL, Bruna EM, Didham RK, Stouffer PC, Gascon C, Bierregaard RO, Laurance SG, Sampaio E (2002) Ecosystem decay of Amazonian forest fragments: a 22-year investigation. Conserv Biol 16:605–618
Li M, Huang C, Zhu Z, Wen W, Xu D, Liu A (2009) Use of remote sensing coupled with a vegetation change tracker model to assess rates of forest change and fragmentation in Missippi, USA. Int J Remote Sens 30:6559–6574
Lillesand TM, Keifer RW (1987) Remote sensing and image interpretation. Willey, New York
Liu C, Frazier P, Kumar L (2007) Comparative assessment of the measures of thematic classification accuracy. Remote Sens Environ 107:606–616
Millennium Ecosystem Assessment (MEA) (2005) Ecosystems and human well-Being: Desertification synthesis. World Resources Institute, Washington DC

Munroe DK, Southworth J, Tucker CM (2004) Modeling spatially and temporally complex land-cover change: the case of western Honduras. Prof Geogr 56:544–559

Naidu LGK, Sidhu GS, Sarkar D, Ramamurthy V (2011) Emerging deficiency of potassium in soils and crops of India. Karnataka J Agric Sci 24(1)

Nandy S, Kushwaha SPS, Dadhwal VK (2011) Forest degradation assessment in the upper catchment of the river Tons using remote sensing and GIS. Ecol Ind 11:509–513

Nelson GC, Harris V, Stone SW (1999) Spatial econometric analysis and project evaluation: modeling land use change in the Darien. Inter-American Development Bank, Environment Division, Sustainable Development Department, Washington, DC, p 33

Nelson GC, Geoghegan J (2002) Deforestation and land use change: sparse data environments. Agric Econ 27(3):201–216

Pan Y, Birdsey RA, Fang J, Houghton R, Kauppi PE, Kurz, WA, Phillips OL, Shvidenko A, Lewis SL (2011a) A large and persistent carbon sink in the world's forests. Science 333(6045):988–993. 1201609

Prasad PN, Subbarayappa CT, Sathish A (2020) Spatial distribution and mapping of available micro nutrient status in relation to soil properties of tomato grown soils of Chikkaballapura District, Karnataka. Int J Plant & Soil Sci, pp 1–10

Pujar KG, Hiremath SC, Pujar AS, Pujeri US, Yadawe MS (2012) Analysis of physicochemical and heavy metal concentration in soil of Bijapur Taluka, Karnataka Sci Revs Chem Commun 2(1):76–79

Ramachandra TV, Kumar U (2011) Characterisation of landscape with forest fragmentation dynamics. J Geogr Inf Syst 3(03):242

Ramachandra TV, Krishnadas G, Setturu B, Kumar U (2012a) Regional Bioenergy planning for sustainability in Himachal Pradesh. India, Journal of Energy, Environment & Carbon Credits 2(1):13–49

Ramachandra TV, Subash Chandran MD, Joshi NV, Pallav Julka, Uttam Kumar, Bharath H Aithal, Prakash Mesta, Rao GR, Vishnu Mukri (2012b) Landslide susceptible zone mapping in Uttara Kannada Central Western Ghats, ENVIS Technical Report 28, Energy & Wetlands Research Group, Centre for Ecological Sciences, Indian Institute of Science, Bangalore 560012

Ramachandra TV, Aithal BH, Sanna DD (2012c) Insights to urban dynamics through landscape spatial pattern analysis. Int J Appl Earth Obs Geoinf 18:329–343

Ramachandra TV, Subash Chandran MD, Sudarshan P Bhat, Bharath H. Aithal, Rao GR, Vishnu M (2013) Status of Forest in Shimoga district, Karnataka, Sahyadri Conservation Series 23, ENVIS Technical Report: 53, ENVIS- Environmental Information System, Centre for Ecological Sciences, Indian Institute of Science, Bangalore

Ramachandra TV, Bharath S, Bharath A (2014a) Spatio-temporal dynamics along the terrain gradient of diverse landscape. J Environ Eng Landsc Manag 22(1):50–63

Ramachandra TV, Subash Chandran MD, Bharath Setturu, Rao GR, Vishnu DM (2014b) Reclamation of mine regions at Bisgod: Approaches and challenges, Sahyadri Conservation Series 43, ENVIS Technical Report 80, CES, Indian Institute of Science, Bangalore 560012, India

Ramachandra TV, Setturu B, Chandran S (2016a) Geospatial analysis of forest fragmentation in Uttara Kannada District, India. For Ecosyst 3(1):1–15

Ranjitha P, Chidanandappa HM, Dhananjaya BC, Prashanth KM (2019) Chemical properties and available zinc status in soils coming under paddy land use cover of Bhadra command, Karnataka. J Pharmacognosy Phytochem 8(2):1220–1226

Ravikumar MA, Patil PL, Dasog GS (2007) Mapping of Nutrients Status Under 48A Distributary in Malaprabha Right Bank Command of Karnataka by Geographic Information System Technique-II: Micro-nutrients. Karnataka J Agric Sci 20(4):735

Riitters KJ, O'Neill R, Wickham JD, Jones B, Smith E (2000) Global-scale patterns of forest fragmentation. Conserv Ecol 4(2):3

Riitters K, Wickham J, O'Neill R, Jones K, Smith E, Coulston J, Wade T, Smith J (2002) Fragmentation of the continental United States forests. Ecosystems 5:815–822

Roy J, Saugier B, Mooney HA (2001) Terrestrial global productivity. Academic Press, London

Sindhu D, Ravikumar AS, Shivakumar, BL (2022) Evaluation of Soil Water Crop Compatibility Using RS and GIS Techniques. In Innovative Trends in Hydrological and Environmental Systems: Select Proceedings of ITHES 2021 (pp. 449–459). Singapore: Springer Nature Singapore

Land Surface Temperature

Aithal BH, Vinay S, Ramachandra TV (2014) Landscape dynamics modeling through integrated Markov, Fuzzy-AHP and cellular automata. In: 2014 IEEE Geoscience and remote sensing symposium. https://doi.org/10.1109/igarss.2014.6947148

Anderson MC, Kustas WP, Norman JM, Hain CR, Mecikalski JR, Schultz L, Gonzalez-Dugo MP, Cammalleri C, d'Urso G, Pimstein A, Gao F (2011) Mapping daily evapotranspiration at field to continental scales using geostationary and polar orbiting satellite imagery. Hydrol Earth Syst Sci 15(1):223–239. https://doi.org/10.5194/hess-15-223-2011

Bharath S, Rajan KS, Ramachandra TV (2013a) Land surface temperature responses to land use land

Cover Dynamics (2013) Geoinformatics & Geostatistics: an overview 01(04). https://doi.org/10.4172/23274581.1000112

Bunai T, Wibowo A (2018) Comparison spatial pattern of land surface temperature with mono window algorithm and split window algorithm: a case study in South Tangerang, Indonesia. IOP Conf Ser Earth Environ Sci 149:012066. https://doi.org/10.1088/17551315/149/1/012066

Chen Z, Xu B, Devereux B (2014) Urban landscape pattern analysis based on 3D landscape models. Appl Geogr 55:82–91. https://doi.org/10.1016/j.apgeog.2014.09.006

Ganasri BP, Dwarakish GS (2015) Study of land use/land cover dynamics through classification algorithms for Harangi catchment area, Karnataka State, India. Aquat Proc 4:1413–1420. https://doi.org/10.1016/j.aqpro.2015.02.183

Grizzetti B, Pistocchi A, Liquete C, Udias A, Bouraoui F, van de Bund W (2017) Human pressures and ecological status of European rivers. Sci Rep 7(1). https://doi.org/10.1038/s41598-017-00324-3

James MM, Mundia CN (2014) Dynamism of land use changes on surface temperature in Kenya: a case study of Nairobi City. Int J Sci Res 3(4):38–41

Koppad AG, Janagowdar BS (2017) Assessment and mapping of soil organic carbon pool through geo-informatics approach in Bhatkal taluka of Uttar Kannada district of Karnataka. Indian Forester 143(7):653–656

Li X, Zhou W, Ouyang Z (2013) Relationship between land surface temperature and spatial pattern of greenspace: What are the effects of spatial resolution? Landsc Urban Plan 114:1–8. https://doi.org/10.1016/j.landurbplan.2013.02.005

Mallick J, Kant Y, Bharath BD (2008) Estimation of land surface temperature over Delhi using Landsat7 ETM+. J Indian Geophys Union 12(3):131–140

Newbold T, Hudson LN, Hill SL, Contu S, Lysenko I, Senior RA, Börger L, Bennett DJ, Choimes A, Collen B, Day J, Purvis A (2015) Global effects of land use on local terrestrial biodiversity. Nature 520(7545):45–50. https://doi.org/10.1038/nature14324

Peng J, Xu Y, Cai Y, Xiao H (2011) Climatic and anthropogenic drivers of land use/cover change in fragile karst areas of southwest China since the early 1970s: a case study on the Maotiaohe watershed. Environ Earth Sci 64(8):2107–2118. https://doi.org/10.1007/s12665-011-1037-5

Ramachandra TV, Bharath S, Chandran S (2016b) Geospatial analysis of forest fragmentation in Uttara Kannada District, India. For Ecosyst 3(1). https://doi.org/10.1186/s40663-016-0069-4

Ramachandra TV, Setturu B, Rajan KS, Subash Chandran MD (2017a) Modelling the forest transition in Central Western Ghats, India. Spat Inf Res 25(1):117–130. https://doi.org/10.1007/s41324-017-0084-8

Ramachandra TV, Vinay S, Bharath S, Shashishankar A (2018a) Focus: ecology and evolution: ecohydrological footprint of a River Basin in Western Ghats. Yale J Biol Med 91(4):431–444
Ramachandra TV, Bharath S (2019a) Global warming mitigation through carbon sequestrations in the Central Western Ghats. Remote Sens Earth Syst Sci 2(1):39–63. https://doi.org/10.1007/s41976-019-0010-z
Ramachandra TV, Bharath S, Vinay S (2019a) Visualisation of impacts due to the proposed developmental projects in the ecologically fragile regions-Kodagu district, Karnataka. Progress Disaster Sci 100038
Szymura TH, Szymura M, Zając M, Zając A (2018) Effect of anthropogenic factors, landscape structure, land relief, soil and climate on risk of alien plant invasion at regional scale. Sci Total Environ 626:1373–1381. https://doi.org/10.1016/j.scitotenv.2018.01.131
Wan Z, Zhang Y, Zhang Q, Li Z (2002) Validation of the land-surface temperature products retrieved from Terra moderate resolution imaging spectroradiometer data. Remote Sens Environ 83(1–2):163–180. https://doi.org/10.1016/s0034-4257(02)00093-7

Vegetation Characteristics

Abbas SZ, Kousar A, Razzaq S, Saeed A, Alam M, Mahmood A (2018) Energy management in South Asia. Energ Strat Rev 21:25–34
Ali MB, Saidur R, Hossain MS (2011) A review on emission analysis in cement industries. Renew Sustain Energy Rev 15(5):2252–2261
Batjes NH (1996) Total carbon and nitrogen in the soils of the world. Eur J Soil Sci 47(2):151–163
Bharath S, Rajan KS, Ramachandra TV (2013b) Land surface temperature responses to land use land cover dynamics. Geoinf Geostat Overvi 1(4)
Blunden J (2017) Long-lived greenhouse gases [in “State of the Climate in 2016”]. Bull Am Meteor Soc 98(8):43–46
Chhabra A, Palria S, Dadhwal VK (2003) Soil organic carbon pool in Indian forests. For Ecol Manag 173(1–3):187–199
Dadhwal VK, Singh S, Patil P (2009) Assessment of phytomass carbon pools in forest ecosystems in India. NNRMS Bullet 41–57
Devagiri GM, Money S, Singh S, Dadhawal VK, Patil P, Khaple A, Devakumar AS, Hubballi S (2013) Assessment of above ground biomass and carbon pool in different vegetation types of south western part of Karnataka, India using spectral modeling. Trop Ecol 54(2):149–165
Dixit AK, Birthal PS (2016) Greenhouse gases emission from livestock production system of India: an actual consumption approach. Indian J Anim Sci 86(11):1331–1336
Do TV, Trung PD, Yamamoto M, Kozan O, Thang NT, Thuyet DV, Thang HV, Phuong NTT, Khuong NV, Cam NV (2018) Aboveground biomass increment and stand dynamics in tropical evergreen broadleaved forest. J Sustain for 37(1):1–14
Eggleston S, Buendia L, Miwa K, Ngara T, Tanabe K (eds) (2006) 2006 IPCC guidelines for national greenhouse gas inventories, vol 5. Institute for Global Environmental Strategies, Hayama, Japan
EMPRI (2015) Karnataka state action plan on climate change 1st Assessment report, Department of Ecology & Environment, Government of Karnataka, The Energy and Resources Institute (TERI), Bengaluru, p 366
Gallardo M, Gómez I, Vilar L, Martínez-Vega J, Martín MP (2016) Impacts of future land use/land cover on wildfire occurrence in the Madrid region (Spain). Reg Environ Chang 16(4):1047–1061
Ghazoul J, Butler RA, Mateo-Vega J, Koh LP (2010) REDD: a reckoning of environment and development implications. Trends Ecol Evol 25(7):396–402
Global Carbon Project (2017) http://www.globalcarbonatlas.org/en/CO2-emissions
Ghose MK, Paul R, Banerjee SK (2004) Assessment of the impacts of vehicular emissions on urban air quality and its management in Indian context: the case of Kolkata (Calcutta). Environ Sci Policy 7(4):345–351

Gupta PK, Sahai S, Singh N, Dixit CK, Singh DP, Sharma C, Tiwari MK, Gupta RK, Garg SC (2004) Residue burning in rice–wheat cropping system: causes and implications. Curr Sci 1713–1717
Guthrie G, Kumareswaran D (2009) Carbon subsidies, taxes and optimal forest management. Environ Resour Econ 43(2):275–293
Gupta D, Singh SK (2012) Greenhouse gas emissions from wastewater treatment plants: a case study of Noida. J Water Sustain 2(2):131–139
Houghton RA, Hackler JL (1999) Emissions of carbon from forestry and land-use change in tropical Asia. Glob Chang Biol 5(4):481–492
IPCC (2000). IPCC special report- scenarios. Cambridge University Press, Cambridge
IPCC (2007) Intergovernmental Panel on Climate Change (IPCC). Climate Change 2007: The Physical Science Basis: Working Group I Contribution to the Fourth Assessment Report of the IPCC. Cambridge University Press, Cambridge, UK 2007. http://www.ipcc.ch/ipccreports/ar4-wg1.htm
Jose VS, Sejian V, Bagath M, Ratnakaran AP, Lees AM, Al-Hosni YA, Sullivan M, Bhatta R, Gaughan JB (2016) Modeling of greenhouse gas emission from livestock. Front Environ Sci 4:27
Kaul M, Mohren GMJ, Dadhwal VK (2010) Carbon storage and sequestration potential of selected tree species in India. Mitig Adapt Strat Glob Chang 15(5):489–510
KFD (2017) Karnataka forest department annual report 2016–2017. Government of Karnataka, India
Lal R (1999) Soil management and restoration for C sequestration to mitigate the accelerated greenhouse effect. Prog Environ Sci 1(4):307–326
Lal R (2004) Soil carbon sequestration impacts on global climate change and food security. Science 304(5677), 1623–1627
Leach M, Scoones I (2013) Carbon forestry in West Africa: the politics of models, measures and verification processes. Glob Environ Chang 23(5):957–967
Lillesand T, Kiefer RW, Chipman J (2015) Remote sensing and image interpretation. Wiley
Liu X, Bae J (2018) Urbanization and industrialization impact of CO2 emissions in China. J Clean Prod 172:178–186
Madhu V, Negandhi D, Wahal AK, Kumar R, Giridhar AK, Kumar A (2014) Revision of rates of NPV applicable for different class/category of forests. Bhopal, India
Meenakshi Bai R, Vidyavathi GY, Yadahalli GS, Rudramurthy HV, Rajexh NL (2018) Land resource characterization and soil classification of Pannur North-3 microwatershed using RS and GIS in Manvi Taluk, Raichur district J Pharm Phytochem 7 (1):671–78
Mundaca G (2017) How much can CO2 emissions be reduced if fossil fuel subsidies are removed? Energy Econ 64:91–104
Murthy IK, Gupta M, Tomar S, Munsi M, Tiwari R, Hegde GT, Ravindranath NH (2013) Carbon sequestration potential of agroforestry systems in India. J Earth Sci Climate Chang 4(1):1–7
Murthy IK, Bhat S, Sathyanarayan V, Patgar S, Beerappa M, Bhat PR, Bhat DM, Ravindranath NH, Khalid MA, Prashant M, Iyer S (2015) Biomass and carbon stock dynamics in tropical evergreen and deciduous forests of Uttara Kannada District, Western Ghats, India. Glob J Sci Front Res H Environ Earth Sci 15(5):21–29
Nair PRK, Nair VD, Kumar BM, Haile SG (2009) Soil carbon sequestration in tropical agroforestry systems: a feasibility appraisal. Environ Sci Policy 12(8):1099–1111
Ni Y, Eskeland GS, Giske J, Hansen JP (2016) The global potential for carbon capture and storage from forestry. Carbon Balanc Manag 11(1):3
Pan Y, Birdsey RA, Fang J, Houghton R, Kauppi PE, Kurz WA, Phillips OL, Shvidenko A, Lewis SL, Canadell JG, Ciais P (2011b) A large and persistent carbon sink in the world's forests. Science 333(6045):988–993
Pandey R, Rawat GS, Kishwan J (2011) Changes in distribution of carbon in various forest types of India from 1995–2005. Silva Lusitana 19(1):41–54
Pongratz J, Reick CH, Houghton RA, House JI (2014) Terminology as a key uncertainty in net land use and land cover change carbon flux estimates. Earth Syst Dyn 5(1):177–195

Rai SN (1986) Ecological studies on four rainforests in Karnataka, India: I. Environment, structure, floristics and biomass. J Ecol 439–454
Rajan K, Natarajan A, Kumar KA, Badrinath MS, Gowda RC (2010) Soil organic carbon–the most reliable indicator for monitoring land degradation by soil erosion. Curr Sci 823–827
Ramachandra TV, Subramanian DK, Joshi NV, Gunaga SV, Harikantra RB (2000a) Domestic energy consumption patterns in Uttara Kannada district, Karnataka state, India. Energy Convers Manag 41(8):775–831
Ramachandra TV, Joshi NV, Subramanian DK (2000b) Present and prospective role of bioenergy in regional energy system. Renew Sustain Energy Rev 4(4):375–430
Ramachandra TV, Chandran MDS, Bhat HR, Dudani S, Rao GR, Boominathan M, Mukri V, Bharath S (2010) Biodiversity, ecology and socio-economic aspects of Gundia River basin in the context of proposed mega hydroelectric power project. Energy and Wetland Research Group, Centre for Ecological Sciences, Indian Institute of Science, CES Technical Report 122, Bangalore, India
Ramachandra TV, Shwethmala (2012) Decentralised carbon footprint analysis for opting climate change mitigation strategies in India. Renew Sustain Energy Rev 16(8):5820–5833
Ramachandra TV, Aithal BH, Sreejith K (2015) GHG footprint of major cities in India. Renew Sustain Energy Rev 44:473–495
Ramachandra TV, Soman D, Naik AD, Chandran MDS (2017b) Appraisal of forest ecosystems goods and services: Challenges and opportunities for conservation. J Biodivers 8(1):12–33
Ramachandra TV, Bharath S, Gupta N (2018b) Modelling landscape dynamics with LST in protected areas of Western Ghats, Karnataka. J Environ Manag 206:1253–1262
Ramachandra TV, Bharath S, Vinay S (2019b) Visualisation of impacts due to the proposed developmental projects in the ecologically fragile regions-Kodagu district, Karnataka. Prog Disaster Sci 3:100038
Ramachandra TV, Bharath S (2019b) Global warming mitigation through carbon sequestrations in the central Western Ghats. Remote Sensing in Earth Systems Sciences 2(1):39–63
Randerson JT, Van der Werf GR, Collatz GJ, Giglio L, Still CJ, Kasibhatla P, Miller JB, White JWC, DeFries, RS, Kasischke ES (2005). Fire emissions from C3 and C4 vegetation and their influence on interannual variability of atmospheric CO2 and δ13CO2. Global Biogeochem Cycles 19(2)
Rao GR, Krishnakumar G, Dudani SN, Chandran MS, Ramachandra TV (2013) Vegetation changes along altitudinal gradients in human disturbed forests of Uttara Kannada, Central Western Ghats. J Biodivers 4(2):61–68
Ravindranath NH, Somashekhar BS, Gadgil M (1997) Carbon flow in Indian forests. Clim Change 35(3):297–320
Ravindranath NH, Ostwald M (2007) Carbon inventory methods: handbook for greenhouse gas inventory, carbon mitigation and roundwood production projects, vol 29. Springer Science & Business Media
Scharlemann JP, Tanner EV, Hiederer R, Kapos V (2014) Global soil carbon: understanding and managing the largest terrestrial carbon pool. Carbon Manag 5(1):81–91
Scholz F, Hasse U (2008) Permanent wood sequestration: the solution to the global carbon dioxide problem. ChemSusChem: Chem Sustain Energy Mater 1(5):381–384
Sheikh MA, Kumar M, Bussman RW, Todaria NP (2011) Forest carbon stocks and fluxes in physiographic zones of India. Carbon Balance Manag 6(1):15
Singh A, Gangopadhyay S, Nanda PK, Bhattacharya S, Sharma C, Bhan C (2008) Trends of greenhouse gas emissions from the road transport sector in India. Sci Total Environ 390(1):124–131
Singhal KK, Mohini M, Jha AK, Gupta PK (2005) Methane emission estimates from enteric fermentation in Indian livestock: Dry matter intake approach. *Current Science*, pp.119–127.
Singh P, Kansal A (2018) Energy and GHG accounting for wastewater infrastructure. Resour Conserv Recycl 128:499–507
Sun J, Peng H, Chen J, Wang X, Wei M, Li W, Yang L, Zhang Q, Wang W, Mellouki A (2016) An estimation of CO2 emission via agricultural crop residue open field burning in China from 1996 to 2013. J Clean Prod 112:2625–2631

System of Environmental Economic Accounting (2012) Central Framework. International Monetary Fund, 2017
Yu S, Wei YM, Guo H, Ding L (2014) Carbon emission coefficient measurement of the coal-to-power energy chain in China. Appl Energy 114:290–300
Zhang B, Tian H, Ren W, Tao B, Lu C, Yang J, Banger K, Pan S (2016) Methane emissions from global rice fields: Magnitude, spatiotemporal patterns, and environmental controls. Global Biogeochem Cycles 30(9):1246–1263

Soil

Chandrakala M, Srinivasamurthy CA, Parama VRR, Bhaskar S, Sanjeev K, Naveen DV (2017) Phosphorus fractions-keys to soil based P management. Int J Current Microbiol Appl Sci 6(11):281–294
Kulkarni V, Ramachandra TV (2006) Environmental management, commonwealth of learning, Canada and Indian Institute of Science, Bangalore, Printed by Capital Publishing Company, New Delhi. (Reprinted in 2009 by TERI Press, New Delhi)
Ramachandra TV (2006) Soil and groundwater pollution from agricultural activities, commonwealth of learning, Canada and Indian Institute of Science, Bangalore, Printed by Capital Publishing Company, New Delhi. (Reprinted in 2009 by TERI Press, New Delhi)
Soils of Karnataka for optimising land use, NBSS Publ.47b (1998). ISBN:81-85460-45-0

Valuation of Ecosystem Services, Karnataka State, India

Abstract Ecosystem services are the contributions of ecosystems to the benefits that are used in economic and other human activities. Further, ecosystem services encompass all forms of interaction between ecosystems and people, including in situ and remote interactions. The supply of an ecosystem service is associated with an ecosystem structure or process or a combination of ecosystem structures and processes that reflect the biological, chemical, and physical interactions among ecosystem components. Ecosystem services are broadly categorised as (i) provisioning services are those ecosystem services representing the contributions to benefits that are extracted or harvested from ecosystems, (ii) regulating and maintenance services are those ecosystem services resulting from the ability of ecosystems to regulate biological processes and to influence climate, hydrological and

T. V. Ramachandra et al., *Natural Capital Accounting and Valuation of Ecosystem Services, Karnataka State, India*, https://doi.org/10.1007/978-981-97-2405-5_3

biochemical cycles, and thereby maintain environmental conditions beneficial to individuals and society, (iii) cultural services are the experiential and intangible services related to the perceived or actual qualities of ecosystems whose existence and functioning contributes to a range of cultural benefits. A range of other benefits, for example, concerning relational and intrinsic values, are not captured in the above categories. India is trying to accelerate economic growth and relax environmental laws, and there is tremendous pressure to divert natural systems to other uses. Hence, there is a pressing need to undertake the natural capital accounting and valuation of the ecosystem services, especially intangible benefits, provided by ecosystems in India. The value of all ecosystem services, including the degradation costs, needs to be understood to develop appropriate policies toward ecosystem conservation and sustainable management. Scientific efforts during the past decade have refined the understanding of ecosystem function and demonstrated the links between functions and the provision of ecosystem services. This knowledge needs to be communicated effectively to decision-makers and the public, which will lead to developing policies that adequately consider the trade-offs between the conservation of ecosystems and natural resources and economic growth. There is an urgent policy need for more comprehensive assessments of the natural capital of ecosystems, which will aid in comparing these aggregate values with the opportunity cost of this land. Policymakers need such information to gain support for conservation funding, engage local communities, and develop market-based instruments for conservation. This necessitates accounting for the natural capital found in the forest ecosystems (among others) and incorporating their economic worth added to the measurement of the wealth of a region. The valuation of ecosystem services of terrestrial ecosystems (forests and agriculture) is implemented district-wise for Karnataka State, India, as per the validated protocol—System of Environmental-Economic Accounting (SEEA EA; SEEA.un.org), which constitutes the statistical framework for natural capital accounting. According to this protocol, ecosystem services are defined as the contributions of ecosystems to the benefits that are used in economic and other human activities. The valuation of ecosystem services (VES) allows for adjusted national accounts, which reflect the output of ecosystem services as well as the depletion of natural resources and the degradation costs (externalized costs of the loss of ecosystem services) of ecosystems in economic terms, which will help raise awareness and provide a quantitative tool to evaluate the sustainability of policies. It provides an unbiased and dependable national framework to value so far unaccounted ecosystem benefits and helps develop meaningful policy interventions. Services of the ecosystem services were quantified considering only the contribution of the ecosystem to the benefit, through the residual value method by taking the gross value of the final marketed good to which the ecosystem service provides input and then deducting the cost of all other inputs, including labour, produced assets, and intermediate inputs. The current chapter focuses on ecosystem services in forest and agricultural ecosystems for 2005 and 2019. Values of 2005 were adjusted through the consumer price index or gross domestic product (GDP) deflator. These values reflect the actual measures of ecosystem services, which could be compared with ecosystem services of 2019. A comparison of the values of services of 2019 with

2005 highlights that there has been a considerable decline in ecosystem services in Karnataka. A 28.5% reduction in provisioning services (51.6% reduction in forest ecosystems), a 21% reduction in regulating services (mainly in forest ecosystems—27.1% reduction), and a 1.9% reduction in cultural services from 2005 to 2019. Ecosystem services were aggregated to compute the Total Ecosystem Supply Value (TESV). This aggregate measure is also referred to as *Gross Ecosystem Product (GEP), which equals the sum of all final ecosystem services* (i.e., *by monetary values of those services) from ecosystem assets.* The TESV of forest and agricultural ecosystems in Karnataka was 3620 billion INR in 2005 (forest ecosystems: 2841 billion INR and agricultural ecosystems: 779 billion INR). However, TESV declined in 2019 to 2912 billion rupees, with forest ecosystems driving this 35% decline in TESV. The TESV was also compared to the GDP of Karnataka, which is about 10128 billion rupees. The TESV of the forest ecosystem is equivalent to 18.1% of the GDP, and the TESV from agriculture ecosystems is equivalent to about 10.6% of the GDP in Karnataka. There has been a 35.4% reduction in the TESV of forest ecosystems from 2005 to 2019, mainly due to the degradation of ecosystems. The decline in the TESV highlights the degradation of forest ecosystem assets from 2005 to 2019, as shown by the reduction of ecosystem extent and ecosystem condition (Ramachandra et al. 2021a, b). The decrease in value is also demonstrated by a fall in the net present value (NPV) of expected future returns of the ecosystem services supplied by forest ecosystem assets. The NPV of the assessed ecosystems based on 2005 ecosystem flows is about 93130 billion INR (forest ecosystem: 73099 billion INR, agriculture ecosystem: 20031 billion INR). However, the NPV of ecosystems in Karnataka, based on 2019 flows, indicates 74938 billion INR (forest ecosystem: 47214 billion INR, agriculture ecosystem: 27,724 billion INR). This highlights that there has been a decline of 35.4% in the asset value of forest ecosystems with the transition of forest ecosystems to croplands or horticulture (agriculture ecosystems), which is correlated to an increase in NPV of agriculture ecosystems by 38%. Ecosystem accounts make the value of ecosystem services visible, allowing them to be internalized into decision-making. This enables an assessment of trade-offs between economic development and environmental conservation and restoration, resulting in better-informed decisions. It also strengthens the economic case for conserving forests in states in India and developing countries where there can be great pressure to relax forest laws and divert forests to non-forest uses without proper consideration of the sustainability of such actions. The ecosystem services computed for Karnataka State also support the viability of markets for ecosystem services. The development of such markets requires additional institutional reforms, such as changes with respect to property rights and reforms in land and labour markets. The main policy challenge of the future concerns is to promote conservation and develop such markets so that those bearing the cost of conservation can be adequately compensated. The current endeavour of the valuation of ecosystem services done in Karnataka State can be replicated in other states so that the accounts can play a vital role in conservation planning and ecosystem-based management across India. This involves (i) inventorying, mapping, and identifying ecologically vital regions and (ii) quantification of ecological services (both direct and indirect) of these regions, which would help in evolving strategies for the conservation of ecosystems to support people's livelihood. Improved knowledge of biophysical links and pre-existing incentive structures experiment would allow the establishment of compensation schemes, which requires:

(i) Strengthening biophysical research on ecosystem services, the loss of which would seem to have the highest economic value potential (e.g., changes in the climatic, hydrologic regime, etc.).
(ii) Inventorying, mapping, and monitoring ecosystems' spatial extent and conditions by applying advanced geoinformatics (spatial technologies with temporal remote sensing data) with artificial intelligence (AI) and machine learning algorithms.
(iii) Promoting valuation studies and internalizing the results in regional policies for revealing current incentives, i.e., the existing distribution of net ecosystem benefits/opportunity costs across stakeholders.
(iv) Developing strategies for developmental activities based on optimal land-use (LU) policies.

The ecosystem management objectives need to be defined by societal demands in relation to the ecology of the ecosystem, especially in the case of ecosystems that are strongly influenced by human use. The key role of science in such human-dominated systems is to provide insights into the social, economic, and environmental benefits of various levels of biodiversity. This can be achieved through innovative ecosystem management options and by effectively communicating the research results to decision-makers and other stakeholders. The economic valuation of ecosystem services and biodiversity can help clarify trade-offs among conflicting environmental, social, and economic goals in the development and implementation of policies.

1 Introduction

Humans depend on the environment for their basic needs, such as food, fuel, minerals, water, air, etc. In developing countries, nearly 80% of the labor force is engaged in agricultural or resource-based activities, contributing significantly to the GDP (World Bank 1998, 2001). Over the years, the dependency on natural resources has led to their degradation and depletion owing to the unsustainable practices involved in their extraction. Burgeoning unplanned development activities to cater to the demands of the increasing population has put tremendous pressure on the natural resources, leading to environmental degradation (Kulkarni and Ramachandra 2009). An increased surge in developmental and technological activities over the last two decades, with no regard to their ecological implications, has led to indiscriminate disposal of wastes (liquid and solid), contributing to the degradation of natural ecosystems. This has resulted in a substantial and largely irreversible loss in the diversity of life on Earth (MEA 2005). And yet, unsustainable utilization of land and other natural resources persists, despite the increasing understanding of the impacts that human activities have on the environment (Euliss Jr et al. 2010). Linkages between the health of the environment and the sustenance of humankind make it imperative to maintain a balance considering the carrying capacity of the environment and the availability of natural resources. Conservation of natural ecosystems has long-term benefits for humans in utilitarian terms through their provision of food, timber, minerals, and a variety of valuable resources that have provided the backbone for economic development. Going beyond utilitarian values, natural ecosystems have also been a source for maintaining gene pools, biodiversity, and other potentially useful factors that are of indirect use to humans. Hence, the intrinsic, anthropocentric, instrumental, and relational values of ecosystems should be considered in the policy design as well as resources exploited for human settlement, food, and energy production.

In this regard, a statistical framing of data on ecosystems plays a vital role in incorporating at least some parts of the wider value of ecosystems as a regular component of decision-making. The SEEA Ecosystem Accounting (SEEA EA) provides such a framework. Adopted by the United Nations Statistical Commission in 2021, the SEEA EA constitutes an integrated and comprehensive statistical framework for organizing ecosystem data, measuring ecosystem services, and tracking ecosystem changes. In addition, the data on ecosystems is linked to information on economic and other activities, as the SEEA EA uses many of the same concepts, definitions, and classifications as the System of National Accounts (SNA). Finally, the SEEA EA enables high-quality and consistent measurement over time by using agreed concepts, definitions, and classifications. Providing relevant time series and trend data on the environment-economy nexus is crucial for effective policy design, decision-making, and evaluation.

The dilemma associated with rapid land-use changes for accommodating the growing demand for natural resources is impacting and degrading the ecosystems (Foley et al. 2005; Ramachandra et al. 2007). The ecosystem service approach systematically captures the full range of environmental impacts and offers a way to understand and deal with the feedback created when ecosystems are used to meet humankind's needs (Rodríguez et al. 2006). The objectives of the current study are to (i) assess the ecosystem services values for the forest, agriculture, and horticulture ecosystem types, district-wise for Karnataka State, India (ii) the computation of the total ecosystem supply value (TESV), and (iii) net present value (NPV) of ecosystem assets. The current chapter focuses on data for the years 2005 and 2019.

It should be noted that the SEEA EA focuses on values of anthropocentric origin—i.e., values that are centered on human beings. Further, the measurement focus of the SEEA EA is on instrumental (is the value attributed to something as a means to achieve a particular end) or use values because these interactions are most readily quantified and because, from a monetary valuation perspective, these values are most readily reflected in monetary terms. From a policy perspective, the focus on anthropocentric, instrumental values may also be considered of high relevance since they concern the types of human interactions with the environment that can place the most pressure on ecosystems (SEEA EA 2021).

The outline of this chapter is as follows: the following section (Sect. 1) defines ecosystem services and accounting for ecosystem services in the context of the SEEA EA. Section 2 explains data sources, and Sect. 3 presents methods adopted for valuation. Section 4 describes the results of ecosystem services accounting for forest ecosystems and agriculture ecosystems. Section 5 concludes with recommendations. Ecosystem-wise services (physical as well as monetary) computed district-wise are presented in Annexure A for forest ecosystems.

1.1 Ecosystem Services

In the SEEA EA, ecosystem services are the contributions of ecosystems to the benefits that are used in economic and other human activities. In this definition, use incorporates direct physical consumption, passive enjoyment, and indirect receipt of services.

An ecosystem services approach to foster an understanding of the relationship between humans and the environment has been emphasized in various initiatives, including The Economics of Ecosystems and Biodiversity initiative (Costanza et al. 1997, 2014; Markandya et al. 2002; MEA 2005; Van der et al. 2010; TEEB 2010a, b; Ten Brink 2011; De Groot et al. 2012, 2017, 2020; Perelet et al. 2014), the Mapping and Assessment of Ecosystems and their Services (MAES) framework (Maes et al. 2013, 2016, 2018, 2020); the Natural Capital Project at Stanford University; the Integrated System for Natural Capital Accounting (INCA) project (Vallecillo et al. 2019); and the Intergovernmental Science-Policy Platform on Biodiversity and Ecosystem Services (IPBES) (Diaz et al. 2015), etc.

Most resource management decisions are influenced by ecosystem services (ESs) entering markets; thus, the non-marketed benefits often remain unaccounted. Both renewable resources (water supply, air quality, etc.) and non-renewable resources (mineral deposits, some soil nutrients, fossil fuels, etc.) are capital assets and provide the backbone for numerous economic activities that account for the development of a region. Yet, traditional national accounts do not include measures of resource depletion or their degradation. GDP, a measure of the current economic well-being of a population based on the market exchange of material well-being, will indicate resource depletion/degradation only through a positive gain in the economy and will not represent the decline in these assets (wealth) at all. Thus, the existing GDP growth percentages used as yardsticks to measure the development and well-being of citizens in decision-making processes are substantially misleading, and yet they are being used (De Groot et al. 2002; Haripriya et al. 2006). GDP cannot be a true measure of the country's sustained economic wealth and cannot be a proxy for understanding its future economic well-being. Quantitative evidence on the economic value of such assets is necessary for most of these services, most of which are not traded in the markets and hence do not have a market value. The monetary valuation of ecosystem services can help in building a better understanding of their influence on well-being and can further facilitate information-driven decisions and policy reforms that align with the Sustainable Development Goals (SDGs). Environmental accounting systems seek to determine a region's environmental and economic assets and can be used to assess whether economic development is consistent with sustainable development or to help ensure optimal use of natural resources and the environment. Recent efforts, especially the System of Environmental-Economic Accounting, Central Framework (SEEA CF), and Ecosystem Accounting (SEEA EA), aim to extend and integrate the national accounts for environmental and ecosystem assets (SEEA 2017; SEEA EA 2021).

Ecosystem services encompass all forms of interaction between ecosystems and people, including in situ and remote interactions. The supply of an ecosystem service is associated with an ecosystem structure or process or a combination of ecosystem structures and processes that reflect the biological, chemical, and physical interactions among ecosystem components. In the SEEA EA, ecosystem services are broadly categorized as (i) provisioning services, which are those ecosystem services representing the contributions to benefits that are extracted or harvested from ecosystems; (ii) regulating and maintenance services, which are those ecosystem services resulting from the ability of ecosystems to regulate biological processes and to influence climate, hydrological and biochemical cycles, and thereby maintain environmental conditions beneficial to individuals and society; and (iii) cultural services, which are experiential and intangible services related to the perceived or actual qualities of ecosystems whose existence and functioning contributes to a range of cultural benefits. There is a range of other benefits, for example, concerning relational and intrinsic values, that are not captured in the above categories.

Ecosystem services serve as the connecting concept between ecosystem assets (contiguous spaces of a specific ecosystem type, i.e., individual ecosystems) and the production and consumption activities as per the SEEA EA. The key concepts of the SEEA EA related to ecosystem services concern (i) the supply of ecosystem services to users; and (ii) the contribution of ecosystem services to benefits (i.e., the goods and services ultimately used and enjoyed by people and society). Further, ecosystem services encompass all forms of interaction between ecosystems and people, including both in situ and remote interactions. A key feature of ecosystem accounting is its capacity to integrate spatially referenced data about ecosystems, i.e., data about the location, size, and condition of ecosystems within a given area and how these are changing over time. Recording these stocks and changes in stocks in a coherent and mutually exclusive manner supports the derivation of indicators. Understanding the size and location of ecosystems also supports the measurement of ecosystem conditions and the quantification and valuation of many ecosystem services, the flows of which will vary from ecosystem to ecosystem.

1.2 Ecosystem Services Accounting—Physical as well as Monetary (Based on Residual Value Techniques)

Ecosystem services contribute to economic welfare in two ways—(i) contributions to the generation of income and wellbeing and (ii) the prevention of damages that inflict costs on society. Both types of benefits are accounted for in policy appraisal with a broader focus on valuing the benefits provided by ecosystems. Policy options that enhance the natural environment are more likely to be considered, demonstrating that investing in natural capital can make economic sense. There is considerable complexity in understanding and assessing the underlying links between a policy, its effects on ecosystems and related services, and valuing its impacts in economic terms. Collaboration between those working in policy, science, and economics disciplines is essential in implementing this approach in practice. The critical importance of the links to scientific analysis, which form the basis for valuing ecosystem services, needs to be recognized. The SEEA EA emphasizes the need to consider the ecosystem as a whole and underlines those changes or impacts on one part of an ecosystem have consequences for the whole system. Therefore, considering the scale and scope of the services to be valued is vital to arrive at any meaningful value.

The key stages in the valuation of ecosystem services in the SEEA EA are (i) setting a scope and baseline through ecosystem extent and condition accounts, (ii) physical quantification of services, and (iii) valuation of ecosystem services, including changes over time. Monetary accounts can further inform a qualitative assessment of the potential impacts of policy options on ecosystem services and quantification of the impacts of policy options on specific ecosystem services, and evaluation of the effects on human welfare.

There is a growing interest in ecosystem services (ESs), and ES conservation management strategies and the valuation of ecosystem services would help equip society with the means to incorporate the values of nature into decision-making at all levels. It also provides a baseline for evaluating management changes. This helps evaluate and prioritize different policies, evaluate potential trade-offs in management decisions, and assess the damages caused by natural disturbances. Apart from these, other benefits are (i) enhanced communication with stakeholders about the economic benefits and costs of potential changes in forest management, as communities' preferences for different ecosystem services may be affected by estimates of economic performance; (ii) a baseline for evaluating management changes. This helps policymakers to take into account the value of ecosystems in development planning and resource allocations and take adequate measures for conservation to ensure the sustenance of the flow of ecosystem services.

The United Nations Statistical Commission (UNSC) endorsed the SEEA-Experimental Ecosystem Accounting (SEEA-EEA in 2013 (System of Environmental-Economic Accounting-Experimental Ecosystem Accounting) as the basis for commencing testing and further development of a common statistical framework for ecosystem accounting. The UNSC also encouraged the use and experimentation of the SEEA-EEA by international and regional agencies (SEEA

2017; SEEA EA 2021). The various research publications from the scientific community on the valuation of ecosystem services have substantially grown to address several challenges and propose common frameworks. The expansion of a worldwide research base with a multidisciplinary scope of ecosystem services is resolving issues that arise in quantification, terminology, classification systems, research methods, and reporting requirements (Polasky et al. 2015; Mengist and Soromessa 2019).

The ecosystem accounts in this report have been developed for Karnataka State, India, as per the SEEA Ecosystem Accounting (SEEA EA) framework.

The objective of the current analysis is to develop the ecosystem services flow accounts in physical and monetary terms, as well as the monetary asset account. The ecosystem service accounts were developed using spatially explicit estimates of the supply of ecosystem services in physical terms and their contributions to benefits in monetary terms for major ecosystems (forests and agriculture). The following set of services is covered:

(i) Provisioning services

- forest ecosystems—timber, bamboo, fodder, fuelwood, non-timber forest produce, fish and other aquatic products provisioning services, medicine, water supply service, and genetic material service for forest ecosystems
- agriculture ecosystems—food (cereals, pulses, oilseeds, vegetables, and commercial crops), fodder, and wood

(ii) Regulating services include global climate regulation services/carbon sequestration, local (micro and meso) climate regulation services, pollination service, soil conservation, groundwater recharge, water purification, waste treatment (for forest ecosystem), carbon fixation, soil carbon, ground water recharge, nitrogen fixation, soil fertility, remediation—organic and inorganic materials, genetic diversity, biological control (for agriculture ecosystem), air filtration services, and

(iii) Cultural services include aesthetic, recreational, spiritual and historical, artistic and cultural, education, scientific and research.

2 Data

Ecosystem extent account: An important foundation for estimating ecosystem services is the ecosystem extent account (Ramachandra et al. 2021a), which is presented in Chap. 1.

Forest ecosystems: Forest ecosystems in Karnataka are managed by the Karnataka Forest Department, and decentralized administration is practiced through (i) forest circles (note: A forest circle is a term used to signify an area containing one or more (usually) demarcated and (usually) protected or resource-managed forests, for administration and coordination, at decentralized levels), and (ii) divisions. The data

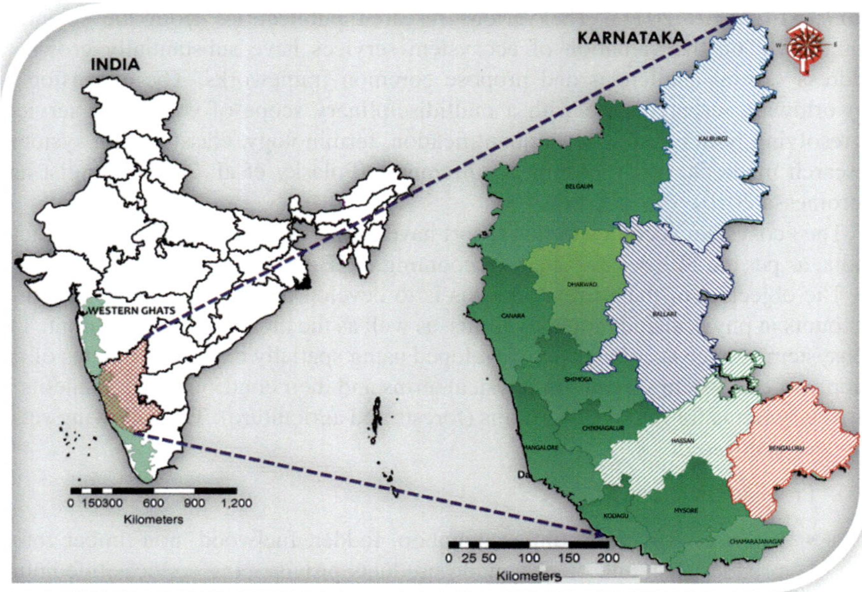

Fig. 1 Karnataka state, India, with forest circles. *Note* Circles shaded in green are part of the Western Ghats (a global biodiversity hotspot)

for provisioning services of forest ecosystems for five years intervals (2001–2005 and 2014–2019) were collected from forest circles. The Karnataka state has 13 forest circles for decentralized administration, and the spatial extent of forest circles is depicted in Fig. 1. The spatial extent of protected areas in each circle and district are listed in Table 1. Global biodiversity hotspot—Western Ghats spread across circles—Canara, Chikmagalur, Shimoga, Mangalore, Kodagu, Chamarajanagar. The state has a protected area network of five national parks (2431.3 km^2), and 21 wildlife sanctuaries (3887.83 km^2), covering nearly 16% of the forest area.

Table 2 lists the data pertaining to the forest ecosystems (with the seigniorage values/residual value of goods used) for computing ecosystem services. Seigniorage value (Haslag 2020) is the revenue received by the government after deducting expenses (costs of labor, produced assets, and intermediate inputs) from the auction price of outputs (benefits). Table 3 lists the data (with source details) used to value agricultural ecosystem services.

Table 1 The districts covered in each forest circle of Karnataka State

	Circle name	District	Protected area-Ha
1	Canara	Uttara Kannada	175,937
2	Mangalore	Dakshina Kannada; Udupi	170,703
3	Kodagu	Kodagu	109,825
4	Chamarajanagar	Chamarajanagar	273,667
5	Belgaum	Belgaum; Bagalkot; Vijayapura	10,973
6	Dharwad	Dharwad; Haveri; Gadag	6,310
7	Shimoga	Shivamoga; Davanagere (Channagiri Taluk); Chikmagalur (Tarikere)	84,976
8	Chikmagalur	Chikmagalur	45,450
9	Mysore	Mysore; Mandya	105,278
10	Kalaburagi	Kalaburagi (Gulbarga); Raichur; Yadgir; Bidar	0
11	Ballari/Bellary	Ballari; Davanagere; Chitradurga; Koppala	4,793
12	Hassan	Hassan; Tumkur	0
13	Bengaluru	Bengaluru (Rural); Bengaluru (Urban); Ramnagara; Kolar; Chikballapur	25,513

Table 2 Data used for computing ecosystem services with the details of data sources

Services	Services and monetary benefits	Source
Ecosystems type: forest ecosystems		
Provisioning services	Provisioning services: the data included (i) service-wise (timber, bamboo, non-timber forest produce, fodder) quantity extracted for the two time periods (2014–2019 and 2001–2005) years and (ii) seigniorage rate as per forest department records for respective provisioning services (timber and other forest goods) for corresponding years	Data pertaining to the provisioning services at the circle level of the supply in physical terms and their benefits in monetary terms were collected for the respective circles from The Karnataka Forest Department offices at respective circles
	Fuelwood	Fuelwood required per person is estimated based on the socio-economic survey carried out in select taluks
	Fish and other aquatic products provisioning services (in streams, reservoir, etc.—which are inseparable parts of the district's forest area)	District wise inland fish catch from districts, Department of Inland Fisheries, GoK https://fisheries.karnataka.gov.in/english Directorate of Economics and statistics, GoK https://des.karnataka.gov.in/english

(continued)

Table 2 (continued)

Services	Services and monetary benefits	Source
	Water supply (domestic, irrigation, industries, hydro-electricity generation)	Land use data (Ramachandra et al. 2021a). Long term meteorological data such as temperature, rainfall, solar radiation were collected from online portals (Worldclim http://www.worldclim.org/), FAO (http://www.fao.org), Climate Research Unit, University of East Anglia http://www.cru.uea.ac.uk, NASA—Climate change and global warming http://climate.nasa.gov/, http://data.giss.nasa.gov/gistemp/, KSNDMC Karnataka (https://www.ksndmc.org/), the Directorate of Economics and Statistics Karnataka (http://des.kar.nic.in/), India Meteorological Department (https://mausam.imd.gov.in/), Food and Agriculture Organisation (http://www.fao.org). Population census for the years 2011 and 2001 was collected from the Census of India (https://censusindia.gov.in/) Livestock data such as Census (http://www.ahvs.kar.nic.in/), water requirements were collected from the Directorate of Economics and Statistics (http://des.kar.nic.in/), District at a Glance, and through public interviews. Agriculture data such as various crops grown, cropping pattern, water requirement at different growth phases were collected from District at a Glance, public interviews, online portals such as Raitamitra (http://raitamitra.kar.nic.in/), iKisan (http://www.ikisan.com), National Food Security Mission (http://www.nfsm.gov.in, Tamil Nadu Agriculture University, etc. and other published literature Field investigations in select stream catchments were carried out for ten months to understand the intra and inter variability of the hydrologic regime in the Central Western Ghats. The information for ungauged streams was compiled from published literature (Ramachandra et al. 2020) Water demand: field survey, review of published literature, daily per capita water requirement (domestic), water use in agriculture—crop-wise, season-wise, and discussion with subject experts Domestic and irrigation water supply—economic values were compiled from farmers and residents Industrial water uses were compiled from publications—Annual reports of Karnataka Power Corporation Ltd, District at a Glance (allocation of water to industries, revenue, and cost)

(continued)

Table 2 (continued)

Services	Services and monetary benefits	Source
	Medicine Various medicinal plants used by the local people were identified, and the value of medicinal plants per unit area of forest area was extrapolated to different types of forests	Public interview and literature review (Ramachandra et al. 2017; De Groot et al. 2020). Benefits are estimated per hectare as per the access and benefit-sharing data (royalty payments from the Karnataka Biodiversity Board and the Medicinal plant conservation authority)
	Genetic material service The economic value of gene-pool conservation in terms of bioprospecting based on (i) number of medicinal plants found in each district; (ii) number of species of conservation importance in each district, and (iii) all species	The estimate is based on all species in the study region and ecosystem extent. Species details obtained from Karnataka Biodiversity Board (https://kbb.karnataka.gov.in/), Medicinal Plants Conservation Authority (https://ayush.karnataka.gov.in/) and genetic resource per hectare as per a case study from India (Verma et al. 2013)
Regulating services	Global climate regulation services/Carbon sequestration data-Ecosystem extent and type of above-ground biomass (AGB), below-ground biomass (BGB), soil carbon, and net primary productivity (NPP)	(i) Land use information, (ii) quantification of AGB through field measurements of girth and height, and sampling of the locations through transect-based quadrat—a survey carried out across forest types (evergreen, deciduous, scrub forests, etc.) in 10 districts
	Soil conservation and soil fertility—data: soil characteristics, land use characteristics, vegetation characteristics, farming practices, topographic effects, etc. Annual rainfall, monthly rainfall, quick flows, historical climate data Bioclimatic variables, long-term weather data, daily rainfall data	Ecosystem extent assessment (Ramachandra et al. 2021a) presented in the first chapter, Ecosystem condition-soil (Ramachandra et al. 2021b; Ma et al. 2019) presented in the second chapter, IMD, GoI (https://mausam.imd.gov.in/), NASA Portal (https://gpm.nasa.gov/data), Worldclim (https://www.worldclim.org/), KSNDMC (www.ksndmc.org)
	Ground water recharge Precipitation, overland flow, infiltration, evapotranspiration, maximum and minimum temperature along with the solar radiation	Overland flow (runoff)—field measurements—four river basins in Uttara Kannada and two river basins in Shimoga districts using a current meter (water velocity measurement—three consecutive days, monthly), IMD, GoI (https://mausam.imd.gov.in/), NASA Portal (https://gpm.nasa.gov/data), Worldclim (https://www.worldclim.org/), KSNDMC (www.ksndmc.org)

(continued)

Table 2 (continued)

Services	Services and monetary benefits	Source
	Water purification	Economic values of water purification and waste treatment are estimated per hectare as per a case study from India (Verma et al. 2013; Ramachandra et al. 2017)
	Pollination service Ecosystem extent, and type Natural forest regeneration and afforestation (replacement) cost	Ecosystem extent based on land use analyses and literature (Ramachandra et al. 2021a). Comparative assessment of natural regeneration of forest patches (with fencing protection from external pressures (Ray et al. 2015) and afforestation cost. The estimates of natural forest regeneration in all forest types are adjusted according to the forest regeneration in plantations (NAP 2009; Ollerton et al. 2011; Hipólito et al. 2019)
	Air filtration services—extent of forest ecosystem Air filtration regulation service values per hectare is based on published literature from India (Ninan and Kontoleon 2016; Joshi and Negi 2011) Compared with values of global studies, adjusted for GDP (PPP) per capita and corresponding currency exchange rate	Ecosystem extent based on land use analyses (Ramachandra et al. 2021a) and Air filtration regulation service, values per hectare based on published literature from India (Ninan and Kontoleon 2016; Joshi and Negi 2011) which are comparable to global studies–global ecosystem service valuation database (ESVD) for tropical forests, and mangroves https://www.es-partnership.org/wp-content/uploads/2020/08/ESVD_Global-Update-FINAL-Report-June-2020.pdf GDP(PPP) per capita for India—https://data.worldbank.org/indicator/NY.GDP.PCAP.PP.CD?locations=IN Currency exchange rate https://www.xe.com/currencyconverter/convert/?Amount=1&From=USD&To=INR
	Local (micro and meso) climate regulation services—extent of forest ecosystem Per hectare local climate regulation service values based on published literature from India (Ghosh 2020; Verma et al. 2007) Compared with global studies adjusted for GDP (PPP) per capita of the country for which values were estimated and corresponding currency exchange rate	Ecosystem extent based on land use analyses (Ramachandra et al. 2021a), per hectare local climate regulation service values based on published literature from India (Ghosh 2020; Verma et al. 2007), which are comparable to values in the global ecosystem (tropical forests, mangroves) service valuation database (ESVD) https://www.es-partnership.org/wp-content/uploads/2020/08/ESVD_Global-Update-FINAL-Report-June-2020.pdf GDP(PPP) per capita for India—https://data.worldbank.org/indicator/NY.GDP.PCAP.PP.CD?locations=IN Currency exchange rate https://www.xe.com/currencyconverter/convert/?Amount=1&From=USD&To=INR

(continued)

Table 2 (continued)

Services	Services and monetary benefits	Source
Cultural services	Aesthetic—National parks, sanctuaries, waterfalls	Karnataka Forest Department (Uttara Kannada, Shimoga, Chikmagalur Dakshina Kannada, and Kodagu districts) (primary survey–entrance fees (park, recreation spots) × the average number of visitors to the park/recreation spots during 2018, 2019, and 2020), and supplemented with the Indian case studies (Ray et al. 2010; Bharath et al. 2017; Ramachandra et al. 2018c), publications—Districts at glance https://des.karnataka.gov.in/english https://kgis.ksrsac.in/kag/
	Spiritual and historic—distribution of sacred groves (relic forests—protected due to belief and customs)	Devotees perform rituals, and the amount is paid for performing rituals (either visiting the grove or in absentia). Also, donating money during birthday celebrations or in the name of elders (or departed soul) is a practice. Data pertaining to the annual collection and expenses were compiled from the administrative/management committees of select groves in Shimoga, Uttara Kannada, and Kodagu districts. The residual method (annual collection for rituals and deducting costs – priest salary and ritual expenses) was used In groves where annual collection details were not available, the travel cost method is used for valuation, considering the number of visitors (visiting groves) for annual rituals, festivals, and other religious activities. This is done through primary surveys of select groves in Uttara Kannada, Shimoga, and Kodagu districts, and supplemented with case-studies from India using the benefit transfer method (Ramachandra et al., 2012, 2016, 2017; Ray et al. 2014b, 2015)
	Tourism and recreational services	The travel cost method (primary survey – benefits to travel operators, entrance fees (park, recreation spots) × the average number of visitors to the park during 2020) is used, supplemented with the Indian case studies. Benefit transfer method—(Ramachandra et al. 2019b; Badola et al. 2017; Gunarekha and Binoy 2017; Sinclair et al. 2020)

(continued)

Table 2 (continued)

Services	Services and monetary benefits	Source
	Education, science, and research	Researchers need to obtain prior permission from the Forest Department to undertake research (and long-term monitoring). Details of the research, duration, project budget (for field research) and research team were compiled from the Karnataka Forest Department This information is supplemented with the data compilation through discussion with researchers and relevant literature of field-based research (ecology, medicinal plants, etc.) (Chandran et al. 2010; Ray and Ramachandra 2010; Gould et al. 2014; Ray et al. 2014a; Dorji et al. 2019; Kreye et al. 2019)

Table 3 Data used for valuing services from agriculture ecosystems

Agriculture—croplands, horticulture			
Services		Variables	Source
Provisioning services	Cereals, pulses, oilseeds, vegetables, and commercial crops (coconut, areca nut, rubber, etc.)	Administrative boundaries (Ecosystem services are evaluated at taluk and district level)	(KSRSAC 2018)
		Crop yield (crop wise yield per unit agriculture area)	(DSO 2019) Public interviews
		Crop area (integrating with land use data to derive actual crop area at the taluk level)	
		Crop production (crop-wise produce at taluk level for valuation of provisioning service)	
		Moderate-resolution satellite data (land use analysis)	(NRSC 2020)
		Virtual earth (Bhuvan and Google earth for land use analysis)	(NRSC 2016; Google 2020)

(continued)

Table 3 (continued)

Agriculture—croplands, horticulture			
		Ecosystem extent	Ecosystem extent based on land use analyses (Ramachandra et al. 2021a)
		Data pertaining to crop yield, produce, net revenue generated fodder production (cattle feed) per unit area, cost of production	Primary survey: public interaction during 2019–2020, data pertaining to revenue, cost, etc.
		Minimum support price (specified by Government of India), and implemented at Mandi (local market) by the Government of Karnataka) was used to determine the monetary value)	(HOPCOMS 2019; AGRICOOP 2020; Coffee Board 2020; Commodities Online 2020; DMI 2020; KMV 2020; MSP 2020)
		Production cost (to determine net revenue from food crops, used for computing residual value of provisioning services)	(EANDS 2020)
		Agriculture revenue (the contribution of the agriculture sector with state gross district domestic product (GDDP))	(DPPMS 2018)
		Government records and Published literature (Economic values for services – Provisioning, Regulating, Cultural)	(Murali 2010; Nayak et al. 2019; De Groot et al. 2020; NAAS 2020)
	Fodder	Type, quantity, likely use, revenue, costs	Public interviews
	Wood	Type (horticulture crop), quantity, likely uses, revenue, cost of harvesting	Public interviews
Regulating services	Air filtration services	Extent of ecosystem, per hectare air filtration services from India	Literature review—case studies from India—ICAR Indian Council of Agriculture Research (http://naasindia.org)

(continued)

Table 3 (continued)

Agriculture—croplands, horticulture			
	Local (micro and meso) climate regulation services	The extent of an ecosystem, climate control services value per hectare based on published literature from India (Ghosh 2020; Verma et al. 2007), compared with based on global studies. These estimates were adjusted for GDP (PPP) per capita of the country for which values were estimated and the corresponding currency exchange rate	Literature review Per hectare local climate regulation service values based on published literature from India (Ghosh 2020; Verma et al. 2007), Compared with global studies adjusted for GDP (PPP) per capita of the country for which values were estimated and corresponding currency exchange rate—global ecosystem service valuation database (ESVD) https://www.es-partnership.org/wp-content/uploads/2020/08/ESVD_Global-Update-FINAL-Report-June-2020.pdf
	Global climate regulation services/ carbon sequestration	The extent of an ecosystem, per hectare carbon sequestration services in agriculture and horticulture	Literature review—case studies from India (Nayak et al. 2019)
	Soil carbon	The extent of an ecosystem, per hectare—soil carbon storage service values	Literature review—case studies from India (Nayak et al. 2019)
	Water flow (ground water recharge)	The extent of an ecosystem, per hectare—groundwater recharge service values	Literature review—case studies from India (Nayak et al. 2019)
	Nitrogen fixation	The extent of an ecosystem, per hectare—nitrogen fixation services values	Literature review—case studies from India (Nayak et al. 2019)
	Soil fertility (NIC 2020)	The extent of an ecosystem, per hectare—soil fertility services values	Literature review—case studies from India (Nayak et al. 2019)
	Remediation—organic and inorganic materials	The extent of an ecosystem, per hectare—remediation services values	Literature review—case studies from India (Nayak et al. 2019)

(continued)

Table 3 (continued)

Agriculture—croplands, horticulture			
	Pollination	The extent of an ecosystem, per hectare—pollination services value	Literature review—case studies from India http://naasindia.org/Policy%20Papers/policy%2094.pdf
	Genetic diversity	The extent of an ecosystem, per hectare—genetic diversity services value	Literature review—case studies from India http://naasindia.org/Policy%20Papers/policy%2094.pdf
	Biological control	The extent of an ecosystem, per hectare—biological control services values	Literature review—case studies from India (Nayak et al. 2019)
Cultural Services	Tourism & recreational	The extent of an ecosystem, per hectare recreation and tourism services	Literature review—case studies from India http://naasindia.org/Policy%20Papers/policy%2094.pdf
	Inspirational, Culture, Art	The extent of an ecosystem, Per hectare inspirational, culture and art services value based on global studies. These estimates were adjusted for GDP (PPP) per capita of the country for which values were estimated and the corresponding currency exchange rate	Literature review (Van Berkel and Verburg 2014; Hirons et al. 2016; Moreno et al. 2018; Cheng et al. 2019) review; Global ecosystem service valuation database (ESVD) https://www.es-partnership.org/wp-content/uploads/2020/08/ESVD_Global-Update-FINAL-Report-June-2020.pdf GDP(PPP) per capita for India—https://data.worldbank.org/indicator/NY.GDP.PCAP.PP.CD?locations=IN Currency exchange rate https://www.xe.com/currencyconverter/convert/?Amount=1&From=USD&To=INR

3 Method

Ecosystem services are accounted through the (i) residual value method, (ii) benefit transfer method, and (iii) biophysical models-InVEST, depending on the availability of data and time constraints.

Residual value method: Provisioning services of ecosystems are accounted for through the residual value (or resource rent) method. The residual value method has been used to estimate the value of an ecosystem service by taking the gross value of the final marketed good (to which the ecosystem service provides input) and then deducting the cost of all non-ecosystem inputs, including labour, produced assets and intermediate inputs (as per SEEA Central Framework, given below).

$$\text{Net return on environmental assets} = \textbf{resource rent} - \text{depletion}$$
$$\textbf{Resource rent} = \textbf{gross operating surplus}$$
$$-\text{consumption of fixed capital (depreciation)}$$
$$-\text{return on produced assets} - \text{labour of self} - \text{employed persons}$$
$$\textbf{Gross operating surplus} = \textbf{Output} - \text{intermediate consumption}$$
$$-\text{compensation of employees}$$
$$-\text{other taxes on production} + \text{other subsidies on production}$$

Economic rent is the surplus value accruing to the extractor or user of an asset calculated after all costs, and normal returns have been considered. The measure of resource rent (i.e., surplus-value of environmental assets) provides a gross measure of the returns to the environmental asset as a direct capital value, giving a reasonable approximation of the market price of the service.

The benefit transfer method or unit value transfer refers to applying economic value estimates from one location to a similar site in another place. Values for ecosystem services at a study site, expressed as a value per unit (usually per unit of area or beneficiary), combined with information on the number of units at the policy site, are used to estimate policy site values. Unit values from the study site are multiplied by the number of units at the policy site. When using the benefit transfer method, unit values are adjusted to reflect differences between the study and policy sites. In this chapter, ecosystem services values are based on case studies from India, which are compared with the global ecosystem service valuation database (ESVD) [https://www.es-partnership.org/wp-content/uploads/2020/08/ESVD_Global-Update-FINAL-Report-June-2020.pdf] and published literature (of case studies from India) considering GDP (PPP) per capita for India (https://data.worldbank.org/indicator/NY.GDP.PCAP.PP.CD?locations=IN) and the currency exchange rate (https://www.xe.com/currencyconverter/convert/?Amount=1&From=USD&To=INR).

InVEST: **InVEST (Integrated Valuation of Ecosystem Services and Trade-offs) is a suite of models used to map and value ecosystem services. It helps explore how changes in ecosystems can lead to changes in the flows of many**

different benefits to people. InVEST returns results in either biophysical terms (e.g., tons of carbon sequestered) or economic terms (e.g., the net present value of that sequestered carbon). InVEST (https://naturalcapitalproject.stanford.edu/software/invest) models are spatially explicit, using maps as information sources and producing maps as outputs.

3.1 Valuation of Forest Ecosystem Services

Provisioning services from forest ecosystem: The provisioning services derived from the forests in this chapter include:

- **Timber**: Forests are the prime source of timber for the local people. Timber includes wood for making furniture, agricultural implements, fencing wood, wood poles, and roof planks. Timber provides revenue and constitutes an essential component of value on forest land properties. Timber is a readily available goods, which has a marketable value from the forests. However, strict rules and regulations are in operation by the State Government to extract timber from the protected forests.

  In many cases, the value of the timber can be several times the value of the land. Timber includes rosewood, teak wood, jungle wood, etc. Timber is mainly prominent in deciduous forests, while it is found in less quantity in evergreen forest patches. Plantation forests primarily consist of timber-producing trees like acacia, teak, etc. Forest-based industries depend on the produce from the forest, which includes roundwood, softwood, matchwood, etc. The data regarding the quantity of timber harvested and seigniorage values (residual value) were obtained from the Karnataka Forest Department. As per the Karnataka Forest Department records, the seigniorage values remain the same for all provisioning services for 2005 and 2019.

  - The ecosystem supply value of Rosewood is 140,017 Rs/Ha/yr (2005) and 140,998 Rs/Ha/yr (2019) based on the data collected across the circles from the Karnataka Forest Department.
  - Teakwood is 79,881 Rs/Ha/yr (2005) and 79,961 Rs/Ha/yr (2019).
  - Eucalyptus wood is 4,304, and 4,265 Rs/Ha/yr for 2005, and 2019 respectively.
  - Other kinds of timber are 4,644 and 4,297 Rs/Ha/yr for 2005, and 2019, respectively.
  - Pulpwood is 3,369 Rs/Ha/yr (2005) and 3381Rs/Ha/yr (2019).
  - Round poles wood is assessed as 4,434, and 4,261 Rs/Ha/yr for 2005, and 2019.
  - Sandalwood is accounted as 4,573, and 4,652 Rs/Ha/yr for 2005, and 2019 respectively.
  - Bamboo for the Karnataka state is assessed as 3,938, and 4,402 Rs/Ha/yr for 2005 and 2019, respectively.

- **Fuelwood**: Fuelwood is the most important forest product in Karnataka. Fuelwood is the energy source for cooking in most parts of rural India, and 50% of the total fuelwood consumed comes from the forests. The yield of fuelwood depends on the ease of access to the forests. The total value of fuelwood includes the value of fuelwood used for domestic purposes, i.e., for cooking and water heating, and also the fuelwood used for various industrial and commercial purposes like jaggery making, areca processing, cashew processing, restaurants, and bakery, parboiling, cremation, etc. Local people collect fallen wood/dry tree branches from the forests. An exploratory survey focusing on fuelwood requirements was initially conducted in various taluks of the agro-climatic zones and through a review of literature based on our earlier work (Ramachandra et al. 2000b). Socio-economic and energy data was collected from randomly selected samples in the Sirsi, Siddapur, Kumta, and Ankola taluks of Uttara Kannada district. The fuelwood required for various purposes (large-scale jaggery making, drying of agro products such as cardamom, etc.) was based on field experiments (Ramachandra et al. 2000a, 2017; Ramachandra and Bharath 2019a). The preliminary results of the survey in households using fuel-efficient stoves in the Sirsi and Kumta taluks showed fuelwood requirements for cooking (kg/person/day) to be 1.80 and 1.78 (summer) and 2.25 and 1.98 (monsoon), respectively. Villagers are permitted to collect fuelwood from the nearby forests, and the forest department collects the collection fee. Based on these data, the forest circle-wise fuelwood requirement is assessed and quantified in physical (tons) and monetary values (million INR). The ecosystem supply value of fuelwood is assessed as 5,097, and 23,623 Rs/Ha/yr for 2005 and 2019 based on the fuelwood consumption data collected for the Karnataka state.
- **Non-timber forest products**: Non-timber forest products play a significant role in the livelihoods of the local communities and often contribute significantly towards the family income of the forest-dependent communities. The Forest Department gains significant revenue (over 50%) from NTFP extractions, and 75–80% of forest export income comes from NTFP exports. Over 275 million of the rural population (27% of India's total) are engaged in collecting NTFP (Bhattacharya and Hayat 2009), resulting in a revenue of 6000 crores to rural communities (Planning Commission 2011). The villagers practice traditional, non-destructive methods of extraction of NTFPs. Due to commercial players, traditional harvesting has been replaced by destructive/excessive harvesting in many places. The data on the harvesting of NTFP was obtained from the Forest Department. The total value of NTFP includes the value of (a) NTFPs extracted by the Forest Department, (b) NTFPs collected by households, and (c) bamboo extracted by the Karnataka Forest Department. Twenty-two varieties of food products derived from the forest were identified, and the value of food extracted per unit area of forest was obtained from literature was extrapolated to the total forest area. Household honey collection, which is a critical provisioning service from forests, was quantified based on the earlier field studies (Ramachandra et al. 2012, 2018a) for all talukas and valued based on the quantity and benefits.

- **Fodder**: Fodder is the most important source of nutrients for livestock. A significant proportion of cattle reared grazes from the forested landscapes. Tree leaves and ground herbage (grass) are collected from the forests to feed the livestock, and fallen leaves are used to prepare compost and mulching in gardens. Leaf litter from the forest floor is collected by the local people and used as cattle beds for some time. It is then transferred to compost pits; once turned into compost, it is then transported to crop fields to replenish soil fertility. The livestock dependent on forests for fodder requirements is estimated, and the cost-adjusted price of fodder is also evaluated, considering the unit market price of the fodder and the cost of collecting fodder. The total value of fodder supplied from the forest was quantified using field data and data from earlier studies (Prasad et al. 1987; Ramachandra et al. 2000b) on herb layer productivity in different types of forests and the extent of different types of forest. Based on these inputs, per hectare value for fodder was assessed, and total values were computed based on market prices, assuming 10% cost factor, with a standard daily fodder requirement of 22 kg/CU/day and the existing livestock. The ecosystem supply value of fodder is 7,736 and 15,476 Rs/Ha/yr for 2005 and 2019, respectively.
- **Fish and other aquatic products provisioning services**: Fish is one of the primary sources of animal protein globally, and inland fishing is an important economic activity in the forest ecosystem. Inland fishing happens in rivers, rivulets, streams, reservoirs, lakes, etc., which are inseparable parts of the district's forest area. Fish are harvested by locals and for commercial purposes in the streams, lakes, and reservoirs of Karnataka. The inland fish quantity harvested for 2005 and 2019 has been compiled from the published reports of the Karnataka Inland Fisheries Department, Government of Karnataka. The revenue generated is quantified by using fishermen's share value per ton. The fishermen's share in consumers' rupee (%) is estimated from the net price received by the fishermen over the price paid by consumers (Aswathy et al. 2014). The fishermen generally receive 56% of the share from the consumers' end price (Willmann et al. 2001; Kumar et al. 2008; Aswathy et al. 2014). Economic values for the ecosystem's contribution were determined (65,000 Rs/Ha/yr for 2005 and 2019) based on the residual value considering revenue and cost of harvesting.
- **Water supply service**: The accessibility and quality of water are intensely influenced by forests, which regulate water flows and control the availability of water resources (Ramachandra et al. 2020). The misconception of evapotranspiration, misinterpretations, and misinformation about vegetation in catchments has resulted in ad-hoc policies (Calder 1996).

  Most of the water resources come from forested catchments. Hydrological services are quantified by the quantity of domestic water utilization, water for irrigation purposes (Ramachandra et al. 2001), water for industrial use, and water used for power generation (hydropower stations and nuclear power stations). Point-based daily rainfall data from various rain gauge stations in and around the study area between 1901 and 2019 were considered for the analyses of spatial and temporal patterns of rainfall (Ramachandra et al. 1999, 2020). Run-off (surface flow and sub-surface flow), infiltration, and groundwater recharge were quantified based on field measurements (Ramachandra et al. 1999, 2020). The field data reveal a correlation between water supply and the quality of forests as follows: (i) moderate overland flow with substantial local recharge in catchments dominated by native vegetation, (ii) higher overland flow, and reduced local water recharge in catchments with vegetation cover <30%. Local water recharge helps in sustaining water in streams and wells during the post-monsoon season. Societal demand depends on the availability of water in streams and wells in the region. Streams are perennial (with 12 months flow) when their catchment is dominated by vegetation (>60%) of native species. This is mainly due to infiltration or percolation in the catchment as the soil is porous with the presence of native species. Diverse microorganisms interact with plant roots, and soil helps in the transfer of nutrients from the soil to plants, and the soil is porous. Analyses of soil samples from the catchments of perennial, intermittent streams reveal that soils in perennial streams catchment have the highest moisture content (61.47–61.57%), higher nutrients (C, N, and K), lower bulk density (0.50–0.57 g/cc). Compared to this, a catchment of intermittent and seasonal streams had higher bulk density (0.87–1.53 g/cc) and relatively lower nutrients. Due to this, water infiltrates and fills the underlying zones, namely saturated zone, and vadose zones, which is crucial for the sustenance of water in the streams during lean seasons. This emphasizes that forest vegetation helps in retarding the water flow in the catchment by allowing infiltration. Contiguous forests of native species moderate the local climate (through transpiration) and also act as a sponge by retaining the water, which is slowly released to the streams during the lean seasons, thereby sustaining the water availability in the catchment to meet biotic needs throughout the year. Water availability for four months is observed in the streams of the degraded catchment, with vegetation cover less than 30%.

Domestic water demand is assessed as the function of water requirement per person per day, population, and season. Water required per person includes water required for bathing, washing, drinking, and other basic needs. Household surveys were conducted with structured questionnaires to understand the agricultural cropping pattern and water needed for various crops in the catchment. Livestock population details were obtained from the district statistics office, and water requirements for different animals were quantified based on the interviews. The crop water requirement for various crops was estimated considering their growth phase and details of the cropping pattern in the catchment (based on the data compiled from household surveys and discussion with the subject experts, and a review of published publications). Water used for electricity generation is compiled from the Annual reports of Karnataka Power Corporation Ltd and the District at a Glance (allocation of water to industries, revenue, and cost). Water supply services are accounted for sectors (domestic, croplands, horticulture, industries, and electricity generation) considering the quantum of water and residual value (revenue and cost details) as per the Water Resources Development Organisation, Government of Karnataka (http://waterresources.kar.nic.in), Department of Minor Irrigation (https://minorirrigation.karnataka.gov.in/english), Karnataka Power Corporation (http://karnatakapower.com/), districts at a glance (kgis.ksrsac.in/kag/), Directorate of Economics and Statistics Karnataka (http://des.kar.nic.in/), etc. The water supply services account for 2,61,360 ₹/Ha/Yr (medium density forests: MDF) to 4,80,315 ₹/Ha/Yr (very dense forests: VDF), which is comparable to the earlier reports (values are within ±10% as per Ravindranath and Ostwald 2008; De Groot et al. 2020).

- **Medicine**: Medicinal plants are a prime source for healing in forest areas, with widespread usage in recent times, even in urban areas. The people of India have an ancient history of using medicinal plants as codified and non-codified healing systems. As per the Botanical Survey of India, about 255 modern medicines are derived from forest medicinal plants. Forest ecosystems in Karnataka are endowed with 1838 species of medicinal plants (http://envis.frlht.org/checklist/karna.pdf). The demand for medicinal plant-based raw materials is growing at the rate of 15–25 percent annually. According to the World Health Organisation (WHO) estimate, the demand for medicinal plants is likely to increase from the current $14 billion a year to $5 trillion in 2050.

  A sampling of medicinal plants has been done through (i) field-based measurements collected across the forests of Karnataka using transect-based quadrat sampling techniques (see below) and (ii) published literature on an inventory of medicinal plants. The study region (Karnataka State) was divided into 2597 grids of $5' \times 5'$ (or 9 km × 9 km) grids corresponding to grids of 1:50000 topographic maps of the Survey of India. Select grids corresponding to agro-climatic zones were chosen for field investigations. The field estimations were done across the varied forest types covering around 424 transects in Uttara Kannada, Shimoga, Chikmagalur, Kodagu, Dakshina Kannada, Udupi, Dharwad, etc. The number of quadrats per transect varied between 3 and 5 depending on species occurrence in the sampling locality. The opportunistic survey was also carried out to list out species not recorded in transect studies to get the total medicinal plant species count. Per hectare population was also calculated for important medicinal plant (trees, shrubs, and herbs) species. The medicinal plants in the region were identified based on the available secondary literature (Rao et al. 2014, 2015; Ramachandra et al. 2015). Medicinal plants used by the local people were identified (Ramachandra et al. 2017), and the value of these medicinal plants per unit area of forest area was computed based on public interviews, discussion with the local experts (with the knowledge of traditional uses of plants) and the review of literature review (Ramachandra et al. 2017; Chanda and Ramachandra 2019a, b). Highly traded medicinal plants such as *Salacia chinensis, Nothapodytes foetida, Embelia ribes, Coscinium fenestratum, Cinnamomum malabathrum, Myristica malabarica, Costus speciosus* and *Garcinia* spp. were found to be well represented in the study area.

Medicinal plants being used at local levels and details of plants, details of use for treating ailments, and likely market value were compiled through public interviews, discussion with the experts, and literature review (Ramachandra et al. 2017; De Groot et al. 2020). Medicinal plants have commercial value (in addition to local uses), and as per the norms (Biodiversity Act 2002, Government of India (http://nbaindia.org/) and norms of access and benefit-sharing as per Article 15 of the Convention on Biological Diversity (CBD)), industries are required to pay the royalty to the government on an annual basis. Medicinal plants harvested (type/species, quantity extracted per season, and royalty/revenue) with the access and benefit-sharing details are obtained from (from Karnataka Biodiversity Board (https://kbbwebportal.karnataka.gov.in/default.aspx), Karnataka Forest Department (https://aranya.gov.in/; http://envis.frlht.org/mpcas) and Medicinal Plants Conservation Authority (https://ayush.karnataka.gov.in/). Medicine services of forest ecosystems were quantified based on the spatial extent of forests, quantity of medicinal plants harvested, and residual value (revenue and costs). The medicinal plants' services of forest ecosystems range from 221 Rs/Ha/Year (MDF) to 445 Rs/Ha/Year (VDF) and are comparable to studies from India (Verma et al. 2013) and international (De Groot et al. 2020) studies (adjusted for GDP (PPP) per capita and the currency exchange rate).

- **Genetic material service**: Forests aid as gene pools and play a significant role in conserving biodiversity, which is being explored for various purposes (medicine, conservation, species richness, biodiversity, etc.) and is increasingly recognized. The economic value of gene-pool conservation in terms of bioprospecting is based on (i) the number of medicinal plants found in each district, (ii) the number of species of conservation importance in each district, and (iii) all species. The estimate is based on all species in the district and ecosystem extent. Species details were obtained from the Karnataka Biodiversity Board (https://kbb.karnataka.gov.in/), Karnataka Forest Department (https://aranya.gov.in/; http://envis.frlht.org/mpcas), and the Medicinal Plants Conservation Authority (https://ayush.karnataka.gov.in/). The genetic material services of forest ecosystems in Karnataka are estimated considering the spatial extent of forest patches (with the distribution of endemic species and species of conservation importance) through the benefit transfer technique (Verma et al. 2013), which ranges from 2,25,856 Rs/Ha/Year (evergreen forests, VDF), 1,79,680 Rs/Ha/Year (evergreen, MDF), 1,09,940 Rs/Ha/Year (moist deciduous) and 67,852 Rs/Ha/Year (dry deciduous) based on studies from India.

Regulating services from forest ecosystems: Forests provide several intangible benefits such as regulating local and global climate, protecting watersheds, controlling soil erosion, nutrient cycling, etc., that are often ignored in policy contexts since these values do not register in conventional markets or are challenging to measure. Valuation studies have uncovered the significance of forest resources and provided a deeper understanding of many ways in which forest resources benefit humankind (De Groot et al. 2002, 2020; Amirnejad et al. 2006; Costanza et al. 2014; Zarandian et al. 2016). However, unlike provisioning services, regulating services pose much more significant challenges in valuation as they are seldom marketed. In the present study, regulating services were quantified through the benefit transfer method (Ramachandra et al. 2000b, 2010, 2017, 2018b; Ramachandra and Bharath 2021). Regulating services considered are:

- **Global climate regulation services/Carbon sequestration**: Forests sequester CO_2, which aids in mitigating the impacts of climate change. However, degradation of forest ecosystems leading to deforestation would lead to loss of carbon sequestration potential. The carbon sequestration potential of Karnataka State's forests was estimated by integrating temporal land-use data with field investigations at the grid level. The study region (Karnataka State) was divided into 2597 grids of $5' \times 5'$ (or 9 km $\times$ 9 km) grids corresponding to $5' \times 5'$ grids of 1:50000 topographic maps of the Survey of India. Carbon sequestration is assessed across the various forest cover types by accounting for the annual increment in above-ground biomass (AGB), below-ground biomass (BGB), soil organic carbon (SOC), and deadwood content. The biomass, annual increment in biomass of various forest types, sequestered carbon, and productivity have been computed using field data integrated with information compiled from various literature listed in Table 4. Carbon sequestration (CO_2 equivalent) is computed by multiplying the carbon sequestration values with the factor 3.67 as per the protocol of The Intergovernmental Panel on Climate Change (Penman et al. 2003).

Table 4 Carbon sequestration based on forest cover types

Index	Forest type	Equation	Quantification
Annual increment in biomass (T/Ha)	Evergreen	(Forestcover) × 10.48	Incremental growth in biomass (Ramachandra et al. 2000a; Pandey et al. 2011; Devagiri et al. 2013)
	Deciduous	(Forestcover) × 13.82	
	Scrub	(Forestcover) × 5.4	
	Plantations	(Extent) × 1.4	
Annual increment in carbon (T/Ha)	All	(Annualincrementinbiomass) × 0.5	Incremental growth in carbon storage
NPP (net primary productivity) or net annual biomass productivity (T/Ha)	Evergreen	(Forestcover) × 3.6	Used to compute the annual availability of woody biomass in the region. (Ramachandra et al. 2000b)
	Deciduous	(Forestcover) × 3.9	
	Scrub	(Forestcover) × 0.5	
	Plantations	(Extent) × 3.6	
Annual increment of soil carbon (T/Ha)	All	(Cover) × 2.5	Annual increment of carbon stored in the soil (Ravindranath et al. 1997; Rajan et al. 2010)

InVEST carbon sequestration model: The InVEST carbon sequestration model estimates the quantity of carbon sequestered from a landscape and values the amount of sequestered carbon over time. It aggregates the biophysical amount of carbon stored in four carbon pools (aboveground living biomass, belowground living biomass, soil, and dead organic matter) based on land use/land cover (LULC) maps. The InVEST 3.9 Carbon model was used to validate the results of 2005 and 2019 field-based estimates. The model considers inputs as land-use maps and a CSV file containing the values of carbon above ground, carbon below ground, soil carbon, and dead carbon concerning each land-use class. Invest also tries to quantify projected sequestration with the input for future land use. The model output summarizes results into raster outputs for sequestration and value as aggregate totals.

Additionally, a REDD+ (reducing emissions from deforestation and forest degradation) scenario can also be included to evaluate the additional future scenario, calculate storage and sequestration, and summarize results. Outputs of the model are expressed as Mg of carbon per pixel. The valuation model estimates the economic value of sequestration as a function of the amount of carbon sequestered, the monetary value of each unit of carbon, a monetary discount rate, and the change in the value of carbon sequestration over time. Thus, valuation can be done in the carbon model for future scenarios.

The land use maps of 2005 and 2019 have been provided as input in the InVEST carbon model to quantify the carbon sequestration across the Karnataka region. InVEST provides carbon sequestration values, which were converted to carbon dioxide (CO_2) equivalent as suggested by IPCC. The value of CO_2 sequestration is 1,17,660 Gg (Giga gram) and 87946 Gg using InVEST for the years 2005 and 2019, compared to the field-based estimates of 1,24,153 Gg; 89,194 Gg. The model's accuracy is around 95%, and 98%, respectively, which depicts good consistency compared with field measurements. The carbon sequestration service value (social cost of CO_2) is computed as a function of the amount of carbon sequestered per year (based on field measurements), the monetary value of each unit of carbon as US$ 80 using the GDP deflator, as per MoSPI (2020).

- **Soil conservation and soil fertility**: Forests also help increase soil fertility through the decomposition of leaves and humus formation. Forests play a key role in tempering droughts as well as floods and protecting against the incidence of landslides. Forests also limit soil/splash erosion, help retain rainwater, maintain soil moisture, intercept and delay high-pressure precipitation, disperse and delay runoff, intercept sediment and protect the surface (Ma et al. 2019). In addition, forests regulate soil fertility through the underground root system, improving soil physical and chemical properties.

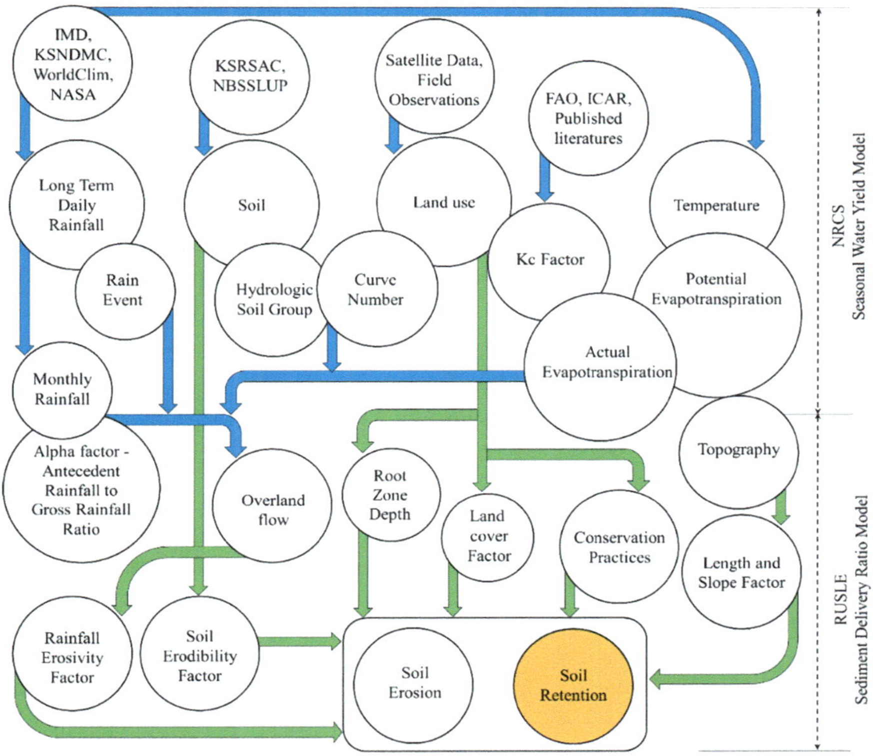

Fig. 2 Procedure for estimating soil retention through the RUSLE module of InVEST with details of input data

The InVEST Sediment Retention model estimates the capacity of a land parcel to retain sediment by using details of geomorphology, climate, vegetative coverage, and management practices. Estimated soil loss and sediment transport of a land parcel are the inputs to the InVEST model, which produces avoided sedimentation as output. The model can also value the landscape in terms of water quality maintenance or avoided reservoir sedimentation and determine how land-use changes may impact the cost of sediment removal. A region's sediment yield and retention characteristics are computed using the Revised Universal Soil Loss Equation (RUSLE) based on the knowledge of soil characteristics, land use characteristics, vegetation characteristics, farming practices, topographic effects, etc. Fig. 2 depicts the method involved in deriving local sediment retention using the RUSLE method. Various factors defining Sedimentation in RUSLE are the R Factor, K factor, L Factor, S Factor, P Factor, and C Factor. R factor is defined as the Rainfall Erosivity Factor. There are numerous methods that incorporate annual rainfall, monthly rainfall, quick flows, etc., to derive the R factor. R Factor is defined based on Quick/Surface flow conditions in the current study and was

Table 5 K factor—Soil erodibility factor

Texture	K	Texture	K
Clayey	0.22	Loamy	0.3
Clayey over Loamy	0.3	Loamy over Sandy	0.11
Clayey over Sandy	0.07	Loamy Skeletal	0.11
Clayey Skeletal	0.22	Rocky Outcrops	0.01
Coarse Loamy	0.3	Sandy	0.02
Fine	0.08	Sandy Skeletal	0.02
Fine Loamy	0.39	Very fine	0.08
Built-up	0.01	Water	0.00

estimated as $R = 81.5 + 0.375 * \text{Quick flow}$. Quick flows were estimated using the Natural Resources Conservation Service (NRCS) method (Seasonal Water Yield Model) in InVEST 3.9.0. Soil erodibility factor varies based on soil texture and depth. Table 5 describes the K factor considered based on published literature. Root zone depth of various vegetation types, Conservation practice factor (CPF), and Land cover factor (LCF) were derived based on land use data and published literature. DEM was used to estimate Length and Slope Factors.

The InVEST sediment yield model integrates various data sets, as described in Table 6, which lists the data used to derive the local sediment retention using the RUSLE method. There are various outcomes[1] with regard to soil retention, but in the current study, sediment retention potential at pixel level was used to derive sediment loss at various land surface features.

Sedimentation outcomes of the model were compared with field experiments carried out at Aghanashini River (https://linkinghub.elsevier.com/retrieve/pii/S0301479717307429) located in Uttara Kannada district, Central Western Ghats. The analysis carried out using InVEST tool showed average sediment yield in the catchment is about 4232 tons per hectare per year i.e., about 1627 cum/ha/year with an average density of 2.6 tons/cum (http://isslup.in/wp-content/uploads/2018/09/Characterisation-Of-Soils-Of-Western-Ghats-In-Dakshina.pdf). In contrast, observed yield in Aghanashini River was in the range of 1367–1567 cum/ha/year. InVEST model was able to predict with an accuracy of 85–96% compared to select field measurements. This calibrated model was used to investigate sediment retention for the year 2020. Prevention of soil erosion varies according to forest types (evergreen, deciduous, scrub), and the quantity varies based on the extent and condition of forests.

[1] Various outcomes with regards to soil retention are (i) total soil loss per pixel in the absence of land-use features equivalent to bare earth, (ii) total amount of sediment exported from each pixel that reaches the drainages/streams, (iii) amount of sediment deposited on the pixels due to retention in upstream resources, (iv) potential soil loss per pixel with original land cover features and (v) sediment retention map at pixel level by comparing soil loss at the pixel level for bare earth and due to land cover features.

Table 6 Data used for assessing soil conservation services

Data and type	Source	Purpose	InVEST	File type
Daily rainfall	• India Meteorological Department • Karnataka State Natural Disaster Monitoring Cell • Tropical Rainfall Measuring Mission • NASA power • NOAA—Climate Data Online • Terrestrial Hydrology Research Group—Princeton University database	Quantification of the number of rainy days in a month (long-term average)	Monthly climate zone factor table—Varies across each month, and across agro-climatic zones. Rainy days are associated with agro-climatic zones across each month	*.csv
		Quantification of antecedent monthly rainfall to annual rainfall	Monthly alpha table-Varies across each month, common values, does not differentiate between agro-climatic zone	*.csv
Monthly rainfall		Quantification of overland flows	Monthly precipitation data stored in the directory	*.tif
Monthly evapotranspiration	• NASA power • Terrestrial Hydrology Research Group—Princeton University database • Food and Agriculture Organisation • Indian Meteorological Department	Quantification of actual evapotranspiration based on land use type and potential (reference) evapotranspiration ET_0 [mm day^{-1}]	Monthly ET_0 data stored in the directory	*.tif
Land use	• Indian Remote Sensing (IRS) satellite data • United States Geological Survey	Quantifications of overland flows, actual evapotranspiration, root zone depth, P factor, and C factor	Land use identified with integer number and titled within the biophysical table	*.tif

(continued)

Table 6 (continued)

Data and type	Source	Purpose	InVEST	File type
Soil	• Karnataka Remote Sensing Application Centre • National Bureau of Soil Studies and Land Use Planning • ICRISAT—Soil Fertility Atlas of Karnataka • Soil Health Card—Department of Agriculture, Cooperation & Farmers Welfare Ministry of Agriculture and Farmers Welfare Government of India • European Soil Data Centre	Assigning HSG (Hydrological Soil Group) based on soil texture and patterns, and based on land use, defining curve numbers Assigning K factor based on soil texture	Soil texture HSG identified as number (HSG A = 1, HSG B = 2, HSG C = 3 and HSG D = 4) K Factor (Table 1) as defined as an image with float values defining soil erodibility factor	*.tif
Kc	• Food and Agriculture Organisation • Published Journal Articles • ICAR-KRISHI • Published Literature	Crop coefficient defining the evaporation coefficient for each month	Month wise crop coefficients for each land use type, stored within the biophysical table	*.csv
Watershed	• NRSC Bhuvan • Karnataka State Remote Sensing Application Centre	Used to define the boundary condition for analysis	Defined with a shape file consisting a single polygon	*.shp
Agroclimatic zones	• University of Agriculture Sciences	Climatic patterns and cropping vary with agroclimatic zones. The number of rainfall days is minimum at arid zones while the coast and ghats have a higher number of rainy days. According to the number of rainy days, the rainfall is spread across the month to define the likely flow patterns	Agroclimatic zones are identified with numbers for each of the raster pixels. The number of rainy days is defined based on agro climatic zone for each month	*.img

(continued)

Table 6 (continued)

Data and type	Source	Purpose	InVEST	File type
R factor	• Published Literature	Defines the rainfall erosivity factor, derived based on quick flow conditions	R factor is identified as float values in an image	*.img
K factor	• Published Literature	Soil erodibility factor, defined by soil texture	K factor is based on soil texture, and defined as float values	*.img
Bio physical table	• ORNL DAAC • NRCS USDA • Published Literature		Quick flow: stores data such as month wise Kc values with respect to land use, curve numbers with respect to soil type and land use Relates Pixel numbers to land use classes Sediment retention: according to the various land use types, stores the sedimentation cover management factor (C Factor), Practice (P Factor), Root Zone depth, Kc factor	*.csv
Topography	• Shuttle Radar Topographic Mission (SRTM)	Defines the flow direction and slope characteristics of terrain. Used to derive L (length) and S (Slope) factors	DEM is directly taken into InVEST tool which automates the calculation of slope and length factors using Moore neighbourhood and pixel dimensions	*.img

(continued)

Table 6 (continued)

Data and type	Source	Purpose	InVEST	File type
Root zone depth	• Published Literature	Vegetation roots		
Max SDR, Borselli k Parameter, Borselli IC0 Parameter			Default values (SDR, k, IC0) provided in InVEST tool has been considered	

InVEST provides the quantum of soil (sediment) retained within the natural forested areas, and considering Rs 48.8 per ton retention of sediments and on the condition of forests (Verma et al. 2013), soil retention services values range from 9064 INR/Ha/yr (dry deciduous, medium-density forests) to 19436 INR/Ha/Yr (evergreen, very dense forests). Analyses of sediment yield based on vegetation type and canopy cover reveal that sediment retention in evergreen forests ranges from 150 tons per hectare (MDF) to 1000 tons per hectare per year (VDF), and soil retention services range from 7320 Rs/ha/year (Open canopy forests) to more than 48800 Rs/Ha/year (VDF) in evergreen forests. Similarly, in deciduous forests, sediment retention ranges between 15 tons per hectare per year to 350 tons per hectare per year (VDF), and soil retention services range from 732 Rs/ha/year (open forests) to 17080 Rs/Ha/year (VDF).

- **Water regulation and groundwater recharge**: Forests in the watershed help retain water supplied from precipitation in underground aquifers just as water is stored in human-made reservoirs. They simplify in increasing the efficient water available, improving water quality, and decreasing water runoff. The quantity of water conserved depends on several parameters: evaporation and runoff rates, interception ratios, tree and forest characteristics, nature and intensity of rainfall, geographic and soil conditions, etc. (Ramachandra et al. 2020). Forests help regulate the hydrological regime locally and sustain water flow in the streams due to their sponge-like effect. Vegetation in the catchment helps in retarding the water flow in the catchment by allowing infiltration.

Water regulation services are the ecosystem contributions to the regulation of river flows and groundwater and lake water tables. They are derived from the ability of ecosystems to absorb and store water, and gradually release water during dry seasons or periods through evapotranspiration and hence secure a regular flow of water. This is recorded as a final or intermediate ecosystem service. Peak flow mitigation services will be supplied together with river flood mitigation services to provide the benefit of flood protection, a final ecosystem service. Field investigations in four river basins of Uttara Kannada district reveal that water infiltrates and fills the underlying zones, namely saturated zone, and vadose zones, in the catchments of streams. The region receives rain for about four months, and the surface run-off during the monsoon is due to the precipitation (after saturation of underlying regions). After the monsoon recedes, the water stored in the vadose regions, and saturated zones flow laterally towards the streams for about 6–8 months (as pipe flow in the post-monsoon period of 4 months and base flow during summer). Water infiltration allows water storage in the saturated and vadose zones, which is crucial for water sustenance in the streams during lean seasons. Catchments with >65% vegetation of native species have perennial streams, higher soil moisture, and groundwater than the catchment dominated by degradation.

InVEST Seasonal Water Yield Model: the model integrates various data sets as described in Table 7. A Natural Resource Conservation Series (NRCS) model or SCS model uses the curve number functions-based soil characteristics, land use characteristics, climate characteristics, etc. Figure 3 depicts the method involved in deriving local water recharge using NRCS curve number method. The method uses the water balance equation to derive the unknown parameter (Precipitation = Overland Flow + Infiltration + Evapotranspiration). The local infiltration parameter defines local water recharge. The NRCS curve number model utilizes various parameters to derive the overland flow; FAO recommends the Penman Monteith/Modified Hargreaves method to derive the potential evapotranspiration (ET_0). In the current study, the Modified Hargreaves method is used to derive the ET_0, and this method uses the maximum and minimum temperatures along with solar radiation. Based on the crop evapotranspiration coefficient (Kc), actual evapotranspiration is derived for land use.

The model uses the NRCS (curve number) method to derive various outcomes with regard to water conservation, which includes (i) base flow per pixel level and upstream contribution, (ii) local recharge at pixel level and upstream contribution, and (iii) quick flow at pixel level and upstream contribution. In the current study, water retention was considered a function of local recharge. Local data sets and field-based observations were used to calibrate flow patterns. The InVEST model was able to predict with an accuracy of over 80%. This calibrated model was used to investigate water retention for 2019 and 2020 (note there is no variation between 2019 and 2020).

Groundwater was the major source of irrigation apart from the canal irrigation system. Water contained in the voids of the geologic materials supports the need for irrigation, domestic and from industries. The groundwater storage available annually is estimated based on area, depth of fluctuation in the groundwater table, and specific yield factors. The groundwater storage in the aquifer depends on the input components such as precipitation, seepage, and return flow (Chatterjee 2011).

InVEST provides the quantum of water recharge within the natural forested areas. The forest of Karnataka locally recharges about 27.2 billion cubic meters of water to the ground, which later flows as base flows. The economic value of groundwater (after deducting costs) is about 262.5 Rs/k cum of water. Based on this, local recharge in evergreen forests ranges between 1000 mm (medium-density forests: MDF) to more than 4000 mm (very dense forests: VDF), and the

Table 7 Data used in InVEST seasonal water yield model for estimating local water recharge services of forest ecosystems

Data and type	Source	Purpose	InVEST	File type
Daily rainfall	• India Meteorological Department • Karnataka State Natural Disaster Monitoring Cell • Tropical Rainfall Measuring Mission • NASA POWER • NOAA—Climate Data Online • Terrestrial Hydrology Research Group—Princeton University Database	Quantification of number of rainy days in a month (long term average)	Monthly climate zone factor table—Varies across each month, and across agroclimatic zones. Rainy days are associated with agroclimatic zones across each month	*.csv
		Quantification of antecedent monthly rainfall to annual rainfall	Monthly alpha table-Varies across each month, common values, does not differentiate between agroclimatic zone	*.csv
Monthly rainfall		Quantification of overland flows and local recharge	Monthly precipitation data stored in directory	*.tif
Monthly evapotranspiration	• NASA POWER • Terrestrial Hydrology Research Group—Princeton University Database • Food and Agriculture Organisation • Indian Meteorological Department	Quantification of actual evapotranspiration based on land-use type and ET_0	Monthly ET_0 data stored in directory	*.tif
Land use	• Indian Remote Sensing Satellite Data • United State Geological Survey	Quantifications of overland flows, local recharge, baseflow, and actual evapotranspiration	Land use identified with integer number and titled within biophysical table	*.tif

(continued)

Table 7 (continued)

Data and type	Source	Purpose	InVEST	File type
Soil	• Karnataka Remote Sensing Application Centre • National Bureau of Soil Studies and Land Use Planning • ICRISAT—Soil Fertility Atlas of Karnataka • Soil Health Card—Department of Agriculture, Cooperation & Farmers Welfare Ministry of Agriculture and Farmers Welfare Government of India • European Soil Data Centre	Assigning HSG (hydrological soil group) based on soil texture and patterns, and based on land use, defining curve numbers	Soil texture HSG identified as number (HSG A = 1, HSG B = 2, HSG C = 3 and HSG D = 4)	*.tif
Kc	• Food and Agriculture Organisation • Published Journal Articles • ICAR-KRISHI • Published Literature	Crop coefficient defining the evaporation coefficient for each month	Month-wise crop coefficients for each land use type, stored within the biophysical table	*.csv
Watershed	• NRSC Bhuvan • Karnataka State Remote Sensing Application Centre	Used to define the boundary condition for analysis	Defined with a shape file consisting a single polygon	*.shp
Agroclimatic zones	• University of Agriculture Sciences	Climatic patterns and cropping vary with agro-climatic zones. The number of rainfall days is minimal at Arid Zones, while the Coast, Ghats have a higher number of rainy days. According to the number of rainy days, the rainfall is spread across the month to define the likely flow patterns	Agro-climatic zones are identified with numbers for each of the raster pixels. The number of rainy days is defined based on the agro-climatic zone for each month	*.img

(continued)

Table 7 (continued)

Data and type	Source	Purpose	InVEST	File type
Biophysical table	• ORNL DAAC • NRCS USDA • Published Literature		Stores data such as month wise KC values with respect to land use Curve numbers with respect to soil type and land use Relates pixel numbers to land use classes	*.csv

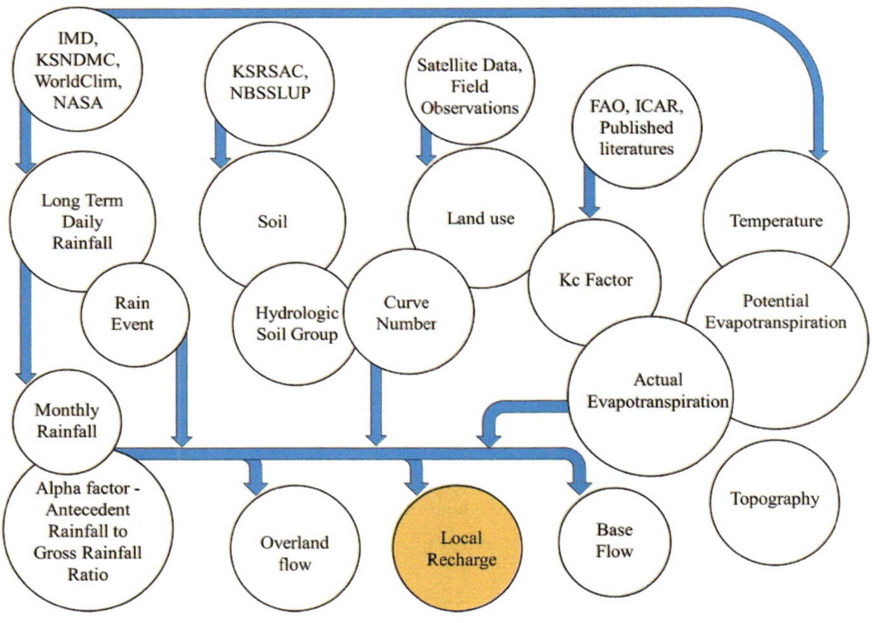

Fig. 3 Procedure to assess groundwater recharge through InVEST

economic value of groundwater recharge services for evergreen forests ranges from 2600 Rs/ha/year (MDF) to more than 5000 Rs/ha/year (VDF). Similarly, local recharge in deciduous forests ranges from 25 mm (MDF) to 1500 mm (VDF). The economic value of groundwater recharge services for deciduous forests ranges from 663 Rs/ha/year to more than 3700 Rs/ha/year.

- **Water purification and waste treatment**: Forests, their diversity, and native vegetation control all hydrological events such as flow, recharge, precipitation, etc. Forest vegetation, diverse root systems, and microorganisms present in soil and water help filter and absorb contaminants and bacteria from the water received from precipitation. In fact, the water obtained from rainfall in forest areas that drips through streams and springs is rich in mineral nutrients and highly valued for its purity and medicinal value (Chatterjee 2011; Terrado et al. 2014; Zawadzka et al. 2019; Ramachandra et al. 2020). Diverse microorganisms interact with plant roots, and soil helps in the transfer of nutrients from the soil to plants which aids in the remediation of water. Soil and water conservation and water purification are interrelated, as these are different services provided by forest ecosystems.

Forest ecosystems have multifunctional potential in terms of water and wastewater treatment, thereby supporting natural processes of ecosystem services supply. Forests aid as sink and support treatment (bioremediation), which depend on the spatial extent, and condition of ecosystems (Zawadzka et al. 2019). Economic values of water purification and waste treatment by forest ecosystems in Karnataka are estimated through the benefit transfer method based on case studies from India. An average of 2,950 Rs/ha/yr was considered to be a water purification service from forests (Verma et al. 2013).

Waste treatment is estimated at 4,716 Rs/ha/yr (Ramachandra et al. 2017; Verma et al. 2013), and values are comparable to the studies across the globe (De Groot et al. 2020) after adjusting for GDP (PPP) per capita for India and currency exchange value.

- **Pollination service**: The natural processes of regeneration of the forest, fruiting, and food production from landscapes are due to pollination and seed dispersal from various agents. Pollination is a key service that governs biological production, and the maintenance of biodiversity. Most flowering plants (approximately 78%) depend on pollinators to reproduce and survive (Ollerton et al. 2011). Pollinator abundance will increase the ecosystem supply value through enhanced production levels by improving qualitative aspects of fruit and seed yields, nutritional content, and general appearance, including fruit size (Balachandran et al. 2017; Porto et al. 2020). This contribution can be well assessed through services such as pollination and seed dispersal (Hipólito et al., 2019), which directly affects the food security of human populations (Porto et al. 2020).

  Pollination services are accounted for through the comparative assessment of the natural regeneration of forest patches (with fencing protection from external pressures Balachandran et al. 2017; Ray et al. 2015) and afforestation cost. The estimates of natural forest regeneration in all forest types are adjusted according to the forest regeneration in plantations (NAP 2009; Ollerton et al. 2011; Hipólito et al. 2019). The economic value of pollination has been estimated in the current study based on the benefit transfer method considering case studies from India (Verma et al. 2013) of natural forest regeneration and its replacement cost (if done artificially) as recommended by the National Afforestation Programme Guidelines (NAP 2009) and based on the success stories of National Beekeeping & Honey Mission (NBHM 2019). The pollination services vary with forest type, and canopy cover and values range from 10167 INR/Ha/Yr (MDF) to 11907 INR/Ha/Yr (VDF) (Verma et al. 2013) with regard to forest regeneration, excluding the contribution to agricultural production. The estimates of natural forest regeneration in all forest types are further adjusted according to the forest regeneration in plantations. The economic value so estimated is limited only to the value of artificially replacing the process of natural forest regeneration and covers partly the economic value of forest succession.

- **Air filtration services**: Forests moderate air quality and reduce pollution that affects human health and well-being, ecosystem health, crops, climate, etc. Forests remove gaseous air pollutants and improve the quality primarily by uptake via leaf stomata (Nowak et al. 2006) and provide clean air by capturing the dust. Densely forested regions can remove air pollution, accounting for around 16%; moderately dense forests can remove up to 4.5%; and sparse cover can remove up to 1%, depending on the meteorological conditions (Nowak et al. 2006). The air quality service value of forest ecosystems ranges from 8,368 INR/Ha/Yr (MDF) to 22,617 INR/Ha/Yr (VDF) based on published literature from case studies in India (Ninan and Kontoleon 2016; Joshi and Negi 2011) which are comparable to tropical forest ecosystems in the global ecosystem service valuation database (ESVD) adjusted for GDP (PPP) per capita for India and the prevailing currency exchange rate (March 2020).
- **Local (micro and meso) climate regulation services**: Forest vegetation plays a crucial role in moderating the local climate. The presence of native vegetation improves the living conditions, regulates ambient atmospheric conditions for the people, and supports economic production. Forests also contribute to local atmospheric climatic moderation, humidity, and rainfall. Local climate regulation services value ranges from 17,933 INR/Ha/Yr (MDF) to 48,468 INR/Ha/Yr (VDF) quantified through benefit transfer method based on published literature from case studies in India (Ghosh 2020; Verma et al. 2007), which are comparable to values for tropical forests in global ecosystem service valuation database (ESVD) adjusted for GDP (PPP) and currency exchange rate as on March 2020 (USD to INR).

Cultural ecosystem services from forest ecosystem: Cultural ecosystem services are the contributions to benefits and well-being people gain from their interactions with different environmental areas, which can be valued in terms of monetary, qualitative, and quantitative methods. Cultural services generally reflect social connections, sensory experiences, symbolic importance, and identity. The forest has a high cultural value; the main reason can be attributed to its aesthetic beauty, recreational benefit, and spiritual and historical (*Kan* forests, which are the sacred groves in the district). Sacred groves are communally protected forest fragments with significant religious connotations (Ramachandra et al. 2010, 2017; Ray et al. 2010). Protected areas in Karnataka are Bannerghatta National Park, Anshi National Park, Bandipur National park, Kudremukh National Park, Arabithittu Wildlife Sanctuary, Bhadra Wildlife Sanctuary, Bhimgad Wildlife Sanctuary, Biligiri Rangaswamy Temple (BRT) Wildlife Sanctuary.

Further, recreational benefits provided by the forest include gaming, trekking, swimming, walking, hunting, etc. The aesthetic beauty of the forest is valuable; the presence of waterfalls and caves adds to the aesthetic value of the district. Education, scientific, and research value provided by the forest is indispensable as the major science and research institutes and organizations set up many long-term ecological monitoring research stations.

- **Aesthetic**—National parks, sanctuaries, waterfalls: Data pertaining to the entry fees (park, recreation spots), and details of visitors were compiled from the forest divisions (Uttara Kannada, Shimoga, Chikmagalur, Dakshina Kannada, and Kodagu districts) of the Karnataka Forest Department. Valuation of aesthetic services is computed considering entrance fees (park, recreation spots) × the average number of visitors to the park/recreation spots during 2018, 2019, and 2020), supplemented with the Indian case studies (Ray et al. 2010; Bharath et al. 2017; Ramachandra et al. 2018c) and travel expenses associated to the travel, based on the address of visitors and considering the connectivity (road/train) and additional details collected from the service providers on revenue during the past three years. The data collected from the service operators include (i) cost (labor, fuel, and maintenance) and (ii) annual revenue. The aesthetic value is computed based on these (travel expenses, entrance fee, maintenance cost, and benefits to service operators), which accounts for 1500 ± 250 ₹/Ha/Yr and an average value of Rs 1500 ₹/Ha/Yr was considered for accounting for the aesthetic cultural services from forest ecosystems in Karnataka.
- **Spiritual and historic**—Distribution of sacred groves (relic forests—protected due to belief and customs): Sacred groves or *kan* forests of Central Western Ghats of Karnataka are climax evergreen forests, preserved through generations as sacred forests by the village communities of Malnadu regions. Patches of forests are dedicated to deities and used by the local communities for worship and cultural activities. In the past, *kans* numbered in the thousands, each *kan* measuring originally from a few hectares to several hundred hectares in area. They were especially characteristic of the traditional land use of Shimoga, Uttara Kannada, and Chikmagalur districts and were equivalent to the devarakadus of Kodagu region. *Kan* forests functioned as important sources of perennial streams and springs used for irrigation crops and domestic needs. However, the curtailment of community rights in the *kans*, in districts of the central Western Ghats, including heavier taxation for collection of forest produce, resulted in the abandonment of many of them, causing various hardships to the villagers (Chandran et al. 2010; Ramachandra et al. 2012, 2016, 2017, 2019a; Ray et al. 2014b, 2015).

  Sacred groves or *devarakadu* in Kodagu district, Karnataka: Sacred groves or *devarakadu* or *kan* forests are culturally protected patches of forests that continue to thrive as a living tradition in Kodagu district. The district has the highest density of groves in the world. The extent of sacred groves in Kodagu is 2,500 ha, i.e., around 2% of the land area in the district, with at least one grove for every 300 ha. The groves are owned by the Forest Department and declared as Protected Forests but are managed by local communities as common property resources. Twenty-four villages in the district have more than ten sacred groves supporting diverse and unique flora and fauna. Kodagu has 1214 sacred groves (Fig. 4) under three taluks. These are community-managed forests protected due to traditional practices and

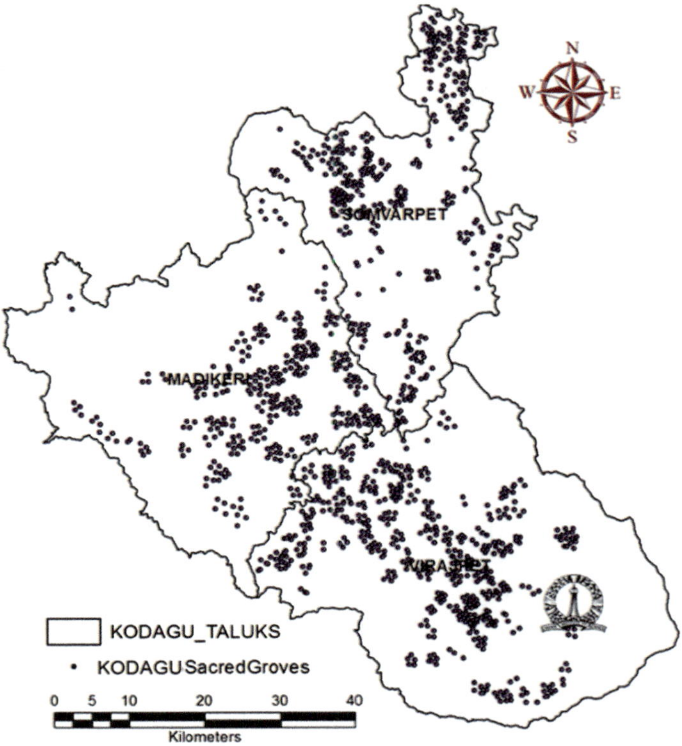

Fig. 4 Distribution of sacred groves/*kans* in Kodagu district

serve as crucial habitats for the conservation of endangered taxa. These groves also act as islands for protecting the biodiversity of rare species in high human-dominated landscapes (Garcia and Pascal 2006; Ormsby and Bhagwat 2010).

The cultural services from the forest ecosystem can be aesthetic, recreational and tourism, spiritual, education, scientific, and research. The spiritual value of the Uttara Kannada district is also high due to the presence of sacred groves, many temples, and pilgrimage centers like Gokarna, Murdeshwar, Dhareshwar, Idagunji, Banavasi, etc. There are 121 deities and unique forms of worship (Fig. 5) practiced by 18 local communities, including Muslims, symbolizing communal unity (Ramachandra et al. 2019a).

The presence of sacred groves is important for cultural services as there are many cultural beliefs associated with the sacred groves in India. Some groves have valuable timber in them but are not harvested for timber due to sacred beliefs. The taluks of Siddapur and Sirsi in Uttara Kannada district have higher cultural values as the region is rich in sacred grooves. The presence of wildlife sanctuaries,

Fig. 5 Various deities in sacred groves

national parks, and groves, in turn, increases the science and educational value of the forest ecosystem.

The valuation of spiritual and historic services is done by collecting primary data pertaining to the amount collected for performing rituals and expenses during 2018, 2019, and 2020. Rituals are performed at the sacred groves, and the amount paid is for performing rituals by devotees (either visiting the grove or in absentia). Every family (irrespective of religion/faith) from the region (residents and those who migrated to cities/abroad) makes donations every year. Also, there is the practice of donating money during birthday celebrations or in the name of elders. Data pertaining to the annual collection and expenses were compiled from the administrative/management committees of select groves in Shimoga, Uttara Kannada, and Kodagu districts. Valuation of spiritual and historical services of sacred groves is done by the residual method (annual collection for rituals and deducting costs – priest salary and ritual expenses).

In groves where annual collection details were unavailable, the travel cost method is used to quantify the service, considering the number of visitors visiting groves for annual rituals, festivals, and other religious activities. This was done through primary surveys of select groves in Uttara Kannada, Shimoga, and Kodagu districts and supplemented with case studies from India that used the benefit transfer method (Ramachandra et al. 2012, 2016, 2017; Ray et al. 2014b, 2015). Based on the field data supplemented with the information from published literature and consultation with the subject experts, the spiritual and historic services provided by forest ecosystems in Karnataka range from 1,200 ₹/Ha/Yr (MDF) to 7,200 ₹/Ha/Yr (VDF).

- **Tourism & recreational services** comprise travel to natural ecosystems for ecotourism, outdoor sports, etc. The recreational sites include the Anashi-Dandeli

Tiger Reserve, Attiveri Bird Sanctuary, and caves in Yana, Kavala, Uluvi, Sintheri, etc. Recreation services are accounted for by considering (i) benefits to travel operators (revenue, costs – labor, fuel, and maintenance of vehicles) and (ii) fees (recreation spots) × the average number of visitors to the park during 2020 – park maintenance charges), (iii) supplemented with the Indian case studies—benefit transfer method (Badola et al. 2017; Gunarekha and Binoy 2017; Ramachandra et al. 2019b; Sinclair et al. 2020). Recreation services ranges from 28,944 ₹/Ha/Yr (MDF) to 2,88,000 (VDF) ₹/Ha/Yr

- **Education, scientific, and research services** of forest ecosystems: Researchers need to get prior permission from the forest department (Forest and Wildlife Research Advisory Committee) to undertake research and monitoring in forests. Details of the research, duration, budget (for field research), and research team details were compiled from the Karnataka Forest Department. This information is supplemented with the data through discussion with researchers and relevant literature on field-based research—ecology, medicinal plants, etc. (Chandran et al. 2010; Ray and Ramachandra 2010; Gould et al. 2014; Ray et al. 2014a; Dorji et al. 2019; Kreye et al. 2019). Based on the data, the education, scientific, and research services provided by forest ecosystems account for 4800 ₹/Ha/Yr.

Figure 6 and Table 8 summarize the method adopted to compute services from forest ecosystems.

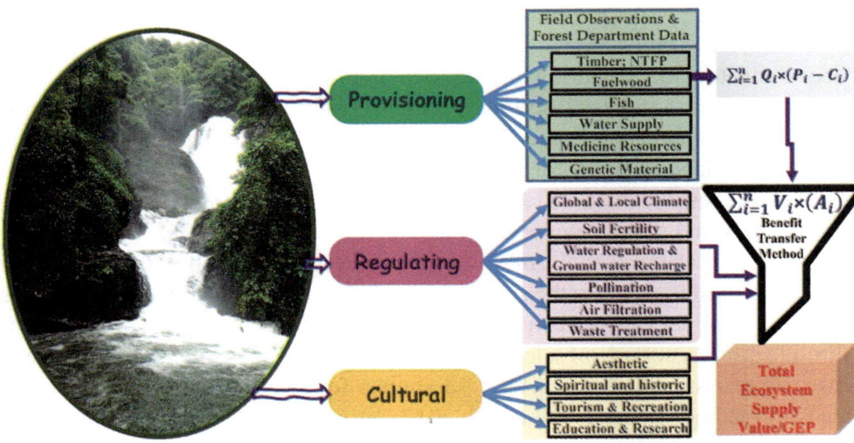

Fig. 6 Method adopted for the valuation of ecosystem services

Table 8 Method for computing goods and services from forest ecosystems

Services	Variables	Data source	Approach
Provisioning services	Timber	Collected from Forest Department, circle wise (the state has 13 circles for the decentralized administration of forests), KFD e-resources: https://aranya.gov.in/aranyacms/English/AnnualReports.aspx; https://aranya.gov.in/aranyacms/downloads/Annual%20Reports/AnnualReportEnglish_19-09-2020_05.06.05.pdf; https://aranya.gov.in/aranyacms/downloads/Annual%20Reports/English%20Annual%20Report%202018-19_28-02-2020_10.58.25.pdf · The ecosystem supply value of rosewood is 140017 Rs/Ha/yr (2005) and 140998 Rs/Ha/yr (2019) based on the data collected across the circles from Karnataka Forest Department · Teak wood is 79881 Rs/Ha/yr (2005) and 79961 Rs/Ha/yr (2019) · Eucalyptus wood is 4304, and 4265 Rs/Ha/yr for 2005, 2019 respectively · Softwood is 2692 Rs/Ha/yr for 2005, 2019 · Other kinds of timber are accounted as 4644, and 4297 Rs/Ha/yr for 2005, 2019 respectively · Pulpwood is 3369 Rs/Ha/yr (2005) and 3381Rs/Ha/yr (2019) · Round poles wood is assessed as 4434, 4261 Rs/Ha/yr for 2005 and 2019, respectively · Sandalwood is accounted as 4573, 4652 Rs/Ha/yr for 2005 and 2019 respectively	**Residual Method** $\sum_{i=1}^{n} Q_i \times (P_i - C_i)$ Where Qi represents quantity, Pi is the price, Ci is the cost involved in the harvest

(continued)

Table 8 (continued)

Services	Variables	Data source	Approach
	Bamboo	The ecosystem supply value of Bamboo for the Karnataka state is assessed as 3938 and 4402 Rs/Ha/yr for 2005 and 2019, respectively	**Residual Method** $\sum_{i=1}^{n} Q_i \times (P_i - C_i)$
	Non-timber forest produce (NTFP)	· The ecosystem supply value of honey is 13177 Rs/Ha/yr (2005) and 13186 Rs/Ha/yr (2019) · Soapnut is 12724, 12977 Rs/Ha/yr for 2005 and 2019, respectively · Cashew nut is 13812 Rs/Ha/yr (2005) and 13945 Rs/Ha/yr (2019) · Tamarind is 14315 and 14346 Rs/Ha/yr for 2005 and 2019, respectively · Rampatri (nutmeg – Myristica malabarica) is 12997 Rs/Ha/yr (2005) and 14436 Rs/Ha/yr (2019) · Murugalu (Kokkum) is 11717 Rs/Ha/yr (2005) and 11740 Rs/Ha/yr (2019)	**Residual Method** $\sum_{i=1}^{n} Q_i \times (P_i - C_i)$ Where Qi represents quantity, Pi is the price, Ci is the cost involved in the harvest
	Fuelwood	Fuelwood required per person is estimated based on a socio-economic survey carried out in select taluks. The ecosystem supply value of fuelwood is assessed as 5097, and 23623 Rs/Ha/yr for 2005 and 2019 based on the fuelwood consumption data collected for the Karnataka state	**Residual Method** $\sum_{i=1}^{n} Q_i \times (P_i - C_i)$
	Fish and other aquatic products	https://des.karnataka.gov.in/english District wise fish catch from districts The ecosystem supply value of fish is 65,000 Rs/Ha/yr based on the fish and other aquatic products from inland aquatic ecosystems	**Residual Method** $\sum_{i=1}^{n} Q_i \times (P_i - C_i)$

(continued)

Table 8 (continued)

Services	Variables	Data source	Approach
	Fodder	Quantity of fodder (estimated forest type-wise) and assuming a 10% cost factor on the market price of fodder, cost-adjusted price of fodder is obtained which is used in the estimation of the economic value of fodder production from forests in each state. Ecosystem supply value from fodder is assessed as 7736 and 15476 Rs/Ha/yr for the years 2005 and 2019, respectively	**Residual Method** $\sum_{i=1}^{n} Q_i \times (P_i - C_i)$
	Water	Quantity of water and price of water with the cost of labor, etc. Sector-wise (industries, residential, irrigation, etc.) water demand, water tariff, revenue, and expenses related to the supply of water (labor, treatment, pumping, etc.). Water services ranges from 2,61,360 ₹/Ha/Yr (MDF), to 4,80,315 ₹/Ha/Yr (VDF), to 2,61,360 ₹/Ha/Yr (MDF), to 4,80,315 ₹/Ha/Yr (VDF)	**Residual Method** $\sum_{i=1}^{n} Q_i \times (P_i - C_i)$
	Medicine	(i) Data of type, quantity, and royalty—received by the government, Karnataka Biodiversity Board, (ii) data of type and quantity extracted by local people from Karnataka Forest Department, (iii) estimates of quantity and type of medicinal plants from medicinal plant conservation authority The medicinal plant services of forest ecosystem based on the primary data range from 221 Rs/Ha/Year (MDF) to 445 Rs/Ha/Year (VDF). They are comparable to studies from India and the global database (EVSD), considering GDP(PPP) per capita for India and exchange rate	**Residual Method** $\sum_{i=1}^{n} Q_i \times (P_i - C_i)$ Supplemented with benefit transfer method—based on studies from India (Verma et al. 2013) and compared with the per hectare values based on EVSD, considering GDP (PPP) per capita for India and exchange rate

(continued)

Table 8 (continued)

Services	Variables	Data source	Approach
	Genetic material	Genetic material service: the economic value of gene-pool conservation in terms of bioprospecting based on (i) number of medicinal plants found in each district; (ii) number of species of conservation importance in each district, and (iii) all species Compared with genetic material values per hectare based on case studies from India (Verma et al. 2013), values are 2,25,856 Rs/Ha/Year (evergreen forests, VDF), 1,79,680 Rs/Ha/Year (evergreen, MDF), 1,09,940 Rs/Ha/Year (moist deciduous), 67,812 Rs/Ha/Year (dry deciduous)	The estimate is based on all species in the study region and ecosystem extent. Species details obtained from Karnataka Biodiversity Board (kbb.karnataka.gov.in), Medicinal plants conservation authority (https://ayush.karnataka.gov.in/) and genetic resource per hectare as per case study from India (Verma et al. 2013)
Regulating services	Global climate regulation—carbon sequestration	Spatiotemporal land use analysis; temporal data—above ground, below ground biomass is estimated based on field data collection across various forest types, integrated with standard literature The value of carbon sequestration services is calculated by considering the social cost of carbon per tonne. The social cost of a tonne of CO_2 is taken as US$ 80 using the GDP deflator (MoSPI 2020). The carbon sequestration from forests depicts the forest circles located in the Western Ghats, which have higher sequestration than other parts of the state due to lower disturbances. Carbon sequestration in forest ecosystems of Karnataka is 124153 Gg/Yr (2005) and 89194 Gg/Yr (2019) due to a decline in the ecosystem spatial extent and conditions (Note: Gg—Gigagram, which is equivalent to 1000 tonnes)	InVEST carbon model, quantity of carbon sequestered annually and the social cost of carbon from MOSPI (MoSPI 2020)

(continued)

Table 8 (continued)

Services	Variables	Data source	Approach
	Soil conservation & soil fertility	Ranges from 7320 Rs/ha/year (Open canopy forests) to more than 48,800 Rs/Ha/year (VDF) in evergreen forests. Similarly, in deciduous forests, it ranges from 732 Rs/ha/year (open forests) to 17080 Rs/Ha/year (VDF) Spatiotemporal land use analysis and meteorological data (rainfall, temperature, evapotranspiration) InVEST provides the quantum of soil (sediment) retained within the natural forested areas and considering Rs 48.8 per ton retention of sediments and on the condition of forests (Verma et al. 2013)	RUSLE, InVEST—quantum of soil (sediment) and valuation based on Benefit transfer method based on case studies from India
	Water regulation and groundwater recharge	2600 Rs/ha/year (MDF) to more than 5000 Rs/ha/year (VDF) in evergreen forests and 663 Rs/Ha/year to more than 3700 Rs/ha/year in deciduous vegetation Based on reference data of groundwater availability combined with the economic value of water compiled from the groundwater authority The economic value of groundwater (after deducting costs) is about 262.5 Rs/k cum of water	InVEST provides the quantum of water recharge within the natural forested areas

(continued)

Table 8 (continued)

Services	Variables	Data source	Approach
	Pollination service	The spatial extent of forest ecosystems and pollination services of forest ecosystems-10167 INR/Ha/Yr (MDF) to 11907 INR/Ha/Yr (VDF) Pollination services are quantified based on the spatial extent of forests, and the economic value of pollination are accounted for through the comparative assessment of natural regeneration of forest patches (with fencing protection from external pressures Ray et al. 2015; Balachandran et al. 2017) and compared with the afforestation cost. These values are comparable to case studies from India (Verma et al. 2013) of natural forest regeneration and its replacement cost (if done artificially) as recommended by the National Afforestation Programme Guidelines (NAP 2009) and based on the success stories of National Beekeeping & Honey Mission (NBHM 2019)	Benefit transfer method $\sum_{I=1}^{n} V_i \times A_i$ Where Vi represents the monetary values per hectare and Ai represents the area
	Water purification	The spatial extent of forest ecosystems and water purification values 2,950 Rs/ ha/yr based on studies from India (Verma et al. 2013)	Benefit transfer method—based on case studies from India
	Waste treatment	The spatial extent of forest ecosystems and waste treatment is estimated at 4716 Rs/ha/yr (Ramachandra et al. 2017; Verma et al. 2013). These values are comparable to the studies across the globe (De Groot et al. 2020) after adjusting for GDP (PPP) per capita for India and currency exchange value	

(continued)

Table 8 (continued)

Services	Variables	Data source	Approach
	Air filtration services	The spatial extent of forest ecosystems and air filtration services of forest ecosystems-8,368 INR/Ha/Yr (MDF) to 22,617 INR/Ha/Yr (VDF) based on published literature from India (Ninan and Kontoleon 2016; Joshi and Negi 2011) which are comparable to tropical forests-global ecosystem service valuation database (ESVD) https://www.es-partnership.org/wp-content/uploads/2020/08/ESVD_Global-Update-FINAL-Report-June-2020.pdf were adjusted for GDP (PPP) per capita of the country for which values were estimated and corresponding currency exchange rate	Benefit Transfer method $\sum_{I=1}^{n} V_i \times A_i$ Where Vi represents the monetary values per hectare and Ai represents the area
	Local (micro and meso) climate regulation services	The spatial extent of forest ecosystems 17,933 INR/Ha/Yr (MDF) to 48,468 INR/Ha/Yr (VDF) based on published literature from India (Ghosh 2020; Verma et al. 2007), which are comparable to values—global ecosystem service valuation database (ESVD) https://www.es-partnership.org/wp-content/uploads/2020/08/ESVD_Global-Update-FINAL-Report-June-2020.pdf adjusted for GDP (PPP) per capita of the country for which values were estimated and corresponding currency exchange rate	

(continued)

Table 8 (continued)

Services	Variables	Data source	Approach
Cultural services	Aesthetic aspects—waterfalls, museums, heritage sites	Karnataka Forest Department (Uttara Kannada, Shimoga, Chikmagalur, Dakshina Kannada and Kodagu districts) (primary survey – entrance fees (aesthetic aspects – waterfalls, museums, heritage sites) × the average number of visitors aesthetic aspects – waterfalls, museums, heritage sites during 2018, 2019, and 2020), and supplemented with the Indian case studies (Ray et al. 2010; Bharath et al. 2017; Ramachandra et al. 2018c), publications—Districts at glance Districts at a Glance https://des.karnataka.gov.in/english https://kgis.ksrsac.in/kag/ Based on these, the value ranges 1500 ± 250 ₹/Ha/Yr, and an average value of Rs 1500 ₹/Ha/Yr was considered	Considering entrance fees (waterfalls, museums, heritage sites) × the average number of visitors to the park/recreation spots during 2018, 2019, and 2020), supplemented with the Indian case studies, and travel expenses associated to the travel, based on the address of visitors and considering the connectivity. Collected additional details from the service providers of revenue during the past three years. The data collected from the service operators include (i) cost (labour, fuel and maintenance) and (ii) annual revenue
	Spiritual and historic	Based on the primary data, spiritual and historic services ranges from 1,200 ₹/Ha/Yr (MDF) to 7,200 ₹/Ha/Yr (VDF) Distribution of sacred groves (relic forests protected under belief) across the state is considered and quantified on the per hectare value-travel cost basis The annual collection and expense data were compiled from administrative/management committees of select groves in Shimoga, Uttara Kannada, and Kodagu districts In groves, where annual collection details were not available, the travel cost method is used to quantify the number of visitors (visiting groves) for annual rituals, festivals, and other religious activities. This is done through primary surveys of select groves in Uttara Kannada, Shimoga, and Kodagu districts and supplemented with case-studies from India—benefit transfer method	Residual method (annual collection for rituals and deducting costs—priest salary and ritual expenses) In groves, where annual collection details were not available, travel cost method and supplemented with case-studies from India—benefit transfer method

(continued)

Table 8 (continued)

Services	Variables	Data source	Approach
	Tourism and recreational	Ranges from 28,944 ₹/Ha/Yr (MDF) to 2,88,000 ₹/Ha/Yr (VDF), based on a primary survey—entrance fees (park, recreation spots) × the average number of visitors to the park during 2018, 2019, and 2020), supplemented with the Indian case studies. Benefit transfer method—(Badola et al. 2017; Gunarekha and Binoy 2017; Ramachandra et al 2017, 2019b; Sinclair et al. 2020)	Travel cost method $\sum_{I=1}^{n} V_i \times A_i$ Where Vi represents the monetary values per hectare and Ai represents the spatial extent of the respective ecosystem and entrance fee with details of visitors
	Education, scientific and research	4800 ₹/Ha/Yr based on the primary data. Details of the (i) nature education programs organized by the Karnataka Forest Department jointly with the non-governmental organisations and universities, (ii) research funding (field research component) research, duration, the project budget (for field research), research team details were compiled from the Karnataka Forest Department This information is supplemented with the data compilation through discussion with researchers, and relevant literature of field-based research (ecology, medicinal plants, etc.) (Chandran et al. 2010; Ray and Ramachandra 2010; Gould et al. 2014; Ray et al. 2014a; Dorji et al. 2019; Kreye et al. 2019)	Based on funding—field research component
Total ecosystem supply value (TESV)	TESV provides the total worth of ecosystem service and is calculated as the sum of provisioning services (*PS*), regulating services (*RS*) and cultural services (*CS*)		$TESV = \sum EVi\ i = 1, 2, 3$ 1: Provisioning, 2: Regulating and 3 Cultural

3.2 Valuation of Services from Agriculture Ecosystems

The revenue due to the services from the agriculture ecosystems is determined through (i) production was calculated for each crop based on the crop area and crop yield per hectare at taluk level; (ii) Minimum support price (MSP) specified by the Government of India, followed by prices at mandi (local market yards, prices fixed by the Government of Karnataka) were used to determine the monetary value; (iii) Regulating services, cultural services, and other provisioning services were obtained based on literature reviews (revenue per unit area is derived from the literature (based in India specific case studies), and for specific parameters where data was not available, international values were used. The method for evaluating the services from the agriculture ecosystem is illustrated in Fig. 7 and Table 9 depicts the revenue of various services per hectare, (iv) Total ecosystem supply value, and (v) NPV.

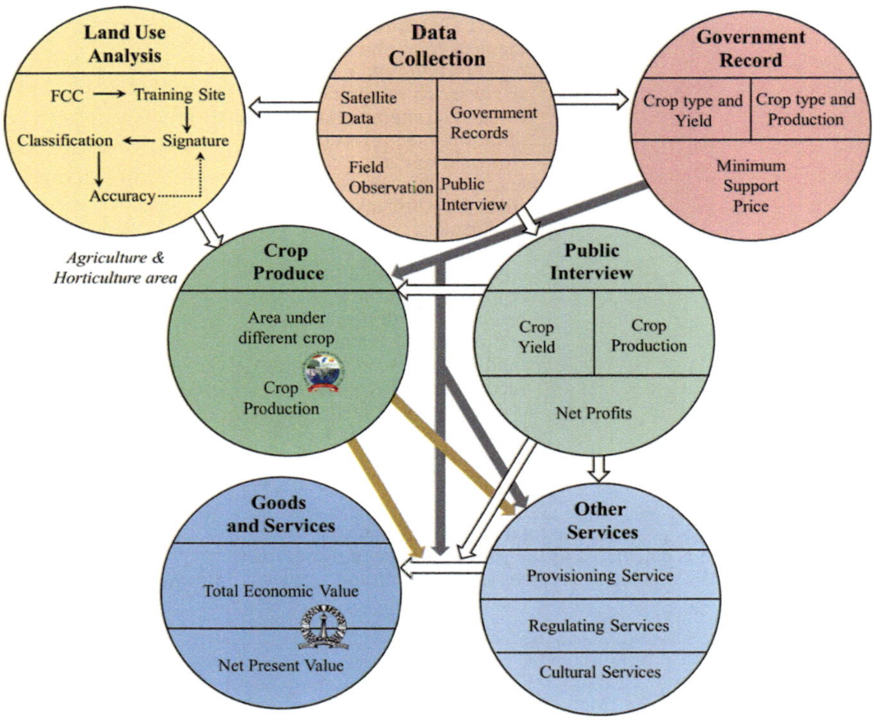

Fig. 7 Method for accounting services in the agriculture (croplands and horticulture) ecosystem

Table 9 Revenue generated per unit area

Services	Variables	Data and source	Approach
Provisioning	Food	MSP—cost of production government records and public interview	**Residual Method** $\sum_{i=1}^{n} Q_i \times (P_i - C_i)$ Where Qi represents quantity, Pi is the price, Ci is the cost involved in the harvest
	Fodder	**3000**–5000 Rs/Acre, 7500–12500 Rs/Ha Public interview, the lowest value is used for accounting	
	Wood	**432**–4000 Rs/Ha, public interview, the lowest value is used for accounting	
Regulating	Air filtration services	3017 Rs per year (Pal 2018)	Benefit transfer method—based on Indian case studies
	Local (micro and meso) climate regulation	720 Rs/Ha/Year, Value is based on global studies based on published literature—global ecosystem service valuation database (ESVD) https://www.es-partnership.org/wp-content/ uploads/2020/08/ESVD_ Global-Update-FINAL-Report-June-2020.pdf adjusted for GDP (PPP) per capita of the country and corresponding exchange rate	Benefit transfer method $\sum_{I=1}^{n} V_i \times A_i$ Where Vi represents the monetary values per hectare and Ai represents the area
	Global climate regulation—carbon sequestration	**Croplands** 36 Rs/Ha/Year Ecosystem extent through land-use analyses, area under crop and type, yield Comparison with the Indian case study (Nayak et al. 2019) Horticulture 5040 Rs/Ha, Ecosystem extent through land-use analyses, area under crop and type, AGB and BGB, Comparison with the Indian case studies (Murali 2010)	InVEST carbon model

(continued)

Table 9 (continued)

Services	Variables	Data and source	Approach
	Soil carbon sequestration	**Agriculture** (croplands) 363.6 Rs/Ha/Year, ecosystem extent based on land use analyses, ecosystem condition (soil), Carbon and Nitrogen (C&N) analyses through elemental analyzer of soil samples (collected from plots representative of agro-climatic regions), Comparison with a case study from India (Nayak et al. 2019) **Horticulture** 14400 Rs/Ha, Comparison with a case study from India (Murali 2010)	CHN elemental analyses of representative soil samples Benefit transfer method InVEST carbon model
	Water flow (groundwater recharge)	**Agriculture (croplands)** 792 Rs/Ha/Year Ecosystem extent, rainfall, etc. Comparison with the study from India (Nayak et al. 2019) **Horticulture** 1224 Rs/Ha/Year Ecosystem extent, rainfall, etc. Comparison with global ecosystem service valuation database (ESVD). Adjusted for GDP (PPP) per capita of the country for which values were estimated and the corresponding exchange rate	InVEST
	Nitrogen fixation	396 Rs/Ha/Year Ecosystem extent based on land use analyses, ecosystem condition (soil) CHN analyses through elemental analyzer of soil samples (collected from plots representative of agro-climatic regions) (Nayak et al. 2019)	C&N elemental analyses of representative soil samples
	Soil fertility (NIC 2020)	2448 Rs/Ha/Year (poor soils) Soil erosion, soil fertility estimated per hectare in terms of monetary values Comparison with (Nayak et al. 2019) 4991 Rs/Ha/Year (good soils) Soil erosion, soil fertility estimated per hectare in terms of monetary values, comparison with http://naasindia.org/Policy%20Papers/policy%2094.pdf	RUSLE, InVEST

(continued)

Table 9 (continued)

Services	Variables	Data and source	Approach
	Remediation—organic and inorganic materials	5760 Rs/Ha/Year (Nayak et al. 2019)	Benefit transfer method (case studies from India)
	Pollination	391 Rs/Ha/Year agriculture (croplands and horticulture) http://naasindia.org/Policy%20Papers/policy%2094.pdf	
	Genetic diversity	12897 Rs/Ha/Year http://naasindia.org/Policy%20Papers/policy%2094.pdf	
	Biological control	115.2Rs/Ha/Year (Nayak et al. 2019)	
Cultural	Tourism & recreational	941 Rs/Ha/Year Travel cost method—Uttara Kannada district (based on the number of visitors visiting farm houses per year, the amount paid and expenses, and benefit to travel operators) and comparison with the case study http://naasindia.org/Policy%20Papers/policy%2094.pdf	Travel cost method
	Inspirational, culture, art	1152 Rs/Ha/Year Values based on global studies based on published literature—global ecosystem service valuation database (ESVD), adjusted for GDP (PPP) per capita of the country for which values were estimated and corresponding exchange rate	$\sum_{I=1}^{n} V_i \times A_i$ Where Vi represents the monetary values per hectare and Ai represents the area
Total ecosystem supply value (TESV) for agriculture (croplands, horticulture) ecosystem	TESV provides the total worth of ecosystem service and is calculated as the sum of provisioning services (*PS*), regulating services (*RS*), and cultural services (*CS*)		$TESV = \sum EVi$ $i = 1, 2, 3$ 1: provisioning, 2: regulating and 3: Cultural

3.3 Total Ecosystem Supply Value [TESV]

The ecosystem services (provisioning, regulating, and cultural) for forest and agriculture ecosystems were then summed across all benefit flows to estimate a total annual flow of value from the respective spatial units.

Temporal comparison of ecosystem services: Monetary values of ecosystem services (provisioning, regulating, cultural services, and TESV) of 2005 and 2019 are compared to understand the changes due to changes in the spatial extent and condition of the ecosystem. The monetary values of 2005 were adjusted to 2019 values by considering the GDP deflator (MoSPI 2020) of an inflation rate of 2.92 times (*Inflation Calculator-Indian Rupee* 2021).

3.4 Net Present Value (NPV) of Ecosystem Assets

The net present value (NPV) *is the value of an asset determined by estimating the stream of income expected to be earned in the future and then discounting the future income back to the present accounting period* (SEEA EA 2021). In ecosystem accounting, it is applied by aggregating the NPV of expected future returns for each ecosystem service supplied by an ecosystem asset. The use of an NPV approach implies that their value will be related to the capacity to supply ecosystem services and how this capacity is expected to change in the future.

NPV is based on the measurement of (i) the ecosystem services supplied by the asset, and the monetary values of these services (ii) the estimation of a pattern of future flows of each ecosystem service, taking into account expected degradation/enhancement and demand (iii) expected future prices for each ecosystem service; (iv) expected institutional arrangements; and (v) expected asset life. In addition, NPV requires a discount rate, which adjusts the value of a stream of future flows to account for time preferences and attitudes to risk. Discount rates are required to convert the expected future ecosystem services flows into a current period estimate of overall value. In this report, a social discount rate (as opposed to an individual discount rate) has been used, as most of the ecosystem services contribute to collective benefits.

Net present value is calculated using Eq. 1 and applied at the level of individual ecosystem services, and the resulting discounted values are aggregated to derive the monetary value of the ecosystem asset.

$$\mathrm{NPV} = \sum_{t=1}^{T} \frac{\mathrm{ES_t}}{(1 + r)^t} \tag{1}$$

where, t—Number of years ranging between 1 and T.

T—Number of years for which this annual benefit from the asset will accrue. This is closely linked to the length of time needed to regenerate the same type and quality of forests. The Hon'ble Supreme Court of India in the Judgement of 26th September 2005 (Page 10, Para 4) has suggested that the basis for calculation of NPV should be the economic value spread over a period of 50 years, which is the period for forest regeneration

ES_t—Ecosystem services at time t

r—Social rate of discount for capital returns. As per the norms in India and the Hon'ble Supreme Court, a social discount rate of 4% is applied for renewables and 2% for non-renewable resources. Hence, for forest ecosystems, 3% is considered, which is the weighted average of renewable and non-renewable.

NPV of ecosystems (forests, agriculture) in Karnataka is computed using TESV-the total value of ecosystem flow based on a social discount rate of 3% and a period of 50 years. The ecosystem service values are determined based on government records, and prices were considered to remain the same with no inflations for 50 years (w.r.t 2019).

4 Results and Discussion

Valuation of ecosystem services entails the computation of spatial extent and conditions of the respective ecosystems, and the quantification of services in physical and monetary units. The spatial extent (Ramachandra et al. 2021a) and the condition of forests (Ramachandra et al. 2021b) assessed district-wise earlier for Karnataka (Ramachandra et al. 2021a, b) were used for computing the value of the provisioning services per hectare. Forests have been reclassified as VDF (very dense forests), MDF (medium dense forests), and open forests (OF) based on the condition (computed through the fragmentation analyses), comparable to the classification approach adopted by the Karnataka Forest Department (based on the forest type and their canopy cover). A similar approach was adopted earlier to compute ecosystem services from forest ecosystems in India (Verma et al. 2013).

The following subsections present (i) the extent of ecosystems in Karnataka, (ii) the valuation of ecosystem services, (iii) computation of TESV: Total ecosystem supply value and NPV of ecosystem assets.

4.1 Assessment of Ecosystem Extent Over Time

An assessment of ecosystem extent was performed, using land use and land cover as proxies. The spatiotemporal land cover/use analysis was carried out from 1985 to 2019 using remote sensing data through a supervised classifier based on the Gaussian maximum likelihood algorithm.

The state witnessed large-scale land-use transitions post-1990s due to globalization and the consequent spurt in industrialization and urbanization, as well as an increase of horticulture crops and the conversion of croplands (cereals, pulses, etc.) to market-based cash crops (coffee, sugarcane, areca nut, etc. with higher economic values), etc.

Temporal land cover/use analyses reveal the decline of forest cover in Karnataka from 1985 to 2019 (Fig. 8). Districts of the Western Ghats region have a higher forest cover than other districts, as depicted in Fig. 8. Currently, 15% of the State's geographical area is forested, compared to 21% in 1985 (Table 10). Large-scale developmental activities such as constructing a series of reservoirs and dams, creating special economic zones, townships, and land conversion for built-up areas have led to the loss of large tracts of forests. The forest cover now is confined to major conservation reserves such as protected areas, national parks, and wildlife sanctuaries. Natural forests show a decline of evergreen forests from 7.5% (1985) to 5.7% (by 2019), moist deciduous forests from 5.7% (1985) to 4.1% (by 2019), and dry deciduous forests from 4.0% (1985) to 2.2% (2019).

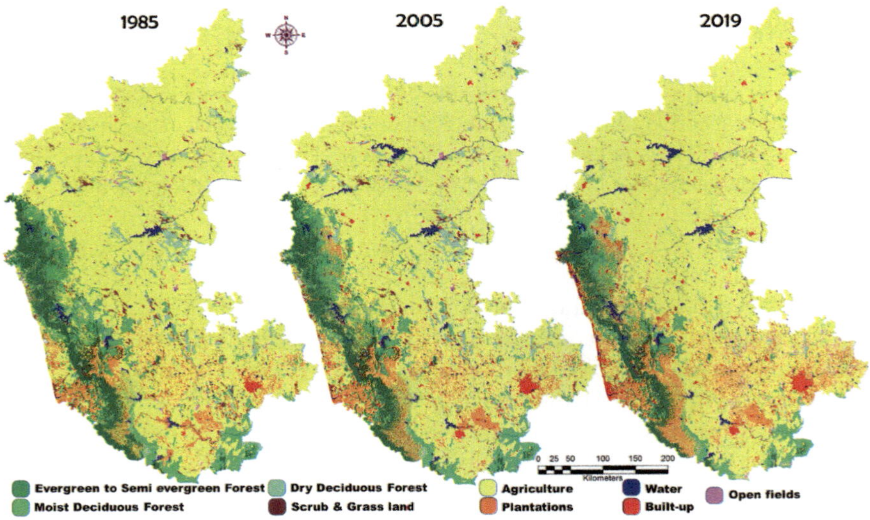

Fig. 8 Land uses from 1985 to 2019 in Karnataka

The abrupt land-use conversion has also resulted in a loss of productive agricultural lands near the cities such as Bengaluru, Mysore, Hubli-Dharwad, Shimoga, etc. Districts such as Kodagu, Uttara Kannada, Bengaluru, Shimoga, Belgaum, Dakshina Kannada, and Chikmagalur have been experiencing large-scale land cover change due to unplanned development activities.

The built-up cover has increased from 0.47 to 3% from 1985 to 2019, causing an impact on agriculture, forests, and lakes. This necessitates sustainable land-use policies to arrest deforestation and abrupt land conversions.

Horticultural areas have increased from 8.8% (1985) to 11.1% (2019), and category-wise land-use dynamics are presented in Table 10. Large-scale monoculture plantations of eucalyptus, rubber, acacia, teak, and areca nuts have increased and now cover 12% of the state. In addition, new urban agglomerations were noticed across cities and major towns such as Bengaluru, Mangalore, Hubli, Hassan, Mysore, etc. These changes are abrupt and have resulted in a disruption in the provision of ecosystem services, affecting the hydrologic regime and the availability of natural resources. While horticultural areas have increased overall, coastal and Tier-1 cities (e.g. Bangalore) and Tier-2 cities (e.g. Mysore, Hubli-Dharwad, Belgaum) are experiencing a loss of agricultural areas in the sub-urban regions with new layouts and satellite towns. The land-use transitions (Table 11) from 1985 to 2019 are computed to understand the probability of changes in each land use type from time t_1 to time t_2. Table 12 summarises the ecosystem extent account for Karnataka, which uses land use—land cover (LULC) classes as proxies for ecosystem type.

The forest ecosystem extent is further disaggregated into very dense forests (VDF) and medium-density forests (MDF) based on the ecosystem condition (fragmentation of forests). Karnataka state has 11,334 km^2 area under VDF (2019), which accounts for 6% of the entire landscape, and MDF covers 12,869 km^2, which accounts for 7% of the state land area (Fig. 9).

Table 13 provides a breakdown of ecosystem types (VDF and MDF categories) for forest circles in Karnataka. Districts such as Uttara Kannada and Kodagu have good VDF coverage compared with other districts. On the other hand, from 2005 to 2019, districts such as Ballari, Bagalkot, Bidar, and Kolar have lost major tracts of MDF.

Table 10 Ecosystem extent—**Karnataka State** (units in km^2 and percentage)—based on temporal remote sensing data analyses

Ecosystem type		Built-up	Crop land	Horticulture	Fallow land	Evergreen forest	Moist deciduous forest	Dry deciduous forest	Scrub \grass lands	Water	Total area
Year	Units										
1985	km^2	904	128468	16790	1678	14293	10960	7622	6733	4344	191791
	%	0.5	67.0	8.8	0.9	7.5	5.7	4.0	3.5	2.3	
2005	km^2	2666	127196	20209	1185	12445	9900	7410	5604	5177	191791
	%	1.4	66.3	10.5	0.6	6.5	5.2	3.9	2.9	2.7	
2019	km^2	5748	127962	21325	2854	10888	7892	4281	4907	5934	191791
	%	3.0	66.7	11.1	1.5	5.7	4.1	2.2	2.6	3.1	
Changes from 1985 to 2019											
1985	km^2	904	128468	16790	1678	14293	10960	7622	6733	4344	191791
2019	km^2	5748	127962	21325	2854	10888	7892	4281	4907	5934	191791
Net change in extent from1985 to 2019											
Extent	km^2	4844	−505	4536	1175	−3405	−3068	−3341	−1826	1590	
	%	535.8	−0.4	27.0	70.0	−23.8	−28.0	−43.8	−27.1	36.6	

Table 11 Ecosystem extent change matrix from 1985 to 2019—Karnataka State (extent in km^2 and percentage)

2019 Ecosystem type / 1985	Units	Built-up	Agriculture-crop lands	Horticulture	Fallow land	Evergreen forest	Moist deciduous forest	Dry deciduous forest	Scrub / Grass lands	Water	Total (opening stock) 1985
Built-up	km^2	859	20	15	2	1	3	1	2	2	904
	%	95.0	2.2	1.7	0.2	0.2	0.4	0.1	0.2	0.3	
Agriculture—Crop lands	km^2	3194	114202	4003	390	940	98	1109	2451	2081	128468
	%	2.5	88.9	3.1	0.3	0.7	0.1	0.9	1.9	1.6	
Horticulture	km^2	775	3344	11661	185	243	258	93	108	122	16790
	%	4.6	19.9	69.5	1.1	1.4	1.5	0.6	0.6	0.7	
Fallow land	km^2	50	843	32	710	1	2	12	13	15	1678
	%	3.0	50.2	1.9	42.3	0.1	0.1	0.7	0.8	0.9	
Evergreen forest	km^2	175	416	2372	151	9097	1402	211	286	182	14293
	%	1.2	2.9	16.6	1.1	63.6	9.8	1.5	2.0	1.3	
Moist deciduous forest	km^2	190	1973	1648	485	388	5581	470	164	61	10960
	%	1.7	18.0	15.0	4.4	0.4	54.1	4.3	1.5	0.6	
Dry deciduous forest	km^2	85	3374	779	458	68	419	2306	87	47	7622
	%	1.1	44.3	10.2	6.0	0.9	5.5	30.3	1.1	0.6	
Scrub/grass lands	km^2	327	3056	701	438	136	113	72	1787	103	6733
	%	4.9	45.4	10.4	6.5	2.0	1.7	1.1	26.5	1.5	
Water	km^2	93	734	114	35	14	16	7	9	3321	4344
	%	2.1	16.9	2.6	0.8	0.3	0.4	0.2	0.2	76.5	
Closing Stock, 2019	km^2	5748	127962	21325	2854	10888	7892	4281	4907	5934	191791
	%	3.0	66.7	11.1	1.5	5.7	4.1	2.2	2.6	3.1	

Table 12 Net additions and reductions in ecosystems in Karnataka State, India

Ecosystem type	Disaggregated ecosystem type	Karnataka				
		Opening stock 1985	Additions to stock	Reduction in stock	Closing stock 2019	Net change (in%) during 1985 to 2019
Built-up land	Built-up	904	4866	45	5725	533.1
	Urban					
	Rural					
	Mining					
	Sub-total 1	**904**	**4866**	**45**	**5725**	**533.1**
Agricultural land	Horticulture	16790	9711	5129	21371	27.3
	Cropland	128468	13760	14317	127910	−0.4
	Fallow Land	1678	6284	968	6994	316.7
	Sub-total 2	**146936**	**29754**	**20414**	**156275**	**6.4**
Forests	Evergreen/ Semi-Evergreen	14293	921	5196	10018	−29.9
	Moist Deciduous	10960	2333	5379	7914	−27.8
	Dry Deciduous	7622	981	5316	3288	−56.9
	Scrub forest	6733	922	4946	2710	−59.8
	Forest Plantation					
	Swamp/ Mangroves					
	Sub-total 4	**39607**	**5158**	**20836**	**23929**	**−39.6**
Grass/ grazing	Grass/grazing					
	Sub-total 5					
Snow and glacier	Snow and Glacier					
	Sub-total 6					
Wetlands/ water bodies	Inland Wetland					
	Coastal Wetland					
	River/stream/ canals					
	Waterbodies	4344	2541	1023	5862	35.0
	Sub-total 7	4344	2541	1023	5862	35.0
Grand total		191791	42319	42319	191791	

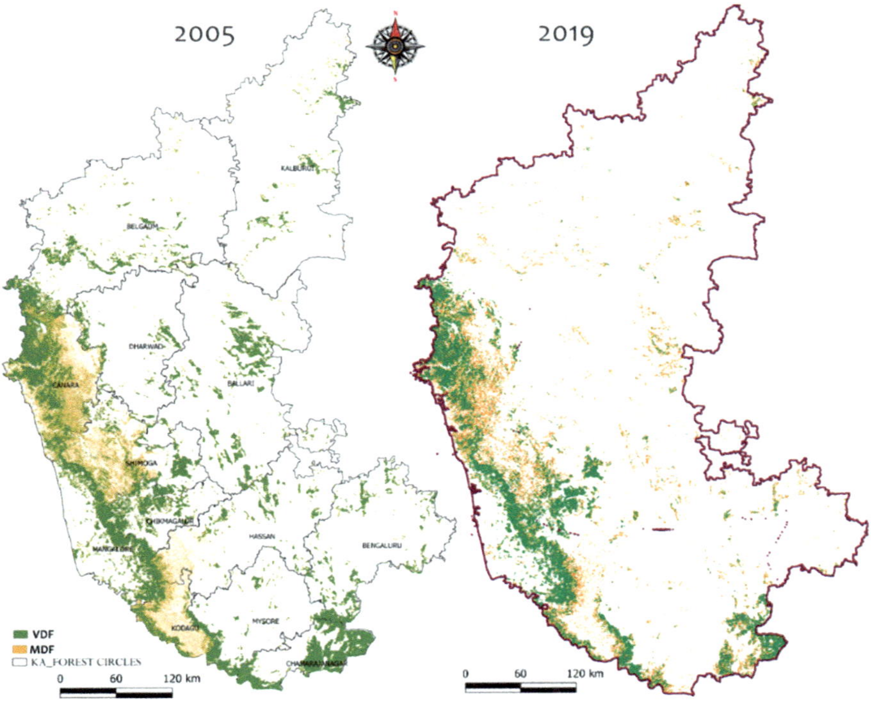

Fig. 9 Very dense forests (VDF) and medium-density forests (MDF) in Karnataka

Table 13 Circle wise forest ecosystem reclassified as VDF and MDF

S.no	Circle	VDF (Ha)		MDF (Ha)	
		2005	2019	2005	2019
1.	Bengaluru	1,44,717	33,696	37,213	63,035
2.	Belgaum	2,01,790	60,393	78,388	87,360
3.	Ballari (Bellary)	2,49,789	13,514	64,122	92,697
4.	Chamarajanagar	2,72,557	96,515	15,222	89,538
5.	Chikmagalur	1,84,516	1,31,484	37,230	64,794
6.	Dharwad	83,942	17,012	25,637	35,280
7.	Kalaburagi (Gulbarga)	79,842	15,575	51,355	38,904
8.	Hassan	1,04,205	44,309	88,492	72,695
9.	Canara	3,78,655	2,56,730	3,10,345	3,32,741
10.	Kodagu	93,764	59,239	1,25,832	1,12,642
11.	Mangaluru	2,60,299	2,23,033	54,209	73,169
12.	Mysore	1,16,604	68,775	19,081	33,795
13.	Shivamogga	1,95,940	1,13,215	2,23,267	1,90,270
Total area		23,66,620	11,33,490	11,30,393	12,86,920

4.2 Valuation of the Ecosystem Services

Ecosystem services and the natural capital stocks in Karnataka State make significant direct and indirect contributions to the district and state economies and human welfare. The evaluation of ecosystem services will aid in formulating policy and legislation that can provide protection and sustainable management of ecosystems to capitalize fully on the most significant ecosystem services. Accounting for ecosystem services in physical terms aims to record, in an accounting structure, the flows of ecosystem services over an accounting period in physical units such as cubic meters and tonnes. Physical quantification commonly focuses on measuring ecosystem structures, processes, and functions, i.e., the supply side of ecosystem service flows (SEEA EA) (SEEA EA 2021).

Forests are multi-functional: they provide an often complex array of goods and services. Forests, both natural and planted, and including trees spread across the terrain, have a critical role in the ecology, aesthetics, and recreational benefits. Tables 14 and 15 present the biophysical ecosystem supply from forests between 2005 and 2019. The flows for each ecosystem service are recorded using a unit of measure that is appropriate for that ecosystem service (for example, cubic meters for timber, kg's/tons for NTFP or fuelwood, etc.).

Recording monetary values for ecosystem services underpins the compilation of two ecosystem accounts: the ecosystem services flow account in monetary terms and the ecosystem asset account. The monetary ecosystem services flow account records the monetary value of flows of ecosystem services based on their exchange values. The data from this account can be used to understand the relative economic significance of different ecosystem services (within the valuation framing of the System of National Accounts), support aggregation of ecosystem services to compare the role of different ecosystem assets, understand changes in monetary value over time, underpin comparison of the inputs of different ecosystem services to different users, and support understanding the role of ecosystem services in different locations (SEEA EA 2021).

Ecosystem services were quantified through the residual value method by taking the gross value of the final marketed good to which the ecosystem service provides input and then deducting the cost of all other inputs, including labour, produced assets, and intermediate inputs (as per the SEEA Ecosystem Accounting). Ecosystem services were computed based on the ecosystem flows in 2005 and 2019. Values of 2005 were adjusted through the consumer price index or GDP deflator; these values reflect the real measures of ecosystem services, which could be compared with ecosystem services of 2019. Tables 16 and 17 provide ecosystem provisioning services in monetary terms for 2005 (at 2019 values) and 2019, respectively.

A comparison of values of services of 2019 with 2005 (at 2019 values) highlights there has been a considerable decline in the provisioning services evident from 42% decline in rosewood, 93% decline in bamboo, NTFP (honey reduced by 97%, tamarind reduced by 75%), 42% decline in fodder and 35% decline in medicine,

Table 14 Ecosystem supply in physical terms, by ecosystem type, 2005

Ecosystem type		Built-up	Crop land	Horticulture	Fallow land	Evergreen forest	Moist deciduous forest		Dry deciduous forest	Scrub\grass land	Water	Total
Year	Units											
2005	Km^2	2,666	127,196		20,209	1,185	12,445	9,900	7,410	5,604	5,177	191,791
Extent	%	1.4	66.3		10.5	0.6	6.5	5.2	3.9	2.9	2.7	100
Ecosystem goods (Physical)—Karnataka												
Rosewood (Cum)							7,536	5,994				13,530
Teak wood (Cum)									3,242			3,242
Eucalyptus (Cum)								8,367	6,263	4,736		19,366
Soft wood (Cum)							4,915	3,910				8,825
Other kinds of timber (Cum)								17,629	13,195	9,979		40,802
Pulpwood (Cum)								1,397	1,046	791		3,234
Round Poles (numbers)								16,176	12,108	9,157		37,441
Sandal wood (Kg)								9,500	7,111	5,378		21,989
Bamboo nos								7,39,718	5,53,668	4,18,725		17,12,112
Canes nos							1,80,407	1,43,514				3,23,921
Firewood ton			15,274		2,427	147	2,14,94,440	1,70,98,832	1,27,98,216	96,78,975		6,10,88,306
Honey ton							2,00,126	1,59,200	1,19,159	90,117		5,68,602
Soap nut ton								1,84,700	1,38,245	1,04,552		4,27,497
Cashew ton								11,194	8,379	6,336		25,909
Tamarind ton								12,978	9,714	7,347		30,039

(continued)

Table 14 (continued)

Ecosystem type Year / Units	Built-up	Crop land	Horticulture	Fallow land	Evergreen forest	Moist deciduous forest		Dry deciduous forest	Scrub\grass land	Water	Total
Ramapatre (nutmeg–*Myristica malabarica*) ton						9,571					9,571
Dalchini ton						5,379	4,279				9,658
Murugalu (kokum or *Garcinia indica*) ton						2859	2275				5,134
Fish ton						57,305	45,586	34,121	25,805	23,838	1,86,655
Fodder k ton		12		19	1	13,749	10,937	8,186	6,191		39,063
Medicine ton						2,72,010	2,16,384	1,61,960	1,22,486		7,72,840
Water million m^3	1,433	68,369		10,863	637	6,689	5,321	3,983	3,012	2,782	1,03,090
Genetic ton						1,33,79,956	1,06,43,757	79,66,691	60,25,012		3,80,15,416
Cereals tons		1,09,39,568			1,01,917						1,10,41,485
Pulses tons		14,13,459			13,168						14,26,627
Oilseeds tons		18,74,861			17,467						18,92,328
Commercial plantation (sugarcane, coffee, tobacco) tons		140,59,607		22,33,801	1,30,984						1,64,24,392
Horticulture (arecanut, coconut)) tons				14,18,616							14,18,616
Fruits tons	30,456	14,53,078		2,30,866	13,537						17,27,937
Vegetables-tons	25,301	11,98,204		1,89,761	10,843						14,24,110

Table 15 Ecosystem supply in physical terms, by ecosystem type, 2019

Ecosystem type		Built-up	Crop land	Horticulture	Fallow land	Evergreen forest	Moist deciduous forest	Dry deciduous forest	Scrub\grass lands	Water	Total area
Year	Units										
2019	Km^2	5,748	1,27,962	21,325	2,854	10,888	7,892	4,281	4,907	5,934	1,91,791
	%	3	66.7	11.1	1.5	5.7	4.1	2.2	2.6	3.1	100
Ecosystem Goods (Physical)—Karnataka											
Rosewood (Cum)						4,591	3,653				8,244
Teak wood (Cum)								3,538			3,538
Eucalyptus (Cum)							10,893	8,153	6,166		25,212
Soft wood (Cum)						4,447	3,538				7,985
Other kinds of timber (Cum)							17,225	12,893	9,751		39,869
Pulpwood (Cum)							2,393	1,791	1,355		5,539
Round Poles (Nos)							17,752	13,287	10,049		41,088
Sandal wood (Kg)							3,786	2,833	2,143		8,762
Bamboo Nos							54,819	41,031	31,031		1,26,881
Canes Nos						19,987	15,900				35,887
Firewood ton			38,320	6,834	861	132,948	1,05,760	79,160	59,867		4,23,921
Honey ton						5,610	4,462	3,340	2,526		15,938
Soap nut ton							36,516	27,331	20,670		84,517
Cashew ton							401	300	227		929
Tamarind ton							3,318	2,484	1,878		7,680
Rampatri (nutmeg–*Myristica malabarica*) ton						4					4

(continued)

Table 15 (continued)

Ecosystem type	Built-up	Crop land	Horticulture	Fallow land	Evergreen forest	Moist deciduous forest	Dry deciduous forest	Scrub\grass lands	Water	Total area
Year / Units										
Dalchini ton					77	62				139
Murugalu (kokum or *Garcinia indica*) ton					73	58				131
Fish ton					57,305	45,586	34,121	25,805	23,838	1,86,655
Fodder k ton		113,412	18,872	1,145	5,913	4,704	3,521	2,662		16,936
Medicine ton					1,88,268	1,49,767	112,098	84,777		5,34,911
Water million m^3	686	32,745	5,202	305	3,203	2,548	1,907	1,442	1,332	49,374
Genetic ton					75,94,703	60,41,588	45,22,037	34,19,905		2,15,78,233
Cereals tons		1,05,74,120		2,35,840						1,08,09,959
Pulses tons		16,39,096		36,558						16,75,654
Oilseeds tons		9,76,167		21,772						9,97,939
Commercial plantation (sugarcane, coffee, tobacco) tons		3,54,33,331	59,05,001	7,90,287						4,21,28,620
Horticulture (arecanut, coconut) tons			34,10,492							34,10,492
Fruits tons	2,23,851	49,83,364	8,30,483	1,11,146						61,48,844
Vegetables-tons	2,41,652	53,72,724	8,94,112	1,20,826						66,29,314

Table 16 Ecosystem supply in monetary terms (million ₹), by ecosystem type, 2005

Ecosystem type		Built-up	Crop land	Horticulture	Fallow land	Evergreen forest	Moist deciduous forest	Dry deciduous forest	Scrub_ grass lands	Water	Total
Year	Units										
2005	km^2	2,666	1,27,196	20,209	1,185	12,445	9,900	7,410	5,604	5,177	1,91,791
	%	1.4	66.3	10.5	0.6	6.5	5.2	3.9	2.9	2.7	21.13
Ecosystem goods (monetary, million ₹)—Karnataka											
Rosewood						₹ 530.9	₹ 422.4				₹ 953.3
Teak wood								₹ 212.0			₹ 212.0
Eucalyptus							₹ 61.3	₹ 45.9	₹ 34.7		₹ 141.9
Soft wood						₹ 6.6	₹ 5.2	₹ 0.0	₹ 0.0		₹ 11.8
Other kinds of timber							₹ 152.7	₹ 114.3	₹ 86.4		₹ 353.4
Pulpwood							₹ 2.2	₹ 1.6	₹ 1.2		₹ 5.0
Round Poles							₹ 5.8	₹ 4.3	₹ 3.3		₹ 13.4
Sandal wood							₹ 142.6	₹ 106.7	₹ 80.7		₹ 330.0
Bamboo							₹ 28.2	₹ 21.1	₹ 15.9		₹ 65.2
Canes						₹ 0.7	₹ 0.5				₹ 1.2
Fuelwood						₹ 17,195.6	₹ 13,679.1	10,238.6	7,743.2		₹ 48,856.4
Honey						₹ 28,043.6	₹ 22,308.7	16,697.7	12,628.1		₹ 79,678.1
Soap nut							₹ 23,947.4	17,924.3	13,555.7		₹ 55,427.3
Cashew							₹ 1,119.4	₹ 837.9	₹ 633.6		₹ 2,590.9
Tamarind							₹ 847.3	₹ 634.2	₹ 479.6		₹ 1,961.1
Ramapatre (nutmeg–*Myristica malabarica*)						₹ 765.7					₹ 765.7

(continued)

Table 16 (continued)

Ecosystem type		Built-up	Crop land	Horticulture	Fallow land	Evergreen forest	Moist deciduous forest	Dry deciduous forest	Scrub_ grass lands	Water	Total
Year	Units										
Dalchini						₹ 33.3	₹ 26.5				₹ 59.8
Murugalu (kokum or *Garcinia indica*)						₹ 1,143.7	₹ 909.9				₹ 2,053.6
Fish						₹ 2,408.6	₹ 1,916.1	1,434.2	1,084.6	1,002.0	₹ 7,845.4
Fodder						₹ 23,550.2	₹ 18,734.2	14,022.3	10,604.7	₹ 0.0	₹ 66,911.4
Medicine						₹ 272.0	₹ 216.4	₹ 162.0	₹ 122.5	₹ 0.0	₹ 772.8
Water		₹ 8,598.0	₹ 4,10,214.1	₹ 65,175.1	₹ 3,821.7	₹ 40,135.8	₹ 31,928.0	23,897.7	18,073.2	₹ 16,696.1	₹ 6,09,941.8
Genetic			₹ 0.0	₹ 0.0	₹ 0.0	₹ 1,33,799.6	₹ 1,06,437.6	₹ 79,666.9	₹ 60,250.1		₹ 3,80,154.2
Cereals			₹ 69,204.2		₹ 644.7						₹ 69,848.9
Pulses			₹ 20,397.4		₹ 190.0						₹ 20,587.4
Oilseeds			₹ 43,023.4		₹ 400.8						₹ 43,424.2
Commercial plantation (sugarcane, coffee, tobacco)			₹ 35,475.4	5,636.4	₹ 330.5						₹ 41,442.2
Horticulture (arecanut, coconut)				73,636.0							₹ 73,636.0
Fruits		₹ 372.7	₹ 17,783.2	2,825.4	₹ 165.7						₹ 21,147.0
Vegetables		₹ 191.1	₹ 9,050.7	1,433.4	₹ 81.9						₹ 10,757.1

Table 17 Ecosystem supply in monetary terms (million ₹), by ecosystem type, 2019

Ecosystem type		Built-up	Crop land	Horticulture	Fallow land	Evergreen forest	Moist deciduous forest	Dry deciduous forest	Scrub\grass lands	Water	Total
Year	Units										
2019	km^2	5,748	1,27,962	21,325	2,854	10,888	7,892	4,281	4,907	5,934	1,91,791
	%	3	66.7	11.1	1.5	5.7	4.1	2.2	2.6	3.1	17.68
Ecosystem goods (monetary, million ₹)—Karnataka											
Rosewood						₹ 307.6	₹ 244.7	₹ 0.0			₹ 552.3
Teak wood								₹ 230.0			₹ 230.0
Eucalyptus							₹ 71.9	₹ 53.8	₹ 40.7		₹ 166.4
Soft wood							₹ 5.5				₹ 12.5
Other kinds of Timber							₹ 144.7	₹ 108.3	₹ 81.9		₹ 334.9
Pulpwood							₹ 3.7	₹ 2.8	₹ 2.1		₹ 8.6
Round Poles							₹ 6.2	₹ 4.6	₹ 3.5		₹ 14.3
Sandal wood							₹ 56.8	₹ 42.5	₹ 32.1		₹ 131.4
Bamboo							₹ 2.1	₹ 1.6	₹ 1.2		₹ 4.8
Canes						₹ 0.1	₹ 0.1				₹ 0.1
Firewood						₹ 20,172.4	₹ 16,047.1	₹ 12,011.0	₹ 9,083.6		₹ 57,314.2
Honey						₹ 785.3	₹ 624.7	₹ 467.6	₹ 353.6		₹ 2,231.3
Soap nut							₹ 4,747.0	₹ 3,553.1	₹ 2,687.1		₹ 10,987.2
Cashew							₹ 40.1	₹ 30.0	₹ 22.7		₹ 92.9
Tamarind							₹ 215.7	₹ 161.4	₹ 122.1		₹ 499.2
Ramapatri						₹ 0.3					₹ 0.3
Dalchini						₹ 24.0	₹ 19.1				₹ 43.1

(continued)

Table 17 (continued)

Ecosystem type		Built-up	Crop land	Horticulture	Fallow land	Evergreen forest	Moist deciduous forest	Dry deciduous forest	Scrub\grass lands	Water	Total
Year	Units										
Murugalu (kokum or *Garcinia Indica*)						₹ 29.2	₹ 23.2				₹ 52.4
Fish						₹ 3,724.8	₹ 2,963.1	₹ 2,217.8	₹ 1,677.3	₹ 1,549.5	₹ 12,132.6
Fodder						₹ 13,748.7	₹ 10,937.1	₹ 8,186.3	₹ 6,191.1		₹ 39,063.2
Medicine						₹ 188.3	₹ 149.8	₹ 112.1	₹ 84.8		₹ 534.9
Water		₹ 4,118.0	₹ 1,96,471.6	₹ 31,215.6	₹ 1,830.4	₹ 19,223.0	₹ 15,291.9	₹ 11,445.8	₹ 8,656.1	₹ 7,996.6	₹ 2,92,130.9
Genetic						₹ 75,947.0	₹ 60,415.9	₹ 45,220.4	₹ 34,199.0		₹ 2,15,782.3
Cereals			₹ 81,497.5		₹ 1,817.7						₹ 83,315.2
Pulses			₹ 28,348.2		₹ 632.3						₹ 28,980.4
Oilseeds			₹ 19,894.0		₹ 443.7						₹ 20,337.7
Commercial (sugarcane, coffee, tobacco)			₹ 97,843.3	₹ 16,305.7	₹ 2,182.2						₹ 1,16,331.3
Horticulture (arecanut, coconut)				₹ 80,605.9							₹ 80,605.9
Fruits		₹ 2,480.9	₹ 55,229.5	₹ 9,204.1	₹ 1,231.8						₹ 68,146.2
Vegetables		₹ 1,481.1	₹ 32,929.1	₹ 5,480.0	₹ 740.5						₹ 40,630.6

which could be attributed to the degradation of forests (extent as well as conditions) in Karnataka during 2005–2019.

Tables 18 and 19 list services by ecosystem type (forest, agriculture, and horticulture) for 2005 (at 2019–2020 values) and 2019, respectively. Table 19 gives a comparison of 2019 ecosystem services values with 2005. There has been a reduction in ecosystem services—28.4% reduction in provisioning services (51.6% reduction in forest ecosystem), 14% reduction in regulating services (mainly in forest ecosystem—27.1% reduction), and 0.2% reduction in cultural services.

4.3 Valuation of the Forest Ecosystem Services—Forest Circle Wise, Karnataka

The forest provisioning services (physical values), area of extraction, and seigniorage value (revenue) for two five-year periods (2001–2005 and 2015–2019) were compiled from the respective forest circles of the Karnataka Forest Department. Averages of five years of goods were used to quantify goods in physical terms for 2005 and 2019. Forests are managed by the Karnataka Forest Department, Government of Karnataka (KFD 2020). The administration and management of forests in Karnataka state are decentralized through forest circles, and there are 13 forest circles in Karnataka. Temporal data of 5 years period helped to account for variability across the study period (years). The seigniorage represents the residual value of the respective goods after deducting the cost involved (harvesting, transportation, etc.). Seigniorage is expressed as revenue received by the government (Haslag 2020) after deducting all expenses from the auction amount of the respective provisioning services.

Provisioning services of forest ecosystems in Karnataka

The provisioning services—timber (rosewood (*Dalbergia latifolia*), teak wood (*Tectona grandis, Terminalia arjuna Terminalia elliptica, Anogeissus latifolia*, etc.), eucalyptus (*Eucalyptus globulus*), soft wood (*Artocarpus hirsutus, Ailanthus excelsa*, etc.), round poles, pulp wood, sandal wood, other timber), bamboo (Bambusa arundinacea, *Dendrocalamus strictus*, etc.), cane (*Calamus neelagiricus, C. vattayila, C. pseudofeanus,* etc.), NTFP (honey, soap nut (*Sapindus saponaria*), cashew nut (*Anacardium occidentale*), tamarind ( ), dalchini (cinnamon-*Cinnamomum malabatrum, Cinnamomum verum*); rampatri (nutmeg—*Myristica malabarica*); murgalu (kokum or *Garcinia indica*)); fuelwood; fodder, etc. extracted (in physical units) with seigniorage values from each circle are provided in Tables 20, 21 and 22, and Figs. 10, 11 and 12 presents circle wise provisioning services. The seigniorage values for 2005 and 2019 are the same for all provisioning services, as per the forest department reports.

Timber: Timber includes rosewood, teak wood, eucalyptus, softwood, round pole, etc. The provisioning services (monetary) of all timber wood products extracted from each circle are assessed considering the respective seigniorage value of individual

Table 18 Ecosystem wise—Provisioning, regulating and cultural services—2005 (Million ₹) (at 2019–2020 values, 2005 values were adjusted to 2019 considering inflation)

Ecosystem type		Built-up	Crop land	Horticulture	Fallow land	Evergreen forest	Moist deciduous forest	Dry deciduous forest	Scrub\grass lands	Water	Total
Year	units										
2005	km^2	2,666	1,27,196	20,209	1,185	12,445	9,900	7,410	5,604	5,177	1,91,791
	%	1.4	66.3	10.5	0.6	6.5	5.2	3.9	2.9	2.7	100
Provisioning, regulating and cultural services (in Million ₹) for forest, agriculture and horticulture ecosystems											
Food (cereal, pulses)			₹ 2,51,411	₹ 39,944	₹ 2,342						₹ 2,93,697
Timber						₹ 712	₹ 566	₹ 424	₹ 321		₹ 2,023
NTFP						₹ 50,167	₹ 39,908	₹ 29,871	₹ 22,590		₹ 1,42,537
Fish						₹ 2,409	₹ 1,916	₹ 1,434	₹ 1,085	₹ 1,002	₹ 7,845
Fuelwood			₹ 262	₹ 42	₹ 2	₹ 17,196	₹ 13,679	₹ 10,239	₹ 7,743		₹ 49,163
Fodder			₹ 1,00,865	₹ 16,026	₹ 940	₹ 23,550	₹ 18,734	₹ 14,022	₹ 10,605		₹ 1,84,742
Medicine						₹ 272	₹ 216	₹ 162	₹ 122		₹ 773
Water						₹ 1,89,899	₹ 1,51,064	₹ 1,13,069	₹ 85,512	₹ 78,996	₹ 6,18,540
Genetic						₹ 1,33,800	₹ 1,06,438	₹ 79,667	₹ 60,250		₹ 3,80,154
Total provisioning (Million ₹)			₹ 3,52,538	₹ 56,011	₹ 3,284	₹ 4,18,004	₹ 3,32,522	₹ 2,48,888	₹ 1,88,228	₹ 79,998	₹ 16,79,473
Air filtration services			₹ 26,941	₹ 4,280	₹ 251	₹ 28,015	₹ 22,286	₹ 16,680	₹ 12,615		₹ 1,11,068
Local (micro and meso) climate regulation services			₹ 6,425	₹ 1,021	₹ 60	₹ 59,655	₹ 47,456	₹ 35,520	₹ 26,863		₹ 1,76,999
Global climate regulation service			₹ 310	₹ 49	₹ 3	₹ 2,79,793	₹ 2,22,576	₹ 1,66,595	₹ 1,25,991		₹ 7,95,318

(continued)

Table 18 (continued)

Ecosystem type		Built-up	Crop land	Horticulture	Fallow land	Evergreen forest	Moist deciduous forest	Dry deciduous forest	Scrub\grass lands	Water	Total
Year	units										
Pollination			₹ 3,489	₹ 554	₹ 33	₹ 13,913	₹ 11,068	₹ 8,284	₹ 6,265		₹ 43,605
Soil erosion						₹ 19,919	₹ 15,846	₹ 11,860	₹ 8,970		₹ 56,595
Soil fertility			₹ 47,784	₹ 7,592	₹ 445	₹ 33,393	₹ 26,564	₹ 19,883	₹ 15,037		₹ 1,46,907
Water purification						₹ 3,167	₹ 2,519	₹ 1,886	₹ 1,426	₹ 1,318	₹ 10,316
Waste treatment						₹ 4,483	₹ 3,567	₹ 2,670	₹ 2,019	₹ 1,865	₹ 14,604
Groundwater						₹ 3,289	₹ 2,616	₹ 1,958	₹ 1,481	₹ 1,368	₹ 10,713
Water flow regulation			₹ 7,068	₹ 1,123	₹ 66						₹ 8,256
Nitrogen fixation			₹ 35,731	₹ 5,677	₹ 333						₹ 41,741
Remediation—organic and inorganic materials			₹ 51,402	₹ 8,167	₹ 479						₹ 60,047
Genetic diversity			₹ 1,15,091	₹ 18,286	₹ 1,072						₹ 1,34,449
Biological control			₹ 1,028	₹ 163	₹ 10						₹ 1,201
Total regulating (Million ₹)			₹ 2,95,270	₹ 46,913	₹ 2,751	₹ 4,45,628	₹ 3,54,497	₹ 2,65,336	₹ 2,00,667	₹ 4,551	₹ 16,15,611
Aesthetic						₹ 1,249	₹ 994	₹ 744	₹ 563		₹ 3,550
Tourism & recreational			₹ 8,397	₹ 1,334	₹ 78	₹ 98,840	₹ 78,627	₹ 58,851	₹ 44,508		₹ 2,90,635
Spiritual						₹ 70	₹ 56	₹ 42	₹ 32		₹ 199
Artistic			₹ 10,280	₹ 1,633	₹ 96	₹ 591	₹ 470	₹ 352	₹ 266		₹ 13,688

(continued)

Table 18 (continued)

Ecosystem type		Built-up	Crop land	Horticulture	Fallow land	Evergreen forest	Moist deciduous forest	Dry deciduous forest	Scrub\grass lands	Water	Total
Year	units										
Education, scientific and research						₹ 5,153	₹ 4,100	₹ 3,068	₹ 2,321	₹ 2,144	₹ 16,786
Total cultural (Million ₹)			₹ 18,678	₹ 2,968	₹ 174	₹ 1,05,903	₹ 84,246	₹ 63,057	₹ 47,688	₹ 2,144	**₹ 3,24,858**
TESV **(Million ₹)**			₹ 6,66,485	₹ 1,05,892	₹ 6,209	₹ 9,69,535	₹ 7,71,265	₹ 5,77,280	₹ 4,36,583	₹ 86,692	₹ **36,19,942**
TESV (Million ₹)			**Agriculture, horticulture**		₹ **7,78,58**	**Forest ecosystem**				₹ **28,41,355**	₹ **36,19,942**

Table 19 Ecosystem wise—provisioning, regulating and cultural services (Million ₹)—2019

Ecosystem type		Built-up	Crop land	Horticulture	Fallow land	Evergreen forest	Moist deciduous forest	Dry deciduous forest	Scrub\grass lands	Water	Total
Year	Units										
2019	km^2	5,748	1,27,962	21,325	2,854	10,888	7,892	4,281	4,907	5,934	1,91,791
	%	3	67	11	2	6	4	2	3	3	100
Provisioning, regulating and cultural services (Million ₹) for forest, agriculture and horticulture ecosystems											
Food (cereal, pulses)			₹ 3,87,216	₹ 64,530	₹ 8,636						₹ 4,60,382
Timber						₹ 565	₹ 409	₹ 222	₹ 254		₹ 1,450
NTFP						₹ 5,414	₹ 3,924	₹ 2,129	₹ 2,440		₹ 13,906
Fish						₹ 3,897	₹ 2,824	₹ 1,532	₹ 1,756	₹ 2,124	₹ 12,133
Fuelwood			₹ 298	₹ 50	₹ 7	₹ 22,313	₹ 16,173	₹ 8,773	₹ 10,056		₹ 57,668
Fodder			₹ 1,05,472	₹ 17,577	₹ 2,352	₹ 15,207	₹ 11,023	₹ 5,979	₹ 6,854		₹ 1,64,464
Medicine						₹ 202	₹ 146	₹ 79	₹ 91		₹ 519
Water						₹ 90,179	₹ 65,365	₹ 35,457	₹ 40,642	₹ 49,148	₹ 2,80,792
Genetic						₹ 81,257	₹ 58,898	₹ 31,949	₹ 36,621		₹ 2,08,726
Total provisioning (Million ₹)			₹ 4,92,986	₹ 82,157	₹ 10,995	₹ 2,19,034	₹ 1,58,763	₹ 86,121	₹ 98,714	₹ 51,272	₹ 12,00,041
Air filtration services			₹ 28,618	₹ 4,769	₹ 638	₹ 20,807	₹ 15,082	₹ 8,181	₹ 9,377		₹ 87,474
Local (micro and meso) climate regulation services			₹ 6,825	₹ 1,137	₹ 152	₹ 44,308	₹ 32,116	₹ 17,421	₹ 19,969		₹ 1,21,929
Global climate regulation service			₹ 3,788	₹ 631	₹ 84	₹ 2,22,348	₹ 1,61,165	₹ 87,424	₹ 1,00,208		₹ 5,75,648
Pollination			₹ 3,706	₹ 618	₹ 83	₹ 10,491	₹ 7,604	₹ 4,125	₹ 4,728		₹ 31,356

(continued)

Table 19 (continued)

Ecosystem type		Built-up	Crop land	Horticulture	Fallow land	Evergreen forest	Moist deciduous forest	Dry deciduous forest	Scrub\grass lands	Water	Total
Year	Units										
Soil erosion						₹ 14,003	₹ 10,150	₹ 5,506	₹ 6,311		₹ 35,969
Soil fertility			₹ 60,426	₹ 10,070	₹ 1,348	₹ 39,408	₹ 28,564	₹ 15,495	₹ 17,760		₹ 1,73,071
Water purification						₹ 2,225	₹ 1,613	₹ 875	₹ 1,003	₹ 1,213	₹ 6,927
Waste treatment						₹ 3,149	₹ 2,283	₹ 1,238	₹ 1,419	₹ 1,716	₹ 9,806
Groundwater						₹ 2,285	₹ 1,656	₹ 898	₹ 1,030	₹ 1,245	₹ 7,115
Water flow regulation			₹ 7,805	₹ 1,301	₹ 174						₹ 9,280
Nitrogen fixation			₹ 3,754	₹ 626	₹ 84						₹ 4,463
Remediation—organic and inorganic materials			₹ 54,601	₹ 9,099	₹ 1,218						₹ 64,918
Genetic diversity			₹ 1,22,255	₹ 20,374	₹ 2,727						₹ 1,45,355
Biological control			₹ 1,092	₹ 182	₹ 24						₹ 1,298
Total regulating (Million ₹)			₹ 2,92,870	₹ 48,807	₹ 6,532	₹ 3,59,024	₹ 2,60,233	₹ 1,41,163	₹ 1,61,805	₹ 4,174	₹ 12,74,610
Aesthetic						₹ 627	₹ 455	₹ 247	₹ 283		₹ 1,612
Tourism & recreational			₹ 8,920	₹ 1,487	₹ 199	₹ 1,09,309	₹ 79,231	₹ 42,979	₹ 49,263		₹ 2,91,388
Spiritual						₹ 71	₹ 51	₹ 28	₹ 32		₹ 181
Artistic			₹ 10,920	₹ 1,820	₹ 244	₹ 439	₹ 318	₹ 173	₹ 198		₹ 14,111
Education, scientific and research						₹ 3,620	₹ 2,624	₹ 1,423	₹ 1,631	₹ 1,973	₹ 11,271
Total cultural (Million ₹)			₹ 19,840	₹ 3,306	₹ 443	₹ 1,14,066	₹ 82,679	₹ 44,849	₹ 51,407	₹ 1,973	₹ 3,18,563

(continued)

Table 19 (continued)

Ecosystem type		Built-up	Crop land	Horticulture	Fallow land	Evergreen forest	Moist deciduous forest	Dry deciduous forest	Scrub\grass lands	Water	Total
Year	Units										
TESV (Million ₹)			₹ 8,05,696	₹ 1,34,270	₹ 17,970	₹ 6,92,124	₹ 5,01,675	₹ 2,72,133	₹ 3,11,926	₹ 57,419	₹ 27,93,213
TESV (Million ₹)		Agriculture, Horticulture			₹ 9,57,936	Forest ecosystems				₹ 18,35,277	₹ 27,93,213

Table 20 Timber extracted from each forest circle of Karnataka during 2005 (based on the averages of 2001–2005 data) and 2019 (based on the averages of 2015–2018), with values in physical units-cum

S.no	Circle name	TIMBER									
		Rosewood (Cum)		Teak wood (Cum)		Eucalyptus (Cum)		Softwood (Cum)		Other kinds of timber (Cum)	
		2005	2019	2005	2019	2005	2019	2005	2019	2005	2019
1	Bengaluru	0	0	0	0	331	561	0	0	854	421
2	Belgaum	46	14	13	15	5	6	0	0	2392	1309
3	Ballari	0	0	0	0	0	0	0	0	387	115
4	Chamarajanagar	0	0	0	0	0	0	0	0	13	77
5	Chikmagalur	324	231	121	212	0	1	133	233	3033	5280
6	Dharwad	4	1	61	85	45	62	0	0	116	93
7	Kalaburagi	0	0	0	0	261	198	0	0	934	708
8	Hassan	37	16	3	3	1894	1556	0	0	180	148
9	Canara	2219	1505	2616	2805	2717	2914	706	758	16682	16715
10	Kodagu	3029	1914	182	163	378	339	174	156	1496	1279
11	Mangalore	50	43	57	77	5861	7911	361	488	2745	3706
12	Mysore	23	14	17	31	5389	9546	0	0	788	1351
13	Shimoga	7798	4506	172	147	2485	2118	7451	6350	11182	8667
Seigniorage value (₹)		67000	67000	6500	65000	6600	6600	1560	1560	8400	8400

Table 21 Provisioning services (pulpwood, round poles, sandalwood, bamboo, canes) extracted from each forest circle of Karnataka during 2005 and 2019, with values in physical units

S.no	Circle name	Pulpwood (Cum)		Round poles (Nos)		Sandalwood (Kg)		Bamboo (Nos)		Canes (Nos)	
		2005	2019	2005	2019	2005	2019	2005	2019	2005	2019
1	Bengaluru	1928	3266	786	1333	38	66	0	0	9150	15500
2	Belgaum	0	0	2337	2605	90	101	318480	0	0	0
3	Ballari	0	0	0	0	0	0	0	0	0	0
4	Chamarajanagar	0	0	0	0	0	0	0	0	0	0
5	Chikmagalur	1306	2273	3675	6396	103	180	26885	14731	0	0
6	Dharwad	0	0	1663	2289	9829	4255	167558	267	1466	2018
7	Kalaburagi	0	0	0	0	0	0	0	0	0	0
8	Hassan	0	0	62	51	4127	3391	0	0	2398	1970
9	Canara	0	0	4824	5173	574	616	489139	85412	272293	10393
10	Kodagu	0	0	2	2	0	0	4820	0	0	0
11	Mangalore	0	0	2261	3052	26	36	17829	21298	33611	6000
12	Mysore	0	0	1720	3048	5263	117	52764	469	3	6
13	Shimoga	0	0	20111	17139	1939	0	634637	4704	5000	0
Seigniorage value (₹)		1560	1560	348	348	15000	15000	38	38	4	4

Table 22 Provisioning services (NTFP—honey, soap nut (*Sapindus saponaria*), cashew nut (*Anacardium occidentale*), tamarind (Tamarindus indica), dalchini (cinnamon–*Cinnamomum malabatrum*, *Cinnamomum verum*); rampatri (nutmeg—*Myristica malabarica*); murgalu (kokum or *Garcinia indica*)) extracted from each forest circle of Karnataka during 2005 and 2019 (in physical tons)

S.no	Circle name	NTFP (tons)													
		Honey		Soapnut		Cashew		Tamarind		Ramapatre		Dalchinni		Murugalu	
		2005	2019	2005	2019	2005	2019	2005	2019	2005	2019	2005	2019	2005	2019
1	Bengaluru	0	0	0	0	1804	67	3967	490	0	0	0	0	0	0
2	Belgaum	0	0	0	0	21282	800	314	35	0	0	0	0	0	0
3	Ballari	0	0	0	0	0	0	139	16	0	0	0	0	0	0
4	Chamarajanagar	33153	9034	868	1000	0	0	0	0	0	0	0	0	0	0
5	Chikmagalur	1109	14	853	101	0	0	38	2	0	0	186	133	574	6
6	Dharwad	0	0	3640	0	0	0	986	75	0	0	9281	0	0	0
7	Kalaburagi	0	0	0	0	0	0	281	13	0	0	0	0	0	0
8	Hassan	0	0	0	0	2508	28	0	0	402	2	0	0	0	0
9	Canara	146	22	15	0	42	11	0	0	14	2	0	0	939	125
10	Kodagu	20775	293	188375	10388	0	0	0	0	8835	0	0	0	3148	0
11	Mangalore	0	0	93937	0	273	23	0	0	320	0	7	6	0	0
12	Mysore	464515	3000	0	0	0	0	4748	150	0	0	184	0	0	0
13	Shimoga	48904	3575	139809	73028	0	0	19566	6899	0	0	0	0	473	0
Seigniorage value (₹)		140000	140000	130000	130000	100000	100000	65000	65000	80000	80000	310000	310000	400000	400000

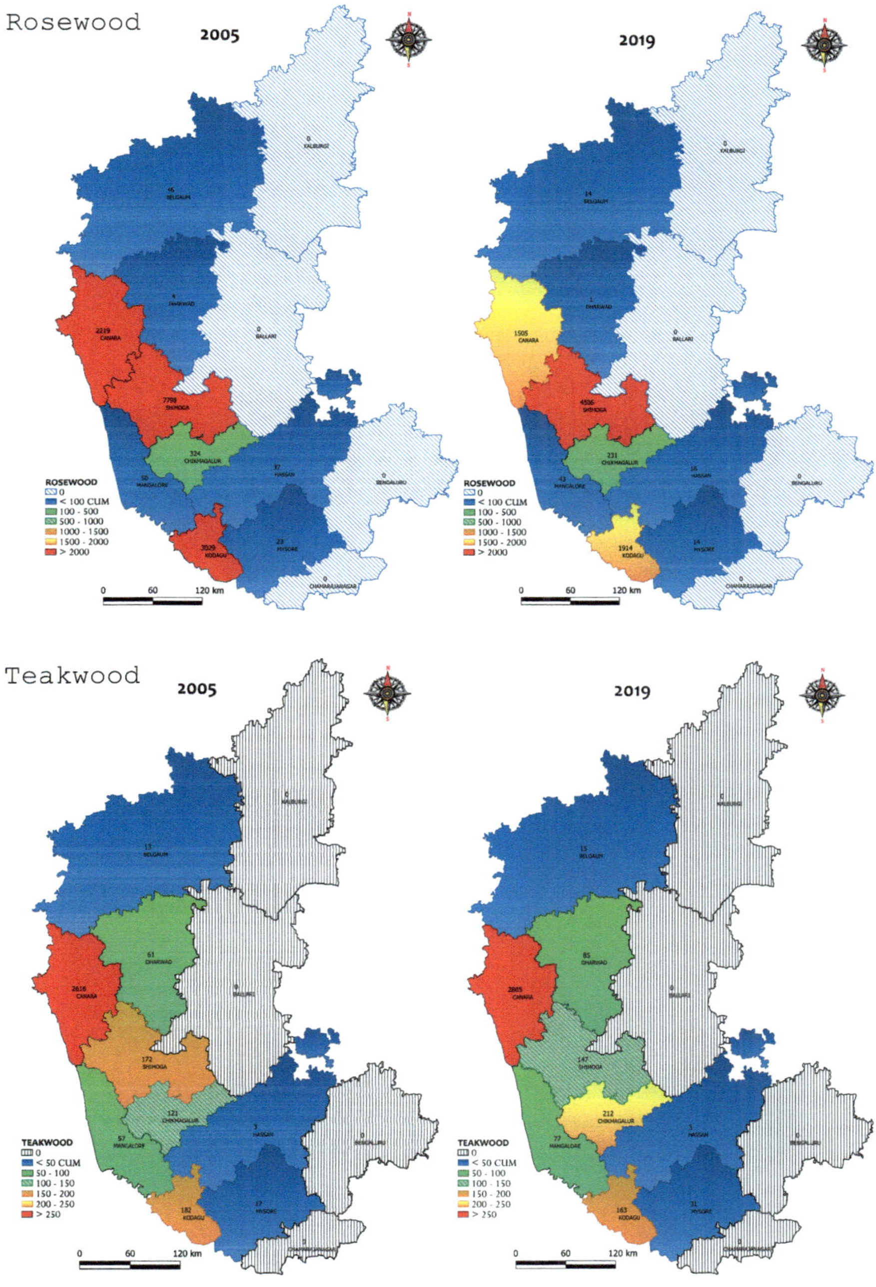

Fig. 10 Timber, bamboo and canes extracted (in the physical units) across forest circles in Karnataka for the years 2005 and 2019

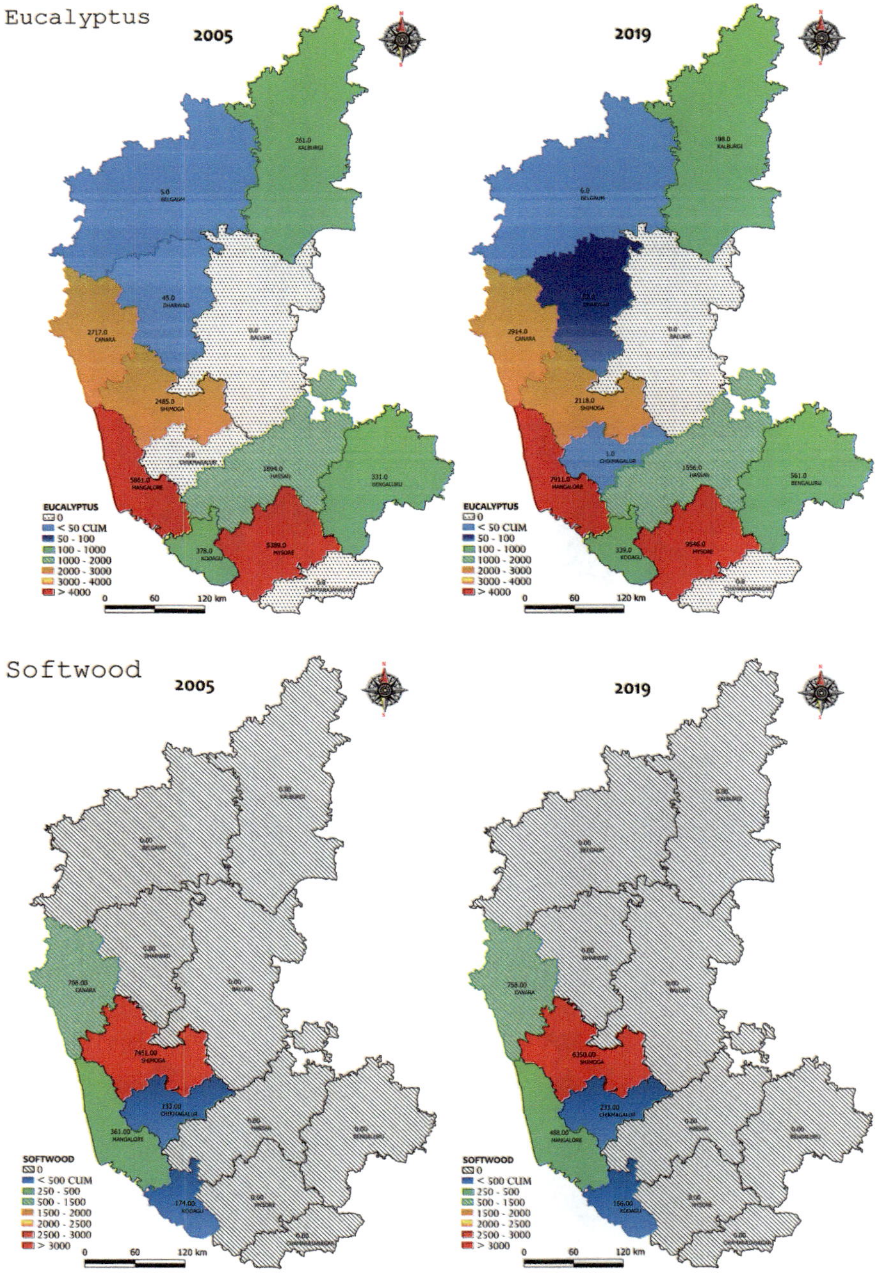

Fig. 10 (continued)

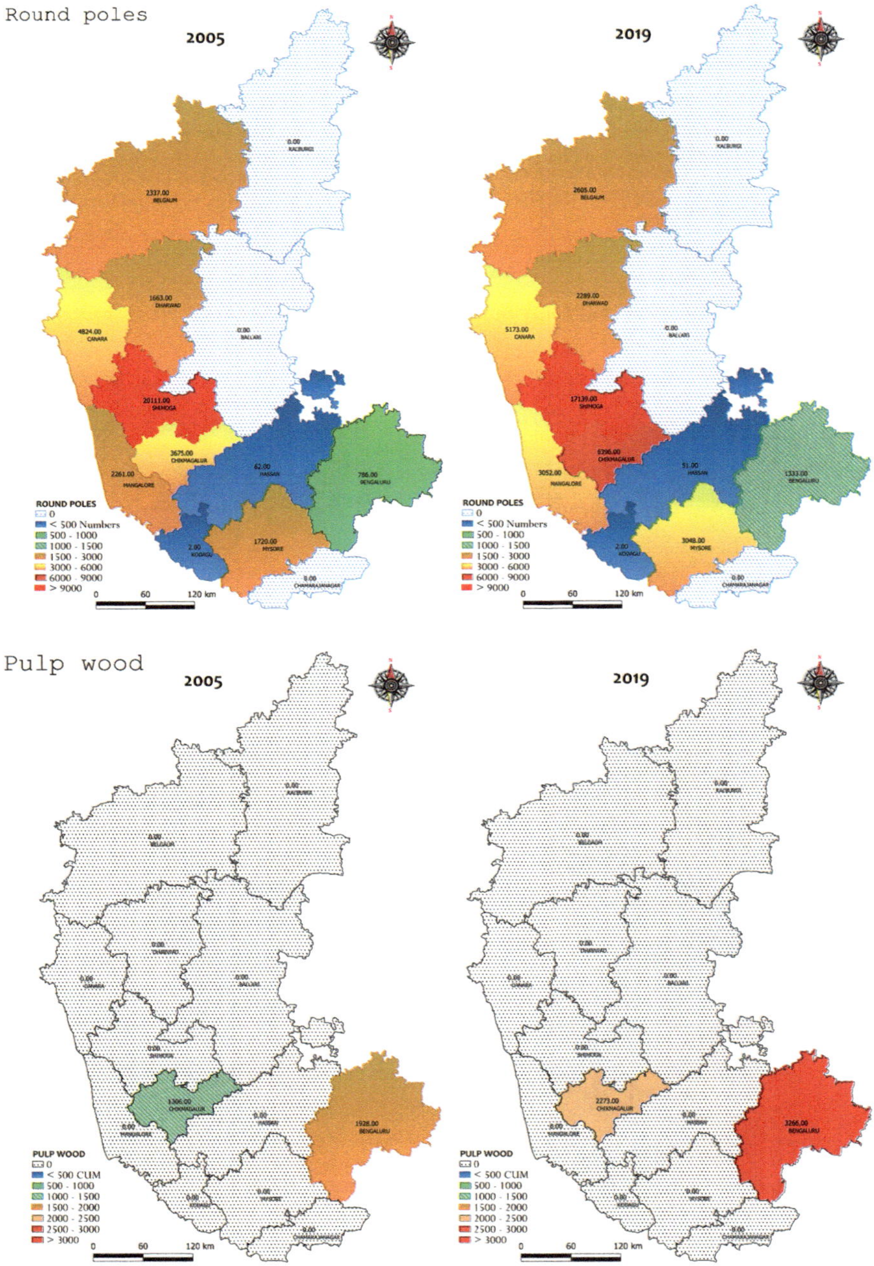

Fig. 10 (continued)

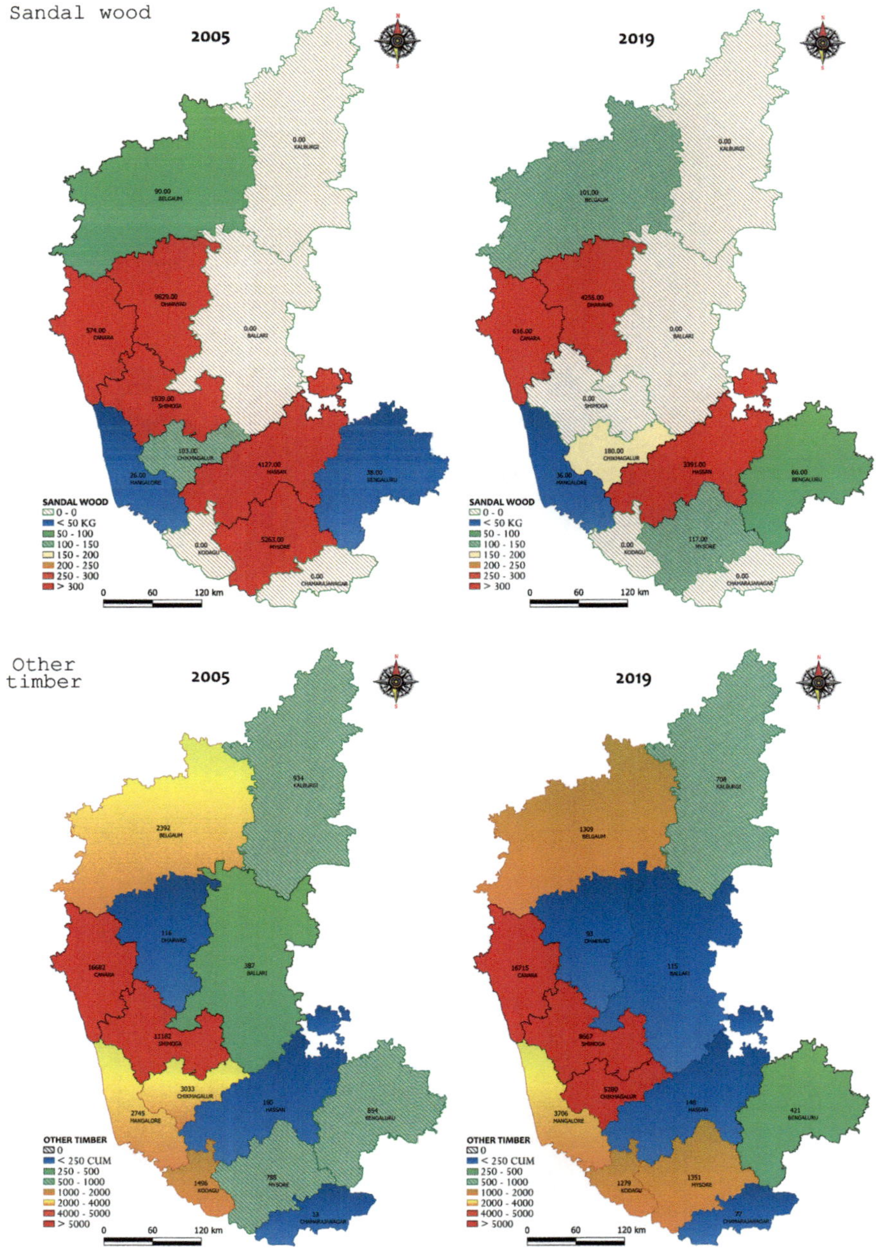

Fig. 10 (continued)

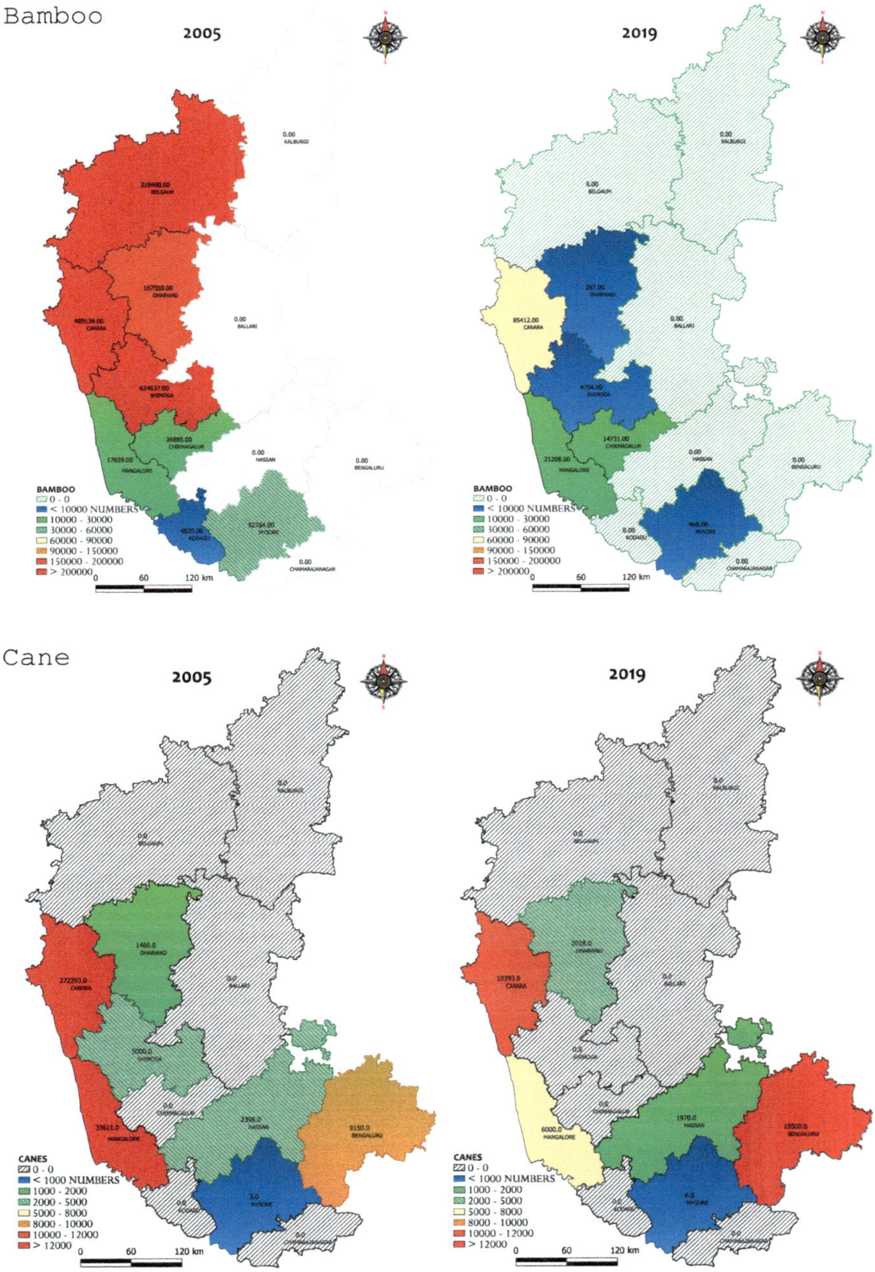

Fig. 10 (continued)

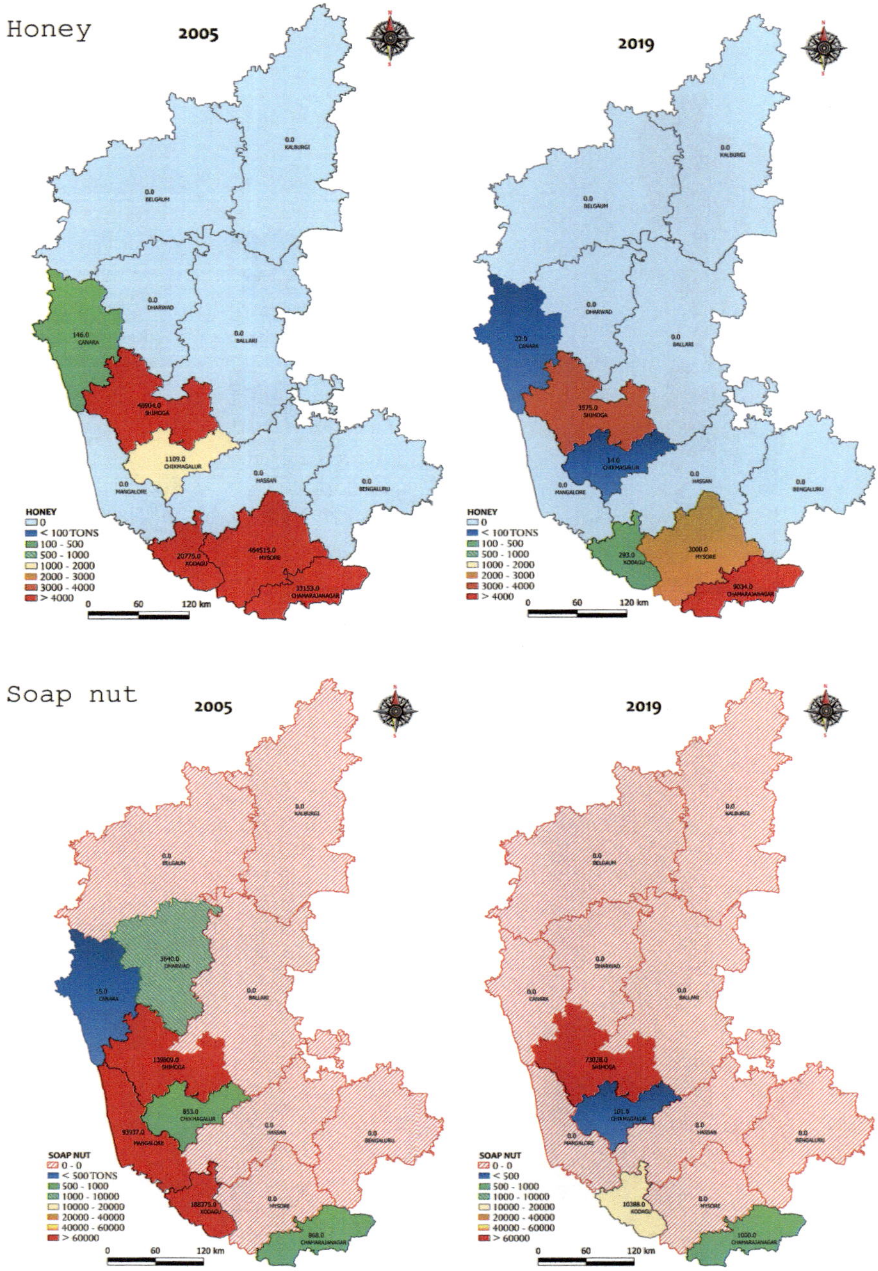

Fig. 11 Quantity of NTFP collected (in physical units) across the circles in Karnataka during 2005 and 2019

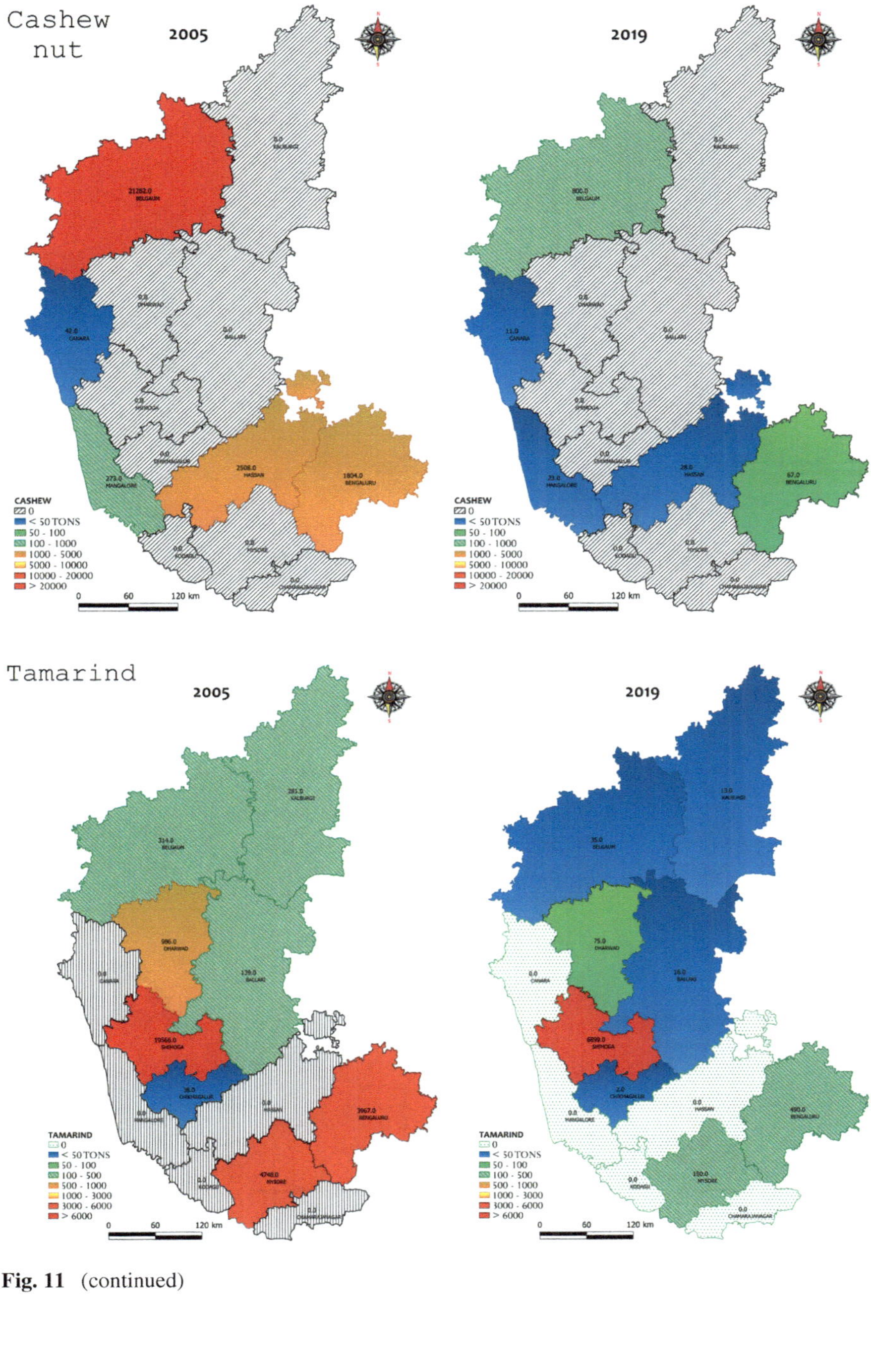

Fig. 11 (continued)

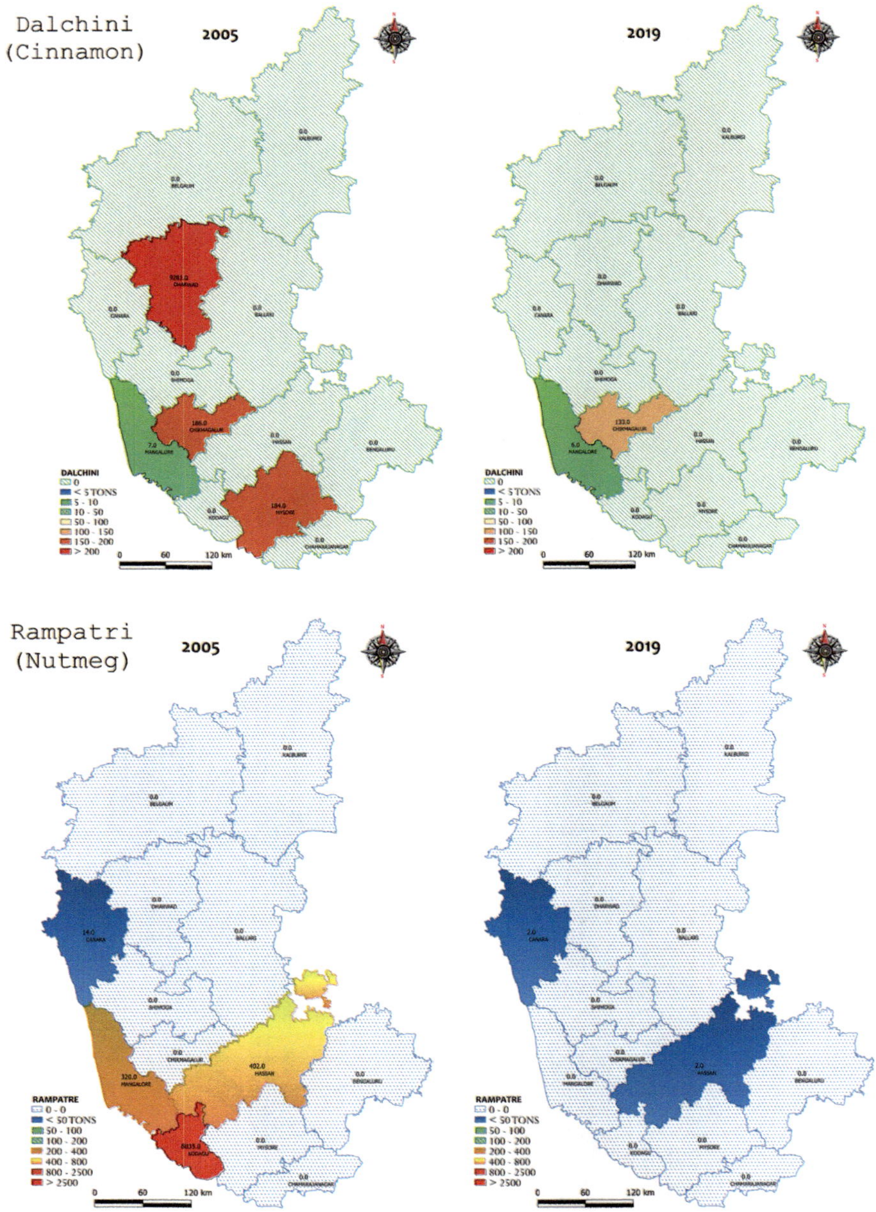

Fig. 11 (continued)

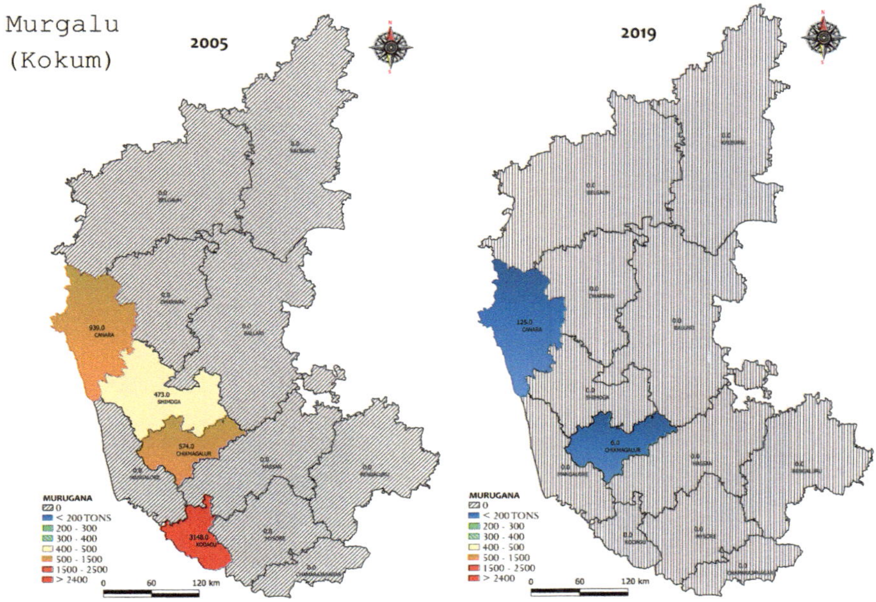

Fig. 11 (continued)

forest goods are listed in Table 23. Based on the extraction area, per hectare values of each timber type are computed, which are listed in Table 24. Figures 12 and 13 depict the spatial variability of the ecosystem value of timber goods in Karnataka. The total provisioning value of timber extracted from Karnataka state is 2,023, and 1,445 billion rupees, respectively, for 2005 and 2019. The rosewood is available only in specific circles, which are part of the Western Ghats. Shimoga, Kodagu, and Canara circles have higher ecosystem supply value as compared with other circles. The ecosystem supply value of rosewood is 140,017 Rs/Ha/yr (2005) and 140,998 Rs/Ha/yr (2019). The maximum value is observed as 240,571 Rs/Ha/yr in the Canara circle, and the minimum value is observed in Dharwad as 60,909 Rs/Ha/yr for the year 2005. In 2019, the maximum value is observed from the Canara circle as 240,656 Rs/Ha/yr and the minimum value as 60,909 Rs/Ha/yr from the Dharwad circle.

Teakwood: Teakwood is extracted in large quantities from Canara, Shimoga, Kodagu, and Chikmagalur circles. The ecosystem supply value of teak wood shows 79,881 Rs/Ha/yr (2005) and 79,961 Rs/Ha/yr (2019) as per the collected data. Maximum and minimum values are observed as 157,744, 41,302 Rs/Ha/yr (2005) for the circles Kodagu, Dharwad respectively, whereas for the year 2019, maximum and minimum values accounted as 158,134, 41,231 Rs/Ha/yr for the circles Kodagu, Dharwad respectively. The average teak wood values are 79,881 and 79,961 Rs/Ha/yr for 2005 and 2019 respectively.

Eucalyptus: The revenue from eucalyptus is high in Bengaluru, and Mysore circles. Sandalwood is extracted in large quantities from Dharwad, Hassan circles.

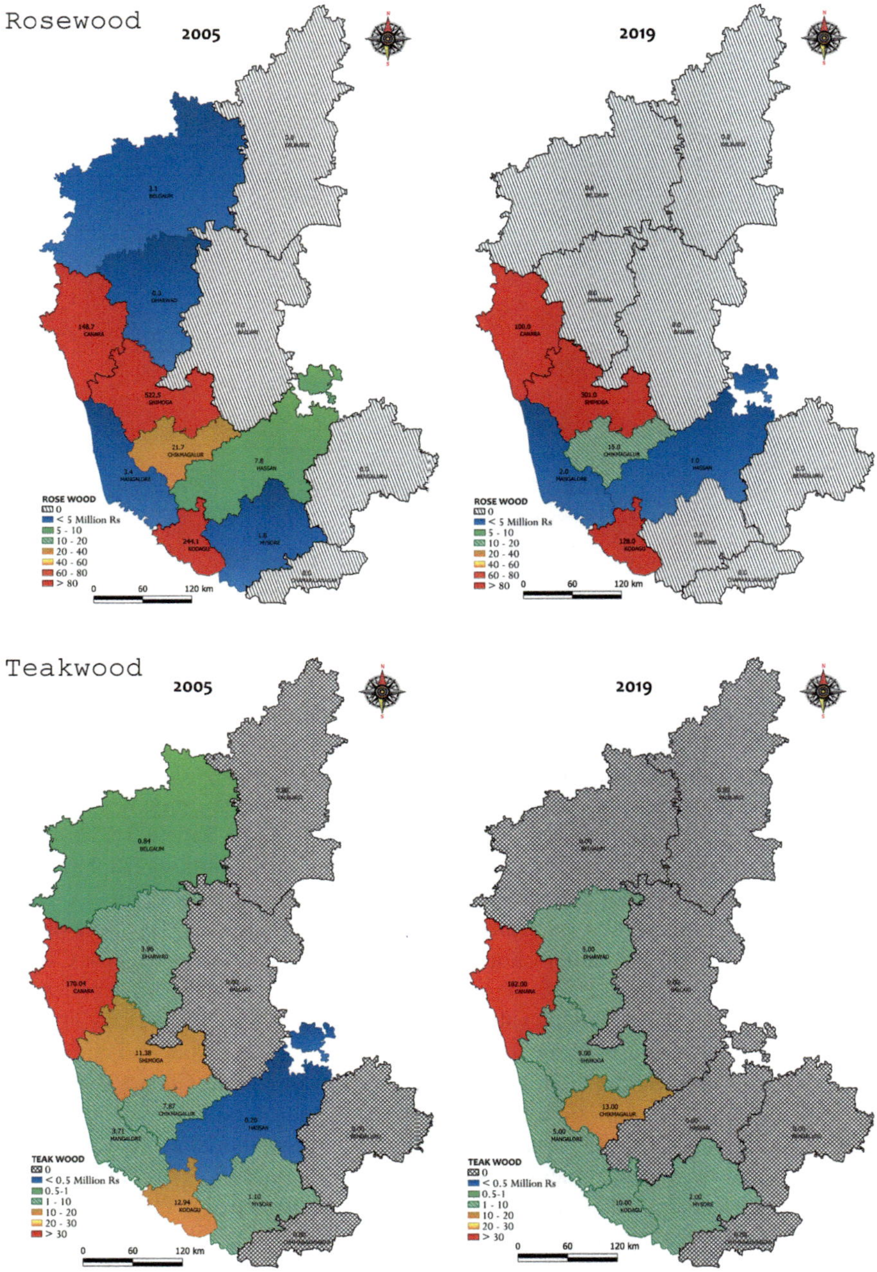

Fig. 12 Ecosystem supply value (INR) of various timber across the circles for 2005 and 2019

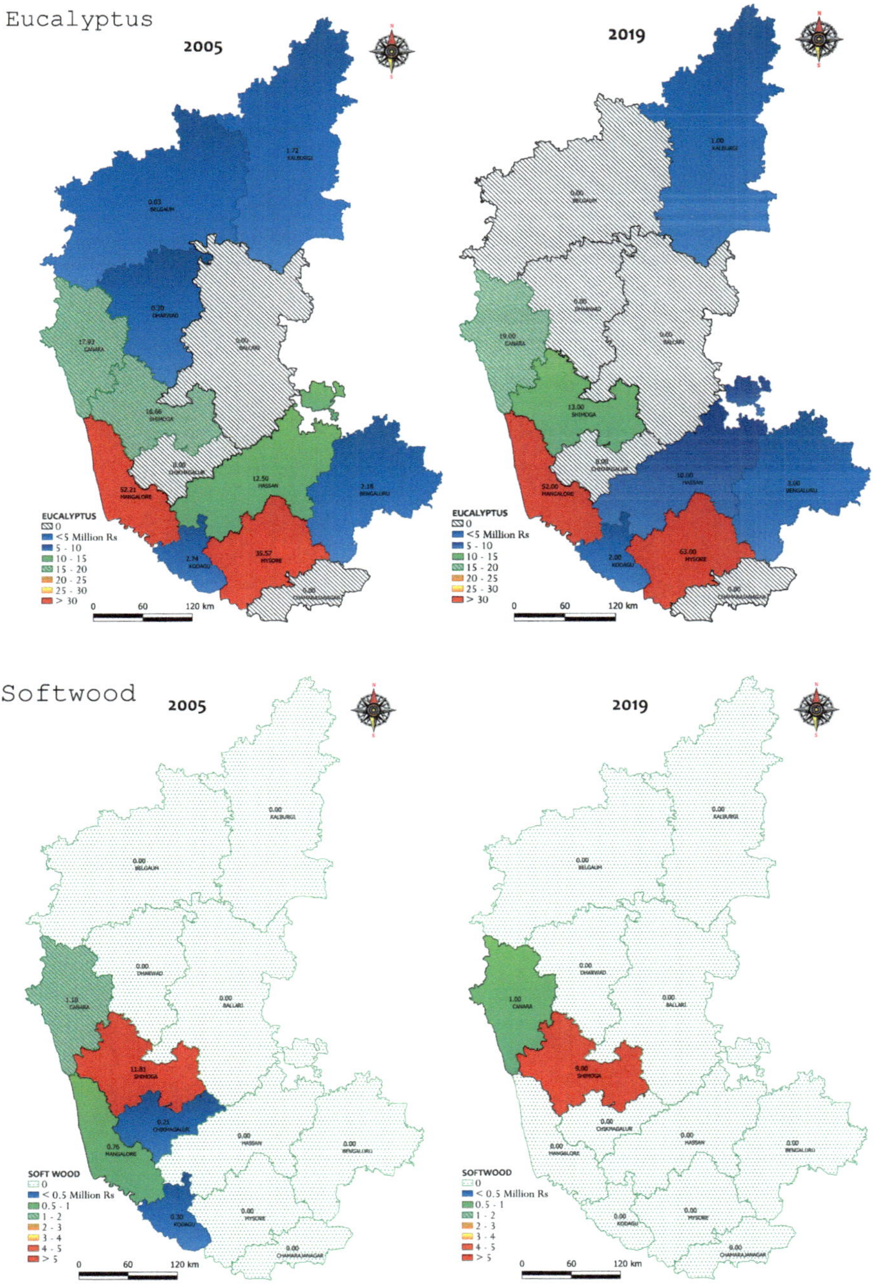

Fig. 12 (continued)

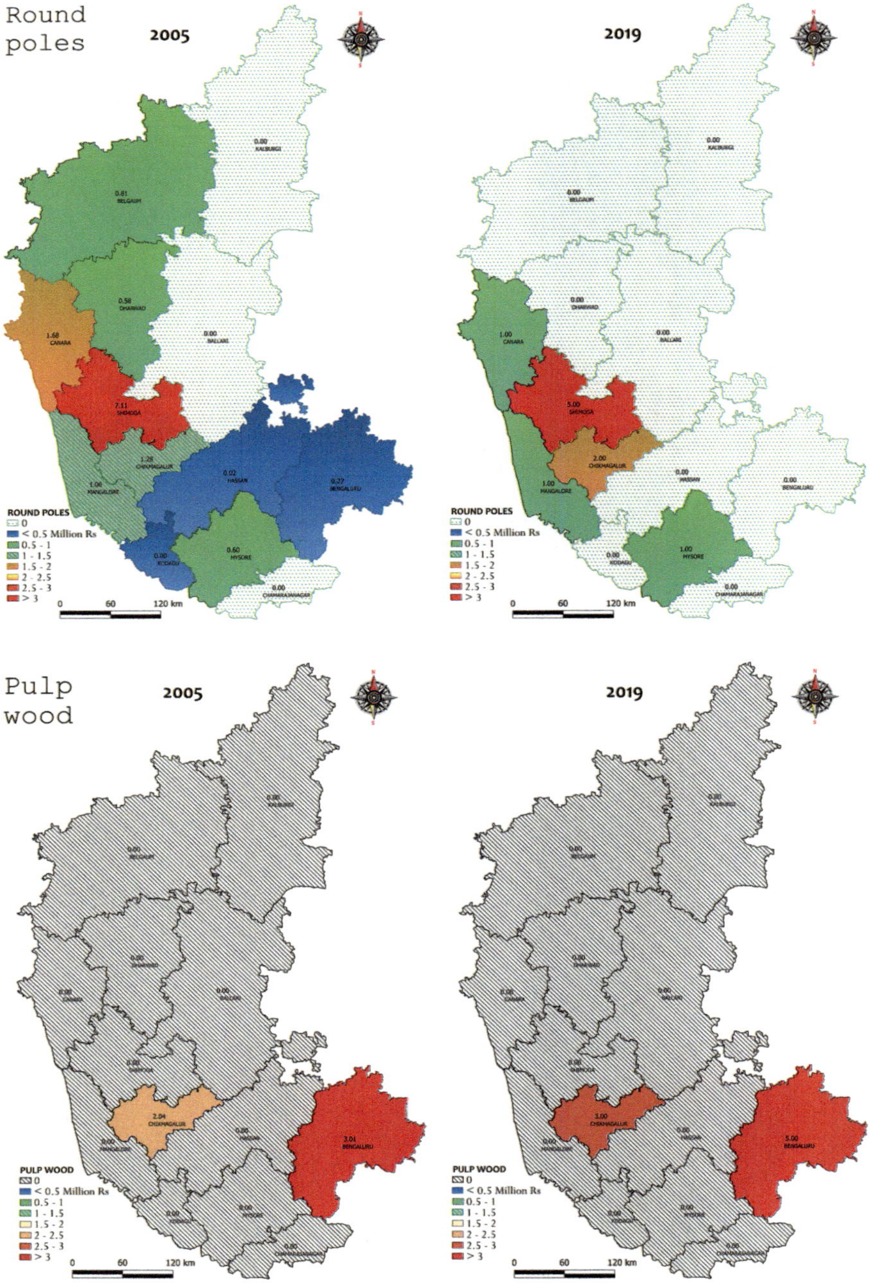

Fig. 12 (continued)

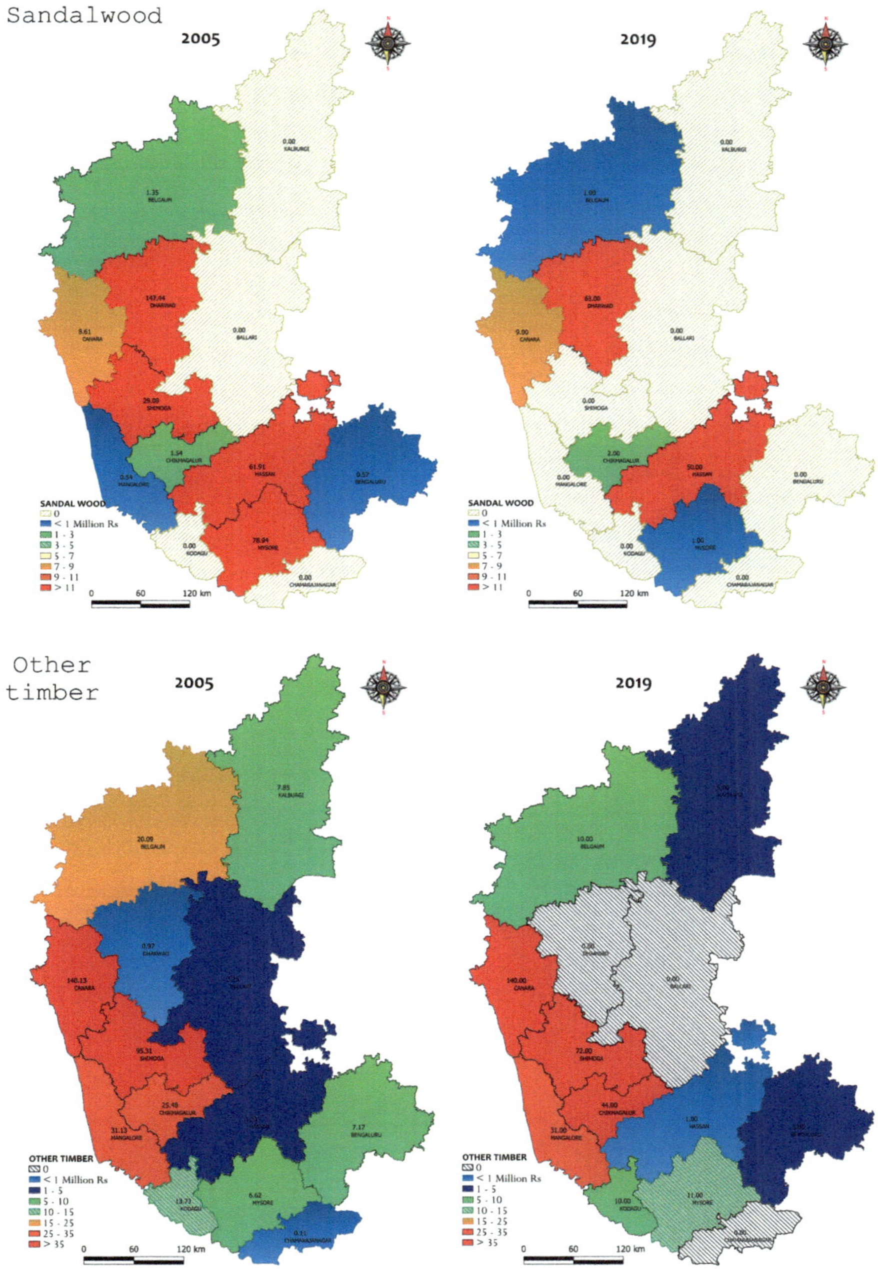

Fig. 12 (continued)

Table 23 District-wise ecosystem supply value (₹-INR) of timber, based on seigniorage values and quantity of timber harvested

S. no	Circle name	TIMBER (Million Rupees)																	
		Rose wood		Teak wood		Eucalyptus		Soft wood		Other kinds of timber		Pulpwood		Round poles		Sandal wood		Total timber	
		2005	2019	2005	2019	2005	2019	2005	2019	2005	2019	2005	2019	2005	2019	2005	2019	2005	2019
1	Bengaluru	0	0	0	0	2	3	0	0	7	3	3	5	0	0	1	0	13	37
2	Belgaum	3	0	1	0	0	0	0	0	20	10	0	0	1	0	1	1	26	63
3	Ballari	0	0	0	0	0	0	0	0	3	0	0	0	0	0	0	0	3	7
4	Chamarajanagar	0	0	0	0	0	0	0	0	0	0	0	0	0	0	0	0	0	0
5	Chikmagalur	22	15	8	13	0	0	0	0	25	44	2	3	1	2	2	2	60	199
6	Dharwad	0	0	4	5	0	0	0	0	1	0	0	0	1	0	147	63	154	375
7	Kalaburagi	0	0	0	0	2	1	0	0	8	5	0	0	0	0	0	0	10	25
8	Hassan	8	1	0	0	13	10	0	0	2	1	0	0	0	0	62	50	84	230
9	Canara	149	100	170	182	18	19	1	1	140	140	0	0	2	1	9	9	488	1428
10	Kodagu	244	128	13	10	3	2	0	0	14	10	0	0	0	0	0	0	274	698
11	Mangalore	3	2	4	5	52	52	1	0	31	31	0	0	1	1	1	0	93	277
12	Mysore	2	0	1	2	36	63	0	0	7	11	0	0	1	1	79	1	125	327
13	Shimoga	522	301	11	9	17	13	12	9	95	72	0	0	7	5	29	0	694	1797

Table 24 Value of various timber (₹/Ha/Yr.) in Karnataka (Forest circle wise/forest type-wise)

S.no	Circle name	Timber type (₹/Ha/Yr.)															
		Rose wood		Teak wood		Eucalyptus		Soft wood		Other kinds of timber		Pulpwood		Round poles		Sandal	
		2005	2019	2005	2019	2005	2019	2005	2019	2005	2019	2005	2019	2005	2019	2005	2019
1	Bengaluru					4284	4280			4278	4278	4272	4295	4274	4286	4318	4285
2	Belgaum	154100	151290	76818	77381	3300	3960			3826	3827			3836	3825	3835	3826
3	Ballari									4277	4293						
4	Chamarajanagar									9100	4293						
5	Chikmagalur	150750	150262	77871	77416		3300	3842	3826	3827	3826	3830	3829	3841	3824	3834	3830
6	Dharwad	60909	60909	41302	41231	4368	4263			4293	4292			4319	4306	4274	4274
7	Kalaburagi					4643	4634			4629	4631						
8	Hassan	125355	126117	65000	65000	9246	9235			9276	9208			10788	8874	9237	9240
9	Canara	240571	240656	156863	156906	1302	1303	1303	1307	1302	1302			1303	1304	1302	1302
10	Kodagu	240303	240597	157744	158134	1302	1302	1307	1308	1302	1302			1392	1547		
11	Mangalore	152273	154892	78830	78203	3826	3827	3864	3864	3826	3826			3834	3834	3857	3857
12	Mysore	124759	134000	65000	65000	9238	9235			9245	9241			9353	9224	9237	9237
13	Shimoga	149148	149235	77381	78320	3827	3827	3828	3828	3826	3826			3828	3828	3826	

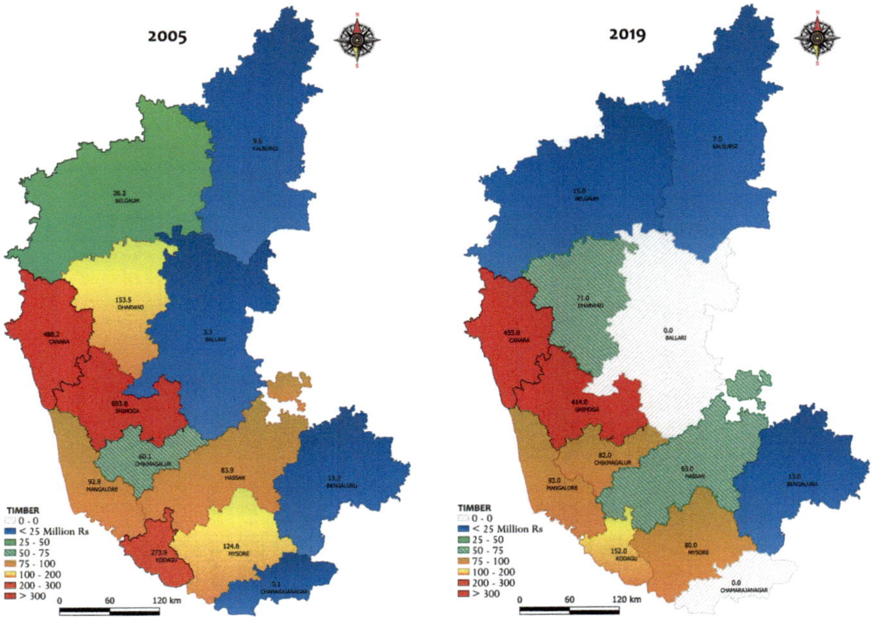

Fig. 13 Aggregated values of timber from forest ecosystems in Karnataka (in a million INR)

These values are comparable to the earlier studies in India (Verma et al. 2013). Overall, the circles such as Belgaum, Mysore, Ballari (Bellary) have lower timber services for 2019 compared to 2005. The ecosystem supply value of Eucalyptus wood is 4,304 and 4,265 Rs/Ha/yr for the data collected for the years 2005, and 2019 respectively. The maximum value observed is 9,246 Rs/Ha/yr in Hassan, and the minimum value in the Canara circle is 1,302 Rs/Ha/yr for the year 2005. In 2019, maximum value is observed from the Hassan circle as 9235 Rs/Ha/yr and the minimum value as 1,302 Rs/Ha/yr from the Kodagu circle.

Softwood: The ecosystem supply value of softwood is 2,692 Rs/Ha/yr as per the collected data for 2005, and 2019. Maximum and minimum values are 3,864 and 1,303 Rs/Ha/yr (2005) for Mangalore, Canara circles, respectively, whereas for the year 2019, maximum and minimum values accounted for 158,134, and 41,231 Rs/Ha/yr for the circles Kodagu, Dharwad respectively.

Other timber: The ecosystem supply value of other kinds of timber is 4,644 and 4,297 Rs/Ha/yr for the data collected for 2005 and 2019, respectively. The maximum value is 9,276 Rs/Ha/yr in Hassan, and the minimum is 1,302 Rs/Ha/yr in Canara, Kodagu circles for 2005. In 2019, the maximum value from the Mysore circle was 9,241 Rs/Ha/yr, and the minimum value was 1,302 Rs/Ha/yr from Canara and Kodagu circles.

Pulpwood: The ecosystem supply value of pulpwood is 3,369 Rs/Ha/yr (2005) and 3,381Rs/Ha/yr (2019) based on the data collected across the circles from the

Karnataka Forest Department. The maximum value is 4,272 Rs/Ha/yr in the Bangalore circle, and the minimum value in Chikmagalur is 3,830 Rs/Ha/yr for 2005. In 2019, the maximum value from the Bangalore circle was 4,295 Rs/Ha/yr, and the minimum value was 3,829 Rs/Ha/yr from the Chikmagalur circle.

Round poles: The ecosystem supply value of round pole wood is assessed as 4,434 and 4,261 Rs/Ha/yr for 2005 and 2019, respectively. The maximum value is 10,788 Rs/Ha/yr in Hassan, and the minimum value in the Canara circle is 1,303 Rs/Ha/yr for the year 2005. In 2019, the maximum value from the Mysore circle was 9,224 Rs/Ha/yr, and the minimum value from the Canara circle was 1,304 Rs/Ha/yr.

Sandalwood: The ecosystem supply value of sandalwood is 4,573, and 4,652 Rs/Ha/yr for 2005, and 2019 respectively. The maximum value is 9,237 Rs/Ha/yr in Hassan, and the minimum value in Canara circles is 1,302 Rs/Ha/yr for 2005. In 2019, the maximum value from Hassan Circle was 9,240 Rs/Ha/yr, and the minimum value was 1,302 Rs/Ha/yr from the Canara Circle.

Bamboo and Cane Production: Timber and biomass of bamboo and cane production have been assessed from the forest inventory. The ecosystem supply value from bamboo and canes was estimated (Table 25, Fig. 14). Circles such as Canara and Mangalore are getting higher revenue from bamboo, and Bengaluru, Canara Dharwad, and Mangalore are getting high revenue from canes. The ecosystem supply value of bamboo has reduced for 2019, which might be attributed to the degradation of the forest ecosystem with the reduced bamboo cover.

Table 25 Value of extracted bamboo and canes across Karnataka

S.no	Circle name	Bamboo (Million Rupees)		Canes (Million Rupees)		Bamboo (₹/Ha/Yr.)		Cane (₹/Ha/Yr.)	
		2005	2019	2005	2019	2005	2019	2005	2019
1	Bengaluru	0	0	0.037	0.1			24400	23846
2	Belgaum	12	0	0	0	3826			
3	Ballari	0	0	0	0				
4	Chamarajanagar	0	0	0	0				
5	Chikmagalur	1	0.6	0	0	3841	3861		
6	Dharwad	6	0	0.006	0	4279	4612	29320	23741
7	Kalaburagi	0	0	0	0	0			
8	Hassan	0	0	0.010	0	0		5642	5629
9	Canara	19	3.2	1.089	0	1302	1303	1304	1341
10	Kodagu	0	0	0	0	1308		0	
11	Mangalore	1	0.8	0.024	0	3836	3836	194	192
12	Mysore	2	0	0	0	9283	8911	5356	5356
13	Shimoga	24	0.2	0.020	0	3827	3886	191	
Seigniorage rate (₹)		38	38	4	4				

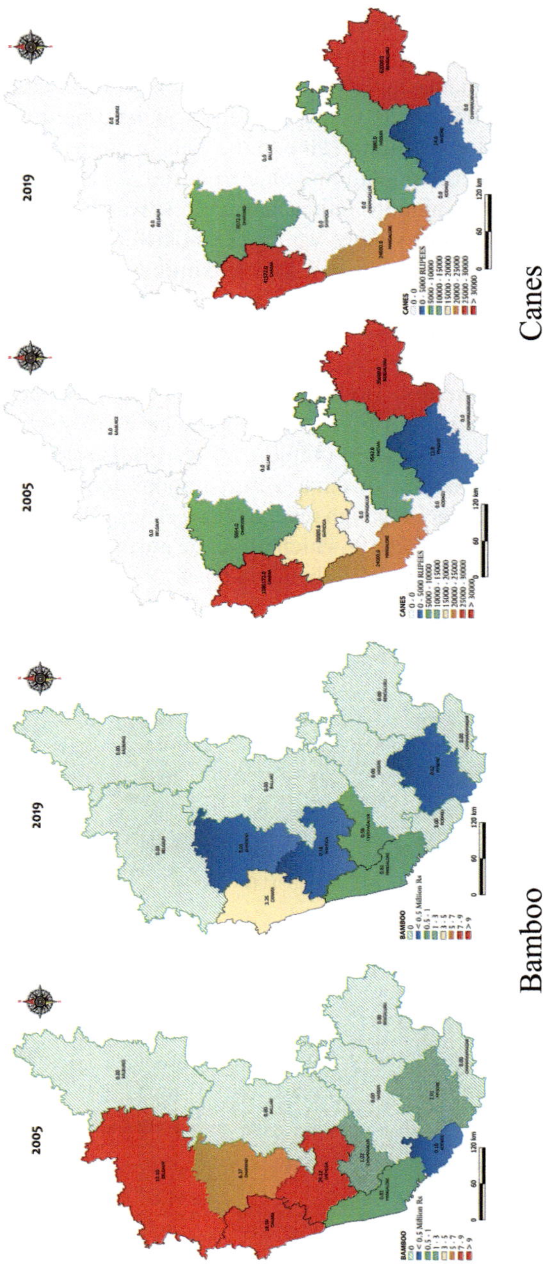

Fig. 14 Ecosystem supply value of bamboo and canes from forests of Karnataka

The ecosystem supply value of Bamboo for the Karnataka state is assessed as 3,938 and 4,402 Rs/Ha/yr for the years 2005 and 2019, respectively. The maximum value is 9,283 Rs/Ha/yr in Mysore, and the minimum value is observed in Canara circles as 1,302 Rs/Ha/yr for 2005. In 2019, the maximum value from the Hassan Circle was 8,911 Rs/Ha/yr, and the minimum value was 1,303 Rs/Ha/yr for the Canara Circle.

Fodder Production: Table 26 lists circle-wise fodder in physical and monetary values per hectare per year. Figure 15 depicts the circle-wise ecosystem value for fodder, which are 66,911 and 39,063 million rupees for 2005 and 2019, respectively. The reduction in fodder availability could be attributed to the degradation of forest ecosystems in the State. Canara, Shimoga, Mangalore, and Chamarajanagar contribute higher compared to the Dharwad circle (least). The ecosystem supply value from fodder for Karnataka state is assessed 7,736, and 15,476 Rs/Ha/yr for the years 2005 and 2019. The maximum value is 8,988 Rs/Ha/yr in Chamarajanagar, and the minimum value in the Kodagu circle is 6,583 Rs/Ha/yr for the year 2005. In 2019, the maximum value witnessed from the Chamarajanagar circle was 20,050 Rs/Ha/yr, and the minimum value from the Ballari circle was 12,779 Rs/Ha/yr.

Non-timber Forest Produce/Non-wood Forest Produce: The quantity of harvested NTFP in physical quantities is assessed and listed category-wise in Table 27 for 2005 and 2019. The spatial variability is depicted in Fig. 16. These are comparable to a study from India (Verma et al. 2013). Table 28 lists NTFP values for various forest types per hectare per year. Figure 16 depicts the ecosystem values from NTFP in a million rupees circle-wise. Ecosystem value from NTFP is estimated as 13,906 million rupees, in which Mysore, Shimoga, and circles show higher values, and

Table 26 Quantity of fodder produced and its ecosystem supply value across Karnataka

S.no	Circle name	Fodder (Ton)		Fodder (MRs)		Fodder INR/ha/yr	
		2005	2019	2005	2019	2005	2019
1	Bengaluru	1658001	548868	3855	1276	8104	13023
2	Belgaum	2580175	1645663	5999	3826	7233	13586
3	Ballari	1988121	449821	4622	1046	7923	12779
4	Chamarajanagar	2722751	2160460	6330	5023	8988	20050
5	Chikmagalur	2210524	1261473	5139	2933	8344	16143
6	Dharwad	922640	317365	2145	738	7985	14089
7	Kalaburagi (Gulbarga)	998633	255515	2322	594	7202	13173
8	Hassan	1807242	1119519	4202	2603	7368	14064
9	Canara	4918160	3220533	11435	7488	6781	15382
10	Kodagu	1511207	1277404	3514	2970	6538	15895
11	Mangalore	2772886	1506332	6447	3502	8320	17671
12	Mysore	1322496	1131299	3075	2630	8395	17808
13	Shimoga	3366242	1907104	7827	4434	7388	17528

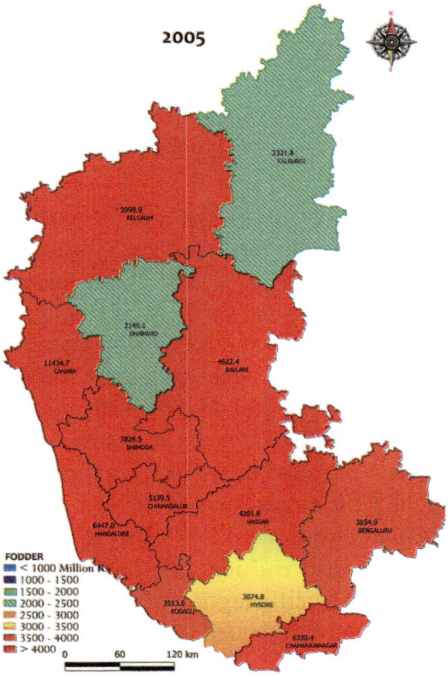

Fig. 15 Ecosystem supply value of fodder

Mangalore, Chikmagalur show lower values. The quantity of NTFP extraction was higher in 2005 than in 2019. The reduction is due to the degradation of the forest cover and the non-availability of labor (discussion with farmers during field investigations) for the sustainable extraction of NTFP.

- The ecosystem supply value of honey is 13,177 Rs/Ha/yr (2005) and 13,186 Rs/Ha/yr (2019) based on the data collected across the circles from the Karnataka Forest Department. The maximum value is 13,411 Rs/Ha/yr in the Chamarajanagar circle, and the minimum value is 13,059 Rs/Ha/yr for 2005 in the Shimoga circle. In 2019, the maximum value of 13,413 Rs/Ha/yr is in the Chamarajanagar circle, and the minimum value is 13,060 Rs/Ha/yr in the Shimoga circle.
- Soapnut is 12,724 and 12,977 Rs/Ha/yr for 2005 and 2019, respectively. The maximum value of 16,757 Rs/Ha/yr is in the Chamarajanagar circle for 2005 and 2019. The minimum value of 11,714 Rs/Ha/yr is in Kodagu, Mangalore, and Shimoga circles for 2005 and Kodagu circle in 2019.
- Cashew nut is 13,812 Rs/Ha/yr (2005) and 13,945 Rs/Ha/yr (2019). The maximum value is 17,029 Rs/Ha/yr in the Bangalore circle, and the minimum in Belgaum is 11,714 Rs/Ha/yr for 2005. In 2019, the maximum value was 17,048 Rs/Ha/yr in the Bangalore Circle, and the minimum value was 11,715 Rs/Ha/yr from the Belgaum Circle.

Table 27 Ecosystem value based on NTFP during 2005 and 2019

S.no	Circle name	NTFP (Million Rupees)															
		Honey		Soapnut		Cashew		Tamarind		Ramapatre		Dalchini		Murugalu		Total	
		2005	2019	2005	2019	2005	2019	2005	2019	2005	2019	2005	2019	2005	2019	2005	2019
1	Bengaluru	0	0.00	0	0.00	64	6.70	179	31.85	0	0.00	0	0.00	0	0.00	242	39
2	Belgaum	0	0.00	0	0.00	749	80.00	14	2.28	0	0.00	0	0.00	0	0.00	763	82
3	Ballari	0	0.00	0	0.00	0	0.00	6	1.04	0	0.00	0	0.00	0	0.00	6	1
4	Chamarajanagar	8089	1264.76	56	130.00	0	0.00	0	0.00	0	0.00	0	0.00	0	0.00	8146	1395
5	Chikmagalur	271	1.96	55	13.13	0	0.00	2	0.13	0	0.00	19	41.23	57	2.40	404	59
6	Dharwad	0	0.00	0	0.00	0	0.00	44	4.88	0	0.00	928	0.00	0	0.00	972	5
7	Kalaburagi	0	0.00	0	0.00	0	0.00	13	0.85	0	0.00	0	0.00	0	0.00	13	1
8	Hassan	0	0.00	0	0.00	88	2.80	0	0.00	4	0.16	0	0.00	0	0.00	92	3
9	Canara	36	3.08	0	0.00	1	1.10	0	0.00	0	0.16	0	0.00	94	50.00	131	54
10	Kodagu	5081	41.02	12318	1350.44	0	0.00	0	0.00	88	0.00	0	0.00	315	0.00	17802	1391
11	Mangalore	0	0.00	6106	0.00	10	2.30	0	0.00	3	0.00	1	1.86	0	0.00	6119	4
12	Mysore	113446	420.00	0	0.00	0	0.00	214	9.75	0	0.00	18	0.00	0	0.00	113678	430
13	Shimoga	11945	500.50	9178	9493.64	0	0.00	886	448.44	0	0.00	0	0.00	47	0.00	22056	10443

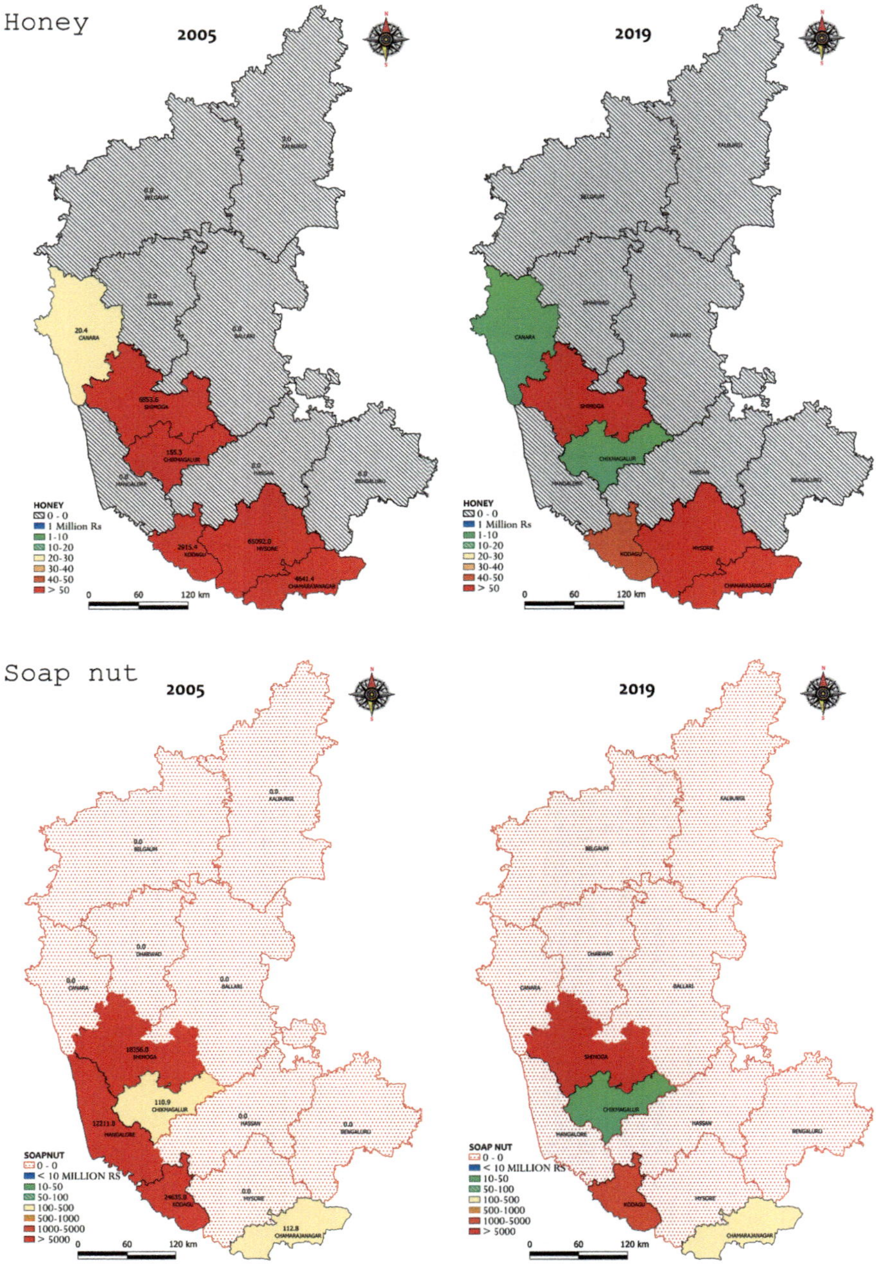

Fig. 16 Ecosystem supply value of from various goods under NTFP category

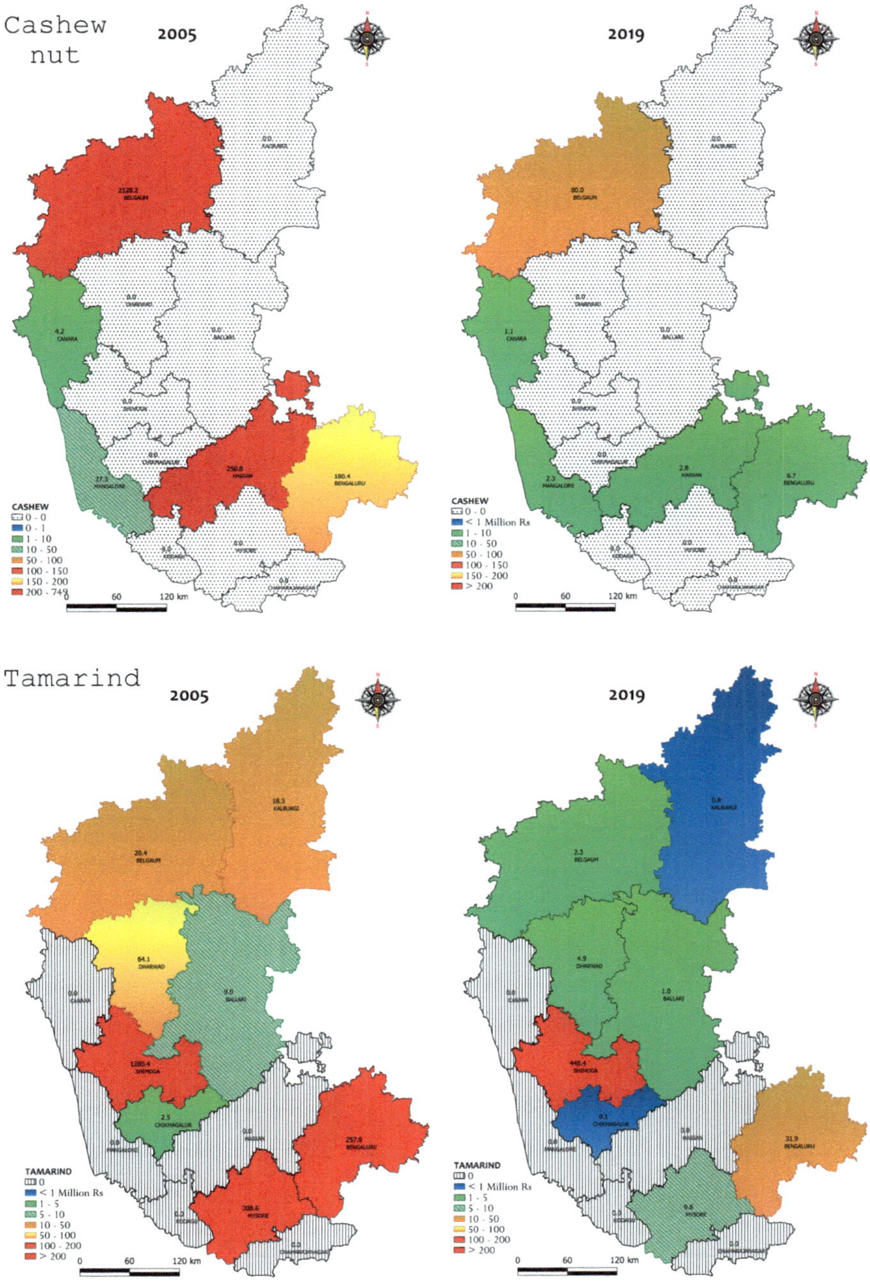

Fig. 16 (continued)

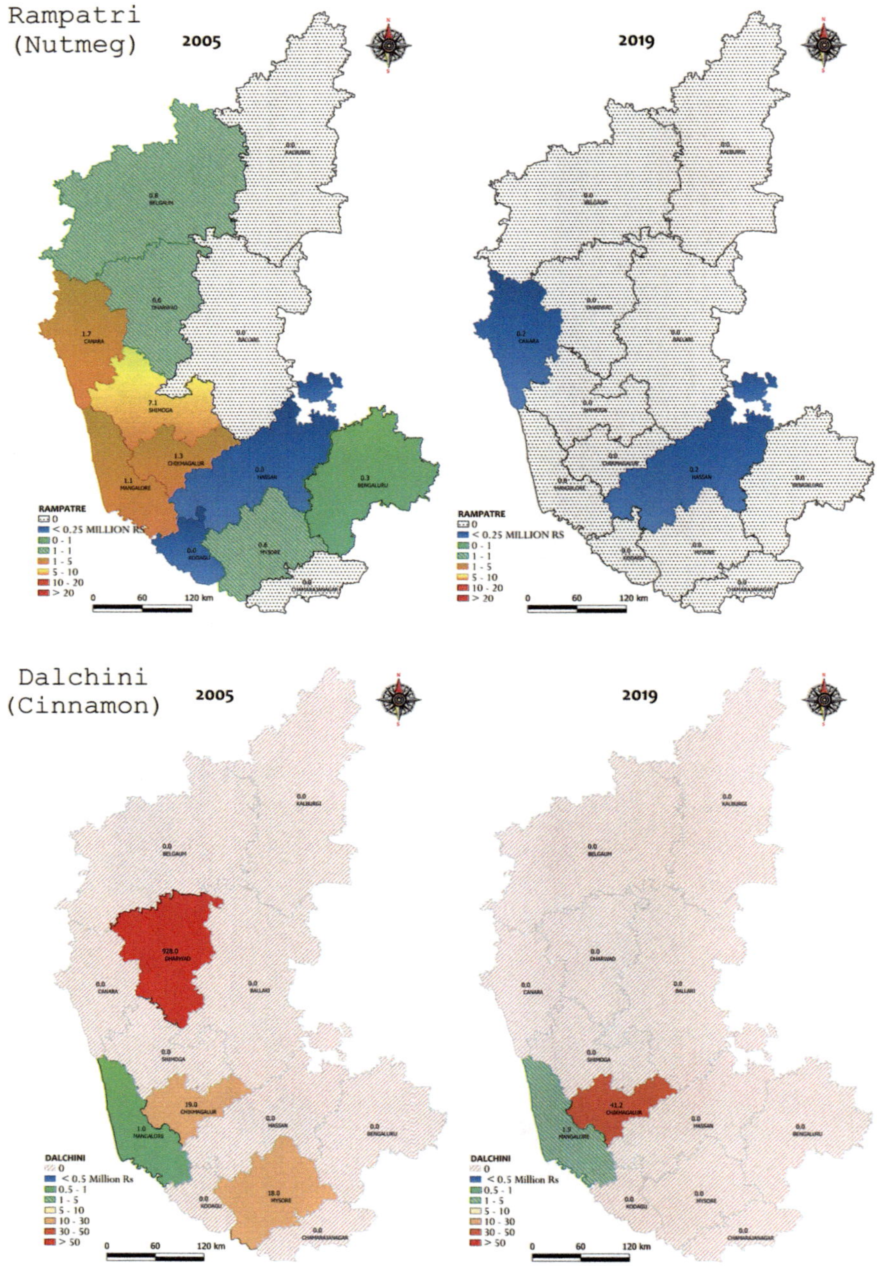

Fig. 16 (continued)

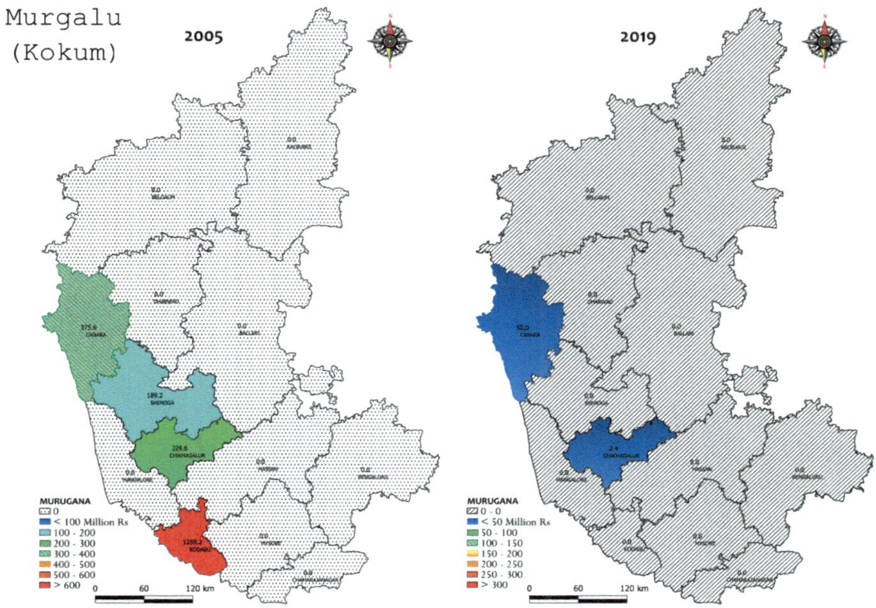

Fig. 16 (continued)

- Tamarind is 14,315 and 14,346 Rs/Ha/yr for 2005 and 2019, respectively. The maximum values of 17,044 and 17,049 Rs/Ha/yr are in the Ballari circle for 2005 and 2019. Minimum values of 11,423, and 11,575 Rs/Ha/yr are in the Kalaburagi circle for 2005 and 2019.
- Rampatri (nutmeg—*Myristica malabarica*) is 12,997 Rs/Ha/yr (2005) and 14,436 Rs/Ha/yr (2019). The maximum value of 16,891 Rs/Ha/yr is in the Hassan circle, and the minimum value is in Canara with 11,667 Rs/Ha/yr for 2005. In 2019, the maximum value of 16,842 Rs/Ha/yr was in the Hassan Circle, and the minimum value was 12,030 Rs/Ha/yr from the Canara Circle.
- Dalchini (Cinnamon) is 13,333 Rs/Ha/yr (2005) and 13,478 Rs/Ha/yr (2019), based on the data collected across the circles from the Karnataka Forest Department.
- Murugalu (Kokkum) is 11,717 Rs/Ha/yr (2005) and 11,740 Rs/Ha/yr (2019). The maximum value is 11,722 Rs/Ha/yr in the Shimoga circle, and the minimum value of 11,714 Rs/Ha/yr for 2005 is in the Canara circle. In 2019, the maximum value observed from the Chikmagalur circle is 11,765 Rs/Ha/yr, and the minimum value of 11,715 Rs/Ha/yr is in the Canara circle.

Overall, Canara, Shimoga, Kodagu, and Chamarajanagar circles provide provisioning services in significant quantities, while Chikmagalur, Belgaum, Hassan, Mangalore, Ballari, Dharwad, and Bengaluru provide moderate amounts, and the least amount is provided by Gulbarga circle.

Table 28 Value of various NTFP per hectare per year (₹/Ha/Yr.)

S.no	Circle name	NTFP (₹/Ha/Yr.)													
		Honey		Soap nut		Cashew		Tamarind		Ramapatri		Dalchini		Murugalu (Kokkum)	
		2005	2019	2005	2019	2005	2019	2005	2019	2005	2019	2005	2019	2005	2019
1	Bengaluru					17029	17048	17027	17032						
2	Belgaum					11714	11715	11726	11727						
3	Ballari							17044	17049						
4	Chamarajanagar	13411	13413	16757	16757										
5	Chikmagalur	13060	13067	11722	11723			11793	11819			13333	13478	11714	11765
6	Dharwad							17033	17045						
7	Kalaburagi							11423	11575						
8	Hassan					16758	16766			16891	16842				
9	Canara	13063	13106			11827	11765			11667	12030			11714	11715
10	Kodagu	13059	13060	11714	11714					11714				11716	
11	Mangalore			11714		11733	12432			11714		13333	13478		
12	Mysore	13411	13412					16758	16810						
13	Shimoga	13059	13060	11714	11715			11713	11714					11722	

Fuelwood

The ecosystem supply values of fuelwood vary with forest type and canopy cover, as described in Table 29. The ecosystem supply of fuelwood is depicted in Fig. 17. The ecosystem supply value from fuelwood amounts to 48,856 (2005) and 57,308 (2019) million rupees. The ecosystem supply value of fuelwood is assessed as 5,097 and 23,623 Rs/Ha/yr for 2005 and 2019 based on the fuelwood consumption data collected for the Karnataka state. The maximum value is 9,366 Rs/Ha/yr in the Canara circle, and the minimum value is observed in Dharwad circles as 1,492 Rs/Ha/yr for the year 2005. In 2019, the maximum value in the Canara circle was 11,499 Rs/Ha/yr, and the minimum value of 1,072 Rs/Ha/yr from the Kalaburagi circle.

Fish and other aquatic products provisioning services

The quantity of the fish harvested in tons and the revenue generated from fish is given in Table 30. The annual revenue from fish accounts for 7,837 (2005) and 12,126 (2019) million rupees. The harvested fish quantity has increased due to improved management and the construction of reservoirs. Spatial variability in the availability of fish and other aquatic products with the ecosystem supply value is evident in Fig. 18.

Table 29 Quantity of fuelwood extracted and its ecosystem supply value across Karnataka

S.no	Circle name	Fuelwood (Tons)		Fuelwood (Million Rupees)	
		2005	2019	2005	2019
1	Bengaluru	3261192	2921359	2609	2337
2	Belgaum	5272025	7786025	4218	6229
3	Ballari	4003969	2444235	3203	1955
4	Chamarajanagar	4889817	7397736	3912	5918
5	Chikmagalur	4276455	5364733	3421	4292
6	Dharwad	1865114	1517578	1492	1214
7	Kalaburagi (Gulbarga)	2227442	1339729	1782	1072
8	Hassan	3916999	5345185	3134	4276
9	Canara	11707192	14373821	9366	11499
10	Kodagu	3731281	5517220	2985	4414
11	Mangalore	5381077	5876867	4305	4701
12	Mysore	2492408	4171073	1994	3337
13	Shimoga	8045491	7587179	6436	6070
Fuelwood INR/ha/yr				5097	23623

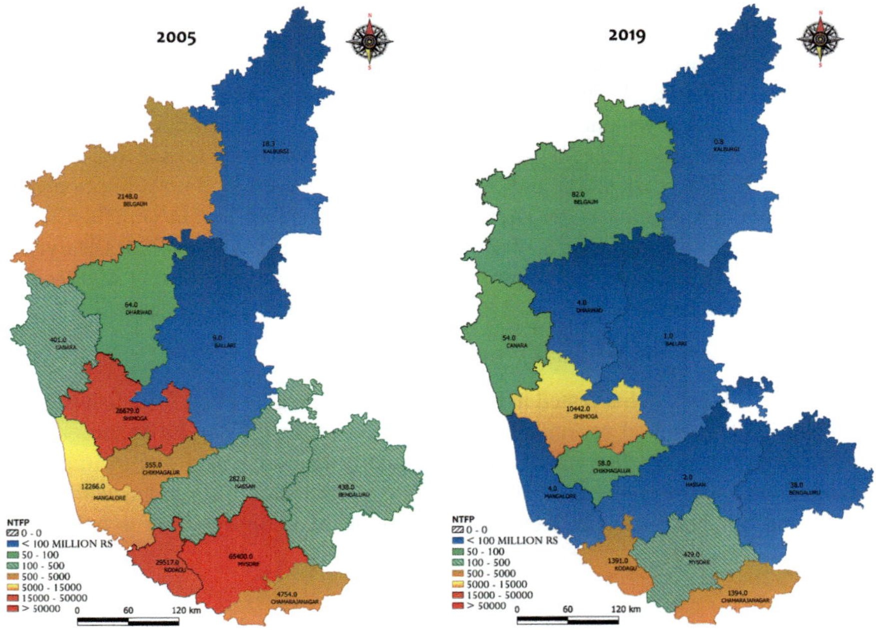

Fig. 17 Ecosystem supply of NTFP (million INR)

Table 30 Fish quantity and ecosystem supply value (physical—tons, monetary—million Rs.)

S.no	Circle name	Fish quantity (Tons)		Ecosystem supply value (Million Rs)	
		2005	2019	2005	2019
1	Bengaluru	14881	15495	967	1007
2	Belgaum	5092	13195	330	857
3	Ballari	20842	23112	1354	1502
4	Chamarajanagar	3454	2017	224	131
5	Chikmagalur	4232	4587	275	298
6	Dharwad	8179	11441	531	743
7	Kalaburagi	10167	27234	660	1770
8	Hassan	22061	18688	1433	1214
9	Canara	5852	20934	380	1360
10	Kodagu	1177	3881	76	252
11	Mangalore	3015	3767	195	244
12	Mysore	14661	24860	952	1615
13	Shimoga	7086	17443	460	1133
Total		120699	186655	7837	12126

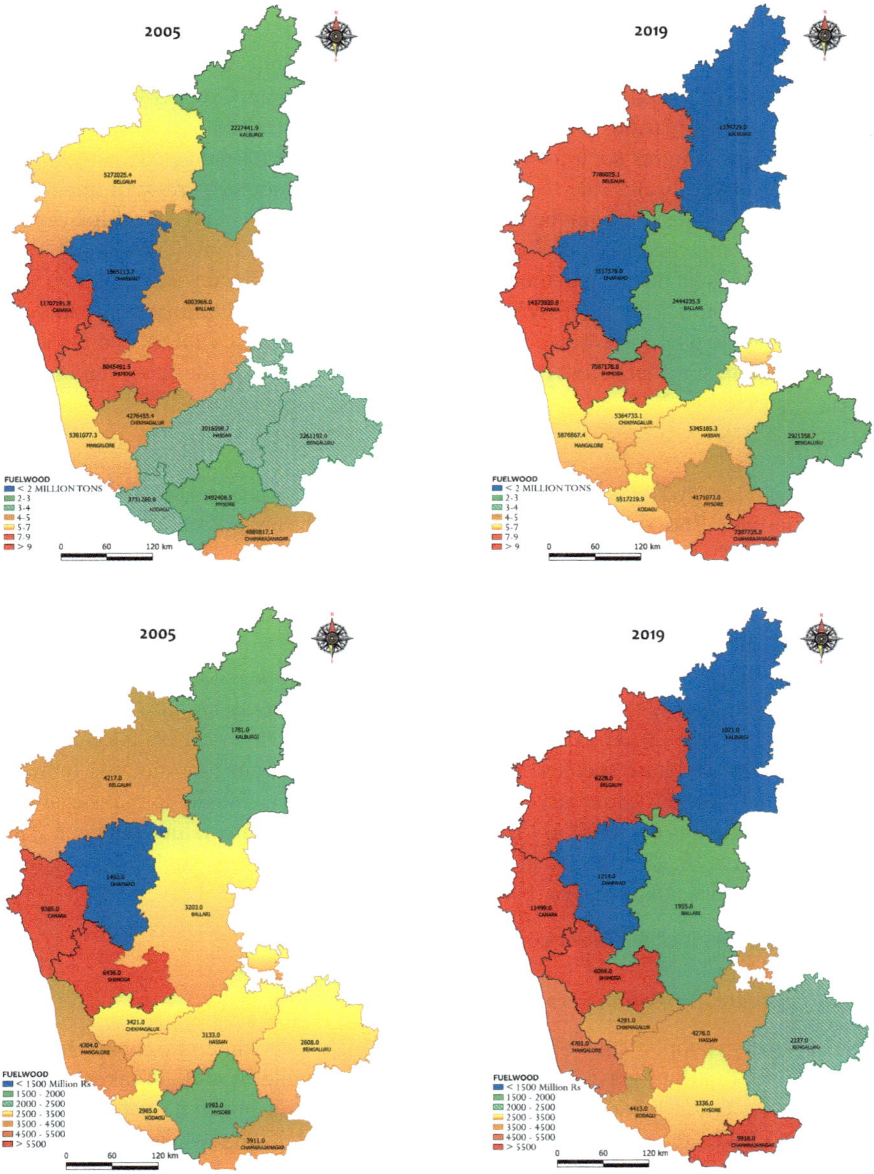

Fig. 18 Quantity of fuelwood extracted and its ecosystem supply value

Water supply

The quantity of water and its service value are assessed and depicted in Fig. 19, and forest circle-wise details are listed in Table 31. The ecosystem supply value of water for the State from various circles accounts for 618,534 (2005) and 280,785 (2019) million rupees. Circles such as Canara, Mangalore, and Chikmagalur have higher values than other circles due to the relatively higher amount of rainfall and better forest cover. Greater losses in the service value depict the direct relationship between deforestation and water supply. Assessment of overland flow (runoff) and local water recharge reveals an increase in the run-off with a decline in local water recharge with land degradation. Local water recharge ensures the water availability in streams and wells during the post-monsoon period. Circles such as Hassan, Ballari, and Dharwad indicate a more significant loss (−86.3, −94.6, and −79.7%) in the ecosystem supply value compared to 2005.

Medicine

Ecosystem value from medicinal plants is assessed and listed circle-wise in Table 32, accounting for 766 million rupees and 514 million rupees for 2005 and 2019, respectively. The spatial variability of the medicinal value of forests is depicted in Fig. 20.

Genetic material service

The ecosystem value of genetic material services varies with forest type and is accounted for through the benefit transfer technique based on case studies from India (Verma et al. 2013), considering the spatial extent of forest patches (with the distribution of endemic species and species of conservation importance). Table 33 lists genetic materials (monetary) estimates by forest circle, and the spatial variability across Karnataka is depicted in Fig. 21. The loss in economic values between 2005 and 2019 points to the change in forest cover in forest circles such as Bengaluru, Belgaum, Ballari and Shimoga, Hassan, and Mysore.

Regulating services of forest ecosystems in Karnataka

Global climate regulation services/carbon sequestration

The ecosystem supply value of carbon sequestration is calculated by considering the social cost of carbon per ton. The social cost of a tonne of CO_2 is taken as US$ 80 using the GDP deflator (MoSPI 2020). The carbon sequestration from forests, as per InVEST, shows that the forest circles located in the Western Ghats have higher sequestration than other parts of the State due to lower disturbances (Fig. 22). Carbon sequestration in forest ecosystems of Karnataka declined from 124153 Gg/Yr (2005) to 89194 Gg/Yr (2019) due to a decline in ecosystem extent and condition. Circles such as Mysore, Ballari, and Kalaburagi have lower carbon sequestration due to deforestation from rapid land-use changes (Table 34, Fig. 23). The ecosystem supply value of carbon sequestration is 794949 and 571138 million rupees for 2005

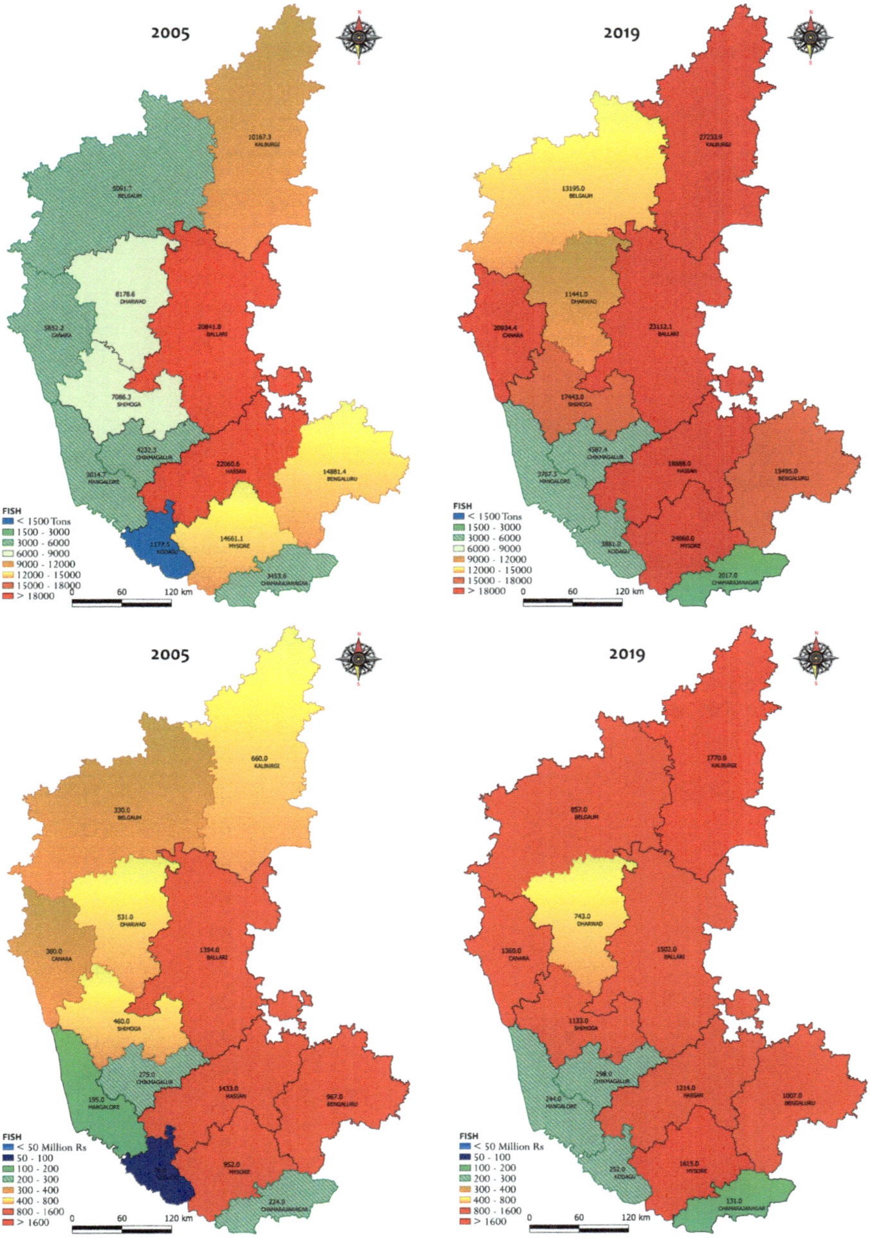

Fig. 19 Fish harvested and its ecosystem supply value

Table 31 Ecosystem value of water supply across forest circles

S.no	Circle name	Water supply (Million cum)		Ecosystem supply value of water (Million Rs)		% Change (2005–2019)
		2005	2019	2005	2019	
1	Bengaluru	1729	1713	37823	8806	−76.7
2	Belgaum	1688	1712	52739	15784	−70.1
3	Ballari	836	860	65284	3532	−94.6
4	Chamarajanagar	299	337	71235	25225	−64.6
5	Chikmagalur	1507	1519	48225	34364	−28.7
6	Dharwad	1392	1409	21939	4446	−79.7
7	Kalaburagi	1996	2011	20867	1681	−91.9
8	Hassan	811	811	27235	3739	−86.3
9	Canara	2072	2077	98965	67098	−32.2
10	Kodagu	1982	1981	24506	12869	−47.5
11	Mangalore	7038	7035	68031	58291	−14.3
12	Mysore	713	719	30475	15361	−49.6
13	Shimoga	2157	2188	51210	29589	−42.2
Total		24219	24371	618534	280785	−54.6

Table 32 Ecosystem value of medicine in 2005 and 2019

S.no	Circle name	Ecosystem supply value of medicine (Million Rs)	
		2005	2019
1	Bengaluru	40	21
2	Belgaum	61	32
3	Ballari	69	23
4	Chamarajanagar	63	41
5	Chikmagalur	49	43
6	Dharwad	24	11
7	Kalaburagi	28	10
8	Hassan	42	19
9	Canara	152	130
10	Kodagu	48	33
11	Mangalore	69	65
12	Mysore	29	20
13	Shimoga	92	66
Total		766	514

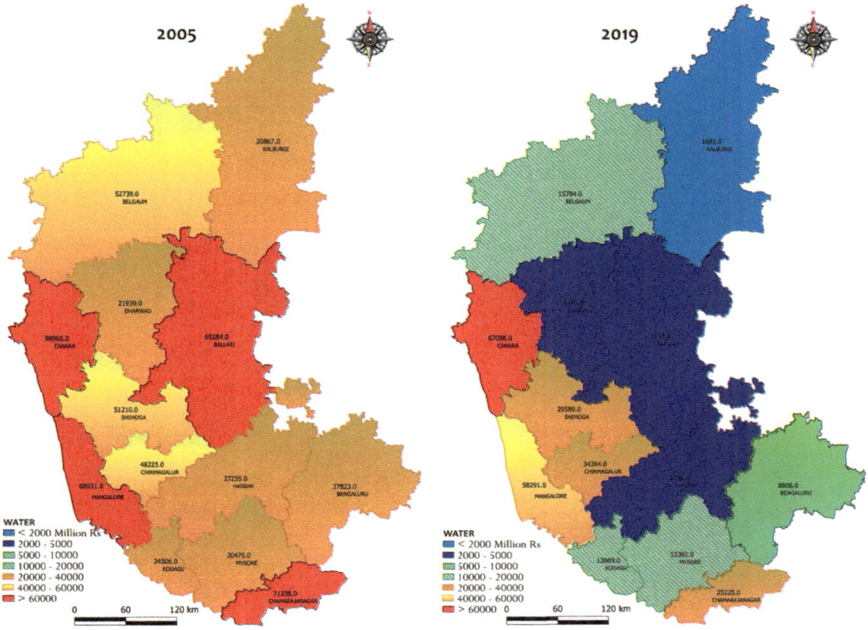

Fig. 20 Ecosystem value of water supply (million INR) in 2005 and 2019

Table 33 Ecosystem supply value of genetic material service

S.no	Circle name	Ecosystem supply value of genetic material service (Million Rs)	
		2005	2019
1	Bengaluru	9819	2286
2	Belgaum	36257	10851
3	Ballari	16948	916
4	Chamarajanagar	28465	10080
5	Chikmagalur	41674	29696
6	Dharwad	8766	1776
7	Kalaburagi	5417	436
8	Hassan	10883	1494
9	Canara	85521	57984
10	Kodagu	21177	11120
11	Mangalore	58790	50373
12	Mysore	12178	6138
13	Shimoga	44254	25570
Total		380149	208725

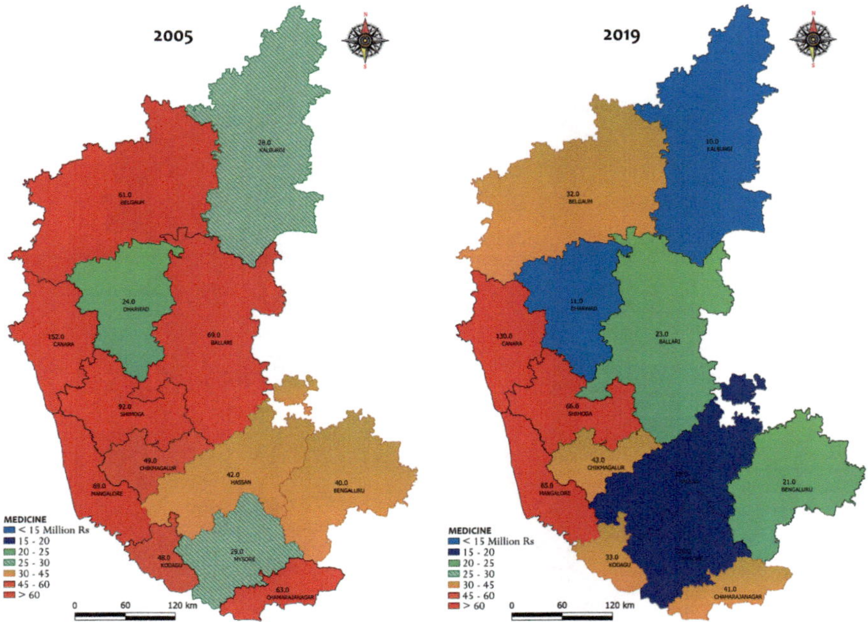

Fig. 21 Ecosystem supply value of medicine across Karnataka

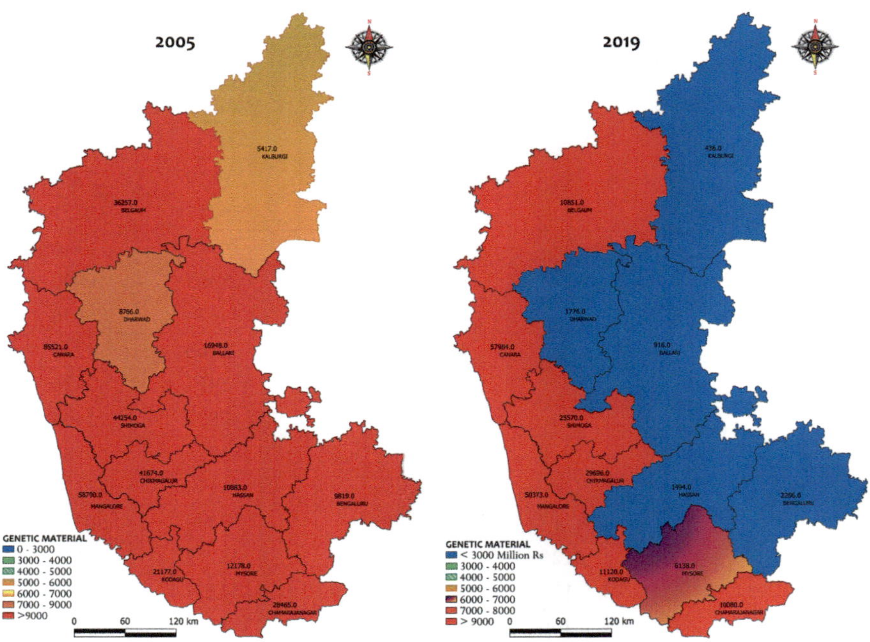

Fig. 22 Genetic material service value across the circles

Table 34 Annual carbon sequestration (CO_2) from forests in Karnataka

S.no	Circle name	Carbon sequestration (CO_2 Gg/Yr)		CO_2 sequestration (Million Rs)	
		2005	2019	2005	2019
1	Bengaluru	5787	3374	37065	21615
2	Belgaum	8320	5026	53275	32187
3	Ballari	6487	2395	41541	15345
4	Chamarajanagar	12066	7934	77250	50800
5	Chikmagalur	9581	8681	61343	55578
6	Dharwad	3285	1699	21039	10887
7	Kalaburagi	2981	1075	19100	6905
8	Hassan	5571	2861	35677	18322
9	Canara	28568	23708	182897	151785
10	Kodagu	8242	5835	52770	37358
11	Mangalore	12752	12174	81650	77946
12	Mysore	5063	3782	32420	24218
13	Shimoga	15450	10650	98922	68192
Total		124153	89194	794949	571138

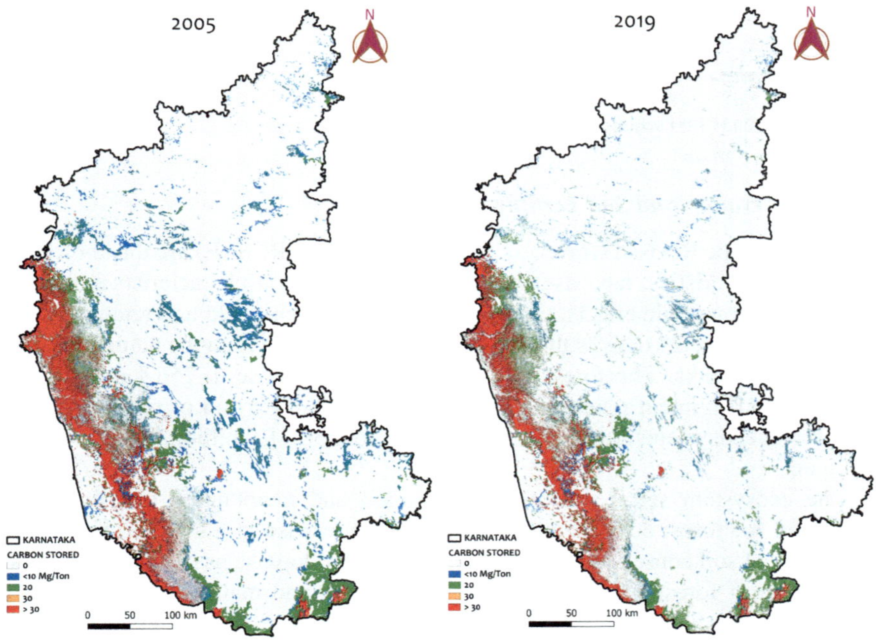

Fig. 23 Carbon sequestration as per InVEST model

and 2019, respectively (Fig. 24 and Table 34). Canara, Mangalore, and Chikmagalur forest circles are contributing larger quantities.

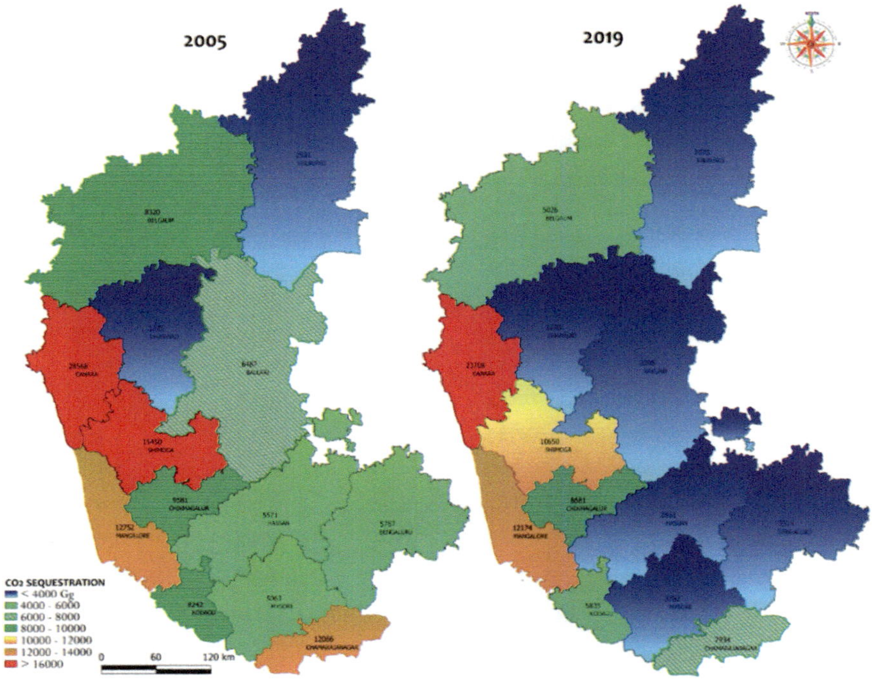

Fig. 24 Carbon (CO_2) sequestration across the circles

Soil Conservation and Soil Fertility

InVEST, and the Revised Universal Soil Loss Equation (RUSLE) use the knowledge of soil characteristics, land use characteristics, vegetation characteristics, farming practices, topographic effects, etc., to derive the sediment yield and retention characteristics. Figure 25 depicts the local soil retention map within the natural forested areas of Karnataka. The forests of Karnataka protect close to 756.4 million tons of sediment (sediment retention). Considering Rs 48.8 per ton of sediment retention, forest ecosystems across the state retain about 36,912 million rupees worth of sediment.

The ecosystem service of soil conservation and maintaining soil fertility is computed for forest ecosystems and is listed circle-wise in Table 35. The spatial variability of soil conservation services of forests, circle-wise, is depicted in Fig. 26. The ability of forest ecosystems to prevent erosion and maintain high fertility is high in Canara, Shimoga, and Mangalore circles due to rich forest cover. With the degradation in the forest cover, soil fertility and erosion prevention have decreased across the circles. Ballari, Belgaum, Kalaburagi, Kodagu, and Hassan were highly impacted due to deforestation.

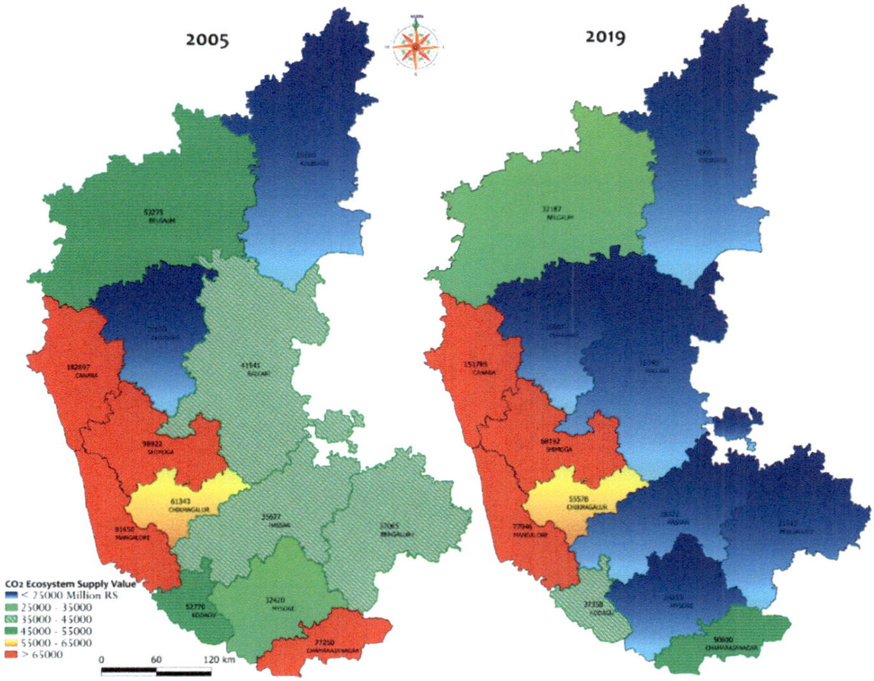

Fig. 25 Ecosystem supply value of CO_2 sequestration across districts in Karnataka

Table 35 Ecosystem supply value of soil conservation & soil fertility services

S.no	Circle name	Soil conservation (million Rs)		Soil fertility (million Rs)	
		2005	2019	2005	2019
1	Bengaluru	2354	1038	2776	3918
2	Belgaum	4482	2212	3954	4351
3	Ballari	4062	1024	3049	4314
4	Chamarajanagar	5951	3255	12785	16074
5	Chikmagalur	4054	3370	3659	4878
6	Dharwad	2118	839	1413	1840
7	Kalaburagi	1576	440	2272	1721
8	Hassan	2251	855	3916	3217
9	Canara	11262	9174	21520	22512
10	Kodagu	3404	2247	10429	9402
11	Mangalore	5740	5254	9953	10792
12	Mysore	2717	1699	5503	6154
13	Shimoga	6615	4554	13641	12046
Total		56856	35969	94870	101227

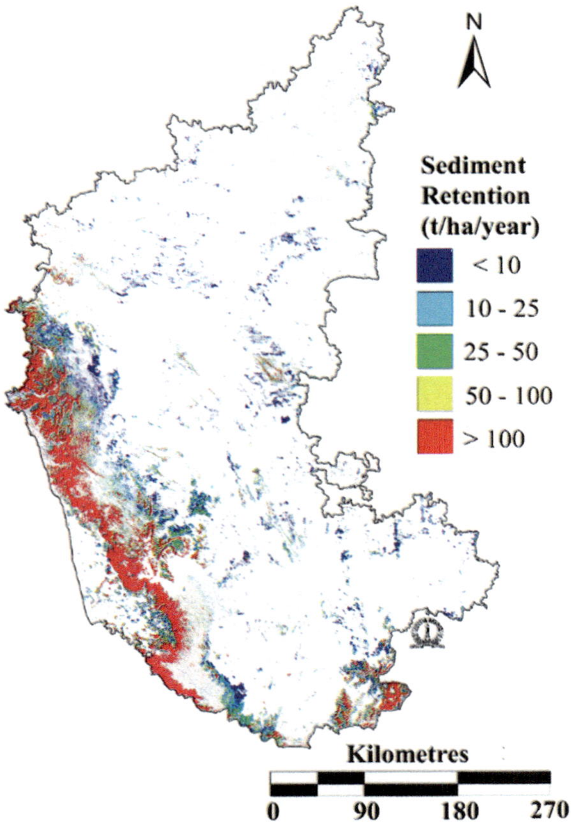

Fig. 26 Sediment retention as per InVEST

Water regulation and groundwater recharge

Figure 27 depicts the local recharge map (seasonal water yield as per INVEST 3.9.0) for the natural forested areas of Karnataka. The forests of Karnataka locally recharge about 27.2 billion cubic meters of water to the ground per year, which later flow as base flows.

The ecosystem supply values across the circles are depicted in Fig. 28 and Table 36, and the value amounts to 7,109 million INR/Yr (2019).

Water Purification: The ecosystem service of water purification ranges from 10,310 INR/Ha/Yr (2005) to 6,921 INR/Ha/Yr (Table 37 and Fig. 29).

Waste treatment: Waste treatment (remediation) amounts to 14,597 million INR/Yr (2005) to 9,799 million INR/Yr (2019). The waste treatment services of the forest ecosystem are given in Table 38 and Fig. 29.

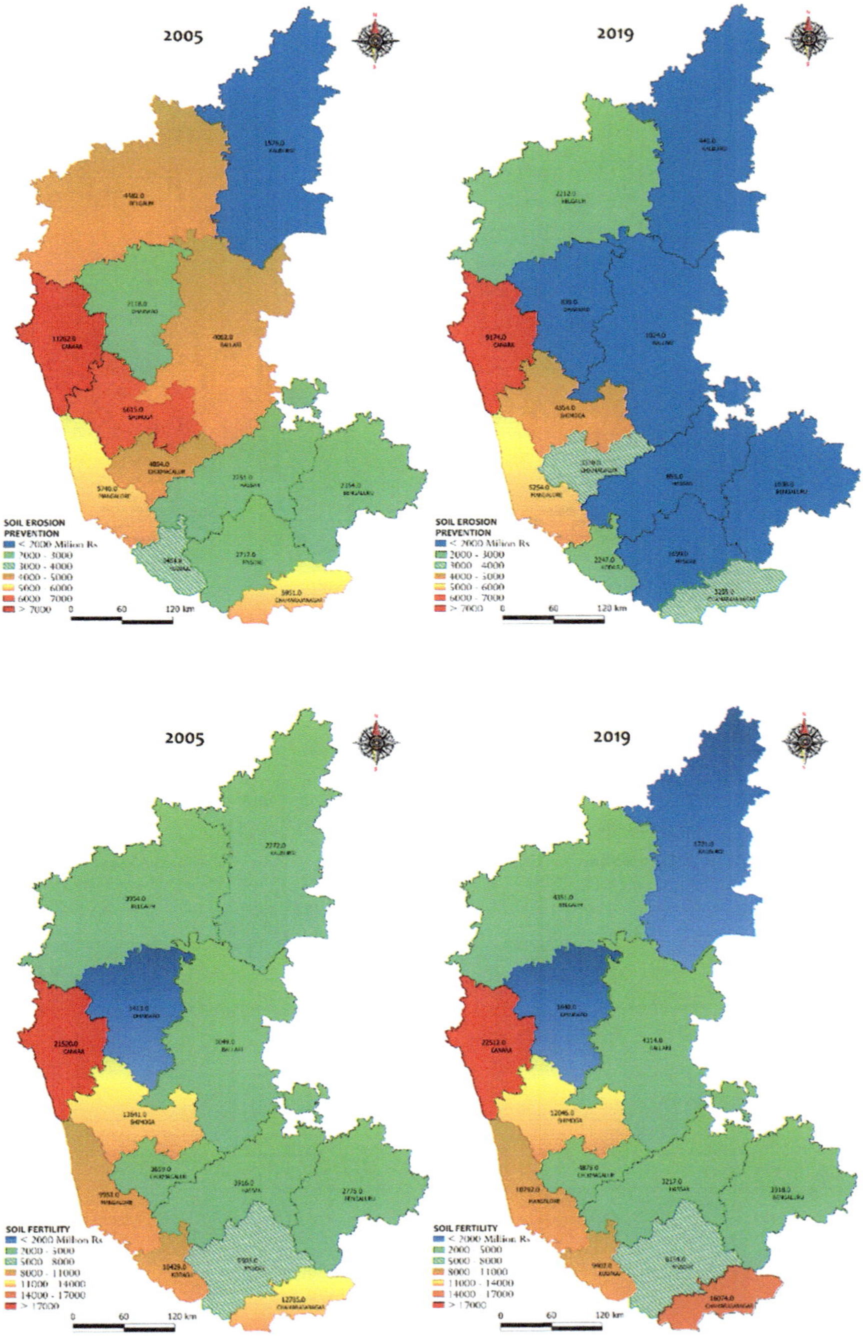

Fig. 27 Ecosystem supply value at circle wise from soil conservation and fertility services

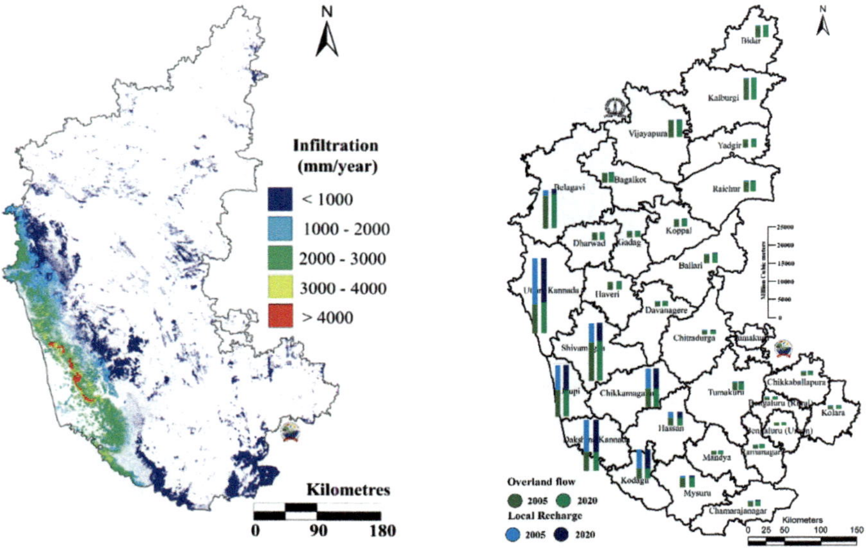

Fig. 28 Local recharge in the natural forested areas of Karnataka

Table 36 Ecosystem services of groundwater recharge across forest circles

S.no	Circle name	Groundwater recharge (million Rs), 2019
1	Bengaluru	145
2	Belgaum	434
3	Ballari	143
4	Chamarajanagar	461
5	Chikmagalur	745
6	Dharwad	121
7	Kalaburagi	61
8	Hassan	120
9	Canara	2000
10	Kodagu	486
11	Mangalore	1166
12	Mysore	238
13	Shimoga	989
Total		7109

Table 37 Forest ecosystem supply value of water purification and treatment services

S.no	Circle name	Water purification (million Rs)		Waste treatment (million Rs)	
		2005	2019	2005	2019
1	Bengaluru	536	1038	759	403
2	Belgaum	826	2212	1170	617
3	Ballari	926	1024	1310	443
4	Chamarajanagar	848	3255	1201	776
5	Chikmagalur	654	3370	926	819
6	Dharwad	323	839	457	218
7	Kalaburagi	387	440	547	189
8	Hassan	568	855	804	363
9	Canara	2032	9174	2877	2461
10	Kodagu	647	2247	917	634
11	Mangalore	927	5254	1313	1236
12	Mysore	400	1699	566	386
13	Shimoga	1236	4554	1750	1254
Total		10310	6921	14597	9799

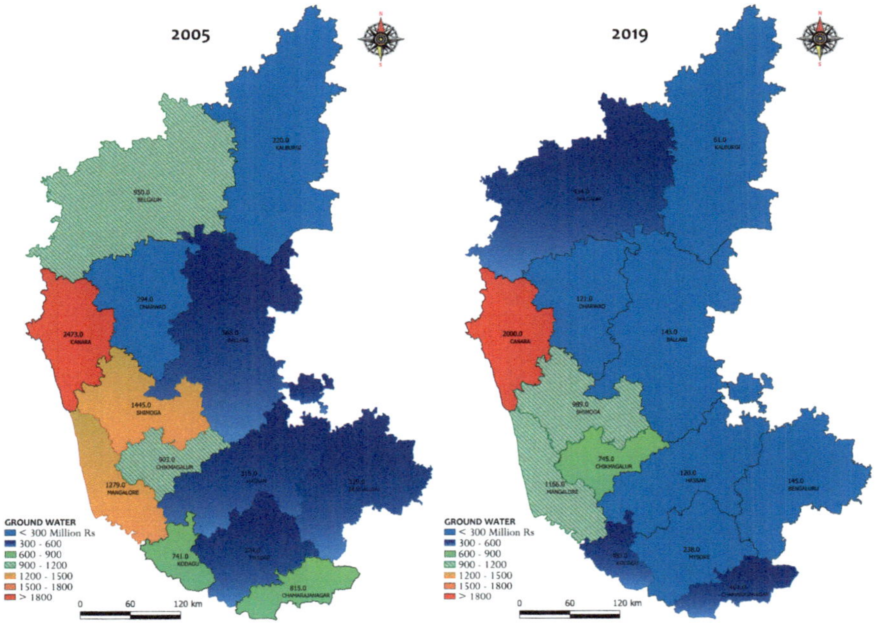

Fig. 29 Ecosystem services of ground-water recharge

Pollination service: The ecosystem supply value of pollination services depicts higher values in Canara, Shimoga, and Mangalore circles, and the State aggregate value is 26,942 million rupees per year (Table 38, Fig. 30).

Air filtration services: The total ecosystem supply value of air quality regulation services is estimated as 79,590 (2005) and 53,440 (2019) million rupees, with Canara, Shimoga, Mangalore, and Kodagu circles contributing a higher share (Table 39). The spatial variability of air quality regulation services is depicted in Fig. 31. The degradation in the forest cover from 2005 to 2019 has resulted in the decline of air filtration services.

Local (micro and meso) climate regulation services: The ecosystem value of moderating climate listed in Table 40 accounts for 169,487 (2005) and 113,807 (2019) million rupees. Canara, Shimoga, Mangalore, and Kodagu circles contribute a higher share due to good forest cover, which moderates the climate (Fig. 32). All forest circles show a decline in local climate regulation ecosystem value due to the loss of forest cover.

Table 38 Circle wise ecosystem supply value from pollination service

S.no	Circle name	Pollination services-2019 (Million Rs/Yr)
1	Bengaluru	983
2	Belgaum	1781
3	Ballari	1079
4	Chamarajanagar	1962
5	Chikmagalur	2337
6	Dharwad	551
7	Kalaburagi	460
8	Hassan	884
9	Canara	7018
10	Kodagu	1808
11	Mangalore	3526
12	Mysore	976
13	Shimoga	3577
Total		26942

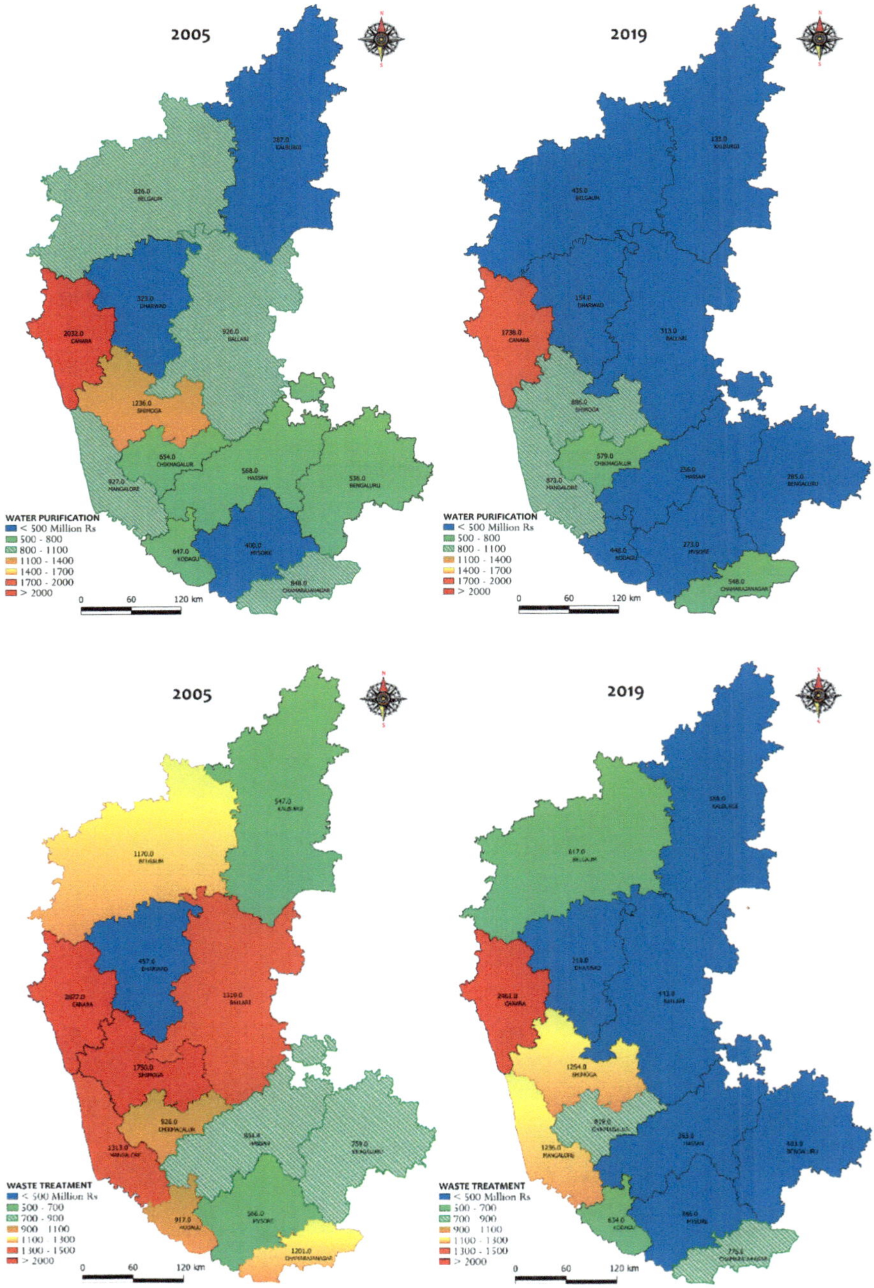

Fig. 30 Ecosystem services of water purification and waste treatment

Table 39 Ecosystem supply value from air filtration service

S.no	Circle name	Air filtration services (Million Rs)	
		2005	2019
1	Bengaluru	4140	2201
2	Belgaum	6377	3363
3	Ballari	7144	2417
4	Chamarajanagar	6550	4234
5	Chikmagalur	5047	4467
6	Dharwad	2494	1190
7	Kalaburagi	2986	1031
8	Hassan	4385	1980
9	Canara	15682	13416
10	Kodagu	4998	3456
11	Mangalore	7158	6741
12	Mysore	3088	2106
13	Shimoga	9541	6838
Total		79590	53440

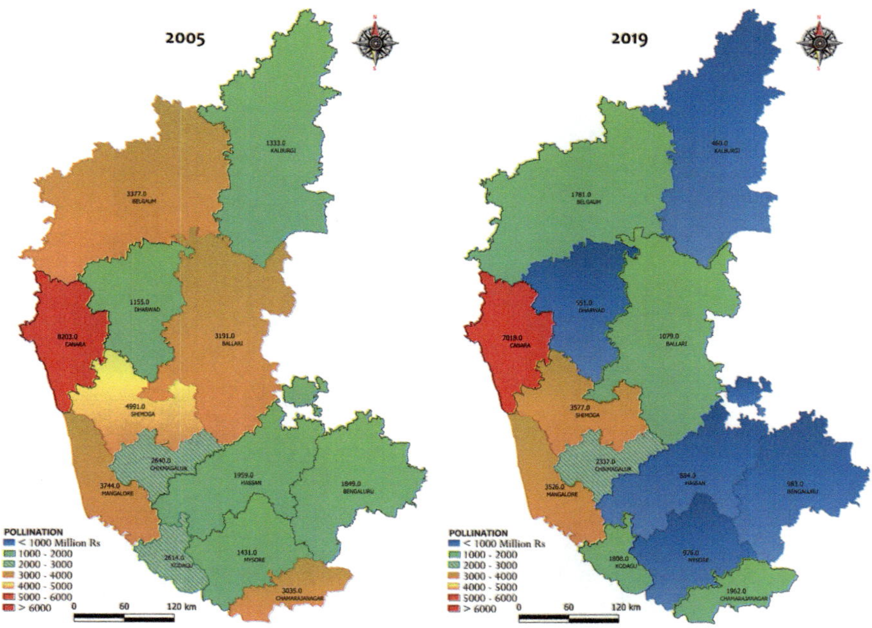

Fig. 31 Ecosystem supply value of pollination service

Table 40 Ecosystem services of local climate regulation

S.no	Circle name	Local climate regulation (million Rs)	
		2005	2019
1	Bengaluru	8817	4688
2	Belgaum	13579	7161
3	Ballari	15214	5147
4	Chamarajanagar	13948	9017
5	Chikmagalur	10747	9513
6	Dharwad	5311	2534
7	Kalaburagi	6358	2197
8	Hassan	9339	4216
9	Canara	33394	28570
10	Kodagu	10643	7361
11	Mangalore	15243	14356
12	Mysore	6576	4486
13	Shimoga	20318	14561
Total		169487	113807

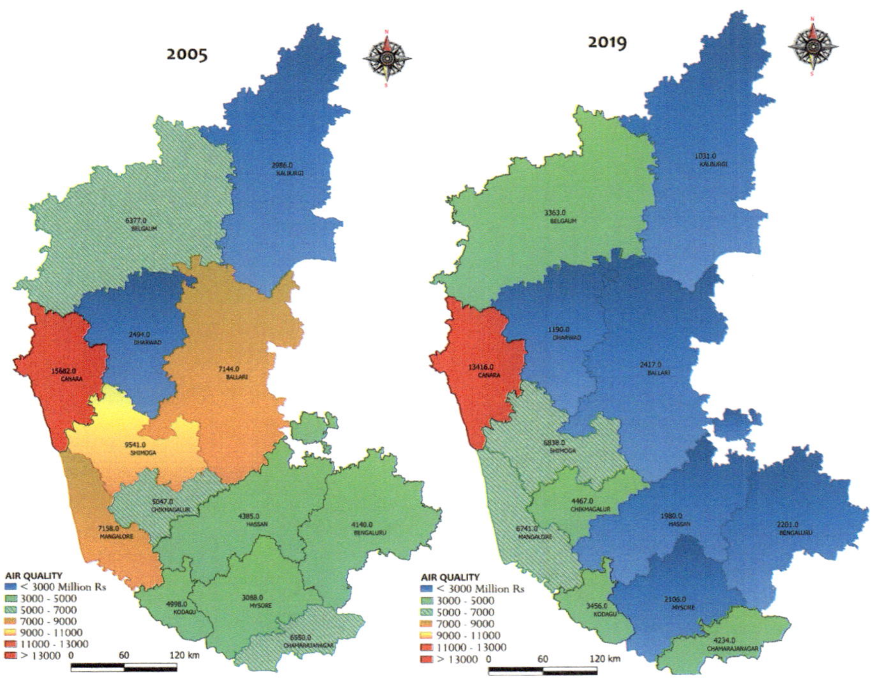

Fig. 32 Air quality regulation services from forest ecosystems in Karnataka

5 Cultural Services from Forest Ecosystems

The ecosystem values of various cultural services are listed in Table 41, and the spatial variability based on ecosystem extent variations is depicted in Fig. 33.

Table 41 Ecosystem supply value of cultural services

S.no	Circle name	Cultural services (million ₹)									
		Aesthetic		Tourism and recreational		Spiritual and historic		Artistic and culture		Education, scientific and research	
		2005	2019	2005	2019	2005	2019	2005	2019	2005	2019
1	Bengaluru	217	50	765	739			87	46	873	464
2	Belgaum	302	90	329	318			134	71	1345	709
3	Ballari	374	20	1380	130			151	51	1507	510
4	Chamarajanagar	408	144	78816	78816			138	89	1381	893
5	Chikmagalur	276	197	13089	13089			106	94	1064	942
6	Dharwad	125	25	189	182			53	25	526	251
7	Kalaburagi	119	9					63	22	630	218
8	Hassan	156	21					92	42	925	418
9	Canara	567	385	50669	50669	38	25	331	283	3307	2829
10	Kodagu	140	73	31629	31629	18	18	105	73	1054	729
11	Mangalore	390	334	49162	49162	26	22	151	142	1510	1422
12	Mysore	174	88	30320	30320			65	44	651	444
13	Shimoga	293	169	24473	24473	110	110	201	144	2012	1442
Total		3541	1605	280821	280777	198	179	1679	1127	16786	11271

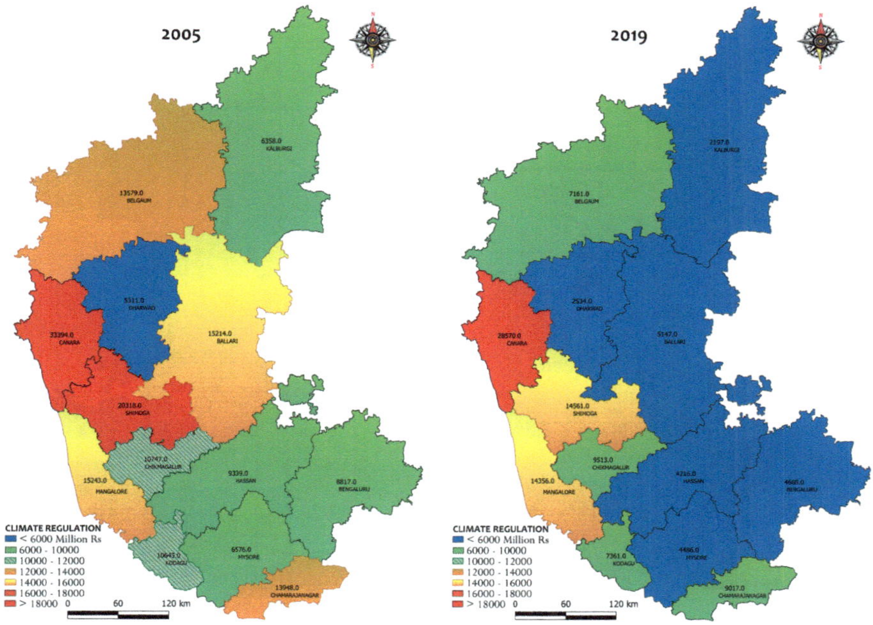

Fig. 33 Local climate regulation service from forests of Karnataka

6 Total Provisioning, Regulating, and Cultural Services of Forest Ecosystems in Karnataka

Provisioning services: The forest circles in the Western Ghats indicate high values for timber, bamboo, fodder, NTFP, genetic resources, medicine, and fuelwood, which are aggregated to compute the total provisioning service of forests (Fig. 34). Relatively higher provisioning services in the few forest circles can be attributed to the presence of rich, intact evergreen forest cover, followed by the transition zones to the east (Deccan plains) and west (Coast) of the Western Ghats (Table 42). The provisioning services of forest ecosystems in Karnataka accounts to 517 (2005) and 531 (2019) billion rupees per year.

Regulating services of forest ecosystems in Karnataka: The total regulating services of forest ecosystems in Karnataka amounts to 1270 (2005) and 926 (2019) billion rupees per year (Table 43). Spatial variations in regulating services across forest landscapes in Karnataka are depicted in Fig. 35. Circles such as Canara, Shimoga, Kodagu, and Mangalore show relatively higher values, emphasizing their forest cover status. The least regulating services were provided in Kalaburagi and Dharwad circles due to the absence of interior/intact forest cover.

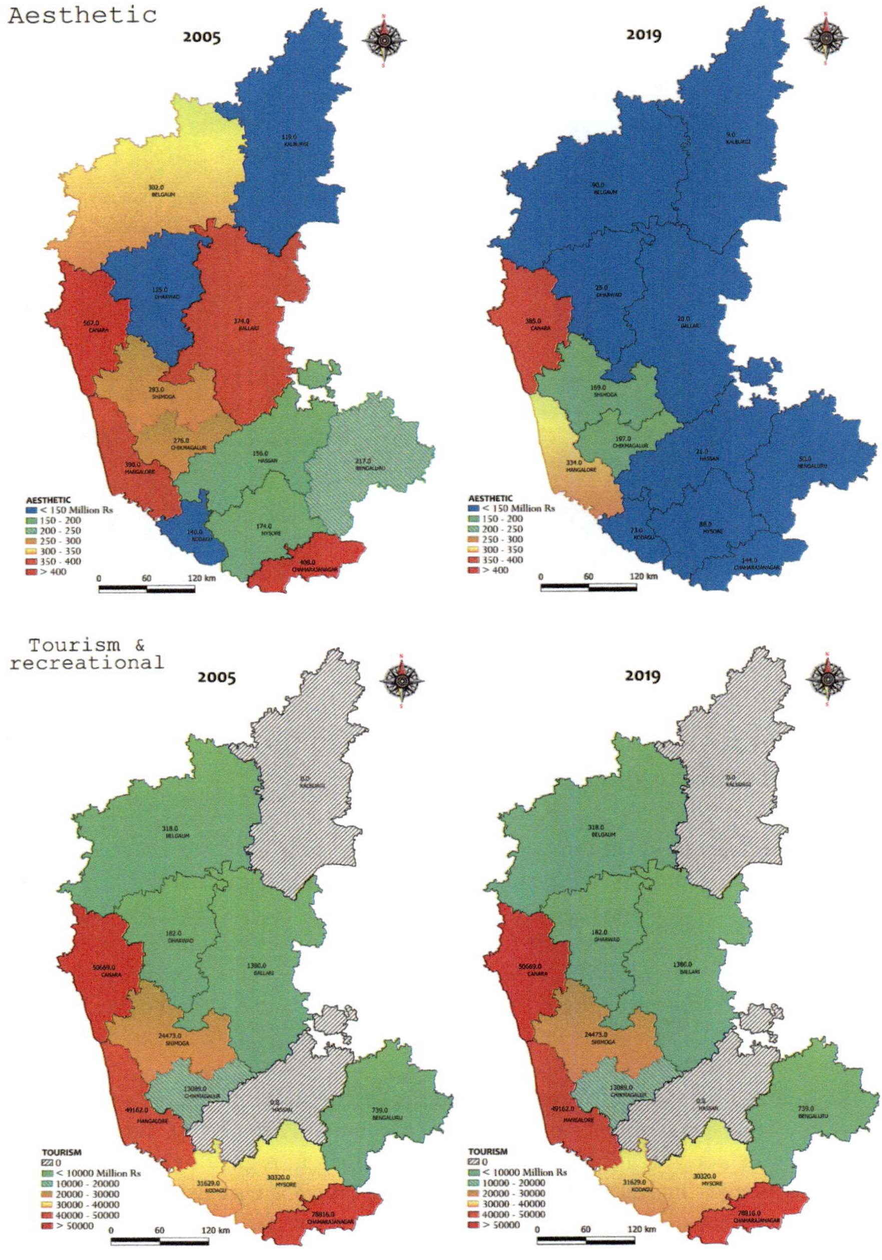

Fig. 34 Cultural services of Karnataka and their ecosystem supply values

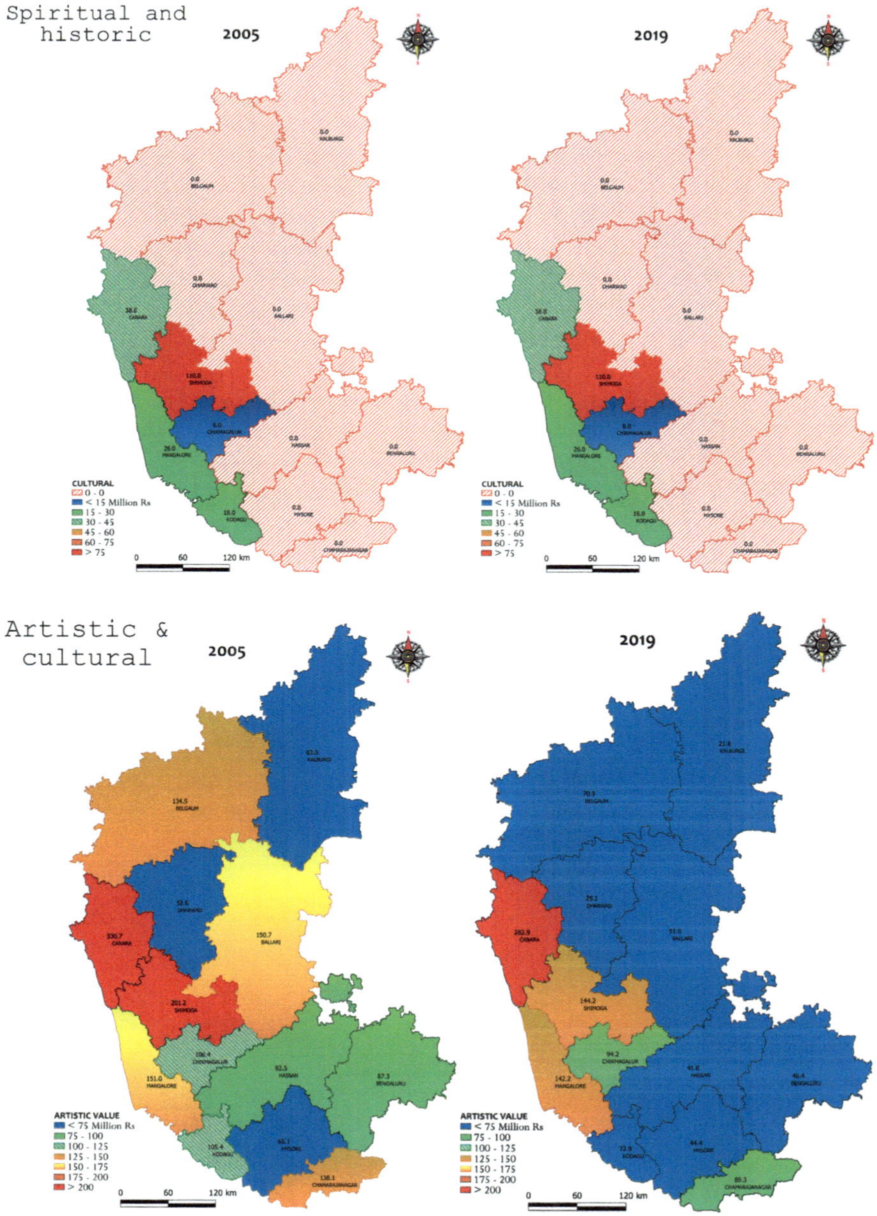

Fig. 34 (continued)

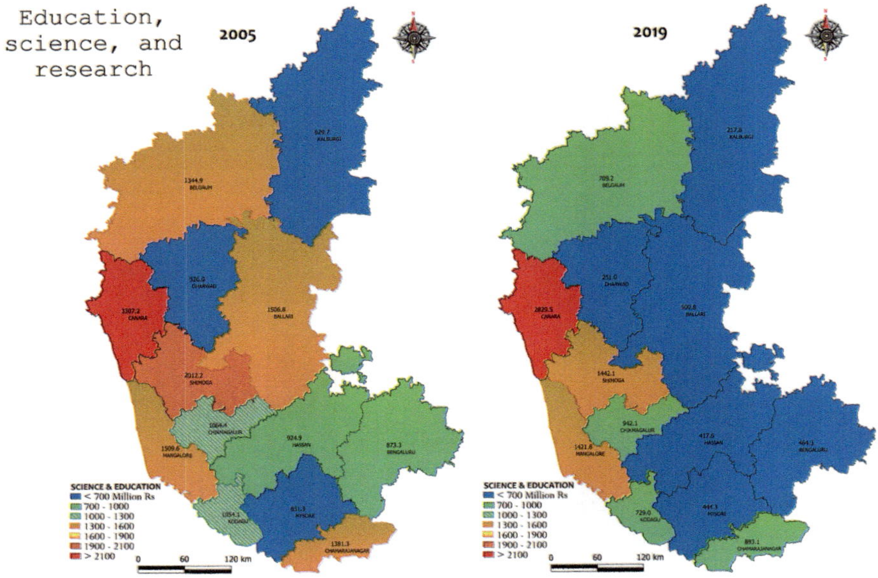

Fig. 34 (continued)

Table 42 Total provisioning services from forest ecosystems in Karnataka

S.no	Circle name	Provisioning services (million Rs)	
		2005	2019
1	Bengaluru	21224	1387
2	Belgaum	30294	28955
3	Ballari	34804	7308
4	Chamarajanagar	46220	38523
5	Chikmagalur	26944	65249
6	Dharwad	13203	7512
7	Kalaburagi	12,261	4500
8	Hassan	16474	7297
9	Canara	56439	129283
10	Kodagu	32361	26364
11	Mangalore	42976	110150
12	Mysore	130886	24535
13	Shimoga	53047	68305
Total		517134	531375

Table 43 Regulating services of forest ecosystem (circle wise) in Karnataka

S.no	Circle name	Regulating services (million Rs)	
		2005	2019
1	Bengaluru	58630	35281
2	Belgaum	87993	52545
3	Ballari	77010	30231
4	Chamarajanagar	122387	87131
5	Chikmagalur	89976	82,289
6	Dharwad	34608	18,338
7	Kalaburagi	34785	13143
8	Hassan	59219	30217
9	Canara	280343	238678
10	Kodagu	87167	63203
11	Mangalore	127012	121896
12	Mysore	53079	40540
13	Shimoga	158464	112901
Total			

Cultural services of forest ecosystems in Karnataka: The total cultural services amount to 303 (2005) and 295 (2019) billion rupees per year (Fig. 36 and Table 44). The circles of the Western Ghats showed higher values in terms of cultural services, primarily spiritual, recreation, and artistic services, emphasizing the intrinsic relation between forests and the culture of the people.

Total Ecosystem Supply Value (TESV) of forest ecosystems in Karnataka

The total ecosystem supply value (TESV) of the forest ecosystem is computed by aggregating all the ecosystem services as given in Eq. 2:

$$\begin{aligned}&\textbf{Total ecosystem supply value (TESV)} =\\&\textbf{Provisioning services} + \textbf{regulating services} + \textbf{cultural services}\end{aligned} \tag{2}$$

The total ecosystem supply value (TESV) of forest ecosystems in Karnataka amounts to 2,894 billion INR/year (2005) and 1,835 billion rupees/year (2019). Provisioning services constitute 44%, regulating services 45%, and cultural services 11% of TESV for 2005 (Fig. 37). Similarly, provisioning services constitute 34%, regulating services 51%, and cultural services 16% of total TESV for the year 2019 (Fig. 37).

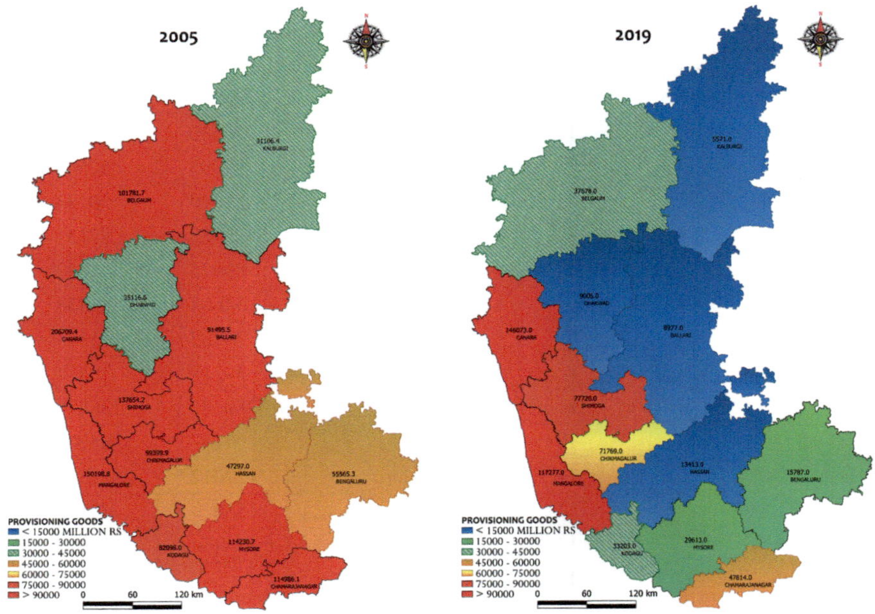

Fig. 35 Provisioning services from forests of Karnataka

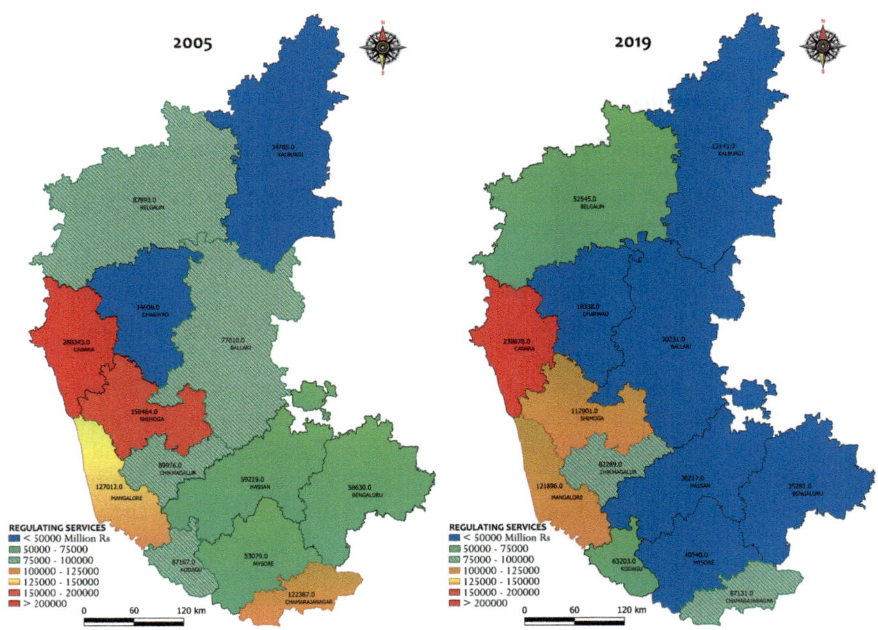

Fig. 36 Regulating services provided by forests of Karnataka

Table 44 Ecosystem supply value of cultural services

S.no	Circle name	Cultural services (million ₹)	
		2005	2019
1	Bengaluru	1942	35281
2	Belgaum	2111	52545
3	Ballari	3412	30231
4	Chamarajanagar	80744	87131
5	Chikmagalur	14543	82289
6	Dharwad	893	18338
7	Kalaburagi	812	13143
8	Hassan	1173	30217
9	Canara	54913	238678
10	Kodagu	32947	63203
11	Mangalore	51239	121896
12	Mysore	31211	40540
13	Shimoga	27090	112901
Total		303030	294967

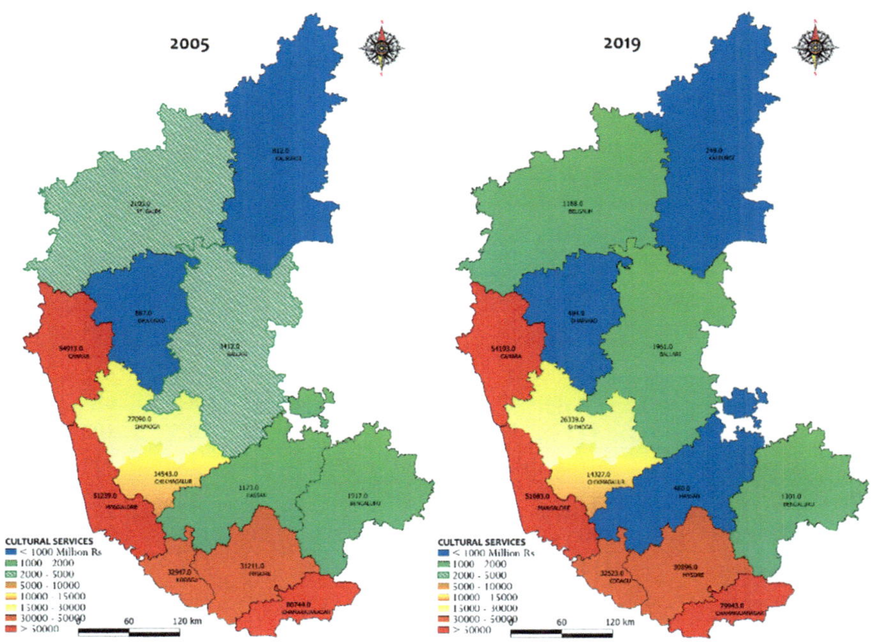

Fig. 37 Cultural services provided by forests of Karnataka

The reduction in TESV and provisioning services is due to the degradation of forests (extent and condition—fragmentation of forests, decline of contiguous intact native forests) from 2005 to 2019. Figure 38 shows that circles such as Canara, Shimoga, Mangalore, Chamarajanagar, Chikmagalur, and Kodagu contribute more to TESV (Table 45). However, the forest ecosystems in circles such as Ballari, Belgaum, Kalaburagi (Gulbarga), Hassan, and Bangalore (Fig. 39) have significantly reduced provisioning and regulating services due to deforestation and forest degradation.

Annexure A provides a district-level assessment of forest ecosystem services (provisioning, regulating, cultural, and TESV).

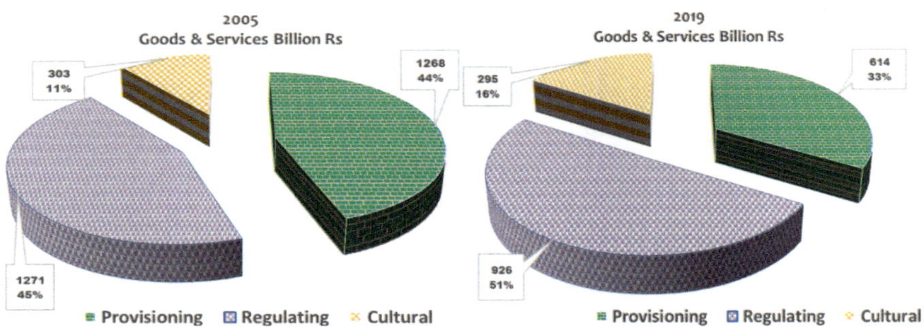

Fig. 38 Share of individual services in TESV

Table 45 Total ecosystem supply value (TESV) of forest ecosystem (circle wise) in Karnataka

S.no	Circle name	Provisioning Billion ₹		Regulating Billion ₹		Cultural Billion ₹		TESV Billion ₹	
		2005	2019	2005	2019	2005	2019	2005	2019
1	Bengaluru	56	16	59	35	2	1	116	52
2	Belgaum	102	38	88	53	2	1	192	91
3	Ballari	91	9	77	30	3	2	172	41
4	Chamarajanagar	115	48	122	87	81	80	318	215
5	Chikmagalur	99	72	90	82	15	14	204	168
6	Dharwad	35	9	35	18	0.9	0.5	71	28
7	Kalaburagi	31	6	35	13	0.8	0.2	67	19
8	Hassan	47	13	59	30	1.2	0.5	108	44
9	Canara	207	146	280	239	55	54	542	439
10	Kodagu	82	33	87	63	33	33	202	129
11	Mangalore	150	117	127	122	51	51	328	290
12	Mysore	114	30	53	41	31	31	199	101
13	Shimoga	138	78	158	113	27	26	323	217
Total		1268	614	1271	926	603	295	2841	1835

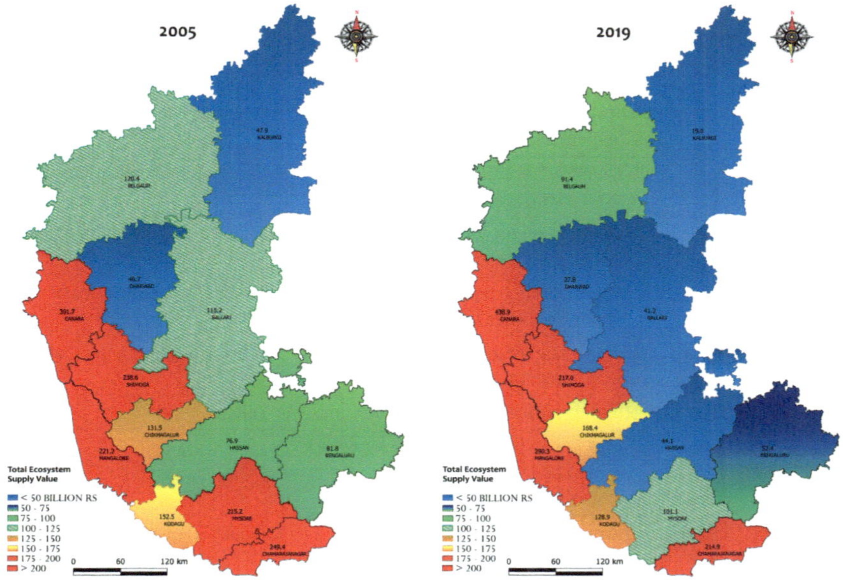

Fig. 39 TESV of forest ecosystem, Karnataka (forest circle wise)

6.1 Valuation of Agriculture (Croplands and Horticulture) Ecosystem Services

Karnataka has diverse cropping patterns across the six major agroecological zones. In addition, numerous irrigation projects have supported growing food crops, and commercial/horticulture crops across the State. Various crops grown in the State are cereals, pulses, oilseeds, fruits, vegetables, commercial crops, horticulture, etc.). Table 46 lists the crops grown in the State according to the Directorate of Economics and Statistics, Government of Karnataka (DES 2017; DSO 2019).

Crop yield per hectare: Crop yield per hectare varies across the State based on the agro-climatic zones and availability of water resources. The yield of some crops is given in Table 47 (based on government records and public interviews in the select ten districts).

Agriculture (cropland, horticulture) ecosystem services at the district level are compiled considering (i) spatial extent—crop-wise, (ii) production as per the statistics from the agriculture department and verified for each crop based on the crop area and crop yield per hectare (iii) Minimum support price fixed by the Ministry of Agriculture, Government of India, and crop-wise cultivation costs and prices at Mandi (crop market set up by the Government of Karnataka) were used to determine the monetary value, (iv) regulating services and cultural services were based on the benefit transfer method through the relevant literature. The spatial extent under each

Table 46 Crop classification as per the Department of Agriculture

Sl.no	Classification	Crops
1	Cereal	Paddy, jowar, bajra, maize, ragi, wheat, minor millets etc.
2	Pulses	Tur dal, green gram, horse gram, black gram, avare (Hyacinth bean), Cowpea, Bengal gram, etc.
3	Oilseeds	Groundnut, sunflower, safflower, Castor, sesamum, soybean, Niger seed, linseed, etc.
4	Fruits	Banana, mango, lemon, pineapple, guava, grapes, sapota, pomegranate, papaya, etc.
5	Vegetables	Potato, tomato, onion, brinjal, beans, cluster beans, chillies, leafy vegetables, other vegetables
6	Commercial crops	Cotton, sugarcane, tobacco, coffee, rubber, coconut, arecanut, cashew, cocoa, cardamom, pepper. Etc
7	Horticulture	Rubber, coffee, cotton, coconut, arecanut, etc.

Table 47 The yield of crops in Karnataka

Crop	Data source	Paddy	Jowar	Bajra	Maize	Ragi	Wheat	Tur Dal	Horse gram
Yield kg/ha	D	1150	674	423	1334	423	501	320	200
	PI	6306	2993	1513	4045	3157	2598	1340	2678
Crop	*Data source*	*Black gram*	*Green gram*	*Avare*	*Cowpea*	*Bengal gram*	*Ground nut*	*Sunflower*	*Safflower*
Yield kg/ha	D	114	108	259	244	450	454	303	329
	PI	695	671	2999	1833	1056	1591	2778	1757
	Data source	*Castor*	*Sesamum*	*Niger Seed*	*Soybean*	*Linseed*	*Cotton*	*Sugarcane*	*Tobacco*
Yield kg/ha	D	256	177	123	179	136	154	69000	738
	PI	1757	1528	1053	1562	626	2400	120000	1165

Note D: Data from the Department of Agriculture, PI: Public interviews

crop with production details at the district level was collated from the government records at the district level (DSO 2019).

Tables 48, 49, 50 and 51 list the spatial extent district-wise under different crops. The area under cultivation of crops (cereals, pulses, fruits, vegetables, oilseeds, commercial crops-sugarcane, coffee, and tobacco) in the State was about 11.5 million hectares, and horticulture is cultivated in 8,03,000 ha.

Cereals: Paddy is cultivated extensively over 1000 km^2 in the coastal districts and districts such as Shimoga, Mysore, Raichur, and Davanagere. Jowar is grown in about 2250 km^2 in Vijayapura (Bijapur), Belgaum, etc. (Fig. 40). Maize is grown in 1200 km^2 in the districts of Davangere, Haveri, and Belgaum. Ragi is grown extensively in the districts of Tumkur and Hassan, encompassing an area of over 1000 km^2. Among the districts, Belgaum has the most significant spatial extent under cultivation of

Table 48 The spatial extent of croplands under cereals (2005—area in hectares)

District	Cereals, 2005—area in hectares								
	Paddy	Jowar	Bajra	Maize	Ragi	Wheat	Barley	Others	Total
Bagalkot	64	155574	50947	50963	0	21202	0	54	278804
Bengaluru Rural	3726	0	0	3123	54283	0	0	0	61132
Belgaum	71341	225605	40727	121537	1960	56811	0	5799	523780
Bellary	66807	88970	18570	53782	5794	848	0	9702	244473
Bengaluru Urban	3503	0	0	724	38454	0	0	26	42707
Bidar	7999	94,797	8912	320	0	7373	0	369	119,770
Vijayapura (Bijapur)	30	244437	97629	19436	0	49116	0	394	411042
Chamarajanagar	16030	27177	619	20119	22210	0	0	64	86219
Chikmagalur	45659	12663	0	2007	57892	6	0	1376	119603
Chikkaballapura	3777	0	40	12842	56383	0	0	252	73294
Chitradurga	4695	33804	7183	44645	43424	658	0	10299	144708
Dakshina Kannada	58838	0	0	0	0	0	0	0	58838
Davanagere	106549	38767	604	164632	23101	316	0	784	334753
Dharwad	23825	57816	0	20723	129	39502	0	3716	145711
Gadag	1343	93659	2015	22621	101	36348	0	917	157004
Gulbaraga	52520	198207	46697	1550	0	11207	0	435	310617
Hassan	53184	5260	0	30071	108854	0	0	319	197688
Haveri	37313	66271	178	133587	1323	783	0	10663	250118
Kodagu	37228	0	0	2875	462	0	0	0	40565
Kolar	3546	0	37	12055	52929	0	0	236	68803
Koppal	62457	63929	73307	16583	0	9762	0	7425	233463
Mandya	91993	1278	0	4116	66654	3	0	0	164044
Mysore	118084	19819	0	18218	79824	0	0	0	235945
Ramanagara	5647	0	0	4734	82277	0	0	0	92658
Raichur	107990	136756	74849	433	0	2746	0	497	323271
Shimoga	123470	607	0	44954	2776	0	0	0	171807
Tumkur	25874	2754	260	10770	173134	0	0	4459	217251
Udupi	62290	0	0	0	0	0	0	0	62290
Uttara Kannada	79144	77	0	442	21	0	0	0	79684
Yadgir	24886	93919	22127	735	0	5310	0	206	147182

Table 49 The spatial extent of croplands under pulses and commercial crops

District	Pulses, 2005—area in hectares				Commercial, 2005—area in hectares			
	Bengal gram	Turdal	Others	Total	Sugarcane	Cotton	Tobacco	Total
Bagalkot	20960	3105	42989	67054	47556	7315	0	54871
Bengaluru Rural	56	2119	10067	12242	870	0	0	870
Belgaum	41421	5517	68035	114973	115849	37659	20707	174215
Bellary	10510	8889	14027	33426	5352	42373	560	48285
Bengaluru Urban	25	1008	6064	7097	25	0	0	25
Bidar	40496	70100	108430	219026	28019	1914	0	29933
Vijayapura (Bijapur)	49658	36948	73566	160172	19166	3043	0	22209
Chamarajanagar	2130	1898	39563	43591	8958	14169	100	23227
Chikmagalur	3038	891	19710	23639	2876	2037	135	5048
Chikkaballapura	74	4078	13969	18122	268	0	0	268
Chitradurga	6295	7746	22573	36614	111	15420	287	15818
Dakshina Kannada	0	0	3120	3120	95	0	0	95
Davanagere	1715	5692	14051	21458	7000	6732	1623	15355
Dharwad	39111	2893	50714	92718	2018	86481	0	88499
Gadag	38988	2784	108500	150272	123	59863	52	60038
Gulbaraga	84154	234573	106398	425124	4657	18917	0	23574
Hassan	2864	2111	41648	46623	6913	1809	7876	16598
Haveri	1877	4375	32606	38858	1865	63400	7	65272
Kodagu	25	0	845	870	0	0	300	300
Kolar	70	3829	13113	17011	252	0	0	252
Koppal	14117	11807	38339	64263	60	29156	0	29216
Mandya	49	1245	37605	38899	28233	5	0	28238
Mysore	983	5309	112724	119016	9535	71938	59482	140955
Ramanagara	86	3212	15258	18556	1318	0	0	1318
Raichur	18483	17583	13817	49883	13	35250	83	35346
Shimoga	39	212	2073	2324	9976	3833	96	13905
Tumkur	518	10787	47383	58688	509	784	75	1368
Udupi	0	0	7684	7684	16	0	0	16
Uttara Kannada	7	105	2182	2294	786	5715	0	6501
Yadgir	39875	111150	50415	201441	2207	8964	0	11171

Table 50 Spatial extent of croplands under oilseeds (2005—area in hectares)

District	Oil seeds									
	Groundnut	Castor seed	Sesamum	Rape seed/ Mustard	Linseed	Soybean	Niger seed	Sunflower	Safflower	Other oil seeds	Total
Bagalkot	22287	2	948	70	3348	9183	226	117258	2857	0	156179
Bengaluru Rural	6802	623	586	558	0	0	652	266	0	0	9488
Belgaum	79533	205	588	254	1750	97168	755	56448	11490	3	248194
Bellary	72964	209	6495	17	10	0	831	88158	5975	284	174943
Bengaluru Urban	203	457	124	645	0	0	655	11	0	0	2095
Bidar	1526	0	8161	10	536	28175	6120	12282	12160	0	68970
Vijayapura (Bijapur)	81452	0	912	0	4338	40	836	307674	5644	0	400896
Chamarajanagar	20904	629	1602	0	0	0	728	5875	0	0	29738
Chikmagalur	4573	537	5953	226	0	0	1115	14909	10	164	27487
Chikkaballapura	25919	153	0	462	0	0	418	2803	0	0	29756
Chitradurga	159559	1315	4699	192	0	0	338	42952	4090	0	213145
Dakshina Kannada	0	0	491	0	0	0	0	0	0	123	614
Davanagere	20746	222	3104	218	0	0	539	14652	151	73	39705
Dharwad	37907	0	441	3	197	22827	408	10052	15581	0	87416
Gadag	57846	102	2960	16	1329	76	632	84106	5142	83	152292
Gulbaraga	40063	47	11552	64	385	139	1321	117971	10425	0	181968
Hassan	2750	2435	4955	408	0	4	4991	10398	0	30	25971
Haveri	25596	133	928	40	40	1253	828	12950	1459	2	43229

(continued)

Table 50 (continued)

District	Oil seeds										
	Groundnut	Castor seed	Sesamum	Rape seed/ Mustard	Linseed	Soybean	Niger seed	Sunflower	Safflower	Other oil seeds	Total
Kodagu	43	0	0	0	0	0	0	0	0	460	503
Kolar	24332	144	0	433	0	0	393	2632	0	0	27933
Koppal	41830	1392	14641	175	2070	0	1452	93507	7295	0	162362
Mandya	4171	1538	13554	0	0	63	2145	48	0	155	21674
Mysore	7623	2373	10493	0	0	18	3673	1214	0	157	25551
Ramanagara	10310	945	889	847	0	0	988	402	0	0	14,380
Raichur	35369	1598	5166	229	160	0	20	201961	7150	19	251672
Shimoga	3573	11	89	28	29	5	112	1427	4	145	5423
Tumkur	139384	4105	886	930	26	3	1604	13257	14	0	160209
Udupi	2006	0	89	0	0	0	0	0	0	323	2418
Uttara Kannada	3342	0	5	0	14	0	0	31	0	14	3406
Yadgir	18984	23	5474	30	182	66	626	55899	4940	0	86223

Table 51 Spatial extent of croplands under fruits (2005—area in hectares)

District	Fruits							
	Mango	Banana	Citrus fruits	Grapes	Pome fruits	Papaya	Other fruits	Total
Bagalkot	158	379	354	520	456	34	1192	3093
Bengaluru Rural	8546	1544	8	533	714	14	98	11457
Belgaum	1119	1148	125	742	471	42	584	4231
Bellary	815	2302	92	17	784	81	2189	6280
Bengaluru Urban	1759	509	0	836	1820	56	28	5008
Bidar	634	225	98	63	158	47	33	1258
Vijayapura (Bijapur)	218	881	8300	4492	245	17	5870	20023
Chamarajanagar	498	1941	103	0	244	0	151	2937
Chikmagalur	2191	1650	415	20	138	0	49	4463
Chikkaballapura	21581	755	42	669	1889	70	271	25277
Chitradurga	1186	1365	222	13	564	261	1071	4682
Dakshina Kannada	1685	3115	55	0	262	82	1381	6580
Davanagere	1777	876	48	1	275	111	229	3317
Dharwad	3615	163	33	0	1587	15	42	5455
Gadag	305	258	54	14	213	2	133	979
Gulbaraga	823	849	843	61	381	33	247	3237
Hassan	2006	3332	480	6	745	53	357	6979
Haveri	889	1150	20	0	337	8	203	2607
Kodagu	44	545	1110	0	76	0	43	1818
Kolar	20258	709	39	628	1774	66	255	23729
Koppal	668	438	70	63	106	120	1219	2684
Mandya	2651	1481	37	0	588	5	83	4845
Mysore	3001	1820	7	0	777	4	73	5682
Ramanagara	12953	2340	11	809	1082	22	149	17366
Raichur	480	22	452	0	133	16	101	1204
Shimoga	2104	4313	36	0	276	23	1810	8562
Tumkur	9658	4530	63	6	500	42	1039	15838
Udupi	1690	1198	79	0	712	57	955	4691
Uttara Kannada	1242	1670	4	0	70	10	1129	4125
Yadgir	390	403	399	29	181	15	117	1534

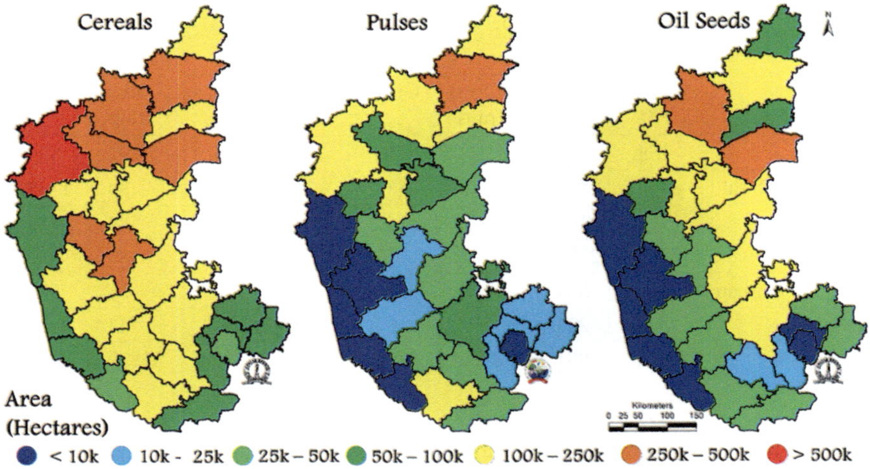

Fig. 40 The spatial extent of croplands under cereals, pulses, oilseeds—2005

cereals, covering 5237 km^2 area, followed by Vijayapura (Bijapur) covering 4110 km^2, Davanagere 3347 km^2, and Raichur with 3232 km^2.

Pulses are grown widely in the arid zones, particularly in the northern belt of Karnataka (Fig. 40). Kalaburagi (Gulbarga) leads in the cultivation of pulses with 4215 km^2 under cultivation, followed by Bidar and Yadgir with more than 2000 km^2 under cultivation. Similar trends are observed in oilseeds (Fig. 40). Vijayapura (Bijapur) has the highest extent under cultivation, i.e., 4008 km^2, followed by Raichur, Belgaum, and Chitradurga, with areas ranging between 2131 and 2517 km^2.

Commercial crops (sugarcane, cotton, rubber, and tobacco) dominate in Belgaum (1742 km^2), followed by Mysore (1409 km^2). Fruits dominate Kolar and Chikkaballapura districts, followed by Vijayapura (Bijapur), with areas ranging from 200 to 250 km^2. Vegetable cultivation dominates the districts of Hassan and Dharwad, with a spatial extent of over 400 km^2.

Horticulture crops (areca nut, coconut, coffee, mango, pomegranate, banana) dominate in the Western Ghats districts (Uttara Kannada, Shimoga, Chikmagalur, Kodagu, and Dakshina Kannada).

The area under cropland—cereal cultivation (single cropping and multi-cropping in croplands) in Karnataka during 2005 was about 115.5 thousand km^2. Belgaum and Vijayapura (Bijapur) encompass the highest extent, i.e., more than ten thousand km^2, followed by Kalaburagi (Gulbarga) with 9481 km^2. Horticulture in the State during 2005 was 8031 km^2, of which Tumkur constituted about 1272 km^2, followed by Chikmagalur and Hassan with 1182 and 1028 km^2, respectively.

Table 52 lists the spatial extent of croplands and horticulture at the district level in 2005 and 2019. From 2005 to 2019, there has been an 11% decline in the area under croplands, from 115 thousand km^2 to 103 thousand km^2 (cereals, pulses, oilseeds, fruits, vegetables, commercial crops). There has been a decline of 18%

Table 52 Spatial extent of croplands under vegetables, Total croplands and horticulture (2005—area in hectares)

District	Vegetable						Total croplands	Horticulture					
	Potato	Tapioca	Sweet potato	Onion	Others	Total		Rubber	Coffee	Cashew	Coconut	Arecanut	Total
Bagalkot	0	0	57	6986	1667	8710	569166	0	1	0	452	2	455
Bengaluru Rural	660	85	8	19	2155	2928	106871	5	80	19	7866	785	8755
Belgaum	5336	0	2164	6967	6660	21127	1087671	0	0	544	599	8	1151
Bellary	0	0	0	3761	2468	6229	514931	0	1	3	1259	32	1295
Bengaluru Urban	204	0	13	90	2838	3145	62818	0	10	0	2498	233	2741
Bidar	47	0	0	441	1887	2375	441386	0	0	30	24	0	54
Vijayapura	0	0	5	5436	1538	6979	1021499	0	0	0	178	0	178
Chamarajanagar	13	0	0	2547	1148	3708	202771	0	686	195	12161	309	13351
Chikmagalur	4051	36	177	5608	2449	12321	310860	709	69526	326	31122	16616	118299
Chikkaballapura	2078	0	0	777	12946	15801	164810	0	0	1181	1084	28	2293
Chitradurga	0	0	0	13134	1317	14451	487294	0	15	9	43354	14498	57876
Dakshina Kannada	0	698	537	0	2370	3605	155525	10113	82	29585	15684	27209	82673
Davanagere	3	0	0	5459	1888	7350	455984	0	20	33	11978	22015	34046
Dharwad	284	0	0	38112	2345	40741	461096	0	0	114	431	11	556
Gadag	0	0	0	30434	1256	31690	552900	0	0	0	622	3	625
Gulbaraga	3	0	21	1302	2343	3669	948687	0	0	0	498	0	498
Hassan	36151	0	68	107	5154	41480	438160	5	38219	16	61098	3483	102821
Haveri	1	0	0	10372	4405	14778	416743	0	0	17	1147	717	1881
Kodagu	0	9	12	0	91	112	134214	1926	83205	2198	1331	1386	90046

(continued)

Table 52 (continued)

District	Vegetable						Total croplands	Horticulture					
	Potato	Tapioca	Sweet potato	Onion	Others	Total		Rubber	Coffee	Cashew	Coconut	Arecanut	Total
Kolar	1950	0	0	730	12153	14833	154714	0	0	1109	1017	26	2152
Koppal	3	0	0	960	2062	3025	495357	0	0	0	344	0	344
Mandya	115	4	14	414	3938	4485	281576	0	85	262	18165	879	19391
Mysore	15	0	0	5	3672	3692	551119	0	26	172	19404	676	20278
Ramanagara	1000	130	12	29	3267	4437	161987	8	121	29	11923	1190	13271
Raichur	0	0	0	1429	1031	2460	664277	0	0	0	441	0	441
Shimoga	0	10	0	20	352	382	240984	881	464	1473	6613	29150	38581
Tumkur	22	28	18	201	1596	1865	582441	0	6	82	110937	16197	127222
Udupi	0	66	439	4	1580	2089	120173	2295	0	19207	14464	5019	40985
Uttara Kannada	0	0	28	77	241	346	117023	6	11	2087	6276	12287	20667
Yadgir	1	0	10	617	1110	1738	449525	0	0	0	236	0	236

in the spatial extent of cereals and a 55% decline in the area under oilseeds, while the area under pulses has increased by 55%, commercial crops by 27%, fruits by 89%, and vegetables by 24%. The area under horticulture has risen by 2%, i.e., from 8.0 to 8.18 km^2. These changes in areas under horticulture, as well as croplands, directly influence revenue. There has been an increase in the spatial extent of cereal crops in Chikkaballapura (33%), Chitradurga (18%), Bellary (16%), Chikmagalur, Dharwad, and Haveri districts. In contrast, the other districts showed a reduction in area under agriculture between 2005 and 2020. Bengaluru Urban district showed a 60% reduction due to urbanization. Similarly, the spatial extent of pulses shows an upward trend from 170 to 272% in Raichur, Vijayapura (Bijapur), and Bagalkot.

Twelve districts among 30 in Karnataka witnessed an increase in spatial extent under pulses. Kodagu, Bengaluru Urban, and Haveri districts have seen over 70% of the area under pulses due to a shift in cropping pattern. Bidar was the only district with a 90% increase in the area under oilseeds, while the rest showed a decline in the area. Bengaluru rural and Dakshina Kannada showed over 90% decline in area under oilseeds. About 26 districts showed an increase in the spatial extent under fruit cultivation, and four districts showed a reduction in area. In particular, Bengaluru urban has lost 48% of its area under fruits. In terms of districts, that increased fruit cultivation, the districts of Dakshina Kannada, Udupi, Kodagu, and Belgaum show over 250% increase in area under fruits, while the districts of Chitradurga, Chamarajanagar, Kolar, Haveri, Uttara Kannada, Koppala, Davanagere, Kalaburagi, Bagalkot, Mysore showed a significant 100–200% increase in area under fruits (Fig. 41).

Similarly, 17 districts showed an increase in the area under vegetable cultivation (Fig. 41). Koppal was the highest, indicating an increase of over 300% in the area under vegetable cultivation. Similarly, Bagalkot, Chamarajanagar, Kodagu, Raichur, Chitradurga, Mandya, Tumkur, Bellary, and Kolar showed a 100–200% increase in

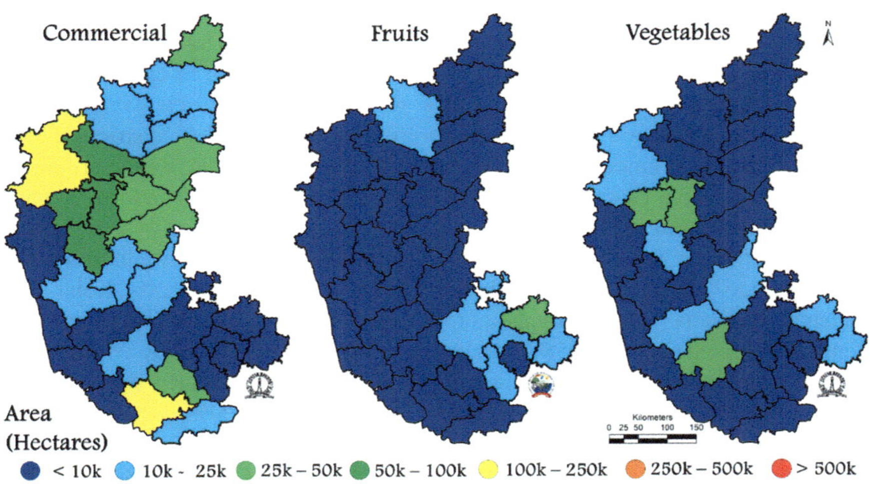

Fig. 41 The spatial extent of commercial crops, fruits, and vegetables—2005

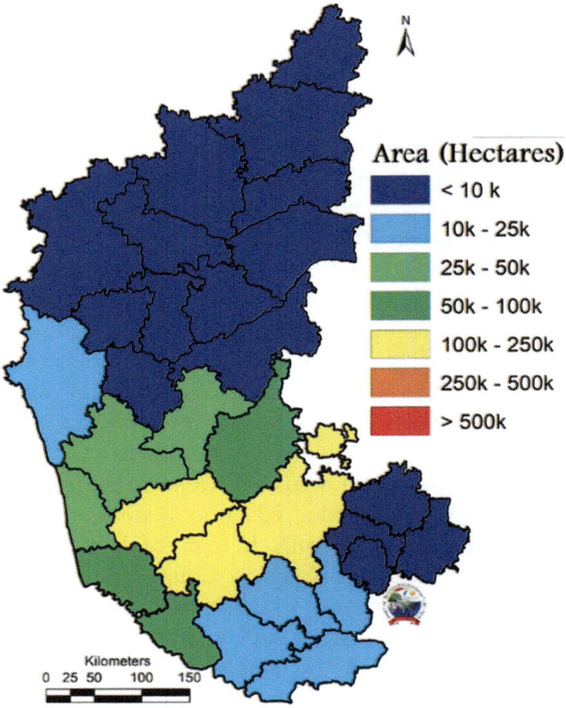

Fig. 42 The spatial extent of horticulture—2005

area under vegetation. On the other hand, Ramanagara, Uttara Kannada, and Dharwad districts lost over 50% of the area under vegetable cultivation.

In terms of commercial crops (sugarcane, cotton, rubber, and tobacco), the districts of Yadgir, Udupi, and Kalaburagi (Gulbarga) saw an increase in extent by 200%. Vijayapura (Bijapur) saw a 176% increase, while about 12 districts witnessed over a 25% increase in area under commercial crops (Fig. 41), and 17 districts showed a decline under commercial crops. Kodagu, Bengaluru Urban, Bengaluru Rural, Kolar, and Dakshina Kannada districts witnessed a decline of 90% in the area under commercial cropping.

Five districts show an increasing trend in agriculture ranging between 5 and 18%, which include Udupi (18%), Chikkaballapura (13%), Kalaburagi (12%), Kolar (8%), Dakshina Kannada (6%) and Bagalkot (5%), while the rest showed declining trends. Bengaluru urban showed a 58% decline in croplands, followed by Mysore with 37%, Bengaluru Rural with 31%, and Tumkur with 30%. Similarly, the area under horticulture has increased in 17 districts (Fig. 42), with a 100% increase in Bidar, Gadag, and Haveri districts. On the other hand, Chamarajanagar, Yadgir, Kalaburagi (Gulbarga), and Bengaluru rural districts witnessed a decline of 70% of the area under horticulture.

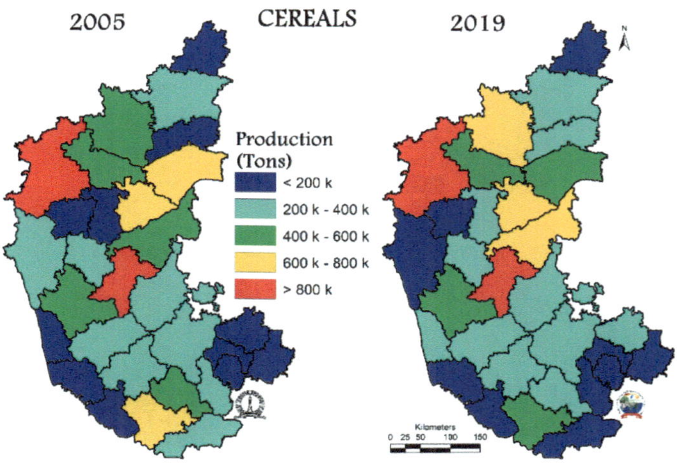

Fig. 43 Cereal production in 2005 and 2019

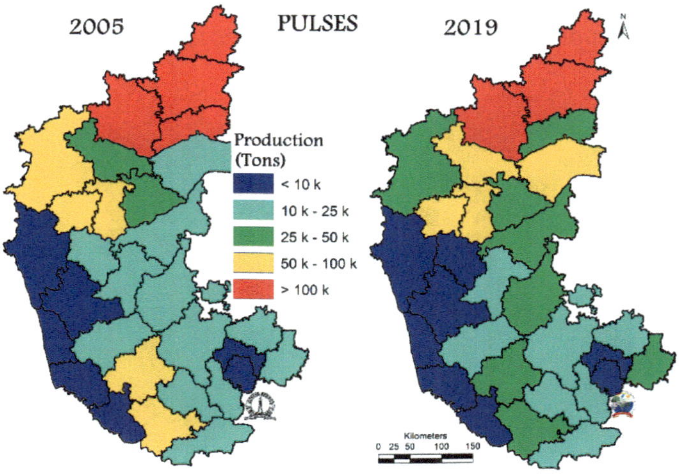

Fig. 44 Pulses production in 2005 and 2019

Crop production: Major crop production (opening and closing stock) is depicted in Figs. 43, 44, 45, 46, 47 and 48 and Tables 53, 54, 55, 56, 57, 58 and 59. The State has produced 11.0 million tonnes of cereals (Fig. 43 and Table 53), 1.4 million tonnes of pulses (Fig. 44 and Table 54), 1.9 million tonnes of oilseeds (Fig. 45 and Table 55), 16.4 million tonnes of commercial crops (Fig. 46 and Table 56), 1.7 million tonnes of fruits (Fig. 47 and Table 57) and 1.4 million tonnes of vegetables, summing up to 33.3 million tonnes of agricultural produce in 2005 (Fig. 48 and Table 58). In addition to this, 1.4 million tonnes of horticulture produce (Fig. 49 and Table 59). District wise

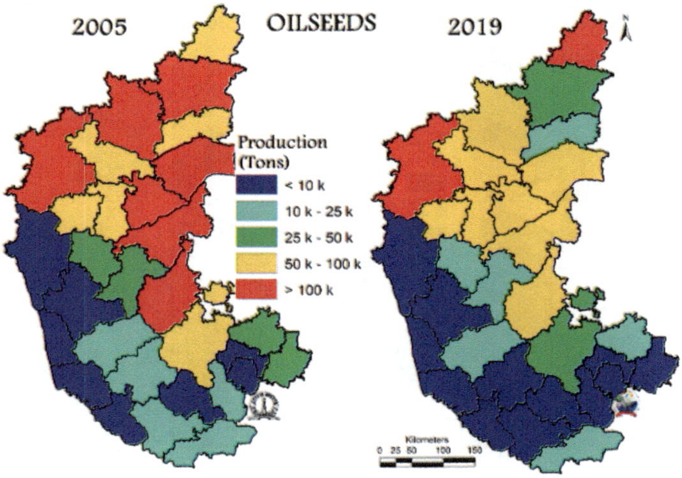

Fig. 45 Oilseed production in 2005 and 2019

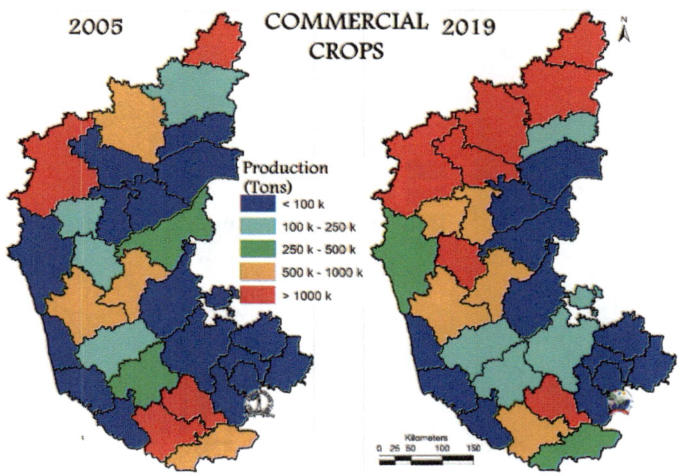

Fig. 46 Commercial crop production in 2005 and 2019

assessment of crops indicates that Davanagere has the highest quantity of cereals, i.e., about 1.4 million tonnes (Table 53), followed by Belgaum with 0.98 million tonnes. Similarly, pulses are predominantly produced in Kalaburagi (Gulbarga), i.e., about 0.32 million tonnes (30% of state produce), followed by Vijayapura (Bijapur), Yadir, and Bidar, contributing over 0.1 million tonnes each (Table 54). Oilseeds are produced in the districts of Belgaum (0.24 million tonnes) and Vijayapura (Bijapur) (0.20 million tonnes). Belgaum contributes to 50% of the total production of all the

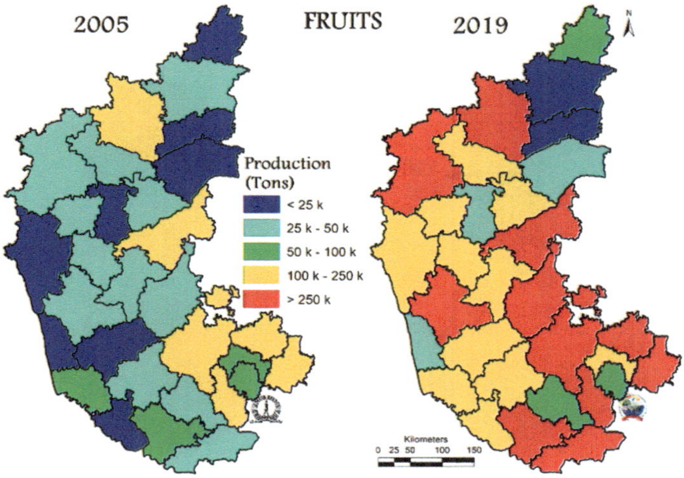

Fig. 47 Fruit production in 2005 and 2019

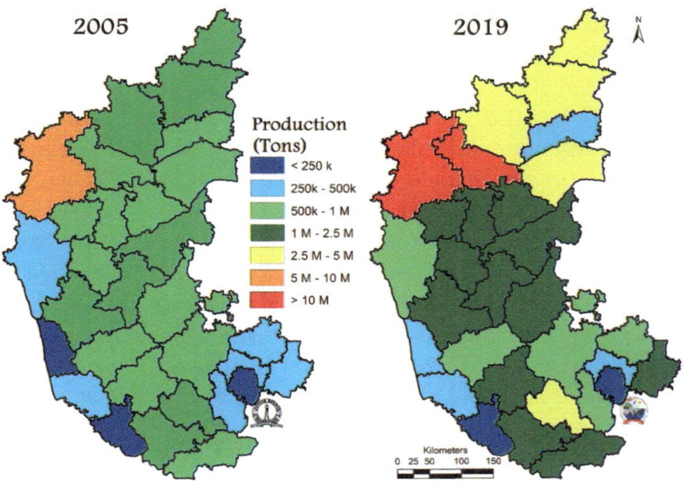

Fig. 48 Total cropland production in Karnataka in 2005 and 2019

commercial crops, i.e., 8.2 million tonnes, followed by Bidar, Mandya and Raichur, contributing over 1 million tonnes each. Fruits are produced the most in the district of Vijayapura (0.2 million tonnes), followed by Chikkaballapura (0.16 million tonnes), Kolar (0.15 million tonnes), and vegetables are produced the most in the districts of Gadag and Hassan, contributing over 0.19 million tonnes each. Aggregation of all produces of the agriculture ecosystem indicates Belgaum produces over 9.7 million tons, followed by Davanagere with 2.2 million tonnes, Bidar and Vijayapura (Bijapur)

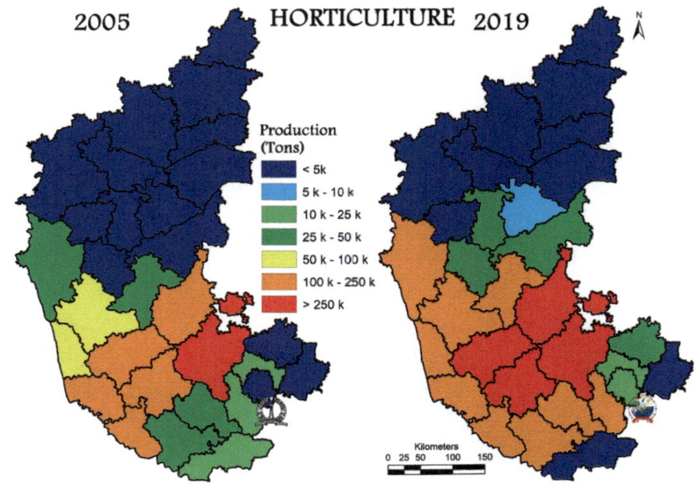

Fig. 49 Total production of horticulture goods

with 1.9 million tonnes each, Mysore and Mandya with 1.8 million tonnes each, and Bellary, Shimoga, and Kalaburagi contributing 1.0 to 1.2 million tonnes each (Table 58). Among all the districts, Tumkur produces 20% of the State's horticulture produce (Table 59), i.e., 0.29 million tonnes, followed by Hassan with 0.19 million tonnes, and Chikmagalur and Kodagu with 0.16 million tonnes each (*Note: each coconut was assumed to be 500 g based on public interviews, this was necessary to convert to standard measuring unit*).

In the last 15 years, Karnataka has witnessed a 102% increase in agricultural production (Table 58) and a 140% increase in horticultural production (Table 59). Oilseed production has declined by 57%, followed by cereals, i.e., about a 2% decline in production. Other crops showed an increase in production between 17 and 366%. Pulses showed an increase of 17%, followed by commercial crops at 157%, fruits at 256%, and vegetables at 366%. The district-wise analysis is presented next.

- Cereals: 12 districts show an increase in production level, including Chikkaballapura 62%, Chitradurga 39% and Gadag 38%. Mandya shows the highest decline, i.e., 43% followed by Bengaluru Urban with 42% (Table 53, Fig. 43).

Table 53 Cereals production (tons)

District	2005	2019
Bagalkot	422610	435628
Bengaluru Rural	154972	123505
Belgaum	980024	927751
Bellary	554970	735662
Bengaluru Urban	136525	78940
Bidar	184925	125111
Vijayapura (Bijapur)	498654	606556
Chamarajanagar	242243	177633
Chikmagalur	239929	277932
Chikkaballapura	185863	301048
Chitradurga	203566	282185
Dakshina Kannada	169924	151184
Davanagere	1399405	1183977
Dharwad	161946	190254
Gadag	165540	229184
Gulbarga	389155	279202
Hassan	313705	349432
Haveri	320579	351813
Kodagu	121956	104055
Kolar	174477	128377
Koppal	604294	765876
Mandya	429152	243529
Mysore	639914	572461
Ramanagara	234894	151361
Raichur	620608	510574
Shimoga	555574	587997
Tumkur	371424	294060
Udupi	179894	215651
Uttara Kannada	200367	183582
Yadgir	184397	245440

Pulses: 12 district shows an increase in production level. Raichur, Vijayapura (Bijapur), and Bagalkot show more than a 135% increase; on the other hand, Yadgiri and Kodagu show over 70% decline (Table 54, Fig. 44).

Table 54 Pulses (tons)

District	2005	2019
Bagalkot	42173	99092
Bengaluru Rural	7058	4132
Belgaum	55926	43716
Bellary	21944	26987
Bengaluru Urban	4786	3194
Bidar	185931	177150
Vijayapura (Bijapur)	102130	276434
Chamarajanagar	23050	21984
Chikmagalur	12679	13900
Chikkaballapura	24504	16451
Chitradurga	22419	35825
Dakshina Kannada	1501	1495
Davanagere	17417	11408
Dharwad	55792	84295
Gadag	73664	94321
Gulbarga	325563	404209
Hassan	99586	41148
Haveri	14258	4383
Kodagu	358	30
Kolar	23003	32390
Koppal	25672	48092
Mandya	15746	12918
Mysore	55870	46339
Ramanagara	10697	13462
Raichur	24008	99749
Shimoga	1301	866
Tumkur	20247	14595
Udupi	3696	2477
Uttara Kannada	1384	963
Yadgir	154265	43648

Oilseeds: Bidar shows a 122% increase in production, followed by Udupi with a 92% increase; the rest of the districts showed a decline in production levels. Bengaluru Urban and Bengaluru Rural showed the highest reduction, i.e., over 92% decline (Fig. 45 and Table 55).

Table 55 Oilseed (tons)

District	2005	2019
Bagalkot	93831	56111
Bengaluru Rural	8188	424
Belgaum	243375	102841
Bellary	167272	88869
Bengaluru Urban	1557	119
Bidar	58914	131020
Vijayapura (Bijapur)	205267	50872
Chamarajanagar	21128	15521
Chikmagalur	17400	14177
Chikkaballapura	29385	17806
Chitradurga	161314	85044
Dakshina Kannada	398	253
Davanagere	36964	14861
Dharwad	74235	69025
Gadag	86367	54002
Gulbarga	123521	38534
Hassan	20384	3899
Haveri	28575	20381
Kodagu	510	71
Kolar	27585	6667
Koppal	128720	80145
Mandya	7111	2727
Mysore	15521	8430
Ramanagara	12410	4181
Raichur	174598	67944
Shimoga	4725	2400
Tumkur	76757	33806
Udupi	2420	4644
Uttara Kannada	5368	3786
Yadgir	5829	1980

- Commercial: 18 districts showed an increase in commercial crop production. Bangalore, Gadag, Udupi show over a tenfold increase, while Chikkaballapura, Kolar, and Kodagu show over a 90% decrease in production (Fig. 46 and Table 56).

Table 56 Commercial crops (tons)

District	2005	2019
Bagalkot	4148	9005885
Bengaluru Rural	0	3868
Belgaum	8247013	15422404
Bellary	354170	86811
Bengaluru Urban	0	0
Bidar	1532236	2187005
Vijayapura (Bijapur)	864909	3437251
Chamarajanagar	595720	370425
Chikmagalur	133483	202810
Chikkaballapura	18320	575
Chitradurga	4684	34235
Dakshina Kannada	0	1490
Davanagere	665263	676818
Dharwad	129803	853624
Gadag	33805	857004
Gulbarga	192068	1878711
Hassan	266060	189286
Haveri	122662	1152447
Kodagu	221	24
Kolar	17197	468
Koppal	7960	80708
Mandya	1384665	3468790
Mysore	1046969	869171
Ramanagara	0	66254
Raichur	12769	0
Shimoga	581993	585453
Tumkur	23958	126075
Udupi	972	14750
Uttara Kannada	92335	421679
Yadgir	91009	134599

Fruit and Vegetables: 27 districts showed an increase in fruit production, and three showed a decline in production. Chamarajanagar, Uttara Kannada, Chikmagalur, Chitradurga showed over 7.5 times increase, while districts, namely Bengaluru urban, Kalaburagi (Gulbarga) and Yadgir, showed an 11–94% decline (Fig. 47 and Table 57). Twenty-six districts showed an increase in vegetable production, while four districts showed a decline in production. About 22 districts showed over a 100% increase in vegetable production of which Koppal, Tumkur, Mysore, Mandya, Bidar, Bengaluru Urban, and Kolar ranged the highest (between 15 and 33-folds).

Table 57 Fruit (tons)

District	2005	2019
Bagalkot	30413	149094
Bengaluru Rural	88900	160955
Belgaum	49000	292565
Bellary	112504	333491
Bengaluru Urban	61848	54868
Bidar	12472	61246
Vijayapura (Bijapur)	201974	443632
Chamarajanagar	31619	359696
Chikmagalur	22877	215946
Chikkaballapura	164121	316560
Chitradurga	44317	391781
Dakshina Kannada	87795	147628
Davanagere	26168	134614
Dharwad	32126	137541
Gadag	7352	38534
Gulbarga	30544	7959
Hassan	39356	172467
Haveri	44904	204,987
Kodagu	20505	108055
Kolar	154066	649409
Koppal	31620	191257
Mandya	35210	80349
Mysore	56738	266896
Ramanagara	134748	359730
Raichur	9292	42644
Shimoga	42652	288330
Tumkur	100011	324760
Udupi	22470	28466
Uttara Kannada	17861	184472
Yadgir	14473	913

- Croplands (aggregating all crops): Overall, 27 districts show an increase in crop production, of which 11 districts show an increase of over 100%, while three show a decline in crop production. The districts of Bagalkot, Haveri, and Kolar show over 250% increase in agricultural production, and Yadgir shows a 13% decline (Fig. 48 and Table 58).

Table 58 Croplands-total production (tons)

District	2005	2019
Bagalkot	662613	10376890
Bengaluru Rural	270310	414777
Belgaum	9703020	17478205
Bellary	1260474	1612653
Bengaluru Urban	207898	203974
Bidar	1979805	2,803709
Vijayapura	1926817	4892366
Chamarajanagar	949889	1165755
Chikmagalur	506937	890255
Chikkaballapura	469555	954677
Chitradurga	538895	1283250
Dakshina Kannada	285036	327440
Davanagere	2209847	2252647
Dharwad	598526	1633937
Gadag	569862	1466936
Gulbaraga	1079674	2617410
Hassan	928967	1033599
Haveri	597256	2375348
Kodagu	145130	214895
Kolar	440788	1545833
Koppal	806742	1541518
Mandya	1877200	3940494
Mysore	1821238	1925672
Ramanagara	409715	660470
Raichur	855399	812537
Shimoga	1187482	1471588
Tumkur	597433	966718
Udupi	220163	283306
Uttara Kannada	318615	797156
Yadgir	511592	446315

Horticulture: 24 districts show increased horticulture production, while six districts show a decline in horticulture production. Gadag and Bidar districts show an over 12-fold increase in production (Fig. 49 and Table 59).

Table 59 Horticulture (tons)

District	2005	2019
Bagalkot	1085	2502
Bengaluru Rural	15373	20573
Belgaum	2008	1642
Bellary	3057	33413
Bengaluru Urban	4816	19568
Bidar	97	1311
Vijayapura	425	2595
Chamarajanagar	17824	1190
Chikmagalur	164620	461123
Chikkaballapura	3515	28820
Chitradurga	109155	523534
Dakshina Kannada	140138	119972
Davanagere	47656	190304
Dharwad	1199	4220
Gadag	1489	48283
Gulbarga	1253	1420
Hassan	193960	403087
Haveri	3763	41,167
Kodagu	160880	107175
Kolar	3300	4993
Koppal	820	8936
Mandya	34253	155408
Mysore	26732	201418
Ramanagara	23301	223501
Raichur	1052	2580
Shimoga	58893	120427
Tumkur	293756	271740
Udupi	68811	219247
Uttara Kannada	34795	190175
Yadgir	594	168

The monetary value of ecosystem goods was evaluated based on the minimum support price (MSP) and relative crop production cost. The minimum support price for crops was obtained from the Price Policy Reports for *rabi* crops (winter crops) and *kharif* crops (monsoon crops), along with other published literature.

Table 60 lists crops, MSP, production cost, and net revenue generated for 2005. Table 61 lists crop-wise MSP in 2019, while Table 62 lists the associated production costs (2019) compiled through public interviews, etc.

Table 60 MSP (Rs/Quintal) and cost of production (CACP 2005a, b; DMI 2020; EANDS 2020; TNAU Agriculture Portal 2020)

Crop	Paddy	Jowar	Bajra	Maize	Ragi	Wheat	Barley	Other Millets
MSP 2005 Rs/Q	600	525	525	540	525	640	495	1172
Cost of production Rs/Q	465	252	317	291	295	431	315	703
Net Rs/ton	1348	2727	2082	2491	2300	2088	1797	4687

Crop	Ground nut	Castor	Sesamum	Mustard	Linseed	Soybean	Niger seeds	Sunflower	Safflower	Others/oilseeds
MSP 2005 Rs/Q	1510	1458	1924	1313	1395	1101	1762	1676	1547	1101
Cost of production Rs/Q	1076	642	734	275	614	661	893	618	497	484
Net Rs/ton	4340	8165	11902	10379	7815	4393	8695	10585	10502	6163

Crop	Pepper	Rubber	Coffee	Cashew	Coconut (Rs/1000 nuts)	Arecanut	Sugarcane (Rs/tons)	Cotton	Tobacco
MSP 2005 Rs/Q	7345	4072	1856	1896	2732	5298	75	1980	2648
Cost of production Rs/Q	2989	1657	755	772	1319	2156	18	1010	1078
Net Rs/ton	43554	24144	11005	11246	14131	31415	569	9702	15700

Crop	Mango	Banana	Lemon	Grapes	Pomegranate	Papaya	Guava
MSP 2005 Rs/Q	928	512	1039	649	2041	297	557
Cost of production Rs/Q	399	220	447	279	878	128	239
Net Rs/ton	5289	2919	5923	3702	11635	1692	3173

Crop	Potato	Tapioca	Sweet Potato	Onion	Green Chillies	Other vegetables	Bengal gram	Tur dal	Others
MSP 2005 Rs/Q	359	359	359	567	675	649	1178	1401	560
Cost of production Rs/Q	101	101	101	328	290	279	489	570	357
Net Rs/ton	2577	2577	2577	2390	3846	3702	6883	8309	2031

Table 61 MSP Rs/Quintal (Rs per 100 kg) (*Fruits Market Price in Bangalore, Karnataka*, 2019; *Krishi Marata Vahini*, 2019; *AgMarknet*, 2020; *Farmers Portal*, 2020; *MSP for Rabi Crops 2019–2020*, 2020; Coffee Board, 2019–2020; Commodities Online, 2019–2020)

Cereals	Paddy	Jowar	Bajra	Maize	Ragi	Wheat	Other millets (barley)		
	1850	2550	2000	1760	3150	1925	3613		
Pulses	Tur Dal	Horse gram	Black gram	Green gram	Avare	Cowpea	Bengal gram	Other pulses	
	5800	2317	4924	7050	4688	4567	4875	2317	
Oil seeds	Groundnut	Sunflower	Safflower	Castor	Sesamum				
	5090	5650	5215	4915	6485				
	Niger seeds	Soybean	Mustard	Linseed	Others/oilseeds				
	5940	3710	4425	4704	3710				
Commercial crops	Cotton	*Sugarcane (Rs/ Tons)	Tobacco	Coffee	*Coconut (Rs/1000 nuts)				
	5335	275	9773	6850	10086				
	Cashew	Coco	Areca	Cardamom	Pepper	Rubber			
	7000	19000	19555	285100	27111	15029			
Fruits	Banana	Mango	Lemon	Pineapple	Guava	Grapes	Sapota	Pomegranate	Papaya
	1380	2500	2800	1000	1500	1750	1000	5500	800
Vegetables	Potato	Tomato	Brinjal	Beans	Cluster Beans	Onion	Green Chillies	Leafy Vegetables	Other Vegetables
	967	532	1058	2000	2200	1527	1818	532	1750

The cost of production for different crops was derived based on public interviews in the districts of Belgaum, Dharwad, Uttara Kannada, Mysore, Mandya, Shimoga, Chitradurga, Davanagere, Tumkur, etc., from December 2019 to April 2020, and on published data by the Department of Agriculture, Farmer Welfare and Directorate of Economics and Statistics, Government of India (EANDS 2020). Table 62 depicts the cost of production for each crop. Across the grades (based on the quality as A1, A2, B1, B2, C1, C2), only A1 was considered in the analysis.

Table 63 depicts the district-wise share of crops (based on MSP and production) in the agriculture ecosystem. Figure 50 illustrates the district-wise contribution of crops in the agriculture ecosystem (based on Table 63). Among the districts in Karnataka, Belgaum contributes about 85.2 billion rupees, of which commercial crops contribute 55.79%, followed by cereals with 20.56%. This is followed by Bagalkot (54.2 billion rupees), and Vijayapura (50.87 billion rupees), while the districts of Chitradurga, Davanagere, and Kalaburagi contributed 43.7–44.3 billion rupees each. The least was contributed by Bengaluru Urban, with revenue of 4.5 billion, of which 51.9% is contributed by cereals, followed by fruits (21.85%). The assessment of dry belts, particularly in northern Karnataka, indicates that pulses are extensively grown, followed by fruits in the south-eastern portions (Kolar and Chikkaballapura). Commercial crops and cereals cropping are practiced in transition

Table 62 Cost of production Rs/Quintal (Rs per 100 kg) in 2019

Cereals	Paddy	Jowar	Bajra	Maize	Ragi	Wheat	Other millets (barley)		
	1434	1225	1207	948	1770	1296	2167		
Pulses	Tur Dal	Horse gram	Black gram	Green gram	Avare	Cowpea	Bengal gram	Other pulses	
	2360	1476	1476	2765	2668	1476	2025	1476	
Oil seeds	Groundnut	Sunflower	Safflower	Castor	Sesamum				
	3627	2081	1675	2162	2473				
	Niger seeds	Soybean	Mustard	Linseed	Others/oilseeds				
	3009	2229	926	2069	1632				
Commercial crops	Cotton	*Sugarcane (Rs/Tons)	Tobacco	Coffee	*Coconut (Rs/1000 nuts)				
	2721	65	3977	2787	4870				
	Cashew	Coco	Areca	Cardamom	Pepper	Rubber			
	2849	7733	7958	116035	11034	6116			
Fruits*	Banana	Mango	Lemon	Pineapple	Guava	Grapes	Sapota	Pomegranate	Papaya
	593	1075	1204	430	645	752	430	2365	344
Vegetables	Potato	Tomato	Brinjal	Beans	Cluster Beans	Onion	Green Chillies	Leafy Vegetables	Other Vegetables
	272	228	454	860	946	882	781	228	752

zones, dry belts with managed water resources, and in the zones close to the Western Ghats.

Crop-wise assessment of revenue generated reveals that Davanagere has the highest revenue in cereal crops (26.1 billion rupees), followed by Belgaum (17.5), and Bellary, Koppal, Mysore, Vijayapura (Bijapur), and Shimoga, range between 10 and 15 billion rupees each. Similarly, Kalaburagi contributed the highest for pulses (22.3 billion), followed by Vijayapura (Bijapur) (15 billion) and Bidar (10 billion rupees). Bidar, Belgaum, Bellary, Koppal, and Chitradurga contributed to oilseeds with 4.3–5.1 billion rupees each. Districts Kolar, Chitradurga, and Vijayapura (Bijapur) contribute to fruits with a 10.4–13.4 billion rupees revenue each. Vegetables contributed between 7.9 billion and 9.6 billion rupees from Bagalkot, Haveri, and Belgaum. Belgaum contributes about 47.5 billion rupees through commercial crops, followed by Bagalkot (21.8), Chikmagalur (24.9 billion rupees), followed by Dakshina Kannada and Shimoga, each with 16 billion rupees.

The annual revenue (in 2019) from the cropland ecosystem in Karnataka State amounts to 866 billion (Fig. 51). Commercial crops contribute to 36.1% (311.8 billion), followed by cereal crops (25.3%, 218.9 billion), fruits (13.8%, 119.5 billion), pulses (10.1%, 87.1 billion), vegetables (9.2%, 79.6 billion) and oilseeds (5.7%, 49.4 billion).

Table 63 District-wise contribution by the agricultural sector (million rupees)

District	Cereals	Pulses	Oil Seeds	Fruits	Vegetables	Commercial	∑Agriculture
Belgaum	17534	2198	4564	5475	7917	47568	85257
Bagalkot	8267	5116	2933	3354	9612	24945	54229
Vijayapura	11116	15097	2737	10442	798	10682	50871
Chitradurga	5938	1895	4371	11481	6598	14067	44351
Davanagere	26147	569	776	2395	2620	11628	44133
Kalaburagi/ Gulbarga	6398	22237	2072	149	115	12797	43767
Chikmagalur	6302	550	796	3602	1951	21848	35048
Mysore	11496	2004	494	4006	1596	14436	34033
Tumkur	7941	600	1723	7710	1508	13639	33120
Bellary	13686	1327	4549	6056	4861	1894	32373
Shimoga	10732	49	134	4212	70	16346	31543
Haveri	6492	213	1025	3227	8520	10497	29975
Koppal	12825	2453	4383	3927	4925	1154	29667
Gadag	4466	4773	2886	676	2962	11327	27090
Dharwad	3644	4247	3133	3044	4330	8297	26694
Bidar	2929	10003	5140	979	1312	6092	26455
Hassan	7286	1880	212	2992	3093	9662	25125
Kolar	3975	1254	341	13466	5621	308	24965
Dakshina Kannada	2797	75	16	1984	304	16846	22022
Mandya	5247	426	153	1291	1725	11513	20355
Raichur	9581	4972	3617	818	1199	26	20212
Udupi	3989	118	237	407	268	14760	19780
Uttara Kannada	3381	44	193	2947	41	12211	18817
Chikkaballapura	6573	886	908	6216	2598	427	17608
Ramanagar	4423	553	213	7399	719	3206	16512
Yadgiri	4602	2588	1008	22	35	5900	14154
Chamarajanagar	3428	697	804	4987	2456	1417	13788
Kodagu	1912	1	4	2230	39	7344	11529
Bangalore Rural	3505	175	22	3058	1186	631	8577
Bangalore Urban	2387	152	6	1005	691	356	4598
Karnataka	218999	87153	49449	119555	79670	311824	866649

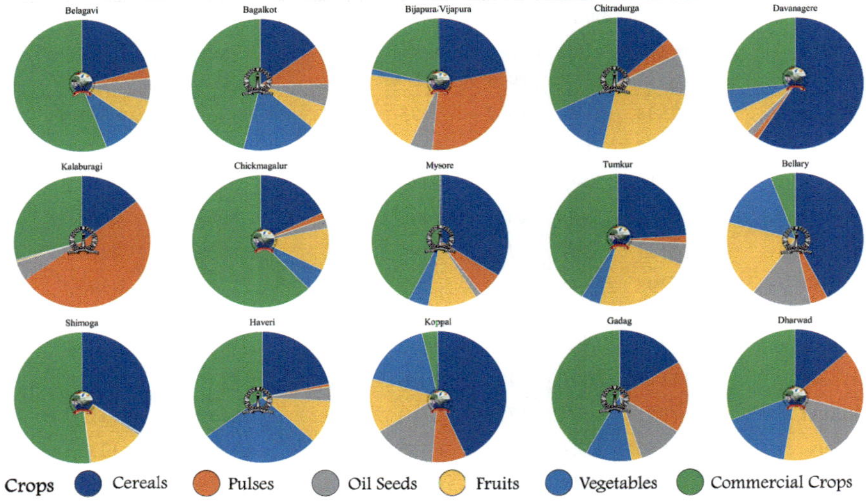

Fig. 50 District wise share of crops (2019), Karnataka

Total revenue (based on the crop production and MSP) generated in a district was compared with the GDDP (Gross District Domestic product) to understand the agriculture sector's contribution to the State economy. Table 64 lists district-wise GDDP obtained from the Economic Survey of Karnataka, 2018–19. Figure 52 indicates district-wise total revenue generated through agriculture, GDDP, and the proportion of revenue generated from the agriculture sector compared to the GDDP. The total GSDP (Gross State Domestic Product) of Karnataka State is 10128 billion rupees. The revenue from croplands is 866 billion rupees, 8.5% of the GSDP. Among districts, Bangalore urban has the highest GDDP of 3535.6 billion rupees (agriculture share is 0.13%), followed by Dakshina Kannada with a GDDP of 587 billion rupees (agriculture contributes 3.75%). Agriculture contributes 27% of GDDP of Vijayapura district, followed by Chitradurga (26%), Koppal (25%), Gadag (25%), Kalaburagi (22%), Davanagere (22%), and Bagalkot (20%).

Ecosystem services of agriculture ecosystem: Table 65 provide details of various ecosystem goods and services and their monetary value across different major crop types.

Table 66 lists ecosystem services district-wise for agriculture and horticulture ecosystems. The agriculture ecosystem services account for 223 billion rupees per year (provisioning services: 106.6 billion rupees, regulating: 110 billion rupees, and cultural service: 6.9 billion rupees). The district-wise assessment indicates that Belgaum has the highest ecosystem services value of 25.1 billion rupees (13.5 billion rupees from provisioning services, 10.89 billion rupees from regulating services, and 0.6 billion rupees from cultural services). Vijayapura (Bijapur) district has an agriculture ecosystem service value of 19.8 billion rupees (8.9 billion rupees from provisioning, 10.2 billion rupees from regulating, and 0.6 billion rupees from cultural

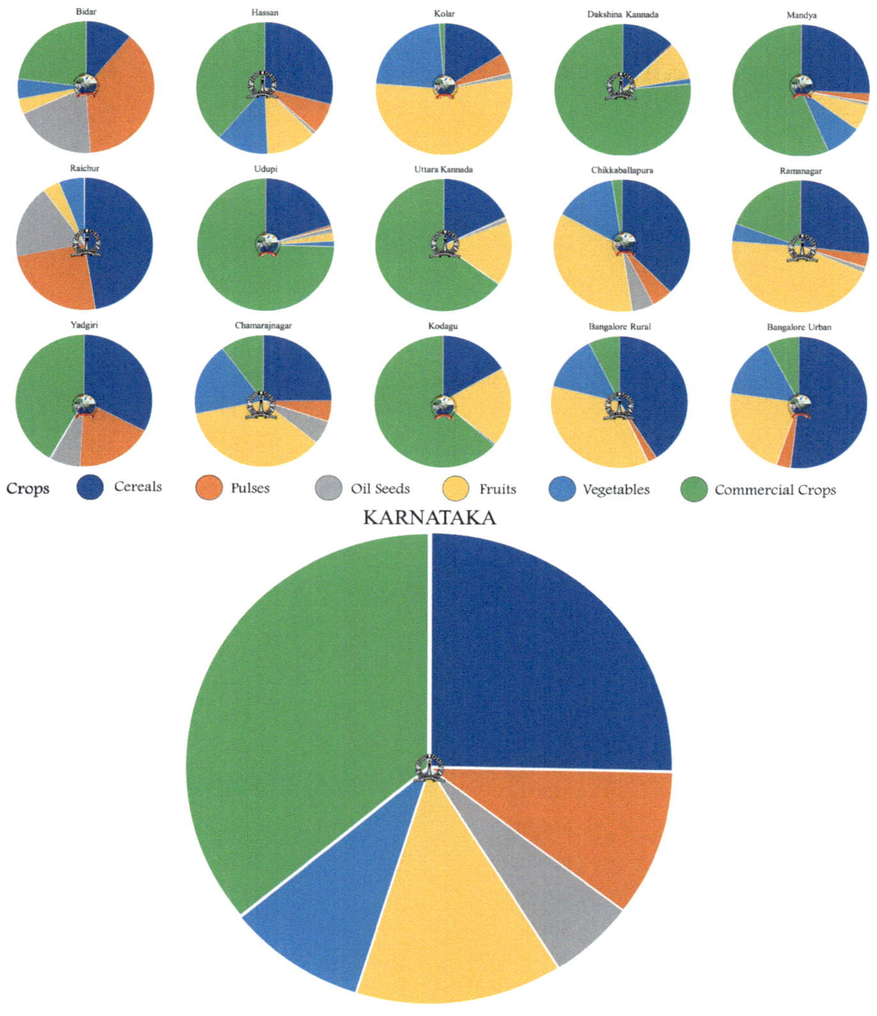

Fig. 51 The relative share of crops in the total revenue of 866 billion (2019)

services). The lowest values are in Bengaluru Urban and Bengaluru Rural districts, with the agriculture ecosystem service value of 1.4 billion Rs each.

Similar to agriculture, the ecosystem services of the horticulture ecosystem amount to 42.9 billion rupees (provisioning services: 34.4 billion rupees, regulating services: 8.0 billion rupees, cultural services: 0.5 billion rupees). A district-wise assessment indicates Tumkur has the highest value of 9.7 billion rupees (8.3 billion rupees from provisioning, 1.2 billion rupees from regulating services, and 0.6 billion rupees from cultural services), followed by Hassan (5.5 billion rupees) and Dakshina

Table 64 District wise GDDP, and revenue from croplands (billion rupees)

District	GDDP	Agriculture	Proportion (%)
Bagalkot	265.5	54.2	20.43
Bengaluru (Urban)	3635.6	4.6	0.13
Bengaluru (Rural)	162.5	8.6	5.28
Belgaum	454.6	85.3	18.76
Bellary	334.8	32.4	9.67
Bidar	144.9	26.5	18.26
Vijayapura	188.1	50.9	27.04
Chamarajanagar	117.0	13.8	11.79
Chikkaballapura	144.4	17.6	12.20
Chikkamagaluru	230.2	35.0	15.23
Chitradurga	169.6	44.4	26.16
Dakshina Kannada	587.2	22.0	3.75
Davanagere	202.1	44.1	21.84
Dharwad	244.7	26.7	10.91
Gadag	109.1	27.1	24.82
Kalaburagi	195.1	43.8	22.44
Hassan	236.4	25.1	10.63
Haveri	155.3	30.0	19.31
Kodagu	61.6	11.5	18.73
Kolar	176.6	25.0	14.14
Koppal	118.9	29.7	24.96
Mandya	267.3	20.4	7.61
Mysuru	352.1	34.0	9.67
Raichur	173.5	20.2	11.65
Ramanagara	159.6	16.5	10.34
Shimoga	300.5	31.5	10.50
Tumkur	385.3	33.1	8.60
Udupi	276.3	19.8	7.16
Uttara Kannada	186.2	18.8	10.10
Yadgir	93.4	14.2	15.16
Karnataka	10128.1	866.6	8.56

Kannada (4.3 billion rupees). Districts with the lowest values include Bidar (3 million rupees), Chikkaballapura (14 million rupees), and Vijayapura (Bijapur) (14 million rupees).

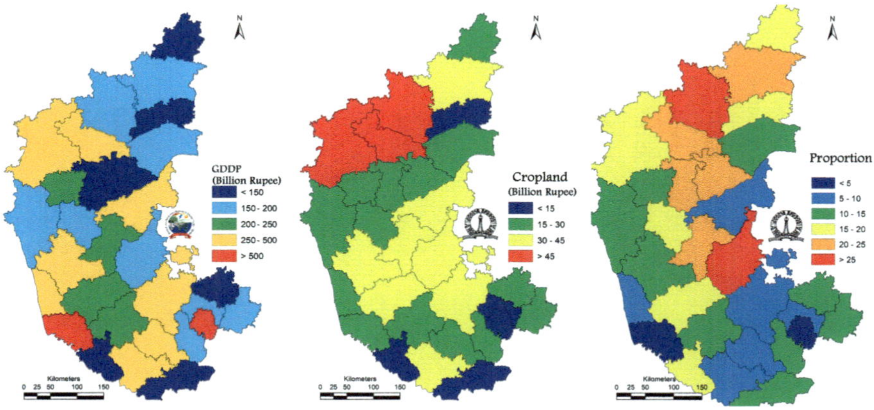

Fig. 52 District wise GDDP, revenue generated from croplands and its relative proportion in the GDDP

Table 65 Agriculture ecosystem goods and services (INR/hectare/Yr) 2005*

	Croplands**	Horticulture	Fruits	Vegetables
Provisioning services				
Food	MSP—cost of production			
Fodder/fiber	3742		2245	2245
Wood		131		
Regulating services				
Air quality	915	915	915	915
Climate	218	218	218	218
Carbon fixation atmosphere	11	1527	3	1
Soil carbon	110	4364	110	110
Water flow regulation	240	371	240	240
Nitrogen fixation	1213	1213	1213	1213
Soil fertility	1512	1512	1512	1512
Remediation—organic and inorganic materials	1745	1745	1745	1745
Pollination	118	118	118	118
Genetic diversity	3908	3908	3908	3908
Biological control	35	35	35	35
Cultural services				
Opportunities for recreation and tourism	285	285	285	285
Inspiration for culture, art and design	349	349	349	349

**Source* Public interviews, government records (CACP 2005a, b; Nayak et al. 2019; CRED 2020; De Groot et al. 2020; DMI 2020; EANDS 2020; NAAS 2020; TNAU Agriculture Portal 2020)
** Services of croplands include all crops except fruits and vegetables

Table 66 Monetary value of agriculture—croplands and horticulture ecosystems services (million Rs)

District	Croplands				Horticulture				TESV (Agriculture) = croplands + horticulture
	Provisioning	Regulating	Cultural	TESV	Provisioning	Regulating	Cultural	TESV	
Bagalkot	4479	5702	361	10542	31	5	0	36	10577
Bengaluru Rural	999	391	25	1414	437	88	6	530	1944
Belgaum	13542	10894	689	25125	45	12	1	57	25182
Bellary	5103	5150	326	10579	87	13	1	100	10680
Bengaluru Urban	831	602	38	1472	137	27	2	166	1638
Bidar	4599	4425	280	9305	2	1	0	3	9307
Vijayapura (Bijapur)	8968	10240	648	19856	12	2	0	14	19870
Chamarajanagar	1979	1899	120	3998	483	134	8	625	4623
Chikmagalur	1739	1931	122	3792	2807	1186	75	4068	7860
Chikkaballapura	1558	789	50	2397	12	2	0	14	2411
Chitradurga	2618	1629	103	4351	3136	580	37	3753	8103
Dakshina Kannada	2167	4306	272	6745	3477	829	52	4358	11103
Davanagere	4105	730	46	4881	1418	341	22	1781	6662
Dharwad	3450	4231	268	7949	32	6	0	38	7986
Gadag	3861	4617	292	8770	42	6	0	49	8819
Gulbaraga	5613	3757	238	9608	35	5	0	41	9649
Hassan	5520	9507	601	15629	4465	1031	65	5561	21190
Haveri	2820	3362	213	6394	109	19	1	129	6524
Kodagu	1811	4160	263	6234	2414	903	57	3374	9608
Kolar	1619	214	14	1847	69	22	1	92	1939

(continued)

Table 66 (continued)

District	Croplands				Horticulture				TESV (Agriculture) = croplands + horticulture
	Provisioning	Regulating	Cultural	TESV	Provisioning	Regulating	Cultural	TESV	
Koppal	3021	1529	97	4647	23	3	0	27	4674
Mandya	3571	4963	314	8848	966	194	12	1173	10021
Mysore	3939	2629	166	6735	756	203	13	972	7707
Ramanagara	1514	593	37	2144	662	133	8	803	2947
Raichur	5033	5323	337	10692	30	4	0	34	10727
Shimoga	1986	1491	94	3571	1792	387	24	2203	5775
Tumkur	4244	6656	421	11321	8380	1276	81	9736	21057
Udupi	1148	2029	128	3306	1549	411	26	1986	5292
Uttara Kannada	2118	4564	289	6971	999	207	13	1219	8190
Yadgir	2660	1780	113	4553	17	2	0	19	4572
Karnatkta	106615	110093	6965	223676	34424	8032	505	42961	266637

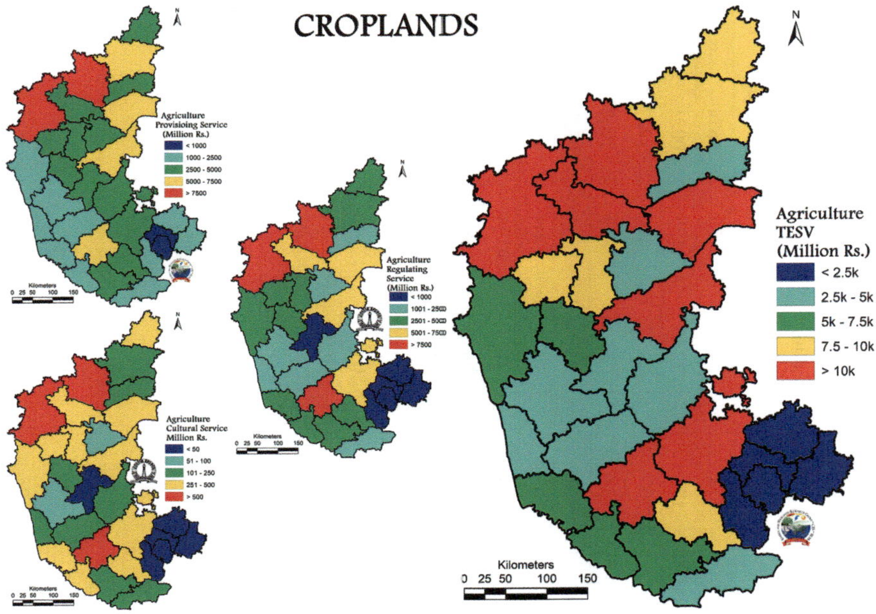

Fig. 53 Services from croplands ecosystem (2005)

Figures 53 and 54 provide the monetary value of agriculture (croplands and horticulture) ecosystem services in 2005. The ecosystem services of agriculture are about 266 billion rupees. Belgaum contributes the highest at 25.1 billion rupees, followed by Hassan (21 billion rupees) and Tumkur (21 billion rupees), and Vijayapura (Bijapur) (19.8 billion rupees).

Valuation of agriculture ecosystem services (2019)

Services of the agriculture ecosystem in Karnataka state with diverse cropping patterns amount to 1,077.6 billion rupees, with 55% from provisioning services, 42% from regulating services, and 3% from cultural services (Fig. 55). The net present value (NPV) of the agriculture ecosystem is 27.72 trillion rupees.

District-wise agriculture ecosystem services are listed in Tables 67, 68, and 69. The annual provisioning services amount to 589 billion rupees (food: 462 billion, fodder:125 billion, and wood 1.5 billion rupees), regulating services amount to 459 billion rupees (air quality: 42 billion, climate regulation: 10 billion, carbon fixation: 18 billion, soil carbon: 54 billion, water flow: 12 billion, nitrogen fixation: 5 billion, soil fertility: 47 billion, remediation—organic and inorganic materials (mineralization of soil nutrients): 80 billion rupees and pollination: 5 billion rupees) and cultural services 29 billion rupees (recreation: 123 billion rupees, culture: 16 billion). TESV of the agriculture ecosystem is 107 billion rupees per year. Based on the annual

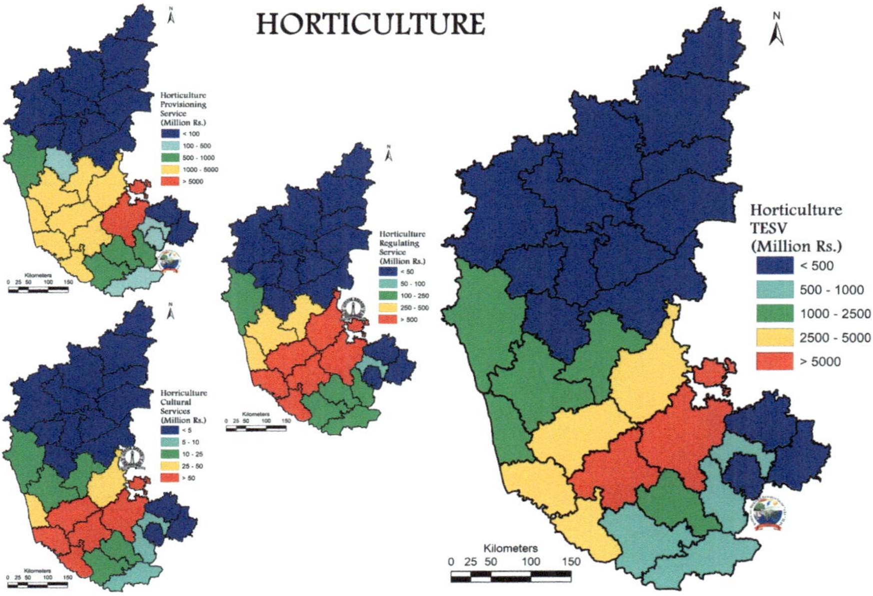

Fig. 54 Services from horticulture ecosystem (2005)

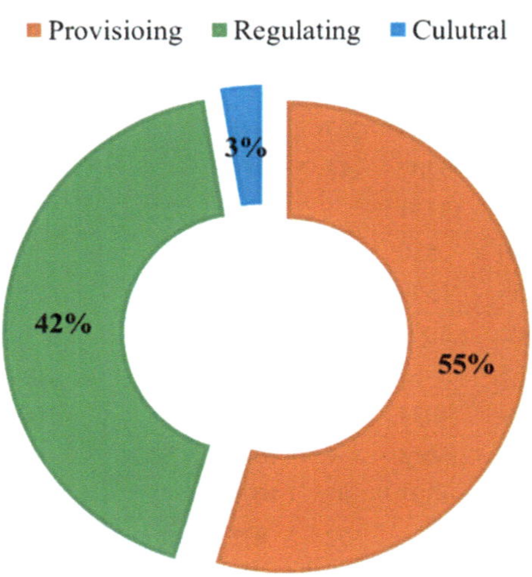

Fig. 55 Agriculture ecosystem services distribution—Karnataka State, 2019

Table 67 Ecosystem services of agriculture ecosystem (million rupees/year), district wise, Karnataka

District	Provisioning services			Regulating services				
	Food	Fodder	Wood	Air quality	Climate regulation	Atmospheric carbon fixation	Soil carbon	Water flow
Bagalkot	33577	7194	1	1805	430	34	253	475
Bangalore Rural	4439	847	9	287	68	106	323	84
Bangalore Urban	2283	292	8	139	33	100	291	45
Belgaum	53017	11982	1	2983	712	44	383	783
Bellary	13620	5865	14	1565	373	186	658	425
Bidar	15761	4704	1	1158	276	20	158	304
Vijayapura	30162	11738	1	2920	696	48	388	767
Chamarajanagar	6865	1585	1	413	99	11	66	109
Chikmagalur	18961	2391	65	1058	252	765	2237	342
Chikkaballapura	8970	2093	0	559	133	7	67	147
Chitradurga	22411	4667	226	2784	664	2653	7684	956
Dakshina Kannada	11981	747	176	1465	349	2056	5894	560
Davanagere	20116	4237	82	1637	390	972	2868	512
Dharwad	13515	5065	2	1286	307	33	205	339
Gadag	14048	5962	21	1635	390	259	869	450
Hassan	13139	3058	227	2375	566	2658	7661	850
Haveri	15505	5207	18	1419	338	224	751	390
Kalaburagi	25491	13134	1	3231	771	46	409	848
Kodagu	6140	530	68	612	146	792	2275	228
Kolar	13556	1590	2	519	124	31	133	138
Koppal	13502	5007	4	1275	304	60	279	338
Mandya	11,935	2562	78	1193	285	922	2689	391
Mysore	16609	4034	87	1614	385	1026	3020	510
Raichur	8779	6643	1	1642	392	32	234	432
Ramanagar	8617	1523	97	1079	257	1131	3267	380
Shimoga	15691	2036	52	875	209	613	1796	282
Tumkur	17411	4011	117	1838	438	1382	4036	600
Udupi	10083	1028	93	928	221	1083	3118	336
Uttara Kannada	9933	1082	82	851	203	962	2772	305
Yadgiri	6233	4588	0	1125	268	14	138	295
Karnataka	462348	125401	1534	42271	10081	18268	54920	12623

Table 68 Ecosystem services of agriculture ecosystem (million rupee/year), district wise, Karnataka

District	Regulating services						Cultural services	
	Nitrogen fixation	Soil fertility	Mineralisation of soil nutrient	Pollination	Genetic diversity	Biological control	Recreation	Culture
Bagalkot	237	2984	3444	234	7711	69	563	689
Bangalore Rural	38	233	548	37	1227	11	90	110
Bangalore Urban	18	113	265	18	594	5	43	53
Belgaum	391	2419	5692	386	12745	114	930	1138
Bellary	205	1269	2986	203	6686	60	488	597
Bidar	152	939	2210	150	4948	44	361	442
Vijayapura	383	4827	5570	378	12,473	111	910	1114
Chamarajanagar	54	683	789	54	1766	16	129	158
Chikmagalur	139	1748	2018	137	4518	40	330	404
Chikkaballapura	73	454	1067	72	2390	21	174	213
Chitradurga	365	4602	5311	361	11892	106	868	1062
Dakshina Kannada	192	1188	2795	190	6257	56	457	559
Davanagere	215	2706	3123	212	6992	62	510	625
Dharwad	169	1043	2453	167	5493	49	401	491
Gadag	214	1325	3119	212	6983	62	510	624
Hassan	312	3926	4531	308	10146	91	740	906
Haveri	186	1150	2707	184	6060	54	442	541
Kalaburagi	424	2620	6165	419	13805	123	1007	1233
Kodagu	80	496	1168	79	2615	23	191	234
Kolar	68	858	990	67	2216	20	162	198

(continued)

Table 68 (continued)

District	Regulating services						Cultural services	
	Nitrogen fixation	Soil fertility	Mineralisation of soil nutrient	Pollination	Genetic diversity	Biological control	Recreation	Culture
Koppal	167	1034	2433	165	5448	49	398	487
Mandya	157	968	2277	155	5098	46	372	455
Mysore	212	2669	3080	209	6897	62	503	616
Raichur	215	1331	3133	213	7014	63	512	627
Ramanagar	141	875	2058	140	4608	41	336	412
Shimoga	115	710	1669	113	3738	33	273	334
Tumkur	241	1490	3507	238	7851	70	573	701
Udupi	122	752	1770	120	3963	35	289	354
Uttara Kannada	112	690	1624	110	3635	32	265	325
Yadgiri	148	913	2147	146	4807	43	351	429
Karnataka	5545	47015	80649	5475	180578	1613	13175	16130

Table 69 Ecosystem services of agriculture ecosystem—district wise TESV and NPV

District	Total services (million rupees/year)				Net present value (billion rupees)			
	Provisioning	Regulating	Cultural	TESV	Provisioning	Regulating	Cultural	NPV
Bagalkot	40771.9	17674.7	1251.4	59697.9	1049.1	454.8	32.2	1536.0
Bangalore Rural	5295.1	2963.1	199.1	8457.3	136.2	76.2	5.1	217.6
Bangalore Urban	2582.5	1621.7	96.4	4300.6	66.4	41.7	2.5	110.7
Belgaum	64999.6	26654.1	2068.4	93722.1	1672.4	685.8	53.2	2411.4
Bellary	19500.0	14616.4	1085.1	35201.5	501.7	376.1	27.9	905.7
Bidar	20464.8	10359.8	802.9	31627.5	526.6	266.6	20.7	813.8
Vijayapura	41900.9	28561.3	2024.1	72486.4	1078.1	734.9	52.1	1865.1
Chamarajanagar	8450.5	4059.8	286.6	12796.9	217.4	104.5	7.4	329.3
Chikmagalur	21416.9	13254.0	733.2	35404.1	551.1	341.0	18.9	910.9
Chikkaballapura	11062.6	4992.2	387.9	16442.7	284.6	128.4	10.0	423.1
Chitradurga	27304.2	37377.0	1929.9	66611.0	702.5	961.7	49.7	1713.9
Dakshina Kannada	12904.6	21001.9	1015.5	34921.9	332.0	540.4	26.1	898.5
Davanagere	24435.3	19688.8	1134.7	45258.8	628.7	506.6	29.2	1164.5
Dharwad	18,581.4	11,541.7	891.4	31,014.4	478.1	297.0	22.9	798.0
Gadag	20030.4	15518.0	1133.3	36681.7	515.4	399.3	29.2	943.8
Hassan	16,424.1	33,423.0	1646.5	51,493.6	422.6	860.0	42.4	1324.9
Haveri	20730.0	13463.0	983.5	35176.5	533.4	346.4	25.3	905.1
Kalaburagi	38625.6	28861.5	2240.3	69727.4	993.8	742.6	57.6	1794.1
Kodagu	6737.0	8516.2	424.4	15,677.6	173.3	219.1	10.9	403.4
Kolar	15148.6	5163.8	359.7	20672.1	389.8	132.9	9.3	531.9
Koppal	18512.7	11553.4	884.1	30950.2	476.3	297.3	22.7	796.3

(continued)

Table 69 (continued)

District	Total services (million rupees/year)				Net present value (billion rupees)			
	Provisioning	Regulating	Cultural	TESV	Provisioning	Regulating	Cultural	NPV
Mandya	14575.4	14179.7	827.4	29582.5	375.0	364.8	21.3	761.2
Mysore	20729.9	19684.2	1119.2	41533.4	533.4	506.5	28.8	1068.6
Raichur	15422.5	14700.1	1138.3	31260.9	396.8	378.2	29.3	804.3
Ramanagar	10236.2	13976.5	747.8	24960.5	263.4	359.6	19.2	642.2
Shimoga	17779.2	10153.5	606.6	28539.3	457.5	261.2	15.6	734.3
Tumkur	21539.9	21690.4	1274.2	44504.4	554.2	558.1	32.8	1145.1
Udupi	11202.8	12447.0	643.1	24292.9	288.2	320.3	16.5	625.1
Uttara Kannada	11097.6	11296.0	589.9	22983.6	285.5	290.6	15.2	591.4
Yadgiri	10820.7	10044.5	780.2	21645.4	278.4	258.4	20.1	556.9
Karnataka	589283.0	459037.2	29305.2	1077625.4	15162.1	11810.9	754.0	27727.0

flow, the net present value (NPV) of the agriculture ecosystem in Karnataka is about 27,727 billion rupees.

Provisioning services: In the current study, provisioning services include food, fodder, and wood-derived from agriculture (croplands and horticulture) ecosystems for 2019 (Table 67). District-wise variability in ecosystem services from the agriculture ecosystem is depicted in Fig. 56.

Belgaum contributes the highest of 64.99 billion rupees/year (Table 67, Fig. 56), followed by Vijayapura (41.9 billion) and Kalaburagi (38.6 billion rupees). The lowest is in Bengaluru Urban (2.5 billion rupees), followed by Bengaluru Rural (5.2 billion rupees).

Regulating services: Regulating services considered (Tables 67 and 68) are air quality, climate regulation, atmospheric carbon fixation, soil carbon, water regulation, nitrogen fixation, soil fertility, remediation—organic and inorganic materials (mineralization of plant nutrients), pollination, genetic diversity, and biological control. Figure 57 depicts the variability in regulating services across districts in Karnataka.

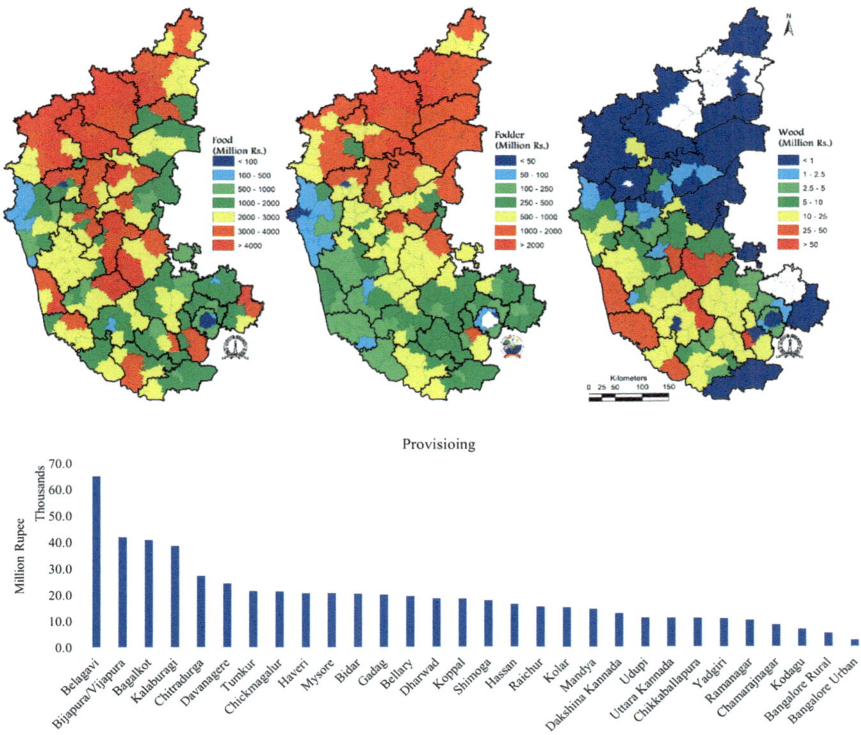

Fig. 56 Provisioning services (food, fodder, wood) during 2019 - Agriculture

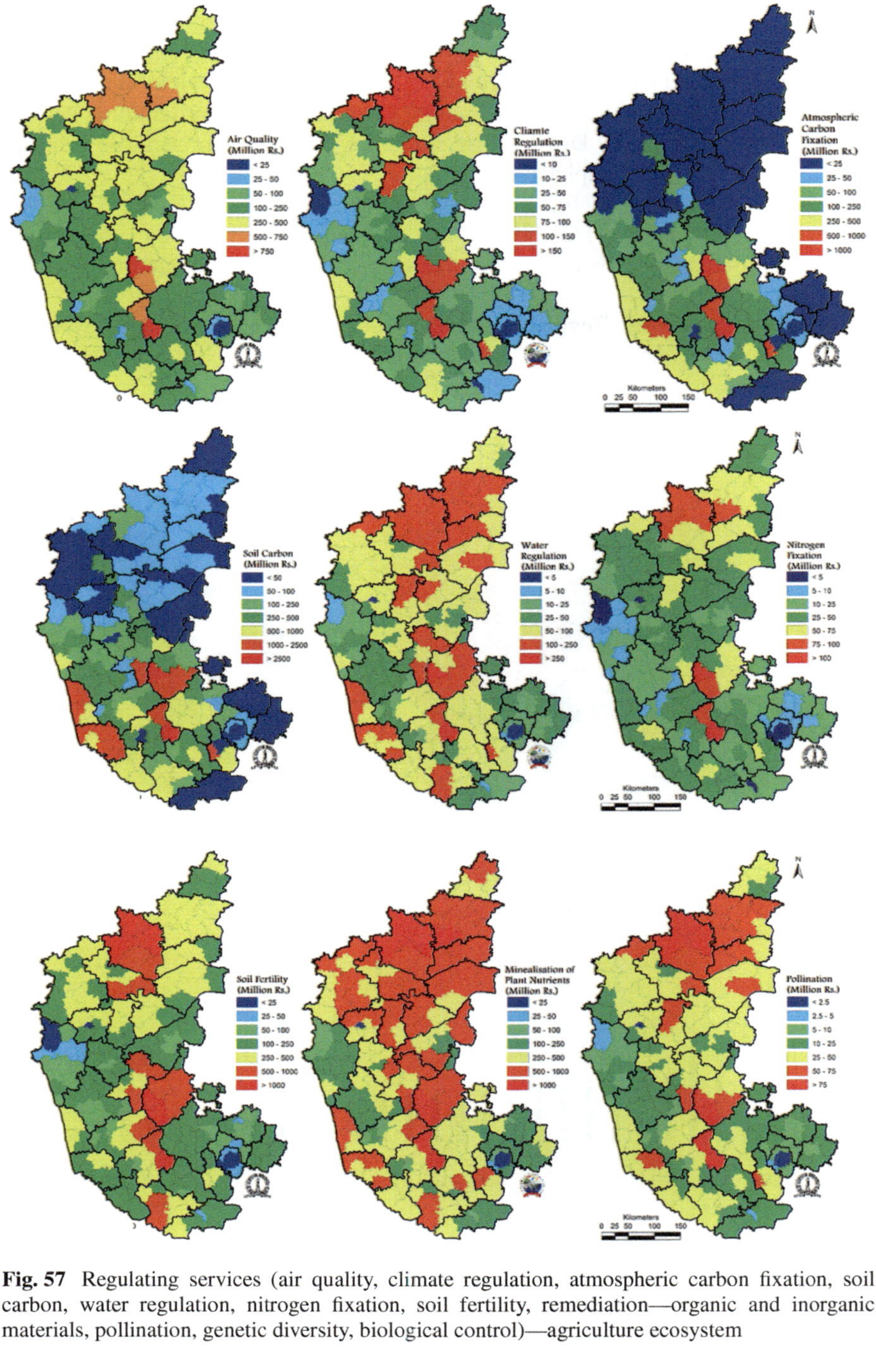

Fig. 57 Regulating services (air quality, climate regulation, atmospheric carbon fixation, soil carbon, water regulation, nitrogen fixation, soil fertility, remediation—organic and inorganic materials, pollination, genetic diversity, biological control)—agriculture ecosystem

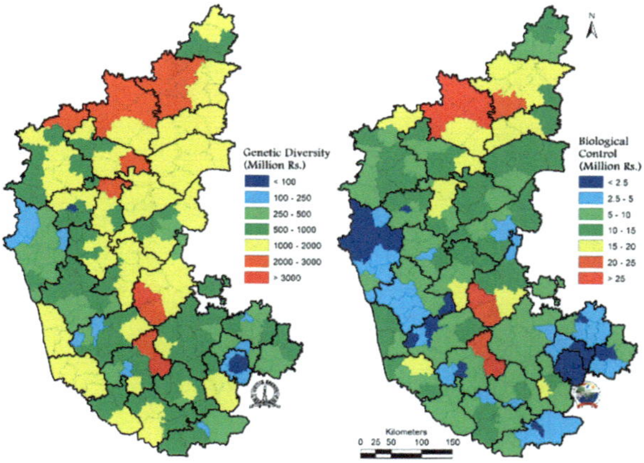

Fig. 57 (continued)

Chitradurga contributes the highest with 37.3 billion rupees/year (961.7 billion rupees—NPV) followed by Hassan with 33.4 billion rupees/year, and Kalaburagi, Vijayapura, and Belgaum with 25–28 billion rupees/year each (Tables 67 and 68, Fig. 57). Bengaluru Urban, Bengaluru rural, Chamarajanagar and Chikkaballapura contributed the lowest, ranging between 1.6 and 4.9 billion rupees each.

Cultural Services: Cultural services (Table 68) include (i) recreation and tourism and (ii) culture and art. The relative share of cultural services is depicted in Fig. 58. Kalaburagi contributed the highest with 2.2 billion rupees (57.6 billion rupees NPV), followed by Belgaum and Vijayapura (Bijapur) with 2 billion rupees/year each (Table 68, Fig. 58). The lowest values were observed in Bengaluru Urban and Bengaluru Rural districts, with 96.4 million rupees/year and 199.1 million rupees/year, respectively.

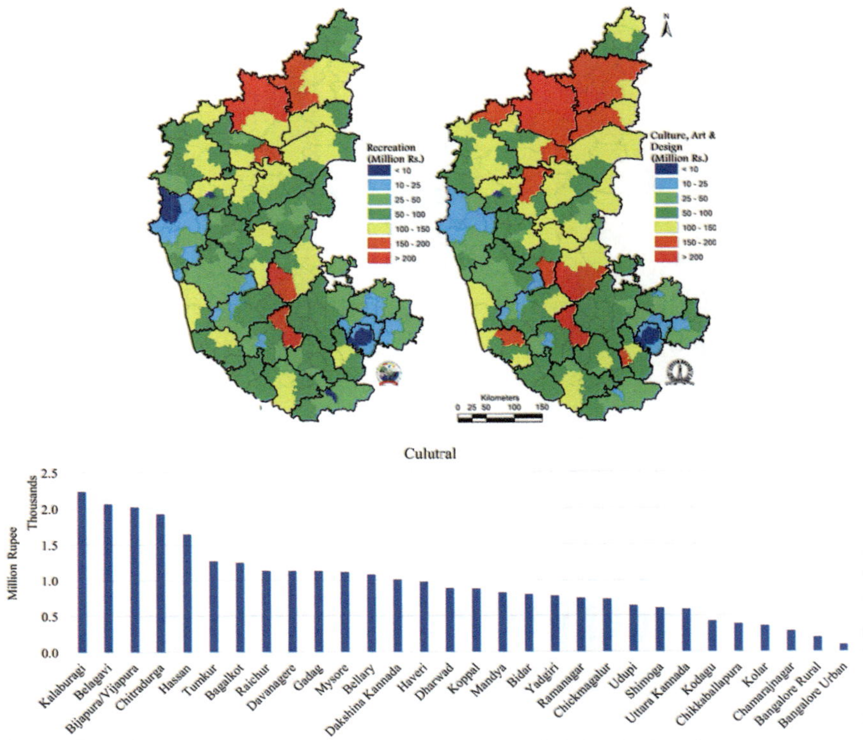

Fig. 58 Cultural services (recreation, culture) from the agricultural ecosystem

Total ecosystem supply value [TESV] of agriculture ecosystem: Provisioning, regulating, and cultural services were aggregated to compute TESV—total ecosystem supply value of agriculture ecosystem in Karnataka State, India, which are listed in Table 69. Belgaum District provides 93.7 Billion Rs./year (2.4 trillion rupees NPV); this was followed by Vijayapura, Kalaburagi, and Chitradurga, with services ranging between 66.6 billion rupees/year to 72.4 billion rupees/year (Fig. 59). The lowest was 4.3 billion rupees/year in the district of Bengaluru Urban, followed by Bengaluru Rural with 8.4 billion rupees/year.

Temporal comparison of agriculture ecosystem services: Monetary values of ecosystem services (provisioning, regulating, cultural services, and TSEV) of 2005 and 2019 are compared to understand the changes due to changes in the spatial extent and condition of the ecosystem. The monetary values of 2005 were adjusted to 2019 by considering the GDP deflator (MoSPI 2020) of an inflation rate of 2.92 times (*Inflation Calculator-Indian Rupee* 2021). Tables 70 and 71 compare goods

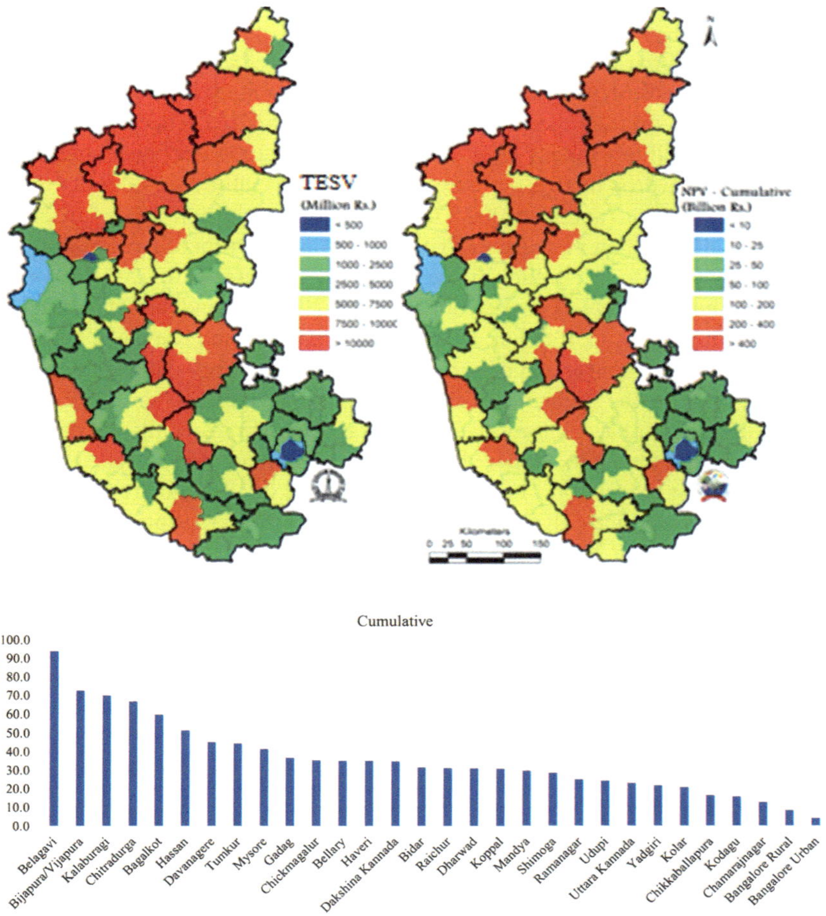

Fig. 59 Agriculture ecosystem TESV—Karnataka, 2019

and services of 2005 (at 2019 price) and 2019 district-wise, for agriculture and horticulture ecosystems.

Croplands Ecosystem: Table 70 indicates that there has been an increase in the provisioning goods of 193 billion rupees (311 billion rupees in 2005 (at 2019 price) increased to 505 billion rupees in 2019, an increase of 62%), a decline in regulating services of 31 billion rupees (321 billion rupees in 2005 declined to 290 billion rupees in 2019, a drop of 10%), a marginal increase in cultural services of 1.5 billion rupees (20.3 billion rupees in 2005 increased to 21.8 billion rupees in 2019, an increase of 8%) and an increase in TESV of 164 billion rupees (653 billion rupees in 2005 increased to 817 billion rupees in 2019, an increase of 25%). However, a TESV decline of 24–80% is noticed during 2005–2019 in Mandya, Bengaluru Urban, Tumkur, Udupi, Hassan, Raichur, Uttara Kannada, Dakshina Kannada, and Kodagu districts, due to the conversion of agriculture lands to commercial plantations, built-up areas, etc. On the other hand, there has been an increase in TESV of 125% in Ramanagara, Kolar, Chitradurga, Kalaburagi (Gulbarga), Chikkaballapura, and Koppal districts, owing to an increase in the spatial extent of cropland.

Horticulture ecosystem: There has been a decline of 19 billion rupees (Table 71) in provisioning services (100 billion rupees in 2005 (at 2019 price), decreased to 80.9 billion rupees in 2019, a decline of 19%), an increase in regulating services of 15 billion rupees (23 billion rupees in 2005 increased to 38 billion rupees in 2019, an increase of 65%), 0.2 billion rupees increase in cultural services (1.5 billion rupees in 2005 increased to 1.7 billion rupees in 2019, an increase of 16%) and decline of 4 billion rupees in TESV (125 billion in 2005 declined to 121 billion rupees in 2019, a drop of 3%). An increase in TESV is noticed in 14 districts, while the rest showed a declining trend. About a 125% increase in TESV is seen in the Uttara Kannada, Gadag, Chikkaballapura, and Bidar districts. In comparison, a 78% decline in Raichur, Chamarajanagar, Bagalkot, Yadgir, and Kalaburagi (Gulbarga) districts was seen in 2005 and 2019.

6.2 TESV—Total Ecosystem Supply Value, GEP—Gross Ecosystem Product, Karnataka [District-Wise] State, India

Ecosystem services (provisioning, regulating, and cultural services) were aggregated to compute the total ecosystem supply value (TESV). The aggregate measure is also referred to as ***gross ecosystem product (GEP), which is equal to the sum of all final ecosystem services (i.e., by economic units) from ecosystem assets*** (Ouyang et al. 2020). The ecosystem monetary asset account also records the changes in the monetary value of ecosystem assets from 2005 to 2019 (accounting period).

Table 72 lists district-wise provisioning services, regulating services, cultural services, and TESV for 2005 (at 2019 monetary values). TESV for Karnataka State

Table 70 Monetary value of goods and services, agriculture ecosystem-comparative assessment 2005 and 2019 (million Rs)

District	Opening stock 2005 (at 2019 price)				Closing stock 2019				% Change			
	Provisioning	Regulating	Cultural	TESV	Provisioning	Regulating	Cultural	TESV	Provisioning (%)	Regulating (%)	Cultural (%)	TESV (%)
Bagalkot	13077	16651	1053	30781	40758	17552	1246	59556	212	5	18	93
Bengaluru Rural	2916	1142	72	4130	4935	2009	156	7099	69	76	116	72
Belgaum	39542	31811	2012	73365	64964	26575	2065	93604	64	−16	3	28
Bellary	14902	15038	951	30891	19291	13066	1015	33372	29	−13	7	8
Bengaluru Urban	2428	1759	111	4298	2377	714	55	3146	−2	−59	−50	−27
Bidar	13430	12921	817	27169	20448	10299	800	31547	52	−20	−2	16
Vijayapura (Bijapur)	26186	29902	1891	57979	41886	28434	2019	72339	60	−5	7	25
Chamarajanagar	5777	5546	351	11674	8399	4002	284	12686	45	−28	−19	9
Chikmagalur	5079	5637	357	11073	9025	5898	419	15342	78	5	17	39
Chikkaballapura	4550	2303	146	6999	10,849	4992	388	16229	138	117	166	132
Chitradurga	7645	4757	301	12704	20,192	11749	834	32775	164	147	177	158
Dakshina Kannada	6329	12572	795	19696	2743	2096	163	5002	−57	−83	−80	−75
Davanagere	11985	2133	135	14253	17209	10373	736	28319	44	386	446	99
Dharwad	10075	12353	781	23210	18505	11376	884	30765	84	−8	13	33
Gadag	11273	13483	853	25609	19713	13296	1033	34042	75	−1	21	33
Gulbaraga	16391	10971	694	28056	38618	28796	2237	69651	136	162	222	148
Hassan	16118	27761	1756	45635	11859	7701	547	20107	−26	−72	−69	−56
Haveri	8234	9817	621	18671	20281	11544	897	32722	146	18	44	75

(continued)

Table 70 (continued)

District	Opening stock 2005 (at 2019 price)				Closing stock 2019				% Change			
	Provisioning	Regulating	Cultural	TESV	Provisioning	Regulating	Cultural	TESV	Provisioning (%)	Regulating (%)	Cultural (%)	TESV (%)
Kodagu	5288	12146	768	18202	2316	1239	96	3651	−56	−90	−87	−80
Kolar	4728	626	40	5394	14965	4919	349	20234	217	686	782	275
Koppal	8821	4466	283	13570	18462	11139	865	30466	109	149	206	125
Mandya	10427	14493	917	25837	13444	5765	448	19657	29	−60	−51	−24
Mysore	11503	7676	486	19665	19240	9831	698	29769	67	28	44	51
Ramanagara	4420	1731	109	6260	15408	14580	1133	31121	249	743	935	397
Raichur	14696	15542	983	31221	8514	3605	280	12399	−42	−77	−72	−60
Shimoga	5799	4354	275	10428	9079	4563	355	13997	57	5	29	34
Tumkur	12393	19436	1229	33058	13762	9079	705	23546	11	−53	−43	−29
Udupi	3352	5926	375	9653	2490	2508	195	5193	−26	−58	−48	−46
Uttara Kannada	6184	13328	843	20355	4626	2470	192	7288	−25	−81	−77	−64
Yadgir	7767	5198	329	13294	10820	10037	780	21636	39	93	137	63
Karnataka	311315	321479	20334	653130	505178	290207	21874	817260	62	−10	8	25

Table 71 Monetary value of goods and services of horticulture ecosystem—comparative assessment 2005 and 2020 (million Rs)

District	Opening stock 2005 (at 2019 price)				Closing stock 2020				% Change			
	Provisioning	Regulating	Cultural	TESV	Provisioning	Regulating	Cultural	TESV	Provisioning (%)	Regulating (%)	Cultural (%)	TESV (%)
Bagalkot	90	13	1	104	13	9	0	22	−85	−34	−56	−79
Bengaluru Rural	1275	256	16	1548	353	120	5	478	−72	−53	−67	−69
Belgaum	131	34	2	167	35	36	2	72	−74	7	−24	−57
Bellary	253	38	2	293	196	115	5	315	−23	202	116	8
Bengaluru Urban	401	80	5	486	198	83	4	285	−51	4	−26	−41
Bidar	6	2	0	8	17	15	1	32	174	826	560	313
Vijayapura (Bijapur)	35	5	0	41	14	9	0	23	−61	74	18	−43
Chamarajanagar	1409	391	25	1825	51	18	1	69	−96	−95	−97	−96
Chikmagalur	8195	3464	219	11878	12379	5949	254	18582	51	72	16	56
Chikkaballapura	35	5	0	40	215	92	4	310	514	1666	1159	667
Chitradurga	9157	1695	107	10959	6913	3000	128	10041	−25	77	20	−8
Dakshina Kannada	10152	2421	153	12725	10034	5154	232	15420	−1	113	52	21
Davanagere	4141	997	63	5201	5196	1956	84	7236	25	96	33	39
Dharwad	92	16	1	110	75	32	1	108	−19	95	39	−1
Gadag	123	18	1	143	298	159	7	464	142	768	519	226
Gulbaraga	104	15	1	119	7	5	0	12	−93	−68	−77	−90
Hassan	13039	3010	190	16240	4370	3546	152	8068	−66	18	−20	−50
Haveri	319	55	3	378	433	186	8	627	36	237	140	66

(continued)

Table 71 (continued)

District	Opening stock 2005 (at 2019 price)				Closing stock 2020				% Change			
	Provisioning	Regulating	Cultural	TESV	Provisioning	Regulating	Cultural	TESV	Provisioning (%)	Regulating (%)	Cultural (%)	TESV (%)
Kodagu	7050	2636	167	9853	4421	7278	328	12027	−37	176	97	22
Kolar	202	63	4	269	183	193	8	384	−9	206	107	43
Koppal	68	10	1	79	47	30	1	78	−31	194	110	−1
Mandya	2822	568	36	3425	1059	658	30	1748	−62	16	−17	−49
Mysore	2208	594	38	2839	1410	775	33	2218	−36	31	−12	−22
Ramanagara	1933	389	25	2346	1633	834	38	2505	−16	115	53	7
Raichur	87	13	1	101	14	9	0	22	−84	−34	−53	−78
Shimoga	5233	1130	71	6434	8667	2029	92	10787	66	80	28	68
Tumkur	24469	3725	236	28429	7681	2168	98	9947	−69	−42	−59	−65
Udupi	4523	1200	76	5799	8644	2578	116	11339	91	115	53	96
Uttara Kannada	2916	605	38	3560	6406	1776	80	8263	120	194	109	132
Yadgir	49	7	0	56	1	1	0	1	−98	−92	−94	−97
Karnataka	100517	23455	1482	125455	80963	38813	1712	121483	−19	65	16	−3

Table 72 District wise ecosystem services—2005 (in 2019 rupees) Million rupees

District	Forest ecosystem			Agriculture croplands and horticulture			Cumulative (Forest and agriculture)			Proportion			Total billion rupees
	Prov	Reg	Cult	Prov	Reg	Cult	Prov	Reg	Cult	Prov (%)	Reg (%)	Cult (%)	TESV
Bagalkot	21928	18958	455	13167	16664	1054	35096	35622	1509	49	49	2	72
Bangalore Rural	7862	8295	275	4191	1398	88	12053	9694	363	55	44	2	22
Bangalore Urban	7504	7918	262	2829	1839	116	10332	9757	379	50	48	2	20
Belagavi	44747	38685	928	39673	31845	2014	84420	70530	2942	53	45	2	158
Bellary	28430	23929	1060	15155	15076	954	43585	39005	2014	52	46	2	85
Bidar	5616	6280	147	13436	12923	817	19052	19203	964	49	49	2	39
Vijayapura (Bijapur)	35106	30350	728	26221	29907	1892	61327	60258	2620	49	49	2	124
Chamarajnagar	114986	122387	80745	7187	5937	376	122173	128324	81120	37	39	24	332
Chikmagalur	114990	107946	17615	13274	9101	576	128264	117047	18191	49	44	7	264
Chikkaballapura	14522	15323	508	4585	2309	146	19107	17632	654	51	47	2	37
Chitradurga	28485	24003	1079	16802	6452	408	45287	30455	1487	59	39	2	77
Dakshina Kannada	86601	73233	29544	16480	14993	948	103082	88225	30492	46	40	14	222
Davanagere	31117	30915	3596	16126	3130	198	47243	34045	3794	56	40	4	85
Dharwad	10897	10739	277	10167	12370	782	21064	23109	1060	47	51	2	45
Gadag	11904	11732	303	11396	13501	854	23300	25233	1157	47	51	2	50
Hassan	18499	23163	459	29157	30772	1946	47656	53935	2406	46	52	2	104
Haveri	12316	12137	313	8553	9872	625	20869	22009	938	48	50	2	44
Kalaburagi	11312	12649	295	16495	10985	695	27806	23635	990	53	45	2	52

(continued)

Table 72 (continued)

District	Forest ecosystem			Agriculture croplands and horticulture			Cumulative (Forest and agriculture)			Proportion			Total billion rupees
	Prov	Reg	Cult	Prov	Reg	Cult	Prov	Reg	Cult	Prov (%)	Reg (%)	Cult (%)	TESV
Kodagu	82098	87168	32948	12338	14783	935	94436	101950	33883	41	44	15	230
Kolar	13621	14373	476	4930	689	44	18551	15062	520	54	44	2	34
Koppal	18693	15733	697	8889	4476	283	27582	20209	980	57	41	2	49
Mandya	50148	23302	13702	13249	15061	953	63397	38363	14655	54	33	13	116
Mysore	64083	29777	17509	13711	8270	523	77794	38047	18032	58	28	13	134
Raichur	8730	9763	228	14783	15555	984	23514	25318	1212	47	51	2	50
Ramanagar	12057	12722	422	6353	2119	134	18409	14841	556	54	44	2	34
Shimoga	106724	122829	20995	11032	5483	347	117756	128313	21342	44	48	8	267
Tumkur	28798	36057	715	36862	23161	1465	65659	59218	2180	52	47	2	127
Udupi	63597	53780	21696	7876	7126	451	71473	60906	22147	46	39	14	155
Uttara Kannada	206709	280344	54914	9101	13933	881	215810	294277	55795	38	52	10	566
Yadgiri	5449	6093	142	7816	5205	329	13265	11298	472	53	45	2	25
Karnataka	1267528	1270583	303034	411834	344933	21819	1679361	1615516	324854	46	45	9	3620

Note Prov: Provisioning services, Reg: regulating services, Cul: Cultural services is about 3,620 billion rupees contributed by provisioning services (1,679 billion rupees, 46%), regulating services (1,615 billion rupees, 45%), and cultural services (324 billion rupees, 9%). Forest ecosystems contribute 2,841 billion rupees (78.5%), while agriculture and horticulture contribute 778 billion rupees (21.5%) in TESV. Figure 60 depicts the district-wise share and the share of provisioning, regulating, and cultural services in TESV of 2005 (in 2019 rupees). Similarly, Table 73 lists district-wise provisioning services, regulating services, cultural services, and TESV for 2019. TESV or GEP for Karnataka state is about 2913 billion rupees contributed by provisioning services (1,203 billion rupees, 41%), regulating services (1,385 billion rupees, 48%), and cultural services (324 billion rupees, 11%). Forest ecosystems contribute 1,835 billion rupees (63%), while agriculture (croplands and horticulture) contributes 1,077 billion rupees (37%) in TESV. Figure 61 depicts the district-wise share and share of provisioning, regulating, and cultural services in TESV of 2019.

Changes in the ecosystem services (provisioning, regulating, and cultural services) are illustrated in Fig. 62. The most significant changes in provisioning services are in Kalaburagi district (48%); the most substantial changes in regulating services are in Chitradurga district (54%); and the most remarkable changes in cultural services are in Kalaburagi (136%). Karnataka State has witnessed a 28% decline in provisioning services, 14% in regulating services, and 20% in TESV. TESV has shown a decreasing trend in the majority of districts (25). District-wise changes in the provisioning, regulating, cultural services, and TESV are listed in Table 74.

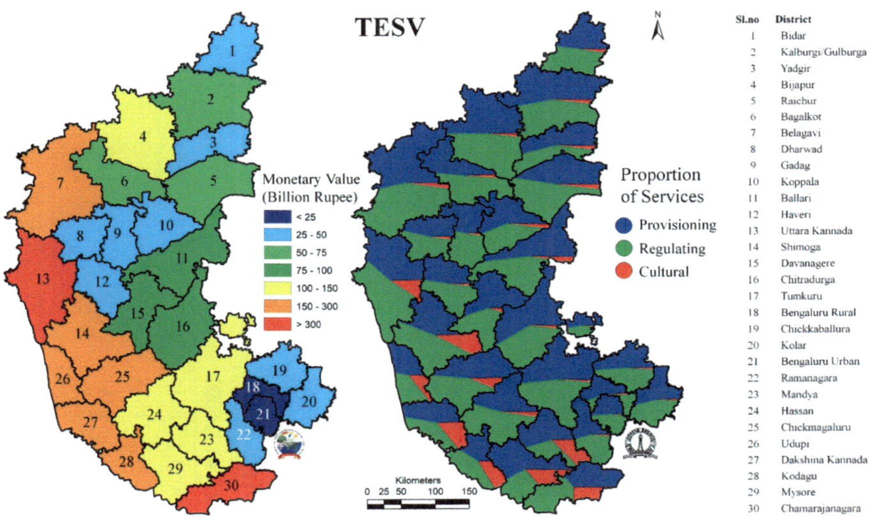

Fig. 60 TESV, with the proportion of services (provisioning, regulating and cultural), 2005

Table 73 District wise goods and services—2019 (million Rs unless otherwise noted)

District	Forest ecosystem			Agriculture ecosystem croplands and horticulture			Cumulative (forest and agriculture)				Proportion			Billion	Billion
	Prov	Reg	Cult	Prov	Reg	Cult	Prov	Reg	Cult	Total	Prov	Reg	Cult	TESV	NPV
Bagalkot	8117	11320	256	40772	17675	1251	48889	28995	1507	79391	62	37	2	79	2043
Bangalore Rural	2233	4991	184	5295	2963	199	7528	7954	383	15865	47	50	2	16	408
Bangalore Urban	2131	4764	175	2582	1622	96	4713	6386	271	11371	41	56	2	11	293
Belagavi	16564	23101	522	65000	26654	2068	81564	49755	2590	133909	61	37	2	134	3445
Bellary	2789	9393	609	19500	14616	1085	22289	24009	1694	47992	46	50	4	48	1235
Bidar	1005	2372	44	20465	10360	803	21470	12732	847	35049	61	36	2	35	902
Vijayapura	12996	18123	410	41901	28561	2024	54897	46684	2434	104015	53	45	2	104	2676
Chamarajnagar	47814	87131	79943	8451	4060	287	56265	91191	80230	227685	25	40	35	228	5858
Chikmagalur	80547	95068	17313	21417	13254	733	101964	108322	18046	228332	45	47	8	228	5875
Chikkaballapura	4126	9220	340	11063	4992	388	15189	14212	728	30129	50	47	2	30	775
Chitradurga	2852	9467	628	27304	37377	1930	30156	46844	2558	79558	38	59	3	80	2047
Dakshina Kannada	67619	70283	29453	12905	21002	1015	80524	91285	30468	202277	40	45	15	202	5205
Davanagere	10175	17758	3260	24435	19689	1135	34610	37447	4395	76452	45	49	6	76	1967
Dharwad	2794	5690	150	18581	11542	891	21375	17232	1041	39648	54	43	3	40	1020
Gadag	3053	6216	164	20030	15518	1133	23083	21734	1297	46115	50	47	3	46	1187
Hassan	5246	11818	188	16424	33423	1646	21670	45241	1834	68746	32	66	3	69	1769
Haveri	3158	6431	169	20730	13463	983	23888	19894	1152	44934	53	44	3	45	1156
Kalaburagi	2026	4779	90	38626	28861	2240	40652	33640	2330	76622	53	44	3	77	1971

(continued)

Table 73 (continued)

District	Forest ecosystem			Agriculture ecosystem croplands and horticulture			Cumulative (forest and agriculture)				Proportion			Billion	Billion
	Prov	Reg	Cult	Prov	Reg	Cult	Prov	Reg	Cult	Total	Prov	Reg	Cult	TESV	NPV
Kodagu	33203	63203	32523	6737	8516	424	39940	71719	32947	144607	28	50	23	145	3721
Kolar	3870	8648	318	15149	5164	360	19019	13812	678	33508	57	41	2	34	862
Koppal	1834	6176	400	18513	11553	884	20347	17729	1284	39360	52	45	3	39	1013
Mandya	13000	17797	13564	14575	14180	827	27575	31977	14391	73944	37	43	19	74	1903
Mysore	16612	22743	17332	20730	19684	1119	37342	42427	18451	98220	38	43	19	98	2527
Raichur	1563	3688	69	15423	14700	1138	16986	18388	1207	36581	46	50	3	37	941
Ramanagar	3425	7655	282	10236	13976	748	13661	21631	1030	36323	38	60	3	36	935
Shimoga	60258	87520	20412	17779	10153	607	78037	97673	21019	196729	40	50	11	197	5062
Tumkur	8167	18398	292	21540	21690	1274	29707	40088	1566	71361	42	56	2	71	1836
Udupi	49657	51613	21629	11203	12447	643	60860	64060	22272	147192	41	44	15	147	3787
Uttara Kannada	146073	238678	54193	11098	11296	590	157171	249974	54783	461928	34	54	12	462	11885
Yadgiri	976	2302	43	10821	10045	780	11797	12347	823	24966	47	49	3	25	642
Karnataka	613883	926346	294955	589283	459037.2	29305.17	1203166	1385383	324260.2	2912809	41	48	11	2913	74946

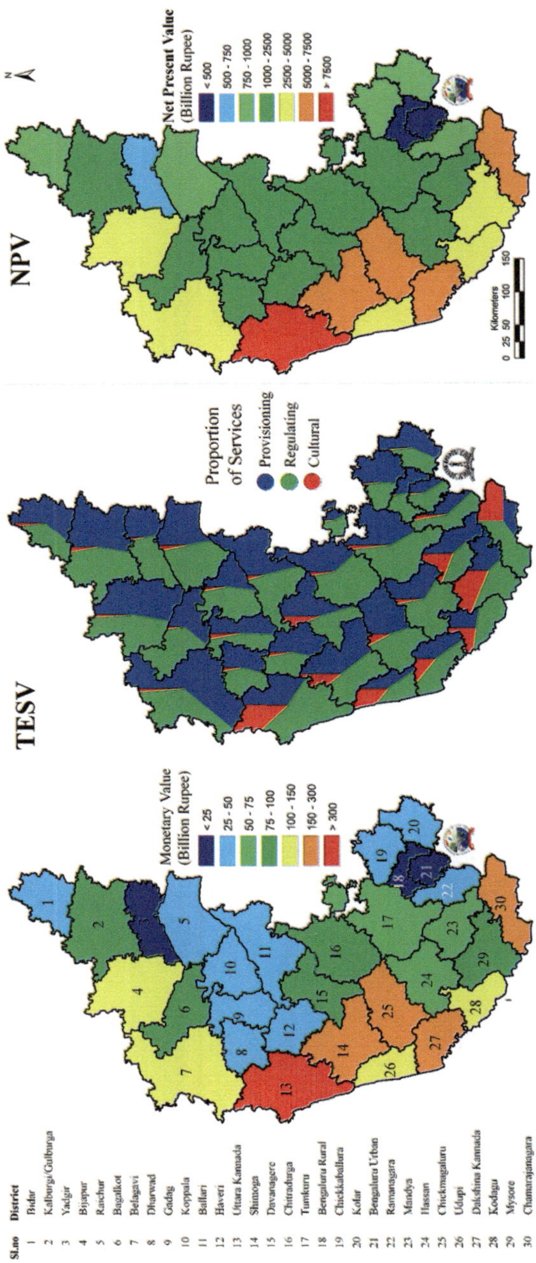

Fig. 61 TESV with the proportion of services (provisioning, regulating and cultural), and NPV 2019–20

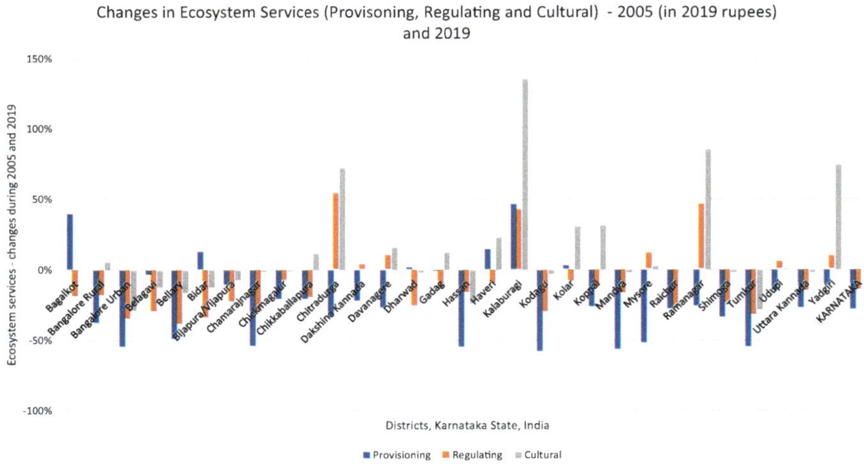

Fig. 62 Changes in ecosystem services (provisioning, regulating and cultural) during 2005 to 2019

Table 75 lists district-wise TESV (forest, agriculture, and total) and GDDP (Gross district domestic product) for Karnataka State, India. The GDP of Karnataka is about 10,128 billion rupees, and estimates of TESV indicate 2,912 billion rupees, of which 1835 billion rupees is derived from forest ecosystems and 1,077 billion rupees from the agricultural ecosystem. The TESV of the forest ecosystem contributes 18.12% of the GDP, and the TESV from agriculture contributes about 10.6% of the GDP in Karnataka.

Figure 63 compares the TESV of 2005 with that of 2019, and Fig. 64 illustrates percentage changes in TESV during 2005 (in 2019 rupees) and 2019. The majority of districts reveal declining TESV, except districts Kalaburagi (48%), Bangalore urban (10%), Ramnagara (6%), Chitradurga (4%), and Haveri (2%), which show an increase in TESV.

Higher NPV values in the Western Ghats districts—Uttara Kannada (NPV: 11,885 billion), Chikmagalur (5,875), Chamarjanagar (5,858), Dakshina Kannada (5,205), Shimoga (5,062), Udupi (3,787), Kodagu (3,721), Belagavi (3,445), and Mysore (2,527) highlight the role of a forest ecosystem with native species of vegetation in supporting rich endemic biodiversity, sustaining water availability during all seasons to meet biotic demands, and supporting the livelihood of people. Understanding these linkages would help the planners/decision-makers with valuable knowledge for integrated ecosystem management. The study highlights the vital ecological function of the Western Ghats, one among 36 global biodiversity hotspots, in sustaining the hydrologic regime and livelihood of local people. Hence, the premium should be towards conserving the forests with native species to sustain water and biotic diversity, which are vital for food security. There is still a chance to restore the lost natural evergreen to semi-evergreen forests through appropriate conservation and management practices in Karnataka State.

Table 74 Changes in the provisioning, regulating, cultural services and TESV (monetary values) in Karnataka from 2005 to 2019

District	Ecosystem services			TESV
	Provisioning (%)	Regulating (%)	Cultural (%)	Billions of rupees (%)
Bagalkot	39	−19	0	10
Bangalore Rural	−38	−18	6	−27
Bangalore Urban	−54	−35	−28	−45
Belagavi	−3	−29	−12	−15
Bellary	−49	−38	−16	−44
Bidar	13	−34	−12	−10
Vijayapura	−10	−23	−7	−16
Chamarajnagar	−54	−29	−1	−31
Chikmagalur	−21	−7	−1	−14
Chikkaballapura	−21	−19	11	−19
Chitradurga	−33	**54**	72	4
Dakshina Kannada	−22	3	0	−9
Davanagere	−27	10	16	−11
Dharwad	1	−25	−2	−11
Gadag	−1	−14	12	−8
Hassan	−55	−16	−24	−34
Haveri	14	−10	23	2
Kalaburagi	**46**	42	**135**	**48**
Kodagu	−58	−30	−3	−37
Kolar	3	−8	30	0
Koppal	−26	−12	31	−20
Mandya	−57	−17	−2	−36
Mysore	−52	12	2	−27
Raichur	−28	−27	0	−26
Ramanagar	−26	46	85	6
Shimoga	−34	−24	−2	−26
Tumkur	−55	−32	−28	−44
Udupi	−15	5	1	−5
Uttara Kannada	−27	−15	−2	−18
Yadgiri	−11	9	74	0
Karnataka	−28	−14	0	−20

Table 75 TESV and share of TESV in GDDP (2019)

District	Revenue billion rupee				Ratio of TESV to GDDP		
	TESV-forest	TESV-agriculture	TESV-total	GDDP	Forest	Agriculture	Total
Bagalkot	19.7	59.7	79.4	265.5	7.4	22.5	29.9
Bangalore Rural	7.4	8.5	15.9	162.5	4.6	5.2	9.8
Bangalore Urban	7.1	4.3	11.4	3635.6	0.2	0.1	0.3
Belagavi	40.2	93.7	133.9	454.6	8.8	20.6	29.5
Bellary	12.8	35.2	48.0	334.8	3.8	10.5	14.3
Bidar	3.4	31.6	35.0	144.9	2.4	21.8	24.2
Vijayapura	31.5	72.5	104.0	188.1	16.8	38.5	55.3
Chamarajnagar	214.9	12.8	227.7	117.0	183.7	10.9	194.6
Chikmagalur	192.9	35.4	228.3	230.2	83.8	15.4	99.2
Chikkaballapura	13.7	16.4	30.1	144.4	9.5	11.4	20.9
Chitradurga	12.9	66.6	79.6	169.6	7.6	39.3	46.9
Dakshina Kannada	167.4	34.9	202.3	587.2	28.5	5.9	34.4
Davanagere	31.2	45.3	76.5	202.1	15.4	22.4	37.8
Dharwad	8.6	31.0	39.6	244.7	3.5	12.7	16.2
Gadag	9.4	36.7	46.1	109.1	8.6	33.6	42.3
Hassan	17.3	51.5	68.7	236.4	7.3	21.8	29.1
Haveri	9.8	35.2	44.9	155.3	6.3	22.7	28.9
Kalaburagi	6.9	69.7	76.6	195.1	3.5	35.7	39.3
Kodagu	128.9	15.7	144.6	61.6	209.3	25.5	234.8
Kolar	12.8	20.7	33.5	176.6	7.3	11.7	19.0
Koppal	8.4	31.0	39.4	118.9	7.1	26.0	33.1

(continued)

Table 75 (continued)

District	Revenue billion rupee				Ratio of TESV to GDDP		
	TESV-forest	TESV-agriculture	TESV-total	GDDP	Forest	Agriculture	Total
Mandya	44.4	29.6	73.9	267.3	16.6	11.1	27.7
Mysore	56.7	41.5	98.2	352.1	16.1	11.8	27.9
Raichur	5.3	31.3	36.6	173.5	3.1	18.0	21.1
Ramanagar	11.4	25.0	36.3	159.6	7.1	15.6	22.8
Shimoga	168.2	28.5	196.7	300.5	56.0	9.5	65.5
Tumkur	26.9	44.5	71.4	385.3	7.0	11.6	18.5
Udupi	122.9	24.3	147.2	276.3	44.5	8.8	53.3
Uttara Kannada	438.9	23.0	461.9	186.2	235.7	12.3	248.1
Yadgiri	3.3	21.6	25.0	93.4	3.6	23.2	26.7
Karnataka	1835.2	1077.6	2912.8	10128.4	18.12	10.64	28.76

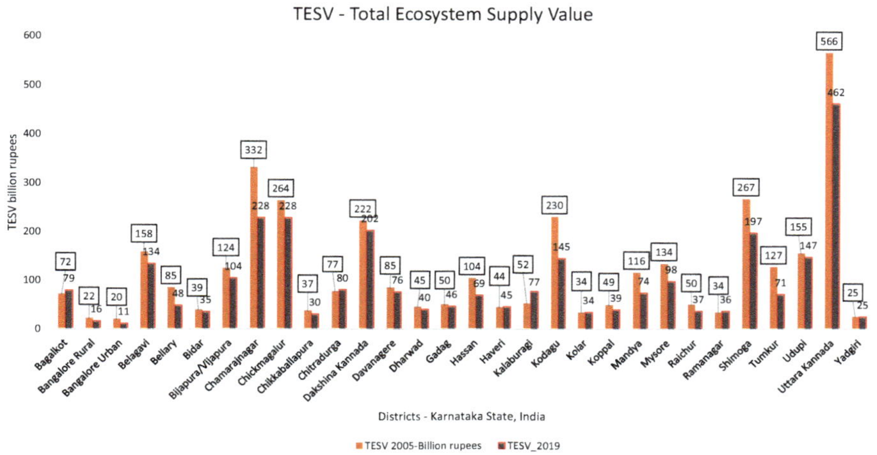

Fig. 63 Comparison of TSEV of 2005 (in 2019 rupees) and 2019

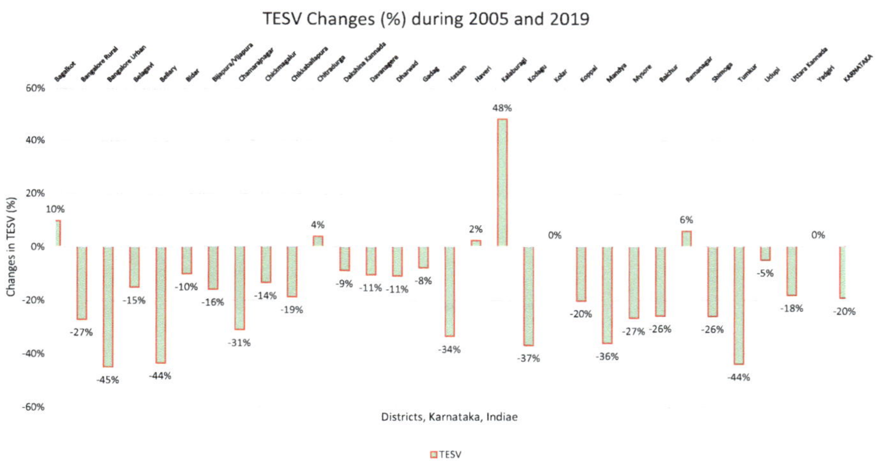

Fig. 64 Percentage changes in TESV—2005 (in 2019 rupees) and 2019

Figure 65 depicts the district-wise share of TESV in GDDP. The district-wise assessment indicates TESV of Bengaluru urban and rural districts are the lowest, amounting to <10% of GDDP. In contrast, forest-rich Western Ghats districts (mainly Uttara Kannada of Canara forest circle, Kodagu of Kodagu forest circle, and Chamarajanagar and Chamarajanagar circle) provide TESV, which is about 200% of the respective district GDDP. The presence of rich forests in these districts contributes to higher TESV, highlighting that TESV share in GDDP (Gross District Domestic Product) is correlated with the extent and conditions of forest ecosystems in the respective districts.

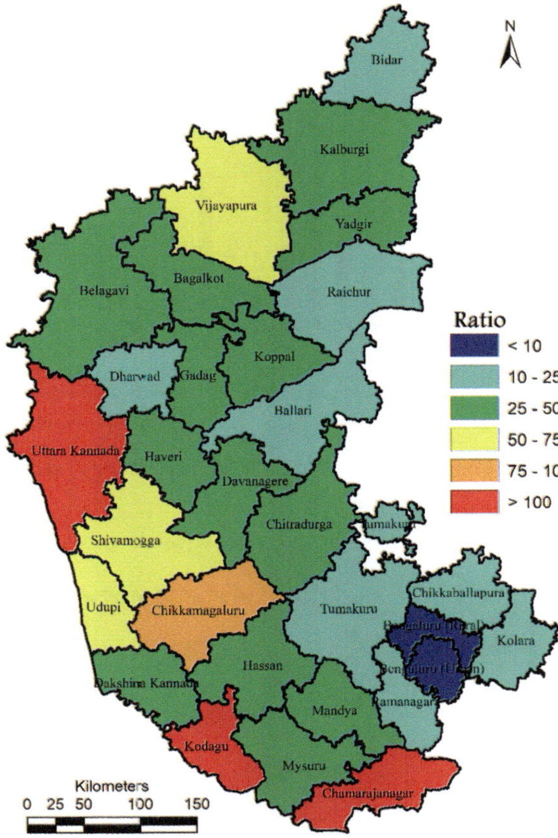

Fig. 65 TESV to GDDP ratio

Table 76 lists TESV or GEP for Karnataka considering forest, and agriculture (croplands and horticulture) ecosystems. The TESV of these ecosystems is 3620 billion rupees in 2005 (forest ecosystem: 2,841 billion rupees and agriculture (croplands and horticulture) ecosystem: 779 billion rupees). Similarly, TESV computed for 2019 indicates 2,912 billion rupees (forest ecosystem: 1,835 billion rupees and agriculture 1,077 billion rupees). There has been a 35.4% reduction in TESV of forest ecosystems from 2005 to 2019, mainly due to the degradation of ecosystems.

The decline of TESV highlights the degradation of forest ecosystem assets from 2005 to 2019 due to the reduction of ecosystem extent and condition. The decrease in value is also demonstrated by a fall in the net present value of expected future returns of the ecosystem services supplied by forest ecosystem assets, as shown through the ecosystem monetary asset account.

Table 76 Comparison of provisioning, regulating, and cultural services and TESV during 2005 (in 2019 rupees) and 2019

Ecosystems	Year	Units	Provisioning	Regulating	Cultural	TESV
Forests	2005	Million ₹	12,67,528	12,70,583	3,03,034	28,41,145
		%	44.6	44.7	10.7	100
Agriculture (croplands and horticulture)		Million ₹	4,11,834	3,44,933	21,819	778,586
		%	52.9	44.3	2.8	100
Total		Million ₹	16,79,361	16,15,516	3,24,854	**36,19,731**
		%	46.4	44.6	9.0	100
Forests	2019	Million ₹	6,13,883	9,26,346	2,94,955	18,35,184
		%	33.5	50.5	16.1	100
Agriculture		Million ₹	5,89,283	4,59,037	29,305	10,77,625
		%	61.2	36.3	2.5	100
Total		Million ₹	12,03,166	13,85,383	3,24,260	**29,12,809**
		%	41.3	47.6	11.1	100

6.3 *Net Present Value (NPV) of Ecosystem Assets*

The ***net present value (NPV)*** of ecosystem assets was determined by considering the stream of income expected to be earned in the future and then discounting the future income back to the present accounting period (SEEA Central Framework, para. 5.11). In ecosystem accounting, it is applied by aggregating the NPV of expected future returns for each ecosystem service supplied by an ecosystem asset. Table 77 provides a monetary asset account (2005–2019). The NPV of accounted ecosystems based on 2005 ecosystem flows is about 93,130 billion INR (forest ecosystem: 73,099 billion INR; agriculture (croplands and horticulture) ecosystem: 20,031 billion INR). The NPV of ecosystems in Karnataka based on 2019 flows is 74,938 billion INR (forest ecosystem: 47,214 billion INR; agriculture ecosystem: 27,724 billion INR). A decline of 35.4% in NPV of forest ecosystems is due to the transition of forest ecosystems to either croplands or horticulture (agriculture ecosystems), which correlates to an increase in NPV of agriculture ecosystems by 23%.

Figure 66 presents the district-wise NPV of forest and agriculture ecosystems based on the 2019 TESV, which totals about **74,938** billion rupees for Karnataka. This shows a decline of 19.5% from 2005, when the NPV of ecosystem assets in Karnataka was **93,130** billion rupees based on 2005 ecosystem service values (in 2019 rupees).

The study reveals that about 63% of TESV and NPV is contributed by the districts of central Western Ghats (Uttara Kannada (11,885 billion rupees), Chikmagalur (5,875 billion rupees), Chamarajnagar (5,858 billion rupees), Dakshina Kannada (5,205 billion rupees), Shimoga (5,062 billion rupees), Udupi (3,787 billion rupees),

Table 77 Monetary asset account (2005–2019)

	Units	Forest ecosystem	Agriculture ecosystem	Total NPV
Opening stock—2005 (at 2019 values)	Billion ₹	**73,099**	**20,031**	**93,130**
Changes (absolute)	Billion ₹	−25,885	7,693	−18,192
Changes	%	−35.4	38.4	−19.5
Provisioning	%	−51.6	43.1	−28.4
Regulating	%	−27.1	33.1	−14.2
Cultural	%	−2.7	34.3	−0.2
Closing stock—2019	Billion ₹	47,214	27,724	74,938

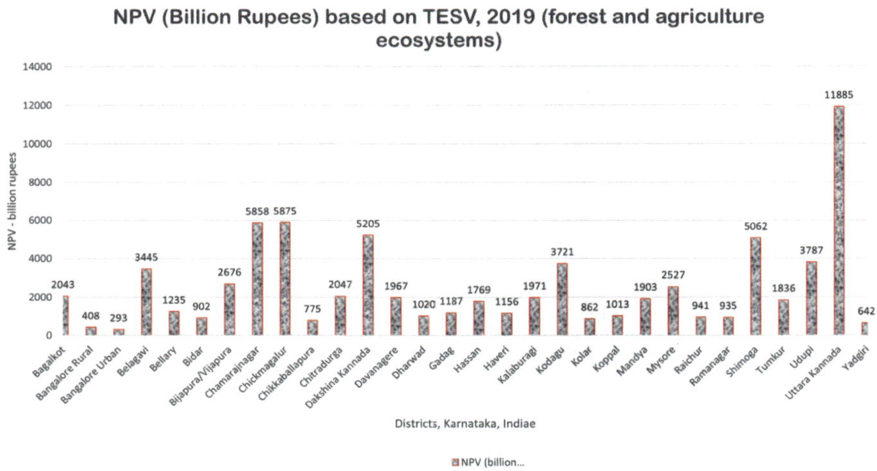

Fig. 66 NPV of ecosystem assets (forest and agriculture) based on TESV of 2019

Kodagu (3,721 billion rupees), Belagavi (3,445 billion rupees), and Mysore (2,527 billion rupees), again reinforcing the critical role of a forest ecosystem with native species of vegetation in providing critical ecosystem services.

7 Conclusion

The compilation of ecosystem extent, service, and asset accounts using the SEEA EA has enabled a thorough analysis of the changes in the provision of ecosystem services in Karnataka between 2005 and 2019. The comparison of the values of goods and services in 2019 with 2005 highlights there has been a considerable reduction in ecosystem services—a 28.5% reduction in provisioning services (including a 51.6% reduction in forest ecosystem), a 21% reduction in regulatory services (mainly in forest ecosystem—27.1% reduction), and a 1.9% reduction in cultural services. In terms of the reductions in provisioning services, these included a 93% decline in bamboo, a decline in NTFP (honey reduced by 97%, tamarind reduced by 75%), a 42% decline in fodder, and a 35% decline in medicine. The large decreases in provisioning and regulatory services can be attributed to the degradation of forests (extent and conditions) in Karnataka from 2005 to 2019.

Ecosystem services were aggregated to compute TESV. This aggregate measure is also referred to as gross ecosystem product (GEP), ***which is equal to the sum of all final ecosystem services (i.e., used by economic units) from ecosystem assets***. The TESV of ecosystems was 3620 billion INR in 2005 (forest ecosystem: 2,841 billion INR and agriculture ecosystem: 779 billion INR). However, the TESV computed for 2019 indicates 2,793 billion INR (forest ecosystem: 1,835 billion INR and agriculture 958 billion INR). While the TESV for agricultural ecosystems increased by 179 billion INR between 2005 and 2019, there was a much larger decrease in TESV for forest ecosystems, which amounted to 1,006 billion INR. This 35.4% reduction in TESV of forest ecosystems can again be attributed to the degradation of ecosystems.

The relationship between the SEEA EA and the System of National Accounts also means that TESV can be compared to GDP. The GDP of Karnataka is about 10,128 billion rupees. Therefore, the TESV of the forest ecosystem is equivalent to 18.1% of the GDP, and TESV from agriculture is equivalent to about 10.6% of GDP in Karnataka. The district-wise assessment indicates that the TESV of Bengaluru's urban and rural districts are the smallest, with <10% of GDDP (Gross District Domestic Product). In contrast, forest-rich Western Ghats districts (mainly Uttara Kannada, Kodagu, and Chamarajanagar) provide TESV, which is about 200% of the respective district GDDP. The presence of rich forests in these districts contributes to higher TESV, highlighting that TESV share in GDDP is correlated with the extent and conditions of forest ecosystems in the respective districts.

The decline of TESV highlights the degradation of forest ecosystem assets from 2005 to 2019 due to the deterioration of ecosystem extent and ecosystem condition. The decrease in value is also reflected in a fall in the NPV of expected future returns of the ecosystem services supplied by forest ecosystem assets. The NPV of forest and agriculture ecosystems based on 2005 ecosystem flows is about 93,130 billion INR (forest ecosystem: 73,099 billion INR; agriculture ecosystem: 20,031 billion INR). Similarly, the NPV of ecosystems in Karnataka based on 2019 flows indicates 74,938 billion INR (forest ecosystem: 47,214 billion INR, agriculture ecosystem: 27,724 billion INR). This indicates a decline of 35.4% in NPV of forest ecosystems, largely due to the transition of forest ecosystems to either croplands or horticulture (agriculture ecosystems). These ecosystem transitions have led to an increase in the NPV of agriculture ecosystems by 23% by 2019.

The drivers behind the land-use change and the decline of forest resources in Karnataka are mainly the expansion of agricultural activities coupled with industrialization and rapid urbanization. However, the increase in the values of agricultural TESV and NPV at the expense of a decrease in the values of forest TESV and NPV points to the need for an adequate assessment of trade-offs in land use policy. Hence, the current study emphasizes the need for the valuation of services of all ecosystems, capitalizing on the advances in geoinformatics, availability of spatial data at regular intervals to estimate the economic value of ecosystems forests, and, in particular, reflect the value of forests in policy decisions.

The overall assessment of the ecological services provides information for prioritizing ecosystems for sustainable land-use practices, promoting off-farm incomes to the dependent communities, restoring degraded sites, biodiversity conservation, water resources, etc., while promoting community-based prudent management of natural resources. There is a need to enhance awareness for the protection of the environment, especially the maintenance of native forest cover, which is crucial for balanced economic and social progress in the country. Over the last few decades, though India has evolved legislation, policies, and programs for environmental protection and conservation of natural resources, these policies and programs have been unevenly implemented. Thus, ecosystem accounting can play a role in two ways. First, ecosystem accounts can help policymakers factor in ecosystem service benefits when making economic policies that impact natural resources and ecosystems. Second, ecosystem accounts, which are regularly compiled, can be used to help monitor the impact of these policies over time and ensure that they are being implemented properly.

Finally, it should be noted that the ecosystem accounts compiled for Karnataka have a large potential to be used for payment for ecosystem services schemes. The Supreme Court of India (2006) directed the national government to set up compensatory payments for the conversion of different types of forested land to non-forest uses and use these payments to improve India's forest cover. The SEEA EA accounts can provide important information on the values of ecosystems and their services, which can help create transparent criteria with which to reward states. Afforestation in the degraded landscape would aid in mitigating changes in the climate due to global warming while sustaining people's livelihood through (i) provision of ecosystem services, (ii) improvements in the crop yield, ii) sustenance of water in the landscape, etc.

8 Recommendations

The ecosystem services computed for Karnataka State support the viability of markets for particular ecosystem services. Developing such markets requires additional institutional reforms, such as changes with respect to property rights and reforms in land and labor markets. Hence, ecosystem services need to be internalized in decision-making, strengthening the economic case for conserving forests in all states in India and developing countries, as there is great pressure to relax forest laws and divert forests to non-forest uses with the illusion of boosting long-term economic growth. The main policy challenge is to promote conservation and develop such markets so that those bearing the cost of conservation are adequately compensated. **The valuation of ecosystem services done in Karnataka State and replicating this exercise in other states will undoubtedly play a vital role in conservation planning and ecosystem-based management in India**. This requires:

(i) Strengthening biophysical research on ecosystem services, with a focus on those that would seem to have the highest economic value potential (e.g., changes in the climatic, hydrologic regime, etc.);
(ii) Inventorying, mapping, and monitoring ecosystems' spatial extent and conditions through the use of advanced spatial technologies with temporal remote sensing data;
(iii) Promoting valuation studies reveals current incentives, i.e., the existing distribution of net ecosystem benefits/opportunity costs across stakeholders, which will aid in internalizing the regional policies, and
(iv) Developing land-use policies considering ecosystem services across different ecosystem types.

The exercise of compiling physical and monetary SEEA EA could be replicated in any region, with a multidiscipline team of experts. It requires (i) all para-state agencies sharing biophysical data, as primary data collection is a time-consuming endeavor, (ii) training programmes and workshops—orientation programs, (iii) hands-on training to enhance the capability of the team to undertake spatial analyses, (iv) collecting biophysical data from the government agencies and in the field, (v) performing data integration and validation, and conducting analyses of the data and interpretation, and (vi) addressing gaps in existing biophysical models (i.e., adapting them to the local context).

Extending this exercise in Karnataka or other states could help evolve strategies to conserve ecosystems to support people's livelihood. As illustrated in this chapter, ecosystem accounts can provide insights into the social, economic, and environmental benefits of various levels of biodiversity that might be achieved under different ecosystem management options at various scales. The economic valuation of forest ecosystem services and biodiversity can help clarify trade-offs among conflicting environmental, social, and economic goals in the development and implementation of policies and to improve management in order to sustain biodiversity. At the same time, there is a need to communicate the research results on these issues more effectively to decision-makers and other stakeholders.

Annexure A Valuation of Forest Ecosystem Services (Provisioning, Regulating and Cultural) at the District Level

Ecosystem services (Provisioning services, regulating services, and cultural services) were computed district-wise by extracting the statistics at the district level by overlaying the district boundaries over the circle-wise layers. Tables 78 to 80 list the provisioning, regulating, and cultural services of 2005. Similarly, Tables 81 to 83 provisioning, regulating, and cultural services during 2019. The spatial variations in the provisioning, regulating, and cultural services district-wise during 2005 and 2019 are depicted in Fig. 67.

Total Ecosystem Supply Value (TESV) is computed by aggregating provisioning services, regulating, and cultural services, which accounts for 2841 billion INR in 2005 and 1835 billion INR in 2019 (Table 84). The decline of 35% in TESV is mainly due to the degradation of the ecosystem (spatial extent as well as conditions. Provisioning services have declined from 1268 (2005) to 614 (2004) billion INR; regulating services declined from 1271 (2005) to 926 (2019) and cultural services from 303 (2005) to 295 (2019). Figure 68 illustrates the variability across the region. Districts such as Uttara Kannada, Chikmagalur, Dakshina Kannada, Shimoga, Kodagu, and Chamarajanagar contribute significantly to TESV. The district-wise assessment indicates Uttara Kannada has the highest TESV, over 439 billion rupees, followed by Chamarajanagar (215 billion), Chikmagalur (193 billion), Shimoga (168 billion), and Dakshina Kannada (167 billion). The lowest was observed in Yadgir, Bidar, Raichur districts.

Table 78 Provisioning services of forest ecosystem (district wise) 2005

S.no	District	Provisioning services (Million Rupees)										
		Timber	Bamboo	Canes	NTFP	Fish	Fuelwood	Fodder	Medicine	Water Supply	Genetic Material	Total
1	Bagalkot	5.6	2.6	0.0	462.9	71.3	908.7	1292.4	13.3	11362.6	7811.6	21928.5
2	Ballari	1.0	0.0	0.0	2.8	420.9	995.3	1436.3	21.6	20285.7	5266.4	28430.0
3	Belgaum	11.5	5.3	0.0	944.6	145.5	1854.2	2637.3	27.2	23186.4	15940.2	44747.0
4	Bengaluru (Rural)	1.9	0.0	0.0	62.0	136.9	369.1	545.4	5.7	5351.5	1389.3	7861.8
5	Bengaluru (Urban)	1.8	0.0	0.0	59.2	130.6	352.3	520.6	5.4	5107.7	1326.0	7503.6
6	Bidar	1.7	0.0	0.0	3.3	119.3	321.7	419.2	5.2	3767.1	978.0	5615.5
7	Chamarajanagar	0.1	0.0	0.0	4754.3	224.5	3911.9	6330.4	63.6	71235.5	28465.9	114986.1
8	Chikkaballapura	3.5	0.0	0.0	114.5	252.8	681.9	1007.5	10.5	9885.2	2566.3	14522.1
9	Chikmagalur	139.1	3.8	0.0	3596.0	327.5	4151.3	6026.3	59.5	54016.8	46673.1	114989.6
10	Chitradurga	1.4	0.0	0.0	19.0	420.4	998.2	1439.6	21.6	20289.6	5295.1	28485.0
11	Dakshina Kannada	53.5	0.5	0.0	7072.8	113.0	2482.1	3717.2	40.1	39225.6	33897.1	86601.4
12	Davanagere	77.5	2.7	0.0	2960.5	285.8	1268.7	1668.7	22.3	16988.9	7844.1	31116.5
13	Dharwad	47.6	2.0	0.0	19.9	165.0	463.0	665.7	7.5	6807.9	2720.4	10896.9
14	Gadag	52.0	2.2	0.0	21.7	180.2	505.8	727.2	8.2	7437.0	2971.8	11903.9
15	Hassan	32.8	0.0	0.0	110.7	560.9	1225.7	1643.5	16.7	10652.5	4256.8	18499.5
16	Haveri	53.8	2.2	0.0	22.5	186.4	523.3	752.3	8.5	7694.3	3074.6	12315.8
17	Kalburgi	3.5	0.0	0.0	6.6	240.3	648.0	844.3	10.5	7588.4	1970.0	11311.7
18	Kodagu	273.9	0.2	0.0	29517	76.5	2985.0	3513.6	48.5	24506.2	21177.2	82098.0
19	Kolara	3.2	0.0	0.0	107.4	237.1	639.6	945.0	9.9	9272.0	2407.1	13621.3
20	Koppal	0.7	0.0	0.0	1.8	276.8	654.4	944.4	14.2	13337.7	3462.6	18692.6
21	Mandya	54.7	0.9	0.0	28711	418.4	875.3	1349.9	13.2	13379.0	5346.3	50148.2

(continued)

Table 78 (continued)

S.no	District	Provisioning services (Million Rupees)										
		Timber	Bamboo	Canes	NTFP	Fish	Fuelwood	Fodder	Medicine	Water Supply	Genetic Material	Total
22	Mysore	69.9	1.1	0.0	36689	534.6	1118.6	1724.9	16.8	17096.6	6831.8	64082.5
23	Raichur	2.7	0.0	0.0	5.1	185.5	500.1	651.6	8.1	5856.7	1520.5	8730.4
24	Ramanagara	2.9	0.0	0.0	95.1	209.9	566.1	836.4	8.7	8206.9	2130.6	12056.6
25	Shimoga	537.4	18.7	0.0	20664	357.4	4988.8	6067.3	71.8	39722.1	34314.9	106723.7
26	Tumkur	51.1	0.0	0.0	172.3	873.1	1907.9	2558.4	25.9	16582.5	6626.4	28797.6
27	Udupi	39.3	0.3	0.0	5194.1	83.0	1822.8	2729.8	29.4	28806.1	24893.0	63597.4
28	Uttara Kannada	488.2	18.6	0.3	401.4	380.4	9365.8	11434	152.3	98965.3	85521.5	206709.4
29	Vijayapura	9.0	4.2	0.0	741.1	114.2	1454.7	2069.1	21.4	18190.9	12505.9	35106.2
30	Yadgir	1.7	0.0	0.0	3.2	115.8	312.1	406.7	5.1	3655.3	949.0	5448.8
Total		2023	65	0.3	142,537	7844	48,852	66,906	772	6,18,460	3,80,133	12,67,528

Table 79 Regulating services of forest ecosystem (district wise) 2005

S.no	District	Regulating services (Million rupees)									
		Air filtration	Local climate regulation	Global climate regulation/ carbon Seq	Pollination	Erosion prevention	Soil fertility	Water purification	Waste treatment	Ground water	Total
1	Bagalkot	1373.9	2925.7	11477.9	727.6	965.8	852.0	178.1	252.1	204.7	18957.8
2	Ballari	2220.1	4727.6	12907.9	991.7	1262.3	947.7	287.7	407.3	176.7	23929.1
3	Belgaum	2803.6	5970.1	23421.7	1484.8	1970.7	1738.7	363.4	514.4	417.7	38685.1
4	Bengaluru Rural	585.9	1247.6	5244.3	261.7	333.1	392.8	75.9	107.5	46.6	8295.5
5	Bengaluru Urban	559.2	1190.8	5005.4	249.8	317.9	374.9	72.5	102.6	44.5	7917.5
6	Bidar	539.1	1147.9	3448.2	240.8	284.7	410.3	69.9	98.9	39.9	6279.6
7	Chamarajanagar	6550.1	13948	77250.3	3035.5	5952.0	12785.1	848.9	1201.8	815.6	122387.4
8	Chikkaballapura	1082.2	2304.5	9687.3	483.4	615.3	725.5	140.3	198.6	86.1	15323.2
9	Chikmagalur	6130.0	13054	72554.8	3206.7	4804.5	5210.2	794.5	1124.7	1067.6	107946.3
10	Chitradurga	2222.8	4733.4	12959.9	993.5	1264.9	955.1	288.1	407.8	177.5	24003.0
11	Dakshina Kannada	4127.4	8789.1	47077.7	2159.2	3310.1	5739.1	534.9	757.3	737.7	73232.5
12	Davanagere	2295.9	4889.0	18167.4	1106.5	1437.5	2041.3	297.6	421.2	258.8	30915.2
13	Dharwad	773.9	1648.1	6528.7	358.7	657.5	438.7	100.3	142.0	91.3	10739.2
14	Gadag	845.5	1800.4	7132.1	391.8	718.2	479.3	109.6	155.1	99.7	11731.6
15	Hassan	1715.5	3653.0	13954.7	766.3	880.8	1531.8	222.3	314.7	123.4	23162.7
16	Haveri	874.7	1862.6	7378.8	405.4	743.1	495.9	113.4	160.5	103.1	12137.4
17	Kalburgi	1085.9	2312.4	6945.8	485.1	573.5	826.5	140.7	199.2	80.3	12649.4
18	Kodagu	4998.2	10643	52770.8	2614.7	3404.7	10429.2	647.8	917.0	741.9	87167.8

(continued)

Table 79 (continued)

S.no	District	Regulating services (Million rupees)									
		Air filtration	Local climate regulation	Global climate regulation/ carbon Seq	Pollination	Erosion prevention	Soil fertility	Water purification	Waste treatment	Ground water	Total
19	Kolara	1015.1	2161.6	9086.3	453.4	577.1	680.5	131.6	186.2	80.8	14372.7
20	Koppal	1459.7	3108.4	8486.9	652.0	830.0	623.1	189.2	267.8	116.2	15733.3
21	Mandya	1355.8	2887.1	14232.8	628.3	1193.1	2416.1	175.7	248.8	164.5	23302.2
22	Mysore	1732.5	3689.3	18187.6	802.9	1524.6	3087.5	224.5	317.9	210.2	29777.0
23	Raichur	838.1	1784.7	5360.8	374.4	442.6	637.9	108.6	153.8	62.0	9762.8
24	Ramanagara	898.5	1913.3	8042.6	401.3	510.8	602.3	116.5	164.8	71.5	12721.6
25	Shimoga	7396.3	15750	76679.9	3869.1	5128.6	10569.6	958.6	1357.0	1120.4	122829.4
26	Tumkur	2670.5	5686.6	21722.9	1192.9	1371.1	2384.5	346.1	490.0	192.2	36056.7
27	Udupi	3031.1	6454.5	34572.4	1585.6	2430.8	4214.6	392.8	556.1	541.8	53779.7
28	Uttara Kannada	15682.3	33395	182897.2	8203.9	11262.1	21520.9	2032.6	2877.3	2473.1	280343.9
29	Vijayapura	2199.6	4683.9	18375.5	1164.9	1546.1	1364.1	285.1	403.6	327.7	30350.3
30	Yadgir	523.1	1113.9	3345.8	233.7	276.2	398.1	67.8	96.0	38.7	6093.2
Total		79,587	1,69,475	7,94,904	39,525	56,589	94,873	10,315	14,601	10,712	127,058

Table 80 Cultural services of forest ecosystem (district wise) 2005

S.no	District	Cultural services (million rupees)					
		Aesthetic	Tourism & recreational	Spiritual	Art & cultural	Education, scientific and research	Total
1	Bagalkot	65.2	70.9	0.0	29.0	289.7	454.8
2	Ballari	116.4	428.9	0.0	46.8	468.2	1060.4
3	Belgaum	133.1	144.7	0.0	59.1	591.2	928.1
4	Bengaluru (Rural)	30.7	108.3	0.0	12.4	123.6	274.9
5	Bengaluru (Urban)	29.3	103.3	0.0	11.8	117.9	262.4
6	Bidar	21.6	0.0	0.0	11.4	113.7	146.7
7	Chamarajanagar	408.8	78816.2	0.0	138.1	1381.3	80744.5
8	Chikkaballapura	56.7	200.0	0.0	22.8	228.2	507.8
9	Chikmagalur	310.0	15864.3	19.2	129.3	1292.7	17615.5
10	Chitradurga	116.4	446.8	0.1	46.9	468.8	1079.0
11	Dakshina Kannada	225.1	28346.1	15.1	87.0	870.4	29543.8
12	Davanagere	97.5	2953.4	12.2	48.4	484.2	3595.7
13	Dharwad	39.1	58.7	0.0	16.3	163.2	277.3
14	Gadag	42.7	64.2	0.0	17.8	178.3	303.0
15	Hassan	61.1	0.0	0.0	36.2	361.8	459.1
16	Haveri	44.2	66.4	0.0	18.4	184.5	313.5
17	Kalburgi	43.6	0.0	0.0	22.9	229.0	295.5
18	Kodagu	140.6	31629.7	18.0	105.4	1054.1	32947.8
19	Kolara	53.2	187.6	0.0	21.4	214.1	476.3
20	Koppal	76.5	282.0	0.0	30.8	307.8	697.2
21	Mandya	76.8	13310.7	0.0	28.6	285.9	13702.0
22	Mysore	98.1	17009.3	0.0	36.5	365.4	17509.3
23	Raichur	33.6	0.0	0.0	17.7	176.7	228.0
24	Ramanagara	47.1	166.0	0.0	18.9	189.5	421.6
25	Shimoga	228.0	18966.1	85.2	156.0	1559.8	20995.1
26	Tumkur	95.2	0.0	0.0	56.3	563.2	714.7
27	Udupi	165.3	20816.5	11.1	63.9	639.2	21696.1
28	Uttara Kannada	568.0	50669.9	38.2	330.7	3307.2	54914.0
29	Vijayapura	104.4	113.5	0.0	46.4	463.9	728.2
30	Yadgir	21.0	0.0	0.0	11.0	110.3	142.3
Total		3,549	280,824	199	1,679	16,784	3,03,034

Table 81 Provisioning services of forest ecosystem (district wise) 2019

S.no	District	Provisioning services (Million rupees)										
		Timber	Bamboo	Canes	NTFP	Fish	Fuelwood	Fodder	Medicine	Water supply	Genetic material	Total
1	Bagalkot	3.2	0.0	0.0	17.7	184.6	1341.8	824.3	6.9	3400.6	2337.8	8117.6
2	Ballari	0.0	0.0	0.0	0.3	466.7	607.5	325.0	7.1	1097.5	284.6	2789.4
3	Belgaum	6.6	0.0	0.0	36.1	376.8	2738.1	1682.1	14.1	6939.2	4770.5	16564.7
4	Bengaluru (Rural)	1.8	0.0	0.0	5.4	142.5	330.7	180.6	3.0	1245.9	323.4	2233.7
5	Bengaluru (Urban)	1.8	0.0	0.0	5.1	136.0	315.6	172.3	2.8	1189.2	308.7	2131.9
6	Bidar	1.3	0.0	0.0	0.2	319.5	193.3	107.2	1.8	303.5	78.7	1005.7
7	Chamarajanagar	0.0	0.0	0.0	1394.0	131.0	5918.0	5023.1	41.0	25225.0	10080.0	47814.0
8	Chikkaballapura	3.4	0.0	0.0	9.9	263.2	610.8	333.5	5.5	2301.5	597.5	4126.0
9	Chikmagalur	129.1	0.6	0.0	1248.0	427.0	4978.1	3435.1	50.5	37697.9	32576.8	80547.0
10	Chitradurga	0.3	0.0	0.0	6.6	466.5	611.2	327.9	7.2	1123.7	308.6	2852.6
11	Dakshina Kannada	53.6	0.5	0.0	2.3	140.7	2710.5	2019.3	37.5	33609.3	29044.0	67619.4
12	Davanagere	45.9	0.0	0.0	1158.3	385.8	1011.8	672.9	11.3	3893.5	2994.6	10174.8
13	Dharwad	22.0	0.0	0.0	1.2	230.6	376.7	229.0	3.4	1379.6	551.1	2794.6
14	Gadag	24.1	0.0	0.0	1.4	251.9	411.5	250.1	3.7	1507.1	602.0	3052.9
15	Hassan	24.6	0.0	0.0	0.8	474.8	1672.5	1018.1	7.4	1462.4	584.4	5246.3
16	Haveri	24.9	0.0	0.0	1.4	260.6	425.8	258.8	3.9	1559.3	622.9	3158.5
17	Kalburgi	2.5	0.0	0.0	0.3	643.7	389.5	216.0	3.6	611.3	158.5	2025.9
18	Kodagu	152.0	0.0	0.0	1391.0	252.0	4413.0	2970.0	33.0	12869.0	11120.0	33203.0
19	Kolara	3.2	0.0	0.0	9.3	246.9	572.9	312.8	5.1	2158.7	560.4	3870.0
20	Koppal	0.0	0.0	0.0	0.2	306.9	399.4	213.7	4.7	721.6	187.1	1834.0
21	Mandya	35.1	0.0	0.0	188.3	709.0	1464.5	1154.7	8.8	6743.6	2694.6	13000.3

(continued)

Table 81 (continued)

S.no	District	Provisioning services (Million rupees)										
		Timber	Bamboo	Canes	NTFP	Fish	Fuelwood	Fodder	Medicine	Water supply	Genetic material	Total
22	Mysore	44.9	0.0	0.0	240.7	906.0	1871.5	1475.6	11.2	8617.4	3443.4	16612.7
23	Raichur	2.0	0.0	0.0	0.2	496.8	300.6	166.7	2.8	471.8	122.4	1563.6
24	Ramanagara	2.8	0.0	0.0	8.2	218.5	507.1	276.9	4.6	1910.7	496.0	3425.5
25	Shimoga	320.7	0.1	0.0	8087.6	878.2	4704.6	3436.9	51.2	22946.5	19829.2	60257.3
26	Tumkur	38.4	0.0	0.0	1.2	739	2603.5	1584.8	11.6	2276.6	909.6	8166.7
27	Udupi	39.4	0.3	0.0	1.7	103.3	1990.5	1482.9	27.5	24681.7	21329.0	49657.6
28	Uttara Kannada	455.0	3.2	0.0	54.0	1360.0	11499.0	7487.7	130.0	67098.0	57984.0	146073.0
29	Vijayapura	5.2	0.0	0.0	28.3	295.6	2148.1	1319.7	11.0	5444.2	3742.7	12995.8
30	Yadgir	1.2	0.0	0.0	0.1	310.0	187.6	104.1	1.8	294.5	76.4	975.9
Total		1445	4.8	0.1	13,900	12,124	57,306	39,062	514	2,80,781	2,08,719	6,13,890

Table 82 Regulating services of forest ecosystem (district wise) 2019

S.no	District	Regulating Services (Million Rupees)									
		Air filtration	Local climate regulation	Global climate regulation/ carbon Seq	Pollination	Erosion prevention	Soil fertility	Water purification	Waste treatment	Ground water	Total
1	Bagalkot	724.5	1542.8	6934.6	383.7	476.6	937.4	93.7	132.9	93.5	11320.6
2	Ballari	751.0	1599.3	4768.1	335.3	318.2	1340.5	97.3	137.7	44.4	9393.5
3	Belgaum	1478.5	3148.2	14150.6	783.0	972.5	1912.9	191.2	271.3	190.8	23100.7
4	Bengaluru (Rural)	311.4	663.3	3058.2	139.1	146.9	554.3	40.3	57.0	20.5	4991.8
5	Bengaluru (Urban)	297.2	633.1	2918.9	132.7	140.2	529.1	38.5	54.4	19.6	4764.4
6	Bidar	186.1	396.6	1246.5	83.0	79.4	310.7	24.0	34.1	11.0	2372.7
7	Chamarajanagar	4234.0	9017.0	50800.0	1962.0	3255.0	16074.0	548.0	776.0	461.0	87131.0
8	Chikkaballapura	575.2	1225.2	5649.1	256.9	271.3	1024.0	74.5	105.3	37.9	9220.8
9	Chikmagalur	5241.6	11162.5	63289.3	2742.2	3885.4	6246.0	679.4	961.1	856.9	95067.9
10	Chitradurga	755.0	1607.7	4816.5	337.5	321.3	1346.5	97.8	138.4	45.2	9467.6
11	Dakshina Kannada	3886.7	8277.4	44942.0	2033.0	3029.3	6222.4	503.4	712.7	672.3	70282.6
12	Davanagere	1177.1	2506.5	10221.2	583.6	682.5	2083.3	152.5	215.8	134.5	17758.4
13	Dharwad	369.3	786.3	3378.3	171.0	260.3	571.0	47.8	67.6	37.5	5690.4
14	Gadag	403.4	859.0	3690.5	186.8	284.4	623.7	52.2	73.9	41.0	6216.3
15	Hassan	774.4	1649.0	7166.4	345.8	334.4	1258.3	100.1	142.0	46.9	11818.9
16	Haveri	417.3	888.7	3818.2	193.2	294.2	645.3	54.0	76.5	42.4	6431.3

(continued)

Table 82 (continued)

S.no	District	Regulating Services (Million Rupees)									
		Air filtration	Local climate regulation	Global climate regulation/ carbon Seq	Pollination	Erosion prevention	Soil fertility	Water purification	Waste treatment	Ground water	Total
17	Kalburgi	374.9	798.9	2511.0	167.3	160.0	625.8	48.4	68.7	22.2	4779.4
18	Kodagu	3456.0	7361.0	37358.0	1808.0	2247.0	9402.0	448.0	634.0	486.0	63203.0
19	Kolara	539.6	1149.2	5298.7	241.0	254.5	960.5	69.9	98.8	35.5	8648.8
20	Koppal	493.8	1051.5	3135.0	220.4	209.2	881.4	63.9	90.5	29.2	6176.2
21	Mandya	924.6	1969.4	10631.9	428.5	745.9	2701.7	119.8	169.5	104.5	17797.4
22	Mysore	1181.4	2516.6	13586.1	547.5	953.1	3452.3	153.2	216.5	133.5	22742.6
23	Raichur	289.4	616.6	1938.0	129.1	123.5	483.0	37.3	53.0	17.1	3688.7
24	Ramanagara	477.6	1017.2	4690.0	213.3	225.2	850.1	61.8	87.4	31.5	7655.3
25	Shimoga	5300.5	11287.1	52866.1	2772.7	3530.2	9335.1	686.8	972.0	766.7	87520.3
26	Tumkur	1205.6	2567.0	11155.6	538.2	520.6	1958.7	155.9	221.0	73.1	18398.1
27	Udupi	2854.3	6078.6	33004.0	1493.0	2224.7	4569.6	369.6	523.3	493.7	51613.4
28	Uttara Kannada	13416.0	28570.0	151785.0	7018.0	9174.0	22512.0	1738.0	2461.0	2000.0	238678
29	Vijayapura	1160.0	2469.9	11101.8	614.3	763.0	1500.7	150.0	212.8	149.7	18123.6
30	Yadgir	180.6	384.8	1209.5	80.6	77.1	301.5	23.3	33.1	10.7	2302.2
Total		53,437	1,13,801	5,71,119	26941	35,959	101,214	6921	9799	7109	926,356

Table 83 Cultural services of forest ecosystem (district wise) 2019

S.no	District	Cultural Services (Million rupees)					
		Aesthetic	Tourism & recreational	Spiritual	Art & cultural	Education, scientific and research	Total
1	Bagalkot	19.39	68.51	0	15.28	152.80	255.95
2	Ballari	6.21	428.80	0	15.84	158.41	609.33
3	Belgaum	39.57	139.80	0	31.18	311.80	522.29
4	Bengaluru (Rural)	7.07	104.56	0	6.57	65.69	184.07
5	Bengaluru (Urban)	6.75	99.80	0	6.27	62.70	175.69
6	Bidar	1.62	0.00	0	3.93	39.29	44.95
7	Chamarajanagar	144.00	78816.00	0	89.31	893.05	79943.00
8	Chikkaballapura	13.07	193.14	0	12.13	121.35	340.02
9	Chikmagalur	216.04	15863.63	16.53	110.55	1105.51	17312.99
10	Chitradurga	6.36	446.70	0.07	15.92	159.25	628.37
11	Dakshina Kannada	192.58	28345.75	12.68	81.98	819.76	29453.36
12	Davanagere	22.21	2953.28	12.20	24.83	248.26	3260.88
13	Dharwad	7.76	56.48	0	7.79	77.89	150.19
14	Gadag	8.47	61.69	0	8.51	85.09	164.07
15	Hassan	8.21	0.00	0	16.33	163.34	187.74
16	Haveri	8.77	63.83	0	8.80	88.03	169.74
17	Kalburgi	3.27	0.00	0	7.91	79.14	90.55
18	Kodagu	73.00	31629.00	18.00	72.90	729.03	32523.00
19	Kolara	12.26	181.16	0	11.38	113.82	318.93
20	Koppal	4.09	281.93	0	10.42	104.16	400.63
21	Mandya	38.63	13310.72	0	19.51	195.07	13563.59
22	Mysore	49.37	17009.28	0	24.93	249.27	17332.41
23	Raichur	2.53	0.00	0	6.11	61.08	69.88
24	Ramanagara	10.85	160.35	0	10.07	100.75	282.29
25	Shimoga	131.06	18965.96	85.20	111.79	1117.88	20412.40
26	Tumkur	12.79	0.00	0	25.43	254.27	292.26
27	Udupi	141.42	20816.25	9.32	60.20	602.01	21629.64
28	Uttara Kannada	385.00	50669.00	25.00	282.95	2829.46	54193.00
29	Vijayapura	31.04	109.68	0	24.46	244.62	409.76
30	Yadgir	1.58	0	0	3.81	38.12	43.62
Total		1605	2,80,775	179	1127	11271	2,94,965

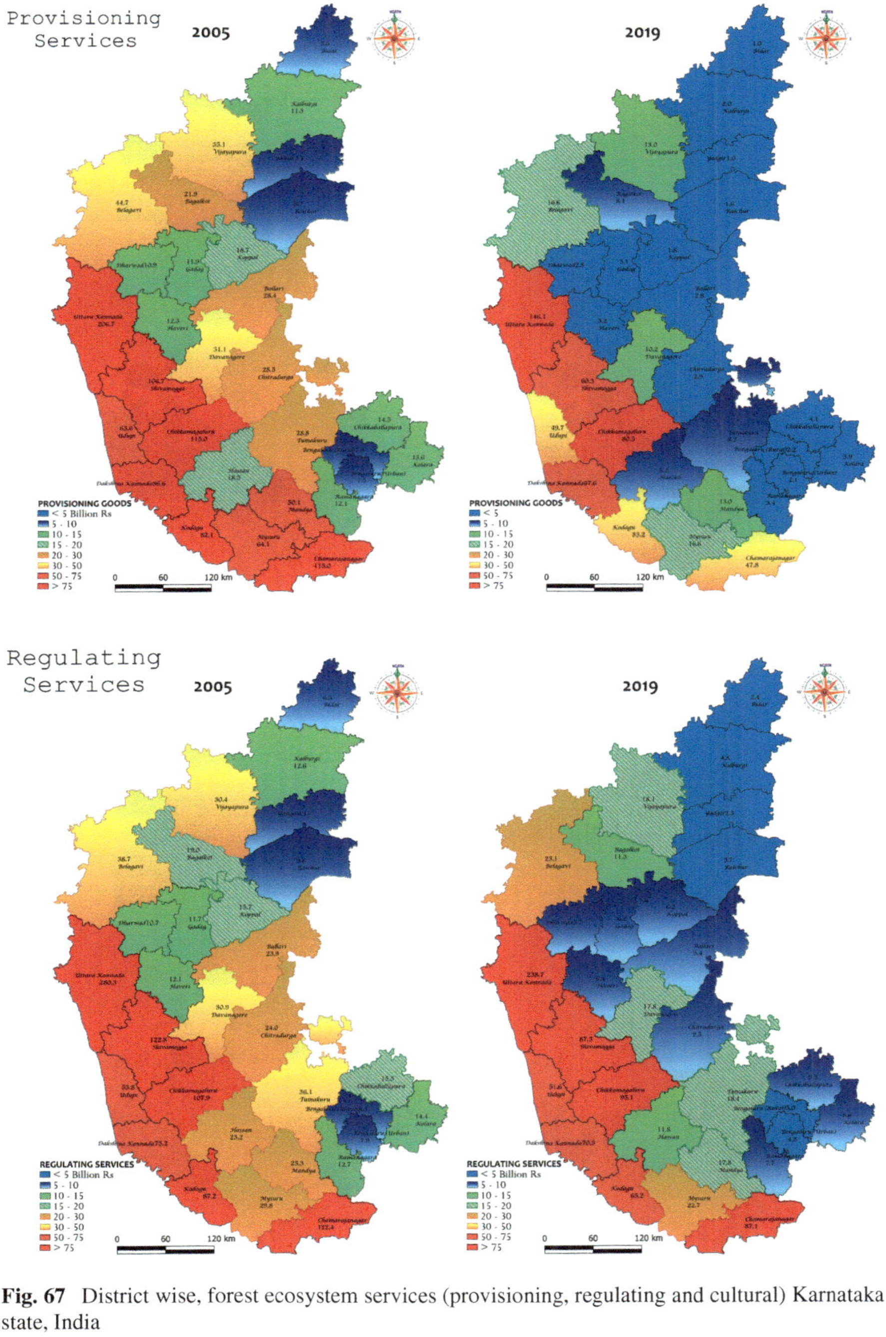

Fig. 67 District wise, forest ecosystem services (provisioning, regulating and cultural) Karnataka state, India

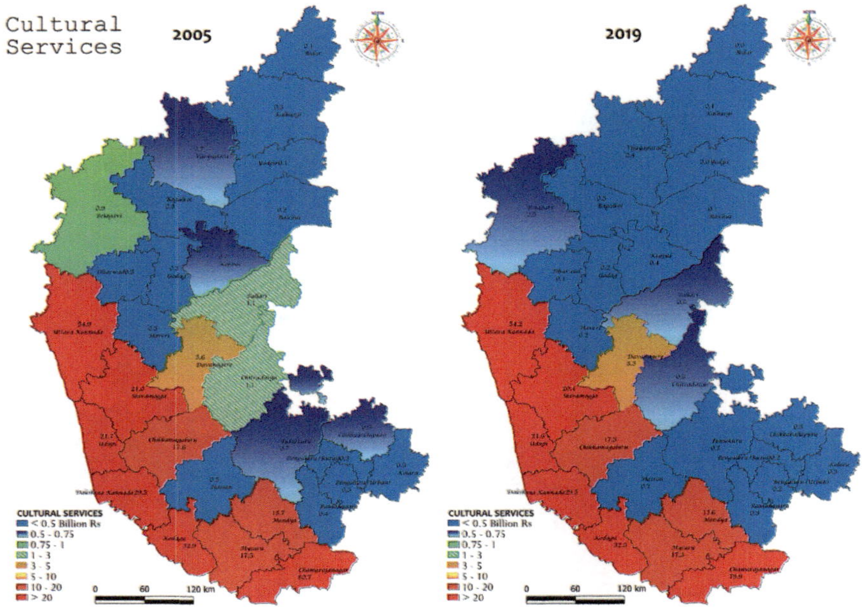

Fig. 67 (continued)

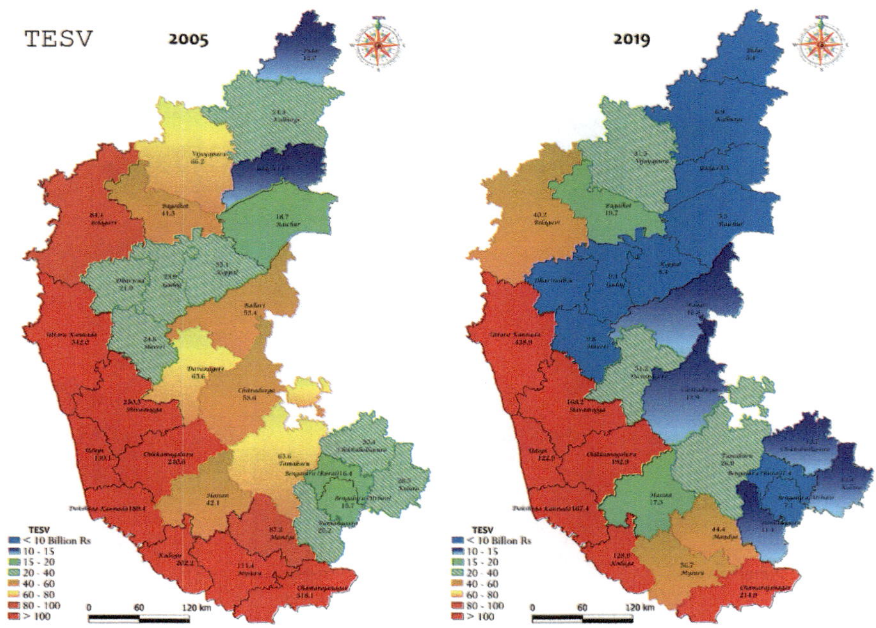

Fig. 68 District-wise TESV (Total Ecosystem Supply Value) of forest ecosystem in Karnataka, India

Table 84 District wise TESV (aggregation of provisioning, regulating and cultural services) of a forest ecosystem in Karnataka in Billion Rs

S.no	District	PROVG_BRS		REGS_BRS		CULTS_BRS		TESV	
		2005	2019	2005	2019	2005	2019	2005	2019
1	Bagalkot	22	8	19	11	0.5	0.3	41	20
2	Ballari	28	3	24	9	1.1	0.6	53	13
3	Belagavi (Belgaum)	45	17	39	23	0.9	0.5	84	40
4	Bengaluru (Rural)	8	2	8	5	0.3	0.2	16	7
5	Bengaluru (Urban)	8	2	8	5	0.3	0.2	16	7
6	Bidar	6	1	6	2	0.1	0.0	12	3
7	Chamarajanagar	115	48	122	87	80.7	79.9	318	215
8	Chikkaballapura	15	4	15	9	0.5	0.3	30	14
9	Chikkamagaluru (Chikmagalur)	115	81	108	95	17.6	17.3	241	193
10	Chitradurga	28	3	24	9	1.1	0.6	54	13
11	Dakshina Kannada	87	68	73	70	29.5	29.5	189	167
12	Davanagere	31	10	31	18	3.6	3.3	66	31
13	Dharwad	11	3	11	6	0.3	0.2	22	9
14	Gadag	12	3	12	6	0.3	0.2	24	9
15	Hassan	18	5	23	12	0.5	0.2	42	17
16	Haveri	12	3	12	6	0.3	0.2	25	10
17	Kalburgi (Gulbarga)	11	2	13	5	0.3	0.1	24	7
18	Kodagu	82	33	87	63	32.9	32.5	202	129
19	Kolar	14	4	14	9	0.5	0.3	28	13
20	Koppal	19	2	16	6	0.7	0.4	35	8
21	Mandya	50	13	23	18	13.7	13.6	87	44
22	Mysuru (Mysore)	64	17	30	23	17.5	17.3	111	57
23	Raichur	9	2	10	4	0.2	0.1	19	5
24	Ramanagara	12	3	13	8	0.4	0.3	25	11
25	Shivamogga (Shimoga)	107	60	123	88	21.0	20.4	251	168
26	Tumakuru (Tumkur)	29	8	36	18	0.7	0.3	66	27
27	Udupi	64	50	54	52	21.7	21.6	139	123
28	Uttara Kannada	207	146	280	239	54.9	54.2	542	439
29	Vijayapura (Bijapur)	35	13	30	18	0.7	0.4	66	32

(continued)

Table 84 (continued)

S.no	District	PROVG_BRS		REGS_BRS		CULTS_BRS		TESV	
		2005	2019	2005	2019	2005	2019	2005	2019
30	Yadgir	5	1	6	2	0.1	0.0	12	3
Total		1268	614	1271	926	303	295	2841	1835

References

AGRICOOP (2020) Farmers Portal. In: Dep. Agric. Coop. Farmers Welf. Minist. Agric. Farmers Welfare, Gov. India

Amirnejad H, Khalilian S, Assareh MH, Ahmadian M (2006) Estimating the existence value of north forests of Iran by using a contingent valuation method. Ecol Econ 58:665–675

Aswathy N, Narayanakumar R, Harshan NK (2014) Marketing costs, margins and efficiency of domestic marine fish marketing in Kerala. Indian J Fish 61:97–102

Badola R, Hussain SA, Dobriyal P et al (2017) Assessment of recreational services of natural landscapes in third world tropics using the travel cost method. Wilderness Wildl Tour 17

Balachandran C, Chandran MDS, Vinay S, Shrikant N, Ramachandra TV (2017) Pollinator diversity and foraging dynamics on monsoon crop of cucurbits in a traditional landscape of South Indian west coast. Biotropia 24(1):16–27

Bharath S, Nimish G, Ramachandra TV (2017) Visualization and prediction of landscape dynamics in the protected areas of Karnataka. IJRET Int J Res Eng Technol 06:2321–7308. https://doi.org/10.15623/ijret.2017.0614011

Bhattacharya P, Hayat SF (2009) Sustainable NTFP management for livelihood and income generation of tribal communities: a case from Madhya Pradesh, India. Non timber For Prod Conserv Manag Policy Trop Ed By Shaanker R Uma, Ankita J Hiremath, Gladwin C Joseph Nitin D Rai Publ by ATTREE Univ Agric Sci Bangalore 21–34

CACP (2005a). https://cacp.da.gov.in/ViewQuestionare.aspx?Input=2&DocId=1&PageId=39&KeyId=347

CACP (2005b). https://cacp.da.gov.in/ViewQuestionare.aspx?Input=2&DocId=1&PageId=40&KeyId=453

Calder IR (1996) Rainfall interception and drop size—development and calibration of the two-layer stochastic interception model. Tree physiology 16(8):727–732. https://academic.oup.com/treephys/article/16/8/727/1668624

Chandran MDS, Rao GR, Gururaja KV, Ramachandra TV (2010) Ecology of the swampy relic forests of Kathalekan from Central Western Ghats, India. Bioremediation Biodivers Bioavailab 4:54–68

Chanda S, Ramachandra TV (2019a) A review on some Therapeutic aspects of Phytochemicals present in Medicinal plants. Int J Pharm Life Sci 10

Chanda S, Ramachandra TV (2019b) Sacred groves—Repository of medicinal plant resources: a review. Res Rev J Ecol 8:12–20

Chatterjee R (2011) Groundwater resources estimation—Case studies from India. J Geol Soc India 77:201–204

Cheng X, Van Damme S, Li L, Uyttenhove P (2019) Evaluation of cultural ecosystem services: a review of methods. Ecosyst Serv 37:100925

Coffee Board (2020) Daily coffee market report. In: Government of India. Ministry of Commerce and Industry

Commodities Online (2020) Mandi rates

Costanza R, D'Arge R, De Groot R et al (1997) The value of the world's ecosystem services and natural capital. Nature. https://doi.org/10.1038/387253a0

Costanza R, De Groot R, Sutton P et al (2014) Changes in the global value of ecosystem services. Glob Environ Chang 26:152–158
Dadhwal VK, Singh S, Patil P (2009) Assessment of phytomass carbon pools in forest ecosystems in India. NNRMS Bull 41–57
De Groot R, Brander L, Van Der Ploeg S et al (2012) Global estimates of the value of ecosystems and their services in monetary units. Ecosyst Serv 1:50–61
De Groot R, Luke B, Solomonides S (2020) Ecosystem services valuation databasE (ESVD) Version June 2020
De Groot RS, Leon B, Costanza R (2017) A short history of the ecosystem services concept. Mapp Ecosyst Serv 31
De Groot RS, Wilson MA, Boumans RMJ (2002) A typology for the classification, description and valuation of ecosystem functions, goods and services. Ecol Econ 41:393–408
DES (2017). https://des.karnataka.gov.in/english
Devagiri GM, Money S, Singh S et al (2013) Assessment of above ground biomass and carbon pool in different vegetation types of south western part of Karnataka, India using spectral modeling. Trop Ecol 54:149–165
Diaz S, Demissew S, Carabias J et al (2015) The IPBES conceptual framework—Connecting nature and people. Curr Opin Environ Sustain 14:1–16
DMI (2020) AgMarknet. In: Directorate of marketing & inspection (DMI), Ministry of Agriculture & Farmers Welfare, Government of India
Dorji T, Brookes JD, Facelli JM et al (2019) Socio-cultural values of ecosystem services from Oak Forests in the Eastern Himalaya. Sustainability 11:2250
DPPMS (2018) Economic Survey of Karnataka 2017–2018
DSO (2019) District Statistics at a Glance, District Statistical Office. Bengaluru
EANDS (2020) Cost of Cultivation/production and related data. In: Director of Economics and Statistics, Department of Agriculture Cooperative & Farmers Welfare, Ministry of Agriculture & Farmers Welfare, Government of India
Euliss Jr NH, Smith LM, Liu S et al (2010) The need for simultaneous evaluation of ecosystem services and land use change
Foley JA, DeFries R, Asner GP et al (2005) Global consequences of land use. Science (80-) 309:570–574
Garcia CA, Pascal JP (2006) Sacred forests of Kodagu: ecological value and social role. In: Cederlof G, Sivaramakrishnan K (eds) Ecological nationalisms: nature, livelihoods, and identities in south Asia. University of Washington Press, Seattle and London, pp 199–229
Ghosh N (2020) Promoting a 'GDP of the Poor': The Imperative of Integrating Ecosystems Valuation in Development Policy, *ORF Occasional Paper No. 239*, March 2020, Observer Research Foundation
Google (2020) Google Earth. http://google.com/earth
Gould RK, Ardoin NM, Woodside U et al (2014) The forest has a story: cultural ecosystem services in Kona, Hawai 'i. Ecol Soc 19
Gunarekha BS, Binoy TA (2017) Community based sustainable tourism development in Karnataka: a study on Mysuru district. Asia Pacific J Res 1:121–126
Haripriya G, Sanyal S, Sinha R, Sukhdev P (2006) Green accounting for Indian States project. TERI Press, New Delhi
Haslag JH (2020) Seigniorage revenue and monetary policy. In: Handbook of monetary policy. Routledge, pp 347–362
Hipólito J, Sousa B dos SB, Borges RC et al (2019) Valuing nature's contribution to people: the pollination services provided by two protected areas in Brazil. Glob Ecol Conserv 20:e00782
Hirons M, Comberti C, Dunford R (2016) Valuing cultural ecosystem services. Annu Rev Environ Resour 41:545–574
HOPCOMS (2019) Fruits market Price in Bangalore, Karnataka
Inflation (2021) Inflation calculator—Indian Rupee. In: Inflat tool. https://www.inflationtool.com/indian-rupee

Joshi J, Negi GCS (2011) Quantification and valuation of forest ecosystem services in the western Himalayan region of India. Int J Biodivers Sci Ecosyst Serv Manag 7(1):2–11. https://doi.org/10.1080/21513732.2011.598134

KFD (2020) Karnataka Forest Department. https://aranya.gov.in/aranyacms/English/Home.aspx

KMV (2020) Krishi Marata Vahini. In: Deparment of Agriculture Marketing Karnataka State Agricultural Marketing Board

Kreye MM, Adams DC, Ghimire R et al (2019) Forest ecosystem services: cultural values

KSRSAC (2018) Karnataka geographical information system. In: Government of Karnataka

Kulkarni V, Ramachandra TV (2009) Environmental management, commonwealth of learning, Canada and Indian Institute of Science, Bangalore

Kumar BG, Datta KK, Joshi PK, Katiha PK, Suresh R, Ravisankar T, Ravindranath K, Menon M (2008) Domestic fish marketing in India–changing structure, conduct, performance and policies. Agric Econ Res Rev 21(conf):345–354

Lal R, Smith P, Jungkunst HF et al (2018) The carbon sequestration potential of terrestrial ecosystems. J Soil Water Conserv 73. https://doi.org/10.2489/jswc.73.6.145A

Ma E, Feng Z, Zheng Y (2019) The effect of forest on soil erosion control based on remote sensing technology. Ekoloji Derg

Maes J, Czúcz B, Keith H et al (2020) A review of ecosystem condition accounts: lessons learned and options for further development

Maes J, Teller A, Erhard M et al (2013) Mapping and assessment of ecosystems and their services. Anal Framew Ecosyst Assess Under Action 5:1–58

Maes J, Teller A, Erhard M et al (2018) Mapping and assessment of ecosystems and their services: an analytical framework for ecosystem condition. Publ Off Eur Union, Luxemb 1–78

Maes J, Zulian G, Thijssen M et al (2016) Mapping and assessment of ecosystems and their services: urban ecosystems. Publ Off Eur Union, Luxemb, p 4

Markandya A, Harou P, Bellu LG et al (2002) Environmental economics for sustainable growth: a handbook for practitioners. Edward Elgar Publishing Ltd.

MEA (2005) Ecosystems and human well-being. Island Press, Washington, DC

Mengist W, Soromessa T (2019) Assessment of forest ecosystem service research trends and methodological approaches at global level: a meta-analysis. Environ Syst Res 8:22. https://doi.org/10.1186/s40068-019-0150-4

Moreno G, Aviron S, Berg S et al (2018) Agroforestry systems of high nature and cultural value in Europe: provision of commercial goods and other ecosystem services. Agrofor Syst 92:877–891

MoSPI (2021) Ecosystem Accounts for India—Report of the NCAVES Project. New Delhi

MoSPI (2020) Forests—The Climate Protectors (Chapter 3). https://mospi.gov.in/documents/213904//649960//1614147361232_b3_ES2_2020.pdf. http://mospi.nic.in/sites/default/files/reports_and_publication/statistical_publication/EnviStats2/ES2_2020_Complete_revised%20on%204_11_2020.pdf

MSP (2020) MSP for Rabi Crops 2019–2020. In: Ministry of Agriculture & Farmers Welfare, Government of India

Murali G (2010) Opportunity to sustain coconut ecosystem services through recycling of the palm leaf litter as vermicompost: Indian scenario (A Technology/Research Note). CORD 26:14. https://doi.org/10.37833/cord.v26i2.130

NBHM (2019). https://nbb.gov.in/default.html

NAAS (2020) Payment for ecosystem services in agriculture. New Delhi

Nair PKR, Nair VD, Kumar BM, Haile SG (2009) Soil carbon sequestration in tropical agroforestry systems: a feasibility appraisal. Environ Sci Policy 12:1099–1111. https://doi.org/10.1016/j.envsci.2009.01.010

NAP (2009) National afforestation programme revised operational guidelines. New Delhi

Nayak AK, Shahid M, Nayak AD et al (2019) Assessment of ecosystem services of rice farms in eastern India. Ecol Process 8:35. https://doi.org/10.1186/s13717-019-0189-1

NIC (2020) Soil Health Maps. In: Deparment of Agriculture Cooperative & Farmers Welfare, Government of India

Ninan KN, Kontoleon A (2016) Valuing forest ecosystem services and disservices—Case study of a protected area in India. Ecosyst Serv 20:1–14. ISSN 2212-0416. https://doi.org/10.1016/j.ecoser.2016.05.001

Nowak DJ, Crane DE, Stevens JC (2006) Air pollution removal by urban trees and shrubs in the United States. Urban forestry and urban greening 4(3–4):115–123. https://www.sciencedirect.com/science/article/pii/S1618866706000173

NRSC (2020) National Remote Sensing Centre-IRS data products. In: Indian Sp. Res. Organ. Government of India

NRSC (2016) Bhuvan-National Remote Sensing Centre. In: Indian Sp. Res. Organ. Government of India

Ollerton J, Winfree R, Tarrant S (2011) How many flowering plants are pollinated by animals? Oikos 120:321–326

Ormsby AA, Bhagwat SA (2010) Sacred forests of India: a strong tradition of community-based natural resource management. Environ Conserv 320–326

Ouyang Z, Song C, Zheng H, Polasky S, Xiao Y, Bateman IJ, Liu J (2020) Using gross ecosystem product (GEP) to value nature in decision making. Proceedings of the National Academy of Sciences 117(25):14593–14601. https://www.pnas.org/doi/10.1073/pnas.1911439117

Pal S (2018) Agriculture and ecosystem services. ICAR-National Institute of Agricultural Economics and Policy Research, New Delhi

Pandey R, Rawat GS, Kishwan J (2011) Changes in distribution of carbon in various forest types of India from 1995–2005 changements dans la distribution du Carbone dans Différents types de Forêts en Inde. Silva Lusit 19:41–54

Penman J, Gytarsky M, Hiraishi T, Krug T, Kruger D, Pipatti R, Buendia L (2003) Good practice guidance for land use, land-use change and forestry. https://www.cabidigitallibrary.org/doi/full/10.5555/20083162304

Perelet R, Mason P, Markandya A, Taylor T (2014) Dictionary of environmental economics. Routledge

Planning Commission (2011) Report of the sub-group II on NTFP and their sustainable management in the 12th five year plan. A Rep Submitt under Plan Comm Work Gr For Nat Resour Manag

Polasky S, Tallis H, Reyers B (2015) Setting the bar: standards for ecosystem services. Proc Natl Acad Sci 112:7356–7361. https://doi.org/10.1073/pnas.1406490112

Porto RG, De Almeida RF, Cruz-Neto O, Tabarelli M, Viana BF, Peres CA, Lopes AV (2020) Pollination ecosystem services: A comprehensive review of economic values, research funding and policy actions. Food Secur 12(6):1425–1442. https://link.springer.com/article/10.1007/s12571-020-01043-w

Prasad SN, Hedge M, Hedge MS (1987) Fuel consumption and conservation method in Urban Centres of Uttara Kannada, CES Technical Report 25. Bangalore

Rai SN, Proctor J (1986) Ecological studies on four rainforests in Karnataka, India. J Ecol 74:439–454

Rajan K, Natarajan A, Anil Kumar KS et al (2010) Soil organic carbon—The most reliable indicator for monitoring land degradation by soil erosion. Curr Sci 99:823–827

Ramachandra TV, Subramanian DK, Joshi NV (1999) Hydroelectric resource assessment in Uttara Kannada district, Karnataka state, India. J Clean Prod 7:195–211

Ramachandra T, Joshi N, Subramanian D (2000a) Present and prospective role of bioenergy in regional energy system. Renew Sustain Energy Rev 4:375–430. https://doi.org/10.1016/S1364-0321(00)00002-2

Ramachandra TV, Subramanian DK, Joshi NV et al (2000b) Domestic energy consumption patterns in Uttara Kannada District, Karnataka State, India. Energy Convers Manag 41:775–831. https://doi.org/10.1016/S0196-8904(99)00151-X

Ramachandra TV, Subramanian DK, Joshi NV (2001) A decision support system for optimal design of hydroelectric projects in Uttara Kannada. Energy Sustain Dev 5:14–31

Ramachandra TV, Chandran MDS, Gururaja KV, Sreekantha (2007) Cumulative environmental impact assessment. Nova Science Publishers, New York, NY (United States)

Ramachandra TV, Chandran MDS, Harish BR et al (2010) Biodiversity, ecology and socio economic aspects of Gundia River basin in the context of proposed Mega Hydro Electric Power Project, CES Technical Report 122. Bengaluru
Ramachandra TV, Chandran MDS, Ananth A et al (2012) Tragedy of the Kan sacred forests of Shimoga District: need for urgent policy interventions for conservation, CES Technical Report: 128,. Bangalore
Ramachandra TV, Chandran MDS, Rao GR et al (2015) Floristic diversity in Uttara Kannada district, Karnataka, Chapter 1, In Biodiversity in India. Regency publications, New Delhi
Ramachandra TV, Chandran MDS, Vinay S et al (2016) Sacred Groves (Kan forests) of Sagara taluk, Shimoga district, Sahyadri Conservation Series: 54, ENVIS Technical Report 102,. Bangalore
Ramachandra TV, Soman D, Naik AD, Chandran MDS (2017) Appraisal of forest ecosystems goods and services: challenges and opportunities for conservation. J Biodivers 8:12–33. https://doi.org/10.1080/09766901.2017.1346160
Ramachandra TV, Chandran MDS, Bharath S et al (2018a) Forest ecosystem: goods and services, ENVIS Technical Report 142, Sahyadri Conservation Series 79,. Bangalore
Ramachandra T, Vinay S, Bharath S, Shashishankar A (2018b) Eco-hydrological footprint of a River Basin in Western Ghats. Yale J Biol Med Inc.
Ramachandra TV, Bharath S, Gupta N (2018c) Modelling landscape dynamics with LST in protected areas of Western Ghats, Karnataka. J Environ Manag 1253–1262
Ramachandra TV, Bharath S (2019a) Global warming mitigation through carbon sequestrations in the Central Western Ghats. Remote Sens Earth Syst Sci 2:39–63. https://doi.org/10.1007/s41976-019-0010-z
Ramachandra TV, Bharath S (2019b) Carbon sequestration potential of the forest ecosystems in the Western Ghats, a global biodiversity hotspot. Nat Resour Res 29:2753–2771. https://doi.org/10.1007/s11053-019-09588-0
Ramachandra T, Bharath S, Vinay S et al (2019a) Grid based monitoring of Natural Resources in the Ecologically Fragile Regions of Kodagu, Karnataka, ENVIS Technical Report 156, Sahyadri Conservation Series 83. Bangalore
Ramachandra TV, Bharath S, Vinay S (2019b) Visualisation of impacts due to the proposed developmental projects in the ecologically fragile regions-Kodagu district, Karnataka. Prog Disaster Sci 3:100038. https://doi.org/10.1016/j.pdisas.2019.100038
Ramachandra TV, Vinay S, Bharath S et al (2020) Insights into riverscape dynamics with the hydrological, ecological and social dimensions for water sustenance. Curr Sci 113891:118
Ramachandra TV, Bharath S (2021) Carbon footprint of Karnataka: accounting of sources and sinks. In: Carbon footprint case studies. Springer, pp 53–92
Ramachandra TV, Bharath S, Vinay S, Bharath HA (2021a) Ecosystem extent account for Karnataka. Bangalore
Ramachandra TV, Bharath S, Vinay S, Bharath HA (2021b) Ecosystem condition account for Karnataka. Bangalore
Rao G, Krishnakumar G, Sumesh Dudani N, Ramachandra TV et al (2013) Vegetation changes along altitudinal gradients in human disturbed forests of Uttara Kannada, Central Western Ghats. J Biodivers 4:61–68
Rao GR, Krishnakumar G, Dudani SN et al (2014) Diversity and Regeneration Aspects of Medicinal Plants at Devimane, Uttara Kannada District, Karnataka, Central Western Ghats. J Biodivers Manag for 8:2
Rao GR, Chandran MDS, Ramachandra TV (2015) Diversity and Regeneration Status of Medicinal Plants in Medicinal Plants Conservation Area (MPCA) at Shirgunji of Uttara Kannada District, Central Western Ghats. MyForest 51:85–99
Ravindranath NH, Ostwald M (2008) Carbon inventory methods: a handbook for greenhouse gas inventory, carbon mitigation and roundwood production projects. Advances in Global Change Research. Carbon Invent methods a Handb Greenh gas Invent carbon Mitig roundwood Prod Proj Adv Glob Chang Res 237–270. https://doi.org/10.1007/978-1-4020-6547-7_10

Ravindranath NH, Somashekhar BS, Gadgil M (1997) Carbon flow in Indian forests. Clim Change 35:297–320. https://doi.org/10.1023/A:1005303405404
Ray R, Ramachandra TV (2010) Small sacred groves in local landscape: are they really worthy for conservation? Curr Sci 98:1178–1180
Ray R, Subash Chandran MD, Ramachandra TV (2010) Ecosystem services from sacred groves of Uttar Kannada: a case study. In: Bangalore: Lake 2010 conference 22nd–24th December
Ray R, Bharath S, Ramachandra TV (2014a) Biodiversity conservation in humanised landscape–challenges and opportunity. In: Lake 2014. EWRG, CES, IISc, Bangalore
Ray R, Chandran MDS, Ramachandra TV (2014b) Biodiversity and ecological assessments of Indian sacred groves. J for Res 25:21–28
Ray R, Chandran MDS, Ramachandra TV (2015) Hydrological importance of sacred forest fragments in Central Western Ghats of India. Trop Ecol 56:87–99
Rodríguez JP, Beard Jr TD, Bennett EM et al (2006) Trade-offs across space, time, and ecosystem services. Ecol Soc 11
SEEA (2017) System of Environmental Economic Accounting 2012: central framework. International Monetary Fund
SEEA EA (2021) System of Environmental-Economic Accounting—Ecosystem Accounting, Final Draft, Version 5,. Department Of Economic And Social Affairs, Statistics Division, United NationsISION, United Nations
Sinclair M, Mayer M, Woltering M, Ghermandi A (2020) Valuing nature-based recreation using a crowdsourced travel cost method: a comparison to onsite survey data and value transfer. Ecosyst Serv 45:101165
Swamy HR (1992) Organic productivity, nutrient cycling and small watershed hydrology of natural forests and monoculture plantations in Chikmagalur District, Karnataka. Wiley, New York, NY (United States)
TEEB (2010a) The economics of ecosystems and biodiversity, ecological and economic foundations. Earthscan, London and Washington
TEEB (2010b) Mainstreaming the economics of nature: a synthesis of the approach, conclusions and recommendations of TEEB. TEEB
Ten Brink P (2011) The economics of ecosystems and biodiversity in national and international policy making. Routledge
Terrado M, Acuña V, Ennaanay D et al (2014) Impact of climate extremes on hydrological ecosystem services in a heavily humanized Mediterranean basin. Ecol Indic 37:199–209
TNAU Agriculture Portal (2020). https://www.agritech.tnau.ac.in/
Vallecillo S, La Notte A, Zulian G et al (2019) Ecosystem services accounts: valuing the actual flow of nature-based recreation from ecosystems to people. Ecol Modell 392:196–211
Van Berkel DB, Verburg PH (2014) Spatial quantification and valuation of cultural ecosystem services in an agricultural landscape. Ecol Indic 37:163–174
Van der P, De Groot RS, Wang Y (2010) The TEEB valuation database: overview of structure, data and results. Wageningen, the Netherlands
Van Do T, Trung PD, Yamamoto M et al (2018) Aboveground biomass increment and stand dynamics in tropical evergreen broadleaved forest. J Sustain for 37:1–14. https://doi.org/10.1080/10549811.2017.1375959
Verma M, Joshi S, Godbole G, Singh A (2007) Valuation of Ecosystem services and forest governance—A scoping study from Uttarakhand, LEAD India, New Delhi
Verma M, Negandhi D, Wahal AK, Kumar R (2013) Revision of rates of NPV applicable for different class/category of forests. AcademiaEdu 143
Willmann R, Boonchuwong P, Piumsombun S (2001) Fisheries management costs in Thai marine fisheries. https://ir.library.oregonstate.edu/concern/conference_proceedings_or_journals/pv63g1109

World Bank (1998) The World Bank annual report 1998 (English). Washington, D.C.
World Bank (2001) The World Bank annual report 2001: Year in review (English). Washington, D.C.
Zarandian A, Baral H, Yavari AR et al (2016) Anthropogenic decline of ecosystem services threatens the integrity of the unique Hyrcanian (Caspian) forests in Northern Iran. Forests 7:51
Zawadzka J, Gallagher E, Smith H, Corstanje R (2019) Ecosystem services from combined natural and engineered water and wastewater treatment systems: going beyond water quality enhancement. Ecol Eng X 2:100006. https://doi.org/10.1016/j.ecoena.2019.100006

Scenario-Based Assessment of Policy Interventions in Karnataka State, India

Abstract Nature functions holistically. A great diversity of biotic species, including humans, have been associated with each other for long periods, and co-evolution is at the centre of all ecosystems. The individual cannot live independently of the living environment, and individuals' actions have an impact on the environment. On the contrary, today's human societies override the integrity of ecosystems, shaking the very foundations of life itself to their detriment. For the lasting well-being of human societies, they need to learn more about the ecosystems. They form a part, assess their complexities and carrying capacity, and modify or derive resources from them without straining their capacity to provide goods and services perpetually. The Karnataka State, India, gifted with enormous natural resources potential, and the mandate of sustainable development based on the foundation of prudent management of ecosystems, is yet to be a reality. Various developmental programmes, which proclaimed to be functioning on sustainability principle, on looking back, have only been fraying the complex web of life, disrupting ecosystems and causing a decline in overall productivity. This is true in four major sectors such as forestry, fisheries, agriculture and hydrology.

In this context, *the sustainable development concept has evolved to balance the ecosystem and its environment* and development. This underlines the need to improve the quality of human life at the same time, which must be within the carrying capacity of the ecosystem. The sound environment is achieved by implementing stipulated procedures related to the environment, and actions adopted and/or enacted by the government or governing bodies. This involves laws, rules, methods, administration, and legislative decisions related to the sustenance of the environment. They address the key issues of quality, control, and mitigation of pollution in the air, water, soil, noise, and issues like conservation and preservation of habitat quality, wildlife, national parks and sanctuaries, forests, and prevention of illegal trafficking of wildlife, etc. These policies are based on the guidelines from international conventions, international statutes, central and state governments, local bodies, and, to a certain extent, from culture and history. Most often, policies consider inputs from interdisciplinary allied subjects like sociology, economics, ecology, natural sciences, geology, etc. In order to have guidelines for regulating and mitigation

T. V. Ramachandra et al., *Natural Capital Accounting and Valuation of Ecosystem Services, Karnataka State, India*, https://doi.org/10.1007/978-981-97-2405-5_4

considering functional components of the environment, environmental policies are being formulated by authorities for;

- preserving, protecting, and improving the quality of the environment,
- protecting human health,
- sustainable utilisation of natural resources prudently and rationally within the carrying capacity of the ecosystem and
- to promote measures to deal with environmental problems.

Sustainable development of a region requires a synoptic ecosystem approach related to the dynamics of natural variability and the effects of human interventions on key indicators of biodiversity and productivity. The concept of cumulative effects of the incremental reduction and erosion of natural systems' integrity from the interactions of developmental activities provides a perspective to redirect impact analysis to deal with the driving causes of unsustainable development. The conservation importance of an area is determined by assessing its ecological values and functions. This necessitates inventorying, mapping and monitoring natural resources to determine viable management strategies. This is achieved through the accounting of ecosystem extent and conditions using temporal remote sensing data coupled with field data.

Integrating environmental dimensions into land-use planning and management processes can greatly contribute to the sustainability of natural resources. Sustainable landscape management requires advanced information on likely land-use changes, which would help in evolving mitigation strategies to sustain dependent communities' livelihoods. This entails modeling and visualization of landscape transitions with temporal-spatial data. Change detection analyses have gained prominence with the availability of spatial data since 1970's. Multiresolution spatial data with advances in modeling techniques would aid in the geo-visualization of likely land uses with the policy decisions, which is vital for sustainable resource management. This would help to examine and statistically define the spatial patterns of LULC changes at a precise interval. Modeling helps identify the most appropriate spatial pattern of future land uses. This information helps assess resource availability and serves as a decision support system (DSS) for managers, planners, and decision-makers. Modeling and visualization will satisfy numerous sustainable development conditions, such as the conservation of biological and cultural diversities through ecosystem protection. Apart from inventorying, mapping, monitoring, and change analyses, modeling, and visualization would empirically interpret the consequences of spatial changes. Predictive statistical and geospatial models used widely for modelling landscape dynamics are logistic regression models, spatial dynamic models, spatial Markov chains (MC), Cellular automata (CA), and multi-criteria decision-making (MCDM) techniques, etc. modeling and visualization of landscape dynamics in Karnataka has been done by considering (i) Business As Usual Scenario-BAU, (ii) Agent based land use transition scenario-ALT, (iii) Reserve Forest Protection scenario-RFP, (iv) Afforestation scenario-AFS and (v) Sustainable Development Policy Scenario -SDP.

The spatiotemporal land use analysis was carried out from 1985 to 2019 using temporal remote sensing data (LANDSAT and IRS) through the Gaussian maximum likelihood algorithm. The districts of the Western Ghats region have higher forest cover than other districts. The State has only 15% of the geographical area under forest compared to 21% in 1985. Post-1990s, the state witnessed large-scale land-use transitions due to industrialization, urbanization, an increase in horticulture crops, conversion from agriculture to market-based crops (higher economic), etc.

BAU depicts a likely increase in built-up area by 11.5% from 3% (2019). The forest cover, and agriculture areas would lose the major tract for the possible expansion of built-up cover. The dry and moist deciduous forest cover was shown higher changes with a loss of 6.5% forest cover. The existing infrastructure and the built-up cover might transform deciduous forests, and agriculture areas along highways across the state. Cities such as Bangalore, Mangalore, Dharwad depicted compact growth, whereas Mysore, Belgaum, Hassan, Tumkur (Tier-2) cities have indicated peri-urban growth.

The ALT scenario incorporates the likely developmental activities - upgrading of urban infrastructure (roads, commercial establishments, information networks, industrial units, sports/recreation), which shows an increase in built-up cover of 15% compared to BAU. ALT scenario presents a more compact urban growth than the BAU, considering that new, high-density residential development occurs inside the cities, major towns, and polycentric urban agglomeration associated with the provision of new transport, industries, and other project investments. Both BAU and ALT scenario projections indicate a significant reduction in forest areas, and agricultural land uses. ALT scenario indicates, only 9% forest cover compared to 15% in 2019. The agriculture area showed a loss of 2%, which might affect food availability in the region. The regions where dry crops are being practiced would witness higher transformation, signifying the agents/driver's role in land use alterations. ALT scenario shows that the higher land use conversion due to transportation, and other projects, which tend to encourage more compact, peri-urban, and mixed land use developments, necessitates effective policies.

The RF scenario assumes the protection of reserve forests in the regions without allowing any developmental activities and expansion of earlier land-uses. The regions covered under reserve forest protection do not favour development and other land-use conversions. The districts that are part of the Western Ghats region will not change compared to BAU and ALT scenarios. The areas with dry deciduous forest cover and not under the reserve forest area protection would experience abrupt growth due to the transportation corridors and urban clusters. The urban area of 2019 is considered another factor influencing land-use change, unlike other scenarios. RF scenario shows a likely built-up cover of 11%, while retaining evergreen forest cover of 5%.

The AF scenario shows the likely improvement in its forest cover due to the plantation activities and protection of degraded forests from uncontrolled land use conversion. The forest cover shows 9.3%, a slight increase compared to BAU and ALT. The built-up cover would limit existing cities and towns by over 11%. The new plantation activities would compensate for the abrupt land-use conversion, especially in the moist, dry deciduous cover regions. The districts such as Bidar, Kolar, Koppala, etc., did not show any improvement due to the least area of forest land availability.

SDP scenario signifies stringent policy implementation initiatives with the least disturbances. The forest cover would remain 11% compared to 17% in 2019, which is the least possible loss compared to all considered scenarios (BAU, ALT, RF, and AF). The urbanization process would restrict the existing tier-1 and tier-2 cities, and the least increase was shown in peri-urban areas. Districts such as Bidar, Gulbarga, Bagalkot, and Kolar might experience higher land use conversions due to existing infrastructure and the absence of any forest policy. Overall, RFP, AF, and SDP scenarios favor the protection of evergreen and moist deciduous forest categories. BAU, and ALT scenarios depict emerging urban clusters, indicating smaller clusters of paved areas are no longer available and tend to agglomerate with bigger ones to form urban centres, satellite towns, and dispersed growth at the periphery of the tier-2 city limits. New constructions might occur in the rural hinterland, which results in the loss of evergreen, moist deciduous cover and agriculture areas. Categories such as plantation, water, and open field categories had similar values across all scenarios. The scenarios would directly help decision-makers/planners reduce the impacts on the environment with stricter growth rules; otherwise, conversion of primary, secondary forest/ moist/dry deciduous forest into urban areas would be favored (BAU, ALT), which would induce ecological imbalances. Another important observation in terms of topography is that construction restrictions in the steep slope areas have played an essential role in restricting urban settlements from further infiltration into forested areas. Collectively, scenarios 3–5 are appropriate for sustainable development without altering the existing edges of agriculture, forests, water bodies, and other protected areas or conservation zones. Additionally, they emphasize a planned growth strategy for a habitable region. Integrated planning through regulated land-use conversion can support self-sufficient towns along newly provided public transport corridors, limiting the abrupt land-use conversion.

The land-use dynamics involving the decline of forests have resulted in a significant change in ecosystem services. The provisioning goods (timber, NTFP, fish, fuelwood, fodder, bamboo, cane, genetic material, etc.) from the forests of Karnataka is about 613 billion INR in 2019. Due to land-use changes with the decline of forests, the provisioning services in 2033 would reduce to 524 billion INR (BAU scenario), 521 billion INR (ALT Scenario), 533 billion INR (RFP Scenario), 530 billion INR (AF Scenario) and 540 billion INR (SDP), which highlights the need to implement afforestation with the reserve forest protection (SDP scenario) to sustain the livelihood through the provision of goods. ALT scenario shows a reduction of 92 billion INR in provisional goods compared to 2019. The SDP scenario has shown improvement in ecosystem services compared to other scenarios, which highlights the role of policy decisions.

Regulating services pertains to services resulting from the ability of ecosystems to regulate biological processes and influence climate, hydrological, and biochemical cycles, thereby maintaining environmental conditions beneficial to individuals and society. Regulating services (Air Quality, Climate Regulation, Carbon Sequestration, Pollination, Soil Erosion, Fertility, Ground Water Recharge, Water Purification, Waste Treatment) quantified district-wise for 2019, and various scenarios of 2033 reveal that the regulating services from forest ecosystems in 2019 are about 926 billion INR compared to 724 billion INR in 2033 (BAU Scenario), 716 billion INR (ALT Scenario), 720 billion INR (RFP Scenario), 725 billion INR (AF Scenario) and 824 billion INR (SDP Scenario), emphasising the need to implement the sustainable development programme (SDP Scenario) to sustain regulating services to support diverse biota on the Earth.

Cultural services of forest ecosystems in 2019 are about 294 billion INR compared to 291 billion INR in 2033 (BAU), 291 billion INR (ALT, RFP, AF Scenarios), and 293 billion INR (SDP Scenario).

The total ecosystem service value of forests in Karnataka is about 1835 billion INR in 2019. However, with the land-use changes leading to degradation of forest ecosystems, the total ecosystem services value would be 1540 billion INR (BAU Scenario), 1530 billion INR (ALT Scenario), 1546 billion INR (RFP Scenario), 1547 billion INR (AF Scenario) and 1658 billion INR (SDP Scenario), emphasizing the need to protect reserve forests and take up afforestation of degraded forest patches to sustain ecosystem goods and services. The relative share of ecosystem services for 2019 and considered policy scenarios. Regulating services dominate (47–51%) compared to the provisioning (32–34%) and cultural services (16–19%).

1 Introduction

The landscape is a mosaic of ecosystem elements, which changes in size, shape, spatial arrangement, and quality of the patches/elements due to the complex, multi-scalar processes that influence the ecosystem's biotic components. The changes in the abiotic and biotic assets of a landscape are referred to as landscape dynamics. Changes in the structure of the landscape will have implications on ecosystem functions and processes. Landscape transitions leading to deforestation in the landscape due to widespread land use land cover (LULC) changes have been acknowledged as prime agents towards contributing to global warming with the enhanced emissions and loss of carbon sequestration potential. Land-use changes in forested landscapes have gained momentum in the middle of the last century due to industrialization, urbanization, and globalization. The forest loss of 2.3 million sq.km was reported globally during the past decade to cater to the burgeoning population's growing demands coupled with unplanned developmental activities. The efforts to restore the landscape through afforestation initiatives are only 0.8 sq.km. LULC changes information provide vital details on natural resources availability, utilization, etc., for evolving appropriate management strategies to ensure the sustainability of natural resources. Thus, LULC changes in a region are the outcome of the complex interaction of policy, economics, management, culture, and environmental factors. Drastic changes in LULC will significantly impact biodiversity, climate, hydrologic regime, biogeochemical dynamics, etc., affecting the sustainability of natural resources. This will profoundly impact land surface climate feedback by altering heat, moisture, and albedo exchange at local, regional, and global scales. Clearing large-scale forests will contribute to releasing the carbon stored in vegetation and soils by altering their physical and chemical properties, which affects the global climate by increasing levels of greenhouse gases in the atmosphere and decreasing evapotranspiration and hydrological cycle. LULC changes have been posing severe challenges to the carrying capacity of the landscape. Mitigation of these impacts entails actions at the local/micro-scale by the preservation of ecosystems. Knowledge of agents in LULC transition with the quantification of landscape dynamics would help frame conservation strategies toward sustainable land uses.

Humans have been altering landscape structures to meet their needs. However, unplanned land-use changes have affected ecological functions; for instance, rapidly changing land cover due to the influence of emerging paved surfaces threatens the sustenance of natural resources, affecting the food and water security of a region. This necessitates land use land cover change (LULCC) analyses to evolve strategies for prudent planning and sustainable management of natural resources and mitigation of changes in the climate. Urban expansions are taking place rapidly, with a higher migration rate from rural areas to fringes and, successively, to core cities. In recent decades, a diversified and growing economy, expanding city limits beyond municipal boundaries to create special economic areas, industrialization, social status, etc., have transformed green spaces into residential or industrial land use. The impacts of the rapid urbanization process on vulnerable natural reserves can be addressed through simulation and geo-visualization of likely changes with robust land-use change models. Challenges in modeling urban growth are manifold since the global level (top-down approach) modifications considering socio-economic and biophysical factors could severely affect local levels. Predicting long-term land-use change with minimum uncertainty is challenging because of the dynamics governing LULCC. The scenario-based modeling helps in designing policies towards human well-being considering geomorphology (environment), connectivity (mobility), facilities, government (development plans), demographic, socio-economic, sustainability (water bodies, green spaces, etc.), and other factors with their effects on LULCC (Wahyudi and Liu 2013). Scenarios may be region-specific or country-specific since enormous factors will affect land-use changes in the future.

Spatially explicit data helps understand the heterogeneity within a landscape to support policy coordination from local to regional scales by establishing a common and agreed set of data and a common framing of the relationship between the environment and economic and human activity within the ecosystem accounting framework. The availability of coherent data supports incorporating environmental data into business and finance sectors' decision-making, complementing the wide range of initiatives underway in those sectors to recognize the importance of ecosystems and biodiversity. Thus, data from ecosystem accounts used with other methods and tools in modeling and scenario analyses aid in policy and decision-making (SEEA EA 2020).

The monetary values recorded in the ecosystem accounts reflect the current use of ecosystems as per the standard protocol of SNA—the system of National Accounts. They are based on the existing management regimes and institutional arrangements regardless of whether the associated patterns of use may be considered (un)sustainable or (in)efficient. However, assessing policy scenarios reflecting alternative management regimes or institutional arrangements for ecosystems in many policy contexts is essential. For example, analyze how certain negative externalities (e.g., pollution) might best be internalized in economic unit decisions. SEEA EA deployed in the application of scenario analysis considers interconnectedness between the natural environment, human societies, and their economies, providing insights into policymaking.

Quantifying scenarios with mathematical simulation models helps create quantitative estimates for various scenarios (e.g., implementing or not implementing a proposed policy) that can inform the policymaking process. The exercise of informing decision-making and using scenarios to assess the outcomes and effectiveness of various policy intervention options is often termed policy scenario analysis. The provision of a standardized approach with consistent and coherent data targeting policy relevance and the involvement of local stakeholders in policy analysis can support the use of accounts, further develop models, and create new models, with the ultimate goal of informing policy decisions. This is achieved through (SEEA EA 2020):

- New knowledge of ecosystem extent and quality in the ecosystem services that benefit communities and human well-being allows for incorporating ecosystem services in social and economic assessments.
- Creating coherent and harmonized accounts allows for developing new models using such data framework.
- Promotion of the use of a systemic approach that assesses (a) the extent of ecosystems, (b) conditions of ecosystems considering the impact of human activity on ecosystems, and (c) models that determine the extent to which ecosystems influence human health and human activity.
- Application of standard approaches for valuation of ecosystem services and assets based on exchange values.
- Improving the analysis performed with sectoral models by introducing physical indicators on ecosystem extent, condition, and services and hence generating a higher degree of realism.
- Generating knowledge on how existing models could be connected better to represent the relations between society, economy, and environment.
- Use of simulations, extending the analysis provided by SEEA, by forecasting or backcasting scenarios.
- Making explicit the importance of site-specific drivers of change, system responses, and impacts, with the use of a spatially explicit analysis that allows users to determine the value of ecosystem services based on the location where these are used (i.e., more explicitly assess demand and supply).

The review of published literature suggests there are well-established models such as Cellular Automata—Markov Chain (CAMC), SLEUTH, CLUE-S, multi-criteria evaluation (MCE), Fuzzy-analytical hierarchical process (AHP) models, agent-based models (ABM), CA-artificial neural network (ANN), etc., which can be incorporated into urban studies to predict future land use trends. These models have been designed both at a local and global scale. Existing modeling techniques such as CA-Markov, constrained-CA, multi-criteria evaluation-CA, and logistic-CA have advantages such as the generation of richer forms of cells at individual class levels (Batty 2005). Still, they fail to calibrate and optimize factors responsible for cell state changes and lack expansion with the intermixing of cells. This chapter explores traditional cellular automata, the Markov chain-based Agent-Based Model (ABM), to visualize realistic urban growth in the future through sustainable development concepts.

2 Review of Land-Use Change Models

Over the last four decades, scientific communities have contributed immensely to urban growth models with a common goal to study land-use dynamics and simulate urban growth using geospatial techniques (Verburg et al., 2002; Sante et al., 2010). Theoretical assumptions, the method followed, spatial, temporal aspects, and geographical extents might vary from each model type, but the outcome of these models is to understand the complex interrelationships between the natural ecosystem and urban environment by observing irreversible heterogeneous patterns of change (Crooks 2010; Li and Liu 2006). Wu and Silva (2010) reviewed urban land dynamics and related models. They highlight the significance of artificial intelligence (AI) and its deep theoretical understanding of urbanization and relative pressure on land use pattern change. Figure 1 shows broad categories of AI systems used in urban models.

2.1 Cellular Automata Model

The origin of cellular automata (CA) based models dates back to the 1940s when Ulam explored the CA framework and started investigating the complex behaviour of systems, its effect, and functional aspects (Ulam 1958; Torrens 2000; Eckhardt, 1987). CA models can be defined as a class of two-dimensional lattices of cells. Each cell representing a land use category tends to change its state over time depending on every cell's local neighborhood. Standard CA applies a "bottom-up" approach. The approach argues that local rules can create complex patterns by running the models in iterations (Batty and Xie 1994). Itami, 1994 states that CA is seen not only as a framework for dynamic spatial modelling but as a paradigm for thinking about complex spatial–temporal phenomena and an experimental laboratory for testing ideas. According to Wolfram (2002), CA can be defined as discrete two-dimensional dynamic systems in which local interactions among components generate global changes in space and time. Torrens (2000) explains that CA is composed of the below four components: Cell space, a number of finite states, neighbourhood, and transition functions or rules. CA has been considered and extensively utilized by the scientific community in the majority of fields, for instance, medicine or biological modeling, forest fire modeling, flood modeling, urban, landscape dynamics, etc.,

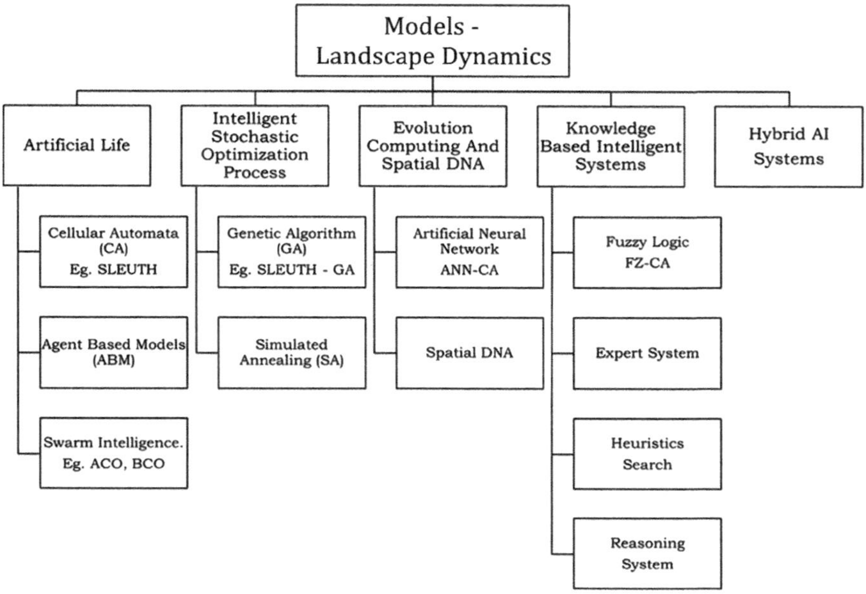

Fig. 1 Different types of landscape dynamics models

due to its simplicity, flexibility, and intuitiveness (Ermentrout and Edelstein 1993; Guidolin et al. 2016; Li and Gong 2016).

The modeling of likely land-use changes can be carried out by the rule-based CA-Markov (Cellular Automata-Markov) technique. Markov's approach has provided information about transition probability between two LU classes with respect to time (t to time t + 1) and transitional area matrix concerning likely land-use changes (extent) of each class. The transitional probability matrix and area matrix are obtained by Eqs. 1 and 2.

$$P = \begin{bmatrix} P_{11} & P_{12} & \dots & P_{1n} \\ \vdots & \vdots & & \vdots \\ P_{n1} & P_{n2} & \dots & P_{nn} \end{bmatrix} \tag{1}$$

where P is the Transitional probability matrix; P_{ij} is the probability of ith land use to convert into jth class during the transition period; n is the number of land use classes.

$$\text{The transition area matrix is obtained by } A = \begin{bmatrix} A_{11} & A_{12} & \dots & A_{1n} \\ \vdots & \vdots & & \vdots \\ A_{n1} & A_{n2} & \dots & A_{nn} \end{bmatrix} \tag{2}$$

where A is the Transitional area matrix; P_{ij} is the sum of the area of ith land use to convert into jth class during the transition period.

CA was used to obtain a spatial context and distribution map. CA aided in simulating and predicting land use changes based on the transitional rules depending on the state of cell changes according to the neighbourhood cells and the current cell's previous state with the local and regional interactions. LU change patterns follow the Markovian random process properties with various constraints that include the average transfer state of LU structure stable, and different LU classes may transform to other land use classes given certain conditions (such as non-transition of urban class to water or vice versa). Thus, Markov was used to derive the land use change probability map for the study region. CA coupled with the Markov chain was then used to predict urban land uses using Eq. 3. CA-Markov incorporates the transitional rules and probabilities collectively and provides better results.

$$L_{(t+1)} = P * L_{(t)} \tag{3}$$

where $L_{(t+1)}$ is the land-use status at time $t + 1$; $L_{(t)}$ is the land-use status at time t.

A contiguity filter of kernel size 5×5 was used to account for the behavior of neighborhood pixels. Past LU data were used for estimating transition probability and area matrices through the Markov model. Then, the Markov-CA model was used to simulate LU, which was compared with the actual LU, and the accuracy of the simulation was assessed through Kappa statistics. The traditional Kappa statistics lack in assessing prediction error with respect to pixel location (Pontius and Millones 2011). The revised kappa statistics were assessed to understand the accuracies of the prediction as compared to traditional kappa (K standard). A revised general kappa is defined as kappa for no ability (Kno), Kquantity and Klocation. The Kquantity and Klocation are able to distinguish clearly between quantification error and location error, respectively (Ahmed et al. 2013).

2.2 Analytical Hierarchical Process

Analytical hierarchical process is used to develop a set of relative weights for a group of factors considered for urban models. The weights are developed by providing a series of pairwise comparisons of the relative importance of factors to the suitability of pixels for the activity being evaluated. AHP process follows steps, including specifying the hierarchical structure, determining the relative importance weights of the factors or criteria and sub-criteria, estimating Eigenvectors or priority vectors, assigning preferred weights of each alternative, and determining the final score (Khwanruthai and Yuji 2011). AHP includes a measure of consistency for the individual comparison matrix of the decision problem. Two factors are considered for pairwise comparison, and judgment is made to assign weights on a scale of 0–9. The consistency index depends on the principal Eigenvector (λ max) value. The consistency ratio (CR) is an approximate mathematical indicator of the consistency of pairwise comparisons. CI and CR are estimated as per Eqs. 4 and 5, respectively.

$$\mathrm{CI} = \frac{\boldsymbol{\lambda}\mathbf{max} - \mathbf{n}}{\mathbf{n} - 1} \tag{4}$$

$$\mathrm{CR} = \frac{\text{CONSISTENCY INDEX}}{\text{RANDOM CONSISTENCY INDEX}} \tag{5}$$

where, $\lambda\max = \sum (\textit{Eigen vector} * \textit{Reciprocal matrix col. sum})$

Saaty, 1980, described that a consistency ratio of 0.10 is acceptable to achieve minimum accuracy. However, lower values are preferred for high accuracy. If CR > 0.1, judgments are not trustworthy because they are too close for the values taken up randomly.

Multi-criteria evaluation (MCE) is a decision support tool for evaluating choices between alternative factors. The basis for a decision is known as a criterion. In a Multi-Criteria Evaluation, an attempt is made to combine a set of criteria to achieve a single composite basis for a decision according to a specific objective. The MCE process is necessary to generate site suitability maps for future land use prediction taken to CA models. Reformed CA approaches such as SLEUTH, agent-based model (ABM-CA), and Genetic Algorithm (GA) (Clarke 2008; Spencer 2009) have their greatest extensibility and adaptability for datasets because of optimization and calibration processes improving the output generated by enhanced self-modified cell transition rules.

2.3 Agent-Based Model

To address issues related to urban growth, such as sprawl, congestion in core cities, and disaggregation, researchers have started adopting bottom-up approaches such as ABM, allowing to simulate and consider individual actions of varied agents by measuring evolving behavior (Crooks and Heppenstall 2012). The aim of the agent-based model (ABM) is to create an environment where interactions are programmed to be self-constrained or in other words, actions can be controlled on their own. Therefore, an agent is autonomous, functioning independently, and interacts socially with other agents in a specified environment while following explicit goals responsible for the agent's behavior (Macal and North 2009; Franklin and Graesser 1996).

Axtell (2000) discusses the need for agents in situations where it is important to adapt and change behaviors, learn, engage in dynamic relationships with other agents, add a spatial component to identify behavior and interactions of agents in the real-world environment, and so on. An autonomous active agent is defined by characteristics that are listed in Table 1 (Holland 1995):

Figure 2 illustrates the basic structure of an agent. The agent interacts with the landscape, leading to four possible relations: the self-influencing behavior of agents,

Table 1 Characteristics of agent-based model

Characteristic	Description
Environment	In a dynamic system, agents operate and interact with other agents and change over time
Sensing capabilities	Defined with respect to agent's sensors. Sensing of an agent may be active or passive, internal or external, depending on the environment
Action	It causes a significant change in the current state and is responsible for behavior in the future. Agents are goal-oriented
Interaction	Agents have extensive ability to communicate with other agents
Drives	Considered as preferences. They can be observed because of the relationship between the functional modules of an agent. The agent may have a single drive or multiple drives
Action selection	Guides future prediction and initiates actions accordingly
Mobility	Agents have the freedom to wander within a specified environment. Hence, taking this advantage, agents are often interspersed and juxtaposed to interact, permitting a vast range of applications
Self-learning	Agents possess a unique characteristic of adaptive or complex adaptive systems where agents alter their state depending on the previous state

the self-influencing state of a landscape at any given time, agents affecting the landscape, and vice versa. Apart from these relations, an agent might also interact with other agents and the environment; meanwhile, the landscape is dependent on external variables or could be influenced by other factors (Crooks and Heppenstall 2012). The major advantages of ABM over traditional modeling techniques are threefold:

(a) Captures emerging spatio-temporal footprint of land-use change;
(b) Provides a domain-specific environment of a region under investigation;
(c) Highly flexible for developing geospatial models in particular.

Apart from these advantages, Bernard (1999) lists a few characteristic changes of the traditional modeling method versus ABM (listed in Table 2). Some challenges need to be addressed when it comes to land-use change models are (1) there is no particular framework available for systematic assessment of land use land cover change prediction and hence its validation considering the decision-making process and authenticity of interactions between various actors in a system, (2) another major challenge lies in the resolution of data, availability, and trade-offs between different resolution. Since modeling results directly depend on various aspects of remote sensing resolution, predicting quantities of change would not be reliable if spatial patterns are not considered under the lens of resolution. Land use change is dependent on land management policy framed by local, regional, national, and global agencies. Therefore, there is a need to integrate these factors along with human–environment interaction to achieve desirable results.

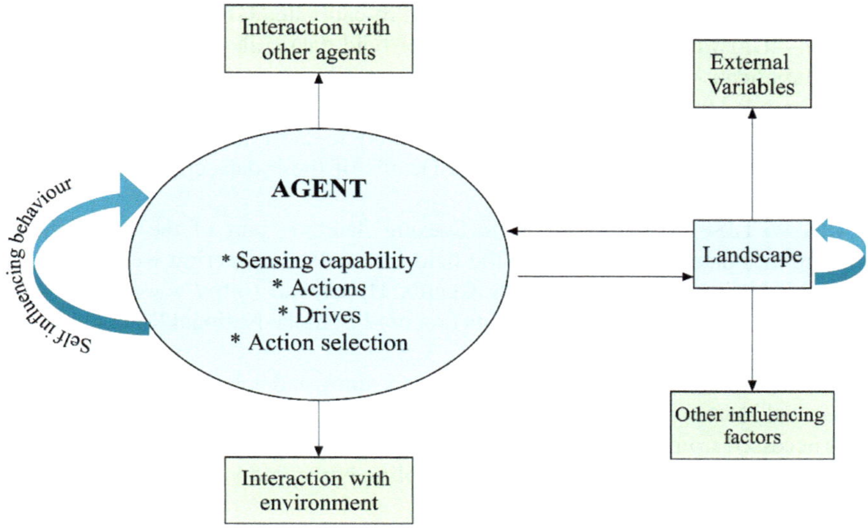

Fig. 2 Structure of an agent (Adapted from Batty 2005)

Table 2 Advantages and comparison of ABM versus traditional modeling method

Traditional modeling	ABM
Deterministic (limited to one feature)	Stochastic (multiple features)
Allocative (top-down approach)	Aggregative (bottom-up approach)
Equation based	Adaptive agents
No explanations	Explanatory in nature
Considers a few parameters	Considers many parameters
Spatially coarse	Spatially explicit
Works in a given environment	Environment is created

3 Data

The process of data acquisition involves primary data collection, such as temporal satellite data, field data, surveys, etc. USGS Earth Explorer (https://earthexplorer.usgs.gov/) has a repository of Landsat data starting from 1984–to date.

Ancillary data include cadastral revenue maps (1:6000), the Survey of India (SOI), topographic maps (1:50,000 and 1:250,000 scales), and a vegetation map of South India developed by the French Institute (1986) of scale 1:250,000. The Topographic maps provided ground control points (GCP's) to rectify remote sensing data and scanned paper maps. The vegetation map of South India (1986) of scale 1:250,000 (Pascal 1986) was digitized to identify various forest cover types and classify RS data of the 1980's. Other ancillary data include land cover maps, administration boundary data, transportation data (road network), etc. Pre-calibrated **GPS** (Global Positioning System—**Garmin GPS**) units were used for field data collection, required for RS data classification, and validation. Various vector layers in the form of point data were compiled from authorized sources, ancillary point, line, and polygon data from virtual online spatial repository sites such as Google Earth (http://earth.google.com) and Bhuvan (http://bhuvan.nrsc.gov.in). Details of these datasets are described in Table 3.

IRS P6 LISS-IV (Indian Remote Sensing Satellite, part of the Indian Space Programme) data concurring with the field data collection period were purchased from the National Remote Sensing Centre, Hyderabad (http://www.nrsc.gov.in). Table 4 provides the details of IRS data procured from the National Remote Sensing Centre (NRSC), Hyderabad.

Agents were delineated in the form of points, lines, and polygons. A list of various agents and sources of data is given in Table 5. The proposed large-scale project details were accessed from the Infrastructure Development Ports & Inland Water Transport Department, Government of Karnataka, weblink https://idd.karnataka.gov.in/.

Table 3 Data used and description

	Data	Description	Source
1	Landsat-5 TM Landsat-7 ETM + Landsat-8 OLI-TIRS	30 m data was collected for the entire Karnataka region. Used to generate maps such as land use, urban, forest, agriculture, etc.	https://earthexplorer.usgs.gov/ https://landsat.gsfc.nasa.gov/ https://landsat.gsfc.nasa.gov/data/more-free-data
2	ASTER DEM	30 m data used to generate slope map for Karnataka region	https://asterweb.jpl.nasa.gov/gdem.asp
3	Open street map (OSM)	Road data updated with classified images (originally vector, rasterized)	https://www.openstreetmap.org/search?query=karnataka#map=7/15.067/78.849
4	City development plans	To create excluded maps or regions restricted from future development	http://www.bdabangalore.org/ http://www.uddkar.gov.in/
5	Google Earth and Bhuvan	Geo-correction, classification and validation. Collection of point, line and polygon data (originally vector, rasterized)	https://www.google.com/earth/ https://bhuvan-app1.nrsc.gov.in/thematic/thematic/index.php
6	Field data—GPS	Geo-correction, training data and validation data, Agents Extraction	

Table 4 Details of IRS data for Karnataka State (2019) with respective path and row

S. no	District	Satellite data path & row	
1	Bagalkot	146,048, 146,049, 145,049	17 scenes of IRS data cover Karnataka state
2	Belagavi	146,048, 146,049	
3	Bellary	145,049, 145,050, 144,049, 144,050	
4	Bengaluru Rural	144,051	
5	Bengaluru Urban	144,051	
6	Bidar	145,057, 145,058	
7	Chamarajanagar	144,052, 144,051	
8	Chikballapur	144,050, 144,051	
9	Chikkamagaluru	145,050, 145,051	
10	Chitradurga	145,050, 145,051, 144,050, 144,051	
11	Dakshina Kannada	146,051, 145,051	
12	Davanagere	145,050	
13	Dharwad	146,049, 146,050	
14	Gadag	145,049, 145,050, 146,049, 146,050	
15	Hassan	145,051, 144,051	
16	Haveri	146,050, 145,050	
17	Kalaburagi	145,048, 144,048	
18	Kodagu	145,051, 145,052	
19	Kolar	143,051	

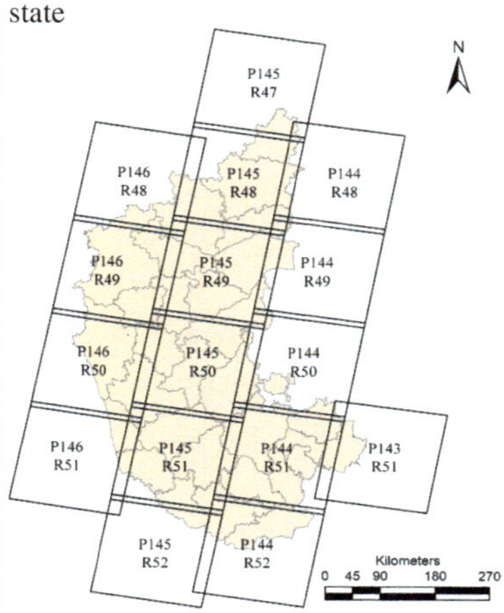

(continued)

Table 4 (continued)

S. no	District	Satellite data path & row
20	Koppal	145,051
21	Mandya	144,051, 144,052
22	Mysuru	144,051, 144,052, 145,051, 145,052
23	Raichur	145,049, 144,049
24	Ramanagara	144,051
25	Shivamogga	146,050, 146,051, 145,050, 145,051
26	Tumakuru	144,050, 144,051, 145,051
7	Udupi	146,050, 146,051, 145,051
28	Uttara Kannada	146,049, 146,050
29	Vijayapura	146,048, 146,049, 145,048, 145,049
30	Yadgir	145,048, 145,049, 144,048, 144,049

4 Method

Chosen scenarios with the modeling exercise's details to visualize likely land uses / land cover in 2033 are listed in Table 6.

Figure 3 outlines the method overview to assess land-use transition in the Karnataka region using CA-ABM modeling technique with various scenarios. The overall method includes three subdomains, namely: (i) Data preparation and image pre-processing (ii) Land use and land cover extraction, and (iii) Hybrid modeling technique.

Table 5 Description of various agents and constraints considered for modeling

S. no	Agent type	Description	Source
1	Power plants	Details of power plant projects in Karnataka with attributes such as power Station, latitude, longitude, type (Hydel, thermal, mini, solar, and wind), and total capacity in Megawatts	http://karnatakapower.com/en/generation/ https://energy.karnataka.gov.in/ https://ksei.gov.in/
2	Ports and harbors	Details of major and minor ports situated along the coastline. The state has 1 major and 12 minor ports functional from the North to the South are Karwar, Belekeri, Tadri, Pavinakurve, Honnavar, Manki, Bhatkal, Kundapur, Hangarkatta, Malpe, Padubidri & Old Mangalore Port	https://kum.karnataka.gov.in/KUM/PDFS/PortPolicy.pdf http://dpal.kar.nic.in/
3	Large scale industries	Details of industries such as iron, steel, aluminum, cement and concrete across the state were digitized as point layers	https://ebiz.karnataka.gov.in/kum/index.aspx https://www.karnatakaindustry.gov.in/
4	Special economic zones	Details of various special economic projects such as information and biotechnology parks, business parks, pharmaceutical, engineering, automobile, and aerospace industry, electronic hardware and textile	https://pib.gov.in/PressReleaseIframePage.aspx?PRID=1578141 https://vtpc.karnataka.gov.in/storage/pdf-files/SEZ_EOU/SEZ%20Status%20Report.pdf https://planning.karnataka.gov.in/
5	Hotels	Details of major hotels of Karnataka	https://www.kstdc.co/
6	Airports	Various types of airports such as international, domestic, airbase, flying school, private and proposed	http://www.ksiidc.com/airstrips.html https://idd.karnataka.gov.in/info-2/Airports++Airstrips+and+Helipad/Airports/en

(continued)

Table 5 (continued)

S. no	Agent type	Description	Source
7	Ecology and environment	Layers related to eco-tourism destinations, lake rejuvenation projects, recreation places, forest regions, and waterfalls	https://aranya.gov.in/ https://parisaramahiti.karnataka.gov.in/ https://www.karnatakatourism.org/ https://www.kstdc.co/ https://empri.karnataka.gov.in/storage/pdf-files/Reports/SoER%20final%20Kannada.pdf
8	Road and the railway line (existing and proposed)	Various types of roads such as NH, SH, MDR, ODR, and sub-arterial road. Existing railway line and proposed railway line connecting major cities of Karnataka	https://transport.karnataka.gov.in/english http://www.kship.in/en/ http://www.shdpkar.in/ http://89.238.162.147/cucpl/About_us.aspx https://kride.in/ https://transportsec.karnataka.gov.in/
9	KIADB and SEIAA industrial layout	Point and polygon digitization of industries location as well as plots sanctioned by the government	http://164.100.133.168/kiadbgisportal/ http://seiaa.karnataka.gov.in/
10	Mines	Demarcating major mines of Karnataka such as granite, coal, metal, etc.	https://karunadu.karnataka.gov.in/dmg/english/Pages/home.aspx https://mines.gov.in/writereaddata/UploadFile/Karnataka.pdf
11	Bus stops	Details of various urban bus stops	https://kgis.ksrsac.in/kgis1/portal.aspx

(continued)

Table 5 (continued)

S. no	Agent type	Description	Source
Constraints			
S. no	Constraints type	Description	
1	Protected areas	National Parks (IUCN Category II), Wildlife Sanctuaries (IUCN Category IV), Community Reserves (IUCN Category V and VI, respectively), Tiger Reserves, Sanctuaries, etc., are declared as protected areas under the Wildlife Protection Act 1972 by the union government on the specific request of state government. Protected areas are prime regions of land scape that protect biological diversity, maintain ecological integrity, and provide livelihoods to local communities. These regions act as an important corridor for wildlife and endemic flora in the Western Ghats of Karnataka, protected by India's Union government	https://aranya.gov.in/aranyacms/English/WildLife.aspx http://www.wiienvis.nic.in/Database/Maps_PAs_1267.aspx https://www.protectedplanet.net/en
2	Reserve forest boundary	The respective state governments earmark reserved forests under the provisions of Section 29 of the Indian forest act 1927 to reduce the pressure on forests. Activities such as hunting, grazing, etc., are banned in reserved forests unless specific orders are issued otherwise	https://aranya.gov.in/aranyacms/English/WildLife.aspx https://parisaramahiti.karnataka.gov.in/
3	Water bodies	Lakes, Reservoirs, and check dams are considered prime source for agriculture and help maintain the ecosystem	https://soinakshe.uk.gov.in/ https://indiawris.gov.in/wris/#/
4	Very dense forests	Interior forest is also known as intact forests which are not having any disturbances. They act as a primary habitat for numerous flora and fauna	Land use analysis, Fragmentation analysis (Ramachandra and Bharath 2021)

Modeling of landscape dynamics has been carried out using a hybrid Fuzzy Analytical Hierarchy Process (AHP) based spatial Markov chains (MC) Cellular automata (CA) (Fuzzy AHP-MCA) technique using temporal data. This technique has advantages than other traditional procedures that integrate agents and distance-based relationships of driving forces in the simulation. Hybrid models assist in identifying criteria (n) and alternatives (m) in interactive decision-making by comparing the relative importance of two elements (criteria or alternatives) i and j for a given pair-wise comparison matrix of the likelihood of events for all possible alternative ranking outcomes. This technique integrates expert knowledge of spatial data to determine each factor's weight, which influences land suitability classification and also lends to evaluating the various policies that influence decision-making. Urbanizing agents,

Table 6 Modeling scenarios

	Method	Datasets	Scope
(i) Business as usual scenario (BAU)	CA Markov Likely, land uses in 2033 will be generated BAU assumes the current development will continue, and evaluates the various agents responsible for the change, and forecasts the future landscape status	Land use transitions (based on temporal, and spatial extent of ecosystems)	Ecosystems in Karnataka state
Policy context			
(ii) Agent-based land-use transition scenario (ALT)	Likely land uses in 2033 will be generated under various scenarios considering (1) Markov Chain transition of base land uses, (2) evaluating the driving factors (agents) and constraints, (3) weightage metric score by fuzzy AHP based estimation and site suitability map generation by MCE, (4) simulation and future prediction of land use by MC-CA algorithms	Various drivers (agents) such as proposed (new) developments by the government, existing and proposed (i) industries, (ii) liner projects, (iii) urbanization, (iv) slope, (v) core built-up areas, (vi) special economic zones (SEZ), etc., responsible for the land-use changes in the neighborhood	All ecosystems, Karnataka state, India
(iii) Reserve Forest Protection (RFP) and stringent conservation of national parks and sanctuaries scenario	Considering protected areas –The modeling approach is the same as (ii)	The spatial extent of reserve forests, national parks, sanctuaries	All ecosystems, Karnataka State, India

(continued)

Table 6 (continued)

	Method	Datasets	Scope
(iv) Afforestation (High conservation) scenario (AF)	Considering afforestation initiatives –Same as (ii)	The spatial extent of afforestation data (during the past decade) and proposed afforestation	All ecosystems, Karnataka state, India
(v) Sustainable development policy	It includes constraints as in scenarios 3 and 4, allowing growth in areas other than forest areas. Though scenarios 3 and 4, the modeling approaches focus on "Ecological protection priority"	This is scenario is more conservative as well as a focused growth scenario This scenario further allows the comparison of development and conservation tradeoffs effectively. This scenario stresses the strict implementation of spatial policies such as not allowing land use change within natural forests, increasing plantation forests in barren lands, degrading woodlands, and protecting water bodies	All ecosystems, Karnataka state, India

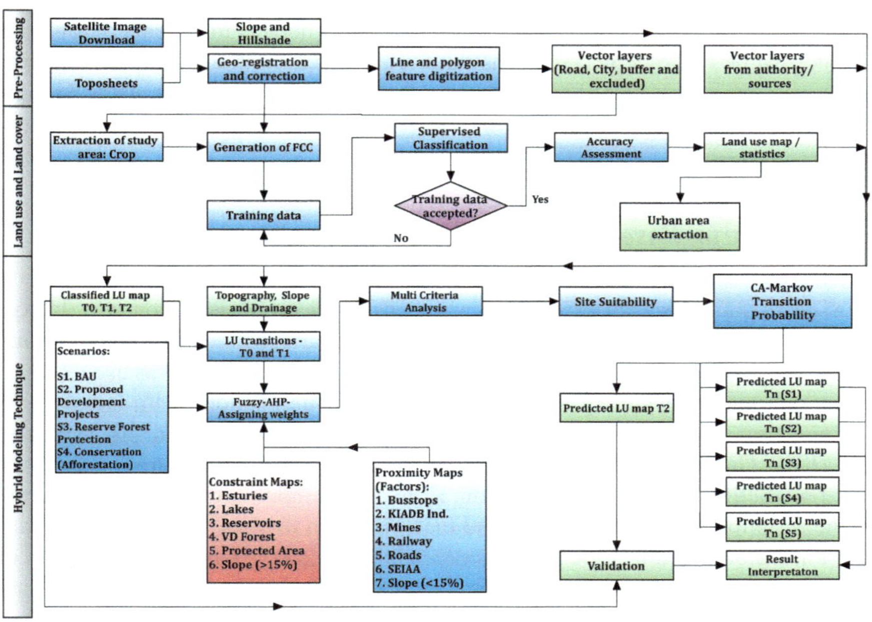

Fig. 3 Method adopted to analyze land use transitions in the Karnataka region

Table 7 Agents and constraints considered for modeling

Agents	Industries, proximities to roads, railway stations, metro stations, educational institutes, religious places, and service infrastructure such as police stations, hotels, hostels, etc.
Constraints	Drainage lines, slope, water bodies, Reserved regions for non-development, Protected areas, catchment areas, etc.

constraints, and classified data were considered as base layers for modeling. Criteria are of two types: Constraints and Factors, which are listed in Table 7. Constraints exclude or constrain areas from being considered for a type of land use. Constraints are represented by Boolean images, where 0 is assigned to areas that are not suitable, and one is assigned to suitable areas. For example, water bodies are not considered suitable for residential/urban development under any condition. Constraints taken into account in this study include water bodies, forest zones, protected areas, catchment areas, etc., based on the city development plan (CDP). Factors are criteria that assign a relative degree of suitability for each location for the land use under consideration. Factors define a degree of suitability and are represented by a fuzzy scale. For example, proximity to roads is a factor that enhances the suitability of land for urban development. Various driving factors used as input for modeling include road and railway networks, bus stops and railway stations, industries, educational institutes, places of religious importance, and service infrastructure such as shopping malls, hospitals, police stations, hotels, etc. Figure 4 depicts the agents of land-use changes, while Fig. 5 depicts the constraints considered for modeling. These factors and constraints were identified by developing attribute information using Google Earth data (https://earth.google.com).

4.1 Scenario-Based Prediction

Four development scenarios were considered based on various complex factors provided by the city development plan for the Karnataka region, 2033. Many researchers have successfully demonstrated the usage of the business as usual (BAU) scenario, which considers past and recent growth patterns along with factors such as population, socio-economic trends, urban density, etc. (Samie et al. 2017; Hamad et al. 2018; Guzman et al. 2020). Another name derived from literature for scenario-based prediction is "what-if" prediction. Here, the simulation will be based entirely on the assumption of certain layers and factors, which implies "what will happen if factor a, factor b, and factor c are considered" (Singh 2003). Knowing the impact of new urban spread and possible placement of new infrastructure, green area conservation, regulation of zones, upgrading existing transport corridors, and related changes in land use categories inside the perimeter of the study region considered will be highly significant for urban planning and municipal authorities. With this context, the four scenarios considered are:

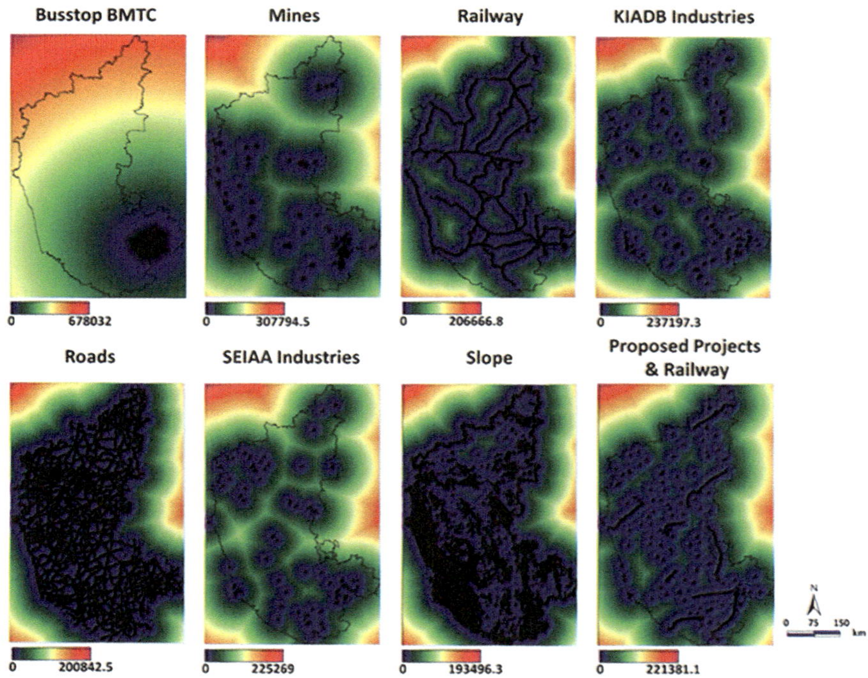

Fig. 4 Distance maps of various urban growth factors

4.1.1 Scenario—1: Business as Usual (BAU)

BAU assumes the historical trend of land-use changes from 1985–2019 will continue without any change in the environmental and economic development policies during 2019–2033. This scenario assumes the current development will continue, evaluate the various agents responsible for the change, and forecast the future landscape status. Further restrictions include protecting existing water bodies such as lakes, reservoirs, and rivers as per the land use map of 2019. There is scope for a major improvement in transport corridors, special economic zones, group housing schemes, and special projects under the Karnataka Housing Act. Other development zones allowed include peri-urban use zones, commercial use zones, manufacturing use zones (industrial layouts), utilities use zones, and multiple-use zones. BAU scenario will provide a benchmark for comparing the consequences of ad-hoc land use policies for the decision-makers.

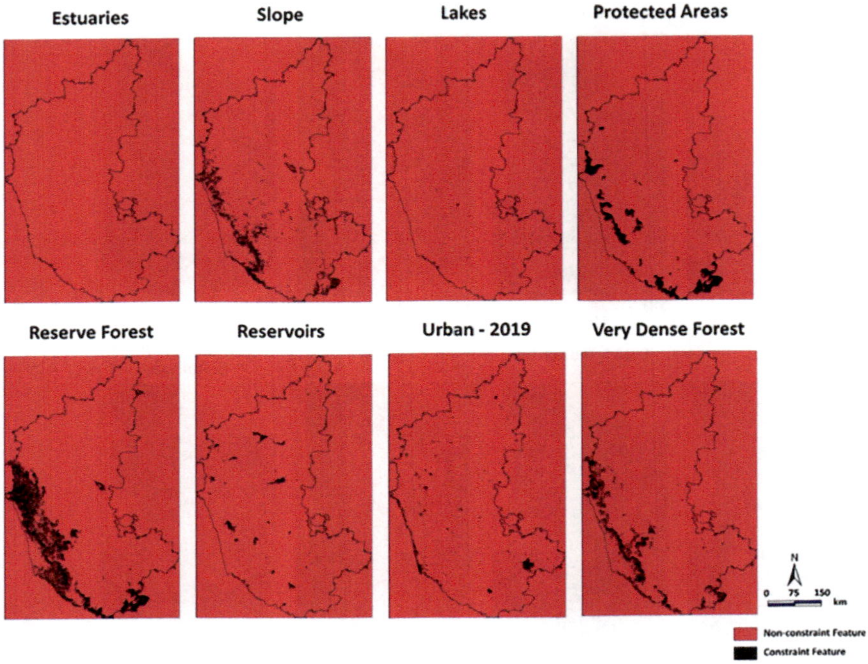

Fig. 5 Constraint maps

4.1.2 Scenario—2: Agent-Based Land-Use Transition Scenario (ALT)

This scenario evaluates the role of various drivers (agents) such as proposed (new) developments by regulatory agencies, existing industries, liner projects, urbanization, slope, core built-up areas, special economic zones (SEZ), etc., responsible for the land-use change on their neighborhood. This scenario emphasizes the priority of economic development by allowing the development to favor economic development by compromising ecological conditions. The scenario weighs individual factors responsible for improving the socio-economic condition and provides the probable land-use change due to these agents based on the multi-criteria decision-making. Unlike scenario 1, it includes proposed railway and industry project layouts as drivers of change.

4.1.3 Scenario—3: Reserve Forest Protection Scenario (RFP)

The National Forest Policy, Government of India, 1988 envisaged a 'forest and tree cover31' of 33% (63 lakh ha) for maintaining the environmental stability apart from meeting the requirements of fuel wood, fodder, minor forest produce and small timber for the rural and tribal populations. Afforestation has been taken up in the State under various State and Centrally Sponsored Schemes and externally funded projects viz., Forestry and Environment Project for Eastern Plains (FEEP) and Karnataka Sustainable Forest Management and Bio-diversity Conservation Project (KSFMBC) funded through loan assistance from Japan Bank for International Co-operation (JBIC). In its report (March 2006), the National Forest Commission recommended that each State have its forest policy within the National Forest Policy's broad parameters, 1988 for sustainable management of its forest. The Government needs to formulate State forest policy on priority. State Forestry Action Programme was drawn up (May 1996) by the Department for a 20-year period 1997–2017 for conservation, rehabilitation, and forestry activities and forest management with people's participation, which had also not been approved by the government. The government, however, introduced (June 2002) a scheme of Joint Forest Planning and Management (JFPM) to promote the participation of people living on the periphery of forest areas and make them joint stakeholders. Under the scheme, Village Forest Committees (VFCs) were to prepare management plans to develop and protect degraded forests.

Forest (Conservation) Act, 1980 (Act) imposes restrictions for the de-reservation of reserved forests and forest land used for non-forest purposes for which approval is required to be obtained from the Central Government, which is given subject to fulfillment of prescribed conditions in each case. The main condition governing such diversion is carrying out 'Compensatory Afforestation' (CA) on equivalent identified non-forest land and twice the area over degraded forests if non-forest land is unavailable or is less available. Departmental instructions provide that CA should be taken up soon after the land is diverted for non-forest purposes by recovering the cost at the prescribed rate in advance from the user agencies. The 'Net Present Value' (NPV) of the existing forests is also recoverable from 31 October 2002. NPV as part of CAMPA charges is to be assessed and recovered at the specified rates of that range, from Rs.5.80 lakh to Rs.9.20 lakh per ha, depending upon the type of forest land and density of the forest.

Reserve Forests are protected under the Indian Forest Act, 1927 (an area duly notified under the provisions of the India Forest Act or the State Forest Acts having full protection) by the state government for conserving endemic flora and fauna. This scenario assumes the policy decision of protecting reserve forests and protected areas from further degradation and allowing growth other than reserve forest areas. This approach helps to evaluate forest protection policy and allows decision-makers to visualize growth by protecting the ecosystem and allowing development in regions other than rich in natural resources. Forest land is to be preserved as such, no exceptions are encouraged. Restriction similar to scenario 1 and further additional areas considered includes special recreation zone (parks, playground, and exhibition ground) and open space conservation as per CDP, all reserved forests (primary, secondary) as notified by the forest department (no other activity other than forestry and afforestation is permitted).

4.1.4 Scenario—4: Afforestation (High Conservation) Scenario (AF)

The Karnataka forest department (KFD) has taken up large-scale planting programs under various schemes to compensate for the deforestation/degradation of forest cover in the state. KFD has prepared comprehensive plans for afforestation across the state and specifies the working plans' sites and extents. After undertaking field surveys, afforestation is managed by drawing up Working Plans (WPs) for 10 years for each territorial division. The WPs contain annual plans for conserving and protecting forest areas, improving degraded forests, and reclamation of mined areas. These also outline the 'working areas' for afforestation in 'blank areas' or areas with a vegetation density of less than 0.25 and 'protection and improvement' in areas with a vegetation density of 0.25 or more. The WPs are to be implemented by formulation of Annual Plans of Operations (APO), which, *inter alia*, contain details of areas to be covered under afforestation and financial requirement thereof.

However, the Government approved the WPs for all 37 divisions and did not project financial requirements for the identified activities. The targets for afforestation (1.70 lakh ha) in the State as fixed in the APOs were short of planned in the WPs (2.73 lakh ha) for the years 2002–07 as budget allocations were insufficient to meet the WP targets. Besides, actual releases were even 29% less during the period. Lower allocation/release of funds for afforestation adversely affected achievements planned in the WPs/APOs. The Department carries afforestation in open and degraded forests and non-forest lands, including government wastelands, institutional lands, canals, and roadside areas. Degraded forests are rehabilitated by undertaking gap planting and cultural operations for natural regeneration in 'working areas' as identified in WPs. Forests are protected from degradation by demarcating the forest areas and taking preventive measures against fire, encroachments, and illegal mining and grazing. As per the Karnataka Forest Department records, Government of Karnataka, the degraded forest area assessed by the Department was 8.63 lakh ha. of the total territorial forest area of 30.04 lakh ha in the State. However, the assessment was made based on 'vegetation density' without assessing the degradation level based on parameters like depletion in soil conditions, moisture regime, nutrient availability, biomass production, and capacity to regenerate. The Government had not formulated any forest policy for the sustainable management of forests. The targets set in Annual Plans of Operations (APO) for afforestation were relatively lower than those planned in the WPs due to insufficient budget allocations. Afforestation was adversely affected due to the absence of quality checks on seeds procured, raising and planting seedlings in deviation from the 'package of practices'/norms, and planting exotic species without site-specific plans.

AF scenario accounts for the region of afforested areas by KFD. It assumes the protection and afforestation activities as a prime variable in the model to account for the policy decision. This approach helps to derive how far the current policy measures can support the development of regions and improve ecosystem services. The area of afforestation is considered based on the data collected from KFD. The afforestation area is considered a constraint for the development or land use conversion. The afforestation polygons were digitized based on the information (spatial extent, location, shape files of the areas chosen for afforestation) provided by KFD using Google Earth (http://karenvis.nic.in/Content/RaisingPlantationofKarnataka_9560.aspx). Protected areas and plantations were treated with high importance—areas beyond the peri-urban zone were treated like an agriculture conservation zone.

4.1.5 Scenario 5: Sustainable Development Policy Scenario (SDP)

This is scenario is more conservative as well as a focused growth scenario. The sustainable development policy scenario ensures (i) the protection of reserve forests and (ii) afforestation and hence includes the constraints same as scenarios 3 and 4 and allows growth in regions other than forest areas. This scenario stresses the strict implementation of spatial policies such as not allowing land use change within natural forests, increasing native species afforestation in barren lands, degraded woodlands, and protecting water bodies. The conservation focus is on holistic "Ecosystem Protection" by integrating reserve forest protection with stringent norms and afforestation of degraded lands (scenarios 3 and 4). This scenario allows the comparison of development and conservation tradeoffs effectively.

5 Results and Discussion

5.1 Assessment of Historical Land-Use Transitions and Simulation

The spatiotemporal land-use analysis was carried out from 1985 to 2019 by using the maximum likelihood algorithm. Districts in the Western Ghats region have a higher forest cover than other districts. Currently, 15% of the State's geographical area is under forest compared to 21% in 1985. Post-1990s the state witnessed large-scale land-use transitions during the post-1990s with globalisaton and consequent spurt in industrialization, urbanization, an increase of horticulture crops, conversion from agriculture to market-based crops (higher economic), etc.

Large-scale developmental activities such as constructing a series of reservoirs and dams, creating special economic zones and townships, and converting land for built-up areas have led to the loss of large tracts of forests. The abrupt land-use conversion has resulted in a loss of productive agricultural lands near cities such as Bengaluru, Mysore, Hubli-Dharwad, Shimoga, etc. The districts such as Kodagu, Uttara Kannada, Bengaluru, Shimoga, Belgaum, Dakshina Kannada, and Chikmagalur have been experiencing large-scale land cover due to unplanned developmental activities. Post-1990s, the state witnessed large-scale land-use transitions due to industrialization, urbanization, an increase in horticulture crops, conversion from agriculture to market-based crops (higher economic), etc. The forest cover is now confined to major conservation reserves such as protected areas, national parks, and wildlife sanctuaries. The built-up cover has increased from 0.47 to 3% from 1985 to 2019, causing an impact on agriculture, forests, and lakes. This necessitates sustainable land-use policies to arrest deforestation and abrupt land conversions.

The built-up area has increased from 0.5% (1985) to 3% (by 2019), and the horticulture area has increased from 8.8% (1985) to 11.1% (2019). Natural forests show the decline evident from the decrease of evergreen forests from 7.5% (1985) to 5.7% (by 2019), moist deciduous forests from 5.7% (1985) to 4.1% (by 2019), and dry deciduous forests from 4.0% (1985) to 2.2% (2019). LU transitions across ecosystems for Karnataka are provided in Table 8. The new urban agglomerations were noticed across the cities and major towns such as Bengaluru, Mangalore, Hubli, Hassan, Mysore, etc. Large-scale monoculture plantations of Eucalyptus, Rubber, Acacia, Teak, and Arecanut have increased and now cover 12% of the state. These changes are abrupt and result in an imbalance of ecosystem services, affecting the hydrologic regime and natural resource availability. The coastal and Tier-1 cities such as Bangalore and Tier-2 cities such as Mysore, Hubli-Dharwad, and Belgaum are experiencing loss of agricultural areas in the sub-urban regions with new layouts and satellite towns (Fig. 6).

The land use transition (Table 8) from 1985 to 2009 is accounted to understand the area of change and probability of change in each land-use type from time t_1 to time t_2 using the Markovian process. The accuracy of the simulated 2019 land use map was assessed by accounting for the consistency of land quantity and spatial position similarity at the pixel level (Table 9). Kappa statistics were evaluated to account for the consistency of land quantity of simulated over actual land use, and the similarity of the spatial position of simulated land use over actual with respect to a certain pixel is also assessed. Table 10 (Fig. 6) lists the simulated land uses of 2019 with area details.

5.2 *Scenario-Based Prediction and Validation*

Distance maps for various urban growth factors (agents) were prepared (Fig. 7), including a common set of maps such as slope, distance to road, proposed railway projects, industries, etc., and five different excluded maps (considered as constraints) for scenarios based on the discussion mentioned above. The site suitability maps were generated to define the suitability of each pixel for the transition to any land-use type. Each pixel in the suitability maps has a value that ranges from 0 to 255, with 0 representing unsuitable and 255 representing the highly suitable area for a particular land use type (Pontius and Malanson 2005).

The distance influence of each agent on specific land use was evaluated, which helped to derive the weights in an effective manner, unlike traditional weighing. Figure 7 shows the factors such as roads, and industries influence on built-up areas. Industries have a higher influence up to 2000 m from urban areas, moderately influence up to 10,000 m, and gradually decrease thereafter. Roads have depicted a higher influence up to 500 m, reducing influence after 1000 m. Figure 8 shows the factors

Table 8 Transitions across LU categories during 1985 to 2019—Karnataka State (Extent in hectares and percentage)

1985	2019 Land use categories										
	Units	Built-up	Agriculture	Horticulture	Fallow land	Evergreen forest	Moist deciduous forest	Dry deciduous forest	Scrub /Grass lands	Water	Total (Opening stock) 1985
Built-up	sq.km	859	20	15	2	1	3	1	2	2	904
	%	95.0	2.2	1.7	0.2	0.2	0.4	0.1	0.2	0.3	
Agriculture	sq.km	3169	114,150	4050	4529	69	119	117	254	2008	128,468
	%	2.5	88.9	3.2	3.5	0.1	0.1	0.1	0.2	1.6	
Plantations	sq.km	775	3344	11,661	185	243	258	93	108	122	16,790
	%	4.6	19.9	69.5	1.1	1.4	1.5	0.6	0.6	0.7	
Openland	sq.km	50	843	32	710	1	2	12	13	15	1678
	%	3.0	50.2	1.9	42.3	0.1	0.1	0.7	0.8	0.9	
Evergreen forest	sq.km	175	416	2372	151	9097	1402	211	286	182	14,293
	%	1.2	2.9	16.6	1.1	63.6	9.8	1.5	2.0	1.3	
Moist deciduous forest	sq.km	190	1973	1648	485	388	5581	470	164	61	10,960
	%	1.7	18.0	15.0	4.4	0.4	54.1	4.3	1.5	0.6	
Dry deciduous forest	sq.km	85	3374	779	458	68	419	2306	87	47	7622
	%	1.1	44.3	10.2	6.0	0.9	5.5	30.3	1.1	0.6	
Scrub_grass lands	sq.km	327	3056	701	438	136	113	72	1787	103	6733
	%	4.9	45.4	10.4	6.5	2.0	1.7	1.1	26.5	1.5	
Water	sq.km	93	734	114	35	14	16	7	9	3321	4344
	%	2.1	16.9	2.6	0.8	0.3	0.4	0.2	0.2	76.5	
Closing stock, 2019	sq.km	5725	127,910	21,371	6994	10,018	7914	3288	2710	5862	191,791
	%	3.0	66.7	11.1	3.6	5.2	4.1	1.7	1.4	3.1	191,791

Table 9 Accuracy of the simulation

Information of allocation	No [n]	Medium [m]	Perfect [p]	Kappa
Perfect [P(x)]	P(n) = 0.3706	P(m) = 0.9541	P(p) = 1	Kno = 0.9003
Perfect stratum [K(x)]	K(n) = 0.3706	K(m) = 0.9541	K(p) = 1	Klocation = 0.9273
Medium grid [M(x)]	M(n) = 0.3559	M(m) = 0.9102	M(p) = 0.9009	KlocationStrata = 0.9273
Medium stratum [H(x)]	H(n) = 0.1	H(m) = 0.3502	H(p) = 0.3598	Kstandard = 0.8619
No [N(x)]	N(n) = 0.1	N(m) = 0.3502	N(p) = 0.3598	

Table 10 Simulated land use for year 2019

Land use categories	Simulated 2019	
	Ha	%
Built-up	5,66,163	2.95
Agriculture	1,17,17,321	61.09
Plantation	27,38,669	14.28
Open fields	1,32,786	0.69
Evergreen forest	12,26,365	6.39
Moist deciduous forest	9,89,814	5.16
Dry deciduous	6,79,597	3.54
Scrub_grass	5,17,639	2.70
Water	6,10,746	3.18
	1,91,79,100	100

that influence forest cover land use. Industries depict high conversion of forests to other land use up to 7000 m thereafter, the influence has reduced, roads have shown higher influence up to 2000 m. This helps in understanding the distance influence of each factor.

The AHP has been used to assign the weights and evaluate their consistency, quantifying the consistency ratio. Scenario-wise weights assigned for each factor are shown in Tables 11, 12, 13 and 14. The consistency ratio of less than 0.1 is achieved for each scenario, and eigenvector weightage is shown for individual factors considered for the scenario. Site suitability of land use transition has been accounted for using MCE, which has been provided as an input for CA-based prediction. The CA has been used to compute projected land use under five different scenarios based on varied inputs. The model was run considering five scenarios, as discussed in the method section. The results of the simulation for the years 2019 and 2033 are illustrated in Figs. 9, 10 and 11 and (Table 15).

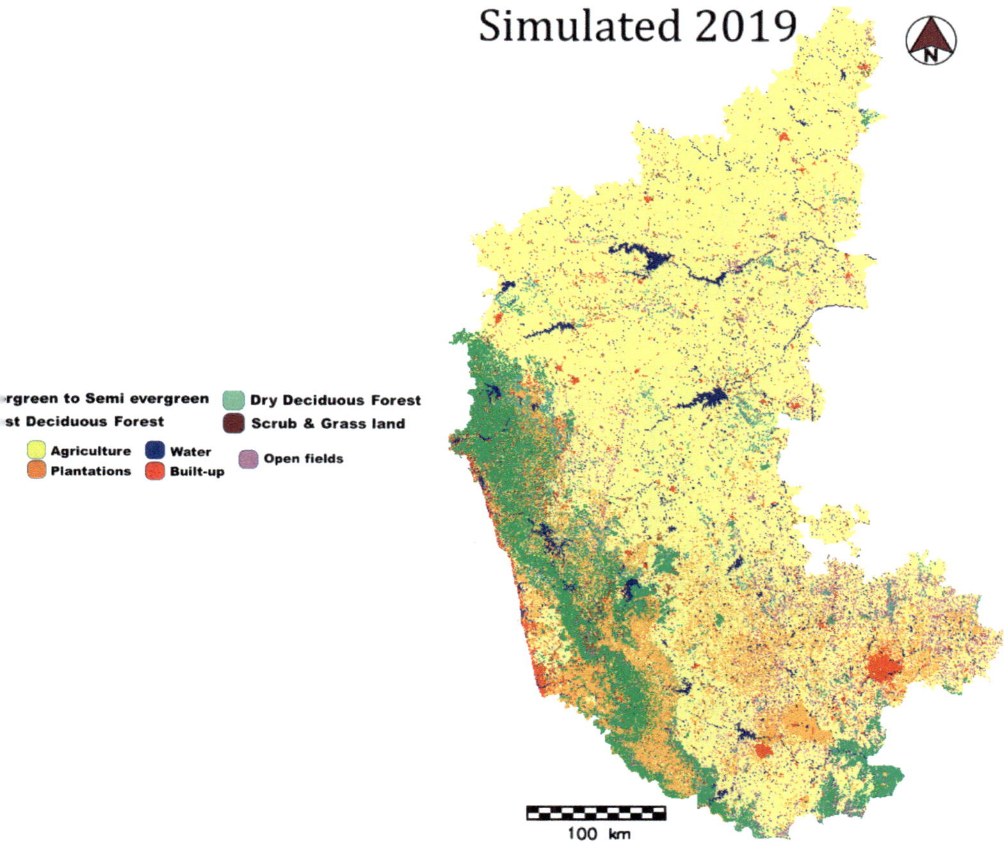

Fig. 6 Simulated land uses in 2019

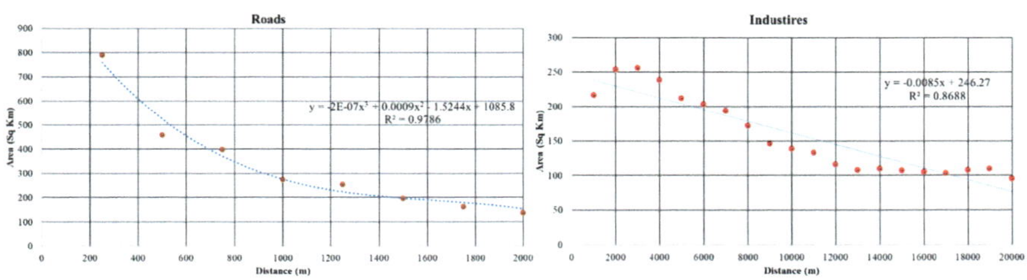

Fig. 7 Distance influence on land use change-built-up

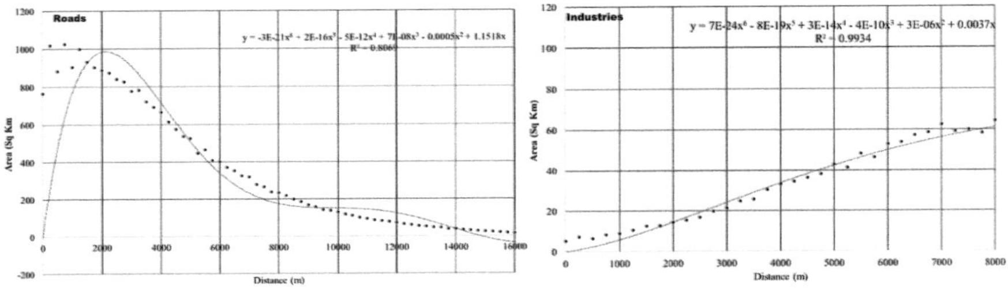

Fig. 8 Distance influence on land use change-forest

BAU can be considered a base case that helps compare or differentiate with alternative growth strategies. BAU depicts the likely increase in the built-up area to 11.5% from 3% compared to 2019 (Table 16, Fig. 12). The forest cover, and agriculture areas would lose the major tract for the likely expansion of built-up cover. The dry and moist deciduous forest cover showed a higher change, as shown in Fig. 9, accounting for a loss of 6.5% of forest cover. Cities such as Bangalore, Mangalore, and Dharwad depicted compact growth, whereas Mysore, Belgaum, Hassan, and Tumkur (Tier-2) cities have indicated peri-urban growth.

Upgrading urban infrastructure (roads, commercial establishments, information networks, industrial units, sports/recreation) would result in a built-up cover of 15% compared to BAU. ALT scenario would result in a more compact urban form than the BAU considering that new, high-density residential development takes place inside the cities, major towns, and polycentric urban agglomeration associated with the provision of new transport, industries, and other project investments (Fig. 9). BAU, ALT scenario projections indicate a significant reduction in forest area, and agricultural land uses were possible. The ALT scenario shows that forest cover would likely be only 9% as compared to 15% in the year 2019 (Table 16, Fig. 12). The agriculture area showed a loss of 2%, which might affect food availability in the region. The regions where dry crops were practiced were shown higher transformation, which signifies the external driver's role in land use alterations. ALT scenario shows that the higher land use conversion due to transportation, and other projects, which tend to encourage more compact, peri-urban, and mixed land use developments, necessitates effective policies.

RF scenario accounts for protecting reserve forests in the regions without altering further for unplanned developmental activities and expansion of earlier land uses. The regions covered under reserve forest protection do not favor development and other land-use conversions. The districts that are part of the Western Ghats region might not affect BAU, and ALT scenarios. The regions with dry deciduous forest cover are not under reserve forest area protection, which might experience abrupt growth due to the transportation corridors and urban clusters. Unlike other scenarios, urban areas of 2019 are considered to be another factor that can influence land-use change. RF scenario shows a likely built-up cover of 11% (Fig. 9), retaining a good amount of evergreen forest cover of 5% (Table 16, Fig. 12). As depicted in Fig. 10,

Table 11 Weights based on AHP for BAU

Factors	BAU								Eigen vector of weight
	Bus stops	KIDAB plots	Mining areas	Railway	Roads	SEIAA_Industries		Slope	
Bus stops	1								0.02
KIDAB plots	9	1							0.32
Mining areas	3	1/3	1						0.06
Railway	5	1/3	3	1					0.10
Roads	9	1/3	3	1	1				0.15
SEIAA Industries	7	1/3	3	3	3	1			0.23
Slope	7	1/3	3	3	1/3	1/3		1	0.12
Less important				Equal	More important				Consistancy ratio: 0.09
Extremely	Very strongly	Strongly	Moderate		Moderate	Strongly	Very strongly	Extremely	Status: Acceptable (Satndard <= 0.1)
1/9	1/7	1/5	1/3	1	3	5	7	9	

Table 12 Weights based on AHP for ALT

Factors	ALT								Eigen vector of weight
	Bus stops	KIDAB plots	Mining areas	Railway	Roads	SEIAA industries	Slope	Proposed projects	
Bus stops	1								0.02
KIDAB plots	9	1							0.22
Mining areas	3	1/7	1						0.04
Railway	5	1/5	1	1					0.07
Roads	7	1/3	3	3	1				0.17
SEIAA industries	5	1/3	3	3	1/3	1			0.11
Slope	3	1/3	3	1/3	1/3	1/3	1		0.06
Proposed projects	7	3	5	3	3	3	3	1	0.30
Less important				Equal	More important				Consistancy ratio: 0.08
Extremely	Very strongly	Strongly	Moderate		Moderate	Strongly	Very strongly	Extremely	Status: Acceptable (Satndard < = 0.1)
1/9	1/7	1/5	1/3	1	3	5	7	9	

Table 13 Weights based on AHP for RFP

Factors	RFP								
	Bus stops	KIDAB plots	Mining areas	Railway	Roads	SEIAA industries	Slope	Urban_2019	Eigen vector of weight
Bus stops	1								0.02
KIDAB plots	9	1							0.28
Mining areas	3	1/7	1						0.05
Railway	5	1/5	1	1					0.12
Roads	7	1/3	3	3	1				0.25
SEIAA industries	5	1/3	3	1	1/3	1			0.10
Slope	3	1/3	3	1/3	1/3	1/3	1		0.06
Urban_2019	5	1/3	3	1/3	1/5	3	3	1	0.12
Less important				Equal	More important				Consistancy ratio: 0.10
Extremely	Very strongly	Strongly	Moderate		Moderate	Strongly	Very strongly	Extremely	Status: Acceptable (Satndard < = 0.1)
1/9	1/7	1/5	1/3	1	3	5	7	9	

Table 14 Weights based on AHP for AF

Factors	AF								
	Bus stops	KIDAB plots	Mining areas	Railway	Roads	SEIAA industries	Slope	Proposed projects	Eigen vector of weight
Bus stops	1								0.02
KIDAB plots	7	1							0.27
Mining areas	3	1/7	1						0.05
Railway	5	1/5	1	1					0.06
Roads	7	1	3	3	1				0.27
SEIAA industries	5	1/3	3	3	1/3	1			0.11
Slope	3	1/3	3	1	1/3	1/3	1		0.07
Plantations	5	1/3	3	3	1/5	3	3	1	0.15
Less important				Equal	More important				Consistancy ratio: 0.07
Extremely	Very strongly	Strongly	Moderate		Moderate	Strongly	Very strongly	Extremely	Status: Acceptable (Satndard < = 0.1)
1/9	1/7	1/5	1/3	1	3	5	7	9	

Table 15 Weights based on AHP for SDP

Factors	SDP								
	Bus stops	KIDAB plots	Mining areas	Railway	Roads	SEIAA industries	Slope	Proposed projects	Eigen vector of weight
Bus stops	1								0.02
KIDAB plots	7	1							0.22
Mining areas	3	1/7	1						0.04
Railway	5	1/7	1	1					0.05
Roads	7	1	3	3	1				0.18
SEIAA industries	7	1/3	3	3	1/3	1			0.12
Proposed projects	3	1/3	3	3	1/3	1/3	1		0.10
Plantations	7	3	5	5	3	3	1	1	0.28
Less important				Equal	More important				Consistancy ratio: 0.09
Extremely	Very strongly	Strongly	Moderate		Moderate	Strongly	Very strongly	Extremely	Status: Acceptable (Satndard < = 0.1)
1/9	1/7	1/5	1/3	1	3	5	7	9	

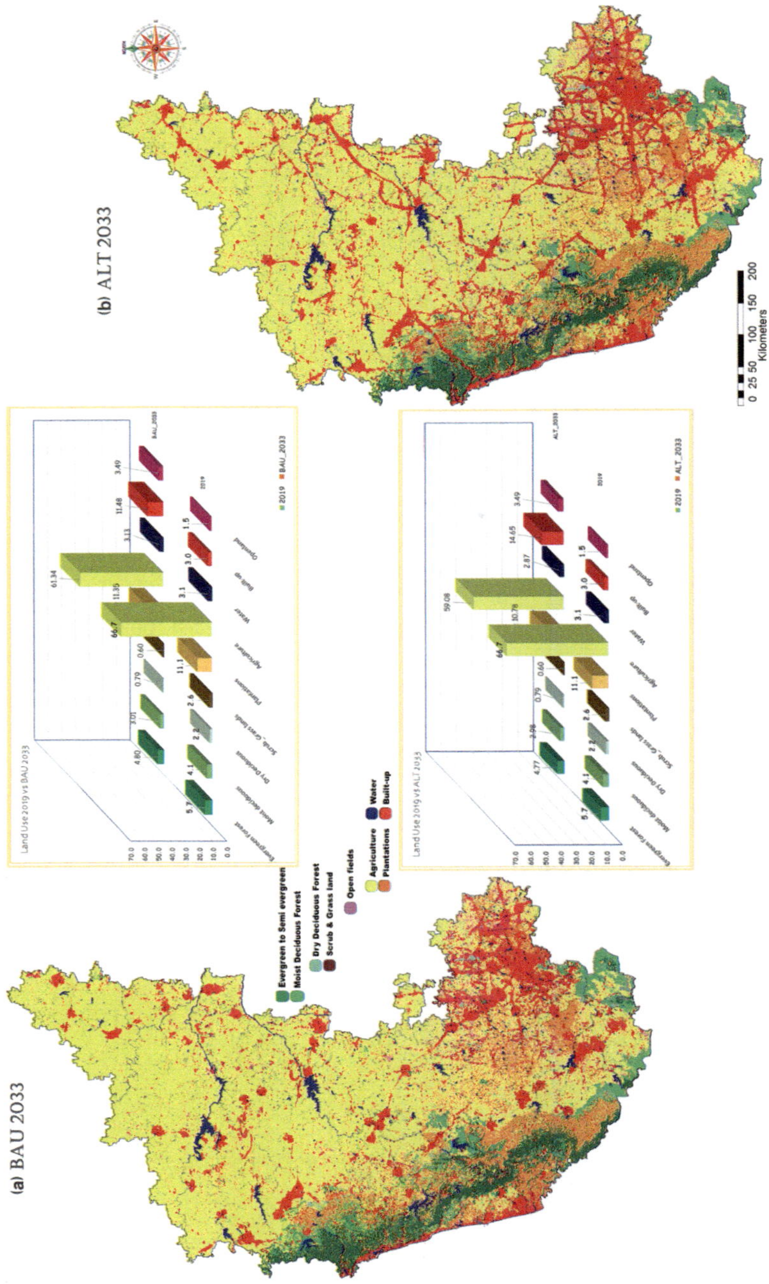

Fig. 9 Projected land use under BAU, ALT scenarios

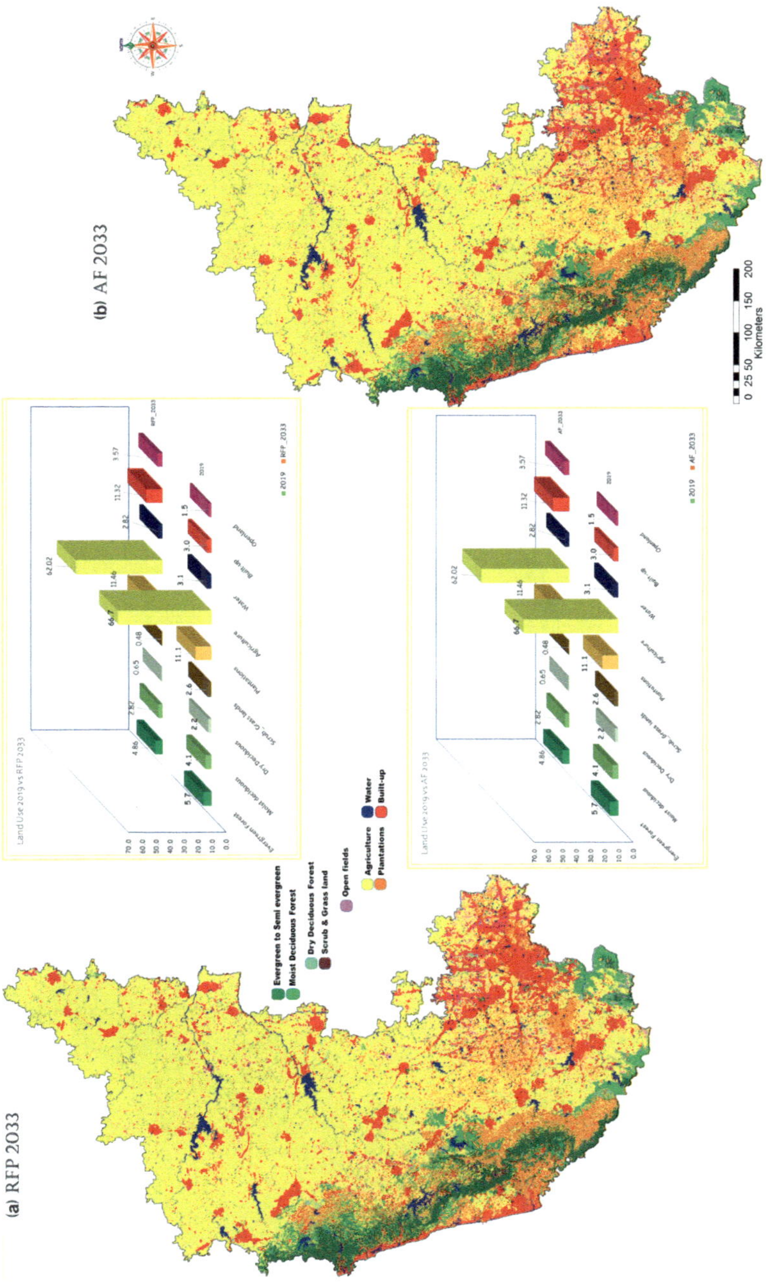

Fig. 10 Projected land use under RFP, AF scenarios

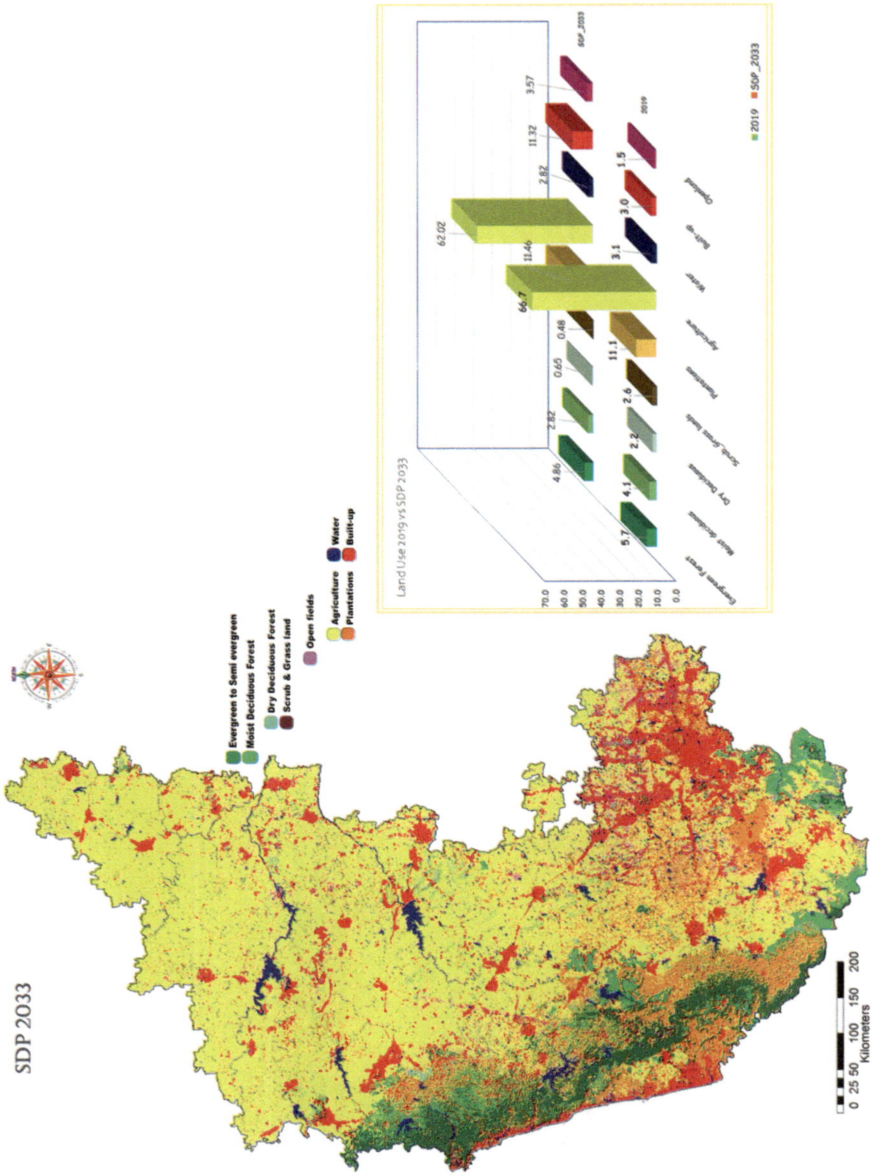

Fig. 11 Projected land use under SDP scenarios

Table 16 Likely land uses in 2033 for various scenarios

Land use categories	BAU_2033		ALT_2033		RFP_2033		AF_2033		SDP_2033	
	Ha	%	Ha	%	Ha	%	Ha	%	Ha	%
Built-up	22,01,515	11.48	28,10,015	14.65	21,70,349	11.32	21,46,203	11.19	20,71,994	10.76
Agriculture	1,17,64,649	61.34	113,30,926	59.08	1,18,94,624	62.02	1,18,58,568	61.83	1,16,67,178	60.88
Plantation	21,77,670	11.35	20,66,713	10.78	21,98,428	11.46	21,69,239	11.31	21,52,053	11.23
Open fields	6,68,824	3.49	6,68,824	3.49	6,85,063	3.57	6,68,824	3.49	6,85,063	3.57
Evergreen forest	9,20,948	4.80	9,15,179	4.77	9,32,811	4.86	9,59,097	5.00	9,72,489	5.07
Moist deciduous forest	5,77,449	3.01	5,72,085	2.98	5,41,381	2.82	5,75,213	3.00	6,66,957	3.48
Dry deciduous	1,52,389	0.79	1,51,217	0.79	1,24,597	0.65	1,56,922	0.82	2,36,863	1.24
Scrub_Grass	1,15,952	0.60	1,14,432	0.60	91,496	0.48	97,938	0.51	1,96,677	1.01
Water	5,99,704	3.13	5,49,709	2.87	5,40,351	2.82	5,47,096	2.85	5,29,826	2.76
Total	1,91,79,100									

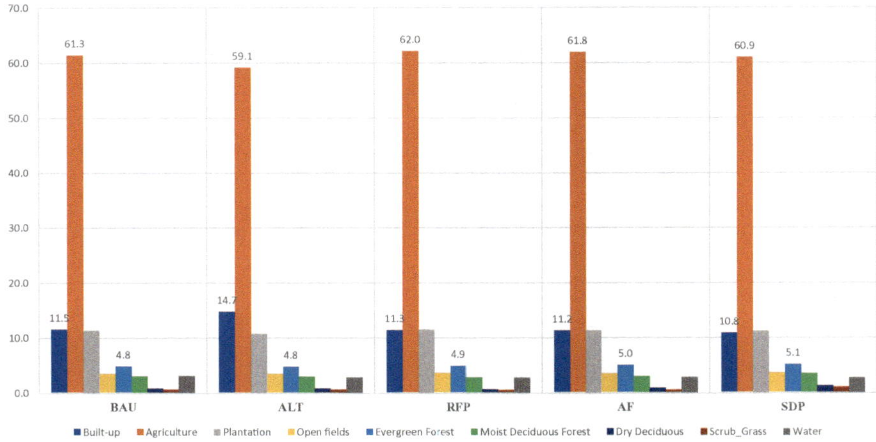

Fig. 12 Projected land uses (in 2033) for modelled scenarios

AF scenario shows the likely improvement in its forest cover due to the plantation activities and protection of degraded forests from uncontrolled land use conversion. The forest cover shows 9.3% (Table 16), a slight increase compared to BAU and ALT. The built-up cover would limit existing cities and towns by over 11%. The new plantation activities would compensate for the abrupt land-use conversion, especially in the moist, dry deciduous cover regions. The districts, such as Bidar, Kolar, Koppala, etc., did not show any improvement due to the least area of forest land availability.

SDP scenario signifies stringent policy implementation initiatives with the least disturbances, and the simulated land uses using this scenario are depicted in Fig. 1. The forest cover would remain at 11% compared to 17% in 2019 (Table 16, Fig. 12), which is the least probable loss compared with all the scenarios. The built-up cover would remain to the existing tier-1 and tier-2 cities; the least increase was shown in peri-urban areas. Districts such as Bidar, Gulbarga, Bagalkot, and Kolar might experience higher land use conversion rates due to existing infrastructure and lack of any forest policy. Overall, RFP, AF, and SDP scenarios favor the protection of evergreen and moist deciduous forest categories. BAU, and ALT scenarios depict the merging of urban clusters, indicating smaller clusters of paved areas are no longer available and tend to agglomerate with bigger ones to form urban centres, satellite towns, and dispersed growth at the periphery of the tier-2 city limits. New constructions might occur in the rural hinterland, which results in the loss of evergreen, moist deciduous cover and agricultural areas. Categories such as plantation, water, and open field had similar values across all scenarios. The scenarios would directly help decision-makers/planners reduce environmental impacts with stricter growth rules. Otherwise, conversion of primary, secondary forest/moist/dry deciduous forest into urban areas would be favored (BAU, ALT), which can induce ecological imbalances. Also, another important observation in terms of statistics is that slope steepness restriction has played an important role in restricting urban settlements from further infiltrating into forested areas. Scenarios 3–5 seem feasible for sustainable development without altering the edges of agriculture, forests, water bodies, and other protected areas or conservation zones. Additionally, they emphasize a planned growth strategy for a habitable city. Integrated planning through regulated land-use conversion can support self-sufficient towns along newly provided public transport corridors, limiting the abrupt land-use conversion.

5.3 *Ecosystem Services with the Simulated Land Uses (of 2033)*

The ecosystem services are estimated using ecosystem asset accounts simulated for 2033, using five scenarios using spatially explicit estimates of the supply of ecosystem services in physical terms and their benefits in monetary terms (Fig. 13). Ecosystem supply accounts are presented in physical terms and monetary terms considering the most broad types of ecosystem services: production of wild biomass (timber, bamboo, fuelwood, etc.), carbon storage and sequestration, pollination, soil conservation, recreation, aesthetic, etc., as listed in Table 17, grouped broadly under-provisioning, regulating and cultural services.

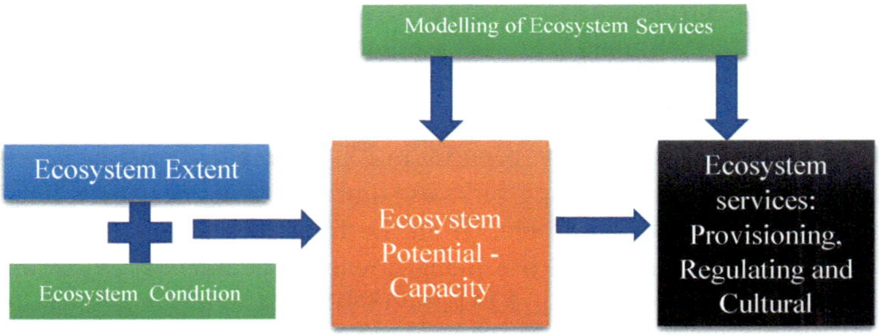

Fig. 13 Ecosystem services for simulated land uses

Table 17 Goods and services valued for forested landscape

Provisioning services	Regulating services	Cultural services
1. Timber	1. Carbon sequestration	1. Recreational (Tourism)
2. Bamboo	2. Carbon storage	2. Spiritual and historic
3. Fodder	3. Pollination and seed dispersal	3. Aesthetic
4. Fuelwood	4. Soil conservation	4. Artistic and culture
5. Non-Timber forest Produce	5. Ground water recharge	5. Science and education
6. Fish	6. Water purification	
7. Genetic material	7. Waste treatment	

- Provisioning services are those ecosystem services representing the contributions to benefits that are extracted or harvested from ecosystems.
- Regulating services are those ecosystem services resulting from ecosystems' ability to regulate biological processes and influence climate, hydrological, and biochemical cycles, thereby maintaining environmental conditions beneficial to individuals and society.
- Cultural services are the experiential and intangible services related to the perceived or actual qualities of ecosystems whose existence and functioning contribute to a
- range of cultural benefits.

Landscape dynamics involve fundamental changes in the structure of ecosystems, changing both extent and condition, which affects the array of ecosystem services evident from the conversion of perennial streams (12 months water availability) to intermittent (6–8 months) or seasonal streams (4 months water) with land degradation leading to deforestation. The analysis addressing land-use changes modeling also attempts to quantify a range of ecosystem services affected by the policy options, such as carbon sequestration, provision of goods, and cultural services. Hence, the accounting implications of different policy options have been studied as part of Natural Capital Accounting in Karnataka, India.

The land-use dynamics involving the decline of forests have resulted in significant changes in ecosystem services. This section presents the provisioning services representing the contributions to benefits extracted or harvested from ecosystems under various policy regimes. The services across various scenarios were evaluated and depicted in Fig. 14 (Fig. 14a provides the ecosystem goods for 2019, while Fig. 14b–f provides the scenarios ecosystem goods). Table 18 lists provision services district-wise for considered policy scenarios in 2033, compared with the provisioning services in 2019. The provisioning services (timber, NTFP, fish, fuelwood, fodder, bamboo, cane, genetic material, etc.) from the forests of Karnataka is about 613 billion INR in 2019. Due to land-use changes with the decline of forests, the provisioning services in 2033 reduced to 524 billion INR (BAU scenario), 521 billion INR (ALT Scenario), 533 billion INR (RFP Scenario), 530 billion INR (AF Scenario) and 540 billion INR (SDP), which highlights the need to implement afforestation with the reserve forest protection (SDP scenario) to sustain the livelihood through the provision of goods. ALT scenario shows a reduction of 92 billion INR in provisional goods compared to 2019. The SDP scenario has shown improvement in ecosystem services compared to other scenarios, highlighting the role of policy decisions.

Regulating services of ecosystems pertains to services resulting from the ability of ecosystems to regulate biological processes and influence climate, hydrological, and biochemical cycles, thereby maintaining environmental conditions beneficial to individuals and society. Table 19 lists regulating services from forest ecosystems, district-wise, for considered policy scenarios in 2033, compared with the provisioning services of 2019. Regulating services (Air Quality, Climate Regulation, Carbon Sequestration, Pollination, Soil Erosion, Fertility, Ground Water Recharge, Water Purification, Waste Treatment) quantified district-wise for 2019 and various

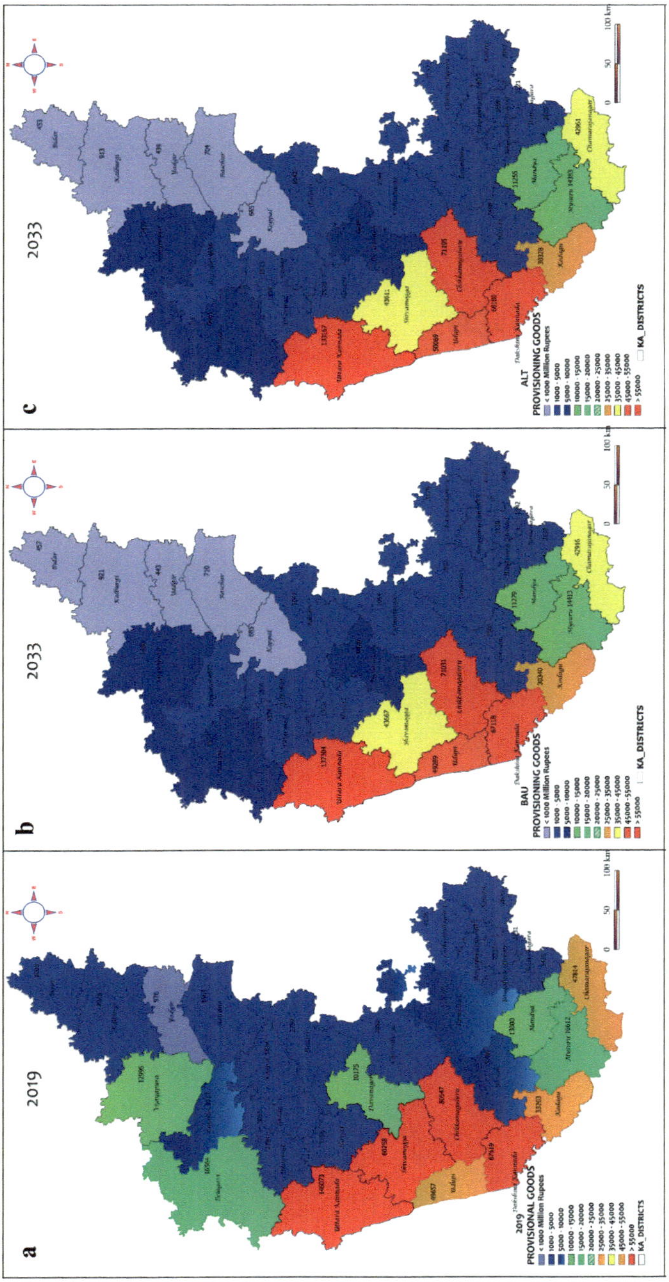

Fig. 14 Provisional goods in million rupees across various scenarios. **a** Provisioning goods in 2019, **b** Provisioning goods in 2033-BAU scenario, **c** Provisioning goods in 2033-ALT scenario

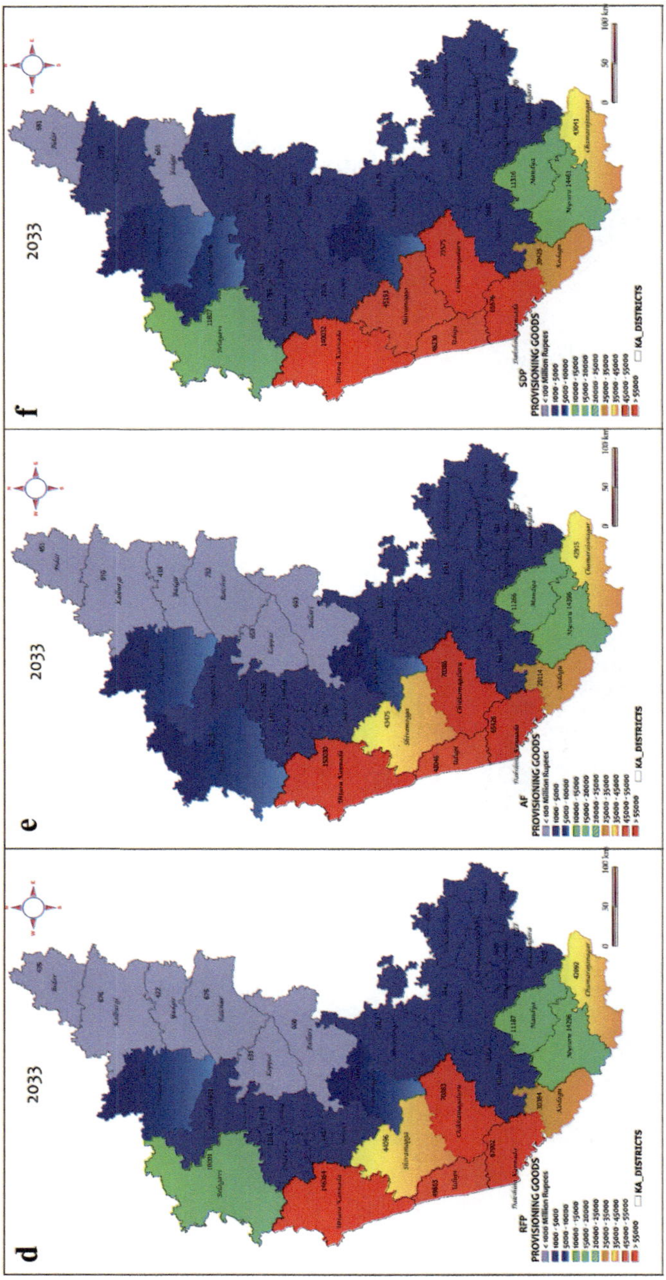

Fig. 14 (continued)

Table 18 Provisioning services (MRs-Million Rupees) from forest ecosystems in 2033 under various scenarios

S. no	District	2019	2033				
		2019_ PROV MRs	BAU_ PROV MRs	ALT_ PROV MRs	RFP_ PROV MRs	AF_ PROV MRs	SDP_ PROV MRs
1	Bagalkot	8117	4672	4636	4901	4712	5786
2	Ballari	2789	1042	1042	960	993	2077
3	Belagavi	16,564	9534	9461	10,001	9615	11,807
4	Bengaluru (Rural)	2233	1720	1698	1600	1631	1645
5	Bengaluru (Urban)	2131	1642	1621	1527	1557	1570
6	Bidar	1005	457	453	435	451	681
7	Chamarajanagar	47,814	42,916	42,961	42,892	42,915	43,041
8	Chikkaballapura	4126	3178	3137	2955	3013	3039
9	Chikkamagaluru	80,547	71,031	71,195	70,883	70,386	73,575
10	Chitradurga	2852	1094	1094	1012	1045	2129
11	Dakshina Kannada	67,619	67,118	68,180	67,902	65,426	65,676
12	Davanagere	10,175	6826	6818	6841	6771	7621
13	Dharwad	2794	1374	1394	1316	1433	1786
14	Gadag	3053	1501	1523	1438	1566	1951
15	Hassan	5246	2506	2480	2212	2257	2800
16	Haveri	3158	1553	1576	1487	1620	2018
17	Kalburgi	2026	921	913	876	910	1373
18	Kodagu	33,203	30,340	30,328	30,384	29,114	30,425
19	Kolara	3870	2981	2943	2772	2826	2850
20	Koppal	1834	685	685	631	653	1365
21	Mandya	13,000	11,279	11,255	11,187	11,266	11,316
22	Mysuru	16,612	14,413	14,383	14,296	14,396	14,461
23	Raichur	1563	710	704	676	702	1060
24	Ramanagara	3425	2639	2605	2453	2502	2523
25	Shivamogga	60,258	43,667	43,611	44,096	43,475	45,193
26	Tumakuru	8167	3902	3861	3443	3514	4359
27	Udupi	49,657	49,289	50,069	49,865	48,046	48,230
28	Uttara Kannada	146,073	137,304	133,167	146,364	150,030	140,032
29	Vijayapura	12,996	7480	7422	7846	7544	9263
30	Yadgir	976	443	439	422	438	661
	Grand total	613,883	524,217	521,654	533,673	530,807	540,313

Table 19 Regulating services (MRs-Million Rupees) from forest ecosystems in 2033 under various scenarios

No	District	2019	2033				
		2019_ REGS MRs	BAU_ REGS MRs	ALT_ REGS MRs	RFP_ REGS MRs	AF_ REGS MRs	SDP_ PROV MRs
1	Bagalkot	11,320	5530	5527	5531	5636	9061
2	Ballari	9393	1550	1551	1391	1494	6445
3	Belagavi	23,101	11,286	11,279	11,287	11,500	18,489
4	Bengaluru (Rural)	4991	3079	3013	2822	2923	3876
5	Bengaluru (Urban)	4764	2939	2876	2694	2790	3700
6	Bidar	2372	397	397	362	411	1576
7	Chamarajanagar	87,131	84,740	84,768	84,703	84,843	85,961
8	Chikkaballapura	9220	5688	5567	5214	5399	7161
9	Chikkamagaluru	95,068	80,686	80,457	82,339	80,270	87,361
10	Chitradurga	9467	1619	1621	1462	1564	6513
11	Dakshina Kannada	70,283	65,770	65,362	65,414	65,131	66,390
12	Davanagere	17,758	10,180	10,172	10,255	10,262	14,311
13	Dharwad	5690	2148	2173	2079	2256	4060
14	Gadag	6216	2346	2374	2271	2464	4436
15	Hassan	11,818	6359	6259	5278	5409	9383
16	Haveri	6431	2428	2456	2349	2550	4589
17	Kalburgi	4779	800	800	731	828	3175
18	Kodagu	63,203	59,325	59,249	59,792	58,951	60,780
19	Kolara	8648	5335	5222	4890	5065	6716
20	Koppal	6176	1019	1020	914	982	4237
21	Mandya	17,797	16,160	16,109	15,980	16,075	16,846
22	Mysuru	22,743	20,650	20,585	20,421	20,541	21,526
23	Raichur	3688	618	617	564	639	2450
24	Ramanagara	7655	4722	4622	4328	4483	5945
25	Shivamogga	87,520	65,117	65,054	66,265	65,906	74,918
26	Tumakuru	18,398	9899	9743	8216	8421	14,607
27	Udupi	51,613	48,299	48,000	48,038	47,830	48,755
28	Uttara Kannada	238,678	196,906	190,822	195,966	201,276	215,273
29	Vijayapura	18,123	8854	8849	8855	9023	14,506
30	Yadgir	2302	385	385	352	399	1529
	Grand total	926,346	724,834	716,929	720,763	725,321	824,575

scenarios of 2033 are listed in Table 20 and depicted district-wise in Fig. 15. Regulating services from forest ecosystems in 2019 is about 926 billion INR compared to 724 billion INR in 2033 (BAU Scenario), 716 billion INR (ALT Scenario), 720 billion INR (RFP Scenario), 725 billion INR (AF Scenario) and 824 billion INR (SDP Scenario), emphasising the need to implement the sustainable development programme (SDP Scenario) to sustain regulating services to sustain diverse biota on the Earth.

Ecosystem Services—Carbon Storage and Sequestration Assessment Through InVEST

The InVEST Carbon Storage and Sequestration model estimates the current amount of carbon stored in a landscape and values the amount of sequestered carbon over time. InVEST aggregates the biophysical amount of carbon stored in four carbon pools (aboveground living biomass, belowground living biomass, soil, and dead organic matter) based on land use/land cover (LULC) maps. With a spatially explicit LU map, the carbon sequestration component of the model estimates the expected change in carbon stocks over time. This portion of the model values the amount of carbon sequestered as an environmental service using additional data on the market value or social cost of carbon.

InVEST carbon model has been used to understand the change in carbon storage and carbon sequestration across the state. The scenario-based carbon storage and sequestrations were made to compare the loss/ gain in carbon content. The economic values corresponding to each scenario have been assessed. Forests sequester large amounts of CO_2 while reducing impacts of climate change and degradation of forests would lead to loss of carbon sequestration potential. Carbon sequestration (CO_2 equivalent) is derived by multiplying the carbon sequestration values with the factor 3.67, as suggested by The Intergovernmental Panel on Climate Change (IPCC 2003). The economic value of carbon sequestration is calculated by considering the social cost of carbon per ton. The social cost of a tonne of CO_2 is considered as US$ 80 using the GDP deflator (MoSPI 2021). The carbon sequestration assessment in forests highlights that districts/forest circles in the Western Ghats have higher carbon and sequestration potential than other parts of the state due to the relatively lesser disturbances (Figs. 16 and 17). Carbon sequestration in forests for 2019 is about 89,194

Table 20 Quantification of carbon sequestration using InVEST

Year	Total carbon sequestration (Gg/yr)	Economic value (Million Rs/yr)	Loss in ES (compared to 2019)
2019	89,194	571,145	–
Projected_2033-BAU	70,559	451,829	119,316
Projected_2033-ALT	69,581	445,561	125,584
Projected_2033-RFP	70,251	449,844	121,301
Projected_2033-AF	70,883	453,889	117,256
Projected_2033-SDP	79,584	509,605	61,540

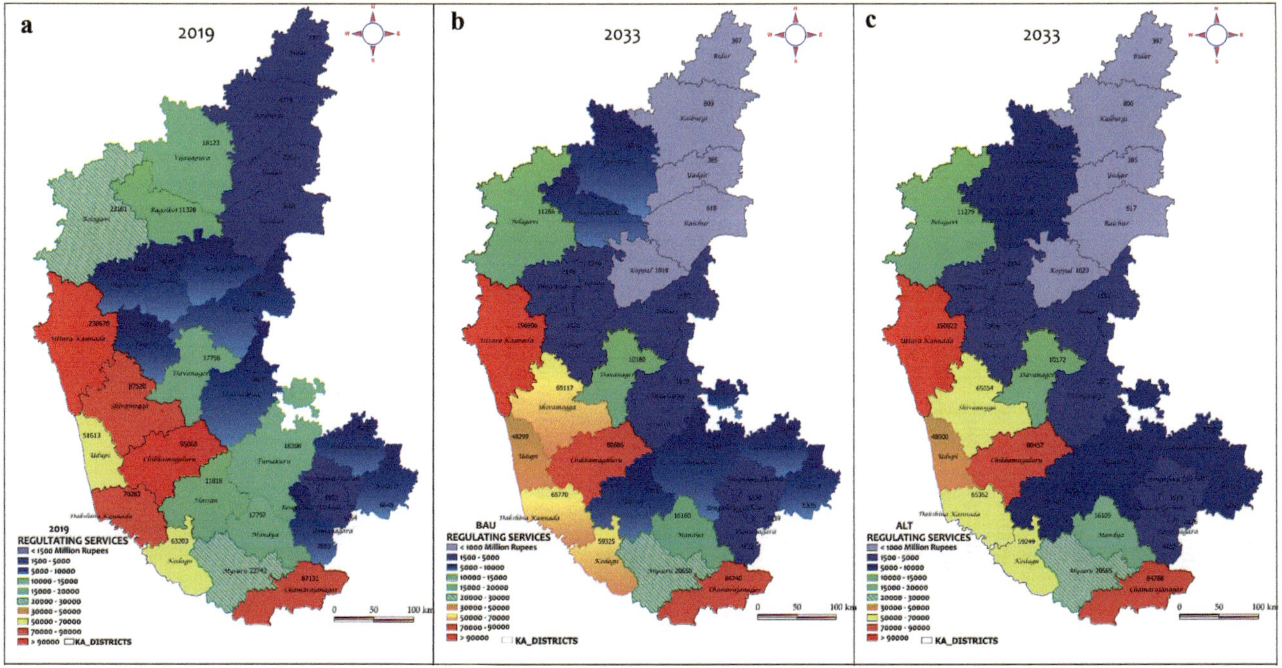

Fig. 15 Regulating services from forest ecosystems in million rupees across various scenarios. **a** Regulating services, district wise Karnataka in 2019 **b** Regulating services, district wise Karnataka in 2033, BAU scenario **c** Regulating services, district wise Karnataka in 2033, ALT scenario **d** Regulating services, district wise Karnataka in 2033, REP scenario e: Regulating services, district wise Karnataka in 2033, AF scenario **f** Regulating services, district wise Karnataka in 2033, SDP scenario

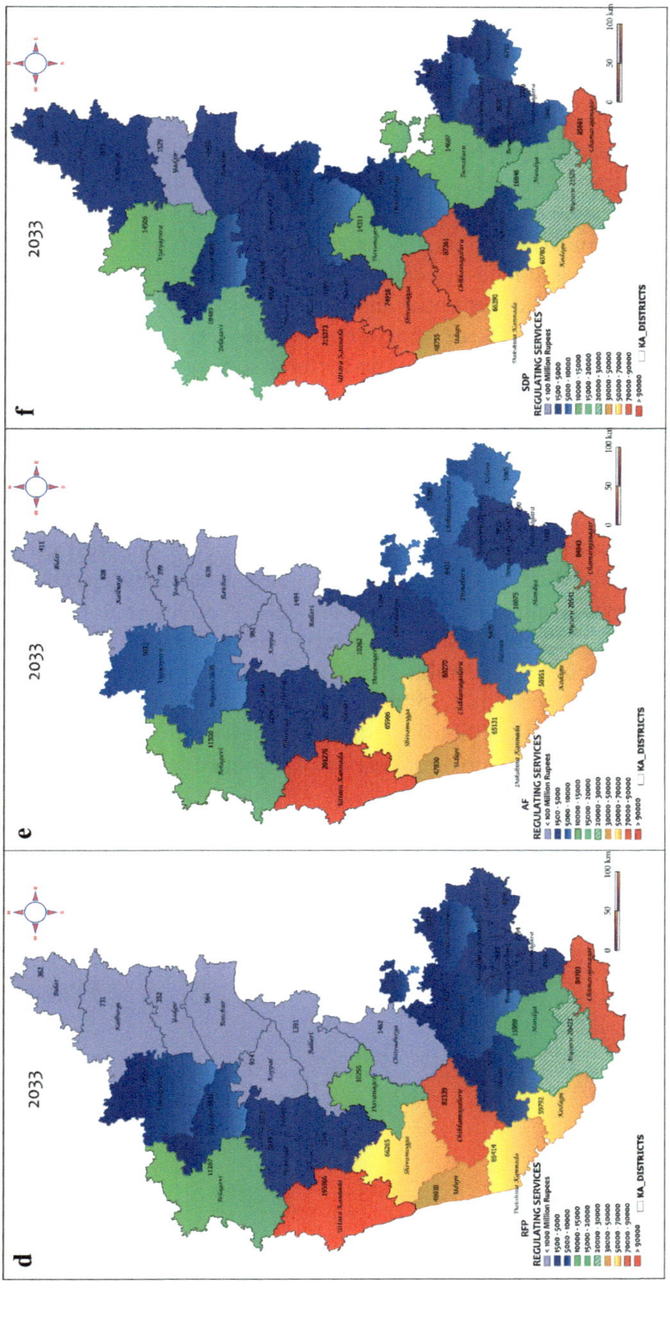

Fig. 15 (continued)

Gg/yr (89 million tonnes), while carbon sequestration potential of forests in 2033 is about 70,559 Gg/yr (BAU), 69,851 Gg/yr (ALT), 70,251 Gg/yr (RFP), 70,883 Gg/yr (AF) and 79,584 Gg/yr (SDP) and details are listed in Table 20. Carbon storage in Karnataka forests for 2019 and 2033 (considering various scenarios) and District-wise details of carbon sequestration and stored carbon are listed in Tables 21 and 22, respectively. Districts such as Bangalore (Urban), Bangalore (Rural), Vijayapura, Bidar, Yadgir, and Gadag have lower carbon storage due to the lower forest cover. The carbon sequestration is least in the districts such as Gadag, Yadgir, Bidar, and Vijayapura. The probable carbon sequestration across various policy scenarios indicates significant loss in BAU and ALT scenarios due to likely land degradation and deforestation with the rapid land-use changes (Table 23).

Cultural services are the experiential and intangible services related to the actual qualities or perceived qualities of ecosystems whose existence and functioning contribute to various cultural benefits. The cultural services of forest ecosystems, considering aesthetic, tourism, spiritual and cultural, artistic, science, and education services, are quantified district-wise for Karnataka State, and results are listed in Table 24 and presented in Fig. 18. Cultural services of forest ecosystems in 2019 is about 294 billion INR compared to 291 billion INR in 2033 (BAU), 291 billion INR (ALT, RFP, AF Scenarios), and 293 billion INR (SDP Scenario).

Table 25 summarises the total ecosystem services value for forest ecosystems in Karnataka State for 2019 and 2033, considering various policy scenarios. The total ecosystem service value of forests in Karnataka is about 1835 billion INR in 2019. However, with the land-use changes leading to degradation of forest ecosystems, the total ecosystem services value would be 1540 billion INR (BAU Scenario), 1530 billion INR (ALT Scenario), 1546 billion INR (RFP Scenario), 1547 billion INR (AF Scenario) and 1658 billion INR (SDP Scenario), emphasizing the need to protect reserve forests and take up afforestation of degraded forest patches to sustain ecosystem goods and services.

Figure 19 depicts the relative share of ecosystem services for 2019 and considered policy scenarios. Regulating services dominate (47–51%) compared to the provisioning (32–34%) and cultural services (16–19%).

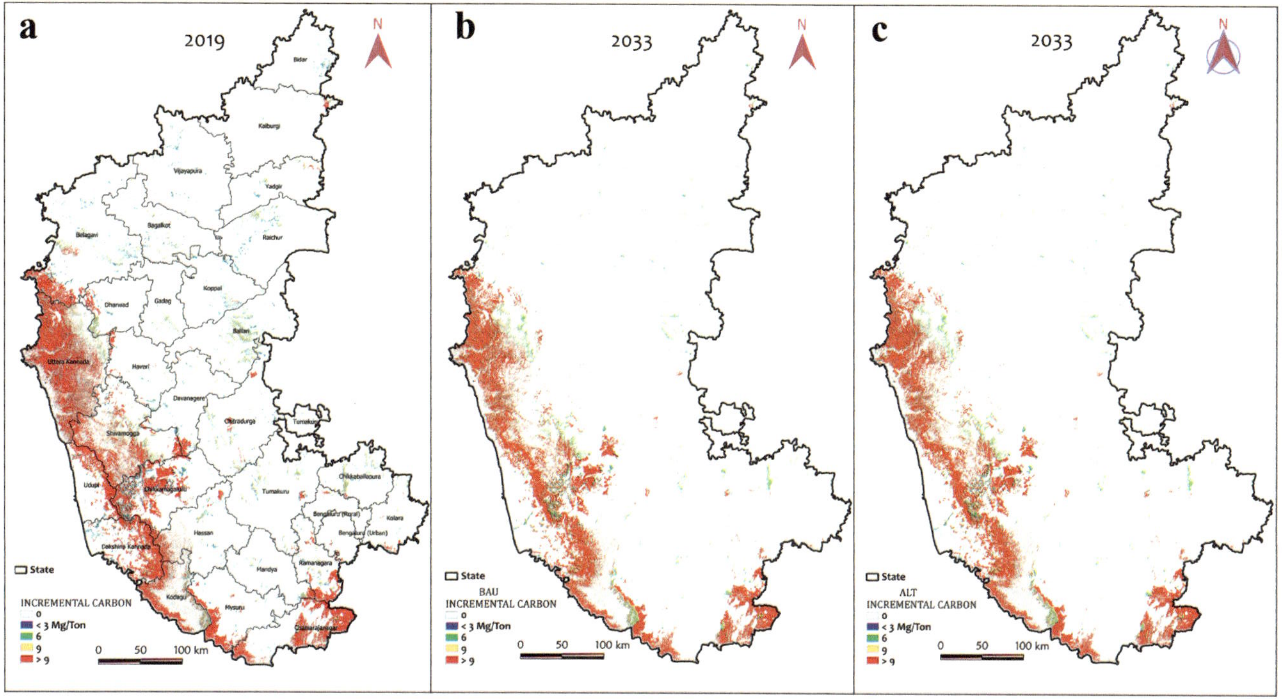

Fig. 16 Carbon sequestration in Karnataka forests from 2019 to 2033 **a** Annual carbon sequestration in forest ecosystems in 2019 **b** Annual carbon sequestration in forest ecosystems in 2033 (BAU scenario) **c** Annual carbon sequestration in forest ecosystems in 2033 (ALT scenario) **d** Annual carbon sequestration in forest ecosystems in 2033 (RFP Scenario) e. Annual carbon sequestration in forest ecosystems in 2033 (AF scenario) **f** Annual carbon sequestration in forest ecosystems in 2033 (SDP scenario)

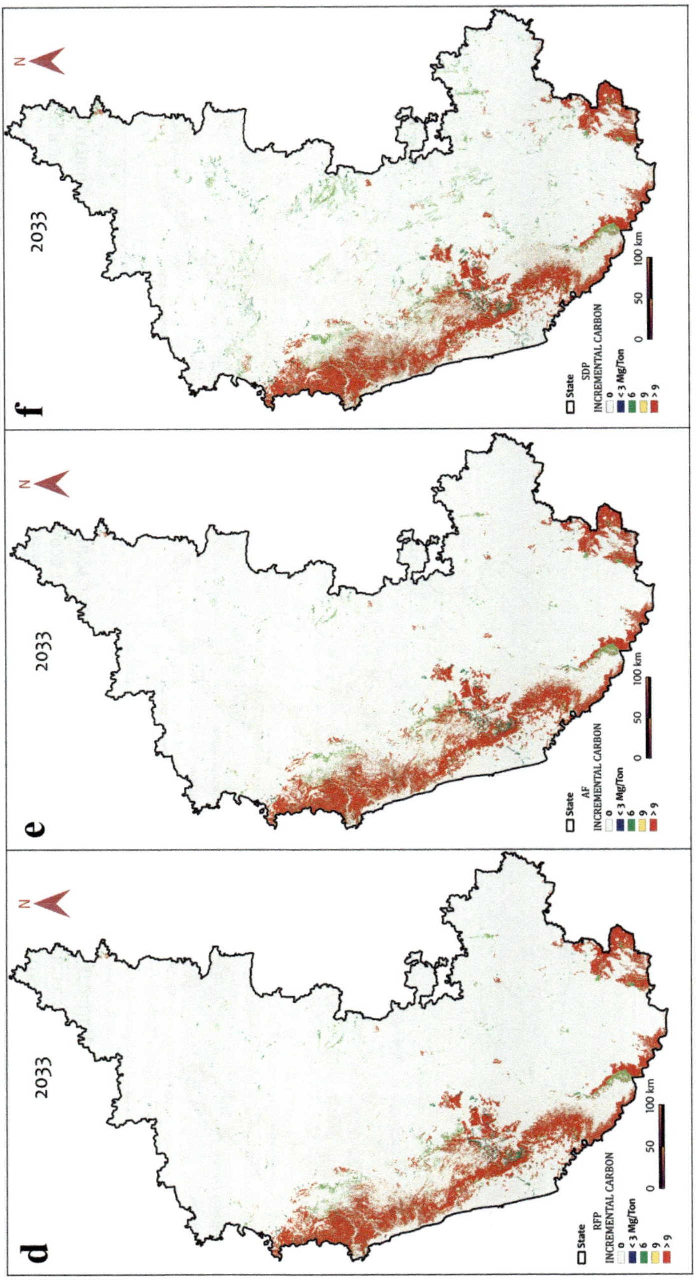

Fig. 16 (continued)

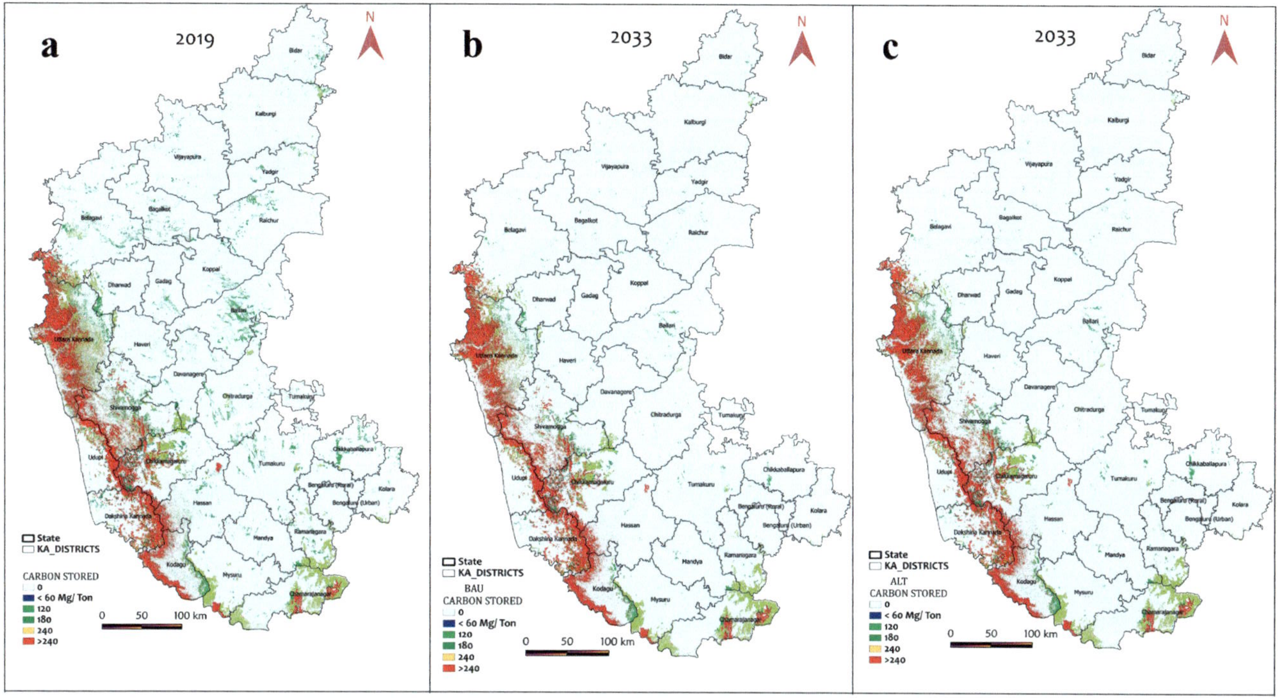

Fig. 17 Carbon storage from 2019 to 2033 for Karnataka **a** Carbon stored in forest ecosystems in 2019 **b** Carbon stored in forest ecosystems in 2033 (BAU scenario) **c** Carbon stored in forest ecosystems in 2033 (ALT scenario) **d** Carbon stored in forest ecosystems in 2033 (RFP Scenario) **e** Carbon stored in forest ecosystems in 2033 (AF scenario) **f** Carbon stored in forest ecosystems in 2033 (SDP scenario)

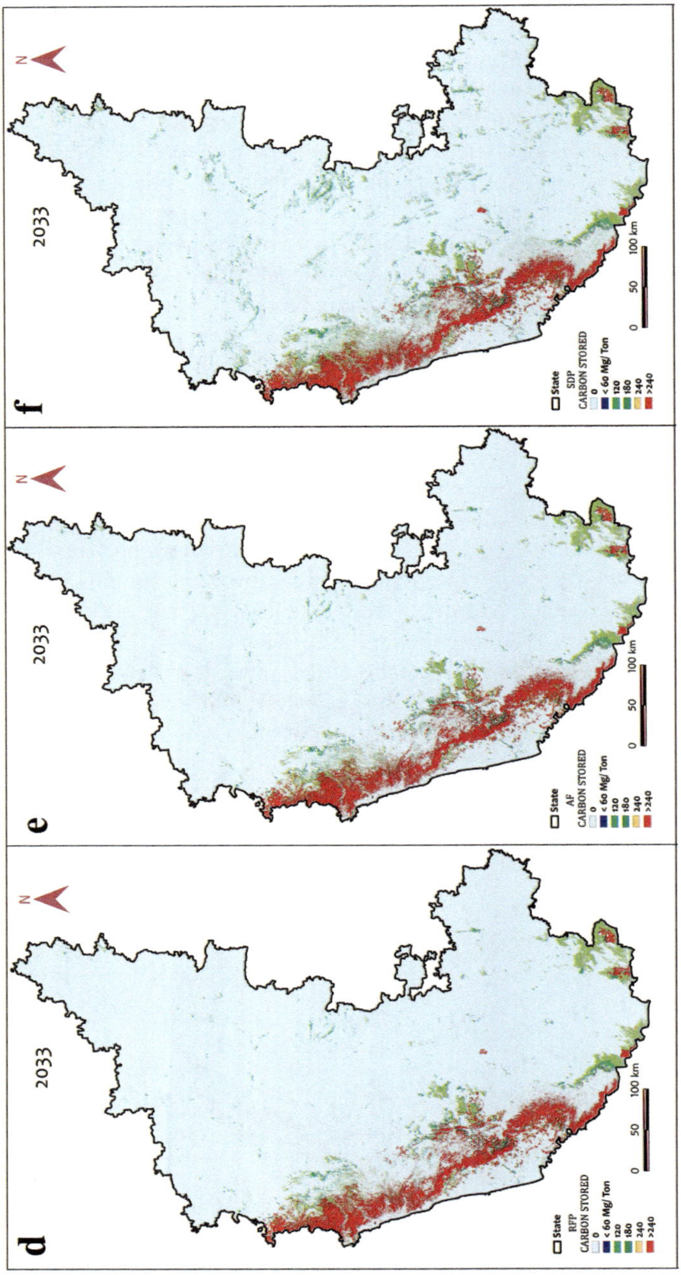

Fig. 17 (continued)

Table 21 Quantification of carbon storage in forest ecosystems (using InVEST)

Year	Total carbon storage (Gg)	Economic value (Million Rs)	Loss in ES (Ref. 2019)
2019	2,440,592	15,625,150	–
Projected_2033-BAU	2,045,774	13,097,455	2,527,696
Projected_2033-ALT	2,021,513	12,942,146	2,683,005
Projected_2033-RFP	2,032,676	13,013,607	2,611,543
Projected_2033-AF	2,045,774	13,097,455	2,527,696
Projected_2033-SDP	2,252,945	14,423,786	1,201,365

6 Conclusions and Recommendations

Integrating environmental dimensions into the land-use planning and management process can significantly contribute to prudent management to ensure the sustenance of natural resources. Sustainable landscape management requires advanced information on likely land use changes, which would help evolve mitigation strategies to sustain the natural resources with dependent communities' livelihoods. This entails inventorying, mapping, monitoring (and change analyses), modeling, and visualization to interpret the consequences of spatial changes empirically. Modeling and visualization of landscape dynamics in Karnataka has been done by considering (i) Business As Usual Scenario-BAU, (ii) Agent-based land-use transition scenario-ALT, (iii) Reserve Forest Protection scenario-RFP, (iv) Afforestation scenario-AFS and (v) Sustainable Development Policy Scenario (SDP).

The spatiotemporal land use analysis was carried out from 1985 to 2019 using temporal remote sensing data (LANDSAT and IRS) through the Gaussian maximum likelihood algorithm. The districts that are part of the Western Ghats region show higher forest cover than other districts. The State now has only 15% of the geographical area under forest compared to 21% in 1985. Post-1990s, the state witnessed large-scale land-use transitions due to industrialization, urbanization, increased horticulture crops, conversion from agriculture to market-based crops (higher economic), etc.

BAU depicts a likely increase in built-up area by 11.5% from 3% (2019). Forests and agriculture areas would lose significant tracts for the likely expansion of built-up cover. The dry and moist deciduous forest cover was shown higher changes with a loss of 6.5% forest cover. The proposed expansion of infrastructure, and built-up cover would transform deciduous forests, and agriculture areas along highways across the state. Cities such as Bangalore, Mangalore, and Dharwad depicted compact growth, whereas Mysore, Belgaum, Hassan, and Tumkur (Tier-2) cities have indicated peri-urban growth.

The ALT scenario incorporates the likely developmental activities—upgrading of urban infrastructure (roads, commercial establishments, information networks, industrial units, sports/recreation), which shows an increase in built-up cover of 15% compared to BAU. ALT scenario presents a more compact urban growth than

Table 22 Change in annual carbon sequestration from 2019 to 2033 and its economic values

S.no	District	2019		BAU		ALT		RFP		AF		SDP	
		Gg	Million Rs	Gg	Million Rs	Gg	Million Rs	Gg	Million Rs	Gg	Million Rs	Gg	Million Rs
1	Bagalkot	337	2162	25	162	25	166	21	140	23	148	213	1368
2	Bellary	1135	7271	187	1203	189	1212	168	1081	190	1222	801	5129
3	Belgaum	4558	29,182	2581	16,529	2574	16,484	2559	16,386	2641	16,913	3786	24,244
4	Bangalore (Rural)	121	780	26	170	24	158	20	134	20	134	72	465
5	Bangalore (Urban)	96	616	65	422	66	426	64	416	66	427	79	508
6	Bidar	150	966	22	142	21	139	18	120	22	143	97	624
7	Chamarajanagar	7934	50,800	7705	49,335	7709	49,359	7703	49,317	7717	49,407	7818	50,054
8	Chikkaballapura	612	3920	253	1623	234	1498	118	757	120	770	378	2425
9	Chikkamagaluru	8681	55,579	7538	48,263	7504	48,042	7759	49,675	7495	47,990	8073	51,686
10	Chitradurga	1026	6574	186	1192	185	1187	167	1070	172	1104	681	4361
11	Dakshina Kannada	7498	48,008	7148	45,764	7014	44,911	7064	45,229	7091	45,398	7107	45,501
12	Davanagere	922	5909	424	2717	425	2724	419	2683	447	2868	690	4422
13	Dharwad	948	6075	418	2680	427	2735	404	2590	442	2831	726	4654
14	Gadag	195	1253	6	41	5	38	5	34	5	34	123	793
15	Hassan	2097	13,430	1397	8950	1396	8943	1235	7912	1260	8069	1862	11,925
16	Haveri	556	3560	220	1411	220	1409	216	1385	237	1518	345	2211
17	Kalburgi	333	2137	91	588	93	599	90	578	107	689	222	1423
18	Kodagu	5835	37,358	5457	34,941	5452	34,907	5497	35,194	5538	35,456	5565	35,633

(continued)

Table 22 (continued)

S.no	District	2019		BAU		ALT		RFP		AF		SDP	
		Gg	Million Rs	Gg	Million Rs	Gg	Million Rs	Gg	Million Rs	Gg	Million Rs	Gg	Million Rs
19	Kolara	343	2197	93	600	90	582	76	488	77	495	214	1376
20	Koppal	234	1501	26	167	25	163	21	136	22	141	149	955
21	Mandya	545	3492	469	3003	464	2974	452	2898	461	2957	511	3273
22	Mysuru	3237	20,727	2972	19,033	2968	19,007	2956	18,926	2969	19,013	3062	19,606
23	Raichur	390	2503	55	352	55	353	50	321	54	348	262	1678
24	Ramanagara	2202	14,102	1703	10,905	1690	10,821	1661	10,639	1790	11,460	1887	12,082
25	Shimoga	9728	62,283	7422	47,519	7410	47,443	7669	49,102	7648	48,968	8407	53,828
26	Tumkur	764	4893	237	1521	219	1406	153	981	162	1038	439	2813
27	Udupi	4676	29,938	4233	27,103	4214	26,981	4212	26,966	4231	27,088	4367	27,962
28	Uttara Kannada	23,708	151,786	19,555	125,196	18,838	120,604	19,436	124,435	19,838	127,007	21,433	137,220
29	Vijayapura	131	844	24	156	25	162	22	143	22	144	87	561
30	Yadgir	202	1300	21	140	20	128	16	107	16	108	128	826
Total		89,194	571,145	70,559	451,829	69,581	445,561	70,251	449,844	70,883	453,889	79,584	509,605

Table 23 Change in carbon storage (Gg) from 2019 to 2033 and its economic values

S. no	District	2019		BAU		ALT		RFP		AF		SDP	
		Gg	Million Rs	Gg	Million Rs	Gg	Million Rs	Gg	Million Rs	Gg	Million Rs	Gg	Million Rs
1	Bagalkot	6648	42,567	463	2967	518	3322	437	2804	463	2967	4202	26,904
2	Bellary	22,364	143,182	3762	24,087	3732	23,897	3327	21,304	3762	24,087	15,771	100,973
3	Belgaum	123,081	787,987	79,943	511,812	78,097	499,990	78,283	501,181	79,943	511,812	107,108	685,724
4	Bangalore (Rural)	2137	13,687	359	2302	430	2753	359	2299	359	2302	1278	8183
5	Bangalore (Urban)	1656	10,605	1129	7234	1131	7241	1098	7033	1129	7234	1359	8703
6	Bidar	3078	19,706	457	2930	446	2856	384	2464	457	2930	1988	12,732
7	Chamarajanagar	160,005	1,024,378	156,221	1,000,156	156,098	999,368	155,984	998,636	156,221	1,000,156	158,002	1,011,555
8	Chikkaballapura	11,952	76,524	2330	14,919	4568	29,246	2290	14,665	2330	14,919	7400	47,381
9	Chikkamagaluru	242,112	1,550,042	213,876	1,369,273	213,108	1,364,351	212,953	1,363,365	213,876	1,369,273	229,186	1,467,285
10	Chitradurga	19,171	122,737	3053	19,550	3305	21,164	2958	18,943	3053	19,550	12,727	81,483
11	Dakshina Kannada	252,137	1,614,224	241,974	1,549,156	238,408	1,526,331	240,747	1,541,306	241,974	1,549,156	244,345	1,564,338
12	Davanagere	16,269	104,157	7602	48,674	7241	46,363	7116	45,561	7602	48,674	12,115	77,566
13	Dharwad	16,956	108,560	7955	50,933	7732	49,507	7303	46,757	7955	50,933	13,054	83,577
14	Gadag	3858	24,704	105	676	116	746	105	676	105	676	2440	15,625
15	Hassan	64,738	414,467	46,086	295,052	48,116	308,051	45,382	290,546	46,086	295,052	59,540	381,186
16	Haveri	9687	62,023	4031	25,811	3746	23,988	3678	23,550	4031	25,811	6017	38,527
17	Kalburgi	6029	38,604	1845	11,814	1615	10,342	1552	9940	1845	11,814	4032	25,814
18	Kodagu	189,356	1,212,290	176,927	1,132,720	181,142	1,159,701	179,463	1,148,952	176,927	1,132,720	184,029	1,178,184

(continued)

Table 23 (continued)

S. no	District	2019		BAU		ALT		RFP		AF		SDP	
		Gg	Million Rs	Gg	Million Rs	Gg	Million Rs	Gg	Million Rs	Gg	Million Rs	Gg	Million Rs
19	Kolara	6089	38,988	1338	8570	1604	10,272	1318	8441	1338	8570	3828	24,513
20	Koppal	4653	29,791	440	2817	509	3262	424	2716	440	2817	2963	18,970
21	Mandya	9647	61,766	7980	51,090	8049	51,537	7821	50,074	7980	51,090	8976	57,468
22	Mysuru	65,130	416,973	60,413	386,775	60,411	386,763	60,185	385,319	60,413	386,775	62,104	397,604
23	Raichur	7883	50,471	1094	7004	1109	7103	1008	6459	1094	7004	5282	33,819
24	Ramanagara	37,633	240,936	30,027	192,243	28,436	182,057	28,540	182,719	30,027	192,243	32,007	204,919
25	Shimoga	279,242	1,787,753	228,065	1,460,109	226,924	1,452,807	223,963	1,433,847	228,065	1,460,109	252,257	1,614,990
26	Tumkur	14,582	93,357	2939	18,819	4098	26,242	2786	17,837	2939	18,819	8476	54,266
27	Udupi	154,643	990,051	143,540	918,966	142,786	914,145	145,849	933,752	143,540	918,966	148,002	947,532
28	Uttara Kannada	703,320	4,502,772	621,034	3,975,959	597,127	3,822,903	616,582	3,947,456	621,034	3,975,959	660,224	4,226,860
29	Vijayapura	2735	17,514	465	2979	524	3358	461	2957	465	2979	1819	11,646
30	Yadgir	3801	24,336	321	2058	387	2480	320	2049	321	2058	2414	15,458
Total		2,440,592	15,625,151	2,045,774	13,097,455	2,021,513	12,942,146	2,032,676	13,013,608	2,045,774	13,097,455	2,252,945	14,423,786

Table 24 Cultural services from forest ecosystems in 2033 under various policy scenarios

S no	District	2019	2033				
		2019_ CULT MRs	BAU_ CULT MRs	ALT_ CULT MRs	RFP_ CULT MRs	AF_ CULT MRs	SDP_ CULT MRs
1	Bagalkot	256	157	157	160	158	215
2	Ballari	609	457	457	454	456	551
3	Belagavi	522	320	320	326	323	439
4	Bengaluru (Rural)	184	152	150	148	148	163
5	Bengaluru (Urban)	175	145	143	141	141	156
6	Bidar	44	7	7	6	7	29
7	Chamarajanagar	79,943	79,906	79,907	79,906	79,907	79,924
8	Chikkaballapura	340	281	278	274	273	302
9	Chikkamagaluru	17,313	17,093	17,094	17,094	17,078	17,186
10	Chitradurga	628	476	476	473	475	570
11	Dakshina Kannada	29,453	29,383	29,395	29,389	29,360	29,384
12	Davanagere	3260	3127	3127	3125	3124	3198
13	Dharwad	150	91	91	89	92	121
14	Gadag	164	99	100	98	101	133
15	Hassan	188	94	92	76	78	143
16	Haveri	169	103	103	101	104	137
17	Kalburgi	90	14	14	13	14	58
18	Kodagu	32,523	32,473	32,472	32,481	32,437	32,500
19	Kolara	318	263	261	257	256	283
20	Koppal	400	300	301	298	300	362
21	Mandya	13,564	13,536	13,535	13,532	13,534	13,547
22	Mysuru	17,332	17,297	17,296	17,293	17,295	17,311
23	Raichur	69	11	11	10	11	45
24	Ramanagara	282	233	231	227	227	251
25	Shivamogga	20,412	20,071	20,070	20,072	20,055	20,200
26	Tumakuru	292	147	143	118	121	223
27	Udupi	21,629	21,578	21,586	21,582	21,561	21,579
28	Uttara Kannada	54,193	53,607	53,533	53,653	53,767	53,837
29	Vijayapura	410	251	251	256	253	344
30	Yadgir	43	7	7	6	7	28
	Grand total	294,955	291,679	291,608	291,658	291,663	293,219

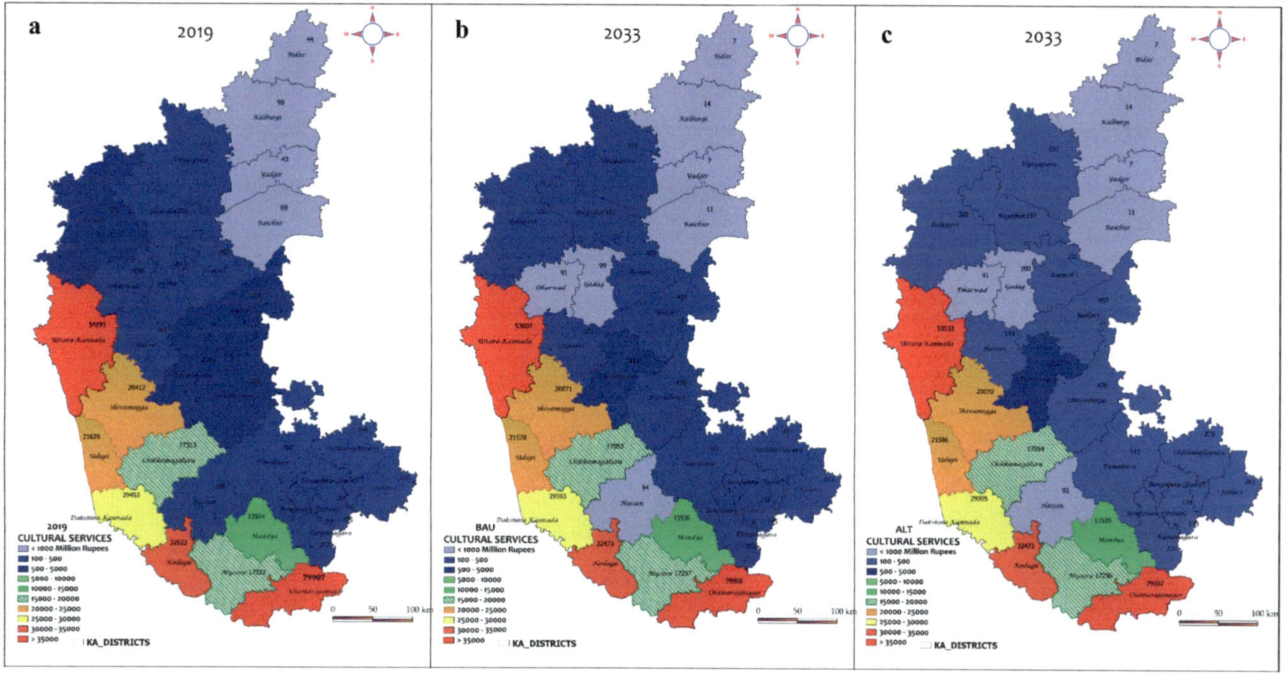

Fig. 18 Cultural Services from forest ecosystem in Karnataka (MRs-Million Rupees) in 2033 across various scenarios. **a** Cultural services from forest ecosystems in 2019, **b** Cultural services from forest ecosystems in 2033, BAU Scenario **c** Cultural services from forest ecosystems in 2033, ALT Scenario **d** Cultural services from forest ecosystems in 2033, RFP Scenario **e** Cultural services from forest ecosystems in 2033, AF Scenario **f** Cultural services from forest ecosystems in 2033, SDP Scenario

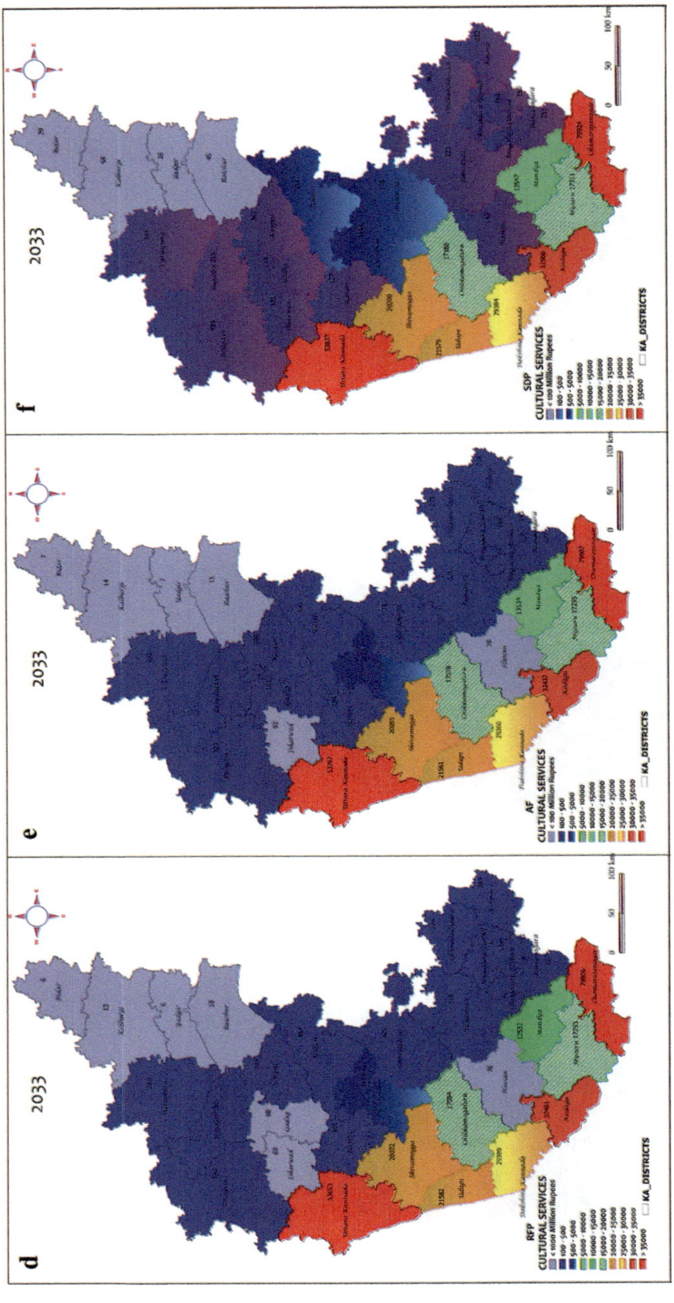

Fig. 18 (continued)

Table 25 Total ecosystem services value of forest ecosystems in Karnataka considering various policy scenarios

Ecosystem: Forests (in billion rupees) Policy scenarios						
Services type	2019	BAU	ALT	RFP	AF	SDP
Provisioning goods	613.90	524.23	521.67	533.69	530.82	540.33
Regulating services	926.36	724.85	716.94	720.78	725.33	824.59
Cultural services	294.97	291.69	291.62	291.67	291.68	293.23
Total	1835.23	1540.77	1530.23	1546.14	1547.83	1658.15

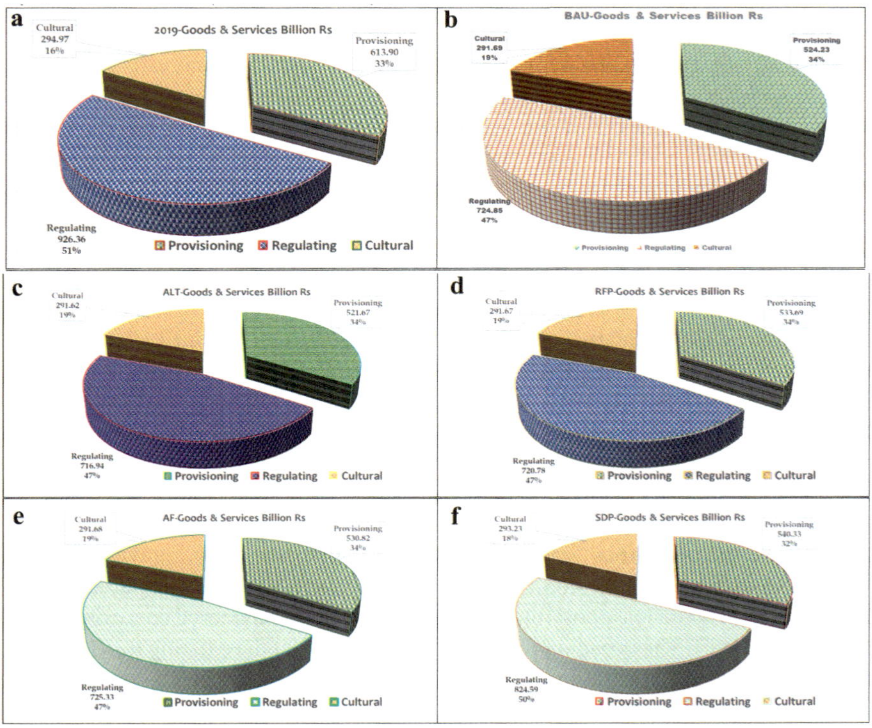

Fig. 19 Relative share of ecosystem services (forest ecosystems), Karnataka State **a** Relative Share of Forest Ecosystem Services, 2019 **b** Relative Share of Forest Ecosystem Services, 2033 (BAU Scenario) **c** Relative Share of Forest Ecosystem Services, 2033 (ALT Scenario) **d** Relative Share of Forest Ecosystem Services, 2033 (RFP Scenario) **e** Relative Share of Forest Ecosystem Services, 2033 (BAU Scenario) **f** Relative Share of Forest Ecosystem Services, 2033 (BAU Scenario)

the BAU, considering that new, high-density residential development occurs inside the cities, major towns, and polycentric urban agglomeration associated with the provision of new transport, industries, and other project investments. BAU and ALT scenario projections indicate a significant reduction in forests and agricultural land uses. ALT scenario shows only 9% forest cover compared to 15% in 2019. The agriculture area showed a loss of 2%, which might affect food availability in the region. The regions where dry crops are practiced would witness higher transformation, signifying the agents/driver's role in land use alterations. ALT scenario shows that the higher land use conversion due to transportation, and other projects, which tend to encourage more compact, peri-urban, and mixed land use developments, necessitates effective policies.

The RF scenario assumes the protection of reserve forests in the regions without allowing any developmental activities or expansion of earlier land use. The regions covered under reserve forest protection do not favor development and other land-use conversions. The districts that are part of the Western Ghats region will not change compared to BAU and ALT scenarios. The regions with dry deciduous forest cover and not under the reserve forest area protection would experience abrupt growth due to the transportation corridors and urban clusters. Unlike other scenarios, the urban area of 2019 is considered to be another factor that can influence land-use change. RF scenario shows likely built-up cover of 11% while retaining evergreen forest cover of 5%.

AF scenario shows the likely improvement in its forest cover due to the plantation activities and protection of degraded forests from uncontrolled land use conversion. The forest cover shows 9.3%, a slight increase compared to BAU and ALT. The built-up cover would limit existing cities and towns by over 11%. The new plantation activities would compensate for the abrupt land use conversion, especially in the moist, and dry deciduous cover regions. Districts such as Bidar, Kolar, Koppala, etc., which have the least forest cover, did not show any improvement.

SDP scenario signifies stringent policy implementation initiatives with the slightest disturbances. The forest cover would remain at 11% compared to 17% in 2019, which is the least possible loss compared to all considered scenarios (BAU, ALT, RF, and AF). The urbanization process would restrict the existing tier-1 and tier-2 cities, and the least increase was shown in peri-urban areas. Districts such as Bidar, Gulbarga, Bagalkot, and Kolar might experience higher land use conversions due to existing infrastructure and the absence of any forest policy. Overall, RFP, AF, and SDP scenarios favor the protection of evergreen and moist deciduous forest categories. BAU, and ALT scenarios depict emerging urban clusters, indicating smaller clusters of paved areas are no longer available and tend to agglomerate with bigger ones to form urban centres, satellite towns, and dispersed growth at the periphery of the tier-2 city limits. New constructions might occur in the rural hinterland, which results in loss of evergreen, moist deciduous cover and agricultural areas. Categories such as plantation, water, and open field had similar values across all scenarios. The scenarios would directly help decision-makers/planners in reducing the impacts on the environment with stricter growth rules; otherwise, conversion of primary, secondary forest/ moist/dry deciduous forest into urban areas would be favored (BAU, ALT), which would induce ecological imbalances. Also, another important observation in terms of topography is that construction restrictions in steep slope areas have played an important role in restricting urban settlements from further infiltrating forested areas. Scenarios 3–5 are appropriate for sustainable development without altering the edges of agriculture, forest, water bodies, and other protected areas or conservation zones. Additionally, they emphasize a planned growth strategy for a habitable region. The integrated planning through regulated land use conversion can support developments such as self-sufficient towns along newly provided public transport corridors, limiting the abrupt land use conversion.

The land use dynamics involving the decline of forests have resulted in significant changes in ecosystem services. The provisioning services (timber, NTFP, fish, fuelwood, fodder, bamboo, cane, genetic material, etc.) from the forests of Karnataka is about 613 billion INR in 2019. Due to land use changes with the decline of forests, the provisioning services in 2033 would reduce to 524 billion INR (BAU scenario), 521 billion INR (ALT Scenario), 533 billion INR (RFP Scenario), 530 billion INR (AF Scenario) and 540 billion INR (SDP), which highlights the need to implement afforestation with the reserve forest protection (SDP scenario) to sustain the livelihood through the provision of goods. ALT scenario shows a reduction of 92 billion INR in provisional goods compared to 2019. The SDP scenario has shown improvement in ecosystem services compared to other scenarios, which highlights the role of policy decisions.

Regulating services of ecosystems pertains to services resulting from the ability of ecosystems to regulate biological processes and influence climate, hydrological, and biochemical cycles, thereby maintaining environmental conditions beneficial to individuals and society. Regulating services (Air Quality, Climate Regulation, Carbon Sequestration, Pollination, Soil Erosion, Fertility, Ground Water Recharge, Water Purification, Waste Treatment) quantified district-wise for 2019 and various scenarios of 2033 reveal that the regulating services from forest ecosystems in 2019 is about 926 billion INR compared to 724 billion INR in 2033 (BAU Scenario), 716 billion INR (ALT Scenario), 720 billion INR (RFP Scenario), 725 billion INR (AF Scenario) and 824 billion INR (SDP Scenario), emphasising the need to implement the sustainable development programme (SDP Scenario) to sustain regulating services to sustain diverse biota on the Earth.

Cultural services of forest ecosystems in 2019 is about 294 billion INR compared to 291 billion INR in 2033 (BAU), 291 billion INR (ALT, RFP, AF Scenarios) and 293 billion INR (SDP Scenario).

The total ecosystem service value of forests in Karnataka is about 1835 billion INR in 2019. However, with the land use changes leading to the degradation of forest ecosystems, the total ecosystem services value would be 1540 billion INR (BAU Scenario), 1530 billion INR (ALT Scenario), 1546 billion INR (RFP Scenario), 1547 billion INR (AF Scenario) and 1658 billion INR (SDP Scenario), emphasizing the need to protect reserve forests and take up afforestation of degraded forest patches to sustain ecosystem goods and services. The relative share of ecosystem services for 2019 and considered policy scenarios. Regulating services dominate (47–51%) compared to the provisioning (32–34%) and cultural services (16–19%).

6.1 Recommendations

Forest resources in Karnataka State have undergone tremendous change and degradation because its value is poorly understood and not considered in the policy-making process. Forests provide various services classified as provisioning, regulating, and cultural services. Most of these services are underestimated or not estimated and are thus undervalued in policy decisions. However, the valuation of regulating services and cultural services is challenging to estimate and thus poses serious challenges to planners and practitioners. As a consequence, the value of these services is often overlooked. Hence, the valuation of the ecosystem services in the income accounting of a region/ nation is essential to make plans and policies more sustainable.

Habitat fragmentation, negligence, conflict of interest, and ineffective restoration/ improvement strategies are major threats. Poor understanding of the complex ecological processes and proper estimation of the ecosystem benefits have often led to the destruction of fragile ecosystems. To improve the scenario, a thorough understanding of the complex ecosystem dynamics and its socio-religious association with community life is important from a conservation and management point of view. Therefore, suggestions are:

- Forests, especially evergreen forests, have high levels of threatened biodiversity, especially belonging to endemic species of plants and animals. Conservation of hydrologically important forests with perennial streams, springs, and swamps must be demarcated on forest management maps for inclusion in the working plans. Conservation activities are implemented mainly by Government agencies, NGOs, and sometimes by communities. However, community participation is often activated by extramural support, which has serious problems in long-term sustainability due to financial limitations. The problem could be mitigated to some extent by awareness generation to raise the interest among people to safeguard its future for their benefit. The premium should be on conserving the remaining fragile ecosystems, which are vital for water security (perenniality of streams), food security (sustenance of biodiversity), and uplift of the local population's livelihoods with carbon credits.
- All forests where globally threatened species of plants and animals occur (figuring in IUCN Red List) should be demarcated along with their buffer zones for maximum care.
- As forests with more endemic tree species have more endemic fishes and frogs in their associated water bodies, the conservation of such water bodies should also include their catchment area forests.
- In the spectre of climatic change that the planet faces with its widespread implications, especially on farming and biodiversity, the need has arisen to increase carbon sequestration in forest areas. There are considerable areas of degraded forests in Karnataka, the biomass of which has to be increased substantially through protection, enrichment, and co-management.
- Realizing that depletion of forests of food resources and human-induced vegetational changes in forests has adverse consequences on wildlife while increasing

crop raids by animals enrichment of secondary forests and poor grade tree plantations with food resources for forest herbivores is highly desirable.

- NTFP collection, which yields only minor revenue to the state, is being overexploited in many forests, destroying the resource itself. We recommend that the Village Forest Committees (VFCs) and other forest dwellers in their respective villages be organized and trained in the scientific harvesting of NTFP, which also serves as medicinal plants.
- Rampant collection of poles, cane, fuelwood, etc., has been taking a heavy toll on forest resources, particularly in the village vicinities. Most of the readily accessible areas with many medicinal plants are more prone to exploitation and converted into scrub and thickets. Even the semi-evergreen and evergreen forests higher up in more inaccessible areas are also being exploited for fuelwood, timber, etc. Many of these forests have thorny thickets as under-growths. Also, bamboo is considered 'poor man's timber', and the villagers may be allowed to harvest it from designated areas for their bonafide use to desist from pole cutting and stake removal from the forests, which destroys lakhs of tree saplings and pole-sized juveniles.
- Carbon sequestration in any given forest is related to forest biomass. Basal area/ha is an index of the forest biomass. Higher carbon sequestration in-stream course/ swamp forests of 211.87 t/ha. In the nine samples from forests away from watercourse areas, carbon sequestration was less at 165.54 t/ha. On a hilltop savannized part, obviously, due to shifting cultivation practices in the past, the carbon sequestration was very poor at only 5.03 t/ha. Numerous hilltops and wind-exposed slopes of the district are in a savannized state with poor biomass, demonstrating the fact clear felling of a rainforest can bring in desertified conditions. Nevertheless, these grassy patches considering forest as a place of tree growth, often also with undergrowth, the savannised forests and several secondary forest samples subjected to ongoing human impacts, or systems recovering from past human impacts, had some of the least basal areas, irrespective of whether they fall in high, moderate or low rainfall areas.
- The quality and quantity of a deciduous forest stand are reflected in its total biomass, of which basal area is an index. Forest fragmentation of high order, shifting cultivation practices in the past, massive conversions into monoculture plantations, clear-felling, and selection felling rampantly practiced in the past are major causes for low basal areas. Compact stretches of forests, especially in areas thinly populated by humans, may be prioritized for developing ideal forests of high stature through special protection and periodical monitoring of natural succession and tree growth progress.
- Fuel extraction, both legal (especially the removal of dead and fallen from interior forests) and illegal by the local population, is instrumental in the degradation of many forests. Energy-efficient stoves, biogas, solar devices, use of agricultural wastes, etc., are to be promoted as fuel in rural areas. Simultaneously, adequate fuelwood/or other alternative fuels should be granted to cottage industries run by potters, lime makers, etc.

- Raising monocultural/mixed tree plantations has to be site-specific. Planting of *Acacia auriculiformis* has to be restricted to rocky or otherwise impoverished terrain and not in lands with good soil resources where native species are to be preferred.
- The practice of afforestation of grassy blanks has to be discontinued unless there is a need for recreation of resource patches (fruit trees and keystone plant resources favoring herbivores' life). These resource patches must be in block planting or linear forms, facilitating corridors for animal movements. A combination of both may also be carried out after planning and deliberations.
- Grassland enrichment is to be thought of for selected grasslands. Very degraded grasslands may be closed to grazing, facilitating a revival of the grasses.
- The aforestation of grassy blanks is to be limited to very unproductive areas only. While selecting tree species for planting, the animal community should be borne in mind.
- Legumes are nitrogen-rich and good as fodder; hence leguminous fodder herbs may be planted in abandoned agricultural fields to promote wildlife. Herbaceous climbers of legumes that provide forage for wildlife may be encouraged experimentally in some of the poor-grade monoculture plantations.
- Natural succession inside monoculture plantations may be directed towards enhancing the food resources for wildlife.
- Herbaceous forage legumes may be considered experimentally for planting along the sides of some of the forest roads.
- The valuation of forest goods and services at the micro-level is expected to explore more effective micro-level planning possibilities. This helps integrate ecosystem services with the region's economic goals while ensuring the sustenance of natural resources and maintaining intergeneration equity.
- Greater interaction with the local community, and forest dwellers and involving them at all stages of planning and implementing forestry programmes run by the Department, and supporting their own planning and implementation of community-based forestry programmes will ensure their resourceful economic and social development. This approach will lead to a sustainable environment, protracted natural resource availability, and holistic development of the region.
- Environmental education can play a vital role in this effort by preparing well sensitized responsible citizenry with a good understanding of various forces operating in the environment and their consequences. Local educational and research institutions should be encouraged to take up documentation of traditional practices, study local and global demands of forest products, research projects to help forest departments enrich local ecology, and farmers shift from non-organic methods to organic farming.
- Energy-efficient stoves, biogas, solar devices, use of agricultural wastes, etc., are to be promoted as fuel in rural areas by creating awareness through schools and panchayats. Energy plantations raised on degraded lands providing the right to the public will help improve the ecological status of the region, provide biomass feedstock for rural bioenergy programmes, and help meet fuelwood demand.

- The cluster-based economic development approach is considered an important aspect of a broader research and economic policy reorientation towards laying the foundations of a microeconomic approach for prosperity and growth. The past decades were under the spell of macroeconomics and the creation of market institutions in developing economies. While there is now a broad consensus on the macroeconomic and legal conditions necessary to achieve economic progress, it is also becoming clearer that these conditions are insufficient. As a new approach to help economies reap the full potential of an improved macroeconomic and legal context, cluster-based efforts have received a lot of attention. Clusters are groups of companies and institutions co-located in a specific geographic region and linked by interdependencies in providing a related group of products and/or services. Because of the proximity among them—both in terms of geography and of activities—cluster constituents enjoy the economic benefits of several types of positive location-specific externalities.
- Financial allocations to restore degraded forest patches are to be made in advance. Budget allocations should be released according to the requirement of afforestation activities that should be planned after adopting a State forest policy and committing funds for it.
- Geo-tagging of afforested locations would help in monitoring the progress of afforestation.
- Seedlings of location-specific native species should be raised in all cases as per the 'package of practices' and planted according to site-specific plans.
- The afforested plots should be maintained for the prescribed period, and the survival rate of plantations beyond the planting year should be recorded with the data available in the public domain.
- Correct assessment of various components of compensatory afforestation (CA) charges should be ensured based on the respective ecosystems' goods and services.

References

Ahmed B, Ahmed R, Zhu X (2013) Evaluation of model validation techniques in land cover dynamics. ISPRS International Journal of Geo-Information 2(3):577–597

Axtell R (2000) Why agents? on the varied motivations for agent computing in the social sciences. Cent Soc Econ Dynamics—Brook Inst 17:1–23

Batty M, Xie Y (1994) From cells to cities. Environment and Planning B 21:31–48

Batty M (2005) Agents, cells, and cities: New representational models for simulating multiscale urban dynamics. Environ Plan A 37(8):1373–1394

Bernard RN (1999) Using adaptive agent-based simulation models to assist planners in policy development: the case of rent control. Santa Fe Institute Working Papers, 7(52):18

Clarke KC (2008) A decade of cellular urban modeling with SLEUTH: Unresolved issues and problems. Plan Support Syst Cities Regs:47–60

Crooks AT (2010) Constructing and implementing an agent-based model of residential segregation through vector GIS. Int J Geogr Inf Sci 24(5):661–675

Crooks AT, Heppenstall AJ (2012) Introduction to agent-based modelling. In: Agent-based models of geographical systems. Springer, Dordrecht, pp 85–105

Eckhardt R (1987) Stan Ulam, John von Neumann, and the monte carlo method. Los Alamos Science, Special Issue, 131–141
Ermentrout GB, Edelstein-Keshet L (1993) Cellular automata approaches to biological modeling. J Theor Biol 160(1):97–133
Franklin S, Graesser A (1996) Is it an agent, or just a program? A taxonomy for autonomous agents. In: Proceedings of the third international workshop on agent theories, architectures, and languages. pp 21–36
Guidolin M, Chen AS, Ghimire B, Keedwell EC, Djordjevic S, Savic DA (2016) A weighted cellular automata 2D inundation model for rapid flood analysis. Environ Model Softw 84:378–394
Guzman LA, Escobar F, Peña J, Cardona R (2020) A cellular automata-based land-use model as an integrated spatial decision support system for urban planning in developing cities: The case of the Bogotá region. Land Use Policy 92:104445
Hamad R, Balzter H, Kolo K (2018) Predicting land use/land cover changes using a CA-Markov model under two different scenarios. Sustainability 10(10):3421
Holland JH (1995) Hidden order: how adapation builds complexity. Basic Books, vol 75
Itami RM (1994) Simulating spatial dynamics: cellular automata theory. Landsc Urban Plan 30:24–47
Khwanruthai B, Yuji M (2011) Site suitability evaluation for ecotourism using GIS & AHP: A case study of Surat Thani Province, Thailand. Procedia Soc Behav Sci 21:269–278
Li X, Gong P (2016) Urban growth models: progress and perspective. Science Bulletin 61(21):1637–1650
Li X, Liu X (2006) An extended cellular automaton using case-based reasoning for simulating urban development in a large complex region. Int J Geogr Inf Sci 20(10):1109–1136
Macal CM, North MJ (2009) Agent-based modeling and simulation. Simulation:86–98
Pascal JP (1986) Explanatory booklet of vegetation maps 1, 2 and 3. Institut Français de Pondichéry, India, pp 88. http://hal.archives-ouvertes.fr/hal-00444285
Pontius JRG, Malanson J (2005) Comparison of the structure and accuracy of two land change models. Int J Geogr Inf Sci 19:243–265
Pontius Jr R.G, Millones M (2011) Death to kappa: birth of quantity disagreement and allocation disagreement for accuracy assessment. Int J Remote Sens, 32(15):4407–4429
Ramachandra TV, Kumar U (2010) Greater Bangalore: Emerging urban heat island. GIS Development 14(1):86–104
Ramachandra TV, Setturu B, Aithal BH (2012a) Peri-urban to urban landscape patterns elucidation through spatial metrics. Int J Eng Res Dev 2(12):58–81
Ramachandra TV, Aithal BH, Sanna DD (2012b) Insights to urban dynamics through landscape spatial pattern analysis. Int J Appl Earth Obs Geoinf 18:329–343
Ramachandra TV, Bharath HA (2016) Bangalore's Reality: towards unlivable status with unplanned urban trajectory. Curr Sci 110(12):2207–2208
Ramachandra TV, Bharath S (2021) Carbon footprint of Karnataka: Accounting of sources and sinks. In: Carbon footprint case studies. Springer, Singapore, pp 53–92
Saaty TL (1980) The analytical hierarchy process: planning, priority setting. McGraw-Hill publication, New York, Resource Allocation
Samie A, Deng X, Jia S, Chen D (2017) Scenario-based simulation on dynamics of land-use-land-cover change in Punjab Province, Pakistan. Sustainability 9(8):1285
Sante I, Garcia AM, Miranda D, Crecente R (2010) Cellular automata models for the simulation of real-world urban processes: A review and analysis. Landsc Urban Plan 96(2):108–122
SEEA EA (2020) System of environmental-economic accounting—ecosystem accounting (SEEA EA). United Nations, White cover publication, pre-edited text subject to official editing. Available at: https://seea.un.org/ecosystem-accounting
Singh AK (2003) Modelling land use land cover changes using cellular automata in a geo-spatial environment. International Institute for Geo-information Science and Earth Observation: Enschede, The Netherlands. Accessed on December 18, 2020. Retrieved from https://pdfs.semanticscholar.org

Spencer D (2009) Cities and complexity: understanding cities with cellular automata, agent-based models, and fractals. J Archit 14(3):446–450
Torrens PM (2000) How cellular models of urban systems work (1. Theory). Paper presented at: Working Paper Series 28, Centre for advanced spatial analysis, University College London, UK
Ulam S (1958) John Von Neumann 1903–1957. Bull Am Math Soc 64(3):1–49
Verburg PH, Soepboer W, Veldkamp A, Limpiada R, Espaldon V, Mastura SSA (2002) Modeling the spatial dynamics of regional land use: The CLUE-S model. Environ Manage 30(3):391–405
Wahyudi A, Liu Y (2013) Cellular automata for urban growth modeling: a chronological review on factors in transition rules. In: 13th international conference on computers in urban planning and urban management
Wolfram S (2002) A new kind of science. Wolfram Media, Champaign, IL
Wu N, Silva EA (2010) Artificial intelligence solutions for urban land dynamics: A review. J Plan Lit 24(3):246–265

Natural Capital Accounting and Valuation of Aquatic Ecosystem Services in Karnataka, India

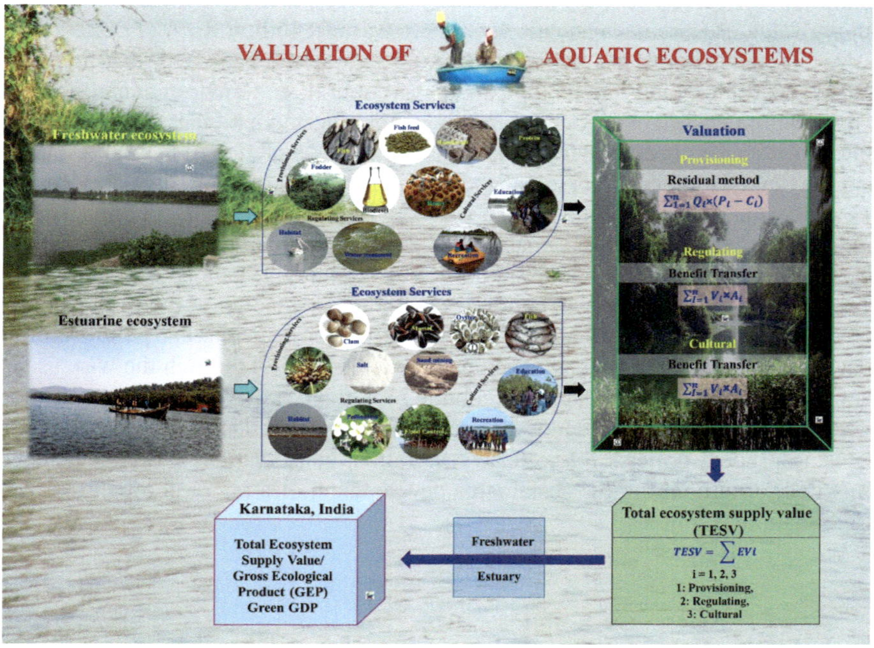

Abstract Wetlands are productive ecosystems providing an array of services that sustain the well-being of dependent biota. The industrialization and globalization era witnessed a spurt in anthropogenic activities, leading to the degradation and decline of fragile ecosystems, affecting the livelihood of the dependent population. This necessitates the conservation of vital ecosystems through sustainable management tenets, which requires an understanding of the livelihood support of ecosystems. Now, many wetlands are degraded due to increasing pollution and a lack of awareness among the public about the economic value of wetland ecosystems. Thus, the current chapter focuses on the valuation of provisioning, regulating, and cultural services through the residual value and resource rent method and the benefit transfer method

T. V. Ramachandra et al., *Natural Capital Accounting and Valuation of Ecosystem Services, Karnataka State, India*, https://doi.org/10.1007/978-981-97-2405-5_5

from aquatic ecosystems in Karnataka, India. The value of provisioning, regulating, and cultural services provided by freshwater lentic ecosystems in Karnataka are 50, 197, and 38 billion rupees/year, respectively. The total ecosystem supply value (TESV) provided by the freshwater ecosystem of Karnataka is 285 billion rupees/ year, and the net present value (NPV) amounted to 7321 billion rupees.

Similarly, the value of services provided by the estuarine ecosystem in Karnataka are 5, 10, and 1 billion rupees/year from the provisioning, regulating, and cultural services, respectively. The TESV provided by the estuarine ecosystem of Karnataka is 16 billion rupees/year, and NPV amounted to 411 billion rupees. The total value of provisioning, regulating, and cultural services considering both freshwater and estuarine ecosystems was 55 billion Rs/yr (183,328 Rs/ha/yr), 207 billion Rs/yr (691,577 Rs/ha/yr), and 39 billion Rs/yr (130,686 Rs/ha/yr), respectively. The total ecosystem supply value of Karnataka aquatic ecosystem was 301 billion Rs/yr (1,005,591 Rs/ ha/yr) and the net present value amounts to 7732 billion rupees. This highlights the importance of wetlands in ecological, social-cultural, and environmental aspects. Appraisal of ecosystem services (ES) allows for adjusted national accounts, which reflect the output of ecosystem services as well as the depletion of natural resources and the degradation costs (externalized costs of the loss of ecosystem services) of ecosystems in economic terms, which will help raise awareness and provide a quantitative tool to evaluate the sustainability of policies toward prudent management and conservation of fragile livelihood supporting ecosystems. The monetary valuation of ecosystem services can help in building a better understanding of their influence on well-being and can further facilitate information-driven decisions and policy reforms that align with the Sustainable Development Goals (SDGs) through the wise use of natural resources. Natural capital accounting research, involving the quantification of services provided by wetlands and insights into economic values, would help in devising prudent policies for wetland restoration, conservation, and management. This will, in turn, aid in the sustainable development of a region through the wise use of natural resources.

1 Introduction

Wetlands are among the most diverse and productive ecosystems, providing numerous ecological, economic, and social benefits to society. Wetlands provide vital ecosystem services and processes, such as nutrient cycling, water purification, reducing pollution, carbon sequestration, groundwater recharge, flood reduction, erosion control, habitats for aquatic flora/fauna, opportunities for education, aesthetics, and recreation (de Groot et al. 2020; Ramachandra et al. 2019; Barbier 2013; Lambert 2003; Costanza et al. 1997). The Ramsar convention on wetlands defined wetlands as "areas of marsh, fen, peat land or water, whether natural or artificial, permanent or temporary, with water that is static or flowing, fresh, brackish or salt, including areas of marine water, the depth of which at low tide does not exceed 6 m". Thus, wetlands refer to locations that are inundated or saturated by surface or ground water at a frequency and duration sufficient to support, and that under normal circumstances do support, a prevalence of vegetation and other biotas typically adapted for life in saturated soil conditions. Wetlands help to conserve biodiversity by providing habitat for fish, plankton, aquatic plants, insects, and crustaceans, as well as feeding and resting areas for water birds (Zhang et al. 2015a). Food chains/food webs describe the structure of communities inhabiting a particular ecosystem, the associated energy and nutrient flows, and the interactions between species (Ramachandra et al. 2005). Wetlands also aid in remediation by removing nutrients (like nitrate, phosphate, etc.) and heavy metals from water (Zedler and Kercher 2005; Ausseil et al. 2015).

Functions of wetlands include the provision of habitat for fish and wildlife, as well as for protecting water quality (remediation), erosion prevention, flood storage, and recreation. Wetlands retard water flows, aid in the percolation of rainwater into the ground (vadose and saturated zones), and reduce downstream soil erosion. Some wetlands, particularly those on floodplains and coastal areas, store excess water during storm events, which helps flood control and protection from storms. Figure 1 illustrates the functions of wetland ecosystems: food production, habitat provision, information provision, and regulation of ecosystem processes (Fig. 1). The food web includes complex, dynamic feeding relationships among the organisms inhabiting wetland environments. The combination of shallow water, high levels of inorganic nutrients, and high rates of primary productivity (the synthesis of new plant biomass through photosynthesis) in many wetlands is ideal for the development of organisms (insects, mollusks, and crustaceans) that form the base of the food web. The primary producers, microalgae, are photosynthetic organisms that provide food and energy to higher trophic levels. Consumers at the next food chain level consume the above-ground live vegetation (herbivore-carnivore food web). At the same time, some utilize the dead plant biomass, which breaks down in the water to form small, nutrient-enriched particles of organic material (detritus). As the plant material continues to break down into smaller and smaller particles, it becomes increasingly enriched (nutritious) due to bacterial, fungal, and protozoan activities. The enriched proteinaceous material, including various microbes (that colonize it), feeds many small aquatic invertebrates and small fish, which then serve as food for larger predatory amphibians, reptiles, fish, birds, and mammals.

The diversity of habitats in a watershed of wetlands is vital for biogeochemical cycling, which involves the biological, physical, and chemical transformations of various nutrients within the biota, soils, water, and air. Wetlands are very important for managing nitrogen, sulfur, and phosphorus. Anaerobic (non-oxygenated) and chemically reduced wetland soils and the muddy sediments of aquatic habitats like estuaries, lakes, and streams support microbes that function in nitrogen and sulfur cycling. Upon death and decay, the nitrogen and sulfur in plant and animal biomass are released through mineralization. Much of this is eventually transformed into gaseous forms and released into the atmosphere, where it once again becomes available to certain plants and their associated nitrogen-fixing bacteria in the soil. This is a major defense for mud since the anaerobic and chemically reducing conditions in

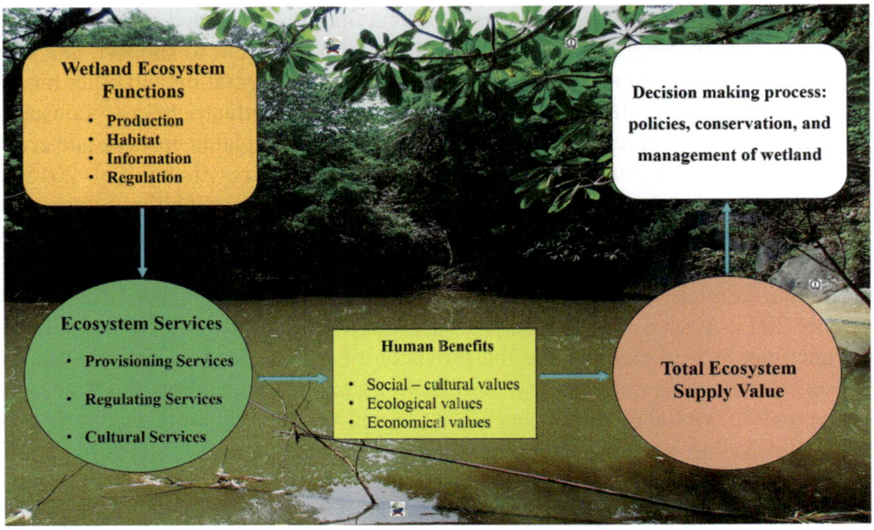

Fig. 1 Wetland ecosystem function and service valuation

the substrate, in conjunction with various microbes, ensure the gaseous release of nitrogen and sulfur. On the other hand, phosphorus does not have a gaseous form, but vascular plants in wetlands transform inorganic forms of phosphorus (that might otherwise be shunted into undesirable algal blooms) into organic forms in their biomass as they grow. Thus, wetlands provide the conditions needed for the removal of both nitrogen and phosphorus from surface water, which improves water quality, including that of drinking water, by intercepting surface runoff and removing or retaining inorganic nutrients, processing organic wastes, and reducing suspended sediments before they reach open water.

The pelagic food chain (phytoplankton at the base) was the longest, with four trophic transfers, compared to the benthic food chain (based on detritus of planktonic origin), which involved three trophic transfers, and the macrophytic detritus-based food chain involved two trophic transfers (Kulkarni and Ramachandra 2006; Peel et al. 2019). The oxygenic photosynthesis by microalgae belonging to cyanobacteria led to an increase in oxygen levels on early Earth, initiating aerobic respiration and leading to the evolution of complex multicellular life forms (Sánchez-Baracaldo and Cardona 2020; Soo et al. 2017). Microalgae provide consumers in the food chain with omega-3 polyunsaturated fatty acids (PUFA) (Ebm et al. 2021). Thus, microalgae provide PUFA, such as eicosapentaenoic acid (EPA) and docosahexaenoic acid (DHA), to other aquatic organisms for their growth (Ruess and Müller-Navarra 2019).

Macrophytes provide food and shelter for other aquatic organisms, serve as fodder for livestock, and have medicinal value used for treating diseases. Macrophytes act as substrata for periphyton (Vadeboncoeur 2006) as well as microhabitats (food and space) for zooplankton (Choi et al. 2014). In aquatic ecosystems, macrophytes aid in nutrient removal, water purification/treatment, and carbon sequestration (Ramachandra et al. 2020a). Fish store nutrients in their tissues, translocate, and excrete dietary nutrients in dissolved forms readily available to primary producers, i.e., microalgae in lakes (Vanni 2002). Fish-feeding activities alter the community structure of phytoplankton, zooplankton, and aquatic insects (Griffiths 2006). Bioturbation of bottom sediments by fish will cause the release of nutrients into overlying surface water in shallow lakes, thus altering the quality of the water.

Wetlands provide numerous ecosystem services, which include provisioning, regulating, and cultural services, and there is a need to quantify the socio-cultural, ecological, and economic value of wetlands (TEEB 2010) for evolving effective wetland conservation and management strategies (Fig. 1). Wetland valuation aids in prudent management, conservation, and sustainable use of wetlands and their resources to improve environmental health and biodiversity, minimize pollution and environmental degradation, and allow decision-makers to develop appropriate policies, and raise financial support for wetland conservation.

2 Ecosystem Services from Wetlands

Ecosystem services are the benefits derived from nature for human wellbeing (Fig. 2). The Millennium Ecosystem Assessment (MEA) and the Common International Classification of Ecosystem Services (CICES) have identified ecosystem services as provisioning, regulating, and cultural services.

Provisioning services include the products/material benefits obtained from ecosystems such as water, food (fish, algae, etc.), fiber, fuel, genetic resources, biochemical, medicines, and pharmaceuticals. Regulating services include the benefits obtained from various ecosystem processes, such as water regulation, climate regulation, water purification, erosion regulation, and flood control. Cultural services include the nonmaterial benefits from ecosystems through spiritual, recreational, aesthetic, and ecotourism experiences (Magalhães Filho et al. 2021; de Groot et al. 2006). These benefits may vary in terms of geographical scale, from local to global (Everard et al. 2019). Different researchers have studied wetland ecosystem services globally (Table 1). The economic value of a wetland depends on catchment (watershed) integrity or characteristics (hydrology, vegetation, and soil), geographical conditions, socio-economic aspects, and the population of the wetland area (Yiran et al. 2014). The monetary value of global wetland ecosystem services (in 2011) is estimated at US$ 47.4 trillion per year, constituting about 43.5% of the total value of all natural biomes (Davidson et al. 2019). The economic value of Songore wetland was estimated at $142,431 (gross financial value) and the wetland income was $20,848 annually (Mahlatini et al. 2020). The total value of ecosystem services from Mabamba Bay wetlands was found to be US$3.6 million/year (Akwetaireho and Getzner 2010).

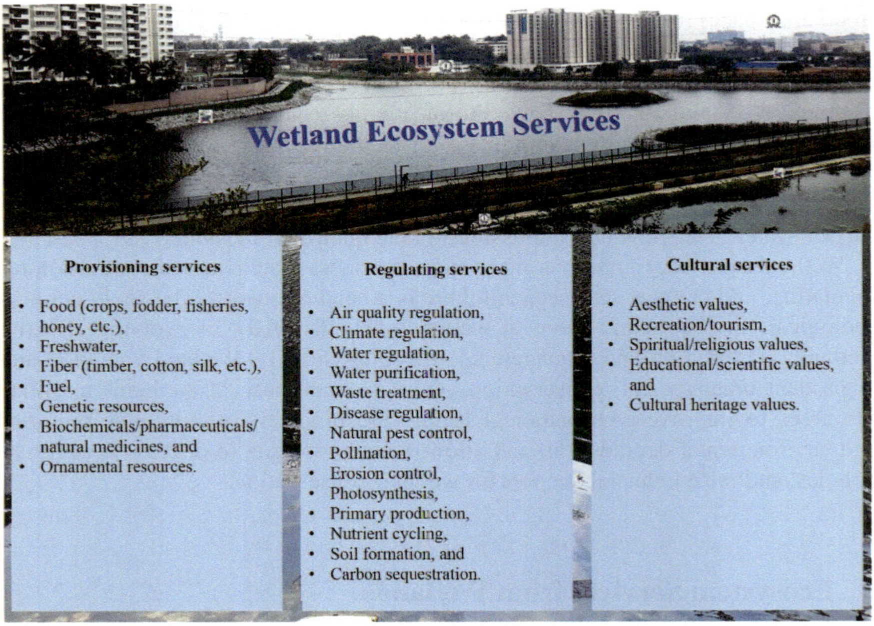

Fig. 2 Ecosystem services from wetlands

In India, freshwater lakes provide various ecosystem services to society. For example, the services provided by Rachenahalli lake in Bangalore are 10,435 Rs/ha/day (Ramachandra et al. 2005). The services provided by freshwater lakes in the Kashmir Himalayas include sources of drinking water, irrigation, agriculture, fishery, navigation, recreation, and socioeconomic development (Najar and Khan 2012). Pandoh Lake water is used to generate hydropower, fishing, and agricultural activities (Anshumali and Ramanathan 2007). Brahmaputra Maun in Muzaffarpur city, Bihar, provides various services to people, such as food (fish), fodder, ornamental resources, traditional medicine, aquaculture, and water for domestic use (Kumar et al. 2013). Tourism is a major source of foreign exchange, income, and employment in countries. Neermahal is the "Lake Palace" in East India, situated in the middle of Rudrasagar Lake, which attracts numerous visitors from India and abroad (Burman et al. 2007). Rewalsar Lake has mythological significance where the pilgrims of Hindu, Sikh, and Buddhist communities perform religious rituals (mass bathing, offering flowers and garlands, offering food for fishes, birds, etc.) during the Baisakhi fair (Harvest festival) in Himachal Pradesh, India (Kashyap et al. 2016). In Pushkar Lake, the festival of Kartik Poornima and the annual cattle fair is held in the month of Kartik (October–November) (Lal and Gardner 2016). According to an Indian wetland study, states such as Karnataka, Gujarat, and Andhra Pradesh are the richest states in terms of wetland wealth, with a worth of 7896 million US$/yr, 7689 million US$/yr, and 7671 million US$/yr respectively (Pandey et al. 2004).

Table 1 Wetland ecosystem services in $/ha/yr

Services	de Groot et al. (2020)	Clarkson et al. (2013)	de Groot et al. (2012)	Costanza et al. (1997)	Russi et al. (2013)	Quintas-Soriano et al. (2016)	Li and Gao (2016)	Zang et al. (2011)	Zhang et al. (2015b)
Food	6,030	614	1111	256	2090	479		26	14
Water	1,934	408	1217	3800	5189	234	3800		2872
Raw materials	1,682	425	358	106	2430		106	5	1
Genetic resources	60		10						
Medicinal resources		99	301						
Ornamental resources		114							
Air quality regulation	34						133	118	
Climate regulation	150	488	65	330	351			1148	65
Moderation of extreme events	13,320	2986			4430				
Regulation of water flows	3,638	5606		15	9369			2345	
Waste treatment	2,043	3015	162,125	4177	4280	175	4177	2376	2562
Erosion prevention		2607	3929	4539		41			
Maintenance of soil fertility		1713	45		4588			112	1
Biological control		948				20	304	326	351
Maintenance of life cycles of migratory species	1,886	1287	10,648	304	917				
Maintenance of genetic diversity	3,427	1168	6490		2554				
Aesthetic information	49	1292			3906	109			
Opportunities for recreation and tourism	2,660	2211	2193	574	3700		574	646	612
Inspiration for culture, art and design	114	700		881	793		881		
Spiritual experience	1								
Information for cognitive development	120								
Existence and bequest values	11,498					8			

Several researchers have integrated various geospatial techniques for wetland monitoring and conservation measures to understand the importance of wetland ecosystem services (Table 1). Geospatial techniques play a significant role in wetland inventory, monitoring, habitat suitability analysis, and the preparation of conservation and management plans (Garg 2015). Geographical Information Systems (GIS) and temporal remote sensing data can help in the mapping of wetland ecosystems and ecosystem services (Langan et al. 2019). GIS-based tools are useful to incorporate public perception into ecosystem assessments (Shoyama and Yamagata 2016). The involvement of stakeholders in ecosystem valuation helps to identify the major relevant services, assess the values of ecosystem services, and discuss trade-offs involved in ecosystem service use (Koko et al. 2020). Various GIS-based mapping, and modelling tools have been designed to assess ecosystem services, include Water-World, Artificial Intelligence for Ecosystem Services (ARIES), Integrated Valuation of Ecosystem Services and Trade-offs (InVEST), Water Evaluation and Planning System (WEAP), Toolkit for Ecosystem Service Site-based Assessment (TESSA) help in policy and decision making (Pandeya et al. 2016). Geoinformatics with geospatial tools are useful in the valuation of wetland ecosystems.

3 Approaches of Ecosystem Service Valuation

In general, market value is used for provisioning services, while non-market values are used for regulating services and cultural services (Table 2). Direct valuation (market prices) was applied to estimate the value of provisioning services, while indirect valuation included the avoided cost method and the replacement cost method (Aulia et al. 2020). The value of regulating services provided by wetlands can be obtained through non-market valuation techniques using travel cost (TC), contingent valuation (CV), or hedonic pricing (HP) methods (Woodward and Wui 2001). The valuation of cultural services is estimated using direct market pricing and travel cost methods (de Groot et al. 2012). The contingent valuation method (CVM) depends on individuals' preferences (willingness to pay for a service) in monetary terms and is used for the economic valuation of ecosystem services, including both use values and non-use values (Vo et al. 2012). The willingness to pay (WTP) and benefit-transfer methods are the two main economic valuation methods (Xu et al. 2020).

The valuation of ecosystem services of lentic aquatic ecosystems (inland wetlands and estuaries) is implemented for Karnataka State, India, as per the validated protocol—System of Environmental-Economic Accounting (SEEA EA; SEEA.un.org), which constitutes the statistical framework for natural capital accounting. According to this protocol, ecosystem services are defined as the contributions of ecosystems to the benefits that are used in economic and other human activities. The valuation of ecosystem services (VES) allows for adjusted national accounts which reflect the output of ecosystem services as well as the depletion of natural resources and the degradation costs (externalized costs of the loss of

Table 2 Various methods adopted for ecosystem services valuation

Wetland	Area	Total value of ecosystem service (ES)	The method used for ecosystem service valuation	References
Rachenahalli Lake, Bengaluru, Karnataka	42.09 ha	10,435 Rs/ha/day	Market prices method and CVM, socio-economy survey and questionnaire	Ramachandra et al. (2005)
Varthur Lake, Bengaluru, Karnataka	220 ha	9,554,000 Rs/220 ha/yr	Market prices method and CVM, socio-economy survey and questionnaire	Ramachandra et al. (2011)
Chan-Ba wetland, China	2533 hm^2	7.7462×10^6¥/a 3.87×10^8¥/a	Market value method, travel cost method, shadow engineering method, ecological method, carbon tax method, afforestation cost method and alternative cost method	Zhou et al. (2018)
Begnas watershed system	49 km^2	3.91 US$ million per year, equivalent to 650.67 US$ per household and 799.79 US$/ha	Market price method, TC method, revealed price method, CVM, and benefit transfer method	Thapa et al. (2020)
Lake and marsh wetlands in China	3.241×10^4 km^2	8.1841×10^{10} (USD)	Value transfer method	Yiran et al. (2014)
35 national nature reserves in 11 coastal provinces/municipalities in China		331.68×10^8 USD/yr	Market price method, avoided cost method, and shadow project method	Li et al. (2020)
Ga-Mampa wetland, South Africa	1 km^2	Provisioning services: 90,000 $/yr		Adekola et al. (2008)
Songore wetland, Zimbabwe	56.25 ha	20,843.31 $/yr	Market price method	Mahlatini et al. (2020)
Rural wetland Letseng-la-Letsie, South Africa	819 ha	Provisioning services: 220·$/ha/yr	Market price method	Lannas and Turpie (2009)

(continued)

Table 2 (continued)

Wetland	Area	Total value of ecosystem service (ES)	The method used for ecosystem service valuation	References
Peri-urban wetland Mfuleni, South Africa	310 ha	Provisioning services: 1765 $/ha/yr	Market price method	
Satajan wetland of North Lakhimpur, Assam, India	95 acre	Provisioning services: Rs 52, 65,600 or 78,591 $/yr	3 types of surveys were conducted: (i) key information interview, (ii) focus group discussion and (iii) household surveys, market cost valuation method and alternate valuation method	Kakoti et al. (2019)
Kalyanthakur para lake, Tripura, Northeast India	7.84 ha	Provisioning services: 26,263.65 US$/yr Cultural services: 2605.68 US$/yr	CVM	Taran et al. (2016)
Kala-oya basin, Sri Lanka	2,870 km^2	Provisioning services: 2972 US$/ha/yr	Participatory appraisal methods and market prices method	Vidanage et al. (2005)
Ghodaghodi wetland, Nepal	2563 ha	674,000 US$/yr	Market price method, net revenue method, value transfer method, and CVM	Aryal et al. (2021)
Koshi Tappu wildlife reserve, Nepal	175 km^2	16 USD million/yr	Market-based and value transfer method	Sharma et al. (2015)

ecosystem services) of ecosystems in economic terms, which will help raise awareness and provide a quantitative tool to evaluate the sustainability of policies. It provides an unbiased and dependable national framework to value so far unaccounted ecosystem benefits and helps develop meaningful policy interventions.

Ecosystem services encompass all interactions between ecosystems and people, including in situ and remote interactions. The supply of an ecosystem service is associated with an ecosystem structure or process or a combination of ecosystem structures and processes that reflect the biological, chemical, and physical interactions among ecosystem components. Ecosystem services are broadly categorised as (i) provisioning services, which are those ecosystem services representing the contributions to benefits that are extracted or harvested from ecosystems; (ii) regulating and maintenance services, which are those ecosystem services resulting from the ability of ecosystems to regulate biological processes and to influence climate, hydrological and biochemical cycles, and thereby maintain environmental conditions beneficial to individuals and society; and (iii) cultural services, which are experiential and intangible services related to the perceived or actual qualities of ecosystems whose existence and functioning contributes to a range of cultural benefits. A range of other benefits, for example, concerning relational and intrinsic values, are not captured in the above categories.

A key feature of ecosystem accounting is its capacity to integrate spatially referenced data about ecosystems, i.e., data about the location, size, and condition of ecosystems within a given area and how these are changing over time. Recording these stocks and changes in stocks in a coherent and mutually exclusive manner supports the derivation of indicators. Understanding the size and location of ecosystems also supports measuring ecosystem conditions and quantifying and valuing many ecosystem services, the flows of which will vary from ecosystem to ecosystem.

The total ecosystem supply value (TESV) is the sum of provisioning, regulating, and cultural services for a particular ecosystem (Fig. 3). In the System of Environmental-Economic Accounting (SEEA 2014), ecosystem assets are generally valued based on the net present value (NPV) of the expected flow of ecosystem services.

There are only a few studies focusing on the valuation of wetland ecosystems in India. The valuation of wetland ecosystems is necessary to protect biodiversity, sustain the use of wetland resources, and conserve wetland. The main objectives of this study were to (a) find out the various aquatic ecosystem services provided by algae, fish, macrophyte, and mangroves; (b) estimate the provisioning services, regulating services, and cultural services provided by wetlands in all the districts in Karnataka; (b) accounting the total ecosystem supply value (TESV) and the net present value (NPV) of wetlands in Karnataka, India.

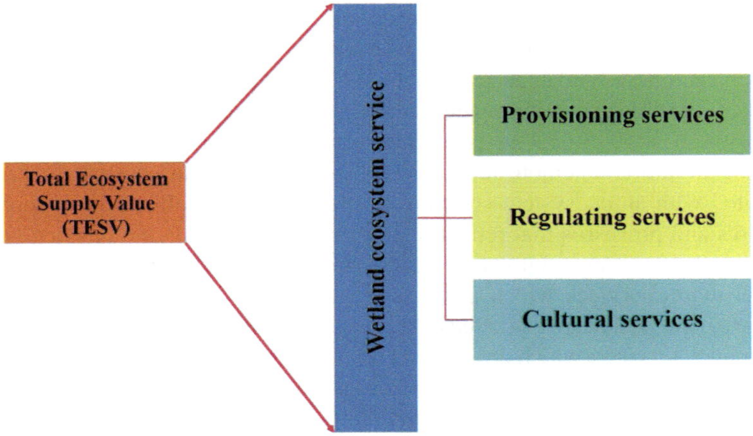

Fig. 3 TESV framework for valuation of ecosystem services

4 Materials and Methods

4.1 Study Area

Karnataka State is endowed with a vast inland water spread that includes lakes, tanks, reservoirs, rivers, and ponds. Karnataka (11° 30′ N and 18° 30′ N latitudes and 74° E and 78° 30′ E longitudes) is the 8th largest state by area and ninth-largest state by population in India. The distribution of wetlands in Karnataka is presented in Fig. 4. The size and district-wise distribution of wetlands in Karnataka is presented in Fig. 5.

Karnataka state is surrounded by the Arabian Sea to the west, Goa to the northwest, Maharashtra to the north, Telangana to the northeast, Andhra Pradesh to the east, Tamil Nadu to the southeast, and Kerala to the southwest. The state is divided into 30 districts, with Bengaluru as the capital city. The total land area of Karnataka is 191,967 km^2 (i.e., 5.83 percent of the total land area of India). Rainfall ranges from 500 mm to over 4000 mm. Agumbe in the Sahyadris receives the second heaviest annual rainfall (7600 mm) in India. Summer temperatures range from 18 °C to 40 °C, whereas winter temperatures range from 14 °C to 32 °C. Ragi, jowar, rice, wheat, sugarcane, coconut, groundnut, and cotton are the major crops grown in Karnataka.

Estuaries in Karnataka

An Estuary is a semi-enclosed coastal body of water that extends to the effective limit of tidal influence, within which seawater entering from one or more free connections with the open sea or any other saline coastal body of water is significantly diluted with freshwater derived from land drainage, and can sustain euryhaline biological species for either part or the whole of their life cycle (Gillanders and Kingsford 2002). The

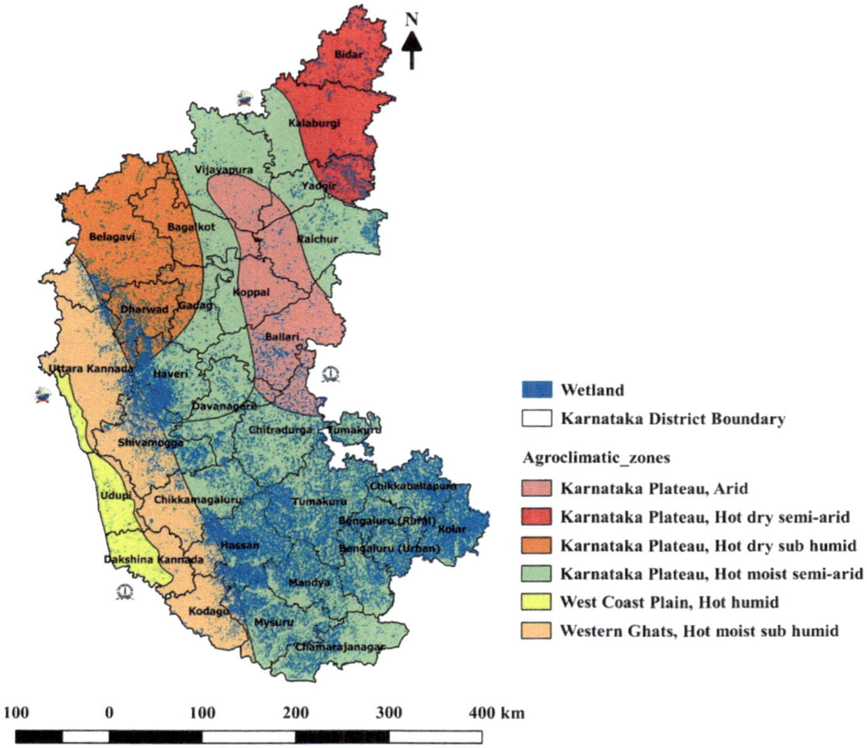

Fig. 4 Wetlands of Karnataka, India

estuary acts as a transitional zone between freshwater and saltwater habitats as well as land and the sea (Thrush et al. 2013).

Estuaries consist of distinct aquatic habitats, comprising seagrass and kelp beds, shellfish beds and coral reefs, hard-bottom communities of sponges and outcrops, soft-bottom communities with mud and sand, rocky inter-tidal zones, fringing mangrove forests, and vegetated marshes/wetlands. Vegetated marshes/wetlands are fundamental estuarine habitats that stabilize shorelines, protect uplands against erosion, and improve water quality by filtering pollutants and trapping fine sediments. The same habitats supply food and shelter for fish, invertebrates, birds, mammals, amphibians, and others. The high productivity of wetlands also plays a vital role in the cycling of food and nutrients within estuaries and in nearby coastal areas (McGlynn 2009). They are the chief habitats for various marine species and are called as the "nurseries of the sea". Estuaries are protected environments supported by mangroves, and organic-rich mudflats contribute to an ideal habitat status for most of the commercially important fish and shellfish species (Bhavan et al. 2021).

Estuaries aid as huge sponges, soaking up excess water during floods and heavy storms. Estuaries stabilize shorelines and protect coastal areas, inland habitats, and

District	Map	Wetland Size (Number)	Wetland size (%)
Bagalkot (Bagalkote) Geographical Area (ha) = 656257.27	 Wetland Area (ha) = 1391.90 (0.21%)	Total Number = 198 < 2.5 ha = 117 2.5–50 ha = 76 > 50 ha = 5	< 2.5 ha = 59.1 2.5–50 ha = 38.4 > 50 ha = 2.5
Ballari (Bellary) Geographical Area (ha) = 845578.94	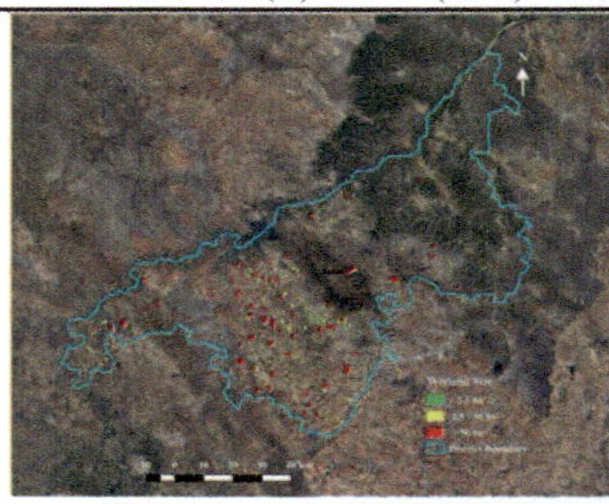 Wetland Area (ha) = 8751.91 (1.04%)	Total Number = 515 < 2.5 ha = 280 2.5–50 ha = 182 > 50 ha = 53	< 2.5 ha = 54.4 2.5–50 ha = 35.3 > 50 ha = 10.3
Bengaluru (Rural) **(Bangalore Rural)** Geographical Area (ha) = 230057.52	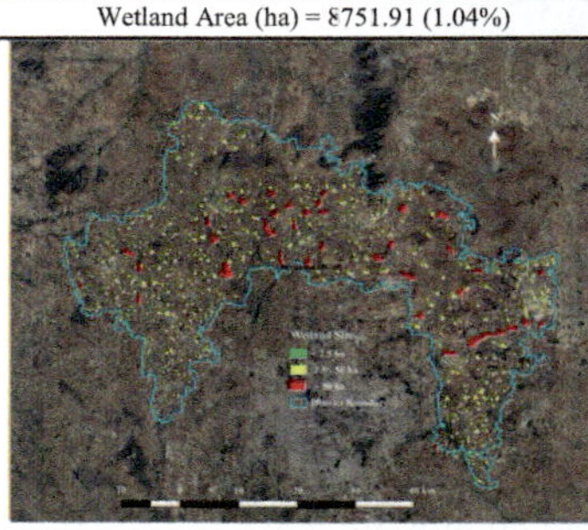 Wetland Area (ha) = 11997.61 (5.22%)	Total Number = 803 < 2.5 ha = 234 2.5–50 ha = 529 > 50 ha = 40	< 2.5 ha = 29.1 2.5–50 ha = 65.9 > 50 ha = 5.0
Bengaluru (Urban) **(Bangalore Urban)** Geographical Area (ha) = 219582.07	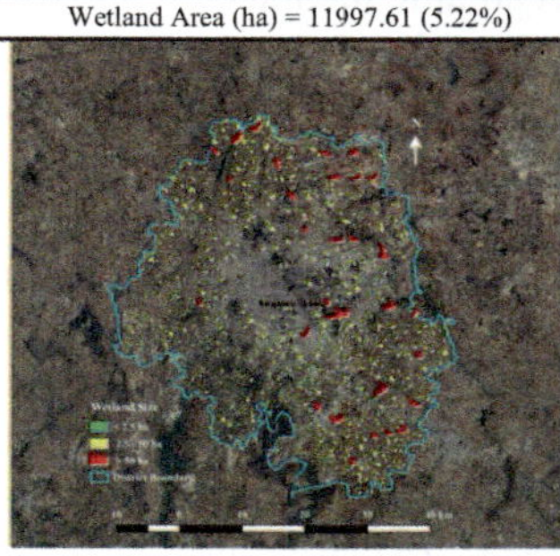 Wetland Area (ha) = 9255.42 (4.22%)	Total Number = 985 < 2.5 ha = 394 2.5–50 ha = 561 > 50 ha = 30	< 2.5 ha = 40.0 2.5–50 ha = 57.0 > 50 ha = 3.0

Fig. 5 Karnataka wetlands: Districtwise distribution, size and numbers

Belagavi (Belgaum) Geographical Area (ha) = 1338197.42	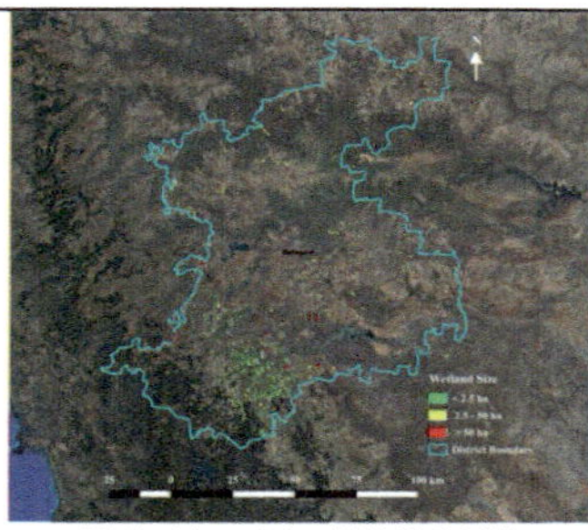 Wetland Area (ha) = 4776.00 (0.36%)	Total Number = 1091 < 2.5 ha = 726 2.5–50 ha = 355 > 50 ha = 10	< 2.5 ha = 66.5 2.5–50 ha = 32.5 > 50 ha = 1
Bidar Geographical Area (ha) = 544874.68	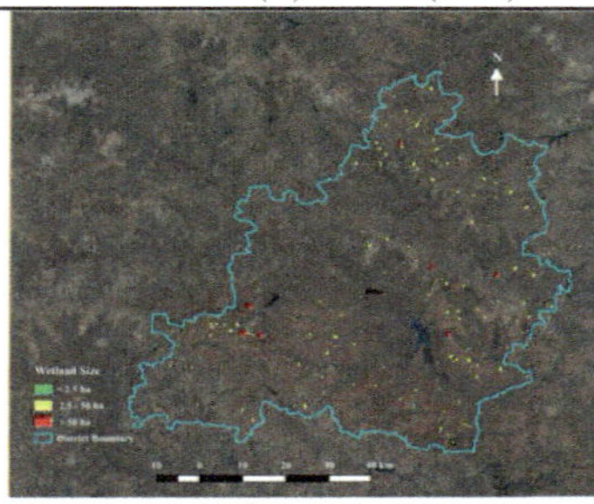 Wetland Area (ha) = 2316.19 (0.43%)	Total Number = 157 < 2.5 ha = 38 2.5–50 ha = 112 > 50 ha = 7	< 2.5 ha = 24.2 2.5–50 ha = 71.3 > 50 ha = 4.5
Chamarajanagar Geographical Area (ha) = 563934.54	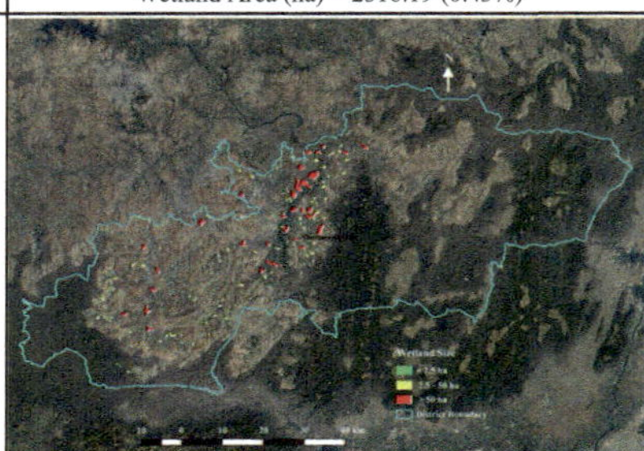 Wetland Area (ha) = 6475.18 (1.15%)	Total Number = 1197 < 2.5 ha = 1020 2.5–50 ha = 143 > 50 ha = 34	< 2.5 ha = 85.2 2.5–50 ha = 12 > 50 ha = 2.8
Chikkaballapura Geographical Area (ha) = 425128.80	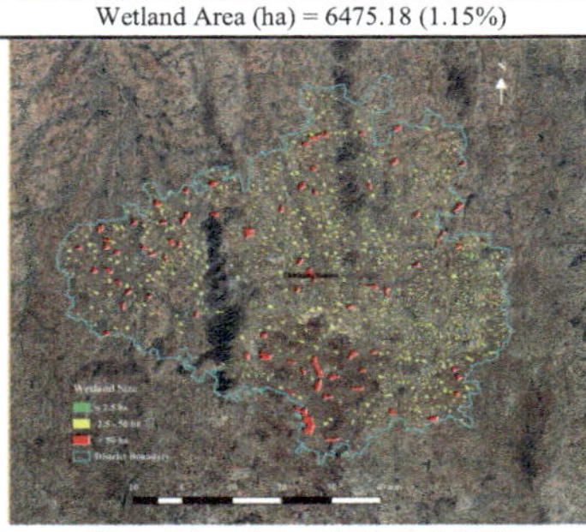 Wetland Area (ha) = 20592.46 (4.84%)	Total Number = 2149 < 2.5 ha = 904 2.5–50 ha = 1175 > 50 ha = 70	< 2.5 ha = 42 2.5–50 ha = 54.7 > 50 ha = 3.3

Fig. 5 (continued)

Chikkamagaluru (Chikmagalur) Geographical Area (ha) = 720981.82	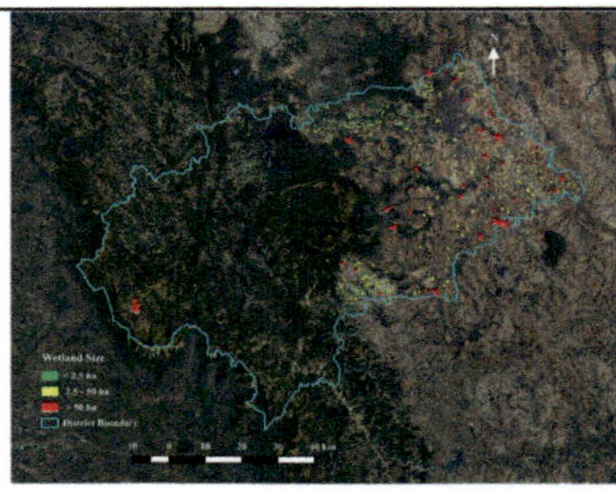 Wetland Area (ha) = 8880.39 (1.23%)	Total Number = 1049 < 2.5 ha = 647 2.5–50 ha = 370 > 50 ha = 32	< 2.5 ha = 61.7 2.5–50 ha = 35.3 > 50 ha = 3
Chitradurga Geographical Area (ha) = 843487.39	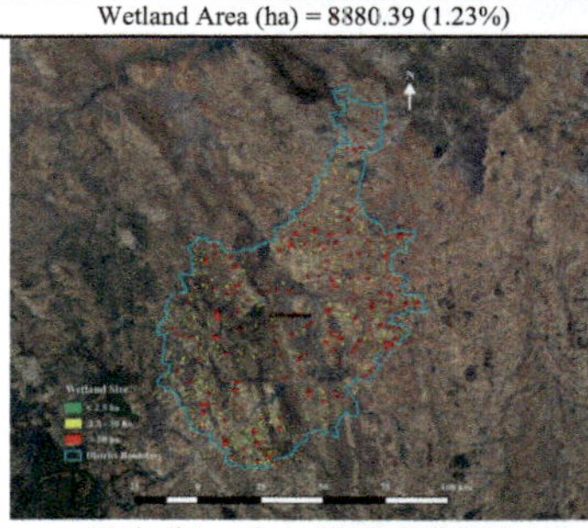 Wetland Area (ha) = 23960.49 (2.84%)	Total Number = 1262 < 2.5 ha = 553 2.5–50 ha = 569 > 50 ha = 140	< 2.5 ha = 43.8 2.5–50 ha = 45.1 > 50 ha = 11.1
Dakshina Kannada Geographical Area (ha) = 484587.84	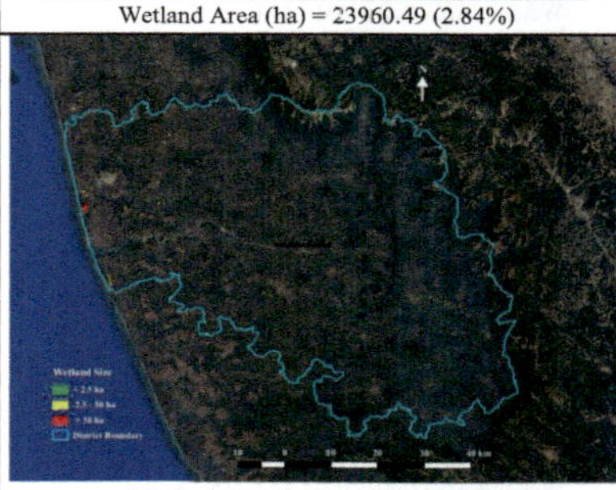 Wetland Area (ha) = 253.72 (0.05%)	Total Number = 59 < 2.5 ha = 49 2.5–50 ha = 9 > 50 ha = 1	< 2.5 ha = 83 2.5–50 ha = 15.3 > 50 ha = 1.7
Davanagere Geographical Area (ha) = 591584.24	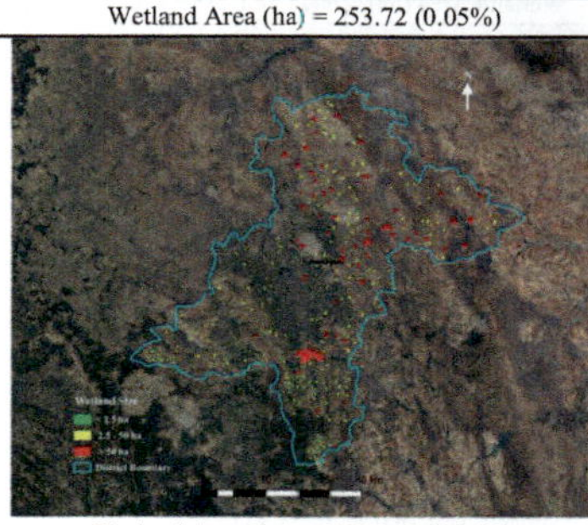 Wetland Area (ha) = 12784.48 (2.16%)	Total Number = 648 < 2.5 ha = 268 2.5–50 ha = 326 > 50 ha = 54	< 2.5 ha = 41.4 2.5–50 ha = 50.3 > 50 ha = 8.3

Fig. 5 (continued)

Dharwad Geographical Area (ha) = 425486.58	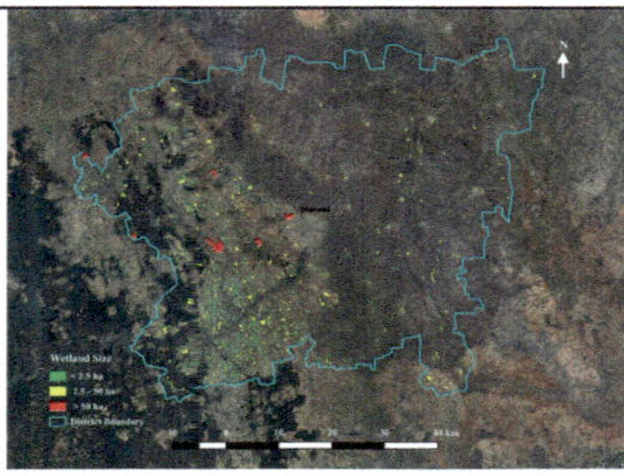 Wetland Area (ha) = 3842.48 (0.9%)	Total Number = 1471 < 2.5 ha = 1198 2.5–50 ha = 267 > 50 ha = 6	< 2.5 ha = 81.4 2.5–50 ha = 18.2 > 50 ha = 0.4
Gadag Geographical Area (ha) = 465532.12	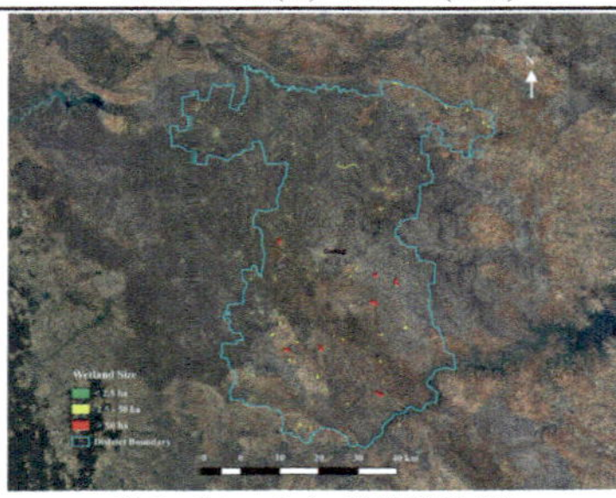 Wetland Area (ha) = 1842.89 (0.4%)	Total Number = 345 < 2.5 ha = 257 2.5–50 ha = 80 > 50 ha = 8	< 2.5 ha = 74.5 2.5–50 ha = 23.2 > 50 ha = 2.3
Hassan Geographical Area (ha) = 681815.33	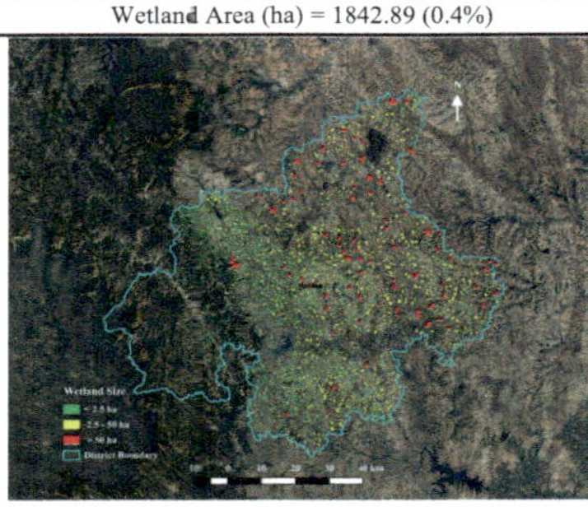 Wetland Area (ha) = 22173.96 (3.25%)	Total Number = 5333 < 2.5 ha = 4082 2.5–50 ha = 1178 > 50 ha = 73	< 2.5 ha = 76.5 2.5–50 ha = 22.1 > 50 ha = 1.4
Haveri Geographical Area (ha) = 481781.94	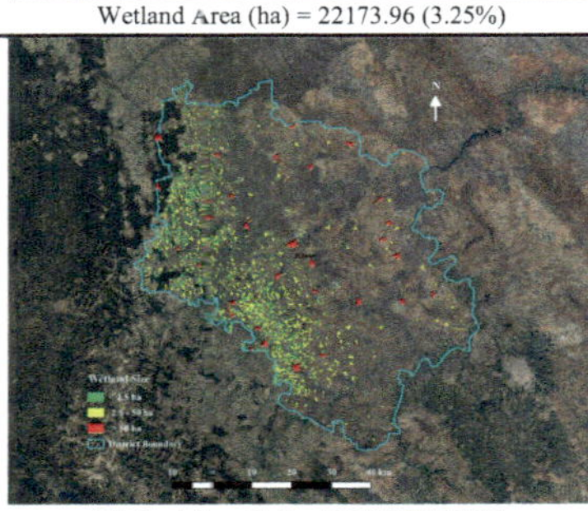 Wetland Area (ha) = 13645.57 (2.83%)	Total Number = 2850 < 2.5 ha = 1805 2.5–50 ha = 1014 > 50 ha = 31	< 2.5 ha = 63.3 2.5–50 ha = 35.6 > 50 ha = 1.1

Fig. 5 (continued)

Kalburgi (Gulbarga) Geographical Area (ha) = 1096696.39	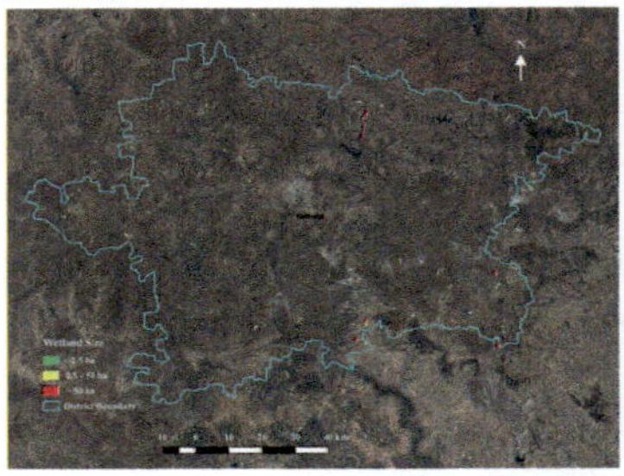 Wetland Area (ha) = 1482.88 (0.14%)	Total Number = 109 < 2.5 ha = 31 2.5–50 ha = 73 > 50 ha = 5	< 2.5 ha = 28.4 2.5–50 ha = 67.0 > 50 ha = 4.6
Kodagu Geographical Area (ha) = 410268.75	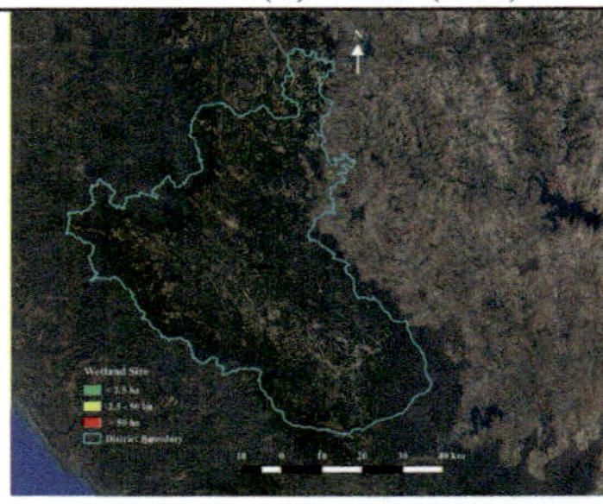 Wetland Area (ha) = 445.14 (0.11%)	Total Number = 447 < 2.5 ha = 421 2.5–50 ha = 26 > 50 ha = 0	< 2.5 ha = 94.2 2.5–50 ha = 5.8 > 50 ha = 0.0
Kolar Geographical Area (ha) = 399027.96	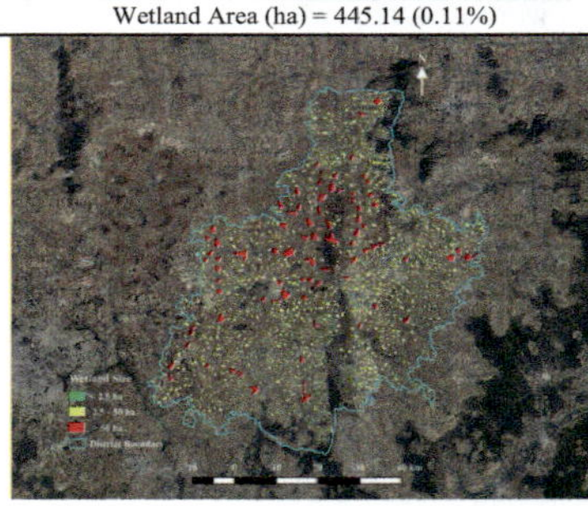 Wetland Area (ha) = 25283.74 (6.34%)	Total Number = 2870 < 2.5 ha = 1197 2.5–50 ha = 1594 > 50 ha = 79	< 2.5 ha = 41.7 2.5–50 ha = 55.5 > 50 ha = 2.8
Koppal Geographical Area (ha) = 557647.83	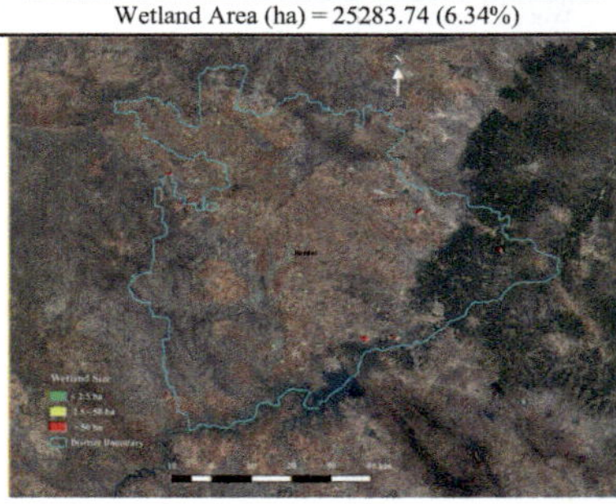 Wetland Area (ha) = 1093.30 (0.2%)	Total Number = 144 < 2.5 ha = 84 2.5–50 ha = 55 > 50 ha = 5	< 2.5 ha = 58.3 2.5–50 ha = 38.2 > 50 ha = 3.5

Fig. 5 (continued)

Mandya Geographical Area (ha) = 494723.01	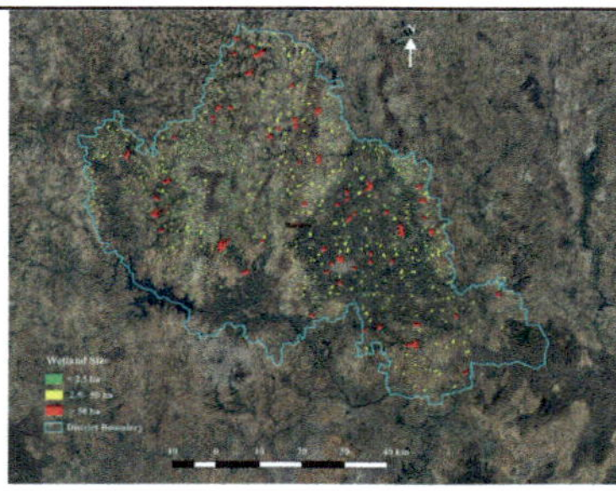 Wetland Area (ha) = 14013.11 (2.83%)	Total Number = 2089 < 2.5 ha = 1429 2.5–50 ha = 606 > 50 ha = 54	< 2.5 ha = 68.4 2.5–50 ha = 29.0 > 50 ha = 2.6
Mysuru (Mysore) Geographical Area (ha) = 631979.77	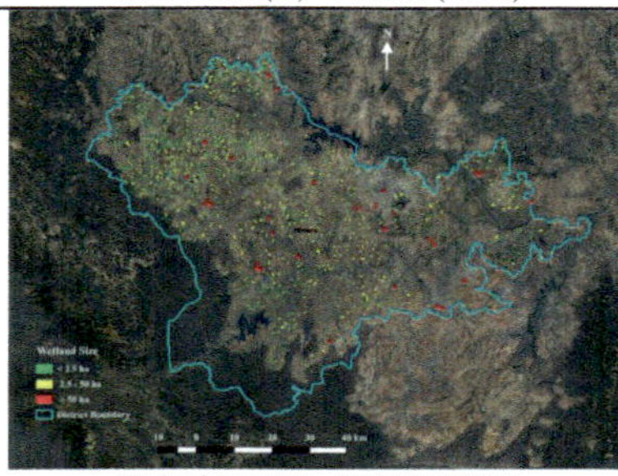 Wetland Area (ha) = 9698.04 (1.53%)	Total Number = 2682 < 2.5 ha = 1931 2.5–50 ha = 727 > 50 ha = 24	< 2.5 ha = 72.0 2.5–50 ha = 27.1 > 50 ha = 0.9
Raichur Geographical Area (ha) = 847033.56	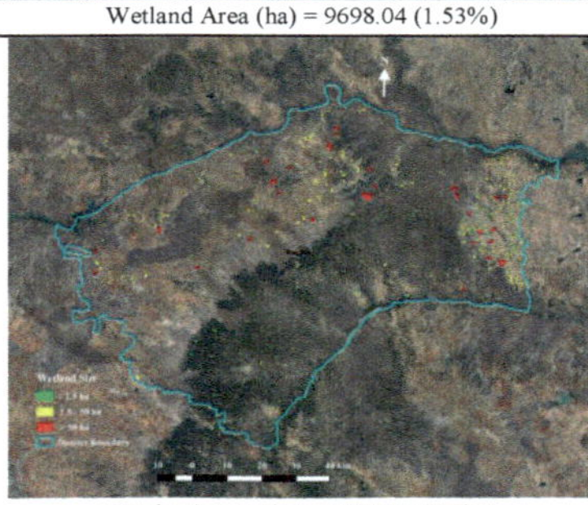 Wetland Area (ha) = 5928.19 (0.7%)	Total Number = 532 < 2.5 ha = 222 2.5–50 ha = 283 > 50 ha = 27	< 2.5 ha = 41.7 2.5–50 ha = 53.2 > 50 ha = 5.1
Ramanagara Geographical Area (ha) = 352688.67	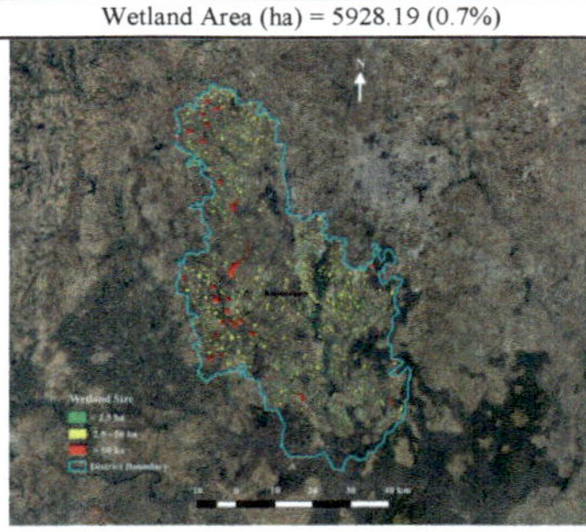 Wetland Area (ha) = 8283.97 (2.35%)	Total Number = 1334 < 2.5 ha = 819 2.5–50 ha = 489 > 50 ha = 26	< 2.5 ha = 61.4 2.5–50 ha = 36.7 > 50 ha = 1.9

Fig. 5 (continued)

Shivamogga (Shimoga) Geographical Area (ha) = 847253.81	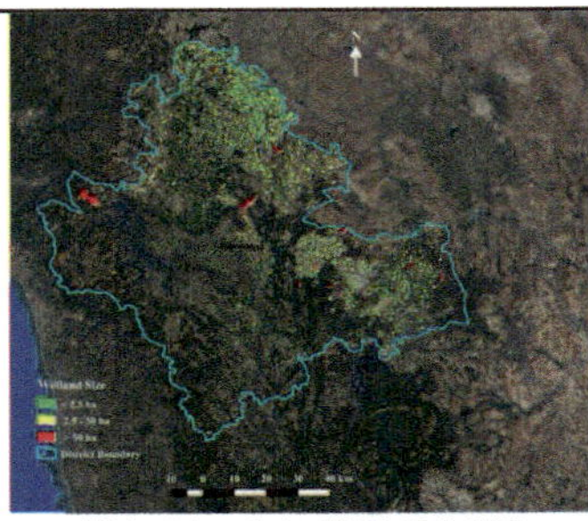 Wetland Area (ha) = 13330.17 (1.57%)	Total Number = 3857 < 2.5 ha = 2643 2.5–50 ha = 1203 > 50 ha = 11	< 2.5 ha = 68.5 2.5–50 ha = 31.2 > 50 ha = 0.3
Tumakuru (Tumkur) Geographical Area (ha) = 1060355.93	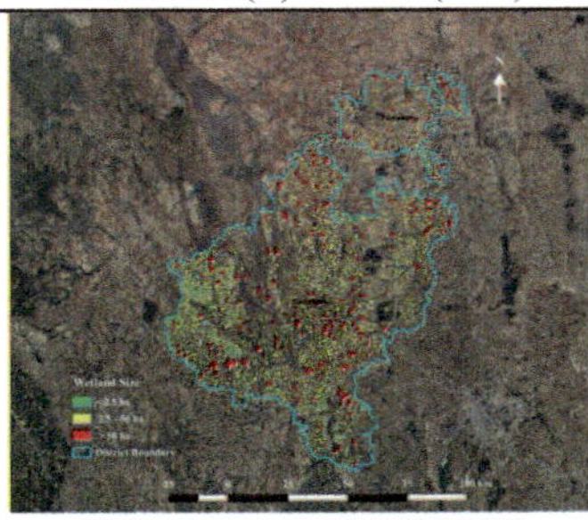 Wetland Area (ha) = 46586.68 (4.39%)	Total Number = 3891 < 2.5 ha = 2051 2.5–50 ha = 1634 > 50 ha = 206	< 2.5 ha = 52.7 2.5–50 ha = 42.0 > 50 ha = 5.3
Udupi Geographical Area (ha) = 357045.21	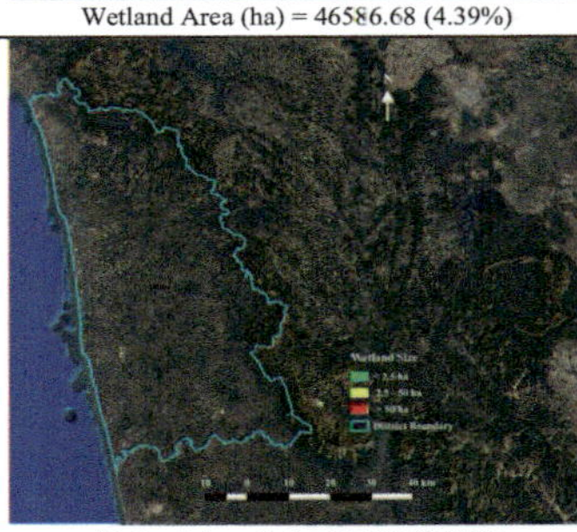 Wetland Area (ha) = 231.95 (0.06%)	Total Number = 79 < 2.5 ha = 55 2.5–50 ha = 24 > 50 ha = 0	< 2.5 ha = 69.6 2.5–50 ha = 30.4 > 50 ha = 0.0
Uttara Kannada Geographical Area (ha) = 1029816.25	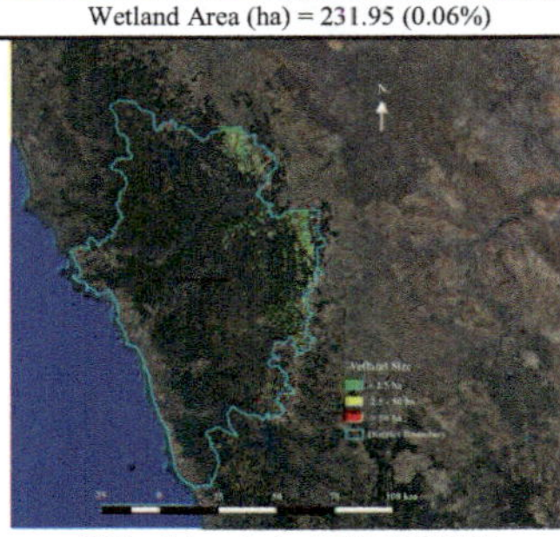 Wetland Area (ha) = 3409.47 (0.33%)	Total Number = 1290 < 2.5 ha = 1032 2.5–50 ha = 252 > 50 ha = 6	< 2.5 ha = 80.0 2.5–50 ha = 19.5 > 50 ha = 0.5

Fig. 5 (continued)

Vijayapura (Bijapur) Geographical Area (ha) = 1050134.43	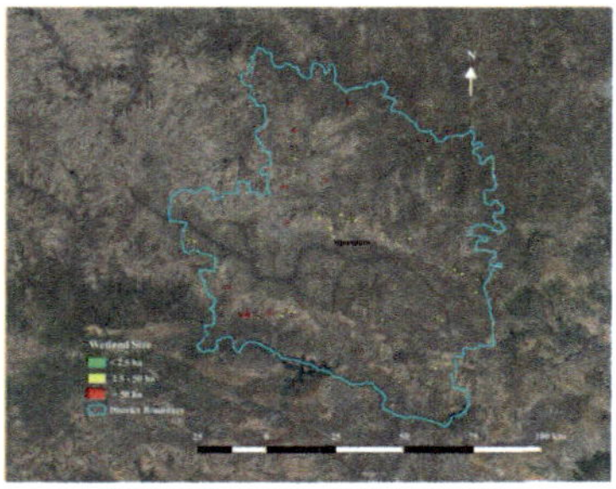 Wetland Area (ha) = 1896.34 (0.18%)	Total Number = 147 < 2.5 ha = 47 2.5–50 ha = 91 > 50 ha = 9	< 2.5 ha = 32.0 2.5–50 ha = 61.9 > 50 ha = 6.1
Yadgir (Yadagiri) Geographical Area (ha) = 528300.05	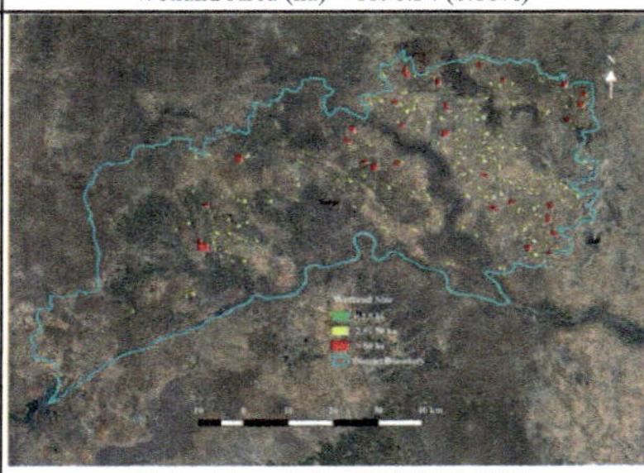 Wetland Area (ha) = 6179.23 (1.17%)	Total Number = 339 < 2.5 ha = 63 2.5–50 ha = 248 > 50 ha = 28	< 2.5 ha = 18.6 2.5–50 ha = 73.2 > 50 ha = 8.3

Fig. 5 (continued)

human communities. Estuaries contribute diverse living and non-living resources, offering employment opportunities, income, amenities, and pleasure to the local community (Ramachandra et al. 2013). Fishing is one of the chief economic activities of the rural coastal communities. Estuaries and inlets are places for harvesting food and fiber (Ramachandra et al. 2013).

Estuaries are the storehouse of mangrove biodiversity, which acts as a protective shield for the coastline apart from providing tremendous other benefits. Mangroves are salt-tolerant forest ecosystems observed usually in tropical and subtropical intertidal regions where they get organic materials from estuarine or oceanic ecosystems (Ramachandra et al. 2013). Mangroves support commercial fisheries through the provision of nurseries, breeding, spawning, and hatching habitats for offshore fisheries and exporting organic matter to the marine environment, producing nutrients for fauna in both the mangroves themselves and in adjacent marine and estuarine ecosystems. Mangroves play an ideal role in shoreline protection, acting as natural barriers, handling the destructive energy of waves, and reducing the impact of cyclones and storm surges (Salem and Mercer 2012). Mangrove supplies forestry products (firewood, charcoal, timber, honey, etc.) and sustain fishery products (fish, prawns, crabs, etc.). Due to their high calorific values, mangrove twigs are used for making charcoal and firewood. Mangrove extracts are used in indigenous medicines. Mangrove swamps act as a trap for the sediments and a sink for the nutrients. The root system

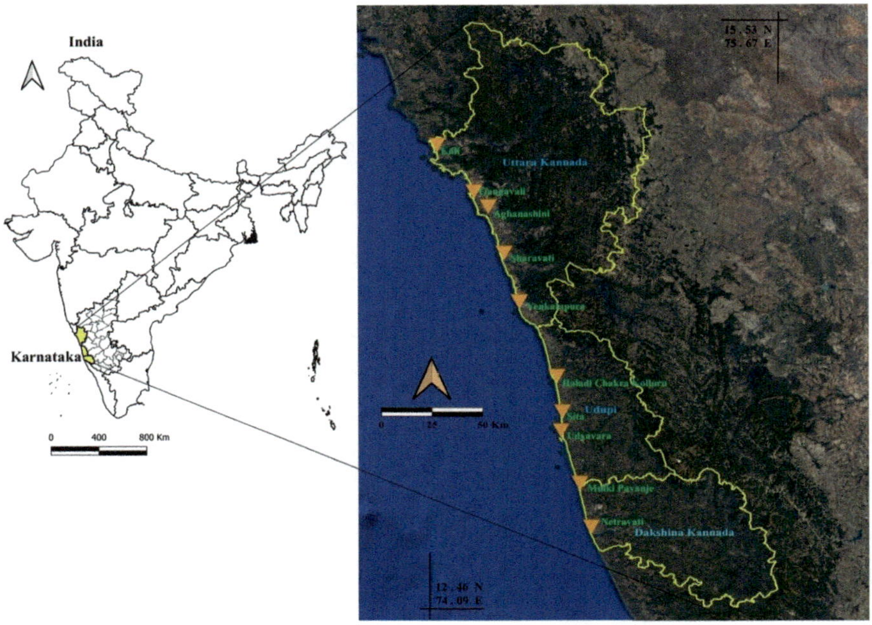

Fig. 6 Estuaries in Karnataka

of plants retains the substrate firm and thus contributes to the lasting stability of the coast (Ramachandra et al. 2013).

This study accounts for ecological services of estuaries on the west coast of Karnataka in Dakshina Kannada, Udupi, and Uttara Kannada districts, which include Mulky-Pavanje, Netravathi, Udyavara, Sita, Kollur-Chakra-Haladi, Kali, Aghanashini, Gangavali, Sharavathi, and Venkatapura estuaries (Fig. 6, Table 3).

Dakshina Kannada District

The district of South Kanara (Dakshina Kannada) is situated on the Western Coast of India. The South Kanara region is separated from the rest of the South Indian peninsula by the towering heights of the Western Ghats. It spreads from the Western Ghats towards the sea to the west. The length of the coastline, which is almost straight but broken at numerous points by rivers, rivulets, creeks, and bays, is 76 nautical miles (140.85 km). The district lies between 12° 27′ and 13° 58′ N, and 74° 35′ and 75° 40′ E. It is about 177 km in length, about 40 km broad at its narrowest, and about 80 km at its widest part. In several respects, it shares the geographical and economic characteristics of its contiguous districts of Coorg, Shivamogga, Chikkamagaluru, etc. The district, which is a long and narrow strip of territory, is generally bounded on the north by the North Kanara district, on the east by the districts of Shivamogga, Chikkamagaluru, Hassan, and Coorg, on the south by the Kasaragod taluk, which is

Table 3 Location, latitude, longitude and area of esturies

Study area	Location	Latitude	Longitude	Area (ha)	Source
Mulki—Pavanje estuary	Dakshina Kannada	12° 70′ N	74° 39′ E	750	(Dakshina Kannada District Gazetteer, 2011)
Netravathi estuary	Dakshina Kannada	12° 50′ N	74° 50′ E	1920	
Udyavara estuary	Udupi	13° 00′ 00″ to 13° 45′ 00″ N	74° 47′ 30″ to 74° 30′ 00″ E	800	(Udupi District Gazetteer, 2012)
Sita estuary	Udupi	13° 33′ 972″N	74° 71′ 056″E	980	
Chakra-Haladi-Kollur estuary	Udupi	13° 64′ 306″N	74° 65′ 861″ E	2550	
Kali estuary	Uttara Kannada	14.816°–14.917° N	74.125°–74.267° E	3240	(Uttara Kannada District Gazetteer, 1985)
Gangavali estuary	Uttara Kannada	14.586°–14.618° N	74.278°–74.385° E	700	
Aghanashini estuary	Uttara Kannada	14.391°–14.585° N	74.304°–74.516° E	4801	
Sharavathi estuary	Uttara Kannada	14.241°–14.385° N	74.418°–74.507° E	1600	
Venkatapura estuary	Uttara Kannada	14.008°–14.085° N	74.502°–74.569° E	200	

at present in Kerala State, and on the west by the Arabian Sea (Dakshina Kannada District Gazetteer, 2011).

Udupi District

Udupi district has a geographical area of 3,880 km^2. Geographically, the Udupi district is separated from the rest of the South Indian Peninsula by the Western Ghats. It spreads from the Western Ghats towards the sea to the west. The newly formed Udupi district has a coastal line that extends for about 100 km and is broken at numerous points by rivers, rivulets, creeks, and bays. The district lies between 13° 05′ and 13° 55′ North latitude and 74° 40′ and 75° 07′ East longitude. The district is bounded by Uttara Kannada district on the north, Shivamogga and Chikkamagaluru districts on the east, Dakshina Kannada on the south, and the Arabian Sea on the west. In its east, the Udupi district is bounded by higher ranges of the Western Ghats. The width of the districts gets narrower as one proceeds from south to north. Swarna, Seetha, Haladi, Chakra, and Kollur are the important rivers of the district. The coast line is generally straight, and the backwater areas are ideal for fisheries and boating (Udupi District Gazetteer, 2012).

Uttara Kannada District

The district is located in the mid-western part of the state. It lies between 13° 55′ to 15° 31′ North latitude, 74° 9′ to 75° 10′ East longitude, and extends over an area of 10,327 km^2, which is 5.37% of the total area of the state. The coast stretches in a long, nearly straight line to the southeast, except for the shallow Karwar and Belekeri bays. Though the coast is unbroken by deep or wide-mouthed estuaries, it is varied and picturesque with rocky islands and capes, stretches of palm-fringed sand beaches, low narrow mouths, and rough bluffs and headlands. The deep, winding valleys, waving wooded hills, and wild background of high peaks stretch behind the coastline boundaries. The district, which is a long and narrow strip of territory, is surrounded by Belgaum district and Goa territory in the north, Dharwad district in the east, Dakshina Kannada, and Shivamogga districts in the south, and the Arabian Sea in the west (Uttara Kannada District Gazetteer, 1985).

5 Water Sample Collection and Water Quality Analysis

Water samples were collected from the lakes of Bengaluru, Karnataka, in clean and sterile one-litre bottles for laboratory analysis of physicochemical parameters. Physicochemical parameters like water temperature (WT); pH; total dissolved solids (TDS); electrical conductivity (EC); and dissolved oxygen (DO) were measured onsite. The analysis of other physicochemical parameters like chemical oxygen demand (COD); biochemical oxygen demand (BOD); total alkalinity (TA); chloride (Cl); total hardness (TH); calcium (Ca); magnesium (Mg); nitrate (N); and orthophosphate (OP) were done in the laboratory according to the standard procedures of APHA AWWA WEF (2005).

Microalgae sample collection and quantification

A plankton net (of 50 μm mesh) was used for microalgae sampling. The samples were transferred to clean and sterile bottles for further identification and quantification. Samples were preserved in 5% neutralized formaldehyde solution. Then, cells were counted on a microscope and identified using standard keys (Desikachary 1959; Prescott 1970; AlgaeBase 2021).

Fish sampling and data collection

Fish sampling was carried out in different lakes in Bengaluru, Karnataka. Fishermen used gill nets, cast nets, and dragnets of varying dimensions to catch fish samples from lakes. The fish samples were identified using standard keys (Jayaram 1999; Talwar and Jhingran 1991; Fishbase 2021).

Macrophyte sampling and data collection

In lakes, quadrats were randomly laid at different locations, and macrophytes were collected (all species in the quadrat were considered). Samples were washed properly to remove adhered sediments and epiphytes, and stored in polythene bags. Then, the samples were identified based on the standard taxonomic literature (Cook 1996).

6 Valuation of Ecosystem Service from Wetlands

The total ecosystem supply value of wetland ecosystems in Karnataka was assessed considering (a) provisioning services, (b) regulating services, and (c) cultural services. The following services were considered for the present study, and this involved data compilation (Tables 4 and 5) from primary (field investigation) and secondary sources (government agencies, published articles in peer-reviewed journals). The study used residual value and resource rent, and the benefit transfer method to estimate the TESV of the total wetlands in Karnataka (Table 6).

Provisioning services include the products obtained from wetland ecosystems such as water, algae, macrophytes, fish, raw materials, genetic resources, medicinal resources, ornamental resources, fuelwood, and irrigation. Regulating services include the benefits obtained from the regulation of ecosystem processes such as air quality regulation, climate regulation, moderation of extreme events, regulation of water flows, waste treatment/water purification, erosion prevention, maintenance of soil fertility, biological control, maintenance of life cycles of migratory species, maintenance of genetic diversity, carbon sequestration, waterborne disease control, pollination, water conservation, flood control, habitat or refugia, and groundwater recharge. Cultural services include the benefits from wetlands through aesthetic information, opportunities for recreation and tourism, inspiration for culture, art and design, spiritual experience, information for cognitive development, existence and bequest values, and education.

7 Data and Source: Freshwater Ecosystem in Karnataka

Data used for computing ecosystem services in freshwater ecosystems in Karnataka are listed in Table 4.

8 Data and Source: Estuaries in Karnataka

Data used for computing ecosystem services in estuarine ecosystems in Karnataka are listed in Table 5.

Table 4 Data used for computing ecosystem services with the details of data sources

Services	Services and monetary benefits	Source
Provisioning services		
Ecosystems type: Wetland ecosystems		
Provisioning services	Microalgae: Data: Microalgal species; microalgal biomass from different wetlands Microalgae samples were collected from the surface water of wetlands by plankton net, identified, and counted under compound microscope	Microalgae species listed through field work to wetlands (of Bengaluru, Uttara Kannada, Mandya; and Kolar district); average microalgal biomass from Bengaluru lakes were considered Ramachandra et al. (2018), Asulabha et al. (2018); Ramachandra et al. (2016), Ramachandra et al. (2015)
	Harvested microalgae from the Bengaluru wetland were dried, and productivity was determined. Microalgal productivity from wetlands was estimated to be 14 $g/m^2/d$ (51.1 t/ha/yr dry biomass). With a harvesting period of 5–7 days (i.e., 52 cycles per year), the total microalgae harvested was 7.28 t/ha/yr It was assumed that out of 7.28 t/ha/yr of dried algal biomass, 6 t of microalgal biomass was used for biodiesel production, 280 kg of microalgal biomass was used for fish feed, and 1 t of biomass for protein production	Ramachandra et al. (2018), Asulabha et al. (2018), Ramachandra et al. (2016), Saranya and Ramachandra (2020)
	Biodiesel • Biodiesel production cost from microalgae was 31.29 Rs/L resulting in a benefit of 25 Rs/L • Lipid content of dried microalgal biomass is around 20–50% • 6 t of microalgal biomass produces 2069.76 L biodiesel giving a benefit of 51,744 Rs	Chisti (2007), Nagarajan et al. (2013), Chisti (2013), Louw et al. (2016), Saranya and Ramachandra (2020)
	Glycerol • During biodiesel production, crude glycerol is produced (10%) as byproduct. Benefit from glycerol amounted to 16 Rs/kg. Benefit from 235.2 kg glycerol was 3763 Rs	Quispe et al. (2013)

(continued)

Table 4 (continued)

Services	Services and monetary benefits	Source
	Feed for Fish • 1 kg of feed for fish will give benefit of 15 Rs/kg. 280 kg of microalgal biomass was used as feed for fish, which gives a benefit of 4200 Rs	Ansari et al. (2020)
	Protein • About 540 kg protein could be obtained from one-ton algal biomass. Benefit from one kg of protein was 94 Rs. Thus, benefit from 540 kg of food (protein) was 50,760 Rs	Asiedu (2015), Mahapatra et al. (2016)
	Macrophyte: Data: Types of macrophytes, biomass. Spatial extent of macrophytes	Macrophyte species listed through field work to wetlands (of Bengaluru, Uttara Kannada, Mandya; and Kolar district) Ramachandra et al. (2018), Ramachandra et al. (2016), Ramachandra et al. (2014), Kiran et al. (2007), Sudarshan Bhat et al. (2017)
	Digitized the area of macrophyte cover in entire wetlands of Karnataka; The total macrophyte cover in Karnataka wetlands is 50432.35 ha	Google Earth and QGIS open software
	Macrophyte is used as food, honey, making handicraft materials, and fodder • Benefits from fodder is 5580 Rs/ha/yr (each person will get a benefit of 1488 Rs/yr for fodder collection) • Benefit from cattle grazing in wetland is estimated to be 1174 Rs/ha/yr • Benefit from food (leaf vegetables like Ipomea, Marsilea, and Amaranthus) is estimated to be 147 Rs/ha/yr • Benefit from handicraft made of aquatic plants is 2310 Rs/ha/yr	Mukherjee (2008), Ramachandra et al. (2005), Ramachandra et al. (2011), Mukherjee (2008), Ramachandra et al. (2005), Ramachandra et al. (2011)
	• The capacity of wetlands to supply honey is 20 to 40 kg/ha/yr. In Karnataka, average benefit from honey is 52 Rs per kg. From wetlands, honey production is estimated to be 2080 Rs/ha/yr	Affek (2018), Tejaswi (2008)

(continued)

Table 4 (continued)

Services	Services and monetary benefits	Source
	Fish: Data: Fish species (location wise), type, quantity	Fish species listed through field work to wetlands (of Bengaluru, and Mandya district) Ramachandra and Sreekantha (2006), Sincy et al. (2018), Fish Base (2021)
	Fish catch data; wetlands of Karnataka studied for determining fish catch, price, and benefit associated with fishing	District wise inland fish catch from districts, Department of Inland Fisheries, GoK https://fisheries.karnataka.gov.in/english Directorate of Economics and statistics, GoK https://des.karnataka.gov.in/english; District at a glance, GoK
	• Total fish catch from Bengaluru wetlands 465 kg/ha/yr. Total fish catch was estimated to be 525 kg/ha/yr from Karnataka fisheries data. So the average fish catch was estimated as 495 kg/ha/yr • The price of fish ranged between 120 and 300 Rs/kg resulting in a benefit of 65 Rs/kg	District at a glance, GoK; Ramachandra et al. (2005), Ramachandra et al. (2011)
	Water: Data: Sector wise water demand, Value of water for domestic consumption, irrigation, livestock use, and industries Compared with values of global studies, adjusted for GDP (PPP) per capita and corresponding currency exchange rate	Ecosystem Services Valuation Database (ESVD), 2020: https://www.es-partnership.org/wp content/uploads/2020/08/ESVD_Global-Update-FINAL-Report-June-2020.pdf GDP(PPP) per capita for India https://data.worldbank.org/indicator/NY.GDP.PCAP.PP.CD?locations=IN Currency exchange rate https://www.thehansindia.com/business/currency-update-today-indian-rupee-against-us-dollar-on-19-november-2021-716019
	Raw materials: Data: Types of raw materials, quantity, value of raw materials like wood, craftwork materials, and construction materials Compared with values of global studies, adjusted for GDP (PPP) per capita and corresponding currency exchange rate	Ecosystem Services Valuation Database (ESVD), 2020: https://www.es-partnership.org/wp-content/uploads/2020/08/ESVD_Global-Update-FINAL-Report-June-2020.pdf GDP(PPP) per capita for India https://data.worldbank.org/indicator/NY.GDP.PCAP.PP.CD?locations=IN Currency exchange rate https://www.thehansindia.com/business/currency-update-today-indian-rupee-against-us-dollar-on-19-november-2021-716019

(continued)

Table 4 (continued)

Services	Services and monetary benefits	Source
	Genetic resources: Data: Genetic resources—types, Value of genetic resources (products for materials science, genes for resistance to plant pathogens and crop pests, and ornamental species) Compared with values of global studies, adjusted for GDP (PPP) per capita and corresponding currency exchange rate	Ecosystem Services Valuation Database (ESVD), 2020: https://www.es-partnership.org/wp-content/uploads/2020/08/ESVD_Global-Update-FINAL-Report-June-2020.pdf GDP(PPP) per capita for India https://data.worldbank.org/indicator/NY.GDP.PCAP.PP.CD?locations=IN Currency exchange rate https://www.thehansindia.com/business/currency-update-today-indian-rupee-against-us-dollar-on-19-november-2021-716019
	Medicinal resources: Data: Fish types, medicinal values of fish species	For checklist of medicinal values of fish species chosen published literature from India (Gogoi and Bora 2020; Borah and Prasad 2017; Borah and Prasad 2016; Gupta and Dey 2017; Teronpi et al. 2012; Chanu et al. 2016; Prakash and Prakash 2021)
	Data: Checklist of aquatic plants, medicinal values	For checklist of medicinal values of aquatic plants chosen published literature from India (Kiran et al. 2007; Taran and Deb 2020; Nandan and Singh 2004; Sarmah et al. 2013; Misra et al. 2012; Sarma and Saikia 2010; Mishra et al. 2016; Chai et al. 2015; Banerjee et al. 2012)
	Data: Value of extraction of medicine from biota (e.g. aquatic plants, and fish) Compared with values of global studies, adjusted for GDP (PPP) per capita and corresponding currency exchange rate	Clarkson et al. (2013) GDP(PPP) per capita for India https://data.worldbank.org/indicator/NY.GDP.PCAP.PP.CD?locations=IN Currency exchange rate https://www.thehansindia.com/business/currency-update-today-indian-rupee-against-us-dollar-on-19-november-2021-716019
	Ornamental resources: Data: considered value of ornamental resources like resources for fashion, handicraft, jewellery, pets, worship, decoration, and souvenirs Compared with values of global studies, adjusted for GDP (PPP) per capita and corresponding currency exchange rate	Clarkson et al. (2013) GDP(PPP) per capita for India https://data.worldbank.org/indicator/NY.GDP.PCAP.PP.CD?locations=IN Currency exchange rate https://www.thehansindia.com/business/currency-update-today-indian-rupee-against-us-dollar-on-19-november-2021-716019

(continued)

Table 4 (continued)

Services	Services and monetary benefits	Source
	Fuelwood: Data: About 10% macrophyte cover area of each wetland is considered as annual fuelwood extraction area Value of fuelwood, and logs from wetland Compared with values of global studies, adjusted for GDP (PPP) per capita and corresponding currency exchange rate	Average value from published Indian literatures (Ramachandra et al. 2005, 2011); other literatures (Schuijt 2002; Zuze 2013; Schuyt and Brander 2008). *In appendix (Table A1) GDP(PPP) per capita for India https://data.worldbank.org/indicator/NY.GDP.PCAP.PP.CD?locations=IN Currency exchange rate https://www.thehansindia.com/business/currency-update-today-indian-rupee-against-us-dollar-on-19-november-2021-716019
	Irrigation: Data: Tank irrigated area for 30 districts in Karnataka as wetland water is used for the cultivation of crops such as paddy, maize and various types of vegetables and fruits	Tank irrigated area from District at a glance (GoK)
	Irrigation service value was estimated as 1826 Rs/ha/yr	Mukherjee (2008)
Regulating services		
Ecosystems type: Wetland ecosystems		
Regulating services	Air quality regulation: Amount of aerosols or chemicals "extracted" - effect on air quality Data: Leaf area index, and NOx-fixation Air quality regulation service value per hectare Compared with values of global studies, adjusted for GDP (PPP) per capita and corresponding currency exchange rate	Ecosystem Services Valuation Database (ESVD), 2020: https://www.es-partnership.org/wp-content/uploads/2020/08/ESVD_Global-Update-FINAL-Report-June-2020.pdf GDP(PPP) per capita for India https://data.worldbank.org/indicator/NY.GDP.PCAP.PP.CD?locations=IN Currency exchange rate https://www.thehansindia.com/business/currency-update-today-indian-rupee-against-us-dollar-on-19-november-2021-716019

(continued)

Table 4 (continued)

Services	Services and monetary benefits	Source
	Climate regulation: Quantity of greenhouse gases, etc., fixed and/or emitted - effect on climate parameters Data: Greenhouse gas-balance (esp. C-fix), DMS production, Land cover characteristics. Climate regulation service value per hectare Compared with values of global studies, adjusted for GDP (PPP) per capita and corresponding currency exchange rate	Ecosystem Services Valuation Database (ESVD), 2020: https://www.es-partnership.org/wp-content/uploads/2020/08/ESVD_Global-Update-FINAL-Report-June-2020.pdf GDP(PPP) per capita for India https://data.worldbank.org/indicator/NY.GDP.PCAP.PP.CD?locations=IN Currency exchange rate https://www.thehansindia.com/business/currency-update-today-indian-rupee-against-us-dollar-on-19-november-2021-716019
	Moderation of extreme events: Data: Ecosystems and living organisms create buffers against natural disasters. They reduce damage from floods, storms, avalanches, landslides, and droughts. The moderation of extreme events service value per hectare Compared with values of global studies, adjusted for GDP (PPP) per capita and corresponding currency exchange rate	Ecosystem Services Valuation Database (ESVD), 2020: https://www.es-partnership.org/wp-content/uploads/2020/08/ESVD_Global-Update-FINAL-Report-June-2020.pdf GDP(PPP) per capita for India https://data.worldbank.org/indicator/NY.GDP.PCAP.PP.CD?locations=IN Currency exchange rate https://www.thehansindia.com/business/currency-update-today-indian-rupee-against-us-dollar-on-19-november-2021-716019
	Regulation of water flows: Data: Quantity of water stored and influence of hydrological regime. The regulation of water flows service value per hectare Compared with values of global studies, adjusted for GDP (PPP) per capita and corresponding currency exchange rate	Ecosystem Services Valuation Database (ESVD), 2020: https://www.es-partnership.org/wp-content/uploads/2020/08/ESVD_Global-Update-FINAL-Report-June-2020.pdf GDP(PPP) per capita for India https://data.worldbank.org/indicator/NY.GDP.PCAP.PP.CD?locations=IN Currency exchange rate https://www.thehansindia.com/business/currency-update-today-indian-rupee-against-us-dollar-on-19-november-2021-716019

(continued)

Table 4 (continued)

Services	Services and monetary benefits	Source
	Waste treatment: Data: Denitrification (kg N/ha/y), Accumulation in plants, - kg –BOD/ha/y, chelation (metal binding). Waste treatment service value per hectare Compared with values of global studies, adjusted for GDP (PPP) per capita and corresponding currency exchange rate	Ecosystem Services Valuation Database (ESVD), 2020: https://www.es-partnership.org/wp-content/uploads/2020/08/ESVD_Global-Update-FINAL-Report-June-2020.pdf GDP(PPP) per capita for India https://data.worldbank.org/indicator/NY.GDP.PCAP.PP.CD?locations=IN Currency exchange rate https://www.thehansindia.com/business/currency-update-today-indian-rupee-against-us-dollar-on-19-november-2021-716019
	Erosion prevention: Data: Erosion prevention service value per hectare Compared with values of global studies, adjusted for GDP (PPP) per capita and corresponding currency exchange rate	Clarkson et al. (2013) GDP(PPP) per capita for India https://data.worldbank.org/indicator/NY.GDP.PCAP.PP.CD?locations=IN Currency exchange rate https://www.thehansindia.com/business/currency-update-today-indian-rupee-against-us-dollar-on-19-november-2021-716019
	Maintenance of soil fertility: Data: Value for Maintenance of soil fertility service value per hectare Compared with values of global studies, adjusted for GDP (PPP) per capita and corresponding currency exchange rate	Clarkson et al. (2013) GDP(PPP) per capita for India https://data.worldbank.org/indicator/NY.GDP.PCAP.PP.CD?locations=IN Currency exchange rate https://www.thehansindia.com/business/currency-update-today-indian-rupee-against-us-dollar-on-19-november-2021-716019
	Biological control: Data: Number and impact of pest control species; number and impact of pollinating species. The biological control service value per hectare Compared with values of global studies, adjusted for GDP (PPP) per capita and corresponding currency exchange rate	Clarkson et al. (2013) GDP(PPP) per capita for India https://data.worldbank.org/indicator/NY.GDP.PCAP.PP.CD?locations=IN Currency exchange rate https://www.thehansindia.com/business/currency-update-today-indian-rupee-against-us-dollar-on-19-november-2021-716019

(continued)

Table 4 (continued)

Services	Services and monetary benefits	Source
	Maintenance of life cycles of migratory species: Data: Wetlands serve as feeding and breeding grounds for migratory species. Maintenance of life cycles of migratory species service value per hectare Compared with values of global studies, adjusted for GDP (PPP) per capita and corresponding currency exchange rate	Ecosystem Services Valuation Database (ESVD), 2020: https://www.es-partnership.org/wp-content/uploads/2020/08/ESVD_Global-Update-FINAL-Report-June-2020.pdf GDP(PPP) per capita for India https://data.worldbank.org/indicator/NY.GDP.PCAP.PP.CD?locations=IN Currency exchange rate https://www.thehansindia.com/business/currency-update-today-indian-rupee-against-us-dollar-on-19-november-2021-716019
	Maintenance of genetic diversity: Data: Genetic diversity (the variety of genes between, and within, species populations) and a gene pool for developing commercial crops and livestock. Maintenance of genetic diversity service value per hectare Compared with values of global studies, adjusted for GDP (PPP) per capita and corresponding currency exchange rate	Ecosystem Services Valuation Database (ESVD), 2020: https://www.es-partnership.org/wp-content/uploads/2020/08/ESVD_Global-Update-FINAL-Report-June-2020.pdf GDP(PPP) per capita for India https://data.worldbank.org/indicator/NY.GDP.PCAP.PP.CD?locations=IN Currency exchange rate https://www.thehansindia.com/business/currency-update-today-indian-rupee-against-us-dollar-on-19-november-2021-716019
	Carbon sequestration: Data: Wetlands are carbon sinks that contribute towards reducing carbon emissions Carbon sequestration services value is calculated by considering the social cost of carbon per tonne. The social cost of a tonne of CO_2 is taken as US$ 86 Field studies from Bengaluru wetlands. Determined carbon sequestration from algae, and macrophyte; quantification of algal biomass and macrophyte biomass	Verma et al. (2014)

(continued)

Table 4 (continued)

Services	Services and monetary benefits	Source
	Microalgae • 1 tC = 3.67 tCO_2 • 1 kg microalgae = 1.83 kg C • Total microalgal biomass: 51.1 t/ha/yr • Total value of carbon sequestration from microalgae: 234,404.05 Rs tCO_2/ha/yr	Baral et al. (2016), Chisti (2008)
	Macrophyte • Total macrophyte biomass: 119,310 t/ha/y • Total C from macrophyte: 124.44 kg/ha/y • Total value of carbon sequestration from macrophyte: 311.92 Rs $kgCO_2$/ha/yr	Kalita (2019)
	Water borne diseases control: Data: Proper management of wetlands with the use of fish that consume mosquito larvae, or of bacterial larvicides that kill them without affecting other organisms Water borne diseases control service values per ha (average of 71, 5726 and 25 Rs/ha/yr.) from Indian literatures	Verma et al. (2001), Kant et al. (2013), Ramachandra et al. (2011)
	Pollination: Data: Value for pollination service per ha (19 Rs/ha/yr.) from literature Value adjusted for GDP (PPP) per capita and corresponding currency exchange rate	Schuyt and Brander (2008) GDP(PPP) per capita for India https://data.worldbank.org/indicator/NY.GDP.PCAP.PP.CD?locations=IN Currency exchange rate https://www.thehansindia.com/business/currency-update-today-indian-rupee-against-us-dollar-on-19-november-2021-716019

(continued)

Table 4 (continued)

Services	Services and monetary benefits	Source
	Water conservation: Data: Water conservation service value per ha (2875 Rs/ha/yr.) from literature Value adjusted for GDP (PPP) per capita and corresponding currency exchange rate	Li and Gao (2016) GDP(PPP) per capita for India https://data.worldbank.org/indicator/NY.GDP.PCAP.PP.CD?locations=IN Currency exchange rate https://www.thehansindia.com/business/currency-update-today-indian-rupee-against-us-dollar-on-19-november-2021-716019
	Flood control: Data: Flood control service values per ha (ranged between 191 to 13,032 Rs/ha/yr.) from literature Compared with values of global studies, adjusted for GDP (PPP) per capita and corresponding currency exchange rate	Kaul et al. (2016), Zuze (2013), Schuyt and Brander (2008). *In appendix (Table A2) GDP(PPP) per capita for India https://data.worldbank.org/indicator/NY.GDP.PCAP.PP.CD?locations=IN Currency exchange rate https://www.thehansindia.com/business/currency-update-today-indian-rupee-against-us-dollar-on-19-november-2021-716019
	Habitat/refugia: Data: Critical breeding, feeding and nesting habitat for populations in wetlands Habitat service values per ha (164 and 3486 Rs/ha/yr.) from literature Compared with values of global studies, adjusted for GDP (PPP) per capita and corresponding currency exchange rate	Schuyt and Brander (2008), Kaggwa et al. (2009) GDP(PPP) per capita for India https://data.worldbank.org/indicator/NY.GDP.PCAP.PP.CD?locations=IN Currency exchange rate https://www.thehansindia.com/business/currency-update-today-indian-rupee-against-us-dollar-on-19-november-2021-716019
	Groundwater recharge: Data: Water filters down from the wetland into an underground aquifer which stores potable water Groundwater service values per ha (192,000 and 238,246 Rs/ha/yr.) from literature. Compared with values of global studies, adjusted for GDP (PPP) per capita and corresponding currency exchange rate	Ramachandra et al. (2019), Kaggwa et al. (2009) GDP(PPP) per capita for India https://data.worldbank.org/indicator/NY.GDP.PCAP.PP.CD?locations=IN Currency exchange rate https://www.thehansindia.com/business/currency-update-today-indian-rupee-against-us-dollar-on-19-november-2021-716019

(continued)

Table 4 (continued)

Services	Services and monetary benefits	Source
Cultural services		
Ecosystems type: Wetland ecosystems		
Cultural services	Aesthetic information: Data: Expressed aesthetic value, e.g., number of houses bordering wetland areas; number of users of "scenic routes" Aesthetic information service value per ha Compared with values of global studies, adjusted for GDP (PPP) per capita and corresponding currency exchange rate	Ecosystem Services Valuation Database (ESVD), 2020: https://www.es-partnership.org/wp-content/uploads/2020/08/ESVD_Global-Update-FINAL-Report-June-2020.pdf GDP(PPP) per capita for India https://data.worldbank.org/indicator/NY.GDP.PCAP.PP.CD?locations=IN Currency exchange rate https://www.thehansindia.com/business/currency-update-today-indian-rupee-against-us-dollar-on-19-november-2021-716019
	Opportunities for recreation and tourism: Data: Value considered for ecotourism, bird-watching, swimming, and outdoor sports Recreation and tourism service values per ha (ranged between 2 and 101,747 Rs/ha/yr.) from literatures Compared with values of global studies, adjusted for GDP (PPP) per capita and corresponding currency exchange rate	Indian literatures (Anitha and Muraleedharan 2006; Dehlavi and Adil 2011; Rao 2018; Verma et al. 2001; Dixit et al. 2016; Mishra 2014; Venkatachalam and Zareena Begam 2016; Prasher et al. 2006) and others (Schuijt 2002; Zuze 2013; Schuyt and Brander 2008; Kaggwa et al. 2009; Baral et al. 2016; Li and Gao 2016; Dehlavi and Adil 2011). *In appendix (Table A3) GDP(PPP) per capita for India https://data.worldbank.org/indicator/NY.GDP.PCAP.PP.CD?locations=IN Currency exchange rate https://www.thehansindia.com/business/currency-update-today-indian-rupee-against-us-dollar-on-19-november-2021-716019

(continued)

Table 4 (continued)

Services	Services and monetary benefits	Source
	Inspiration for culture, art and design: Data: Presence of culturally important landscape features or species; Number of people "using" ecosystems for cultural heritage and identity Inspiration for culture, art and design service values per ha (ranged between 302 and 13,985 Rs/ha/yr.) from literatures Compared with values of global studies, adjusted for GDP (PPP) per capita and corresponding currency exchange rate	Indian literatures (Ramachandra 2016; Ramachandra et al. 2019) and others (Zuze 2013; Schuyt and Brander 2008; Kaggwa et al. 2009). *In appendix (Table A4) GDP(PPP) per capita for India https://data.worldbank.org/indicator/NY.GDP.PCAP.PP.CD?locations=IN Currency exchange rate https://www.thehansindia.com/business/currency-update-today-indian-rupee-against-us-dollar-on-19-november-2021-716019
	Spiritual experience: Data: Number of people who attach religious significance to ecosystems; number of books, paintings, etc., using ecosystems as inspiration Spiritual experience service value per ha Compared with values of global studies, adjusted for GDP (PPP) per capita and corresponding currency exchange rate	Ecosystem Services Valuation Database (ESVD), 2020: https://www.es-partnership.org/wp-content/uploads/2020/08/ESVD_Global-Update-FINAL-Report-June-2020.pdf GDP(PPP) per capita for India https://data.worldbank.org/indicator/NY.GDP.PCAP.PP.CD?locations=IN Currency exchange rate https://www.thehansindia.com/business/currency-update-today-indian-rupee-against-us-dollar-on-19-november-2021-716019
	Information for cognitive development: Data: Cognitive development resulting from wetland organisms. Information for cognitive development service value per ha Compared with values of global studies, adjusted for GDP (PPP) per capita and corresponding currency exchange rate	Ecosystem Services Valuation Database (ESVD), 2020: https://www.es-partnership.org/wp-content/uploads/2020/08/ESVD_Global-Update-FINAL-Report-June-2020.pdf GDP(PPP) per capita for India https://data.worldbank.org/indicator/NY.GDP.PCAP.PP.CD?locations=IN Currency exchange rate https://www.thehansindia.com/business/currency-update-today-indian-rupee-against-us-dollar-on-19-november-2021-716019

(continued)

Table 4 (continued)

Services	Services and monetary benefits	Source
	Existence and bequest values: Data: Expressed (through, e.g., donations and voluntary work) or stated preference for nature protection for ethical reasons Existence and bequest values service value per ha Compared with values of global studies, adjusted for GDP (PPP) per capita and corresponding currency exchange rate	Ecosystem Services Valuation Database (ESVD), 2020: https://www.es-partnership.org/wp-content/uploads/2020/08/ESVD_Global-Update-FINAL-Report-June-2020.pdf GDP(PPP) per capita for India https://data.worldbank.org/indicator/NY.GDP.PCAP.PP.CD?locations=IN Currency exchange rate https://www.thehansindia.com/business/currency-update-today-indian-rupee-against-us-dollar-on-19-november-2021-716019
	Education: Data: Value for use of wetlands for scientific and education Education service value per ha (48,000, 34,660, and 4773 Rs/ha/yr) Compared with values of global studies, adjusted for GDP (PPP) per capita and corresponding currency exchange rate	Indian literature (Ramachandra 2016; Ramachandra et al. 2019) and Li and Gao (2016) GDP(PPP) per capita for India https://data.worldbank.org/indicator/NY.GDP.PCAP.PP.CD?locations=IN Currency exchange rate https://www.thehansindia.com/business/currency-update-today-indian-rupee-against-us-dollar-on-19-november-2021-716019

Table 5 Data used for computing ecosystem services from estuaries with the details of data sources

Ecosystem services	Variables	Data sources	Approach
Provisioning sources	Fishery resources (Fish, Clams, Oysters and Mussels) provide the major livelihood opportunities of the estuarine dependent communities in the coastal villages	Indian case studies—Ramachandra et al. (2013) CMFRI Annual Report, 2014–15 (http://eprints.cmfri.org.in/id/eprint/10461); CMFRI Annual Report, 2017–18 (http://eprints.cmfri.org.in/id/eprint/13044); CMFRI Annual Report, 1999–2000 (http://eprints.cmfri.org.in/id/eprint/6336); CMfRI Annual Report, 2010–11 (http://eprints.cmfri.org.in/id/eprint/8676); Indian Journal of Fisheries, 1985 (http://eprints.cmfri.org.in/id/eprint/407)	*Residual Value Method* $\sum_{i=1}^{n} Q_i(P_i - C_i)$ Where Qi represents quantity, Pi is the price, Ci is the cost involved in the harvest
	Agricultural activities (Coconuts and Gazani paddy)—which provide a significant source of income for the local communities	Indian case study—Ramachandra et al., 2013; Google earth pro is used for estimating total area (ha) of coconut plantation http://earth.google.com	*Residual Value Method* $\sum_{i=1}^{n} Q_i(P_i - C_i)$ Where Qi represents quantity, Pi is the price, Ci is the cost involved in the harvest
	Mining activity (Sand production)—which provides economic benefits to the estuarine dependent communities in the coastal villages	Indian case study—(Ramachandra et al. 2013)	*Residual Value Method* $\sum_{i=1}^{n} Q_i(P_i - C_i)$ Where Qi represents quantity, Pi is the price, Ci is the cost involved in the harvest

(continued)

Table 5 (continued)

Ecosystem services	Variables	Data sources	Approach
	Mangrove products (charcoal, fodder, timber and thatch)—which provides major livelihoods to the local communities	Indian case study—Ramachandra et al. 2013 Google earth pro software is used to estimate total area (ha) of Mangrove plantation http://earth.google.com	*Residual Value Method* $\sum_{i=1}^{n} Q_i(P_i - C_i)$ Where Qi represents quantity, Pi is the price, Ci is the cost involved in the harvest
	Salt production—which provides main livelihood option to the local communities	Indian article https://www.whatshot.in/bangalore/producing-brown-salt-since-300-years-this-tiny-village-is-karnatakas-rann-of-kutch-c-28641	*Residual Value Method* $\sum_{i=1}^{n} Q_i(P_i - C_i)$ Where Qi represents quantity, Pi is the price, Ci is the cost involved in the harvest
Regulating services	Gas regulation	Gas regulation service value per hectare is based on the Indian case study (Ramachandra et al. 2019)	Benefit transfer method

(continued)

Table 5 (continued)

Ecosystem services	Variables	Data sources	Approach
	Climate regulation	Climate regulation service values per hectare are based on Indian case study (Ramachandra et al. 2019) Compared with the values of International case studies (de Groot et al. 2020; Costanza et al. 2012) Adjusted for GDP (PPP) per capita for India (https://data.worldbank.org/indicator/NY.GDP.PCAP.PP.CD?locations=IN) and corresponding currency exchange rate https://www.xe.com/currencyconverter/	Benefit transfer method
	Disturbance regulation	Disturbance regulation service values per hectare is based on Indian case study (Ramachandra et al. 2019) Compared with the values of International case study (Costanza et al. 2012) Adjusted for GDP (PPP) per capita for India (https://data.worldbank.org/indicator/NY.GDP.PCAP.PP.CD?locations=IN) and corresponding currency exchange rate https://www.xe.com/currencyconverter/	Benefit transfer method

(continued)

Table 5 (continued)

Ecosystem services	Variables	Data sources	Approach
	Soil fertility	A soil fertility service value per hectare is based on International case study (de Groot et al. 2020) Adjusted for GDP (PPP) per capita for India (https://data.worldbank.org/indicator/NY.GDP.PCAP.PP.CD?locations=IN) and corresponding currency exchange rate https://www.xe.com/currencyconverter/	Benefit transfer method
	Water regulation	Water regulation service values per hectare is based on Indian case study (Ramachandra et al. 2019) Compared with the values of International case studies (Taye et al. 2021; de Groot et al. 2020) Adjusted for GDP (PPP) per capita for India (https://data.worldbank.org/indicator/NY.GDP.PCAP.PP.CD?locations=IN) and corresponding currency exchange rate https://www.xe.com/currencyconverter/	Benefit transfer method
	Air quality	Air quality service value per hectare is based on International case studies (Taye et al. 2021; de Groot et al. 2020) Adjusted for GDP (PPP) per capita for India (https://data.worldbank.org/indicator/NY.GDP.PCAP.PP.CD?locations=IN) and corresponding currency exchange rate https://www.xe.com/currencyconverter/	Benefit transfer method

(continued)

Table 5 (continued)

Ecosystem services	Variables	Data sources	Approach
	Erosion control	Erosion control service value per hectare is based on the Indian case study (Ramachandra et al. 2019) Compared with the values of International case studies (de Groot et al. 2020; Costanza et al. 2012) Adjusted for GDP (PPP) per capita for India (https://data.worldbank.org/indicator/NY.GDP.PCAP.PP.CD?locations=IN) and corresponding currency exchange rate https://www.xe.com/currencyconverter/	Benefit transfer method
	Nutrient cycling	Nutrient cycling service value per hectare is based on the Indian case study (Ramachandra et al. 2019) Compared with the values of International case study (Costanza et al. 2012) Adjusted for GDP (PPP) per capita for India (https://data.worldbank.org/indicator/NY.GDP.PCAP.PP.CD?locations=IN) and corresponding currency exchange rate https://www.xe.com/currencyconverter/	Benefit transfer method

(continued)

Table 5 (continued)

Ecosystem services	Variables	Data sources	Approach
	Pollination	Pollination service value per hectare is based on the International case study (de Groot et al. 2020) Adjusted for GDP (PPP) per capita for India (https://data.worldbank.org/indicator/NY.GDP.PCAP.PP.CD?locations=IN) and corresponding currency exchange rate https://www.xe.com/currencyconverter/	Benefit transfer method
	Carbon sequestration	Carbon sequestration service value per hectare is based on the Indian case study (Ramachandra et al. 2019) Compared with the values of International case study (Salem and Mercer 2012) Adjusted for GDP (PPP) per capita for India (https://data.worldbank.org/indicator/NY.GDP.PCAP.PP.CD?locations=IN) and corresponding currency exchange rate https://www.xe.com/currencyconverter/	Benefit transfer method
Cultural services	Recreation services	Recreation services value per hectare is based on the Indian case study (Ramachandra et al. 2019) Compared with the values of International case studies (de Groot et al. 2020; Costanza et al. 2012) Adjusted for GDP (PPP) per capita for India (https://data.worldbank.org/indicator/NY.GDP.PCAP.PP.CD?locations=IN) and corresponding currency exchange rate https://www.xe.com/currencyconverter/	Benefit transfer method

(continued)

Table 5 (continued)

Ecosystem services	Variables	Data sources	Approach
	Aesthetic information	Aesthetic Information services value per hectare is based on the Indian case study (Ramachandra et al. 2019) Compared with the value of International case study (de Groot et al. 2020) Adjusted for GDP (PPP) per capita for India (https://data.worldbank.org/indicator/NY.GDP.PCAP.PP.CD?locations=IN) and corresponding currency exchange rate https://www.xe.com/currencyconverter/	Benefit transfer method
	Education, scientific and research services	Education, scientific and research services value per hectare is based on the Indian case study (Ramachandra et al. 2019)	Benefit transfer method

Table 6 Method for computing goods and services from wetland ecosystems

Services	Variables	Data source	Approach
Provisioning services			
Provisioning services	Microalgae (PS_1)	Microalgae services: for biofuel production, serve as food (protein) and feed for fish • Biodiesel (Bi): 51,744 Rs/ha/yr • Glycerol (Gl): 3763 Rs/ha/yr • Feed for fish (Ff): 4200 Rs/ha/yr • Food (Protein) (P): 50,760 Rs/ha/yr Benefits from Microalgae (BMA) $BMA = \sum Bi + Gl + Ff + P$ $PS_1 = BMA * A$ Where, A = area of wetlands in districts of Karnataka (in ha)	Residual Method

(continued)

Table 6 (continued)

Services	Variables	Data source	Approach
	Macrophyte (PS_2)	Macrophyte services: as food, honey, making handicraft materials, and fodder. Total macrophyte cover in Karnataka wetlands is 50432.35 ha • Fodder (Fr): 5580 Rs/ha/yr • Honey (Ho): 2080 Rs/ha/yr • Grazing (G): 1174 Rs/ha/yr • Food (F): 147 Rs/ha/yr • Handicraft material (Hm): 2310 Rs/ha/yr Benefits from Macrophyte (BM) $BM = \sum Fr + Ho + G + F + Hm$ $PS_2 = BM * A_m$ Where, A_m = Macrophyte covered area of wetlands in districts of Karnataka (in ha)	Residual Method

(continued)

Table 6 (continued)

Services	Variables	Data source	Approach
	Fish (PS_3)	Fish as food • Benefit: 65 Rs/kg • Average fish catch: 495 kg/ha/yr • Benefit from Fish (BF): 32,175 Rs/ha/yr Benefits from Fish (BF) $PS_3 = BF * A$ Where, $A =$ Area of wetlands in districts of Karnataka (in ha)	Residual Method
	Water (PS_4)	Value for water service per hectare: 15,359 Rs/ha/yr $PS_4 = 15359 * A$ Where, $A =$ Area of wetlands in districts of Karnataka (in ha)	Benefit transfer method based on case studies (ESVD, 2020)
	Raw materials (PS_5)	Value for raw materials per hectare: 13,358 Rs/ha/yr $PS_5 = 13358 * A$ Where, $A =$ Area of wetlands in districts of Karnataka (in ha)	Benefit transfer method based on case studies (ESVD, 2020)

(continued)

Table 6 (continued)

Services	Variables	Data source	Approach
	Genetic resources (PS_6)	Value for genetic resources per hectare: 476 Rs/ha/yr $PS_6 = 476 * A$ Where, A = Area of wetlands in districts of Karnataka (in ha)	Benefit transfer method based on case studies (ESVD 2020)
	Medicinal resources (PS_7)	Value for medicinal resources per hectare: 786 Rs/ha/yr $PS_7 = 786 * A$ Where, A = Area of wetlands in districts of Karnataka (in ha)	Benefit transfer method based on case study (Clarkson et al. 2013)
	Ornamental resources (PS_8)	Value for ornamental resources service per hectare: 905 Rs/ha/yr $PS_8 = 905 * A$ Where, A = Area of wetlands in districts of Karnataka (in ha)	Benefit transfer method based on case study (Clarkson et al. 2013)

(continued)

Table 6 (continued)

Services	Variables	Data source	Approach
	Fuelwood (PS_9)	Fuelwood extraction from-10% of macrophyte covered area of wetlands in districts of Karnataka Value for fuelwood service per hectare: 5833 Rs/ha/yr $PS_9 = 5833 * A_f$ Where, A_f = Area of fuelwood in wetlands in districts of Karnataka (in ha)	Benefit transfer method based on case studies from India (Ramachandra et al. 2005, 2011)
	Irrigation (PS_{10})	Data on irrigation and agricultural crops grown in each district of Karnataka obtained from government reports (District at a Glance) Value for irrigation service per hectare: 1826 Rs/ha/yr $PS_{10} = 1826 * A_i$ Where, A_i = Tank irrigated area in districts of Karnataka (in ha)	Benefit transfer method based on case studies from India (Mukherjee 2008)
Regulating services			

(continued)

Table 6 (continued)

Services	Variables	Data source	Approach
Regulating services	Air quality regulation (RMS_1)	Value for air quality regulation service per hectare: 270 Rs/ha/yr $RMS_1 = 270 * A$ Where, A = Area of wetlands in districts of Karnataka (in ha)	Benefit transfer method based on case studies (ESVD 2020)
	Climate regulation (RMS_2)	Value for climate regulation service per hectare: 1191 Rs/ha/yr $RMS_2 = 1191 * A$ Where, A = Area of wetlands in districts of Karnataka (in ha)	Benefit transfer method based on case studies (ESVD 2020)
	Moderation of extreme events (RMS_3)	Value for moderation of extreme events service per hectare: 105,781 Rs/ha/yr $RMS_3 = 105781 * A$ Where, A = Area of wetlands in districts of Karnataka (in ha)	Benefit transfer method based on case studies (ESVD 2020)
	Regulation of water flows (RMS_4)	Value for regulation of water flow service per hectare: 28,891 Rs/ha/yr $RMS_4 = 28891 * A$ Where, A = Area of wetlands in districts of Karnataka (in ha)	Benefit transfer method based on case studies (ESVD 2020)

(continued)

Table 6 (continued)

Services	Variables	Data source	Approach
	Waste treatment (RMS_5)	Value for waste treatment service per hectare: 16,225 Rs/ha/yr $RMS_5 = 16225 * A$ Where, A = Area of wetlands in districts of Karnataka (in ha)	Benefit transfer method based on case studies (ESVD 2020)
	Erosion prevention (RMS_6)	Value for erosion prevention service per hectare: 20,704 Rs/ha/yr $RMS_6 = 20704 * A$ Where, A = Area of wetlands in districts of Karnataka (in ha)	Benefit transfer method based on case study (Clarkson et al. 2013)
	Maintenance of soil fertility (RMS_7)	Value for maintenance of soil fertility service per hectare: 13,604 Rs/ha/yr $RMS_7 = 13604 * A$ Where, A = Area of wetlands in districts of Karnataka (in ha)	Benefit transfer method based on case study (Clarkson et al. 2013)
	Biological control (RMS_8)	Value for biological control service per hectare: 7529 Rs/ha/yr $RMS_8 = 7529 * A$ Where, A = Area of wetlands in districts of Karnataka (in ha)	Benefit transfer method based on case study (Clarkson et al. 2013)

(continued)

Table 6 (continued)

Services	Variables	Data source	Approach
	Maintenance of life cycles of migratory species (RMS_9)	Value for maintenance of life cycles of migratory species service per hectare: 14,978 Rs/ha/yr $RMS_9 = 14978 * A$ Where, A = Area of wetlands in districts of Karnataka (in ha)	Benefit transfer method based on case studies (ESVD 2020)
	Maintenance of genetic diversity (RMS_{10})	Value for maintenance of genetic diversity service per hectare: 27,216 Rs/ha/yr $RMS_{10} = 27216 * A$ Where, A = Area of wetlands in districts of Karnataka (in ha)	Benefit transfer method based on case studies (ESVD 2020)

(continued)

Table 6 (continued)

Services	Variables	Data source	Approach
	Carbon sequestration (RMS_{11})	Microalgae Total value of carbon sequestration from microalgae: 234,404.05 Rs tCO_2/ha/yr Macrophyte Total value of carbon sequestration from macrophyte: 311.92 Rs $kgCO_2$/ha/yr Value for carbon sequestration service per hectare (summation of carbon sequestration from microalgae and macrophyte): 234,716 Rs/ha/yr $RMS_{11} = 234716 * A$ Where, A = Area of wetlands in districts of Karnataka (in ha)	Benefit transfer method based on case studies from India (Verma et al. 2014; Kalita 2019; Baral et al. 2016)
	Water borne disease control (RMS_{12})	Value for water borne disease control service per hectare: 1941 Rs/ha/yr $RMS_{12} = 1941 * A$ Where, A = Area of wetlands in districts of Karnataka (in ha)	Benefit transfer method based on case studies from India (Verma et al. 2001; Kant et al. 2013; Ramachandra et al. 2011)

(continued)

Table 6 (continued)

Services	Variables	Data source	Approach
	Pollination (RMS_{13})	Value for pollination service per hectare: 19 Rs/ha/yr $RMS_{13} = 19 * A$ Where, A = Area of wetlands in districts of Karnataka (in ha)	Benefit transfer method based on case study (Schuyt and Brander 2008)
	Water conservation (RMS_{14})	Value for water conservation service per hectare: 2875 Rs/ha/yr $RMS_{14} = 2875 * A$ Where, A = Area of wetlands in districts of Karnataka (in ha)	Benefit transfer method based on case study (Li and Gao 2016)
	Flood control (RMS_{15})	Value for flood control service per hectare: 7053 Rs/ha/yr $RMS_{15} = 7053 * A$ Where, A = Area of wetlands in districts of Karnataka (in ha)	Benefit transfer method based on case studies from India (Kaul et al. 2016), Zuze (2013), Schuyt and Brander (2008)
	Habitat/refugia (RMS_{16})	Value for habitat/refugia service per hectare: 1825 Rs/ha/yr $RMS_{16} = 1825 * A$ Where, A = Area of wetlands in districts of Karnataka (in ha)	Benefit transfer method based on case studies (Schuyt and Brander 2008; Kaggwa et al. 2009)

(continued)

Table 6 (continued)

Services	Variables	Data source	Approach
	Groundwater recharge (RMS_{17})	Value for Groundwater recharge service per hectare: 215,123 Rs/ha/yr $RMS_{17} = 215123 * A$ Where, A = Area of wetlands in districts of Karnataka (in ha)	Benefit transfer method based on case studies from India (Ramachandra et al. 2019), Kaggwa et al. (2009)
Cultural services			
Cultural services	Aesthetic information (CS_1)	Value for aesthetic information service per hectare: 389 Rs/ha/yr $CS_1 = 389 * A$ Where, A = Area of wetlands in districts of Karnataka (in ha)	Benefit transfer method based on case studies (ESVD 2020)
	Opportunities for recreation and tourism (CS_2)	Value for opportunities for recreation and tourism service per hectare: 12,111 Rs/ha/yr $CS_2 = 12111 * A$ Where, A = Area of wetlands in districts of Karnataka (in ha)	Benefit transfer method based on case studies from India (Anitha and Muraleedharan 2006; Dehlavi and Adil 2011; Rao 2018; Verma et al. 2001; Dixit et al. 2016; Mishra 2014; Venkatachalam and Zareena Begam 2016; Prasher et al. 2006; Schuijt 2002; Zuze 2013; Schuyt and Brander 2008; Kaggwa et al. 2009; Baral et al. 2016; Li and Gao 2016; Dehlavi and Adil 2011)

(continued)

Table 6 (continued)

Services	Variables	Data source	Approach
	Inspiration for culture, art and design (CS_3)	Value for inspiration for culture, art and design service per hectare: 3346 Rs/ha/yr $CS_3 = 3346 * A$ Where, $A =$ Area of wetlands in districts of Karnataka (in ha)	Benefit transfer method based on case studies from India (Ramachandra 2016; Ramachandra et al. 2019; Zuze 2013; Schuyt and Brander 2008; Kaggwa et al. 2009)
	Spiritual experience (CS_4)	Value for spiritual experience service per hectare: 8 Rs/ha/yr $CS_4 = 8 * A$ Where, $A =$ Area of wetlands in districts of Karnataka (in ha)	Benefit transfer method based on case studies (ESVD 2020)
	Information for cognitive development (CS_5)	Value for information for cognitive development service per hectare: 953 Rs/ha/yr $CS_5 = 953 * A$ Where, $A =$ Area of wetlands in districts of Karnataka (in ha)	Benefit transfer method based on case studies (ESVD 2020)

(continued)

Table 6 (continued)

Services	Variables	Data source	Approach
	Existence and bequest values (CS_6)	Value for existence and bequest values service per hectare: 91,312 Rs/ha/yr $CS_6 = 91312 * A$ Where, A = Area of wetlands in districts of Karnataka (in ha)	Benefit transfer method based on case studies (ESVD 2020)
	Education (CS_7)	Value for education service per hectare: 29,144 Rs/ha/yr $CS_7 = 29144 * A$ Where, A = Area of wetlands in districts of Karnataka (in ha)	Benefit transfer method based on case studies from India (Ramachandra 2016; Ramachandra et al. 2019; Li and Gao 2016)

9 Method

The SEEA Central Framework provides concepts, definitions, and classifications to support integrated accounting for physical flows (natural inputs from and residual flows to the environment such as water, energy, air emissions, and solid waste); environmental transactions and transfers (e.g., environmental taxes, environmental subsidies, and environmental protection expenditure); and individual environmental assets (e.g., mineral and energy resources, timber, fish, land, soil, and water).

Residual value and resource rent methods: The residual value and resource rent methods estimate the value of an ecosystem service by taking the gross value of the final marketed good to which the ecosystem service provides input and then deducting the cost of all other inputs, including labour, produced assets, and intermediate inputs (SEEA 2021).

The Residual value method used for provisioning services is computed as per the Eq. (1).

$$\sum\nolimits_{1=1}^{n} Q_i \times (P_i - C_i) \tag{1}$$

where Qi represents quantity, Pi is the price, Ci is the cost involved in the harvest.

Benefit transfer method: The benefit transfer method involves transferring monetary values of ecosystem services from previous studies or literature that focused on a different region to our area of interest (Ramachandra et al. 2019; Adekola 2007).

The Benefit transfer method (Eq. 2).

$$\sum_{I=1}^{n} V_i \times A_i \tag{2}$$

where Vi represents the monetary values per hectare and Ai represents the area.

Ecosystem accounting through benefit transfer method requires adjusting GDP for the value of the PPP (purchasing power parity). This adjustment converts nominal GDP into a number, which helps in easy comparison between countries with different currencies. The per capita income gap between the countries is reduced under PPP exchange rates. The exchange rate of the US dollar to the Indian rupee is now Rs 74.22. Based on PPP, the GDP per capita of India is projected at 7333 international dollars in 2021. GDP per capita (US$ PPP) for the US is projected at 68,309.

Macrophyte

Macrophyte covered ~18% of the Karnataka wetland area. The total macrophyte cover in Karnataka wetlands is 50432.35 ha (Fig. 7, Table 7). The total benefit from macrophyte was computed considering fodder, honey production, food, grazing, and handicraft making through the residual value method for calculating provisioning services.

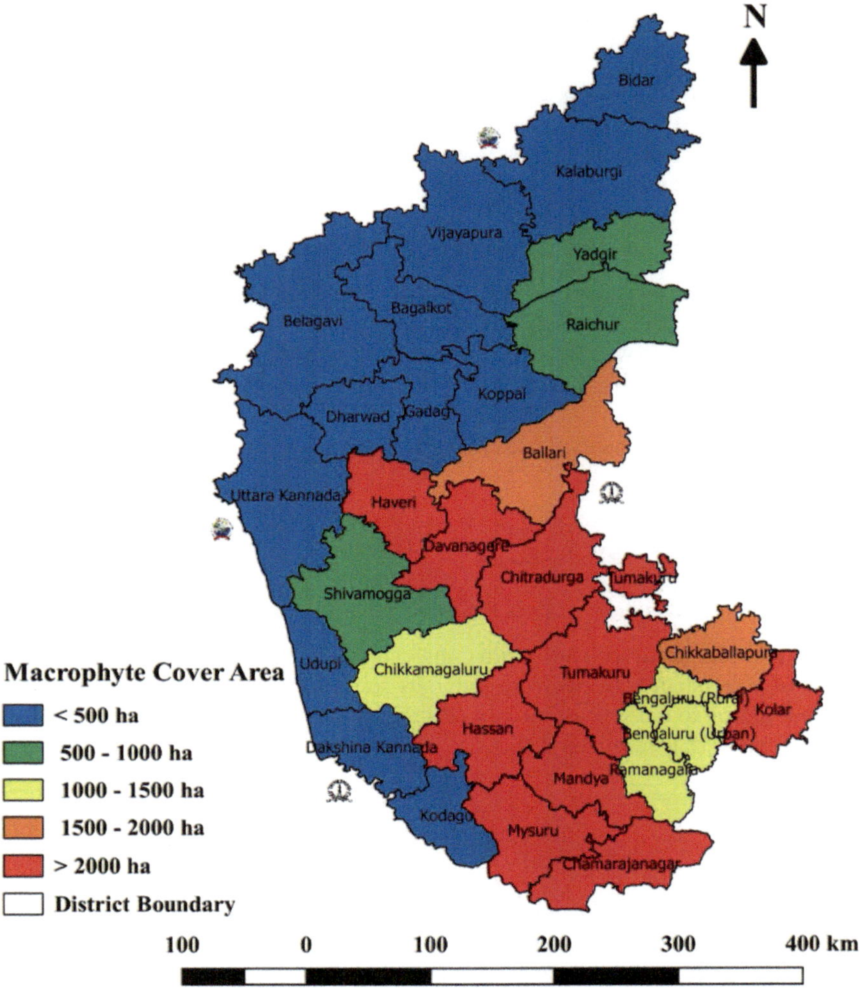

Fig. 7 Macrophyte covered area in wetlands of districts of Karnataka

Fuelwood

Wetland trees provide materials such as timber and fuel wood. The households near the lakes collect fuelwood as cooking energy. Fuelwood quantity is quantified considering 10% of the macrophyte spread area in wetlands (Table 8).

Table 7 Macrophyte cover district-wise in Karnataka

District	Macrophyte cover (ha)
Bengaluru urban	1316.63
Bengaluru rural	1448.9
Chamarajanagar	3032.63
Mysuru	3052.41
Bidar	137.59
Raichur	656.74
Yadgir	949.19
Dharwad	284.54
Belagavi	200.61
Gadag	348.18
Vijayapura	29.82
Kalaburgi	171.98
Shivamogga	865.73
Bagalkot	103.86
Dakshina Kannada	15.06
Chikkamagaluru	1330.86
Chikkaballapura	1741.72
Kolar	2733.96
Ramanagara	1253.06
Udupi	41.99
Uttara Kannada	316.46
Mandya	3968.29
Chitradurga	5734.12
Davanagere	2599.28
Kodagu	86.49
Hassan	4291.89
Haveri	2266.65
Koppal	263.74
Ballari	1782.32
Tumakuru	9407.65

Irrigation

Wetland water is used for irrigation of agriculture and horticulture fields. The data on irrigation (Fig. 8, Table 9) and agricultural crops grown in each district of Karnataka was obtained from government reports (District at a Glance, GoK).

Table 8 Fuelwood—harvested (spatial extent in hectare)

Sl. No	District	Fuelwood area in ha
1	Bagalkot	10.39
2	Ballari	178.23
3	Bengaluru Rural	144.89
4	Bengaluru Urban	131.66
5	Belagavi	20.06
6	Bidar	13.76
7	Chamarajanagar	303.26
8	Chikkaballapura	174.17
9	Chikkamagaluru	133.09
10	Chitradurga	573.41
11	Dakshina Kannada	1.51
12	Davanagere	259.93
13	Dharwad	28.45
14	Gadag	34.82
15	Hassan	429.19
16	Haveri	226.67
17	Kalaburgi	17.20
18	Kodagu	8.65
19	Kolar	273.40
20	Koppal	26.37
21	Mandya	396.83
22	Mysuru	305.24
23	Raichur	65.67
24	Ramanagara	125.31
25	Shivamogga	86.57
26	Tumakuru	940.77
27	Udupi	4.20
28	Uttara Kannada	31.65
29	Vijayapura	2.98
30	Yadgir	94.92

In this study, the residual value method and the benefit transfer method is used for ecosystem service valuation. Values used for the valuation of ecosystem service are listed in Table 10.

The values adopted for the valuation of ecosystem services are shown in Table 10. For wetland valuation, provisioning service was calculated by Eq. (3). PSn is the total provisioning service district wise. n is Karnataka's 1–30 districts, i is the various provisioning services ($i = 1$ to 10 services), An is the wetland area of each district,

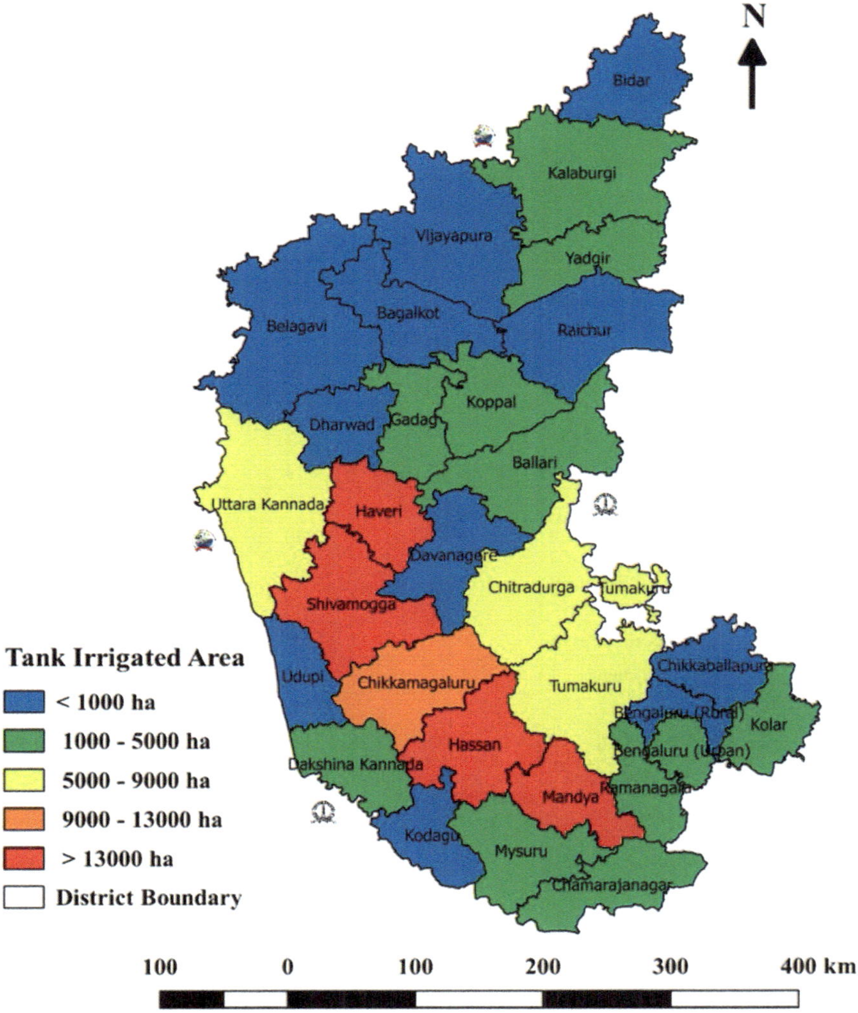

Fig. 8 Tank irrigated area in districts of Karnataka

but the area is different for the macrophyte (considered macrophyte cover area), irrigation (considered tank irrigated area), and fuelwood service (considered 10% of the macrophyte area). The regulating service (*RS/RMS*) is calculated by Eq. (4), and the cultural service (*CS*) is calculated by Eq. (5). The total ecosystem supply value (*TESV*) is calculated using Eq. (6), which is the sum of provisioning, regulating, and cultural services. The Eq. (7) is used to calculate the *NPV* (net present value) of a wetland ecosystem (SEEA 2014, 2021).

Table 9 Tank irrigated area in hectare

Sl. No	District	Tank irrigated area in ha
1	Bagalkot	839
2	Ballari	1139
3	Bengaluru rural	624
4	Bengaluru urban	1449
5	Belagavi	307
6	Bidar	688
7	Chamarajanagar	1646
8	Chikkaballapura	251
9	Chikkamagaluru	9124
10	Chitradurga	5041
11	Dakshina Kannada	2172
12	Davanagere	914
13	Dharwad	238
14	Gadag	1379
15	Hassan	17,407
16	Haveri	20,768
17	Kalaburgi	1069
18	Kodagu	171
19	Kolar	3368
20	Koppal	2984
21	Mandya	14,654
22	Mysuru	4335
23	Raichur	789
24	Ramanagara	1688
25	Shivamogga	48,572
26	Tumakuru	5377
27	Udupi	453
28	Uttara Kannada	6155
29	Vijayapura	945
30	Yadgir	3525

$$PS_n = \sum_{i=1}^{10} PS_i \times A_n \quad (3)$$

$$RS_n = \sum_{i=1}^{17} RS_i \times A_n \quad (4)$$

Table 10 Values used for wetland ecosystem valuation

Symbol	Services	Unit value (Rs/ha/yr)
Provisioning services		
PS1	Algae	110,467
PS2	Macrophyte	11,291
PS3	Fish	32,175
PS4	Water	15,359
PS5	Raw materials	13,358
PS6	Genetic resources	476
PS7	Medicinal resources	786
PS8	Ornamental resources	905
PS9	Fuelwood	5833
PS10	Irrigation	1826
Regulating services		
RMS1	Air quality regulation	270
RMS2	Climate regulation	1191
RMS3	Moderation of extreme events	105,781
RMS4	Regulation of water flows	28,891
RMS5	Waste treatment	16,225
RMS6	Erosion prevention	20,704
RMS7	Maintenance of soil fertility	13,604
RMS8	Biological control	7529
RMS9	Maintenance of life cycles of migratory species	14,978
RMS10	Maintenance of genetic diversity	27,216
RMS11	Carbon sequestration	234,716
RMS12	Water borne disease control	1941
RMS13	Pollination	19
RMS14	Water conservation	2875
RMS15	Flood control	7053
RMS16	Habitat/refugia	1825
RMS17	Groundwater recharge	215,123
Cultural services		
CS1	Aesthetic information	389
CS2	Opportunities for recreation and tourism	12,111
CS3	Inspiration for culture, art and design	905
CS4	Spiritual experience	8
CS5	Information for cognitive development	953
CS6	Existence and bequest values	91,312
CS7	Education	29,144

$$CS_n = \sum_{i=1}^{7} CS_i \times A_n \quad (5)$$

$$TESV = \sum_{n=1}^{30} PS_n + RMS_n + CS_n \quad (6)$$

$$NPV = \sum_{t=1}^{T} \frac{R_t}{(1+r)^t} \quad (7)$$

where, NPV = net present value; R = net cash flow (TESV) in year t; T = discount period (50 year); and r = discount rate (3%).

The NPV reflects the monetary value of an asset on the basis of its cash flow during a discount period and a discount rate. Macrophyte spatial extent is quantified using Google Earth, and QGIS open-source GIS software is used for visualization of ecosystem services provided by wetlands.

10 Results and Discussion

10.1 Water Quality of Lakes

The water quality of aquatic ecosystems depends on the climate, geology, topography, and soil properties of a region. In the case of Bengaluru lakes in Karnataka, 79% of lakes belong to class E, 19% belong to classes D and E, and 2% belong to class A as per CPCB standards (Fig. 9). The loss and degradation of wetlands in Karnataka occurred due to urbanization, industrial development, agriculture expansion, water extraction for domestic and industrial usage, encroachment, and pollution of water bodies due to domestic sewage as well as industrial effluents (Ramachandra et al. 2020b, 2021; Ramachandra and Aithal 2016; Ravikumar et al. 2013). The sustained inflow of untreated domestic sewage and industrial effluents enhances nutrients in water bodies, which allows invasive exotic species to proliferate, leading to a decline of native species, which alters the community structure and composition of aquatic biota (Ramachandra et al. 2014).

Degradation of wetlands occurs due to urbanization, industrialization, agricultural intensification, water extraction for domestic use and industries, and pollution (Singh et al. 2013). The physical, chemical, and geomorphological variables of an aquatic ecosystem strongly influence freshwater biodiversity (Oikonomou and Stefanidis 2020). The current study highlights that among Bangalore wetlands, 24 lakes are less polluted (with low ionic as well as nutrient content), 22 lakes are moderately polluted, and 34 lakes are highly polluted (with high ionic, organic, and nutrient content) in Bengaluru, Karnataka (Fig. 10).

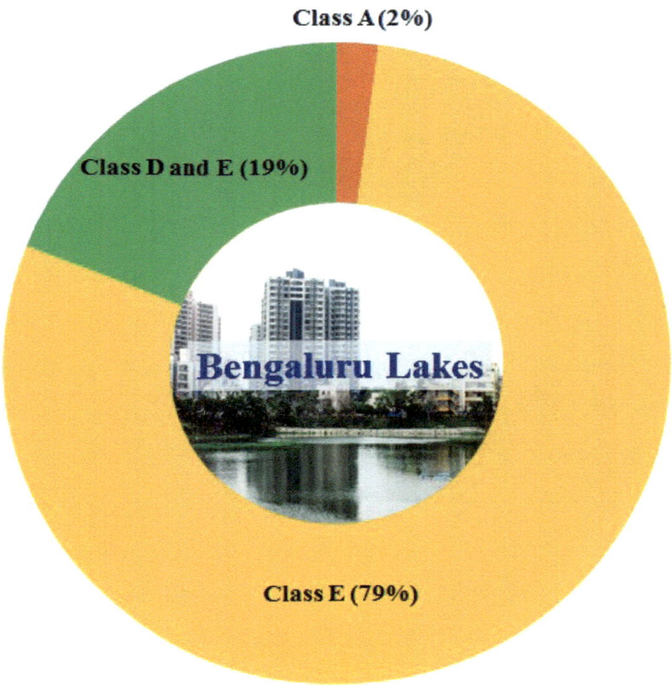

Fig. 9 Classification of lakes as per CPCB standards

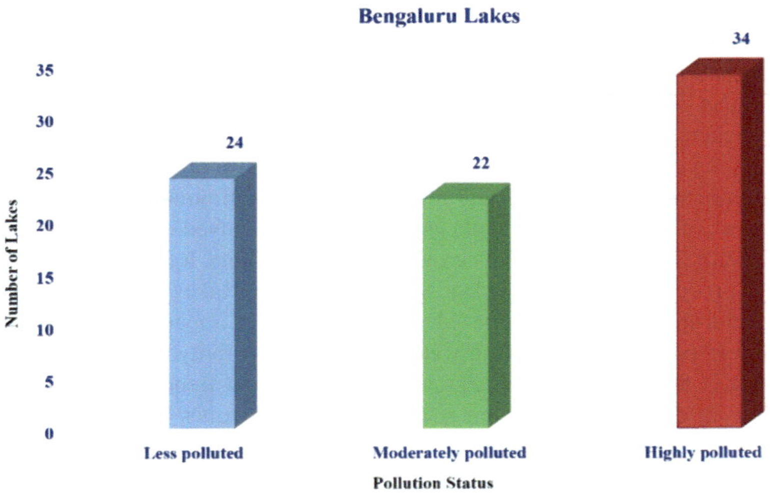

Fig. 10 Pollution status of lakes in Bengaluru (based on water quality data)

Fig. 11 Sewage fed lakes in Bengaluru

The study on 105 wetlands in Bengaluru district (Fig. 11) revealed that about 90% of the lakes were under threat due to (i) the sustained inflow of untreated domestic sewage and industrial effluents, (ii) dumping of organic wastes, (iii) dumping of construction and demolition wastes, (iv) encroachment of lake bed and stormwater drains (Ramachandra et al. 2016), (v) narrowing and concretising storm water drains, and (vi) dumping of waste (solid and liquid waste) in stormwater drains.

Nutrient enrichment in water leads to eutrophication, stimulating the profuse growth of invasive algae and aquatic plants, resulting in dissolved oxygen depletion (Tang et al. 2020), and affected the native species of aquatic biota. It was observed that benthic algal production is higher in clear water than in turbid lakes (McGowan et al. 2005). High densities of cyanobacteria (blue-green algae) can impair lake ecosystem services such as poor-quality drinking water, recreational activity, and biodiversity (Schallenberg et al. 2013). This is aggravated by the recent trends of changes in climate affecting ecosystems and their related services (Marquès et al. 2013). The alarming decline in the quality and quantity of wetland resources will likely affect the livelihoods of various stakeholders who depend directly or indirectly on these wetland resources.

Wetland loss and conversion to other land use with unplanned urbanization due to the expansion of built-up areas, agricultural extension, water abstraction, and flow modification (Pal and Talukdar 2018). Unplanned urbanization, industrialization, and waste disposal practices impose negative effects on wetland ecosystem services (Alam et al. 2017). The major threats to wetlands are eutrophication, sewage, solid waste, overfishing, detergent usage, and agricultural sources (Biswasroy et al. 2011). Over-exploitation of wetland resources and anthropogenic activities deteriorated the Kuttanad wetland ecosystem, causing an ecological imbalance (Rao and Balasubramanian 2019). Even though inhabitants depended on Loktak Lake's bioresources such as fishes, prawns, eels, vegetable items, fuelwood, thatching materials, handicraft materials, medicinal plants, and oysters for livelihood and income generation, various anthropogenic pressures resulting in pollution and ecological degradation led to the poor socio-economic condition of the people (Laishram and Dey 2013). Unplanned developmental activities drivers such as habitat loss, habitat degradation, and fragmentation, overharvesting of biological resources, illegal hunting and poaching, monoculture plantation, agricultural intensification, loss of genetic resources, shifting cultivation with insufficient fallow period, human-wildlife conflict, invasive alien species, and atmospheric pollution have an adverse impact on conservation and sustainable use of biodiversity and ecosystem services (Xu et al. 2019). The driving forces for the decline in ecosystem services such as food supply (rice production, and fish production), and biodiversity in China's Poyang Lake and Bangladesh's Tanguar Haor wetland from 2000 to 2012 were land use policies, population increase, and economic development (Sun et al. 2017).

The problems of encroachment of lakes and wastewater discharges into lakes with an increase in urbanization need to be addressed with strict enforcement of acts, and regulations (Merugu and Seetharaman 2013). Reducing nutrient input into lakes from point and nonpoint sources and banning anthropogenic activities is necessary to protect lakes (Kumar et al. 2006).

11 Ecosystem Services Provided by Microalgae

Microalgae are ubiquitous photosynthetic microorganisms that generally thrive in freshwater and saltwater ecosystems. They have chlorophyll and produce atmospheric oxygen through the process of photosynthesis, which converts water and carbon dioxide into carbohydrates with the help of solar energy. Wetland primary production is influenced by a variety of factors such as vegetation type, hydrology, climate, geographic location, light intensity, water temperature, and the nutrient status of water and sediment. Microalgae provide many essential ecosystem services and are widely used for commercial applications as food and nutritional supplements (Asulabha et al. 2018). Recently, several studies have been carried out on the utilization of microalgae for carbon dioxide sequestration and wastewater remediation. Microalgae help in bioremediation (purification of water/wastewater) since they have the ability to accumulate nutrients, heavy metals, inorganic and organic toxic substances, pesticides, and other pollutants in their tiny cells. Microalgae help in the reduction of greenhouse gas (GHG) emissions, wastewater treatment, nutrient removal/recycling, and the production of carbon–neutral biofuels (Silva et al. 2021).

Microalgae biomass (Table 11, Fig. 12) composed of carbohydrates, lipids, and proteins has been widely used in industries to produce fuel (biodiesel, bioethanol, methane, biobutanol, and biogas), feed (spirulina, and chlorella powder), biofertilizers, and medicines (pharmaceuticals and nutraceuticals).

Some species of microalgae produce bioactive compounds and commercially valuable pigments such as chlorophyll, phycobiliproteins, carotenoids, and astaxanthin (Khan et al. 2018a). For example, *Chlorella* sp., which contains large amounts of proteins, carbohydrates, lipids, vitamins, micronutrients, and carotenoids, is directly used in animal feed. Microalgae are rich sources of proteins and can produce 2.5 to 7.5 tons/ha/yr of proteins (Khan et al. 2018a). In order to produce 160 MT/day of protein commercially, 336 MT/day of dry algae (with 54 wt.% protein) is required (Asiedu 2015). Polyunsaturated fatty acids, including eicosapentaenoic acid (EPA)

Table 11 Review of algal biomass productivity

References	Area ha	Cultivation	Biomass productivity	Lipid content (%)	USD per kg algal biomass	USD per L algal lipid or biodiesel
Amer et al. (2011)	500	Pond	24 g m^{-2} day^{-1}	40		4–7.50
Davis et al. (2011)	1950.58	Pond	25 g m^{-2} day^{-1}	25		2.25
Norsker et al. (2011)	100	Pond	21 t ha^{-1} $year^{-1}$		6.57	
Delrue et al. (2012)	333.3	Pond	20–30 g m^{-2} day^{-1}	20–50		1.3–2.53
Nagarajan et al. (2013)	400	Pond	30 g m^{-2} day^{-1} or 60 g m^{-2} day^{-1}	50		0.42–0.97

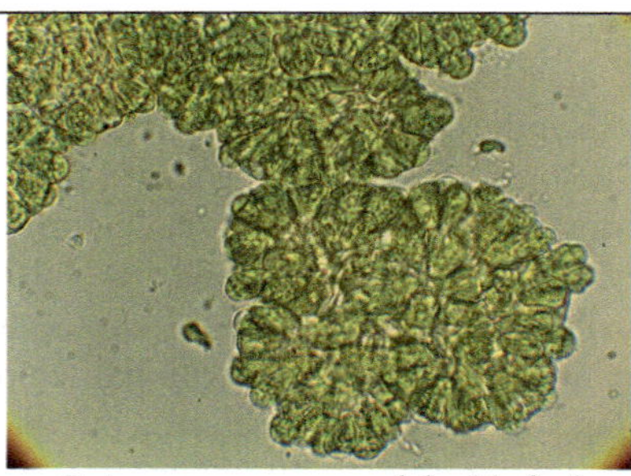

Botryococcus sp. used for biofuel production

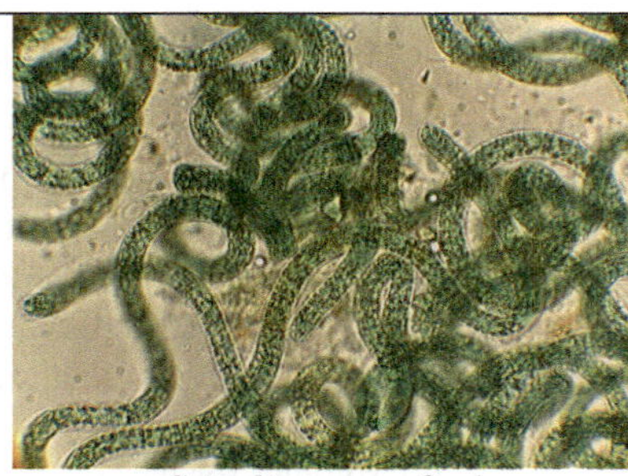

Spirulina sp. as feed

Fig. 12 Microalgae images under the microscope

and docosahexaenoic acid (DHA), extracted from microalgae have been marketed by numerous algae-based or biotechnological companies worldwide.

Microalgae is the most preferred source for biofuel production due to several advantages: (i) higher productivity than other crops; (ii) higher photosynthesis efficiency (3–8%) than terrestrial plants (0.5%), (iii) shorter doubling time; (iv) higher lipid content; (v) no competition with resources needed for food supply; (vi) ability to grow in wastewaters and nutrient removal; (viii) reduced land requirements; (ix) ability for CO_2 sequestration and; (x) ability to accumulate other products of commercial value (Silva et al. 2021; Amit et al. 2021). The total oil content of third-generation feedstocks (biofuels from microalgae) was higher than the first-generation (biofuels from edible food crops such as rapeseed, oil seeds, sunflower, soybean, potato, barley, wheat, and sugarcane) and second-generation (biofuels from non-edible crops like jatropha, jojoba, pongamia, and lignocellulosic biomass) feedstocks (Chhandama et al. 2021). Microalgal biodiesel production begins with the harvesting of microalgal biomass, followed by drying and lipid extraction, and then the transesterification process. During the transesterification process, the fatty acids of one molecule triacylglyceride (TAG) react with alcohol to form fatty acid methyl esters (FAMEs) and glycerol (Schlagermann et al. 2012). Techniques for harvesting microalgae biomass include filtration, centrifugation, flocculation, and flotation (Tan et al. 2020). Table 12 lists the cost and benefits of microalgae byproducts.

Microalgae play a significant role in wastewater treatment and produce valuable biomass in large quantities that can be used for biofuel production (Randrianarison and Ashraf 2017). Microalgae-based biofuels (third-generation biomass energy) are a renewable and sustainable source of energy required to meet the global energy demand for the transportation sector. Microalgal biodiesel is an alternative to reduce the unsustainable utilization of fossil fuels that cause global warming. Microalgal biodiesel showed reduced emissions of carbon monoxide (CO), unburned hydrocarbon (HC), and oxide of nitrogen (NOx) compared to petroleum diesel (Ali et al. 2017). Microalgal biodiesel can displace petroleum-derived diesel since it has properties similar to petroleum-derived diesel and can be directly used in conventional diesel engines (Chisti 2008). The calorific value of biodiesel is about 39 MJ/kg. Pure biodiesel can be used directly in a diesel engine or it may be blended with conventional diesel before use (Pandey et al. 2018). About 20 kg of wet de-oiled biomass can produce one cubic meter of biogas/day, which is equivalent to generating about 6 kWh of power. This is sufficient to light three 100-W incandescent bulbs for 20 h (Ali et al. 2017). Biofuels derived from algae biomass include hydrogen, methane, syngas (gaseous), ethanol, jet fuel, diesel, butanol (liquid), and charcoal (solid fuel) through different methods such as photobiological method, gasification, fermentation, anaerobic digestion, liquefaction, transesterification, and pyrolysis (Chowdhury and Loganathan 2019; Aliyu et al. 2021). Biodiesel complies with the three key aspects of sustainability, i.e., economic (lower production cost and high profit), environmental (lower environmental load), and social (ensure food security, livelihood, and employment) sustainability (Mizik and Gyarmati 2021). The provisioning service provided by microalgae from the wetlands of Karnataka amounts to 1,10,467 Rs/ha/yr (Table 13).

Table 12 Costs and benefits of algae derived byproducts

Algal Product	Cost	Selling price	Benefit	References
Biodiesel	0.42 $/L or 31.29 Rs/L	56 Rs/L	25 Rs/L	Nagarajan et al. (2013), Chisti (2013), Louw et al. (2016)
Glycerol			16.39 Rs/kg	Quispe et al. (2013)
Protein	$2.86/kg	$4.13/kg	$ 1.27/kg	Asiedu (2015)
Algal fuel	$0.45/L to $7.93/L	$1.32/L to $5.94/L	0.87$/L to 1.99 $/L	Gu et al. (2020)
Biodiesel	0.33 €/L			Amit et al. (2021)
	0.33 €/L	0.75 €/L	0.42 €/L	Branco-Vieira et al. (2020)
Biofuel	$0.44/L to $8.76/L			Chowdhury and Loganathan (2019)
Biofuel	$ 0.43/L ($ 1.64/gal) $ 7.92/L ($ 30.00/gal)	$ 1.82/L ($ 7/gal)		Roles et al. (2021)

Table 13 Provisioning service provided by microalgae

Microalgae	Provisioning services (Rs/ha/yr)
Biodiesel	51,744
Glycerol	3763
Protein/Food	50,760
Fishmeal	4200
Total benefit	110,467

12 Ecosystem Services Provided by Fish

Fisheries sector provides livelihoods, income, and nutritious food to society (Sincy et al. 2018). Fish excretion contributes to nutrient recycling and nutrient translocation, influencing microalgal biomass production and community composition in aquatic ecosystems (Williamson et al. 2018). Physical habitat variables (stream channel width and depth, and organic matter) and water quality variables (conductivity, pH, water temperature, chlorophyll, turbidity, and dissolved oxygen) play a decisive role in the fish species assemblage (Vieira and Tejerina-Garro 2020). The vertical distribution of fish in lakes is determined by water temperature and oxygen stratification. Vertical migration of fish along with diel periodicity helps in avoiding predation and reducing competition (Perivolioti et al. 2020). Fish for their growth, development, and reproduction, require long-chain polyunsaturated fatty acids (PUFA), such as eicosapentaenoic acid (EPA 20:5n-3), docosahexaenoic acid (DHA, 22:6n-3), and arachidonic acid (AA, 20:4n-6) (Sargent et al. 1999). Figure 13 shows some of the commercially important fish species in Karnataka.

Ctenopharyngodon idella feeds on numerous aquatic weeds, such as species of Lemna, Potamogeton, Elodea, Ceratophyllum, Vallisneria, Hydrocharis, and Myriophyllum (Lancar and Krake 2002), and has been useful in weed control. Introducing planktivorous fish alters plankton communities in food webs (Izaguirre et al. 2018). Fish rich in protein, essential fatty acids, and micronutrients (Fe, Zn, Ca, and vitamin A) constitute an important component of the human diet and serve as medicine (Table 14). Crude protein (% ww) content in *Catla catla*, *Cirrhinus mrigala*, *Clarias batrachus*, and *Labeo rohita* ranges from 16.2 ± 0.51, 15.5 ± 0.51, 16.4 ± 0.31, and 15.9 ± 0.41 respectively. The DHA and EPA (mg/100 g wet wt.) content in freshwater fish species *Amblypharyngodon mola*, *Puntius sophore*, *Tor putitora* was found as 133.3 ± 6.8 and 94.6 ± 5.4; 161.7 ± 7.6 and 303.8 ± 8.3; 115.5 ± 12.2 and 201.9 ± 13.1 respectively (Mohanty et al. 2016). The blood, bile, fat, and meat of eel fish were used as traditional medicine to treat obesity, alcoholism, gout, mental disorders, and during pregnancy (Svanberg and Locker 2020). The Indian major carps

Fish as food source | Fish in wetland

Fig. 13 Commercially important fish species

Table 14 Fish species being used for medicinal purpose

Species	Part used	Medicinal use for curing diseases	Application
Amblypharyngodon mola	Whole fish	Premenstrual pain	Fish is boiled and consumed
		Pox, pain, asthma	Cooked fish
Anabas testudineus	Whole fish	Dysmenorrhoea	Fish head, Tal tree leaf (*Borassus flabellifer*), and chilly are boiled together and eaten
Anguilla bengalensis	Fats	Rheumatoid-arthritis	Fat is applied and massaged into the affected parts to relieve pain
	Whole body	Piles and meningitis	Fish is cooked with *Stellaria media* and orally taken as soup once daily for 2 weeks
	Body mucous	Burns	Mucous is applied to the affected parts
Bagarius bagarius	Fat	Rheumatism, gout and joint pain	Fat is applied to the joints and massaged
Chaca chaca	Whole body	Polio	Dried fish is cooked with spices and taken orally
Channa gachua	Whole body	Gall bladder stone	Flesh is cooked and consumed
		Abdominal pain	Boiled fish is consumed
Channa punctatus	Whole fish	Tuberculosis	The fish is boiled with black pepper and consumed
		Diarrhoea	Boiled fish is consumed
	Bile	Malaria	Bile is taken thrice a day till recovery
Channa stewartii	Whole body	Diabetes, pain, high pressure	Boiled in water and consumed
Clarias batrachus	Whole body	Body ache, wound healing	Cooked with black pepper and consumed
		Small pox	Cooked fish is consumed to cure small pox
		Weakness after delivery	Boiled fish is taken regularly
Eutropiichthys vacha	Flesh	Tuberculosis, brain health	Used as curry two times a week for one month
Heteropneustes fossilis	Whole body	Pain, wound healing	Cooked with black pepper and prescribed to eat

(continued)

Table 14 (continued)

Species	Part used	Medicinal use for curing diseases	Application
		Anemia	Boiled fish taken as a tonic
	Brain	Sting by the fish itself	Brain is consumed raw when stung by the fish, as analgesic
Hilsa ilisha	Body oil	Arthritis, cracked heels, scurvy	Used as ointment for hands, heels and legs
Labeo pangusia	Flesh	Weakness after delivery	Boiled fish is consumed regularly
Labeo pangusia	Bile	Stomach ache	Bile is taken orally to relieve stomach pain
Labeo rohita	Gall bladder (Bile)	Gastric problem, gastric ulcer, intestinal cancer	The fish is ground with water and consumed orally
Mystus bleekeri	Whole body	Dysentery	Fish is boiled with *Portulaca oleracea*, and curry taken 2 times daily for 3 days
Mystus sp.	Whole fish	Small pox	Cooked fish is consumed to cure small pox
Notopterus notopterus	Whole fish	Delivery pain, abdominal pain	The burned fish is cooked with mustard oil and black pepper and consumed
Osteobrama cotio cotio	Whole body	Ringworm infection	Sundried fish is crushed into a powder and consumed
Puntius sp.	Whole body	Eye problem	Fish is cooked and consumed
		Blood purifier	Fish is cooked with bamboo shoots and taken orally to purify the blood
		Common cold	Fish is cooked with chilly and consumed
	Head	Night blindness	The cooked head part of fish is taken regularly
		Memory	A boiled head of the fish is taken to increase memory
Wallago attu	Head	Liver tonic	Boiled head is consumed regularly to improve liver function

(continued)

Table 14 (continued)

Species	Part used	Medicinal use for curing diseases	Application
	Barbels	Treatment of diarrhoea	Roasted barbel is mixed with powdered tender leaves of *Psidium guajava*
Xenentodon cancila	Whole fish	Joint pain, swelling	Cooked and consumed. Spine and bone is used to prick out the clotted blood

Source Gogoi and Bora (2020), Borah and Prasad (2017), Borah and Prasad (2016), Gupta and Dey (2017), Teronpi et al. (2012), Chanu et al. (2016), Prakash and Prakash (2021)

(IMC) *Labeo rohita*, *Catla catla*, *Cirrhinus mrigala*, and the exotic carps *Cyprinus carpio*, *Ctenopharyngodon idella*, and *Hypothalmichthys molitrix* with high market value fetches good revenue to fishermen in India (Dasgupta and Panigrahi 2014). Fishing tourism (relating to fishing activity for tourist and educational services) helps local development by integrating socio-economic and cultural activities. Hence, sustainable management of resources, with habitat maintenance/conservation, aids in sustaining the fish species, which provides employment and income to the dependent fishing communities (Tursi et al. 2015). In the present study, the value of the provisioning service provided by fish from wetlands in Karnataka accounts to 32,175 Rs/ha/yr.

Fish in wetlands provide income and employment to society (Table 15). The revenue from fish products from the Sundarbans amounted to US$ 158,368 in 2014–2015 (Islam and Hossain 2017). The reservoirs of small and medium categories had a fish yield of 200 kg/ha/yr, and those under the large category had a fish yield of 100 kg/ha/yr (Rao et al. 2002). Stocking of fingerlings and adopting improved management practices in reservoirs (Sarkar et al. 2018) would enhance fish production. Overexploitation, including killing juvenile and brood fish in floodplain wetland fisheries (Sarkar and Borah 2018), has been a major factor responsible for the depletion of fish stock in wetlands.

13 Ecosystem Services Provided by Macrophyte

Aquatic plants are generally referred to as macrophytes, which are either emergent, submergent, or floating types. The ecological significance of macrophytes includes providing protective cover for fish, acting as substrate for aquatic invertebrates, producing oxygen, and serving as food for some fish and other wildlife.

Macrophyte provides numerous ecosystem services to society, including food production, medicine, water purification/treatment, carbon sequestration, and recreational opportunities. The charophytes (Chara, Nitella, and Nitellopsis) play a vital role in maintaining a clear-water state (Poikane et al. 2018). *Ceratophyllum demersum* and *Potamogeton pectinatus* are dominant in nutrient-enriched water (Stefanidis et al. 2019). Submerged macrophytes provide food and refugia for aquatic fauna ranging from microscopic crustaceans to large fish (Gopal 2016). Symbiotic association between macrophyte (Azolla) and algae belonging to cyanobacteria (*Anabaena* sp.) aid in fixing atmospheric nitrogen (Slembrouck et al. 2018). Macrophytes (Fig. 14) provide (i) food for other aquatic organisms, (ii) fodder for livestock, (iii) medicine for treating animal and human diseases, (iv) fiber, (v) green manure, (vi) industrial raw materials (manufacture of essential oils), and (vii) pesticide for killing pests/insects, and some species are useful as ornamental plants (Zhang et al. 2014). Macrophytes like *Alternanthera sessilis*, *Eleocharis dulcis*, and tubers of *Colocasia esculenta* are used as vegetables, while *Fimbristylis dichotoma*, *Cyperus iria*, and *C. pangorie* are used for making mats (Rao et al. 2008). Table 16 lists the commercially important aquatic macrophytes with use values.

Table 15 Economic value of fishing from wetlands

Water body	Area (ha)	Value (USD/yr)	References
Kuttanad wetlands, Kerala, India	162,125	82,949,309	Rao (2018)
Bhoj wetland, Bhopal, India	3229	113,103.4	Verma et al. (2001)
Wular lake, Jammu & Kashmir, India	13,000	686,166.7	Kaul et al. (2016)
Nakivubo Urban wetland, Uganda	529	3300	Schuijt (2002)
Hadejia Nguru wetland, Nigeria	350,000	3,465,100	
Chilwa wetland, Malawi	240,000	18,675,500	
Zambezi basin wetland, Southern Africa	2,982,900	78,620,700	
Chilwa wetland, Malawi	240,000	18,675,478	Zuze (2013)
That Luang wetland, Laos	2000	1,092,092	
Muthurajawela wetland, Sri Lanka	6000	64,904	
Mfolozi flood plain, South Africa	20,886	149,543.8	
Muthurajawela wetland, Sri Lanka	3068	64,904	Schuyt and Brander (2008)
Whangamarino wetland, New Zealand	10,320	10,518	
Rachenahalli lake, Bangalore, Karnataka, India	42.09	7449	Ramachandra et al. (2005)
Varthur lake, Bangalore, Karnataka, India	220	51,642	Ramachandra et al. (2011)
Pong Dam wetland, India	15,500	54,160	Prasher et al. (2006)
Ulluru tank, Karnataka, India	3.99	4384	Ramachandra and Sreekantha (2006)
Kaspadi tank, Karnataka, India	1.7	1047	
Nagara tank, Karnataka, India	3.58	2835	

Grasses like *Saccharum spontaneum* and *Sporobolus maderaspatanus* aid in mitigating floods and soil erosion. Macrophytes are used as human food, fodder (*Cynodon dactylon*), carpet grass (*Axonopus compressus*), medicine, roofing materials, and as raw materials for making household products like mats, ropes, toys, and hats (Kalita 2019; Sukumaran and Jeeva 2011). Some species of Nymphoides are used for treating injuries, jaundice, dysentery, convulsions, and snake bites (Khan et al. 2018b). Aquatic plants store carbon and enhance the carbon sequestration capacity of wetlands. The total carbon sequestered by the twelve macrophytes of Hokersar wetland was 0.85 kg C/m^2/yr, which varied with season and habitat (Lolu et al. 2019). Table 17 shows the beneficial uses of different species of macrophyte. In the present study, the value of the provisioning service provided by macrophytes in the wetlands of Karnataka was 11,291 Rs/ha/yr (Table 17).

Fig. 14 Commercially important aquatic macrophytes

Table 16 Value of commercially important aquatic macrophytes (INR/kg)

Macrophytes – Scientific name	Common name	INR/kg
Euryale ferox	Prickly waterlily	300
Trapa natans	Water chestnut	170
Colocasia esculenta	Taro/elephant ear	30
Nelumbo nucifera	Lotus	230
Centella asiatc	Indian pennywort	250
Ipomoea aquatc	Water spinach	25
Acorus calamus	Sweet flag	72
Bacopa monnieri	Brahmi	150
Eclipta alba	Bhringraj	150

Source Raut et al. (2020)

Table 17 Benefits of macrophytes

Species	Family	Uses
Alternanthera philoxeroides	Amaranthaceae	Consumed as a vegetable, used as animal feed. The juice of younger twigs is consumed to treat worm problems, and its paste is applied to cattle wounds
Azolla pinnata	Azollaceae	Used to treat urinary problems
Bacopa monnieri	Plantaginaceae	Its juice is used to cure fevers
Ceratophyllum demersum	Ceratophyllaceae	Use to relieve pain from insect and scorpion bites; fruit is fed to domestic ducks
Cynodon dactylon	Poaceae	It is a good source of food for aquatic molluscans. This plant is used to cure skin diseases. People utilizes this plant for rituals. Juice is used in the treatment of piles and on cuts and wounds
Cyperus corymbosus	Cyperaceae	Its stem is used for making mats
Cyperus iria and *Cyperus difformis*	Cyperaceae	The plant is a source of fiber, fuel, and food
Eichhornia crassipes	Pontederiaceae	It is used as a fertilizer, to reduce oil in lake water, and as manure in potato fields
Euryale ferox	Nymphaeaceae	Used as a vegetable, it has medicinal value
Hydrilla verticillata	Hydrocharitaceae	Used as fish food, aquarium plants, and fertilizer. It is used by fish farmers to breed *Cyprinus carpio* since their eggs are attached to this plant
Ipomoea aquatica	Convolvulaceae	Its young leaves and stems are used as vegetables and serve as animal feed
Ludwigia adscendens	Onagraceae	Aerial parts are used to make poultices for treating skin problems. Its shoots are used as vegetables
Ludwigia perennis	Onagraceae	The plant is boiled with coconut oil and applied externally to reduce fever
Monochoria vaginalis	Pontederiaceae	Used as cattle feed, liquid extracts from the plant improved the seed viability rate and increased the seedling vigour of some crops
Nymphaea pubescens	Nymphaeaceae	Young stems and unopened flower buds can be boiled and served as a vegetable
Nymphaea rubra	Nymphaeaceae	Flowers are used for religious purposes. Its rhizomes are used for curing dyspepsia and dysentery. It is also cultivated as an ornamental plant
Nymphaea stellata	Nymphaeaceae	Used for the treatment of diabetes, liver disorders, inflammation, urinary disorders, menstruation problem
Nymphoides indica	Menyanthaceae	Used to reduce fevers, its young leaves, stems, fruits, and unopened flower buds can be boiled and served as a vegetable

(continued)

Table 17 (continued)

Species	Family	Uses
Pistia stratiotes	Araceae	Its root is used as a pain killer for hemorrhoids. Its leaf juice is boiled with coconut oil and used for treating skin diseases. Its juice mixed with sugar is recommended for curing asthma and cough
Polygonum barbatum	Polygonaceae	Roots and seeds are used for medicinal purposes
Polygonum hydropiper	Polygonaceae	Medicinal value
Salvinia cucullata	Salviniaceae	Used as fodder
Spirodela polyrrhiza	Araceae	Plants are used as organic manures with high nitrogen content (6–7%); used as fodder for cattle; plants are also consumed by herbivorous fish, ducks, goose, and swans; a source of animal feed due to high protein content; used for wastewater purification (phytoremediation)
Trapanatans var. bispinosa	Trapaceae	Used as edible fruit, raw fruits are used for treating diarrhoea and dyspepsia
Utricularia aurea	Lentibulariaceae	Edible as it is rich in nutrients and is used to treat urinary diseases
Vallisneria spiralis	Hydrocharitaceae	Used as fish food, it helps to clean the water and is used for ornamental purposes
Wolffia globosa	Araceae	Leaves are cooked as vegetables

Source Kiran et al. (2007), Taran and Deb (2020), Nandan and Singh (2004), Sarmah et al. (2013), Misra et al. (2012), Sarma and Saikia (2010), Mishra et al. (2016), Chai et al. (2015), Banerjee et al. (2012)

14 Ecosystem Services from Wetlands in Karnataka

Wetlands play several vital roles, including the cycling of water, carbon, and nutrients; food and fiber production; water purification; regulation of flows; habitat for aquatic organisms; and tourism and recreation services (Fig. 15).

The spatial distribution of wetlands in Karnataka is provided in Fig. 16. The spatial analyses reveal that about 61% of wetlands had an area of <2.5 ha, 36% of wetlands had an area of 2.5–50 ha, and 3% of wetlands had an area of >50 ha.

Accounting of provisioning services shows that the major share is from microalgae (57%), fish (17%), water provision (8%), raw materials (7%), macrophytes (6%), and other services (5%). Other services in the provisioning service include genetic resources, medicinal resources, ornamental resources, and fuelwood. The percentage composition of each service under the regulating service is carbon sequestration (33%), groundwater recharge (31%), moderation of extreme events (15%), other services (10%), regulation of water flows (4%), erosion prevention (3%), and maintenance of genetic diversity (4%). Other services include air quality regulation, climate

Fig. 15 Wetland ecosystem services in Karnataka

Crop cultivation near wetland | Temple near wetland

Performing religious rituals/pooja | Recreation

Lake view apartments (Hedonic values) | Habitat for migratory birds

Recreation | Ornamental use

Wetland survey | Wetland research

Fig. 15 (continued)

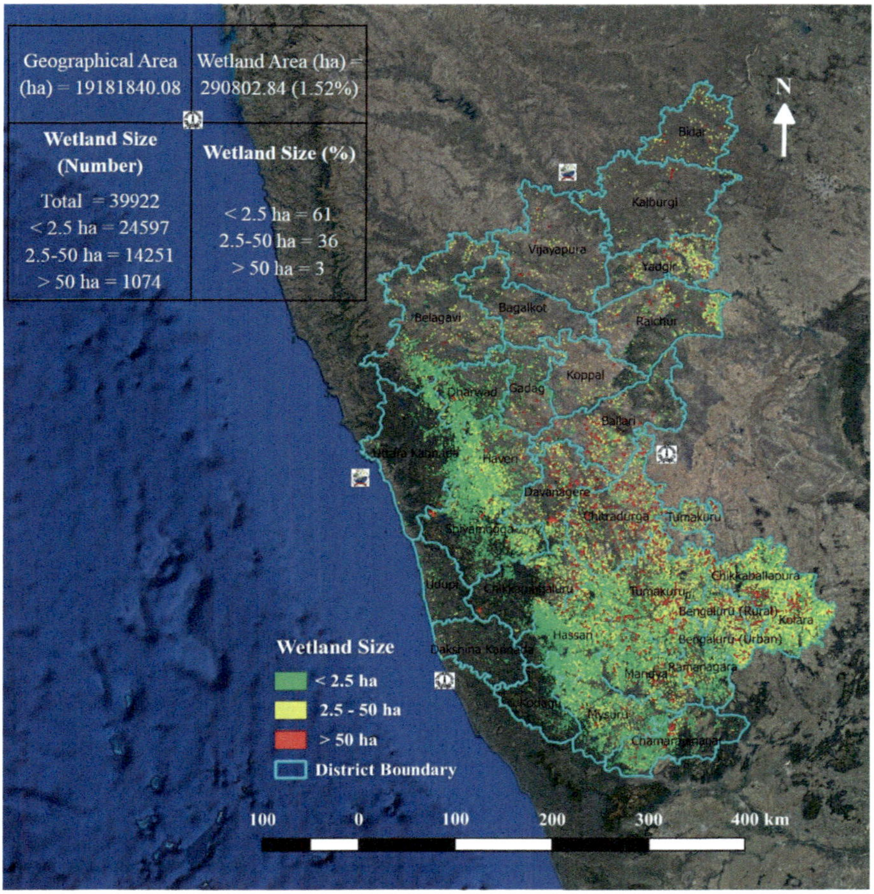

Fig. 16 Classification of wetlands of Karnataka based on size

regulation, waste treatment/water purification, maintenance of soil fertility, biological control, maintenance of life cycles of migratory species, waterborne disease control, pollination, habitat or refugia, water conservation, and flood control. In the case of cultural services, existence, and bequest values (68%), have the highest contribution, followed by education (21%), recreation and tourism (9%), and other services (2%). Other services in cultural service include aesthetic information, inspiration for culture, art, and design, spiritual experience, and information for cognitive development (Fig. 17).

Figure 18 illustrates the share of ecosystem services—provisioning, regulating, and cultural services contributed about 18%, 69%, and 13% to the total ecosystem supply value.

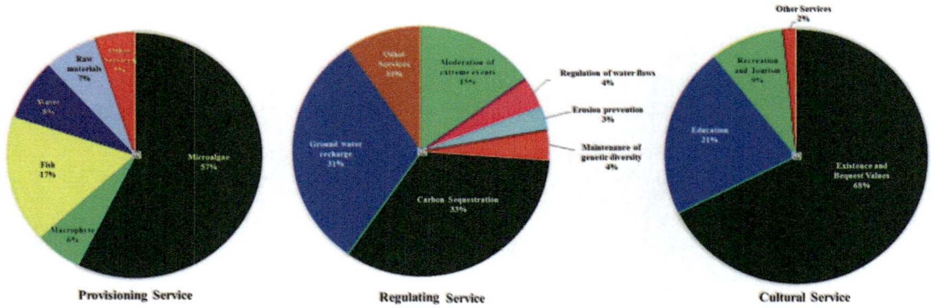

Fig. 17 Contribution of different freshwater wetland ecosystem services

Fig. 18 Contribution of ecosystem services provided by freshwater wetlands of Karnataka

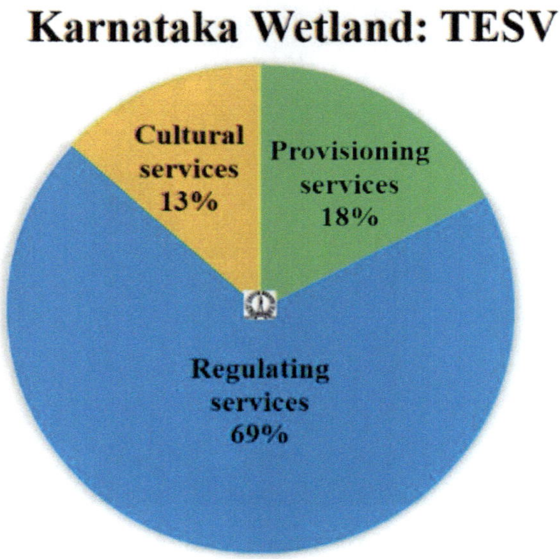

The value of provisioning, and regulating services provided by freshwater wetlands in Karnataka was 49.70 and 196.89 billion Rs/yr, respectively. Figures 19 and 20 depict the district-wise share of provisioning and regulating services, respectively. The value of provisioning services for the Ga-Mampa wetland is about $90,000 per year (Adekola et al. 2008). The value of Satajan wetland and bird sanctuary provisioning services in Assam was Rs 52,65,600 or $78,591 when wild edible plants, fodder, firewood, fish, fresh water, and wild medicinal plants were considered (Kakoti et al. 2019). The total provisioning value of the Kalyanthakur para lake considering aquatic plants, fishes, molluscs, crabs, domestic uses, collection of timber, and fuelwood was US$ 26,263.65/year, while total cultural value accounted to US$ 2605.68/year (Taran et al. 2016). The average net monetary value of wetlands provisioning services when material collection, fishing, crop production, hunting, and logging are considered is $11,508 per household per year (Adekola et al. 2015).

Recreational benefits from the wetlands of Karnataka include scenic beauty, swimming, boating, botanical gardening, an amusement park, and picnic spots. For example, Lalbagh Lake, situated in the famous Lalbagh Botanical Gardens of south Bangalore, is a prime attraction for nature lovers, bird watchers, and tourists as it supports a variety of flora and fauna. Thonnur Lake in Mysore is an attractive spot for bird-watching, boating, and swimming. A beautiful butterfly park has been built over a small island within the lake. The Regional Museum of Natural History, which is also located on the banks of Karanji Lake, provides information on the natural

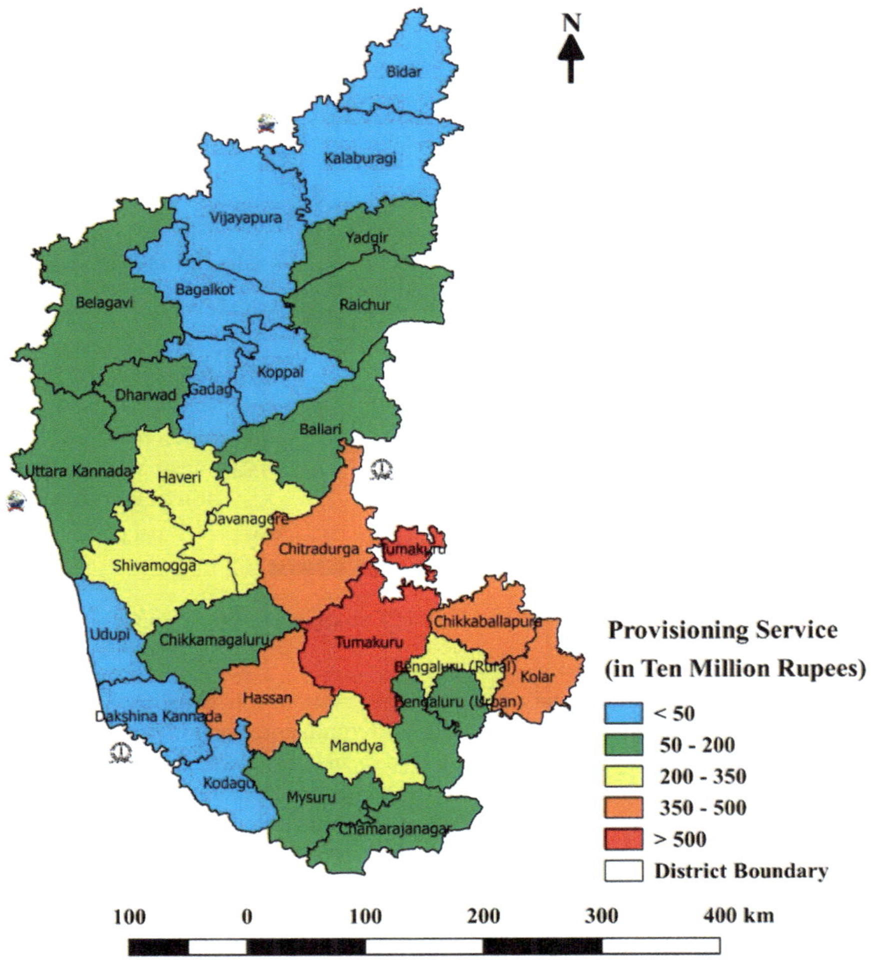

Fig. 19 Provisioning service of freshwater wetlands of Karnataka

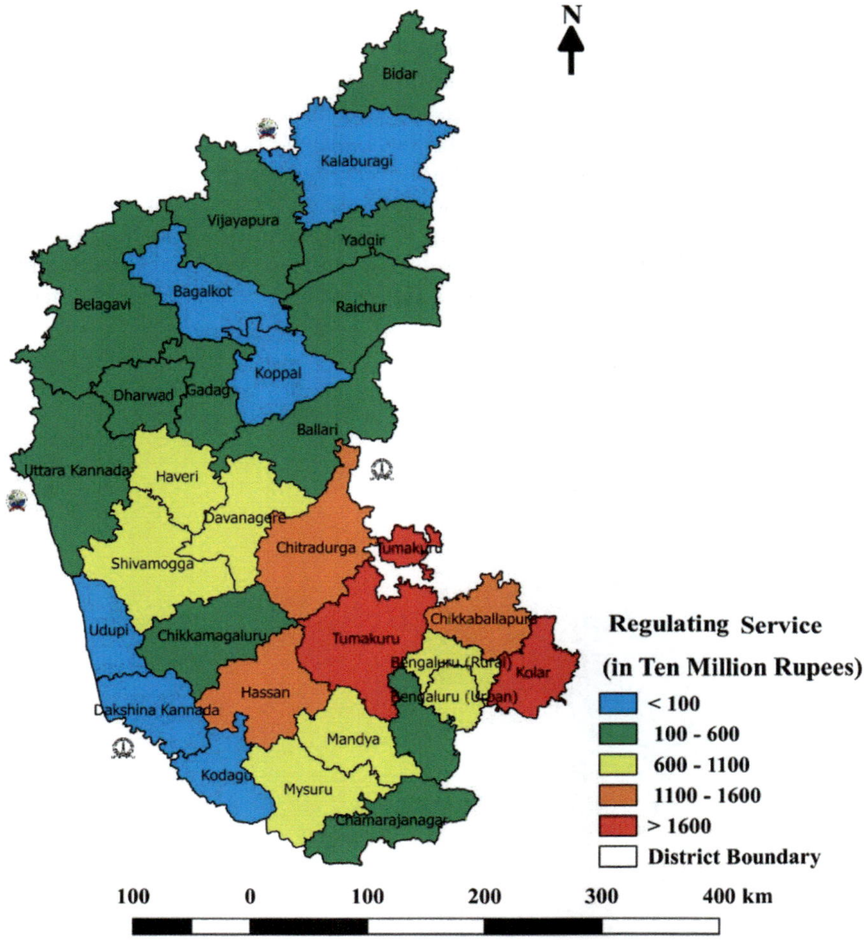

Fig. 20 Regulating service of freshwater wetlands of Karnataka

environment of South India and on nature conservation. Lakes in Karnataka have cultural significance, which is evident from cultural practices at Pampa Sarovar, a sacred lake in the Koppal district near Hampi in Karnataka. A special pooja is conducted in the Honnamana kere (Honnama lake) in Kodagu district during the Gowri festival, during which people offer bagina (puja items along with flowers and bangles placed in a bamboo sieve) to the lake and visitors pray for good rain. Ganesh Chaturthi is mainly celebrated in Karnataka, and the practice involves immersing idols (Ganesha) in Kalyani near the lakes. Visitors come to the lakes during festivals such as Durga puja, Jagadhatri puja, Lakshmi puja, and Ganesh Chaturthi to perform puja (Bhattacharya et al. 2014; Bengani et al. 2020).

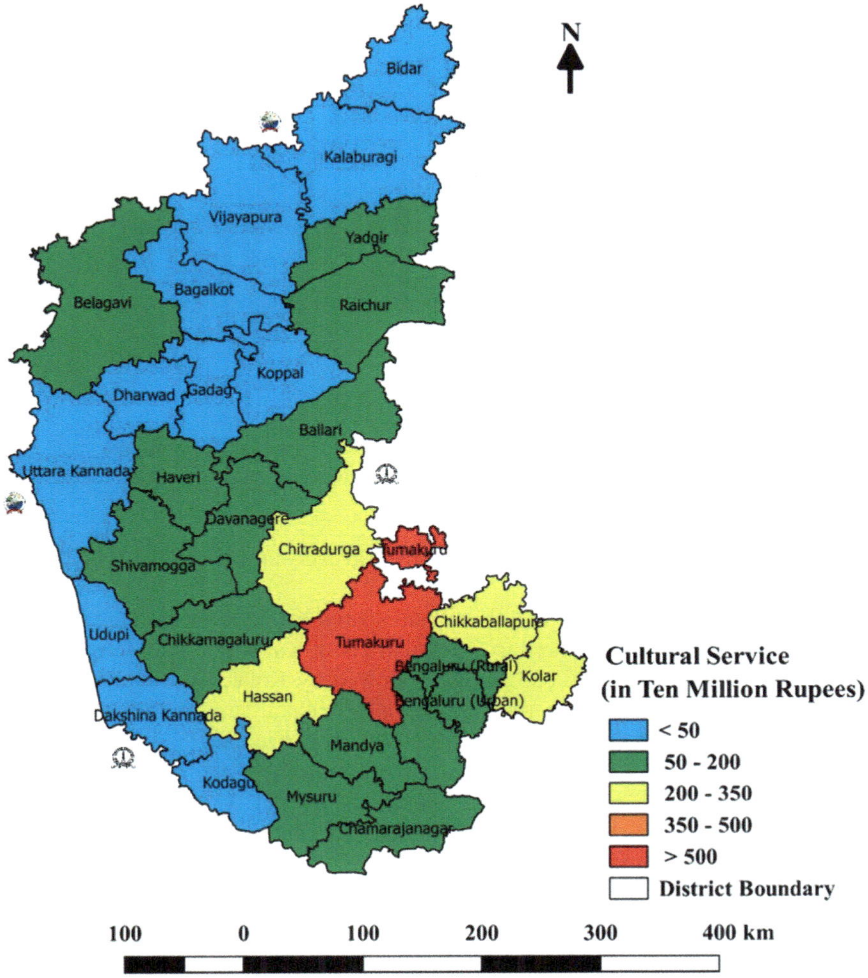

Fig. 21 Cultural service of freshwater wetlands of Karnataka

Cultural services provided by freshwater wetlands in Karnataka amount to 37.93 billion Rs/yr (Fig. 21). The annual economic value of the cultural ecosystem service of the Pateira de Fermentelos freshwater wetland in Portugal is estimated at 3087 €/ha/yr (Roebeling et al. 2016). About 90% of people are willing to pay (WTP) for recreational activities in the Kanibrazan wetland, with an average estimate of 38,217 Rials per person (Zarandi et al. 2019). In 2014–2015, the revenue from tourism (96,949 native and 3868 foreign tourists) amounted to US$ 144,832 from the Sundarbans (Islam and Hossain 2017). The average WTP for the recreational

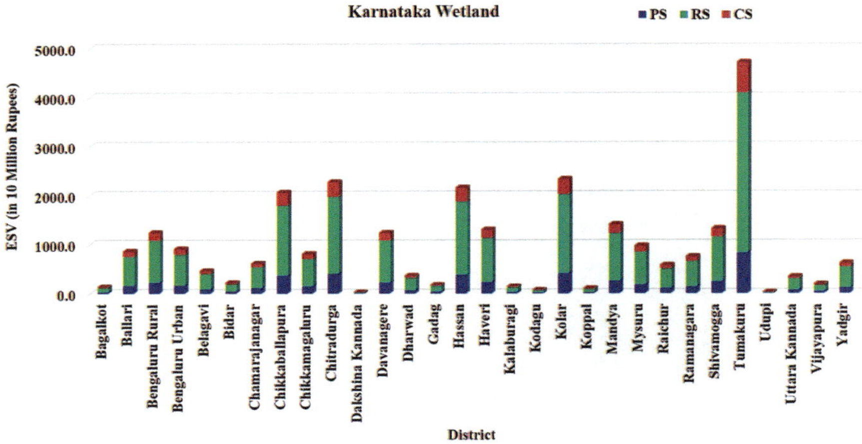

Fig. 22 District-wise ecosystem supply value of freshwater wetlands of Karnataka

benefits provided by Pilikula Lake through the travel cost method (TCM) is 238 INR (Jala and Nandagiri 2015).

Tumakuru district has the largest wetland spatial extent with the highest values for ecosystem services compared to other districts. It was found that Tumakuru has the highest wetland area (46,633.36 ha), followed by Kolar, Chitradurga, Hassan, Chikkaballapura, Mandya, Shivamogga, etc. In Karnataka, Tumakuru district had the highest ecosystem service potential, whereas Udupi district had the lowest values of ecosystem services (Fig. 22). The value of provisioning, regulating, and cultural services provided by wetlands in the Tumakuru district was 8214, 32,641, and 6287 million Rs/yr, respectively. Udupi district had the lowest values for wetland ecosystem services compared to other districts in Karnataka because of its smaller wetland area (223.63 ha). The value of provisioning, regulating, and cultural services provided by wetlands in the Udupi district was 40, 157, and 30 million Rs/yr, respectively.

15 Total Ecosystem Supply Value (TESV) and Net Present Value (NPV) of Freshwater Wetlands in Karnataka

Total ecosystem supply value is the summation of provisioning, regulating, and cultural services provided by the freshwater wetland in Karnataka. The total ecosystem supply value (TESV) of freshwater wetlands in Karnataka was 284.52 billion Rs/yr, with the highest value recorded in Tumakuru district at 47.14 billion Rs/yr (Fig. 23). Total ecosystem supply value differs with the various ecosystem

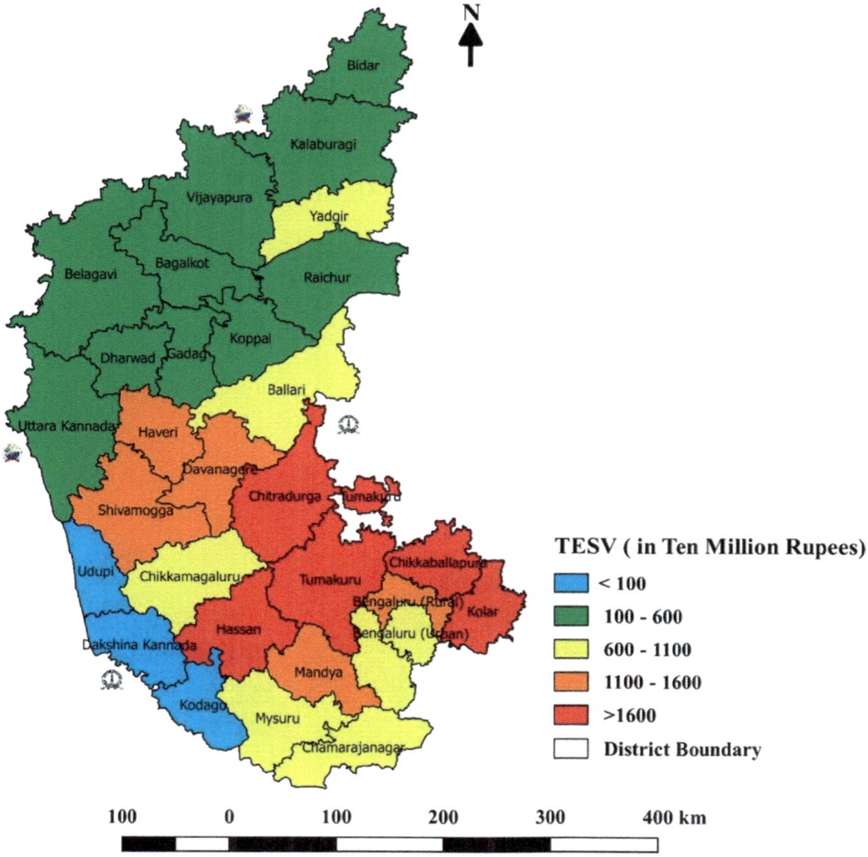

Fig. 23 TESV of freshwater wetlands of Karnataka

services considered and the ecosystem conditions of each wetland. The annual economic value of Ghodaghodi wetland is estimated to be about USD 6,74,000, with the highest contribution from use value (96%) compared to the non-use value (4%) of the wetland (Aryal et al. 2021). Overall, wetlands dominated by vegetation (13,852 $/ha/yr) and open water (4,042 $/ha/yr) had a total ecosystem service value of 17,894 $/ha/yr (Patterson et al. 2018). The average total economic value for food, water provisioning, and wastewater treatment in the Boeng Cheung Aek wetland is 30.12 million USD/yr, ranging from 15.71 to 48.96 million USD/yr (Ro et al. 2020).

NPV of freshwater wetlands in Karnataka is 7320.6 billion rupees. The highest NPV value was recorded in the Tumakuru district, with 1212.9 billion rupees. NPV ranged from 450 to 650 billion rupees in districts like Chitradurga, Chikkaballapura,

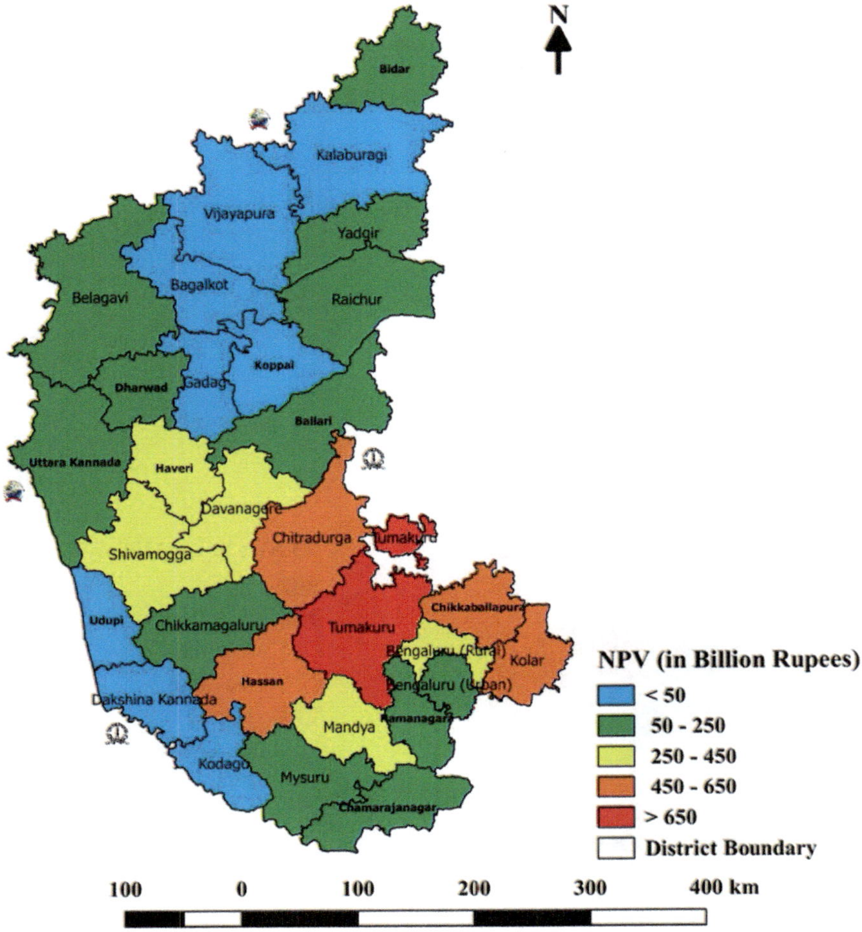

Fig. 24 NPV of freshwater wetlands of Karnataka

Hassan, and Kolar, whereas it ranged from 250 to 450 billion rupees in Davanagere, Haveri, and Shivamogga (Fig. 24). In earlier studies, the net present value of revenue (benefits) earned from Sukhna Lake in Chandigarh city during the last ten years was estimated at Rs. 451 million (Chaudhry et al. 2013). The net present value of three mangrove ecosystems in Thailand ranges from 10,158 to 12,392 US$/ha (Barbier 2011).

Table 18 Total ecosystem supply value of Karnataka freshwater wetlands

Services	Details	Value
	Wetland: Total area (ha) based on grid	281,299.5
Provisioning service	Total Rs/yr (in billion Rupees)	49.70
	Production Rs/ha/yr (in Lakhs)	1.8
	% distribution	18
Regulating service	Total Rs/yr (in billion Rupees)	196.89
	Production Rs/ha/yr (in Lakhs)	7
	% distribution	69
Cultural service	Total Rs/yr (in billion Rupees)	37.93
	Production Rs/ha/yr (in Lakhs)	1.3
	% distribution	13
TESV	Total Rs/yr (in billion Rupees)	284.52
	Production Rs/ha/yr (in Lakhs)	10.1
NPV	NPV in Billion Rupees	7320.6

Wetlands are fundamental to the economic, social, and cultural well-being of the population in India. The total benefit from wetland fisheries, fodder collection, and jute retting was 5.51 lakh Rs/yr, 2.23 lakh Rs/yr, and 1.24 lakh Rs/yr, respectively, from Kalobaur beel (Mukherjee 2008). In the case of Koshi Tappu Wildlife Reserve, the total net benefit value from wetland fodder was estimated at USD 42,51,919 (Sharma et al. 2015). In Karnataka, wetlands cover an area of 281,299.5 ha. Karnataka freshwater wetlands provided a provisioning service worth 1.8 lakh rupees per hectare per year, a regulating service worth 7 lakh rupees per hectare per year, and a cultural service worth 1.3 lakh rupees per hectare per year. The TESV of Karnataka freshwater wetlands is 284.52 billion Rs/yr, and the NPV amounted to 7320.6 billion rupees (Table 18). Wetland restoration would improve wetland habitats, the wetland ecosystem's functioning and biodiversity, and would also improve water quality (Ramachandra et al. 2020b).

16 Estuarine Ecosystem in Karnataka

16.1 Provisioning Services

Provisioning services from the estuarine ecosystem include fishery (fish, clam, oyster, and mussel), agricultural products (saline paddy and coconut), mining products (sand), mangrove resources, and salt production. The residual value method is used for valuing provisioning services.

Estuarine Fishery

The fishery sector provides the major livelihood opportunities of the estuarine-dependent communities in the coastal villages. It includes the common estuarine fishes, clams, oysters, mussels, etc. The market price of fishery resources and the quantity obtained for each category of fish resources are given in Table 19.

The annual income generated from the utilization of estuarine fish resources (fishes, clams, oysters, and mussels) from Mulki—Pavanje estuary is 18,61,52,000 Rs/year. The annual income from estuarine fish resources is 42,18,32,000 Rs/year in the Netravathi estuary, 18,51,58,000 Rs/year in Udyavara estuary, 36,35,10,600 Rs/year in Sita estuary, 56,71,76,800 Rs/year in Haladi—Chakra—Kollur estuary, 33,48,97,140 Rs/year in Kali estuary, 5,00,01,740 Rs/year in Gangavali estuary, 2,43,31,56,920 Rs/year in Aghanashini estuary, 45,65,09,200 Rs/year in Sharavathi estuary, and 1,48,30,040 Rs/year in Venkatapura estuary (Fig. 25).

Agricultural Products

The estuarine belt of Dakshina Kannada, Udupi, and Uttara Kannada districts supports saline-tolerant paddy (gazani) and coconut cultivation. The market price and total quantity of coconut and paddy production are given in Table 20. The annual income provided from coconut plantations is 7,84,350 Rs/year in the Mulki-Pavanje estuary, 19,98,640 Rs/year in the Netravathi estuary, 8,35,200 Rs/year in Udyavara estuary, 10,23,120 Rs/year in Sita estuary, and 26,62,200 Rs/year in Haladi-Chakra-Kollur estuary. The annual income from coconut and paddy plantations is 1,50,07,500 Rs/year in Kali estuary, 51,68,430 Rs/year in Gangavali estuary, 1,83,09,420 Rs/year in Aghanashini estuary, 80,80,560 Rs/year in the Sharavathi estuarine belt, and 10,36,264 Rs/year in Venkatapura estuary. Variations in the ecosystem benefits from paddy and coconut plantations are depicted in Fig. 26.

Mining Product

Mining and dredging activities are prevalent at a significant level in the estuaries of Dakshina Kannada, Udupi, and Uttara Kannada districts. The quantity and price of mining products from estuaries are given in Table 21.

Mining provides an annual income of 540 Rs/year in Mulki—Pavanje estuary, 1,382 Rs/year in Netravathi estuary, 598 Rs/year in Udyavara estuary, 764 Rs/year in Sita estuary, 1,989 Rs/year in Chakra—Haladi—Kolluru estuary, 7,896 Rs/year in Kali estuary, 532 Rs/year in Gangavali estuary, 476 Rs/year in Aghanashini estuary, 672 Rs/year in Sharavathi estuary, and 106 Rs/year in Venkatapura estuary (Fig. 27).

Table 19 Estuarine fisheries

Estuary	Area of Estuary (ha)	Resource	Total fish catch (ton)	Price/Ton	Cost/Ton	Residual value (Price - Cost)	Total benefit Rs/year	Total benefits Rs/ha/year
Mulki-Pavanje estuary	750	Clam	304	2,40,000	1,44,000	96,000	29,184,000	38,912
		Oyster	36	1,40,000	84,000	56,000	2,016,000	2688
		Mussel	6	4,80,000	2,88,000	1,92,000	1,152,000	1536
		Fish	769	5,00,000	3,00,000.00	2,00,000.00	153,800,000	205,066.6667
		Total					186,152,000	248,202.6667
Netravathi estuary	1920	Clam	292	2,40,000	1,44,000	96,000	28,032,000	14,600
		Oyster	0	1,40,000	84,000	56,000	0	0
		Mussel	0	4,80,000	2,88,000	1,92,000	0	0
		Fish	1969	5,00,000	3,00,000.00	2,00,000.00	393,800,000	205,104.1667
		Total					421,832,000	219,704.1667
Udyavara estuary	800	Clam	195	2,45,000	1,47,000	98,000	19,110,000	23,887.5
		Oyster	24	1,45,000	87,000	58,000	1,392,000	1740
		Mussel	0	4,85,000	2,91,000	1,94,000	0	0
		Fish	820	5,02,000.00	3,01,200	2,00,800.00	164,656,000	205,820
		Total					185,158,000	231,447.5
Sita estuary	980	Clam	1565	2,45,000	1,47,000	98,000	153,370,000	156,500
		Oyster	124	1,45,000	87,000	58,000	7,192,000	7338.77551
		Mussel	5.9	4,85,000	2,91,000	1,94,000	1,144,600	1167.959184
		Fish	1005	5,02,000.00	3,01,200	2,00,800.00	201,804,000	205,922.449
		Total					363,510,600	370,929.1837

(continued)

Table 19 (continued)

Estuary	Area of Estuary (ha)	Resource	Total fish catch (ton)	Price/Ton	Cost/Ton	Residual value (Price - Cost)	Total benefit Rs/year	Total benefits Rs/ha/year
Haladi-Chakra-Kollur estuary	2550	Clam	383	2,45,000	1,47,000	98,000	37,534,000	14,719.21569
		Oyster	75	1,45,000	87,000	58,000	4,350,000	1705.882353
		Mussel	0	4,85,000	2,91,000	1,94,000	0	0
		Fish	2616	5,02,000.00	3,01,200	2,00,800.00	525,292,800	205,997.1765
		Total					567,176,800	222,422.2745
Kali estuary	3240	Clam	55	2,50,000	1,50,000	1,00,000	55,00,000	1,697
		Oyster	0.459	1,50,000	90,000	60,000	27,540	8.5
		Mussel	20	4,90,000	2,94,000	1,96,000	39,20,000	1,209
		Fish	1624	5,01,000.00	3,00,600.00	2,00,400.00	32,54,49,600	1,00,447
		Total					33,48,97,140	1,03,363
Gangavali estuary	700	Clam	11	2,50,000	1,50,000	1,00,000	11,00,000	1,571
		Oyster	0.069	1,50,000	90,000	60,000	4,140	5
		Mussel	0	4,90,000	2,94,000	1,96,000	0	0
		Fish	244	5,01,000.00	3,00,600.00	2,00,400.00	4,88,97,600	69,853
		Total					5,00,01,740	71,431
Aghanashini estuary	4801	Clam	76	2,50,000	1,50,000	1,00,000	76,00,000	1,583
		Oyster	0.642	1,50,000	90,000	60,000	38,520	8
		Mussel	28	4,90,000	2,94,000	1,96,000	54,88,000	1,143
		Fish	12,076	5,01,000.00	3,00,600.00	2,00,400.00	2,42,00,30,400	5,04,067
		Total					2,43,31,56,920	5,06,802

(continued)

Table 19 (continued)

Estuary	Area of Estuary (ha)	Resource	Total fish catch (ton)	Price/Ton	Cost/Ton	Residual value (Price - Cost)	Total benefit Rs/ year	Total benefits Rs/ ha/year
Sharavathi estuary	1600	Clam	10	2,50,000	1,50,000	1,00,000	10,00,000	625
		Oyster	0	1,50,000	90,000	60,000	0	0
		Mussel	0	4,90,000	2,94,000	1,96,000	0	0
		Fish	2273	5,01,000.00	3,00,600.00	2,00,400.00	45,55,09,200	2,84,693
		Total					45,65,09,200	2,85,318
Venkatapura estuary	200	Clam	2	2,50,000	1,50,000	1,00,000	200,000	1000
		Oyster	0.014	1,50,000	90,000	60,000	840	4
		Mussel	0	4,90,000	2,94,000	1,96,000	0	0
		Fish	73	5,01,000.00	3,00,600.00	2,00,400.00	1,46,29,200	73,146
		Total					1,48,30,040	74,150

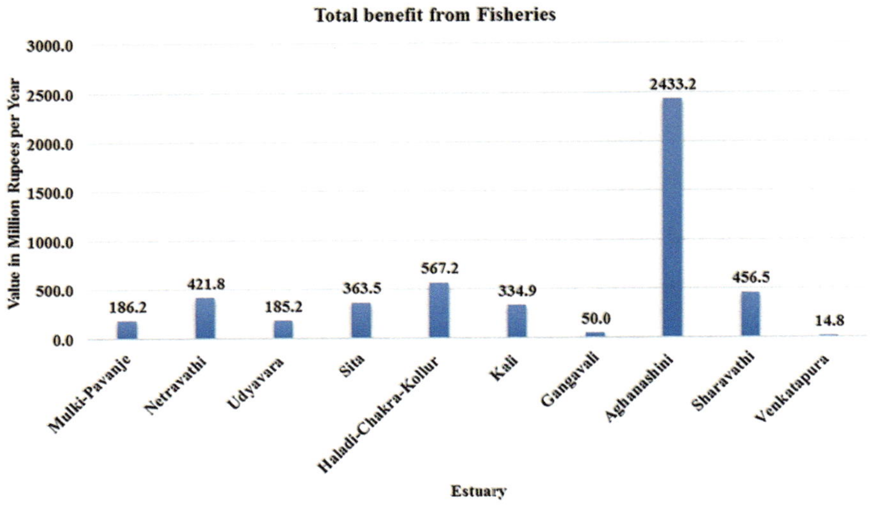

Fig. 25 Total benefit from fisheries in estuaries of Karnataka

Mangrove Products

The mangrove forest is being used for fodder, timber, firewood, and construction activities by the local inhabitants. The market price and quantity of mangrove resources are given in Table 22. The annual income generated from mangrove product extraction is 8,30,290.5 Rs/year in Mulki-Pavanje estuary, 21,25,891.2 Rs/year in Netravathi estuary, 8,85,788 Rs/year in Udyavara estuary, 10,85,090.3 Rs/year in Sita estuary, 2,823,328.75 Rs/year in Haladi-Chakra-Kollur estuary, 2,090,780 Rs/year in Kali estuary, 2160 Rs/year in Gangavali estuary, 2,002,340 Rs/year in Aghanashini estuary, 21,02,180 Rs/year in Sharavathi estuary, and 4,14,780 Rs/year in Venkatapura estuary (Fig. 28).

Salt Production

The market price and quantity of salt production are given in Table 23. The annual income generated by salt production in the Aghanashini estuary is 31,20,000 Rs/year.

Total Provisioning Service

Table 24 lists estuary-wise provisioning services and the total provisioning service amounts to 18,77,67,180 Rs/year in Mulki—Pavanje estuary, 42,59,57,913 Rs/year in Netravathi estuary, 18,68,79,586 Rs/year in Udyavara estuary, 36,56,19,574 Rs/year in Sita estuary, 57,26,64,317 Rs/year in Haladi—Chakra—Kollur estuary,

Table 20 Paddy and coconut from estuaries in Karnataka west coast

Estuary	Area of estuary (ha)	Area of coconut plantation (ha)	Annual production (tons)	Price/ Ton	Cost/ Ton	Residual value (Price - Cost)	Total benefit Rs/ yr
Mulki-pavanje	750	20.23	67.5	29,050	17,430	11,620	7,84,350
Netravathi	1920	16.9	172	29,050	17,430	11,620	19,98,640
Udyavara	800	15.7	72	29,000	17,400	11,600	8,35,200
Sita	980	26.12	88.2	29,000	17,400	11,600	10,23,120
Haladi-Chakra-Kollur	2550	49.5	229.5	29,000	17,400	11,600	26,62,200
Kali	Gazani	3240	2,036	18,000	10,800	7,200	1,46,59,200
	Coconut		30	29,025	17,415	11,610	348,300
	Total						1,50,07,500
Gangavali	Gazani	700	713	18,000	10,800	7,200	51,33,600
	Coconut		3	29,025	17,415	11,610	34,830
	Total						51,68,430
Aghanashini	Gazani	4801	2443	18,000	10,800	7,200	1,75,89,600
	Coconut		62	29,025	17,415	11,610	7,19,820
	Total						1,83,09,420
Sharavathi	Gazani	1600	0	18,000	10,800	7,200	0
	Coconut		696	29,025	17,415	11,610	80,80,560
	Total						80,80,560
Venkatapura	Gazani	200	143	18,000	10,800	7,200	10,29,600
	Coconut		0.574	29,025	17,415	11,610	6,664
	Total						10,36,264

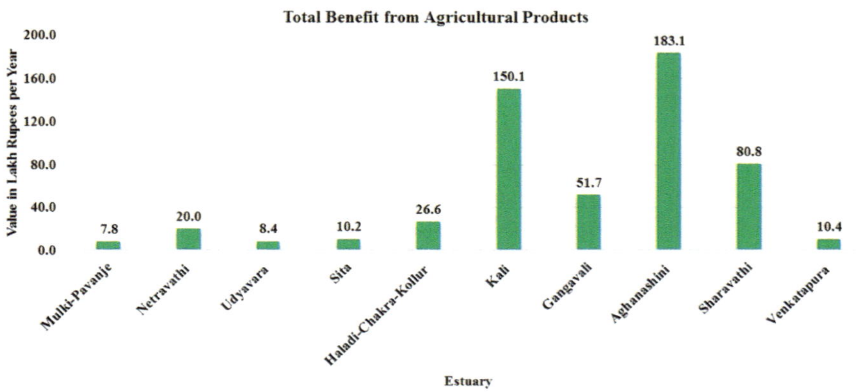

Fig. 26 Total benefit from paddy and coconut products in estuaries of Karnataka

Table 21 Income generated from mining activities

Estuary	Area (ha)	Component	Annual production (in tonnes)	Price/ton	Cost/ton	Residual value (Price - Cost)	Total benefit Rs/year
Mulki-pavanje	750	Sand	22.5	60	36	24	540
Netravathi	1920	Sand	57.6	60	36	24	1,382
Udyavara	800	Sand	23	65	39	26	598
Sita	980	Sand	29.4	65	39	26	764
Haladi-Chakra-Kollur	2550	Sand	76.5	65	39	26	1,989
Kali	3240	Sand	282	70	42	28	7,896
Gangavali	700	Sand	19	70	42	28	532
Aghanashini	4801	Sand	17	70	42	28	476
Sharavathi	1600	Sand	24	70	42	28	672
Venkatapura	200	Sand	3.8	70	42	28	106

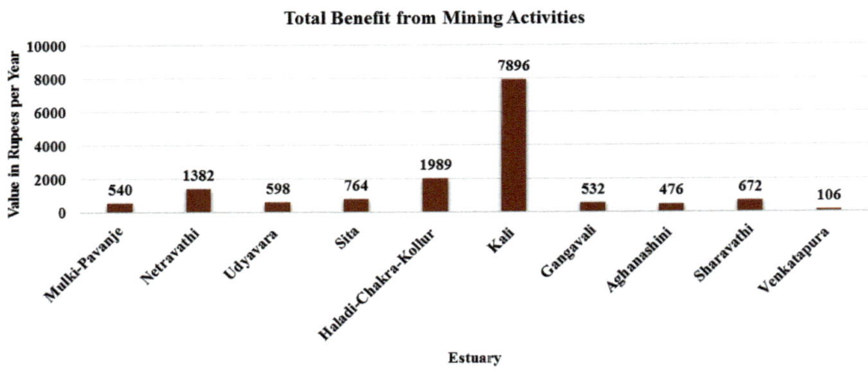

Fig. 27 Total benefit from mining activities in estuaries of Karnataka

35,20,03,316 Rs/year in Kali estuary, 5,51,72,862 Rs/year in Gangavali estuary, 2,45,65,89,156 Rs/year in Aghanashini estuary, 46,66,92,612 Rs/year in Sharavathi estuary, and 1,62,81,190 Rs/year in Venkatapura estuary.

Table 22 Estimation of annual income from mangrove product harvesting

Estuary	Area (ha)	Area of mangrove plantation (ha)	Mangrove product	Total mangrove production (in tons)	Price/ton	Cost/ton	Residual value (Price - Cost)	Total benefit Rs/ Year
Mulki-Pavanje estuary	750	16.03	Fodder	3337	602	361	241	804,217
			Timber	26	152	91	61	1586
			Charcoal	7.5	152	91	61	457.5
			Thatch	30	2,002	1,201	801	24,030
			Total					830,290.5
Netravati estuary	1920	4.8	Fodder	8544	602	361	241	2,059,104
			Timber	67.2	152	91	61	4099.2
			Charcoal	19.2	152	91	61	1171.2
			Thatch	76.8	2,002	1,201	801	61,516.8
			Total					2,125,891.2
Udyavara estuary	800	0.5	Fodder	3560	601	360	241	857,960
			Timber	28	151	90	61	1708
			Charcoal	8	151	90	61	488
			Thatch	32	2,001	1,200	801	25,632
			Total					885,788
Sita estuary	980	6.5	Fodder	4361	601	360	241	1,051,001
			Timber	34.3	151	90	61	2092.3
			Charcoal	9.8	151	90	61	597.8
			Thatch	39.2	2,001	1,200	801	31,399.2
			Total					1,085,090.3

(continued)

Table 22 (continued)

Estuary	Area (ha)	Area of mangrove plantation (ha)	Mangrove product	Total mangrove production (in tons)	Price/ton	Cost/ton	Residual value (Price - Cost)	Total benefit Rs/ Year
Haladi-Chakra-Kollur estuary	2550	10.2	Fodder	11,347	601	360	241	2,734,627
			Timber	89.25	151	90	61	5444.25
			Charcoal	25.5	151	90	61	1555.5
			Thatch	102	2,001	1,200	801	81,702
			Total					2,823,328.75
Kali estuary	3240	100	Fodder	7,920	600	360	240	19,00,800
			Timber	148	150	90	60	8,880
			Charcoal	45	150	90	60	2,700
			Thatch	223	2,000	1,200	800	178,400
			Total					2,090,780
Gangavali estuary	700	43.2	Fodder	0	600	360	240	0
			Timber	36	150	90	60	2160
			Charcoal	0	150	90	60	0
			Thatch	0	2,000	1,200	800	0
			Total					2160
Aghanashini estuary	4801	101.6	Fodder	7200	600	360	240	1,728,000
			Timber	215	150	90	60	12,900
			Charcoal	64	150	90	60	3840
			Thatch	322	2,000	1,200	800	2,57,600
			Total					2,002,340

(continued)

Table 22 (continued)

Estuary	Area (ha)	Area of mangrove plantation (ha)	Mangrove product	Total mangrove production (in tons)	Price/ton	Cost/ton	Residual value (Price - Cost)	Total benefit Rs/ Year
Sharavathi estuary	1600	5	Fodder	8640	600	360	240	20,73,600
			Timber	23	150	90	60	1,380
			Charcoal	0	150	90	60	0
			Thatch	34	2,000	1,200	800	27,200
			Total					21,02,180
Venkatapura estuary	200	8.64	Fodder	1728	600	360	240	4,14,720
			Timber	1	150	90	60	60
			Charcoal	0	150	90	60	0
			Thatch	0	2,000	1,200	800	0
			Total					4,14,780

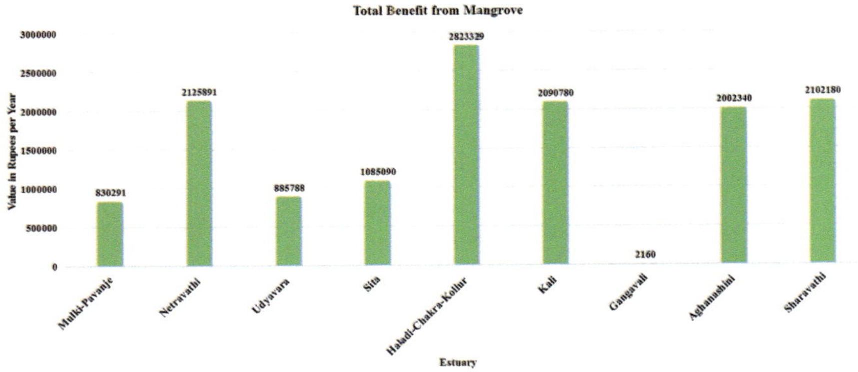

Fig. 28 Total benefit from mangrove products in estuaries of Karnataka

Table 23 Salt production in estuarine catchment

Estuary	Area (ha)	Item	Quantity extracted (ton)	Price (Rs/ ton)	Cost (Rs/ ton)	Residual value (Price - Cost)	Income generated Rs/year
Kali	3240	Salt	0	0	0	0	0
Gangavali	700	Salt	0	0	0	0	0
Aghanashini	4801	Salt	12,000	650	390	260	31,20,000
Sharavathi	1600	Salt	0	0	0	0	0
Venkatapura	200	Salt	0	0	0	0	0

17 Regulating Services

Regulating services provided by estuarine ecosystems include gas regulation, climate regulation, disturbance regulation, soil fertility, water regulation, air quality, erosion control, nutrient cycling, pollination, and carbon sequestration are given estuary wise in Table 25. The regulating services provided by estuarine ecosystems are computed through the benefit transfer method considering the spatial extent of an estuary.

The regulating services amount to 41,86,91,347 Rs/year in Mulki—Pavanje estuary, 1,06,55,86,339 Rs/year in Netravathi estuary, 44,35,22,429 Rs/year in Udyavara estuary, 54,44,08,037 Rs/year in Sita estuary, 1,41,53,25,562 Rs/ year in Haladi—Chakra—Kollur estuary, 1,81,44,55,680 Rs/year in Kali estuary, 39,60,21,325 Rs/year in Gangavali estuary, 2,67,99,94,740 Rs/year in Aghanashini estuary, 88,77,87,490 Rs/year in Sharavathi estuary, and 11,24,61,485 Rs/year in Venkatapura estuary.

Table 24 Total provisioning services from estuaries of Dakshina Kannada, Udupi and Uttara Kannada districts

Estuary	Mulki - Pavanje	Netravathi	Udyavara	Sita	Haladi Chakra Kollur estuary
Area (in ha)	750	1,920	800	980	2,550
Fishery Resources (Rs/year)	186,152,000	421,832,000	185,158,000	363,510,600	567,176,800
Agriculture activities (Rs/year)	7,84,350	19,98,640	8,35,200	10,23,120	2,662,200
Mining activity (Rs/year)	540	1382	598	764	1989
Mangrove products (Rs/year)	830,290	2,125,891	885,788	1,085,090	2,823,328
Total provisioning service value (Rs/year)	18,77,67,180	42,59,57,913	18,68,79,586	36,56,19,574	57,26,64,317
Estuary	Kali estuary	Gangavali estuary	Aghanashini estuary	Sharavathi estuary	Venkatapura estuary
Area (ha)	3,240	700	4,801	1,600	200
Fishery resources (Rs/year)	33,48,97,140	5,00,01,740	2,43,31,56,920	45,65,09,200	14,830,040
Agricultural activities (Rs/year)	1,50,07,500	51,68,430	1,83,09,420	80,80,560	1,036,264
Sand	7,896	532	476	672	106
Mangrove products (Rs/year)	20,90,780	2,160	20,02,340	21,02,180	4,14,780
Salt (Rs/year)	0	0	31,20,000	0	0
Total provisioning service value (Rs/year)	35,20,03,316	5,51,72,862	2,45,65,89,156	46,66,92,612	1,62,81,190

Table 25 Regulating services from estuaries of Dakshina Kannada, Udupi and Uttara Kannada districts

Estuary		Mulki-Pavanje estuary Rs/yr	Netravathi estuary Rs/yr	Udyavara estuary Rs/yr	Sita estuary Rs/yr	Chakra-Haladi-Kollur Rs/yr
Total area of mangroves (ha)		**16.03**	**7.3**	**0.5**	**6.5**	**10.2**
Total Area of estuary (ha)		**750**	**1920**	**800**	**980**	**2550**
Regulating services	*Average value* Rs/ha/year					
Gas regulation	9,600	72,00,000	1,84,32,000	76,80,000	94,08,000	2,44,80,000
Climate regulation	4,710	35,32,500	90,43,200	37,68,000	46,15,800	1,20,10,500
Disturbance regulation	32,556	5,21,872	2,37,658	16,278	2,11,614	3,32,071
Soil fertility	31,072	4,98,084	2,26,825	15,536	2,01,968	3,16,934
Water regulation	73,726	1,181,827	5,38,199	36,863	4,79,219	7,52,005
Air quality	20,968	3,36,117	1,53,066	10,484	1,36,292	2,13,873
Erosion control	65,061	4,87,95,750	12,49,17,120	5,20,48,800	6,37,59,780	16,59,05,550
Nutrient cycling	4,74,916	35,61,87,000	91,18,38,720	37,99,32,800	46,54,17,680	1,21,10,35,800
Pollination	19,203	3,07,824	1,40,181	9,601	1,24,819	1,95,870
Carbon sequestration	8,133	1,30,371	59,370	4,066	52,864	82,956
Total Regulating services value (Rs/year)		41,86,91,347	1,06,55,86,339	44,35,22,429	54,44,08,037	1,41,53,25,562
Estuary		Kali (Rs/year)	Gangavali (Rs/year)	Aghanashini (Rs/year)	Sharavathi (Rs/year)	Venkatapura (Rs/year)
Total area of mangroves (ha)		**100**	**43.2**	**101.6**	**5.0**	**8.64**

(continued)

Table 25 (continued)

Estuary		Kali (Rs/year)	Gangavali (Rs/year)	Aghanashini (Rs/year)	Sharavathi (Rs/year)	Venkatapura (Rs/year)
Total area of estuary (ha)		**3240**	**700**	**4801**	**1600**	**200**
Regulating services	*Average value* (Rs/ha/year)					
Gas regulation	9,600	3,11,04,000	67,20,000	4,60,89,600	1,53,60,000	19,20,000
Climate regulation	4,710	1,52,60,400	32,97,000	2,26,12,710	75,36,000	9,42,000
Disturbance regulation	32,556	32,55,600	14,06,419	33,07,689	1,62,780	2,81,283
Soil fertility	31,072	31,07,200	13,42,310	31,56,915	1,55,360	2,68,462
Water regulation	73,726	73,72,600	31,84,963	74,90,561	3,68,630	6,36,992
Air quality	20,968	20,96,800	9,05,817	21,30,348	1,04,840	1,81,163
Erosion control	65,061	21,07,97,640	4,55,42,700	31,23,57,861	10,40,97,600	1,30,12,200
Nutrient cycling	4,74,916	1,53,87,27,840	33,24,41,200	2,28,00,71,716	75,98,65,600	9,49,83,200
Pollination	19,203	19,20,300	8,29,569	19,51,024	96,015	1,65,913
Carbon sequestration	8,133	8,13,300	3,51,345	8,26,312	40,665	70,269
Total regulating service value (Rs/year)		1,81,44,55,680	39,60,21,325	2,67,99,94,740	88,77,87,490	11,24,61,485

18 Cultural Services

The cultural services provided by estuaries include science and education, recreation, and aesthetic information, which are accounted for through the benefit transfer method considering the spatial extent, and estuary-wise culture services are listed in Table 26.

The cultural services amount to 4,82,67,750 Rs/year in Mulki-Pavanje estuary, 12,35,65,440 Rs/year in Netravathi estuary, 5,14,85,600 Rs/year in Udyavara estuary, 6,30,69,860 Rs/year in Sita estuary, 16,41,10,350 Rs/year in Haladi-Chakra-Kollur estuary, 20,85,16,680 Rs/year in Kali estuary, 4,50,49,900 Rs/year in Gangavali estuary, 30,89,77,957 Rs/year in Aghanashini estuary, 10,29,71,200 Rs/year in Sharavathi estuary, and 1,28,71,400 Rs/year in Venkatapura estuary.

Ecosystem services provided by estuaries in Karnataka are depicted in Fig. 29, which highlights that Aghanashini estuary provides the highest value compared to other estuaries in Karnataka. The share of provisioning, regulating, and cultural services of estuaries in Karnataka is depicted in Fig. 30, which shows that 32% by provisioning services, 61% by regulating services, and 7% by cultural services. The highest contribution is from regulating services. Provisioning services is 27%, regulating services is 65%, and cultural services is 8% in the Dakshina Kannada district. Uttara Kannada district provides provisioning services 34%, regulating services 59%, and cultural services 7%, whereas Udupi district provides provisioning services 30%, regulating services 63%, and cultural services 7%.

19 Ecosystem Services of Estuaries in the West Coast, Karnataka, India (Districtwise Analyses)

Dakshina Kannada

The total provisioning services provided by estuaries in Dakshina Kannada district ranged from 18,77,67,180 Rs/year (Mulki-Pavanje estuary) to 42,59,57,913 Rs/year (Netravathi estuary). The annual value per hectare of Mulki-Pavanje estuary is 2,50,356 Rs/ha/year, and Netravathi estuary is 2,21,853 Rs/ha/year.

Regulating services value provided by estuaries of Dakshina Kannada ranged from 41,86,91,347 Rs/year (Mulki-Pavanje estuary) to 1,06,55,86,339 Rs/year (Netravathi estuary).

Cultural services value ranged from 4,82,67,750 Rs/year (Mulki-Pavanje estuary) to 12,35,65,440 Rs/year (Netravathi estuary).

Udupi District

The total provisioning services provided by estuaries in the Udupi district ranged from 18,68,79,586 Rs/year (Udyavara estuary) to 57,26,64,317 Rs/year (Chakra-Haladi-Kollur estuary). The Sita estuary gives a value of 36,56,19,574 Rs/year. The

Table 26 Cultural services from estuaries of Dakshina Kannada, Udupi and Uttara Kannada districts

Estuary		Mulki-Pavanje estuary Rs/year	Netravathi estuary Rs/year	Udyavara estuary Rs/year	Sita estuary Rs/year	Chakra-Haladi-Kollur Rs/year
Total area of estuary (ha)		**750**	**1920**	**800**	**980**	**2550**
Cultural services	*Average value* Rs/ha/year					
Science and education	34,660	2,59,95,000	6,65,47,200	2,77,28,000	3,39,66,800	8,83,83,000
Recreation	21,603	1,62,02,250	4,14,77,760	1,72,82,400	2,11,70,940	5,50,87,650
Aesthetic information	8,094	60,70,500	1,55,40,480	64,75,200	79,32,120	2,06,39,700
Total cultural service value (Rs/year)		4,82,67,750	12,35,65,440	5,14,85,600	6,30,69,860	16,41,10,350
Estuary		*Kali* (Rs/year)	*Gangavali* (Rs/year)	*Aghanashini* (Rs/year)	*Sharavathi* (Rs/year)	*Venkatapura* (Rs/year)
Total area of estuary (ha)		**3240**	**700**	**4801**	**1600**	**200**
Cultural services	*Average value* (Rs/ha/year)					
Science and education	34,660	11,22,98,400	2,42,62,000	16,64,02,660	5,54,56,000	69,32,000
Recreation	21,603	6,99,93,720	1,51,22,100	10,37,16,003	3,45,64,800	43,20,600
Aesthetic	8,094	2,62,24,560	56,65,800	3,88,59,294	1,29,50,400	16,18,800
Total cultural service value (Rs/year)		20,85,16,680	4,50,49,900	30,89,77,957	10,29,71,200	1,28,71,400

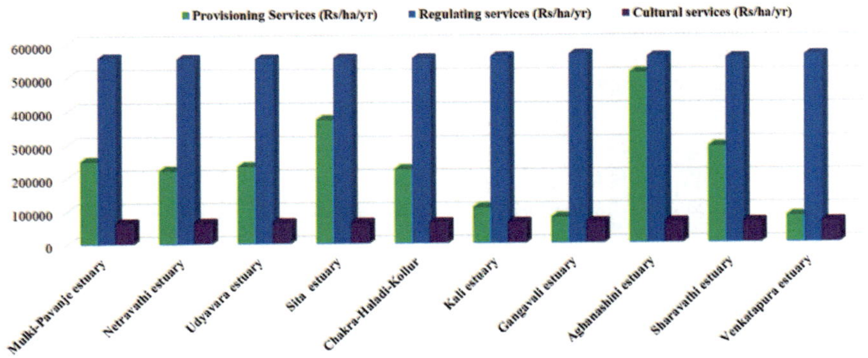

Fig. 29 Ecosystem services provided by estuaries of Karnataka

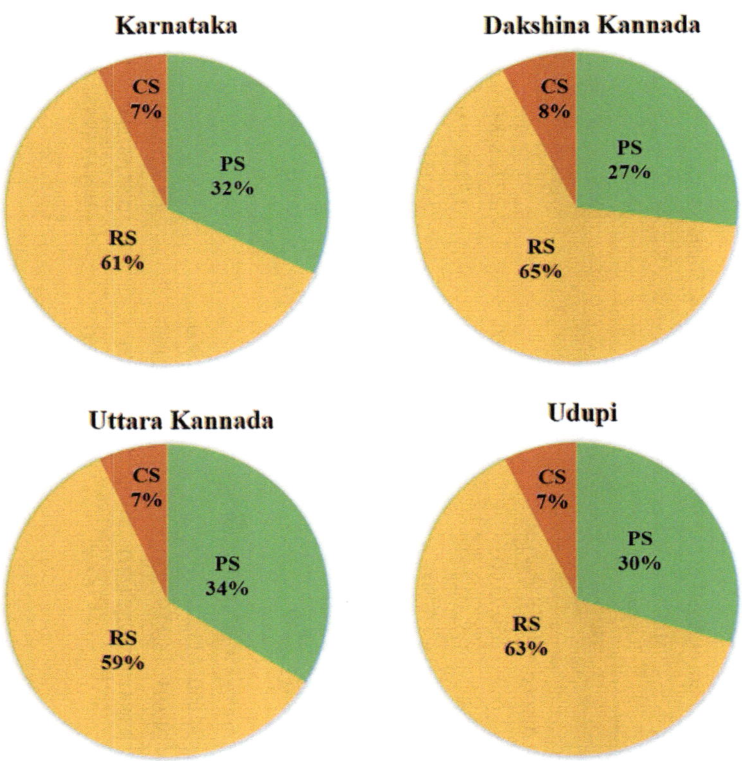

Fig. 30 District wise ecosystem services from estuaries of Karnataka

annual value per hectare of the Udyavara estuary is 2,33,599 Rs/ha/year, Sita estuary is 3,73,081 Rs/ha/year and Chakra-Haladi-Kollur estuary is 2,24,574 Rs/ha/year.

Regulating services provided by Udyavara estuary ranged from 44,35,22,429 Rs/year (Udyavara estuary) to 1,41,53,25,562 Rs/year (Chakra-Haladi-Kollur estuary) and Sita estuary gives the regulating services of 54,44,08,037 Rs/year.

Cultural services range from 5,14,85,600 Rs/year (Udyavara estuary) to 16,41,10,350 Rs/year (Chakra-Haladi-Kollur estuary), and Sita estuary provides cultural services of 6,30,69,860 Rs/year.

Uttara Kannada District

The total provisioning services provided by estuaries in Uttara Kannada district ranged from 1,62,81,190 Rs/year (Venkatapura estuary) to 2,45,65,89,156 Rs/year (Aghanashini estuary). The total provisioning services provided by the Kali estuary is 35,20,03,316 Rs/year, the Gangavali estuary is 5,51,72,862 Rs/year, and the Sharavathi estuary is 46,66,92,612 Rs/year. The annual value per hectare of Kali estuary is 1,08,642 Rs/ha/year, Gangavali estuary is 78,818 Rs/ha/year, Aghanashini estuary is 5,11,682 Rs/ha/year, Sharavathi estuary is 2,91,682 Rs/ha/year, and Venkatapura estuary is 81,405 Rs/ha/year.

Regulating services provided by the estuaries of Uttara Kannada district range from 11,24,61,485 Rs/year (Venkatapura estuary) to 2,67,99,94,740 Rs/year (Aghanashini estuary). The regulating services provided by Kali estuary is 1,81,44,55,680 Rs/year, Gangavali estuary is 39,60,21,325 Rs/year, and Sharavathi estuary is 88,77,87,490 Rs/year.

Cultural services provided by estuaries of Uttara Kannada district range from 1,28,71,400 Rs/year (Venkatapura estuary) to 30,89,77,957 Rs/year (Aghanashini estuary). The cultural services provided by Kali estuary are 20,85,16,680 Rs/year, Gangavali estuary is 4,50,49,900 Rs/year, and Sharavathi estuary is 10,29,71,200 Rs/year.

20 Total Ecosystem Supply Value (TESV)

Total ecosystem supply value is the aggregation of provisioning, regulating, and cultural services. The total ecosystem supply value (TESV) calculated for ten estuaries in Dakshina Kannada, Udupi, and Uttara Kannada districts is given in Table 27.

The total ecosystem supply value is 65,47,26,277 Rs/year in the Mulki-Pavanje estuary, 1,61,51,09,692 Rs/year in the Netravathi estuary, 68,18,87,615 Rs/year in Udyavara estuary, 97,30,97,471 Rs/year in Sita estuary, 2,15,21,00,229 Rs/year in Haladi-Chakra-Kollur estuary, 2,37,49,75,676 Rs/year in Kali estuary, 49,62,44,087 Rs/year in Gangavali estuary, 5,44,55,61,853 Rs/year in Aghanashini estuary, 1,45,74,51,302 Rs/year in Sharavathi estuary, and 14,16,14,075 Rs/year in Venkatapura estuary (Fig. 31).

Table 27 Total ecosystem supply value of estuaries of Dakshina Kannada, Udupi and Uttara Kannada

Estuary	Mulki-Pavanje estuary	Netravathi estuary	Udyavara estuary	Sita estuary	Chakra-Haladi-Kollur
Area (in ha)	750	1,920	800	980	2,550
Provisioning services (Rs/yr)	18,77,67,180	42,59,57,913	18,68,79,586	36,56,19,574	57,26,64,317
Regulating services (Rs/yr)	41,86,91,347	1,06,55,86,339	44,35,22,429	54,44,08,037	1,41,53,25,562
Cultural services (Rs/yr)	4,82,67,750	12,35,65,440	5,14,85,600	6,30,69,860	16,41,10,350
Total ecosystem supply value (Rs/yr)	65,47,26,277	1,61,51,09,692	68,18,87,615	97,30,97,471	2,15,21,00,229
Total ecosystem supply value (Rs/ha/yr)	8,72,968	8,41,202	8,52,359	9,92,956	8,43,960
Estuary	*Kali estuary*	*Gangavali estuary*	*Aghanashini estuary*	*Sharavathi estuary*	*Venkatapura estuary*
Area (ha)	3240	700	4801	1600	200
Provisioning services (Rs/yr)	352,003,316	55,172,862	2,456,589,156	466,692,612	16,281,190
Regulating services (Rs/yr)	1,81,44,55,680	39,60,21,325	2,67,99,94,740	88,77,87,490	11,24,61,485
Cultural services (Rs/yr)	208,516,680	4,50,49,900	30,89,77,957	10,29,71,200	1,28,71,400
Total ecosystem supply value (Rs/yr)	2,37,49,75,676	49,62,44,087	5,44,55,61,853	1,45,74,51,302	14,16,14,075
Total ecosystem supply value (Rs/ha/yr)	7,33,017	7,08,920	11,34,255	9,10,907	7,08,070

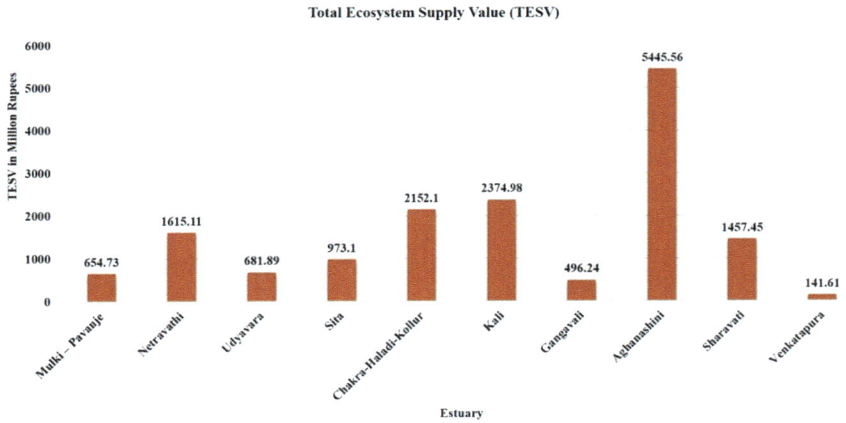

Fig. 31 TESV from estuaries of Karnataka

Estuary-wise, TESV is depicted in Fig. 32, which reveals that the Aghanashini estuary (13%) of Karnataka has the highest contribution, followed by Sita (12%), Sharavathi (11%), etc.

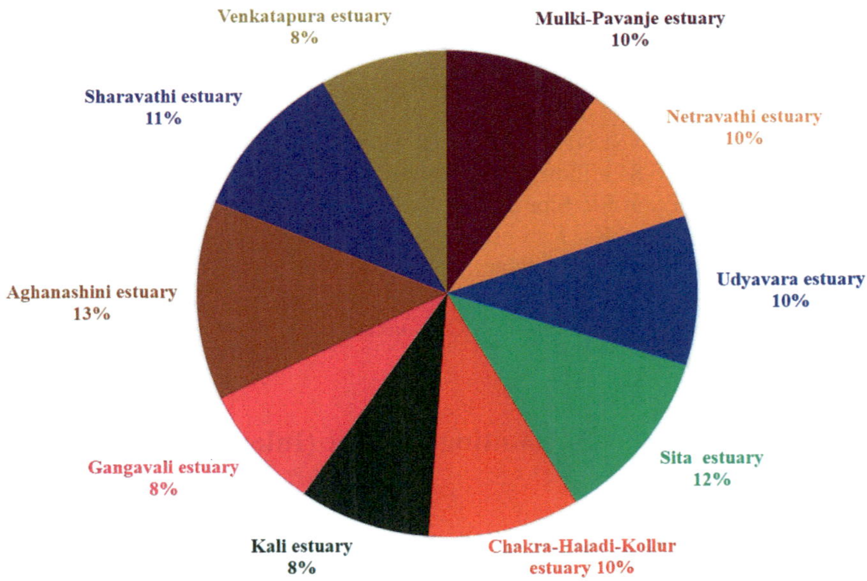

Fig. 32 Contribution of estuaries of Karnataka towards TESV

Table 28 Net Present Value (NPV) of estuaries of Dakshina Kannada, Udupi and Uttara Kannada districts

District	Estuary	Net present value	Total (District wise)
Dakshina Kannada	Mulki–Pavanje	16,84,40,77,455	
	Netravathi	41,55,17,65,535	
	Total		58,39,58,42,990
Udupi	Udyavara	17,54,28,54,482	
	Sita	25,03,47,81,326	
	Haladi–Chakra–Kollur	55,36,68,67,381	
	Total		97,94,45,03,189
Uttara Kannada	Kali	61,10,07,61,718	
	Gangavali	12,76,68,22,001	
	Aghanashini	1,40,09,74,25,235	
	Sharavathi	37,49,57,03,901	
	Venkatapura	3,64,32,91,146	
	Total		2,55,10,40,04,001
Karnataka state	Total net present value of all selected estuaries	4,11,44,43,50,180	

21 Net Present Value (NPV)

NPV of estuary assets is computed by considering annual flows of ecosystem services (TESV), 3% social rate of discount for capital returns, and a time period of 50 years. Table 28 lists district-wise NPV of estuaries. The analysis reveals that NPV of Mulki-Pavanje estuary is 16.84 billion Rs, Netravathi estuary is 41.55 billion Rs, Udyavara estuary is 17.54 billion Rs, Sita estuary is 25.03 billion Rs, Chakra-Haladi-Kollur estuary is 55.36 billion Rs, Kali estuary is 61.10 billion Rs, Gangavali estuary is 12.76 billion Rs, Aghanashini estuary is 1,40 billion Rs, Sharavathi estuary is 37.49 billion Rs/year, and Venkatapura estuary is 3.64 billion Rs. Figure 33 depicts the NPV of estuaries in the west coast of Karnataka.

22 Framework for Accounting Aquatic (Inland, Estuaries) Ecosystems

The framework for the valuation of aquatic ecosystems considering both freshwater and estuarine ecosystems is outlined in Fig. 34. The provisioning, regulating, and cultural services with the total value of ecosystem services for inland and estuaries are presented in Table 29. Figures 35, 36 and 37 present districtwise provisioning, regulating, and cultural services of aquatic ecosystems in Karnataka.

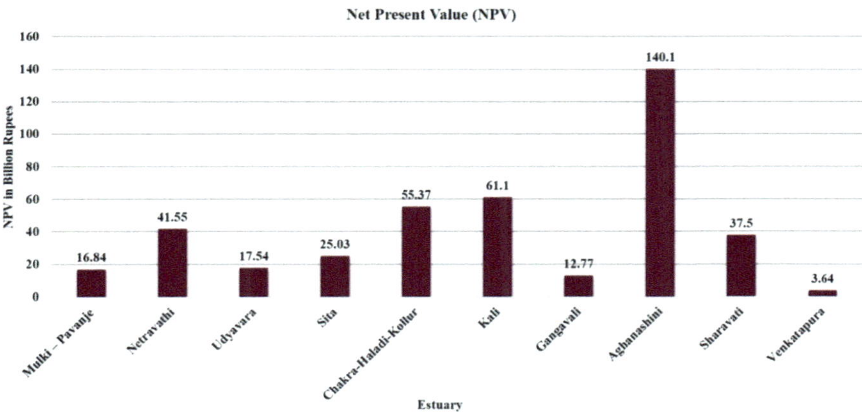

Fig. 33 NPV of estuaries in Karnataka State, India

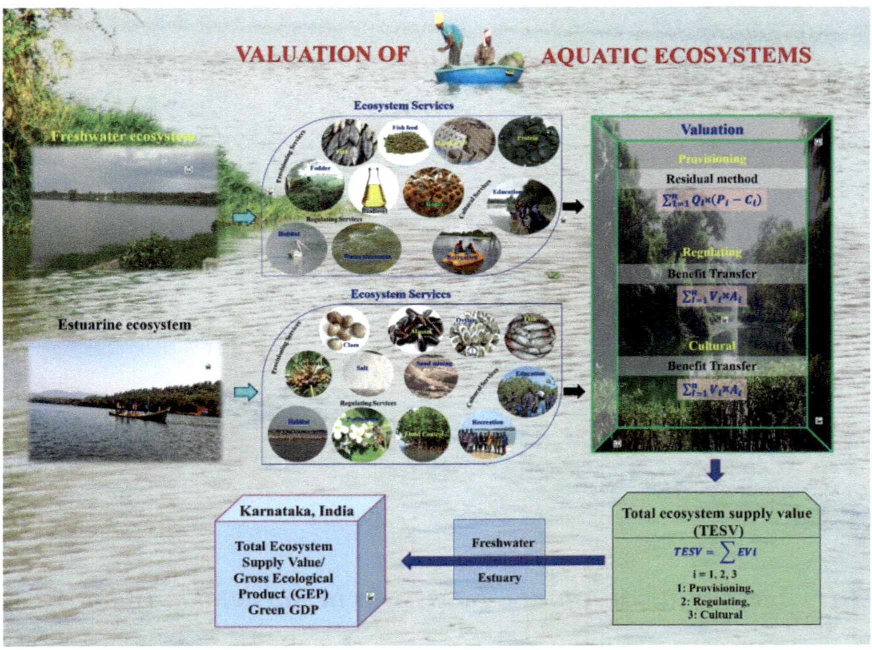

Fig. 34 Framework for valuation ecosystem services of aquatic ecosystems in Karnataka

Table 29 Ecosystem services of aquatic ecosystem in Karnataka

Ecosystem	Provisioning services (Rs/ha/yr)	Regulating services (Rs/ha/yr)	Cultural services (Rs/ha/yr)	TESV (Rs/ha/yr)
Freshwater	1,76,681	6,99,941	1,34,822	10,11,444
Estuarine	2,89,928	5,57,451	64,357	9,11,736
Aquatic	1,83,328	6,91,577	1,30,686	10,05,591

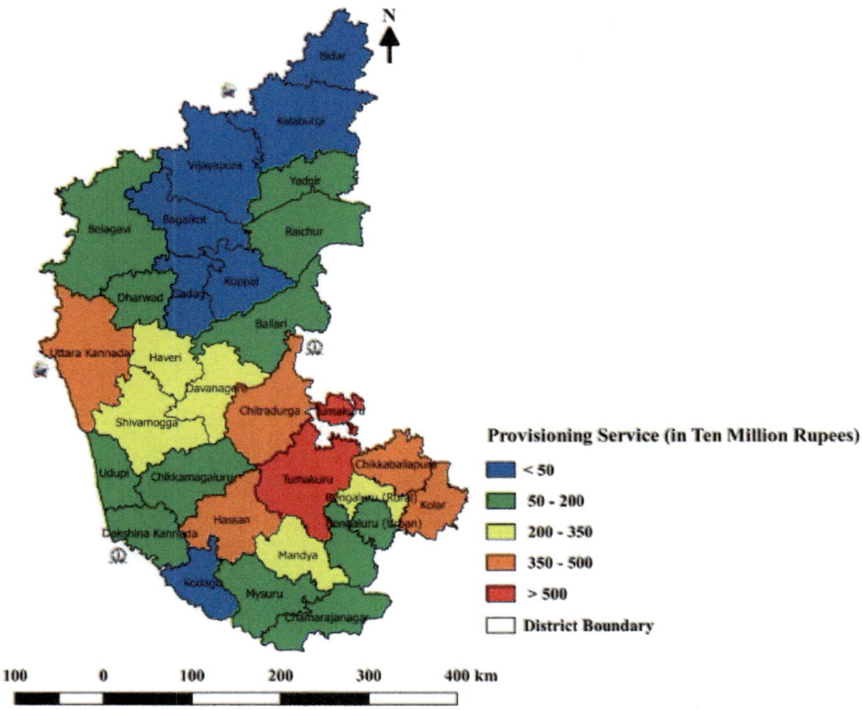

Fig. 35 Provisioning service of aquatic ecosystems in Karnataka

22.1 *Accounting of Aquatic Ecosystem in Karnataka, India*

Provisioning service ranges from 129 to 8214 million rupees annually (Fig. 35). In Karnataka, the aquatic ecosystem provides a total provisioning service value of 55 billion rupees per year.

Regulating service ranges from 515 to 32,641 million rupees per year (Fig. 36). In Karnataka, the aquatic ecosystem provides a total regulating service value of 207 billion rupees per year.

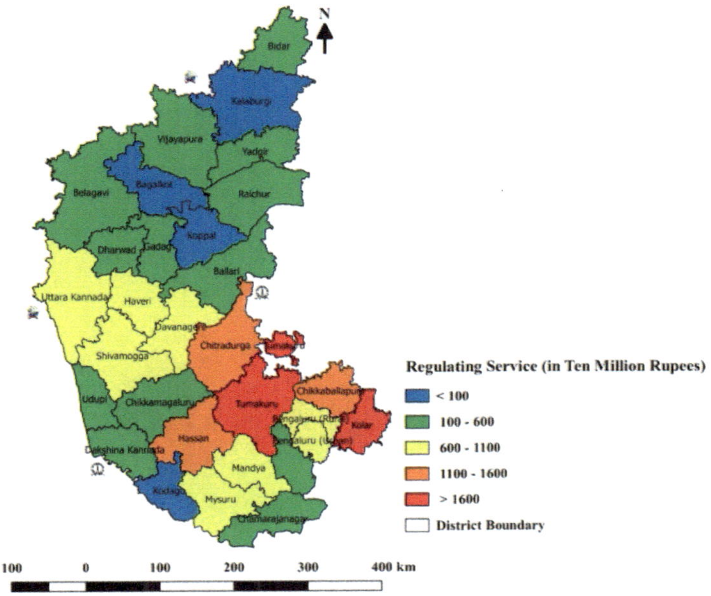

Fig. 36 Regulating service of aquatic ecosystems in Karnataka

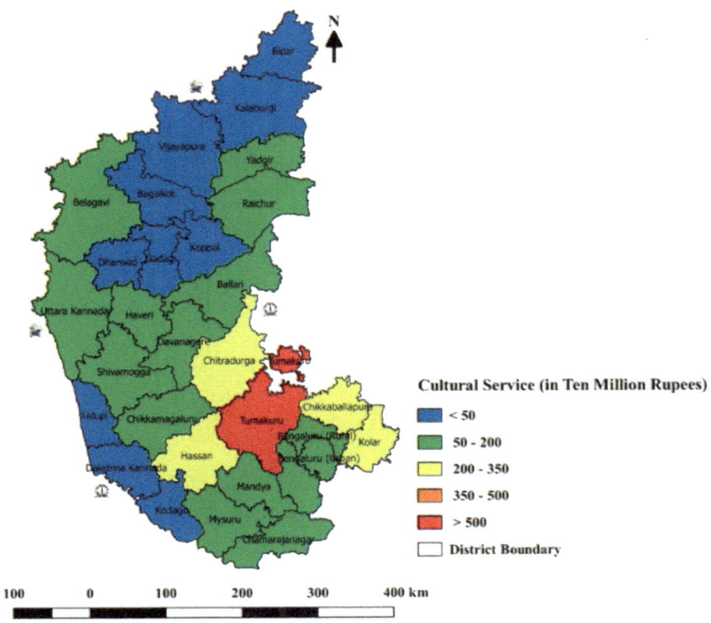

Fig. 37 Cultural service of aquatic ecosystems in Karnataka

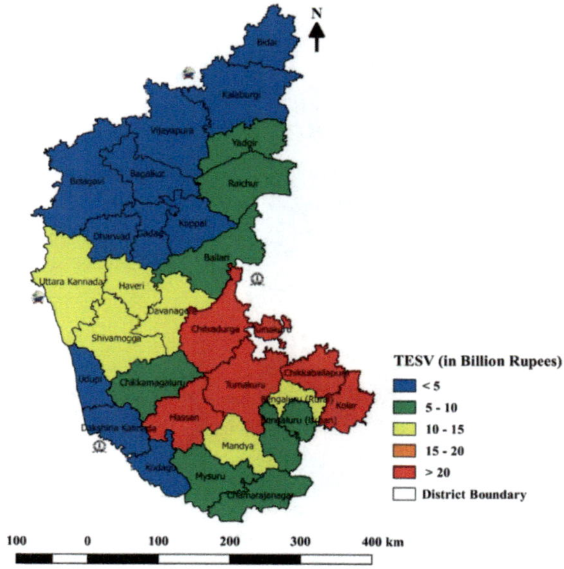

Fig. 38 TESV of aquatic ecosystems in Karnataka

Cultural services range from 99 to 6287 million rupees annually (Fig. 37). In Karnataka, the aquatic ecosystem provides a total regulating service value of 39 billion rupees per year.

Natural capital accounting and valuation of ecosystem services reveal that the aquatic ecosystem in Karnataka provides TESV of 301 billion rupees per year. Kodagu has the lowest TESV, whereas Tumakuru has the highest TESV. The total ecosystem supply value ranges from 0.7 to 47.1 billion rupees per year (Fig. 38).

Net present value of the aquatic ecosystem ranges from 19.1 to 1212.9 billion rupees (Fig. 39). In Karnataka, the NPV of aquatic ecosystem assets amounts to 7732 billion rupees.

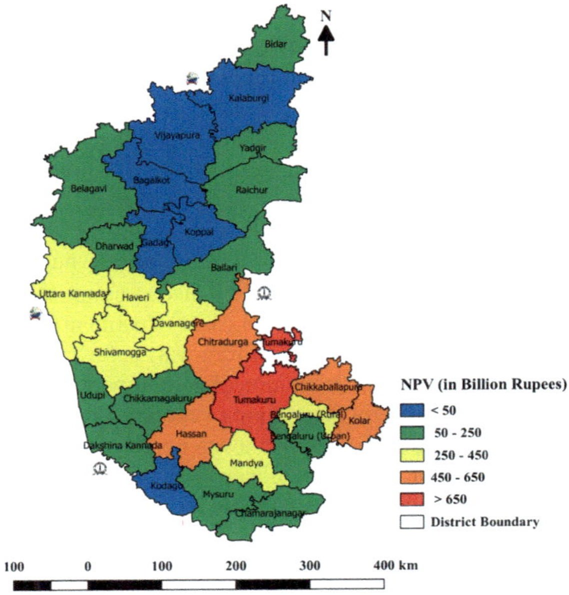

Fig. 39 NPV of aquatic ecosystems in Karnataka

23 Conclusion

Natural capital accounting and valuation of ecosystem services reveal significant direct and indirect contributions to the district and state economies and human welfare. Evaluating ecosystem services would aid in formulating policy and legislation that can provide protection and sustainable management of aquatic ecosystems to fully capitalize on the most significant ecosystem services. Among the districts of Karnataka, the Udupi district had the lowest values of ecosystem services, whereas the Tumakuru district had the highest ecosystem service potential. The value of provisioning, regulating, and cultural services ranged from 40–8214 million Rs/yr, 157–32,641 million Rs/yr, and 30–6287 million Rs/yr, respectively. The total ecosystem supply value of Karnataka wetlands was 285 billion Rs/yr, while the NPV value of Karnataka wetlands was 7321 billion rupees. The valuation of wetland services underlines the fact that wetlands are highly productive and economically viable ecosystems, provides crucial information on wetland resources and services, which will help policymakers and stakeholders with wetland conservation/management, and stresses the need for sustainable and wise use of wetland resources.

NPV of estuaries of Dakshina Kannada district is 58.39 billion Rs. Similarly, the NPV of three estuaries in Udupi district is 97.94 billion Rs, and five estuaries in Uttara Kannada district is 255 billion Rs.

The total value of provisioning, regulating, and cultural services considering both freshwater and estuarine ecosystems are 1,83,328 Rs/ha/yr, 6,91,577 Rs/ha/yr, and 1,30,686 Rs/ha/yr, respectively. The total ecosystem supply value of Karnataka aquatic ecosystem was 10,05,591 Rs/ha/yr. Aquatic ecosystems and their resources need to be conserved to sustain ecosystem services, which benefit millions of local people. This entails (i) regular monitoring of water quality by involving a network of local education institutions (schools, colleges), (ii) documenting diversity of aquatic species, (iii) regular desilting or dredging of wetlands to enhance water storage capacity (and also the silt provides valuable nutreints to agricultural lands in the command area of wetlands, (iv) discouraging the disposal of solid waste in the catchment, drains and in wetlands, (v) treatment of liquid waste through constructed wetlands (for optimal nutrient removal) before allowing to aquatic ecosystems, (vi) maintaining riparian vegetation of native species, (vii) protecting flood plains (buffer regions) integrity by maintaining vegetation cover, (viii) regulating the introduction of exotic species, (ix) public awareness through seminars, workshops, and media, (x) encouraging research on wetlands, educational institutions can adopt a lake and monitor it regularly, (xi) involving local stakeholders in the maintenance and sustainable management of wetlands and (xii) constituting an advisory committee of experts from diverse subjects (ecologists, hydrologists, social scientists) domain including environmental scientists and policy makers for wetland protection.

Considering the ecological significance (breeding area, habitat) there shall be no activity that would alter the physical and chemical integrity of estuaries in a region upto 500 m from HTL. Regions that get affected by tides up to 500 m from the high tide line (HTL) and the land area between the low tide line (LTL) and the high tide line are coastal regulation zone (CRZ). Ecologically sensitive and vulnerable areas need to be mapped for conservation, with no development zone in each estuary.

The Environment Ministry came up with the Coastal Regulation Zone rules (CRZ rules) in February 1991, under the Environment Protection Act in 1986. The rules were notified in 2011. In 2018, the government issued the Coastal Regulation Zone Notification 2018 to remove restrictions on building, streamlining the clearance process and increasing tourism in the coastal areas.

As per the CRZ rules, the coastal areas of creeks, seas, bays, rivers, and backwaters that get affected by tides up to 500 m from the high tide line (HTL) and the land area between the low tide line (LTL) and the high tide line are declared as coastal regulation zone (CRZ). The CRZ rules and regulations have been formulated with the objective of protecting coastal ecosystems, by regulating human and industrial activities near the coastline. CRZ norms (https://mczma.gov.in/sites/default/files/CRZ%20Notification%202019.pdf) prohibits major developmental activities such as new industries, disposal of hazardous materials, large-scale fish processing units, land reclamation, discharge of untreated waste, ports, mining of sand, altering sand dunes, disposal of plastic waste, etc.

Wetlands are to be conserved and managed in accordance with the principle of 'wise use' as per Wetlands Authority (Wetlands (Conservation and Management) Rules 2017, Government of India) with stringent restrictions on the (i) conversion for non-wetland uses, including encroachment of any kind; (ii) discharge of untreated wastes and effluents from industries, cities, towns, villages and other human settlements; (iii) construction of a permanent nature except for boat jetties within fifty meters from mean high flood level.

Appendix

See Tables A1, A2, A3 and A4.

Table A1 Provisioning services—Fuelwood

Water body	Area (ha)	Value (USD/yr)	Value per unit area (USD/ha/y)	Adjustment for GDP (PPP) per capita	Value in INR	References
Hadejia Nguru wetland, Nigeria	350,000	1,601,700	4.58	0.49	36	Schuijt (2002)
Lake Chiuta, Malawi	31,000	28,561.71	0.92	0.10	7	Zuze (2013)
Muthurajawela wetland, Sri Lanka	6000	82,530	13.76	1.47	109	
Muthurajawela wetland, Sri Lanka	3068	82,530	26.9	2.88	214	Schuyt and Brander (2008)
Rachenahalli lake, Bengaluru, Karnataka, India	42.09	113,218	2690	287.83	21,363	Ramachandra et al. (2005)
Hebbal lake, Bengaluru, Karnataka, India	60.7	139,014	2290	245.03	18,186	
Varthur lake, Bengaluru, Karnataka, India	220	25,284	115	12.31	913	Ramachandra et al. (2011)
Average value					**5833**	

Table A2 Regulating services—Flood control

Water body	Area (ha)	Value (USD/yr)	Value per unit area (USD/ha/y)	Adjustment for GDP (PPP) per capita	Value in INR	References
Wular lake, Jammu & Kashmir, India	13,000	17,500,000	1346	144.02	10,689	Kaul et al. (2016)
That Luang wetland, Laos	2000	2,842,000	1421	152.05	11,285	Zuze (2013)
Muthurajawela wetland, Sri Lanka	6000	5,033,800	839	89.77	6663	
Mfolozi flood plain, South Africa	20,886	504,605.76	24	2.57	191	
Muthurajawela wetland, Sri Lanka	3068	5,033,800	1641	175.59	13,032	Schuyt and Brander (2008)
Whangamarino wetland, New Zealand	10,320	601,037	58	6.21	461	
Average value					**7053**	

Table A3 Cultural services—Opportunities for recreation and tourism

Water body	Area (ha)	Value ($/yr)	Value per Unit Area ($/ha/y)	Adjustment for GDP (PPP) per capita	Value in INR	References
Jagadishpur Ramsar site, Nepal	19,600	90,769.5	5	0.54	40	Baral et al. (2016)
Lakeside wetland Park, Hefei, China	1072	13,734,310.34	12,812	1370.88	101,747	Li and Gao (2016)
Athirapally-Vazhachal, Kerala, India	133	1,123,646	8448	903.94	67,090	Anitha and Muraleedharan (2006)
Keenjhar lake, Pakistan	14,000	42,200,000	3014	322.50	23,936	Dehlavi and Adil (2011)
Kuttanad wetlands, Kerala, India	162,125	3,609,831.029	22	2.35	175	Rao (2018)
Bhoj wetland, Bhopal, India	3229	1,114,229	345	36.92	2740	Verma et al. (2001)
Little Rann of Kachchh, Gujarat, India	357,000	4,600,000	13	1.39	103	Dixit et al. (2016)
Zambezi Basin wetland, Southern Africa	2,982,900	813,800	0.27	0.03	2	Schuijt (2002)
Muthurajawela wetland, Sri Lanka	6000	54,743	9	0.96	71	Zuze (2013)
Mfolozi flood plain, South Africa	20,886	167,505.72	8	0.86	64	
Pantanal wetland, Brazil	13,800,000	423,640,000	31	3.32	246	Schuyt and Brander (2008)
Muthurajawela wetland, Sri Lanka	3068	54,743	18	1.93	143	
Whangamarino wetland, New Zealand	10,320	2,022,720	196	20.97	1557	

(continued)

Table A3 (continued)

Water body	Area (ha)	Value ($/yr)	Value per Unit Area ($/ha/y)	Adjustment for GDP (PPP) per capita	Value in INR	References
Wetlands in Bushenyi	18,300	8,985,300	491	52.54	3899	Kaggwa et al. (2009)
Hussain Sagar lake, Hyderabad, India	550	210,929	384	41.09	3050	Mishra (2014)
Ousteri wetland, Puducherry, India	800	95,333	119	12.73	945	Venkatachalam and Zareena Begam (2016)
Pong Dam wetland, India	15,500	133,547	9	0.96	71	Prasher et al. (2006)
Average value					**12,111**	

Table A4 Cultural services—Inspiration for culture, art and design

Ecosystem	Area (ha)	Value ($/yr)	Value per Unit Area	Adjustment for GDP (PPP) per capita	Value in INR	References
Mfolozi flood plain, South Africa	20,886	803,484.42	38 $/ha/yr	4.07	302	Zuze (2013)
Pantanal wetland, Brazil	13,800,000	1,144,490,000	83 $/ha/yr	8.88	659	Schuyt and Brander (2008)
Uttara Kannada, Central Western Ghats, India			480 Rs/ha/yr		480	Ramachandra (2016)
Wetlands in Bushenyi	18,300	32,226,300	1761 $/ha/yr	188.43	13,985	Kaggwa et al. (2009)
Aghanashini, Karnataka, India			1305 Rs/ha/yr		1305	Ramachandra et al. (2019)
Average value					**3346.2**	

References

Adekola O (2007) Economic valuation and livelihood analysis of the provisioning services provided by Ga-Mampa wetland South Africa (Doctoral dissertation, Wageningen University and Research Centre), pp 1–200

Adekola O, Mitchell G, Grainger A (2015) Inequality and ecosystem services: the value and social distribution of Niger Delta wetland services. Ecosyst Serv 12:42–54. https://doi.org/10.1016/j.ecoser.2015.01.005

Adekola O, Morardet S, de Groot R, Grelot F (2008) The economic and livelihood value of provisioning services of the Ga-Mampa wetland, South Africa. In: 13th IWRA world water congress, pp 1–24

Affek AN (2018) Indicators of ecosystem potential for pollination and honey production. Ecol Ind 94:33–45. https://doi.org/10.1016/j.ecolind.2017.04.001

Akwetaireho S, Getzner M (2010) Livelihood dependence on ecosystem services of local residents: a case study from Mabamba Bay wetlands (Lake Victoria, Uganda). Int J Biodiversity Sci, Ecosyst Serv Manag 6(1–2):75–87. https://doi.org/10.1080/21513732.2010.521139

Alam MZ, Carpenter-Boggs L, Rahman A, Haque MM, Miah MRU, Moniruzzaman M, Qayum MA, Abdullah HM (2017) Water quality and resident perceptions of declining ecosystem services at Shitalakka wetland in Narayanganj city. Sustain Water Quality Ecol 9:53–66. https://doi.org/10.1016/j.swaqe.2017.03.002

AlgaeBase (2021). https://www.algaebase.org/

Ali M, Sultana R, Tahir S, Watson IA, Saleem M (2017) Prospects of microalgal biodiesel production in Pakistan—A review. Renew Sustain Energy Rev 80:1588–1596. https://doi.org/10.1016/j.rser.2017.08.062

Aliyu A, Lee JGM, Harvey AP (2021) Microalgae for biofuels: a review of thermochemical conversion processes and associated opportunities and challenges. Bioresour Technol Reports 15:1–19. https://doi.org/10.1016/j.biteb.2021.100694

Amer L, Adhikari B, Pellegrino J (2011) Technoeconomic analysis of five microalgae-to-biofuels processes of varying complexity. Biores Technol 102:9350–9359. https://doi.org/10.1016/j.biortech.2011.08.010

Amit, Dahiya D, Ghosh UK, Nigam PS, Jaiswal AK (2021) Food industries wastewater recycling for biodiesel production through microalgal remediation. Sustainability 13(15):1–17. https://doi.org/10.3390/su13158267

Anitha V, Muraleedharan PK (2006) Economic valuation of ecotourism development of a recreational site in the natural forests of southern Western Ghats. Kerala Forest Research Institute, Peechi–680, pp 1–116

Ansari FA, Nasr M, Guldhe A, Gupta SK, Rawat I, Bux F (2020) Techno-economic feasibility of algal aquaculture via fish and biodiesel production pathways: A commercial-scale application. Sci Total Env 704:135259

Anshumali, Ramanathan AL (2007) Seasonal variation in the major ion chemistry of Pandoh lake, Mandi district, Himachal Pradesh, India. Appl Geochem 22(8):1736–1747. https://doi.org/10.1016/j.apgeochem.2007.03.045

APHA, AWWA, WEF (2005) Standard methods for the examination of water and wastewater. 21st Edition. American Public Health Association, Washington, DC

Aravindh P, Nair DR, Harikumar S (2019) Conservation of Kole wetlands–willingness to pay approach. Indian J Econ Dev 7(11):1–10

Aryal K, Ojha BR, Maraseni T (2021) Perceived importance and economic valuation of ecosystem services in Ghodaghodi wetland of Nepal. Land Use Policy 106:1–12. https://doi.org/10.1016/j.landusepol.2021.105450

Asiedu ANY (2015) Techno-economic analysis of protein concentrate produced by flash hydrolysis of microalgae. Master of Science (MS), thesis, Civil/Environmental Engineering, Old Dominion University, pp 1–75. https://doi.org/10.25777/tjyb-xv07

Asulabha KS, Sincy V, Jaishanker R, Ramachandra TV (2018) Algal diversity in urban lakes of Vrishabhavathi valley, Greater Bangalore. In: LAKE 2018: conference on conservation and sustainable management of riverine ecosystems. The 11th Biennial Lake Conference, Alva Campus, Vidyagiri, Moodbidri, D.K. Dist., Karnataka, India

Aulia AF, Sandhu H, Millington AC (2020) Quantifying the economic value of ecosystem services in oil palm dominated landscapes in Riau Province in Sumatra Indonesia. Land 9(6):1–24

Ausseil AG, Jamali H, Clarkson BR, Golubiewski NE (2015) Soil carbon stocks in wetlands of New Zealand and impact of land conversion since European settlement. Wetlands Ecol Manag 23(5):947–961

Banerjee S, Kar D, Banerjee A, Palit D (2012) Utilization of some aquatic macrophytes in Borobandh—A lentic water body in Durgapur, West Bengal, India: implications for socio-economic upliftment of local stakeholder. Indian J Appl Pure Biol 27(1):83–92

Baral S, Basnyat B, Khanal R, Gauli K (2016) A total economic valuation of wetland ecosystem services: an evidence from Jagadishpur Ramsar site, Nepal. Sci World J 2016:1–10. https://doi.org/10.1155/2016/2605609

Barbier EB (2011) Wetlands as natural assets. Hydrol Sci J 56(8):1360–1373. https://doi.org/10.1080/02626667.2011.629787

Barbier EB (2013) Valuing ecosystem services for coastal wetland protection and restoration: progress and challenges. Resources 2(3):213–230. https://doi.org/10.3390/resources2030213

Bengani R, Ujjania NC, Sangani K, Lovely R (2020) Idol Immersion and its consequences on water quality of Tapi River, Surat (Gujarat). Int J Adv Res Biol Sci 7(10):137–144. https://doi.org/10.22192/ijarbs

Bhattacharya S, Bera A, Dutta A, Ghosh UC (2014) Effects of idol immersion on the water quality parameters of Indian water bodies: environmental health perspectives. Int Lett Chem, Phys Astron 39:234–263. https://doi.org/10.18052/www.scipress.com/ILCPA.39.234

Bhavan SG, Rakshit N, Lal DM, Ingole B, Rivonkar P, Naik G, Mayekar T, Bhanudasrao CE (2021) Ecosystem modelling to understand the trophic dynamics and ecosystem health of a small tropical Indian estuary. Eco Inform 66:101429

Bilanovic D, Andargatchew A, Kroeger T, Shelef G (2009) Freshwater and marine microalgae sequestering of CO_2 at different C and N concentrations—Response surface methodology analysis. Energy Convers Manag 50:262–267

Biswasroy M, Samal NR, Roy PK, Mazumdar A (2011) Watershed management with special emphasis on fresh water wetland: a case study of a flood plain wetland in West Bengal, India. Global NEST J 13(1):1–10

Borah MP, Prasad SB (2016) Ethnozoological remedial uses by the indigenous inhabitants in adjoining areas of Pobitora wildlife sanctuary, Assam. India. Int J Pharm Pharmaceut Sci 8(4):1–7

Borah MP, Prasad SB (2017) Ethnozoological study of animals based medicine used by traditional healers and indigenous inhabitants in the adjoining areas of Gibbon Wildlife Sanctuary, Assam, India. J Ethnobiol Ethnomed 13(1):1–13. https://doi.org/10.1186/s13002-017-0167-6

Bošnjaković M, Sinaga N (2020) The perspective of large-scale production of algae biodiesel. Appl Sci 10(22):8181. https://doi.org/10.3390/app10228181

Branco-Vieira M, Mata TM, Martins AA, Freitas MAV, Caetano NS (2020) Economic analysis of microalgae biodiesel production in a small-scale facility. Energy Rep 6:325–332. https://doi.org/10.1016/j.egyr.2020.11.156

Burman PD, Cajee L, Laloo DD (2007) Potential for cultural and eco-tourism in North East India: a community-based approach. WIT Trans Ecol Environ 102:715–724. https://doi.org/10.2495/SDP070692

Campbell PK, Beer T, Batten D (2011) Life cycle assessment of biodiesel production from microalgae in ponds. Biores Technol 102(1):50–56. https://doi.org/10.1016/j.biortech.2010.06.048

Chai TT, Ooh KF, Quah Y, Wong FC (2015) Edible freshwater macrophytes: a source of anticancer and antioxidative natural products—A mini-review. Phytochem Rev 14(3):443–457

Chandrasekhar V, Murali Gopal S, Vidhyavathi A, Jayanthi C, Sathy R, Gopal N (2020) Recreational value of Vembanad Lake in Kerala using individual travel cost method. Int J Curr Microbiol Appl Sci 9(6):3280–3291. https://doi.org/10.20546/ijcmas.2020.906.391
Chanu TA, Teron R, Singh SK (2016) Ethnomedicinal use of certain fish species by ethnic groups of Bishnupur district in Manipur, NE India. World J Environ Biosci 5(2):1–5
Chaudhry P, Bhargava R, Sharma MP, Tewari VP (2013) Conserving urban lakes for tourism and recreation in developing countries: a case from Chandigarh, India. Int J Leisure Tour Market 3(3):267–281
Chhandama MVL, Satyan KB, Changmai B, Vanlalveni C, Rokhum SL (2021) Microalgae as a feedstock for the production of biodiesel: a review. Bioresour Technol Reports 15:1–13. https://doi.org/10.1016/j.biteb.2021.100771
Chisti Y (2007) Biodiesel from microalgae. Biotechnol Adv 25(3):294–306
Chisti Y (2008) Biodiesel from microalgae beats bioethanol. Trends Biotechnol 26:126–131
Chisti Y (2013) Constraints to commercialization of algal fuels. J Biotechnol 167(3):201–214. https://doi.org/10.1016/j.jbiotec.2013.07.020
Choi JY, Jeong KS, Kim SK, La GH, Chang KH, Joo GJ (2014) Role of macrophytes as microhabitats for zooplankton community in lentic freshwater ecosystems of South Korea. Eco Inform 24:177–185
Chowdhury H, Loganathan B (2019) Third-generation biofuels from microalgae: a review. Curr Opinion Green Sustain Chem 20:39–44. https://doi.org/10.1016/j.cogsc.2019.09.003
CICES. The common international classification of ecosystem services. European Environment Agency. https://www.cices.eu. Accessed 13 July 2021
Clarkson BR, Ausseil AGE, Gerbeaux P (2013) Wetland ecosystem services. In: Ecosystem services in New Zealand: conditions and trends. Manaaki Whenua Press, Lincoln, pp 192–202
CMFRI Annual Report 1999–2000. http://eprints.cmfri.org.in/id/eprint/6336
CMFRI Annual Report 2010–2011. http://eprints.cmfri.org.in/id/eprint/8676
CMFRI Annual Report 2014–2015. http://eprints.cmfri.org.in/id/eprint/10461
CMFRI Annual Report 2017–2018. http://eprints.cmfri.org.in/id/eprint/13044
Cook CDK (1996) Aquatic and wetland plants of India. Oxford University Press, New Delhi
Costanza DP, Badger JM, Fraser RL, Severt JB, Gade PA (2012) Generational differences in work-related attitudes: a meta-analysis. J Bus Psychol 27(4):375–394
Costanza R, d'Arge R, De Groot R, Farber S, Grasso M, Hannon B, Limburg K, Naeem S, O'neill RV, Paruelo J, Raskin RG (1997) The value of the world's ecosystem services and natural capital. Nature 387(6630):253–260
Dasgupta S, Panigrahi AKR (2014) Studies on the effect of aquatic pollution on ichthyofaunal diversity of the East Kolkata wetlands. Int J Res Appl, Nat Soc Sci 2(4):145–152
Davidson NC, Van Dam AA, Finlayson CM, McInnes RJ (2019) Worth of wetlands: revised global monetary values of coastal and inland wetland ecosystem services. Mar Freshw Res 70(8):1189–1194. https://doi.org/10.1071/MF18391
Davis R, Aden A, Pienkos PT (2011) Techno-economic analysis of autotrophic microalgae for fuel production. Appl Energy 88:3524–3531. https://doi.org/10.1016/j.apenergy.2011.04.018
De Groot R, Brander L, Solomonides S (2020) Update of global ecosystem service valuation database (ESVD). FSD report No 2020-06 Wageningen, The Netherlands, pp 1–58
De Groot R, Brander L, Van Der Ploeg S, Costanza R, Bernard F, Braat L, Christie M, Crossman N, Ghermandi A, Hein L, Hussain S (2012) Global estimates of the value of ecosystems and their services in monetary units. Ecosyst Serv 1(1):50–61
De Groot RS, Stuip MAM, Finlayson CM, Davidson N (2006) Valuing wetlands: guidance for valuing the benefits derived from wetland ecosystem services, Ramsar Technical Report No. 3/CBD Technical Series No. 27. Ramsar Convention Secretariat, Gland, Switzerland & Secretariat of the Convention on Biological Diversity, Montreal, Canada. ISBN 2-940073-31-7
Dehlavi A, Adil IH (2011) Valuing the recreational uses of Pakistan's wetlands: an application of the travel cost method. Working Paper-58, South Asian Network for Development and Environmental Economics (SANDEE), Kathmandu, pp 1–38

Delrue F, Setier PA, Sahut C, Cournac L, Roubaud A, Peltier G, Froment AK (2012) An economic, sustainability, and energetic model of biodiesel production from microalgae. Biores Technol 111:191–200. https://doi.org/10.1016/j.biortech.2012.02.020

Desikachary TV (1959) Cyanophyta. Indian Council for Agricultural Research, New Delhi

Devi SS (2017) Loktak lake and ecotourism prospects. Int J Dev Res 7(8):14576–14580

Dixit AM, Bandyopadhyaya S, Kumar L, Bedamatta S (2016) Economic valuation of landscape level wetland ecosystem and its services in Little Rann of Kachchh, Gujarat. In: The economics of ecosystems and biodiversity india initiative. GIZ India, pp 1–160

Ebm N, Guo F, Brett MT, Bunn SE, Kainz MJ (2021) Polyunsaturated fatty acids in fish tissues more closely resemble algal than terrestrial diet sources. Hydrobiologia 848(2):371–383. https://doi.org/10.1007/s10750-020-04445-1

Elaigwu AM, Oladele AH, Umaru J (2019) Protein, energy and micronutrient of five different fishes from Tiga reservoir, Nigeria. Asian J Fisheries Aquatic Res 3(2):1–9

ESVD (2020). https://esvd.net

Everard M, Kangabam R, Tiwari MK, McInnes R, Kumar R, Talukdar GH, Dixon H, Joshi P, Allan R, Joshi D, Das L (2019) Ecosystem service assessment of selected wetlands of Kolkata and the Indian Gangetic Delta: multi-beneficial systems under differentiated management stress. Wetlands Ecol Manag 27(2):405–426. https://doi.org/10.1007/s11273-019-09668-1

Fishbase (2021). https://www.fishbase.in/search.php

Garg JK (2015) Wetland assessment, monitoring and management in India using geospatial techniques. J Environ Manag 148:112–123. https://doi.org/10.1016/j.jenvman.2013.12.018

Gillanders BM, Kingsford MJ (2002) Impact of changes in flow of freshwater on estuarine and open coastal habitats and the associated organisms. Oceanogr Mar Biol 40:233–309

Gogoi C, Bora M (2020) Zoo-therapeutic practices among the deori tribes of Dhemaji district, Assam, India. Int J Fauna Biol Stud 7(4):196–198

Gopal B (2016) Should 'wetlands' cover all aquatic ecosystems and do macrophytes make a difference to their ecosystem services? Folia Geobot 51(3):209–226

Griffiths D (2006) The direct contribution of fish to lake phosphorus cycles. Ecol Freshw Fish 15:86–95. https://doi.org/10.1111/j.1600-0633.2006.00125.x

Gu W, Li H, Zhao P, Yu R, Pan G, Gao S et al (2014) Quantitative proteomic analysis of thylakoid from two microalgae (*Haematococcus pluvialis* and *Dunaliella salina*) reveals two different high light-responsive strategies. Sci Rep 4:1–12. https://doi.org/10.1038/srep06661

Gu X, Yu L, Pang N, Martinez-Fernandez, JS, Fu X, Chen S (2020) Comparative techno-economic analysis of algal biofuel production via hydrothermal liquefaction: One stage versus two stages. Appl Energy 259:114115. https://doi.org/10.1016/j.apenergy.2019.114115

Gupta T, Dey M (2017) Ichthyotherapy: Use of fishes as medicine by ethnic Karbi people of Assam, India. European J Pharmaceut Med Res 4(10):341–343. https://doi.org/10.1186/s13002-020-00410-3

https://www.ramsar.org/sites/default/files/documents/pdf/lib/lib_rtr03.pdf

Hua K, Cobcroft JM, Cole A, Condon K, Jerry DR, Mangott A, Praeger C, Vucko MJ, Zeng C, Zenger K, Strugnell JM (2019) The future of aquatic protein: implications for protein sources in aquaculture diets. One Earth 1(3):316–329

Islam MM, Hossain MM (2017) Community dependency on the ecosystem services from the Sundarbans mangrove wetland in Bangladesh. In: Wetland Science. Springer, New Delhi, pp 301–316. https://doi.org/10.1007/978-81-322-3715-0_16

Izaguirre I, Lancelotti J, Saad JF, Porcel S, Marinone MC, Roesler I, del Carmen Dieguez M (2018) Influence of fish introduction and water level decrease on lakes of the arid Patagonian plateaus with importance for biodiversity conservation. Global Ecol Conserv 14:1–17. https://doi.org/10.1016/j.gecco.2018.e00391

Jala, Nandagiri L (2015) Evaluation of economic value of Pilikula lake using travel cost and contingent valuation methods. Aquatic Procedia 4:1315–1321

Jayaram KC (1999). The Freshwater Fishes of India, Pakistan, Bangladesh, Burma and Sri Lanka—A Handbook, Zoological Survey of India, Kolkata, pp 1–551

Kaggwa R, Hogan R, Hall B (2009) Enhancing wetlands' contribution to growth, employment and prosperity. Environment and Natural Resources Report Series, pp 1–73
Kakoti D, Phukan MM, Devi NB (2019) Assessment of provisioning ecosystem services of Satajan wetland and bird sanctuary, Lakhimpur district, Assam, India. Environmentalism 4(1):124–136
Kalita G (2019) Biomass and rate of carbon sequestration by aquatic macrophytes in Deepor (Beel) wetland—A Ramsar site. Int J Basic Appl Res 9(1):492–506
Kant R, Haq S, Srivastava HC, Sharma VP (2013) Review of the bioenvironmental methods for malaria control with special reference to the use of larvivorous fishes and composite fish culture in central Gujarat, India. J Vector Borne Dis 50(1):1–12
Kashyap R, Verma KS, Brahmi MK (2016) Appraisal of water quality of Rewalsar Lake during baisakhi fair (harvest festival) in Himachal Pradesh, India. J Hill Agric 7(1):90–95. https://doi.org/10.5958/2230-7338.2016.00017.3
Kaul R, Masoodi A, Rasool A, Murty MN, Kishwan J (2016) Economic feasibility of willow removal from Wular Lake, Jammu & Kashmir. In: The economics of ecosystems and biodiversity India initiative. GIZ India, pp 1–52
Khan MI, Shin JH, Kim JD (2018a) The promising future of microalgae: current status, challenges, and optimization of a sustainable and renewable industry for biofuels, feed, and other products. Microb Cell Fact 17(1):1–21. https://doi.org/10.1186/s12934-018-0879-x
Khan ZR, Chowdhury NS, Sharmin S, Sohrab MH (2018b) Medicinal values of aquatic plant genus Nymphoides grown in Asia: a review. Asian Pac J Trop Biomed 8(2):113–119
Kiran BR, Hamsa A, Puttaiah ET (2007) Aquatic and marshy plants and their economic importance in Bhadra reservoir project region, Karnataka. Nat Environ Pollut Technol 6(3):429
Koko IA, Misana SB, Kessler A, Fleskens L (2020) Valuing ecosystem services: stakeholders' perceptions and monetary values of ecosystem services in the Kilombero wetland of Tanzania. Ecosyst People 16(1):411–426. https://doi.org/10.1080/26395916.2020.1847198
Kulkarni V, Ramachandra TV (2006) Environmental Management. The Energy and Resources Institute (TERI), Bangalore 560 012
Kumar M, Prusty AK, Jeniffer PN (2013) Ecosystem (provisioning) services and resource utilization pattern of an oxbow lake in Muzaffarpur urban Agglomeration, Bihar. The Ecoscan 3:289–295
Kumar V, Rai SP, Singh O (2006) Water quantity and quality of Mansar Lake located in the Himalayan Foothills, India. Lake Reservoir Manag 22(3):191–198. https://doi.org/10.1080/07438140609353896
Laishram J, Dey M (2013) Bio-resource utilization and socio-economic conditions of the people living in Ithing and Karang island villages of Loktak Lake, Manipur, India. Int J Bio-Resour Stress Manag 4(2):132–136
Lal D, Gardner JJ (2016) Effect of anthropogenic activities on Indian pilgrimage sites—A case study of Pushkar Lake. Arch Appl Sci Res 8(6):1–7
Lambert A (2003) Economic valuation of wetlands: an important component of wetland management strategies at the river basin scale. Conservation Finance Guide, Washington, pp 1–10
Lancar L, Krake K (2002) Aquatic weeds and their management. In: International commission on irrigation and drainage, pp 1–71
Langan C, Farmer J, Rivington M, Novo P, Smith JU (2019) A wetland ecosystem service assessment tool; Development and application in a tropical peatland in Uganda. Ecol Ind 103:434–445. https://doi.org/10.1016/j.ecolind.2019.04.019
Lannas KS, Turpie JK (2009) Valuing the provisioning services of wetlands: contrasting a rural wetland in Lesotho with a peri-urban wetland in South Africa. Ecol Soc 14(2)
Li T, Gao X (2016) Ecosystem services valuation of lakeside wetland park beside Chaohu Lake in China. Water 8(7):1–19. https://doi.org/10.3390/w8070301
Lolu AJ, Ahluwalia AS, Sidhu MC, Reshi ZA (2019) Carbon sequestration potential of macrophytes and seasonal carbon input assessment into the Hokersar wetland, Kashmir. Wetlands 39(3):453–472. https://doi.org/10.1007/s13157-018-1092-8
Louw TM, Griffiths MJ, Jones SM, Harrison ST (2016) Techno-economics of algal biodiesel. Algae biotechnology. Springer, Cham, pp 111–141

Magalhães Filho LNL, Roebeling PC, Bastos MI, Rodrigues W, Ometto GC (2021) A global meta-analysis for estimating local ecosystem service value functions. Preprints 1–19. https://doi.org/10.20944/preprints202106.0497.v1

Mahapatra DM, Chanakya HN, Ramachandra TV (2016) Algae derived single-cell proteins: economic cost analysis and future prospects. In Protein Byproducts, pp 275–301. Academic Press

Mahlatini P, Hove A, Maguma LF, Chemura A (2020) Using direct use values for economic valuation of wetland ecosystem services: a case of Songore wetland, Zimbabwe. Geojournal 85(1):41–51. https://doi.org/10.1007/s10708-018-9947-3

Marquès M, Bangash RF, Kumar V, Sharp R, Schuhmacher M (2013) The impact of climate change on water provision under a low flow regime: a case study of the ecosystems services in the Francoli river basin. J Hazard Mater 263:224–232

Mathew GM, Raina D, Narisetty V, Kumar V, Saran S, Pugazhendi A, Sindhu R, Pandey A, Binod P (2021) Recent advances in biodiesel production: challenges and solutions. Sci Total Environ 794:1–15. https://doi.org/10.1016/j.scitotenv.2021.148751

McGlynn CA (2009) Native and invasive plant interactions in wetlands and the minimal role of invasiveness. Biol Invasions 11(8):1929–1939

McGowan S, Leavitt PR, Hall RI, Anderson NJ, Jeppesen E, Odgaard BV (2005) Controls of algal abundance and community composition during ecosystem state change. Ecology 86(8):2200–2211

MEA (2003) Ecosystems and human well-being: a framework for assessment. Island Press, Washington, USA, pp 1–266

Merugu CS, Seetharaman R (2013) Comparative analysis of land use and lake water quality in rural and urban zones of south Chennai, India. Environ Dev Sustain 15(2):511–528. https://doi.org/10.1007/s10668-012-9391-3

Mishra N, Panda T, Pradhan BK, Rout SD, Mohanty RB, Kishor A, Singh RR (2016) Indigenous knowledge in utilization of wetland plants of Bhadrak district, Odisha, India. Indian J Nat Prod Resour 7(1):82–89

Mishra PP (2014) Potential benefits and earnings from improving the Hussain Sagar Lake in Hyderabad: a combined revealed and stated preference approach. In: South Asian network for development and environmental economics, pp 1–40

Misra MK, Panda A, Sahu D (2012) Survey of useful wetland plants of South Odisha, India. Indian J Tradit Knowl 11(4):658–666

Mizik T, Gyarmati G (2021) Economic and sustainability of biodiesel production—A systematic literature review. Clean Technol 3(1):19–36. https://doi.org/10.3390/cleantechnol3010002

Mohanty BP, Ganguly S, Mahanty A, Sankar TV, Anandan R, Chakraborty K, ... Sridhar N (2016) DHA and EPA content and fatty acid profile of 39 food fishes from India. Biomed Res Int 1–14. https://doi.org/10.1155/2016/4027437

Mukherjee S (2008) Economic valuation of a wetland in West Bengal, India. In: International Water Management Institute (IWMI)-TATA Water Policy Research Programme, pp 1–14

Nag SK, Nandy SK, Roy K, Sarkar UK, Das BK (2019) Carbon balance of a sewage-fed aquaculture wetland. Wetlands Ecol Manag 27(2):311–322. https://doi.org/10.1007/s11273-019-09661-8

Nagarajan S, Chou SK, Cao S, Wu C, Zhou Z (2013) An updated comprehensive techno-economic analysis of algae biodiesel. Biores Technol 145:150–156. https://doi.org/10.1016/j.biortech.2012.11.108

Najar IA, Khan AB (2012) Assessment of water quality and identification of pollution sources of three lakes in Kashmir, India, using multivariate analysis. Environ Earth Sci 66(8):2367–2378. https://doi.org/10.1007/s12665-011-1458-1

Nandan KB, Singh CB (2004) Useful macrophytes in Kawar Lake, North Bihar, India. Indian J Forest 27(3):241–244

Norsker NH, Barbosa MJ, Vermuë MH, Wijffels RH (2011) Microalgal production—A close look at the economics. Biotechnol Adv 29:24–27. https://doi.org/10.1016/j.biotechadv.2010.08.005

Oikonomou A, Stefanidis K (2020) α-and β-diversity patterns of macrophytes and freshwater fishes are driven by different factors and processes in lakes of the unexplored Southern Balkan biodiversity hotspot. Water 12(7):1984
Pal S, Talukdar S (2018) Drivers of vulnerability to wetlands in Punarbhaba river basin of India—Bangladesh. Ecol Ind 93:612–626. https://doi.org/10.1016/j.ecolind.2018.05.043
Pandey A, Lee DJ, Chang JS, Chisti Y, Soccol CR (eds) (2018). Biomass, biofuels, biochemicals: biofuels from algae. Elsevier
Pandey JS, Joseph V, Kaul SN (2004) A zone-wise ecological-economic analysis of Indian wetlands. Environ Monit Assess 98(1):261–273
Pandeya B, Buytaert W, Zulkafli Z, Karpouzoglou T, Mao F, Hannah DM (2016) A comparative analysis of ecosystem services valuation approaches for application at the local scale and in data scarce regions. Ecosyst Serv 22:250–259. https://doi.org/10.1016/j.ecoser.2016.10.015
Patterson M, Richardson J, Hardy D, Smith H (2018) The real economics of the adaptation to climate change on the Tahamata dairy farm—Assessing future scenarios from an integrated economic production and ecosystem services valuation approach. Horowhenua Coastal Climate Change Research Team, pp 1–72
Peel RA, Hill JM, Taylor GC, Weyl OL (2019) Food web structure and trophic dynamics of a fish community in an ephemeral floodplain lake. Front Environ Sci 7:192. https://doi.org/10.3389/fenvs.2019.00192
Perivolioti TM, Frouzova J, Tušer M, Bobori D (2020) Assessing the fish stock status in lake Trichonis: a hydroacoustic approach. Water 12(6):1–19
Poikane S, Portielje R, Denys L, Elferts D, Kelly M, Kolada A, Mäemets H, Phillips G, Søndergaard M, Willby N, van den Berg MS (2018) Macrophyte assessment in European lakes: diverse approaches but convergent views of 'good' ecological status. Ecol Ind 94:185–197
Prakash S, Prakash S (2021) Ethnomedicinal use of fishes by tribal communities in India: a review. Pharma Innov J 10(5):1315–1321
Prasher RS, Negi YS, Kumar V (2006) Valuation and management of wetland ecosystem (a case study of Pong Dam in Himachal pradesh). In: Man & Development, pp 77–92
Prescott GW (1970) How to Know Freshwater Algae. 2nd ed., W.C. Brown. Dubuque, Iowa
Quintas-Soriano C, Martín-López B, Santos-Martín F, Loureiro M, Montes C, Benayas J, García-Llorente M (2016) Ecosystem services values in Spain: a meta-analysis. Environ Sci Policy 55:186–195. https://doi.org/10.1016/j.envsci.2015.10.001
Quispe CA, Coronado CJ, Carvalho JA Jr (2013) Glycerol: production, consumption, prices, characterization and new trends in combustion. Renew Sustain Energy Rev 27:475–493
Ramachandra TV (2016) Valuation of goods and services from forests ecosystem of Uttara Kannada, Central Western Ghats. ENVIS Bull Himalayan Ecol 24:1–25
Ramachandra TV, Rajinikanth R, Ranjini VG (2005) Economic valuation of wetlands. J Environ Biol 26(2):439–447
Ramachandra TV, Sreekantha R (2006) Conservation values of wetlands in Sharavathi River Basin. Pollut Res 25(1):61–66
Ramachandra TV, Aithal BH (2016) Bengaluru's reality: towards unlivable status with unplanned urban trajectory. Curr Sci 110(12):2207–2208
Ramachandra TV, Alakananda B, Rani A, Khan MA (2011) Ecological and socio-economic assessment of Varthur wetland, Bengaluru (India). J Environ Sci Eng 53(1):101–108
Ramachandra TV, Asulabha KS, Lone AA (2014) Nutrient enrichment and proliferation of invasive macrophytes in urban lakes. J Biodivers 5(1–2):33–44
Ramachandra TV, Mahapatra DM, Bhat SP, Joshi NV (2015) Biofuel production along with remediation of sewage water through algae. Algae and environmental sustainability 33–51
Ramachandra TV, Asulabha KS, Sincy V, Bhat SP, Aithal BH (2016) Wetlands: treasure of Bangalore, ENVIS Technical Report 101, Energy & Wetlands Research Group, CES, Indian Institute of Science, Bangalore - 560012
Ramachandra TV, Raj RK, Aithal BH (2019) Valuation of Aghanashini estuarine ecosystem goods and services. J Biodivers 10(1–2):45–58. https://doi.org/10.31901/24566543.2019/10.1-2.093

Ramachandra TV, Sincy V, Asulabha KS (2020a) Efficacy of rejuvenation of lakes in Bengaluru, India. Green Chem Technol Lett 6(1):14–26. https://doi.org/10.18510/gctl.2020.613

Ramachandra TV, Sudarshan P, Vinay S, Asulabha KS, Varghese S (2020b) Nutrient and heavy metal composition in select biotic and abiotic components of Varthur wetlands, Bangalore, India. SN Appl Sci 2(8):1–14

Ramachandra TV, Asulabha KS, Sincy V (2021) Phosphate loading and foam formation in urban lakes. G P Globalize Res J Chem 5(1):33–52

Ramachandra TV, Sincy V, Asulabha KS, Mahapatra DM, Bhat SP, Aithal BH (2018) Optimal treatment of domestic wastewater through constructed wetlands. J Biodivers 9(1–2):81–102. https://doi.org/10.31901/24566543.2018/09.1-2.077

Ramachandra TV, Subash Chandran MD, Joshi NV, Raj RK, Mesta PN, Dudani SN (2013) Valuation of estuarine ecosystem, Uttara Kannada district, Karnataka, Sahyadri Conservation Series 27

Randrianarison G, Ashraf MA (2017) Microalgae: a potential plant for energy production. Geol, Ecol Landsc 1(2):104–120. https://doi.org/10.1080/24749508.2017.1332853

Rao CS (2018) Economic analysis of wetland ecosystems and their conservation and management: the case of Kuttanad coastal wetland ecosystem of Kerala, Doctoral dissertation, TNAU, Coimbatore

Rao DK, Das AK, Ramakrishniah M, Karthikeyan M (2002) Primary production and fishery potential of reservoirs of Karnataka. Indian J Fish 49(2):123–133

Rao GR, Mesta DK, Chandran MS, Ramachandra TV (2008) Wetland flora of Uttara Kannada. In: Environmental education for ecosystem conservation. Capital Publishing Company, New Delhi, pp 140–152

Rao SC, Balasubramanian R (2019) Conserving the Kuttanad wetlands: stakeholder preferences of management alternatives. Agric Econ Res Rev 32(2):199–208. https://doi.org/10.5958/0974-0279.2019.00032.6

Raut S, Gupta N, Everard M, Singh IS (2020) Commercially and medicinally significant aquatic macrophytes: Potential for improving livelihood security of indigenous communities in northern Bihar, India. J Threatened Taxa 12(13):16819–16830

Ravikumar P, Mehmood MA, Somashekar RK (2013) Water quality index to determine the surface water quality of Sankey tank and Mallathahalli lake, Bangalore urban district, Karnataka, India. Appl Water Sci 3(1):247–261. https://doi.org/10.1007/s13201-013-0077-2

Ro C, Sovann C, Bun D, Yim C, Bun T, Yim S, Irvine KN (2020) The economic value of peri-urban wetland ecosystem services in Phnom Penh, Cambodia. IOP Conference Series: Earth and Environmental Science 561(1):1–8

Roebeling P, Abrantes N, Ribeiro S, Almeida P (2016) Estimating cultural benefits from surface water status improvements in freshwater wetland ecosystems. Sci Total Environ 545:219–226. https://doi.org/10.1016/j.scitotenv.2015.12.063

Roles J, Yarnold J, Hussey K, Hankamer B (2021) Techno-economic evaluation of microalgae high-density liquid fuel production at 12 international locations. Biotechnol Biofuels 14(1):1–19. https://doi.org/10.1186/s13068-021-01972-4

Ruess L, Müller-Navarra DC (2019) Essential biomolecules in food webs. Front Ecol Evol 7:1–18. https://doi.org/10.3389/fevo.2019.00269

Russi D, ten Brink P, Farmer A, Badura T, Coates D, Förster J, Kumar R, Davidson N (2013) The economics of ecosystems and biodiversity for water and wetlands. In: IEEP, London and Brussels; Ramsar Secretariat, Gland, pp 1–84

Salem ME, Mercer DE (2012) The economic value of mangroves: a meta-analysis. Sustainability 4:359–383

Sánchez-Baracaldo P, Cardona T (2020) On the origin of oxygenic photosynthesis and cyanobacteria. New Phytol 225(4):1440–1446. https://doi.org/10.1111/nph.16249

Saranya G, Ramachandra TV (2020) Life cycle assessment of biodiesel from estuarine microalgae. Energy Convers Manag X 8:100065

Sargent J, Bell G, McEvoy L, Tocher D, Estevez A (1999) Recent developments in the essential fatty acid nutrition of fish. Aquaculture 177(1–4):191–199

Sarkar UK, Borah BC (2018) Flood plain wetland fisheries of India: with special reference to impact of climate change. Wetlands Ecol Manag 26(1):1–15

Sarkar UK, Sandhya KM, Mishal P, Karnatak G, Lianthuamluaia, Kumari S, Panikkar P, Palaniswamy R, Karthikeyan M, Mol SS, Das BK (2018) Status, prospects, threats, and the way forward for sustainable management and enhancement of the tropical Indian reservoir fisheries: an overview. Rev Fish Sci Aquacult 26(2):155–175

Sarma SK, Saikia M (2010) Utilization of wetland resources by the rural people of Nagaon district, Assam. Indian J Tradit Knowl 9(1):145–151

Sarmah BP, Baruah D, Bakalial B (2013) Wetland medicinal plants in floodplains of Subansiri and Ranga river of Lakhimpur district, Assam, India. Asian J Plant Sci Res 3(3):54–60

Sayre R (2010) Microalgae: the potential for carbon capture. Bioscience 60(9):722–727. https://doi.org/10.1525/bio.2010.60.9.9

Schallenberg M, de Winton MD, Verburg P, Kelly DJ, Hamill KD, Hamilton DP (2013) Ecosystem services of lakes. In: Ecosystem services in New Zealand: conditions and trends. Manaaki Whenua Press, Lincoln, pp 203–225

Schlagermann P, Göttlicher G, Dillschneider R, Rosello-Sastre R, Posten C (2012) Composition of algal oil and its potential as biofuel. J Combust 2012:1–14. https://doi.org/10.1155/2012/285185

Schuijt K (2002) Land and water use of wetlands in Africa: economic values of African wetlands. IIASA Interim Report, pp 1–48. https://core.ac.uk/download/pdf/33898193.pdf

Schuyt K, Brander L (2008) The economic values of the world's wetlands. In: Prepared with support from the Swiss Agency for the Environment, Forests and Landscape (SAEFL). Gland/Amsterdam, pp 1–32

SEEA, EEA (2021) System of environmental-economic accounting: ecosystem accounting, Final draft. United Nations, New York, pp 1–362

SEEA, EEA (2014) System of environmental-economic accounting 2012: experimental ecosystem accounting. United Nations, New York, pp 1–198

Shah MR, Lutzu GA, Alam A, Sarker P, Chowdhury MK, Parsaeimehr A, Liang Y, Daroch M (2018) Microalgae in aquafeeds for a sustainable aquaculture industry. J Appl Phycol 30(1):197–213. https://doi.org/10.1007/s10811-017-1234-z

Sharma B, Rasul G, Chettri N (2015) The economic value of wetland ecosystem services: evidence from the KoshiTappu Wildlife Reserve, Nepal. Ecosyst Serv 12:84–93

Shoyama K, Yamagata Y (2016) Local perception of ecosystem service bundles in the Kushiro watershed, Northern Japan—Application of a public participation GIS tool. Ecosyst Serv 22:139–149. https://doi.org/10.1016/j.ecoser.2016.10.009

Silva MI, Gonçalves AL, Vilar VJ, Pires J (2021) Experimental and techno-economic study on the use of microalgae for paper industry effluents remediation. Sustainability 13(3):1–29. https://doi.org/10.3390/su13031314

Sincy V, Asulabha KS, Jaishanker R, Ramachandra TV (2018) Ichthyo-diversity in sewage-fed lentic ecosystems of Bangalore. In: International conference on water resources (ICWR-2018), 15-17 March 2018, Department of Environmental Sciences & Department of Geology University of Kerala, Kariavattom, Trivandrum, India

Singh PK (2019) Comparative study of composite fish culture and local practices of fish culture in Surguja district of Chhattisgarh. J Krishi Vigyan 7(2):36–39. https://doi.org/10.5958/2349-4433.2019.00007.2

Singh S, Nair SS, Gupta AK (2013) Ecosystem services for disaster risk reduction: a case study of wetland in East Delhi region, India. Global J Human Soc Sci, Geogr, Geo-Sci, Environ Disaster Manag 13(4):37–47

Slembrouck J, Samsudin R, Pantjara B, Sihabuddin A, Legendre M, Caruso D (2018) Choosing floating macrophytes for ecological intensification of small-scale fish farming in tropical areas: a methodological approach. Aquat Living Resour 31:1–9. https://doi.org/10.1051/alr/2018017

Soo RM, Hemp J, Parks DH, Fischer WW, Hugenholtz P (2017) On the origins of oxygenic photosynthesis and aerobic respiration in cyanobacteria. Science 355(6332):1436–1440

Stefanidis K, Sarika M, Papastegiadou E (2019) Exploring environmental predictors of aquatic macrophytes in water-dependent Natura 2000 sites of high conservation value: results from a long-term study of macrophytes in Greek lakes. Aquat Conserv Mar Freshwat Ecosyst 29(7):1133–1148

Sudarshan Bhat, Mahesh MK, Ramachandra TV (2017) Macrophytes of Bangalore Wetlands. ENVIS technical report 126, energy & wetlands research group, CES, Indian Institute of Science, Bangalore

Sukumaran S, Jeeva S (2011) Angiosperm flora from wetlands of Kanyakumari district, Tamilnadu, India. Check List 7(4):486–495

Sun C, Zhen L, Miah MG (2017) Comparison of the ecosystem services provided by China's Poyang Lake wetland and Bangladesh's Tanguar Haor wetland. Ecosyst Serv 26:411–421. https://doi.org/10.1016/j.ecoser.2017.02.010

Svanberg I, Locker A (2020) Ethnoichthyology of freshwater fish in Europe: a review of vanishing traditional fisheries and their cultural significance in changing landscapes from the later medieval period with a focus on northern Europe. J Ethnobiol Ethnomed 16(1):1–29

Talwar PK, Jhingran AG (1991) Inland Fishes of India and Adjacent Countries, Oxford & IBH, New Delhi, 1 & 2, 1158

Tan JS, Lee SY, Chew KW, Lam MK, Lim JW, Ho SH, Show PL (2020) A review on microalgae cultivation and harvesting, and their biomass extraction processing using ionic liquids. Bioengineered 11(1):116–129. https://doi.org/10.1080/21655979.2020.1711626

Tang X, Li R, Han D, Scholz M (2020) Response of eutrophication development to variations in nutrients and hydrological regime: a case study in the Changjiang River (Yangtze) Basin. Water 12(6):1–19

Taran M, Deb S (2020) Utilization pattern of macrophytes in Rudrasagar Lake, a Ramsar site in India. Appl Ecol Environ Sci 8(4):179–186. https://doi.org/10.12691/aees-8-4-6

Taran M, Deb S, Roy J (2016) Physico-chemical characteristics, biodiversity assessment and economic valuation of Kalyanthakur Para lake: a community based lake of Tripura, North-East India. Lakes, Reservoirs Ponds 10(2):96–109

Taye FA, Folkersen MV, Fleming CM, Buckwell A, Mackey B, Diwakar KC, Le D, Hasan S, Saint Ange C (2021) The economic values of global forest ecosystem services: a meta-analysis. Ecol Econ 189:107145

TEEB (2010) The economics of ecosystems and biodiversity. Ecological and Economic Foundations Earthscan London

Tejaswi PB (2008) Non-timber forest products (NTFPs) for food and livelihood security: an economic study of tribal economy in Western Ghats of Karnataka, India, pp 1–90. (No. 138–2016–2006)

Teronpi V, Singh HT, Tamuli AK, Teron R (2012) Ethnozoology of the Karbis of Assam, India: use of ichthyofauna in traditional health-care practices. Anc Sci Life 32(2):99–103. https://doi.org/10.4103/0257-7941.118547

Thapa S, Wang L, Koirala A, Shrestha S, Bhattarai S, Aye WN (2020) Valuation of ecosystem services from an important wetland of Nepal: a Study from Begnas watershed system. Wetlands 40(5):1071–1083

Thrush SF, Townsend M, Hewitt JE, Davies K, Lohrer AM, Lundquist C, Cartner K, Dymond J (2013) The many uses and values of estuarine ecosystems. In: Ecosystem services in New Zealand–conditions and trends. Manaaki Whenua Press, Lincoln, New Zealand, pp 226–237

Tursi A, Maiorano P, Sion L, D'Onghia G (2015) Fishery resources: between ecology and economy. Rendiconti Lincei 26(1):73–79. https://doi.org/10.1007/s12210-014-0372-3

Vadeboncoeur Y (2006) Substratum as a driver of variation in periphyton chlorophyll and productivity in lakes. J N Am Benthol Soc 25(2):379–392

Vanni MJ (2002) Nutrient cycling by animals in freshwater ecosystems. Annu Rev Ecol Evol Syst 33:341–370. https://doi.org/10.1146/annurev.ecolsys.33.010802.150519

Venkatachalam L, Zareena Begam I (2016) Economic valuation of ecosystem services: a case study of Ousteri wetland, Puducherry. In: The Economics of Ecosystems and Biodiversity India Initiative. GIZ India, pp 1–103
Verma M, Bakshi N, Nair RP (2001) Economic valuation of Bhoj wetland for sustainable use. Indian Institute of Forest Management, Bhopal, pp 1–227
Verma M, Negandhi D, Wahal AK, Kumar R, Kinhal GA, Kumar A (2014) Revision of rates of NPV applicable for different class/category of forests. Indian Institute of Forest Management, Bhopal, India, pp 1–165
Vidanage S, Perera S, Kallesoe M (2005) The value of traditional water schemes: small tanks in the Kala Oya Basin, Sri Lanka. IUCN Water, Nature and Economics Technical Paper No. 6, IUCN, pp 1–76
Vieira TB, Tejerina-Garro FL (2020) Relationships between environmental conditions and fish assemblages in tropical savanna headwater streams. Sci Rep 10(1):1–12. https://doi.org/10.1038/s41598-020-59207-9
Walton MEM, Vilas C, Cañavate JP, González-Ortegón E, Prieto A, van Bergeijk SA, Green AJ, Librero M, Mazuelos N, Le Vay L (2015) A model for the future: ecosystem services provided by the aquaculture activities of Veta la Palma, Southern Spain. Aquaculture 448:382–390
Williamson TJ, Vanni MJ, González MJ, Renwick WH, Bremigan MT, Conroy JD (2018) The importance of nutrient supply by fish excretion and watershed streams to a eutrophic lake varies with temporal scale over 19 years. Biogeochemistry 140(2):233–253. https://doi.org/10.1007/s10533-018-0490-6
Woodward RT, Wui YS (2001) The economic value of wetland services: a meta-analysis. Ecol Econ 37(2):257–270
Xu J, Badola R, Chettri N, Chaudhary RP, Zomer R, Pokhrel B, Hussain SA, Pradhan S, Pradhan R (2019) Sustaining biodiversity and ecosystem services in the Hindu Kush Himalaya. In: The Hindu Kush Himalaya assessment. Springer, Cham, pp 127–165. https://doi.org/10.1007/978-3-319-92288-1_5
Xu X, Chen M, Yang G, Jiang B, Zhang J (2020) Wetland ecosystem services research: a critical review. Global Ecol Conserv 22:1–10. https://doi.org/10.1016/j.gecco.2020.e01027
Yang F, Hanna MA, Sun R (2012) Value-added uses for crude glycerol—A byproduct of biodiesel production. Biotechnol Biofuels 5(1):1–10. https://doi.org/10.1186/1754-6834-5-13
Yiran Z, Demin Z, Zhenguo N, Fengjiao X (2014) Valuation of lake and marsh wetlands ecosystem services in China. Chin Geogra Sci 24(3):269–278. https://doi.org/10.1007/s11769-013-0648-z
Zang S, Wu C, Liu H, Na X (2011) Impact of urbanization on natural ecosystem service values: a comparative study. Environ Monit Assess 179(1):575–588. https://doi.org/10.1007/s10661-010-1764-1
Zarandi MTP, Abesht A, Abedi S, Ahangari I (2019) The estimation of economic value of wetland ecosystem protection and recreational services: case study of the Kanibrazan international wetland. J Mater Environ Sci 10(11):1172–1184
Zedler JB, Kercher S (2005) Wetland resources: status, trends, ecosystem services, and restorability. Annu Rev Environ Resour 30:39–74. https://doi.org/10.1146/annurev.energy.30.050504.144248
Zhang L, Wu B, Yin K, Li X, Kia K, Zhu L (2015a) Impacts of human activities on the evolution of estuarine wetland in the Yangtze Delta from 2000 to 2010. Environ Earth Sci 73(1):435–447
Zhang P, He L, Fan X, Huo P, Liu Y, Zhang T, Pan Y, Yu Z (2015b) Ecosystem service value assessment and contribution factor analysis of land use change in Miyun County, China. Sustainability 7(6):7333–7356
Zhang Y, Xu H, Chen H, Wang F, Huai H (2014) Diversity of wetland plants used traditionally in China: a literature review. J Ethnobiol Ethnomed 10(1):1–19
Zhou B, Zhu JW, Lu P, Huang ZQ (2018) The service function value assessment analysis of urban wetland ecosystem—A case study of Xi'an Chan-Ba wetland. IOP Conference Series: Earth and Environmental Science 191(1):1–9
Zuze S (2013) Measuring the economic value of wetland ecosystem services in Malawi: a case study of lake Chiuta wetland. MSc. Thesis in Iwrm, University of Zimbabwe, pp 1–93

GPSR Compliance

The European Union's (EU) General Product Safety Regulation (GPSR) is a set of rules that requires consumer products to be safe and our obligations to ensure this.

If you have any concerns about our products, you can contact us on ProductSafety@springernature.com

In case Publisher is established outside the EU, the EU authorized representative is:

Springer Nature Customer Service Center GmbH
Europaplatz 3
69115 Heidelberg, Germany

Batch number: 10583311

Printed by Printforce, the Netherlands